Bacterial Polysaccharides

Current Innovations and Future Trends

Edited by Matthias Ullrich

School of Engineering and Science
Jacobs University Bremen
Bremen
Germany

Caister Academic Press

Caister Academic Press
Norfolk, UK

www.caister.com

British Library Cataloguing-in-Publication Data
A catalogue record for this book is available from the British Library

ISBN: 978-1-904455-45-5

Printed and bound in Great Britain

Contents

Contributors

Christina M. Baker
Department of Entomology & Plant Pathology
Oklahoma State University
Stillwater, OK
USA

christy.baker@okstate.edu

Anke Becker
Institut für Biologie III
Freiburg University
Freiburg
Germany

anke.becker@biologie.uni-freiburg.de

Carol L. Bender
Department of Entomology & Plant Pathology
Noble Research Center
Oklahoma State University
Stillwater, OK
USA

carol.bender@okstate.edu

José Antonio Bengoechea
Program Infection and Immunity
Fundación Caubet-Cimera Illes Balears
Consejo Superior de Investigaciones Científicas (CSIC)
Bunyola
Spain

bengoechea@caubet-cimera.es

Ute Bertsche
Eberhard-Karls-Universität Tübingen
Fakultät für Biologie
Institut für Mikrobiologie
Lehrstuhl für Mikrobielle Genetik
Tübingen
Germany

ute.bertsche@uni-tuebingen.de

Suresh Bhamidi
Department of Microbiology
Immunology & Pathology
College of Veterinary Medicine & Biomedical Sciences
Colorado State University
Fort Collins, CO
USA

bsuresh@colostate.edu

David Corbett
Faculty of Life Sciences
University of Manchester
Manchester
UK

david.r.corbett@manchester.ac.uk

Michael Gänzle
University of Alberta
Department of Agricultural Food and Nutritional Science
Edmonton, AB
Canada

mgaenzle@ualberta.ca

Karuppasamy Geetha
Department of Genetics
Centre for Excellence in Genomic Sciences
School of Biological Sciences
Madurai Kamaraj University
Madurai
India

geetgene@gmail.com

Klaus Geider
Julius Kühn Institut für Pflanzenschutz in Obst- und Weinbau
Dossenheim
Germany

k.geider@bba.de, klaus.geider@jki.bund.de

Joanna B. Goldberg
Department of Microbiology
University of Virginia
Charlottesville, VA
USA

jbg2b@virginia.edu

Paramasamy Gunasekaram
Department of Genetics
Centre for Excellence in Genomic Sciences
School of Biological Sciences
Madurai Kamaraj University
Madurai
India

pguna@eth.net

Johannes Hübner
Division of Infectious Disease
Department of Medicine
University Hospital Freiburg
Freiburg
Germany

johannes.huebner@uniklinik-freiburg.de

Kimberly K. Jefferson
Department of Microbiology & Immunology
Virginia Commonwealth University
Richmond, VA
USA

kkjefferson@vcu.edu

Elizaveta Krol
Institut für Biologie III
Freiburg University
Freiburg
Germany

elizaveta.krol@biologie.uni-freiburg.de

Dennis L. Kasper
Harvard Medical School
Channing Laboratory
Brigham and Women's Hospital
Boston, MA
USA

dennis_kasper@hms.harvard.edu

Michael R. McNeil
Department of Microbiology, Immunology & Pathology
College of Veterinary Medicine & Biomedical Sciences
Colorado State University
Fort Collins, CO
USA

mmcneil@colostate.edu

Ryoma Nakao
Department of Bacteriology
National Institute of Infectious Diseases
Toyama
Shinjuku-ku
Tokyo
Japan

ryoma73@nih.go.jp

Alejandro Penaloza-Vazquez
Department of Entomology & Plant Pathology
127 Noble Research Center
Oklahoma State University
Stillwater, OK
USA

apenalo@okstate.edu

Jeyaprakash Rajendhran
Department of Genetics
Centre for Excellence in Genomic Sciences
School of Biological Sciences
Madurai Kamaraj University
Madurai
India

jrajendhran@gmail.com

Narayanan Rajnish
Department of Genetics
Centre for Excellence in Genomic Sciences
School of Biological Sciences
Madurai Kamaraj University
Madurai
India

knrmku@gmail.com

Clara G. de los Reyes-Gavilán
Instituto de Productos Lácteos de Asturias
Consejo Superior de Investigaciones Científicas
(IPLA-CSIC)
Villaviciosa
Asturias
Spain

greyes_gavilan@ipla.csic.es

Ian S. Roberts
Faculty of Life Sciences
University of Manchester
Manchester
UK

ian.s.roberts@manchester.ac.uk

Patricia Ruas-Madiedo
Instituto de Productos Lácteos de Asturias
Consejo Superior de Investigaciones Científicas
(IPLA-CSIC)
Villaviciosa
Asturias
Spain

ruas-madiedo@ipla.csic.es

Roy R. B. Russell
Oral Biology
School of Dental Sciences
Newcastle University
Newcastle upon Tyne
UK

r.r.russell@newcastle.ac.uk

Nuria Salazar
Instituto de Productos Lácteos de Asturias
Consejo Superior de Investigaciones Científicas
(IPLA-CSIC)
Villaviciosa
Asturias
Spain

nuriasg@ipla.csic.es

Michael S. Scherman
Department of Microbiology, Immunology &
Pathology
College of Veterinary Medicine & Biomedical Sciences
Colorado State University
Fort Collins, CO
USA

michael.scherman@colostate.edu

Clarissa Schwab
University of Alberta
Department of Agricultural, Food and Nutritional
Science
Edmonton, AB
Canada

cschwab@ualberta.ca

Mikael Skurnik
Department of Bacteriology and Immunology
Haartman Institute
University of Helsinki
Finland

mikael.skurnik@helsinki.fi

Abhishek Srivastava
School of Engineering and Science
Jacobs University Bremen
Bremen
Germany

a.srivastava@jacobs-university.de

Christian Theilacker
Div. of Infectious Disease / Department of Medicine
University Hospital Freiburg
Freiburg
Germany

christian.theilacker@uniklinik-freiburg.de

Erin B. Troy
Harvard Medical School
Channing Laboratory
Brigham and Women's Hospital
Boston, MA
USA

erin_troy@student.hms.harvard.edu

Matthias S. Ullrich
School of Engineering and Science
Jacobs University Bremen
Bremen
Germany

m.ullrich@jacobs-university.de

Rudi F. Vogel
Lehrstuhl für Technische Mikrobiologie
Technische Universität München
Freising
Germany

rudi.vogel@wzw.tum.de

Florian W. Waldherr
Lehrstuhl für Technische Mikrobiologie
Technische Universität München
Freising
Germany

florian.waldherr@wzw.tum.de

Alan D. Welman
Fonterra Research Centre
Fonterra Cooperative Group Ltd.
Palmerston North
New Zealand

alan.welman@fonterra.com

Darran J. Wigelsworth
Harvard Medical School
Channing Laboratory
Brigham and Women's Hospital
Boston, MA
USA

dwigelsworth@partners.org

Daria Zhurina
School of Engineering and Science
Jacobs University Bremen
Bremen
Germany

d.zhurina@jacobs-university.de

Preface

This book compiles 19 comprehensively written reviews summarizing our current molecular knowledge, innovations, and future trends in research, infection medicine, and industrial application of bacterial polysaccharides. Diverse polysaccharide macromolecules are synthesized by bacteria using a rich arsenal of distinct pathways and function as cell wall components or storage units, to counteract harsh environmental conditions, as masking agents, or as part of the matrix by which bacterial cells reside in sessile life style. Over the past few years, it has become clear that there are unifying themes in bacterial polysaccharide synthesis, regulation, and function – and those are summarized herein. The gathered expert opinions are organized in seven book parts devoted to individual scientific topics and to specific research or application problems. Regardless of the individual topic, overlapping themes such as biofilm formation, coupling of polysaccharide synthesis with intracellular metabolic functions, or the embedding of regulatory pathways controlling polysaccharide synthesis in global – often virulence-associated – regulatory hierarchies can be found in each of the 19 chapters.

Unlike other microbiological traits, bacterial polysaccharides link primary metabolism with extracellular function, thus acting at the interface between bacterium and host, and therefore biosynthesis needs to be tightly controlled at the level of transcription due to their high demand for cellular energy. Bacterial cells 'invest' in polysaccharide synthesis without immediately 'knowing' the beneficial outcome of this synthesis since many of those macromolecules are simply secreted or produced outside of the cell. Consequently, understanding the regulatory links between intracellular energy conservation, polymer synthesis and modification, and the external ecological functions becomes increasingly important for us to better benefit from or to more efficiently combat biological effects mediated by bacterial polysaccharides. The first decade of this new century is an enormously exciting time for researchers. It has been fully appreciated for quite some time that a sessile life is presumably the dominant way of bacteria thriving – be it on a surface, such as our teeth, or the pipelines in biotechnology – within various organs and tissues of eukaryotic hosts, inside dirt, or on any marine aggregate floating in the oceans. At the same time it has become clear that not only exopolysaccharides but also the non-watery composite of biofilm matrices is of tremendously diverse origin. Consequently, researchers are beginning to understand that polysaccharides might be – at least in part – waste disposal storage sites for later recycling, which evolutionarily have become powerful tools, connectors, or barriers for microbe–microbe and microbe–host interactions.

Part I is devoted to the role of polysaccharides as indispensable components of the bacterial cell envelope. The bacterial peptidoglucan matrix, its variable composition, and how it is influenced by stress are summarized in Chapter 1 by Ute Bertsche (University of Tübingen). Bacterial lipopolysaccharide synthesis, its genetics, and regulation with special emphasis on the systems in *Yersinia* pathogens are subjects of the review article provided by Mikael Skurnik (University of Helsinki) and Jose A. Bengaechea (CSIC Bunyola, Spain). In the third chapter of this section, Suresh Bhamidi, Michael S. Scherman, and Michael R. McNeil (Colorado State University) focus on the mycobacterial cell wall polymer, arabinogalactan, its structure and synthesis, as well as its potential to act as a therapeutic target.

Parts II and III deal with polysaccharides from medically important bacteria of the respiratory, digestive, or urogenital tracts with a special emphasis on dental pathogens. David Corbett and Ian S. Roberts (University of Manchester) summarize the current molecular knowledge of the genetics and transcriptional regulation of enterobacterial and streptococcal pathogens' polysaccharides. In her chapter, Joanna B. Goldberg (University of Virginia) highlights the potential of anti-polysaccharide strategies for the development of novel vaccines to fight infectious diseases such as those caused by *Pseudomonas aeruginosa*. The particular roles of bacterial polysaccharides in pathogen recognition by the various 'instruments' of the human immune system, and therapies directed towards these are expertly reviewed by Darran J. Wigelsworth, Erin B. Troy, and Dennis L. Kasper (Harvard Medical School). Nowhere in the digestive tract do bacteria experience such a wealth of sugar nutrients as in the oral cavity. In their chapters, Ryoma Nakao (National Institute of Infectious Disease, Tokyo) and Roy B. Russell (University of Newcastle), dissect the synthesis and composition of polysaccharides from periodontopathic bacteria, as well as the intimate relationship between polysaccharide synthesis, attachment, and biofilm formation within the oral cavity.

Although it is clear from any of the chapters that the bacterial sessile life style and polysaccharide synthesis form a multifaceted functional unit, the particular importance of biofilms in pathogenic and environmental bacteria is specifically highlighted in Part IV. Christian Theilacker and Johannes Hübner (Freiburg University Hospital) provide an in-depth focus on polysaccharide-mediated biofilm formation in Gram-positive pathogens and – in this context – concentrate on the cell wall-associated carbohydrate polymers. Their chapter is followed by a complementary chapter written by Kimberly K. Jefferson (Harvard Medical School), who systematically dissects in detail the role and the molecular regulation of the synthesis of poly-N-acetylglucosamine as a mediator of bacterial biofilm formation.

Part V is devoted to diverse polysaccharides produced by plant-associated bacteria, the 'number 1 source of metabolizable carbohydrates in nature. Elizaveta Krol and Anke Becker (University of Bielefeld) comprehensively review current innovations and future trends in research on surface polysaccharides as fitness factors of rhizospheric nitrogen-fixing bacteria and highlight the importance of specific polysaccharides as molecular signals and mediators of the plant–microbe symbiosis. The chapter by Klaus Geider (Federal Research Institute Dossenheim, Germany) gives an expert point of view on the structure, biosynthesis, and regulation of capsular polysaccharides of the fire blight pathogen, *Erwinia amylovora*. Two chapters, one by Alejandro Penaloza-Vazquez, Christina M. Baker, and Carol L. Bender (Oklahoma State University), and the other by Abhishek Srivastava, Daria Zhurina, and Matthias S. Ullrich (Jacobs University Bremen), dissect the three major polysaccharides produced by the plant pathogen, *Pseudomonas syringae*. The former deals with osmoregulated periplasmic glucans and alginate while the latter gives insights into the function and regulation of the extracellular enzyme forming levan.

Polysaccharide formation in lactic acid bacteria is the major subject comprehensively dealt with in Part VI of this book. In their review, Michael Gänzle and Clarissa Schwab (University of Alberta) focus on the genetic organization, transcriptional regulation, and ecological functions of homo- and heteropolysaccharides produced by lactic acid bacteria. Convincingly complementing this, Patricia Ruas-Madiedo, Nuria Salazar, and Clara G. de los Reyes-Gavilan (Instituto de Productos Lacetos, Spain) provide an in-depth dissection of the mechanisms of synthesis and the nature of the enzymes involved in production of lactic acid bacterial polysaccharides.

The final part, Part VII, covers our current knowledge and future aspects of the commercial use, food industry applications, and detrimental effects in bioethanol synthesis of bacterial polysaccharides. Consequently, the chapters in this section provide valuable insights into the application-associated aspects of research on bacterial exopolymers. Florian Waldherr and Rudi Vogel (Technical University Munich) focus on the commercial exploitation of homo-exopolysaccharides in dairy and non-dairy food systems but also refer to other groups of food. The nutritional and functional benefits of exopolysaccharides from lactic acid bacteria is reviewed by Alan D. Welman (Fonterra Research Centre, New Zealand) in his interdisciplinary chapter. This part of the book is completed by the chapter authored by K. Geetha, K.N. Rajnish, J. Rajendhran, and P. Gunasekaran (Madurai Kamaraj University, India), who highlight the impact of bacterial exopolysaccharides on the efficiency of biofuel production by *Zymomonas mobilis*.

What makes this book particularly interesting to read is not only the enormous wealth of novel experimental data and scientific conclusions obtained from state-of-the-art technologies but also the comprehensive collection of different perspectives. The content of this book impressively demonstrates that current innovations in research on, and biotechnological applications of, bacterial polysaccharides can be obtained only by taking an interdisciplinary approach. The biosynthesis, structure, regulation and function of polysaccharides can be fully understood only if molecular details, genome-, transcriptome- and proteome-wide perspectives, ecological aspects and systems biology approaches are considered simultaneously in future work. Hence, this book on bacterial polysaccharides will give a sound basis for future research directions and high-efficiency applications of bacterial polysaccharides in medicine, the food industry, and renewable energy production.

Matthias S. Ullrich
March, 2009

Part I

Bacterial Cell Wall Structures and Polysaccharides

The Polysaccharide Peptidoglycan and how it is Influenced by (Antibiotic) Stress

1

Ute Bertsche

Abstract
The peptidoglycan or murein sacculus is the stress-bearing structure of bacterial cells. It consists of glycan strands cross-linked by peptide bridges. Even though studies on murein have a very long tradition, it is not yet known how the glycan strands are actually arranged. However, the chemical fine structure and the muropeptide composition of different Gram-negative and Gram-positive bacteria have been investigated in detail. This chapter discusses *Escherichia coli* and *Staphylococcus aureus* as representatives for both Gram forms. During cell growth the stress-bearing structure has to be elongated and/or divided by the insertion of new and elimination of old material without loosing its strength. Therefore multienzyme complexes containing both murein synthases and murein hydrolases have been postulated. Peptidoglycan biosynthesis is the target for many antibiotics such as β-lactams, D-cycloserine and glycopeptide antibiotics such as vancomycin. The last part of this chapter gives an overview of different bacterial strategies of bacteria for coping with antibiotic and osmotic stress.

Introduction

In almost all eubacteria the cytoplasmic membrane is surrounded by a bag-shaped macromolecule: the peptidoglycan or murein sacculus. In Gram-negative bacteria it is located within the periplasm between the cytoplasmic membrane and the outer membrane, while in Gram-positive bacteria the peptidoglycan forms the outermost part of the cell. Eubacteria known not to contain peptidoglycan are *Planctomyces*, mycoplasmas, *Chlamydiae* and *Orientia* (*Rickettsia*) *tsutsugamushi*. Peptidoglycan has been studied for several decades and the chemical composition has long been solved. As the term 'peptidoglycan' suggests, it consists of polysaccharides (glycan strands) cross-linked by peptide moieties. The sacculus can be isolated as a whole and viewed under the electron microscope. Its shape corresponds exactly to the form of the original cell (Weidel and Pelzer, 1964). Unfortunately, to date it has not been possible to actually visualize the fine structure of this macromolecule, resulting in controversial discussions about the orientation of the glycan strands relative to the rod axes.

The peptidoglycan sacculus has to be elongated and divided during bacterial growth. As it is a stress-bearing structure several models have been described for a safe enlargement and separation. Some of these models are described here. The existence of multienzyme complexes for temporal and spatial control is also discussed.

As the peptidoglycan sacculus is a distinct feature of bacteria it is a target for several different kinds of bacteriolytic antibiotics. To cope with these stress factors, different resistance mechanisms have evolved, some of them changing the structure of the murein sacculus. An overview is given in the last section of this chapter.

The role of peptidoglycan

The main purpose of the peptidoglycan sacculus is to maintain bacterial shape and to counteract the internal pressure of the bacterial cell, which is approximately 3–5 atm in Gram-negative bacteria

(Koch, 1983) and up to 25 atm in Gram-positive bacteria (Kunin and Rudy, 1991). This is reflected by the thickness of the peptidoglycan sacculus. Experimental evidence suggests that the sacculus is mainly single layered in Gram-negatives while Gram-positives have up to 40 layers of peptidoglycan.

Peptidoglycan also serves as an anchor for proteins. In Gram-negatives the only protein known to be covalently attached to the peptidoglycan is Braun's lipoprotein (Lpp), which links the sacculus to the outer membrane (Braun and Rehn, 1969; Braun, 1975). Approximately one-third of the Lpp is covalently bound to the α-carboxyl-group of meso-diaminopimelic acid (m-A_2pm) of the stem peptide by the ε-amino group of the Lys at the Lpp C-terminus (Braun and Wolff, 1970) (Fig. 1.1). The other two-thirds are freely associated with the outer membrane (Hirashima *et al.*, 1973; Braun, 1975). Covalent binding is achieved by an L,D-transpeptidase reaction catalysed by three different proteins: ErfK, YcfS and YbiS, of which the latter seems to convey the main transpeptidase activity (Magnet *et al.*, 2007). During stationary growth the abundance of the bound form increases (Pisabarro *et al.*, 1985).

In Gram-positive bacteria, proteins, capsular polysaccharides (discussed in detail in the chapters 'Regulation of Capsule expression in Bacteria' and 'Extracellular polysaccharides and biofilms in oral streptococci'), and teichoic acids are covalently and non-covalently associated with peptidoglycan. These molecules are responsible for bacteria–host interactions and virulence. The covalent attachment of proteins is mediated by sortases (Mazmanian *et al.*, 1999), which recognize a specific cell wall sorting signal (CWS) located in the C-terminus of the attached protein (Marraffini *et al.*, 2006; Dramsi *et al.*, 2008). *S. aureus* contains two different sortases: SrtA, recognizing the CWS 'LPXTG', anchors at least 21 proteins to peptidoglycan including protein A (Spa), fibronectin-binding proteins (Fnbp) A and

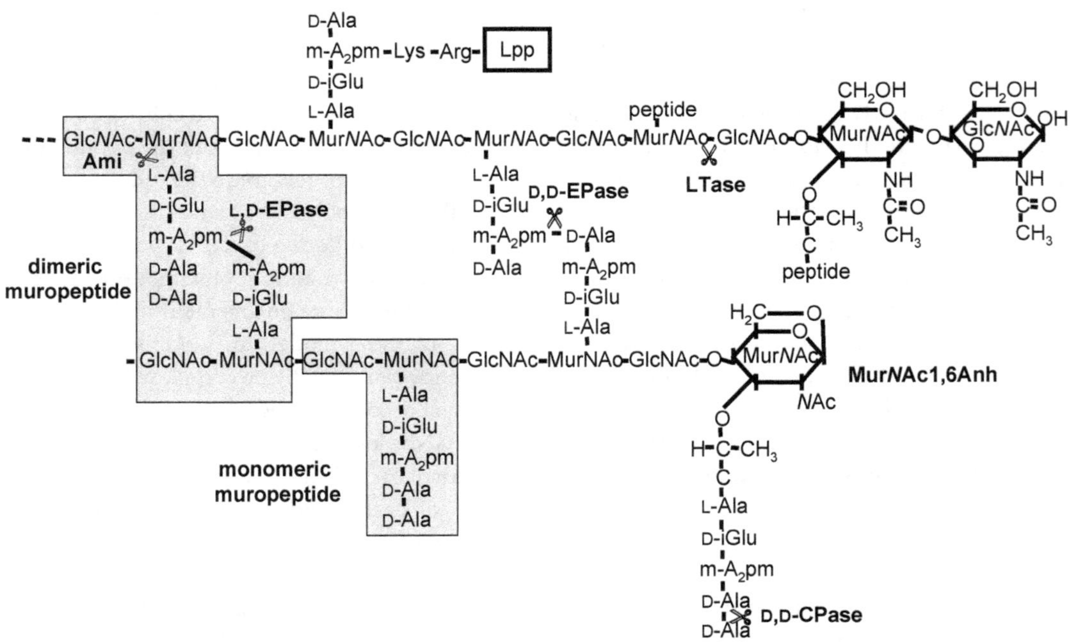

Figure 1.1 The murein structure of Gram-negative bacteria (*E. coli*). In Gram-negative bacteria, the peptide moieties are directly cross-linked either between m-A_2pm and D-Ala (D,D-cross-linkage) or between two m-A_2pm (L,D-cross-linkage). In addition, the surface protein Braun's lipoprotein (lpp) can be bound to m-A_2pm. The ✂ shows the cleavage sites of the murein hydrolases. Cleavage by LT results in the formation of an intramolecular 1,6-anhydro-bond (1,6Anh) at the Mur*N*Ac. The two grey boxes highlight a monomeric and a dimeric muropeptide structure respectively. LT, lytic transglycosylase, CP, carboxypeptidase, EP, endopeptidase, Ami, *N*-acetylmuramyl-L-alanine amidase.

B, clumping factor (Clf) A and B, and collagen adhesion protein (Cna), all of which are responsible for the manifestation of infections. SrtA directly anchors the proteins to the murein precursor molecule lipid II in a two-step transacetylation reaction, thus forming an amide bond between threonine (Thr) of the LPXTG-motif and glycine (Gly) at position five of the pentaglycine-bridge (Fig. 1.2). In many Gram-positive bacteria this pathway is universal (Marraffini *et al.*, 2006). The second sortase of *S. aureus* is SrtB, whose only substrate is the NPQTN-containing protein IsdC (Mazmanian *et al.*, 2002). This iron-uptake protein is attached to non-cross-linked Gly_5 of mature peptidoglycan by an amide bond between Thr and Gly (Marraffini and Schneewind, 2005).

In other Gram-positive bacteria sortases of the C-family polymerize fimbriae (Wu and Fives-Taylor, 2001) and pili (Ton-That and Schneewind, 2003) and anchor them to the murein sacculus (Dramsi *et al.*, 2008). Sortases of the D-family play a role in developmental processes, e.g. during sporulation of *Bacillus anthracis* (Marraffini and Schneewind, 2006) and mycelium formation in *Streptomyces coelicolor* (Claessen *et al.*, 2003; Elliot *et al.*, 2003). The covalent amide bond is always formed between the Thr and the Gly_5 (Dramsi *et al.*, 2008).

Chemical and physical features of the cell wall

The thickness of the peptidoglycan sacculus

The peptidoglycan sacculus can be isolated as a whole and reflects the shape of the bacterium. Isolated sacculi were studied in detail, focussing on their thickness, elasticity, molecular composition, and the length of the glycan chains.

In *Escherichia coli*, a representative of the Gram-negative bacteria, the thickness of the peptidoglycan sacculus was determined to be between 1.5 nm and 15 nm depending on the procedure applied. However, these measurements might be influenced by dehydration and fixation, or they might just represent the line of the contrasting metal (Vollmer and Höltje, 2004; Vollmer *et al.*, 2008a). Small angle-neutron scattering, which

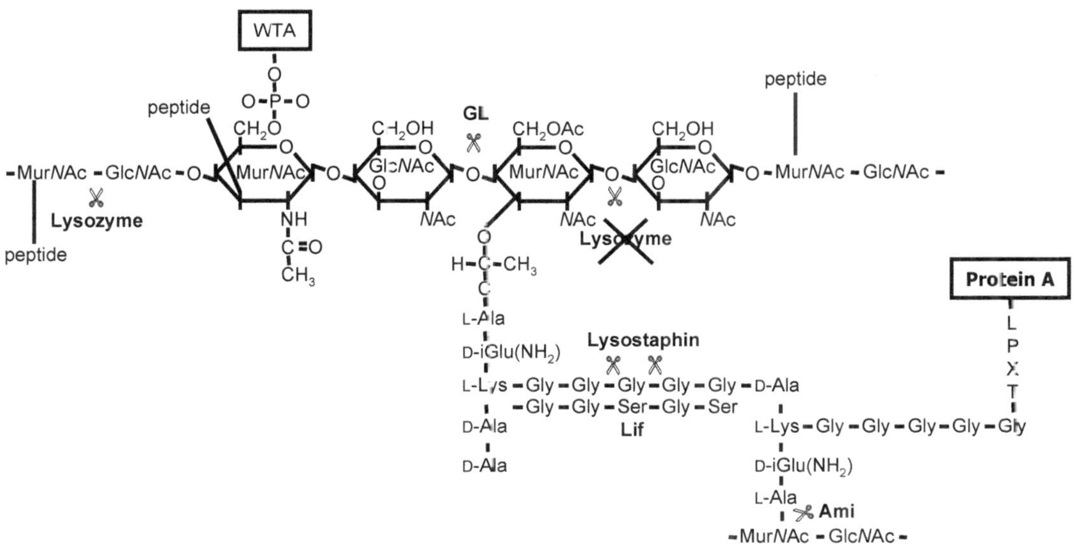

Figure 1.2 The murein structure of Gram-positive bacteria (*S. aureus*). The peptide moieties of Gram-positive glycan strands are cross-linked indirectly via interpeptide bridges. They consist of five Gly residues protruding from the L-Lys to D-Ala. In lysostaphin resistant strains two Gly residues are exchanged by Ser due to Lif (lysostaphin immunity factor) activity. Surface proteins such as protein A are covalently bound to the pentaglycine bridge via their LPXTG-motive by sortases. The wall teichoic acid (WTA) is covalently bound to some C_6-atoms of Mur*N*Ac. Strains carrying an O-acetyl group (OAc) at position C_6 of Mur*N*Ac are lysozyme resistant. The ✂ symbol shows the cleavage sites of the murein hydrolases. GL, *N*-acetylglucosaminidase, Ami, *N*-acetylmuramyl-L-alanine amidase.

allows the measurement of fully hydrated purified peptidoglycan sacculi without staining, showed that peptidoglycan thickness is not uniform in cells of the exponential phase. Rather, 20 to 25% of the surface was up to 7 nm thick, indicating a triple layer, while the remaining 75% was only 2.5 nm thick, probably reflecting a single layer (Labischinski *et al.*, 1991). Atomic force microscopy (AFM) determined rehydrated sacculi of *E. coli* to be about 6 nm thick, while sacculi of *Pseudomonas aeruginosa* were found to be only half as thick (Yao *et al.*, 1999). In Gram-positive bacteria the peptidoglycan sacculus is assumed to be multilayered and up to 50 nm in thickness (Jordan *et al.*, 2008).

A 'periplasmic space' in Gram-positive bacteria

Cryo-transmission electron microscopy (cryo-TEM) was utilized to study the cell wall of Gram-positive bacteria. This revealed an 'inner-wall-zone' (IWZ) or periplasmic space of low density (LD) and an 'outer-wall-zone' (OWZ) of higher density (HD) in *Staphylococcus aureus*. This 'periplasmic space' is presumably composed of soluble components. The thickness of the OWZ, containing the peptidoglycan and teichoic acids, was determined to be between 15 and 30 nm thick, depending not only on the species but also on the growth phase and growth conditions (Matias and Beveridge, 2006). At the septum five alternating zones were observed: two HD zones sandwiched between three LD zones with a tight contact between the HD zones and the membrane of the tip of the septum. It is most likely that the new peptidoglycan is synthesized here and inserted into the existing cell wall by murein synthases and murein hydrolases. These could act in a multienzyme complex (Matias and Beveridge, 2007) as was proposed for *E. coli* (Höltje, 1996) (see *Elongation and division of the peptidoglycan sacculus*). The newly synthesized peptidoglycan at the septum seems to be split in halves, but is still connected by a LD zone bridge, protecting the newly synthesized septal murein against the autolytic activity of the murein hydrolases working at the tip of the septum. Once the septum is closed and the cell separated the bridge thins out and the HD zones thicken to withstand the internal turgor pressure (Matias and Beveridge, 2007).

The chemical fine structure of peptidoglycan

The chemical fine structure of the peptidoglycan of various organisms has been studied in great detail. Peptidoglycan consists of glycan strands composed of alternating β-1→4-linked *N*-acetylglucosamine (Glc*N*Ac) and *N*-acetylmuramic acid (Mur*N*Ac), which are cross-linked by short peptides. In the context of peptidoglycan one unit of Glc*N*Ac and Mur*N*Ac is defined as the disaccharide (DS) (Rogers *et al.*, 1980). Digestion of peptidoglycan with a muraminidase such as lysozyme results in muropeptides. These are peptides of different length and composition, which are covalently linked to DS (Fig. 1.1). As muraminidases hydrolyse only the glycosidic bond but does not affect the peptide bond between two stem peptides, these peptides remain cross-linked, thus resulting in polymeric compounds. The muropeptide composition of different *E. coli* and *S. aureus* strains has been analysed by HPLC under various conditions (Glauner, 1988; Glauner *et al.*, 1988; Kohlrausch and Höltje, 1991; de Jonge *et al.*, 1992). An overview of the muropeptide composition, as well as the degree of cross-linkage is given in Table 1.1. A typical feature of the murein of Gram-negative bacteria is disaccharides without a reducing end group. Instead, the muramic acid residue possesses an internal 1 to 6 anhydro linkage (Mur*N*Ac16Anh), which is formed by the action of lytic transglycosylases. These enzymes catalyse an intramolecular glycosyl transferase reaction instead of transferring the glycosyl moiety onto H_2O (Thunnissen *et al.*, 1995). These anhydro-groups are found only at the end of the glycan chains (Höltje *et al.*, 1975) and account for 3–6% of all muropeptides (Glauner *et al.*, 1988). So far it is clear neither when these anhydro bonds are formed nor what they are used for. They could either be formed at the stage of lipid I (see *Formation of lipid II*) or be the result of the action of lytic transglycosylases on continuously polymerized glycan strands. The anhydro bond might be a specific marker for murein turnover and induces β-lactam resistance (Höltje, 1998) (see *Resistance against antibiotics*).

Another Gram-negative feature are compounds containing an additional Lys-Arg residue at the stem peptide, which shows that they were

Table 1.1 Muropeptide composition[1] of *E. coli* and *S. aureus*

E. coli W7 (calculated from Kohlrausch and Höltje, 1991)			*S. aureus* RN2677 (methicillin-susceptible strain; MSSA) (de Jonge *et al.*, 1992)		
Grown with penicillin G	–	641 µg/ml	Grown with methicillin	–	0.06 µg/ml
Monomers	49.6	45.9	Monomers	19.7	28.4
Ala-Glu-m-A_2pm	4.2	9.8	Ala-Glx-Lys-Ala-Ala	0.6	0.5
Ala-Glu-m-A_2pm-Ala	37.9	28.5	Ala-Glx-Lys-(Gly_4)-Ala-(Gly_2)	5.7	5.0
Ala-Glu-m-A_2pm-Ala-Ala	0.19	0.12	Ala-Glx-Lys-(Gly)-Ala-Ala	1.9	2.1
Ala-Glu-m-A_2pm(-Lys-Arg)	4.5	4.5	Ala-Glx-Lys-(Gly_5)-Ala-Ala	8.9	17.3
			Ala-Glx-Lys-(Ala)-Ala-Ala	1.7	2.1
			Ala-Glx-Lys-(Ala-Gly_4)-Ala-Ala	0.9	1.4
Dimers	43.3	47.1	Dimers	17.4	23.0
Tetra-Tetra	31.5	23.9	Gly_5-interpeptide bridge	1.7	1.3
Tetra-Tri-m-A_2pm	0.6	3.5	Gly_5-interpeptide bridge + free Gly_5	14.8	19.8
			Gly_5-interpeptide bridge + free Ala-Gly_4	0.9	1.9
Trimers	6.5	6.5	Trimers	10.6	11.7
			Tetramers	7.2	6.4
Monoanyhdro-compounds	3.8	6.7	Pentamers	4.4	3.7
Dianhydro-compounds	0.07	0.24	Hexamers	3.9	3.0
			Heptamers	3.6	2.3
			Octamers	2.9	1.7
			Nonamers	2.5	1.3
			> Nonamers	17.8	8.1
Degree of cross-linkage	26.7	28.5	Degree of cross-linkage	60.5	48.1
D,D-cross-linkage	24.1	23.0			
L,D-cross-linkage	1.0	4.1			

[1]Muropeptide composition was determined by HPLC. The values are area percentages. For *E. coli*, only the most prominent compounds are considered. For *S. aureus* only those compounds with a definitely assigned muropeptide structure are presented and used for calculation of the degree of cross-linkage, which was calculated as follows: $(1/2 \times \Sigma\text{dimers} + 2/3 \times \Sigma\text{trimers} + 9/10 \times \Sigma\text{oligomers})/\Sigma(\text{dimers} + \text{trimers} + \text{oligomers})$. The description of the muropeptides reflects the composition of the stem peptide. Amino acids branching off the stem peptide are denoted in parentheses.

previously bound to Braun's lipoprotein (Braun, 1975) (Fig. 1.1). Like all cell wall attached proteins, Braun's lipoprotein has been enzymatically cleaved off by trypsin during murein preparation for analysis.

Features of the glycan strands

The length of the glycan strands covers a wide range. In *E. coli,* the mean value lies between 25 and 35 DS units. However, approximately one-third of the glycan strands are actually up to over 80 DS long (Harz *et al.*, 1990). Glycan strands of *S. aureus* mainly consist of 3–10 DS units. However, longer chains with a length of 23–26 DS units have been discovered (Boneca *et al.*, 2000). Pathogenic staphylococci show an additional acetyl-group on position C6-OH of about 40% of the MurNAc, rendering them resistant towards lysozyme, an enzyme which serves as one of the first host defence mechanisms (Bera *et al.*, 2005; Bera *et al.*, 2006).

The DS units of the glycan strands probably form a helical structure with the stem peptides protruding outwards in all directions and forming angles of about 90° to each other (Labischinski *et al.*, 1985; Koch, 2000). However, a recent NMR study using the smallest possible unit, a tetrasaccharide, came to the conclusion that the periodicity per helix turn is only three DS units (Meroueh *et al.*, 2006).

Arrangement of the glycan strands

It is still a matter of debate how the glycan strands are actually arranged. If one pictures the glycan strands as running around the bacterial cell horizontally (the layered model) and assumes four DS units per helix, this would suggest that every second stem peptide lies in the same plane and would be able to form cross-links with the stem peptide of an adjacent glycan strand. (Verwer *et al.*, 1978, 1980) This is in accordance with the cross-linkage observed: in *E coli* the mainly single-layered sacculus is 40% and 60% cross-linked depending on the strain and growth conditions (Glauner *et al.*, 1988; Quintela *et al.*, 1995). On the other hand, in Gram-positive *S. aureus*, which contains interpeptide bridges added to the stem peptide, the stem peptides can protrude further than only to the next adjacent glycan strand. In fact, cross-linking in this organism is up to 90%. Cross-linking can also occur between stem peptides of different layers (Wilkinson, 1997). However, this also depends on the strain and growth conditions.

The peptidoglycan sacculus is not a rigid structure, but can actually shrink and expand. Isolated sacculi can be stretched up to 300% without damage (Koch and Woeste, 1992), and they are two to three times more elastic in the long axis than in the short axis (Yao *et al.*, 1999). This is in accordance with the above described distribution of the fixed glycan strands (Labischinski *et al.*, 1979) and flexible peptides (Barnickel *et al.*, 1979).

As suggested by Koch, the planar glycan strands of *E. coli* follow a zig-zagged line rather than a straight one. This would result in the formation of a hexagonal structure – the so-called 'tessera' – as the smallest pore (Koch, 1998). However, the murein sacculus in not perfect, meaning that in *E. coli* sometimes less than 50% of all stem peptides are part of cross-links. The presence of short glycan strands results in pores, which are larger than the average size of a tessera. Actually, in a living cell there must be pores within the expanded sacculus which are permeable for proteins of up to 100 kDa to allow the transport of proteins and protein complexes, which are inserted into the outer membrane. The murein sacculus of *E. coli* can therefore be regarded as a molecular sieve which keeps everything over 100 kDa within the cell (Vazquez-Laslop *et al.*, 2001).

The mean number of glycan chains per *E. coli* cell (2–4 μm long and 0.75–1 μm in diameter) is about 1.2×10^5 with about 30 DS units each, resulting in 3.5×10^6 single DS units (Wientjes *et al.*, 1991) with a length of 1 nm and a width of 1.1 nm. The length of a peptide bridge is about 4.1 nm (Dmitriev *et al.*, 2003). Based on these numbers, 70–120 glycan chains are necessary to completely surround the cell once, and 500 to 1000 parallel glycan chains will cover the whole length of the cell (Vollmer and Höltje, 2004).

However, another model for the peptidoglycan structure was proposed a number of years ago. This model envisages the glycan strands as oriented vertical to the cell surface and is known as the 'Scaffold Model' (Dmitriev *et al.*, 1999; Dmitriev *et al.*, 2003; Dmitriev *et al.*, 2004). However, this model hardly agrees with most of the above mentioned data. Particularly the length of the glycan chains of *E. coli* cannot be fitted into this model. A detailed discussion of this question was written a while ago (Vollmer and Höltje, 2004). The scaffold model might be more easily applied to *S. aureus*, which has shorter glycan chains and a thicker peptidoglycan sacculus.

Nevertheless, to date it has not been possible to obtain an actual picture of the fine structure of the murein sacculus, even though much research was devoted to this issue. Atomic force microscopy with *S. aureus* cells showed an old cell wall with a loosely arrayed network containing low-bond fibrils and holes, resembling a gel-like matrix. However, it was impossible to determine the orientation of the glycan strands (Touhami *et al.*, 2004). NMR studies on *E. coli* cells confirmed that glycan strands are more rigid than the peptide moieties. An interaction between added proteins and glycan chains was also found, but no conclusion on the peptide structure could be formed (Kern *et al.*, 2008).

Biosynthesis of the peptidoglycan sacculus

Formation of lipid II

The biosynthesis of peptidoglycan begins in the cytoplasm with the formation of lipid II, the substrate for the murein synthases (Fig. 1.3). This pathway involves 12 consecutive steps. Peptidoglycan biosynthesis is essentially the same in both Gram-negative and Gram-positive bacteria. Starting with fructose-6-P, this sugar is converted into UDP-Glc*N*Ac and further into UDP-Mur*N*Ac. UDP-Glc*N*Ac is also part of other cell wall macromolecules namely lipopolysaccharides and the enterobacterial common antigen in Gram-negatives (see chapter on 'Synthesis and membrane translocation of lipid-linked precursors of cell-surface polysaccharides') as well as teichoic acids in Gram-positives (discussed in detail in chapter 'Composition and functional role of polysaccharides in Gram-positive biofilm infections). UDP-Mur*N*Ac serves as substrate for the enzymes responsible for the addition of the stem peptide, which consists of five amino acids. Highly specific cytoplasmic ADP/peptide-forming enzymes (MurC to F), which require ATP/Mg^{2+}, catalyse stem peptide formation (van Heijenoort, 2001).

Even though there are some variations in the composition of the stem peptide depending on the organism, it usually consists of L-Ala bound as the first amino acid to the lactyl group of UDP-Mur*N*Ac. Some species, however, contain Gly or L-Ser instead of L-Ala. The second compound is D-iGlu, which in most Gram-positives and some mycobacteria is later converted into D-iGln (Vollmer *et al.*, 2008a). In bacteria there are two sorts of enzymes capable of D-Glu formation (van Heijenoort, 2007): a cofactor-independent glutamate racemase, which converts L-Glu into D-Glu and in the case of *E. coli* absolutely requires the presence of UDP-Mur*N*Ac-L-Ala (Doublet *et al.*, 1994; Doublet *et al.*, 1996), and a D-amino acid aminotransferase, which synthesizes

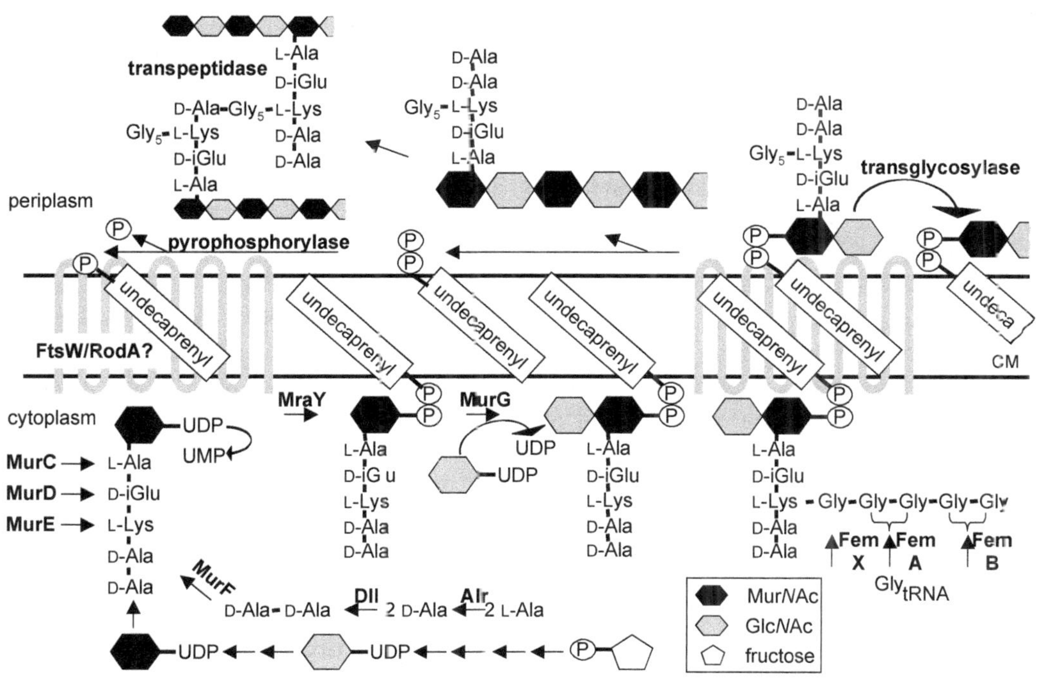

Figure 1.3 Murein biosynthesis of *S. aureus*. Cell wall biosynthesis begins in the cytoplasm with the sequential addition of amino acids of the peptide moiety to UDP-activated Mur*N*Ac. This precursor is then translocated to undecaprenol forming lipid I. Transfer of Glc*N*ac forms lipid II, which is completed by a pentapeptide bridge synthesized by the Fem proteins. This molecule is then translocated across the cytoplasmic membrane by an unknown mechanism. Within the periplasm, glycan strands are polymerized by transglycosylases and adjacent glycan strands are cross-linked by transpeptidases.

D-amino acids from a ketoacid precursor with D-Ala as amino donor (Tanizawa *et al.*, 1989; Fotheringham *et al.*, 1998). It was proposed that D-Glu is mainly formed by a glutamate racemase while the transaminase reaction provides D-amino acids for secondary metabolites, such as the exopolypeptides in some bacilli (van Heijenoort, 2007).

The α-carboxyl-group of the second amino acid forms an amide bond with the ε-amino-group of the third amino acid, which is always a diamino acid. In most Gram-negatives as well as in Bacilli and Mycobacteria it is meso-2,6-diaminopimelic acid (m-A_2pm), while Gram-positives mostly contain L-Lys at position three. Position three is the most variable amino acid within the stem peptide. The amino acids of position four and five are always added as a dipeptide. In all bacteria the fourth amino acid is D-Ala, as is the fifth amino acid in most species. Exceptions are *Enterococcus gallinarum* (D-Ser), *Lactococcus casei* and enterococci with an acquired resistance against vancomycin, which contain D-Lac at position five (Vollmer *et al.*, 2008a) (see also *Control of peptidoglycan biosynthesis by multienzyme complexes*). D-Ala is synthesized by an alanine racemase from L-Ala (Alr) (Wood and Gunsalus, 1951). This enzyme contains two negatively charged residues (Asp 164 and Glu 165) and two positively charged residues (Arg 280 and Arg 300), which have recently been shown to be of importance for substrate orientation (Wu *et al.*, 2008). The racemase reaction is followed by D-Ala-D-Ala synthesis catalysed by the specific ATP-dependent amide bond forming ligase(s) Ddl (Ito and Strominger, 1962; Neuhaus, 1962). This dipeptide is then added to the third amino acid by Mur F (van Heijenoort, 2001). This crucial part of lipid II biosynthesis can be inhibited by the antibiotic D-cycloserine, whose cytotoxic target was recently suggested to be the Ddl-enzyme (Halouska *et al.*, 2007).

The resulting UDP-Mur*N*Ac-pentapeptide is a hydrophilic molecule, which cannot be transported across the cytoplasmic membrane. Therefore, the next steps are synthesized at the cytoplasmic membrane, where UDP-Mur*N*Ac-pentapeptide is coupled to undecaprenyl phosphate (C_{55}-P) (Anderson *et al.*, 1965; Struve and Neuhaus, 1965) by the transferase MraY, a transmembrane protein (Ikeda *et al.*, 1991). This results in the formation of lipid I. MurG, a second translocase (Salmond *et al.*, 1980) transfers the Glu*N*Ac moiety of UDP-Glc*N*Ac to lipid I (C4 hydroxyl group), thereby forming a β-1→4-linked disaccharide called lipid II, the precursor for peptidoglycan biosynthesis (Bouhss *et al.*, 2008).

Most Gram-positives and some Gram-negatives contain so called interpeptide bridges, which are bound to the third amino acid of the stem peptide and consist of one to seven amino acids, such as Gly, L-Ala, L- or D-Ser, D-Asn, D-Asp, L- or D-Glu. This bridge is synthesized by the so called 'branching enzymes' (Vollmer *et al.*, 2008a). The function of these enzymes has been investigated in recent years. For example, Fem-transferases (factor essential for the expression of methicillin resistance) incorporate glycine and L-amino acids from aminoacyl-tRNAs into the interpeptide bridge (Berger-Bächi *et al.*, 1989; Mainardi *et al.*, 2008). D-amino acids, on the other hand, are inserted as activated acyl phosphates by proteins of the ATP-grasp family, which for example catalyse reactions such as the ATP-dependent ligation of a carboxyl group to an amino or imino nitrogen (Galperin and Koonin, 1997). Which precursor is used for the addition of the interpeptide bridge depends on the bacterial species. In *S. aureus* lipid II is the only substrate for the branching enzymes. FemX inserts the first Gly, Fem A the Gly residues 2 and 3. Residues 4 and 5 are then added by FemB (Schneider *et al.*, 2004). On the other hand UDP-Mur*N*Ac-pentapeptide is the substrate for FemX of *Weisella viridescens* while *Enterococcus faecalis* FemX uses both lipid II and UDP-Mur*N*Ac-pentapetide (Vollmer *et al.*, 2008a).

Peptidoglycan biosynthesis takes place on the outside of the cytoplasmic membrane in a compartment called the periplasm, which describes the region between the inner (cytoplasmic) membrane and the outer membrane in Gram-negative bacteria. The existence of a periplasmic space was recently also proposed for Gram-positives and was suggested to consist mostly of soluble low-density molecules located between the cytoplasmic membrane and the multilayered peptidoglycan (Matias and Beveridge, 2006). Therefore, lipid II, the substrate of the murein synthases must be transported across the cytoplasmic membrane. The enzyme(s) catalysing this reaction are not known. However, two proteins have been proposed as acting as so called 'flippases': FtsW and RodA, both containing 10 membrane-spanning regions (Lara and Ayala, 2002). Both proteins belong to the SEDS-family, proteins for 'shape, elongation, division and sporulation' (Ishino *et al.*, 1982; Ikeda *et al.*, 1989; Henriques *et al.*, 1998). In *E. coli* the gene for FtsW lies within the same operon as the murein synthase penicillin-binding protein (PBP)3 gene, also known as FtsI (Hara *et al.*, 1997) and the genes for RodA and PBP2 are also adjacent to each other (Asoh *et al.*, 1983). This suggests a role for RodA during cell elongation and FtsW during cell division, as PBP2 and PBP3 are specific for elongation and division respectively (Spratt, 1975; Uehara and Park, 2008). Co-transcription of FtsW and PBP3 has also been shown for other organisms (Margolin, 2000). However, in *S. aureus* none of the RodA/FtsW homologues is located near one of the known PBPs.

Synthesis of cross-linked glycan strands

The polymerization of glycan strands is catalysed by transglycosylases (TG), which are either monofunctional proteins like MtgA of *E. coli* (Di Berardino *et al.*, 1996) and MgtA of *S. aureus* (Wang *et al.*, 2001), or are part of the bifunctional transglycosylase/transpeptidase (TG/TP) penicillin-binding proteins. The transpeptidase domain is responsible for cross-link formation between the peptide side chains of adjacent glycan strands. These proteins were named after their binding function to beta-lactams such as penicillin, which inhibits TP activity. The β-lactam moiety has structural similarity to the D-Ala-D-Ala of the stem peptide. Its binding to the TP domain leads to the irreversible formation of a penicillinoyl–enzyme complex.

Monofunctional class B PBPs lacking a TG domain also exist in addition to these class A PBPs. Both belong to the high-molecular-weight (HMW) PBPs. Furthermore, there are several low-molecular- weight (LMW) PBPs, which are mainly murein hydrolases, while some function as β-lactamases. All these enzymes are membrane bound proteins, the active sites of which are located within the periplasm (Ghuysen, 1991; Goffin and Ghuysen, 1998; Sauvage *et al.*, 2008; Vollmer and Bertsche, 2008). The nomenclature of the PBPs existing in different species is quite confusing, as they were named according to their size after separation by SDS-PAGE. This means for example, that PBP2 in *E. coli* is a monofunctional transpeptidase while PBP2 of *S. aureus* is the only bifunctional PBP present in this species. An overview of all known PBPs in these two species and their amounts per cell is given in Table 1.2.

From crystal structures of PBP2 of *S. aureus* (Lovering *et al.*, 2007) and PBP1A of *Aquifex aeolicus* (Yuan *et al.*, 2007) as well as from experiments with various substrates (Fraipont *et al.*, 2006; Perlstein *et al.*, 2007), it was concluded that the growing glycan chain acts as the glycosyl donor and lipid II is the glycosyl acceptor. This reaction is catalysed by deprotonation of the 4-OH nucleophile of GlcNAc by a glutamate residue within the active site of the TG domain. A second glutamate stabilizes the leaving lipid moiety (Sauvage *et al.*, 2008). New glycan strands are synthesized as rather long polymers (50–60 DS) and are then rapidly processed to a length of about 25 to 30 DS (Glauner and Höltje, 1990).

The penicillin-binding module (PB-module) harbours three conserved motifs specific for acyltransferases: motif [S*X_2K] containing the active site Ser residue (S*), motif II [(S/Y)X(N/C)] and motif III [(K/H)(T/S)G] with X being any amino acid (Goffin and Ghuysen, 1998; Goffin and Ghuysen, 2002; Vollmer and Bertsche, 2008).

During the transpeptidation reaction, the D-Ala-D-Ala-bond of the donor peptide is first cleaved by a PBP enzyme, thus forming an enzyme-substrate intermediate between the conserved Ser-residue of the SX_2K-motif and the NH-group of the D-Ala on position 4. Thereby, the terminal D-Ala is released. In a second step, a new peptide bond is formed between the peptidyl moieties of the donor and the acceptor (Terrak *et al.*, 1999). In *E. coli* the last D-Ala from stem peptides which have not been cross-linked is released by PBP5, a D,D-carboxypeptidase (Broome-Smith *et al.*, 1988). Deletion of the encoding gene results in a branched phenotype (de Pedro *et al.*, 2003b) and the accumulation of pentapeptides within the murein sacculus.

Table 1.2 Murein synthases of *E. coli* and *S. aureus*

Class	Protein	MW [kDa]	Gene	Intracellular localization	No. per cell[1]	Function	Reference
E. coli (MC4100)							
	MtgA	27	*mgt* (*mtgA*)	Septation site (Derouaux *et al*., 2008)	ND	TG	Di Berardino *et al*. (1996)
HMW A	PBP1A	95	*ponA* (*mrcA*)	ND	221 ± 20	TG/TP	Ishino *et al*. (1980), Broome-Smith *et al*. (1985)
	PBP1B	93	*ponB* (*mrcB*)	Lateral wall and septation site (Bertsche *et al*., 2006)	127 ± 13	TG/TP	Nakagawa *et al*. (1984)
	PBP1C	70	*pbpC*		ND	TG/TP	Schiffer and Höltje)1999)
HMW B	PBP2	67	*pbpA* (*mrdA*)	Lateral wall and septation site (den Blaauwen *et al*., 2003)	120 ± 14	TP for cell elongation	Asoh *et al*. (1983), Ishino *et al*. (1986)
	PBP3	60	*ftsI* (*pbpA*)	Septation site (Weiss *et al*., 1997)	132 ± 17	TP for cell division	Ishino and Matsuhashi (1981)
S. aureus (RN450M, methicillin-resistant *S. aureus* (MRSA) strain)							
	Mgt		*mgt*	ND	ND	TG	Terrak and Nguyen-Distèche (2006)
HMW A	PBP2	73	*pbpB* (*pbp2*)	Septation site (Pinho and Errington, 2005)	449 ± 97	TG/TP	Murakami *et al*. (1994), Hackbarth *et al*. (1995)
HMW B	PBP1	80	*pbpA*	Septation site (Pereira *et al*., 2007)	166 ± 35	TP	Pucci *et al*. (1997)
	PBP3	70	*pbpC*	Unknown	197 ± 60	TP	Pinho *et al*. (2000)
	PBP2A or PBP2'	74	*mecA*	Unknown	824 ± 379	TP for β-lactam resistance	Hartman and Tomasz (1984), Song *et al*., 1987)
LMW (TP)	PBP4	46	*pbpD* (*pbp4*)	Unknown	299 ± 61	TP for secondary cross-linking	Domanski and Bayles (1995)

[1]The number of molecules per cell has been determined by direct measurement (Dougherty *et al*., 1996; Pucci and Dougherty, 2002). TG, transglycosylase; TP, transpeptidase; HMW, high molecular weight; LMW, low molecular weight; ND, not determined.

The most common type of cross-linkage occurs between the diamino acid at position three, the acyl acceptor, and the D-Ala at position four, the acyl donor (Schleifer and Kandler, 1972; Vollmer *et al.*, 2008a). It can either be direct, as it is the case for most Gram-negatives (Fig. 1.2), or through the already mentioned interpeptide bridge (most Gram-positives, Fig. 1.3). In *Micrococcus luteus* an unknown enzyme is responsible for the detachment of a large portion of stem peptides [L-Ala-γ-(α-D-Glu-Gly)-L-Lys-D-Ala] from Mur*N*Ac. One or several of these detached stem peptides join 'head-to-tail' and form a peptide cross-bridge between positions 3 and 4 (Ghuysen *et al.*, 1968).

In phytopathogenic corynebacteria there is a further type of cross-linkage formed between the α-carboxyl-group of D-Glu at position 2 (acyl acceptor) and the carboxyl group of D-Ala at position 4 (acyl donor). This involves a diamino acid (e.g. ornithine) as interpeptide bridge (Vollmer *et al.*, 2008a). Cross-links of the 4→3 type are transformed into a 3→3 type by an L,D-transpeptidase during stationary growth of *Mycobacterium tuberculosis*. This might contribute to a 'dormant' state, which is able to survive antibiotic treatment for several months (Lavollay *et al.*, 2008). To a certain extend these L,D-cross-links are also formed between two m-A_2pm in *E. coli* (Glauner, 1988) especially under antibiotic stress (Kohlrausch and Höltje, 1991). (Fig. 1.2). After many years of research the two L,D-transpeptidases catalysing this reaction have been identified as YcbB and YnhG. These enzymes and the L,D-transpeptidases, which add Braun's lipoprotein to m-A_2pm, also appear to hydrolyse D-Ala and replace it with Gly (Magnet *et al.*, 2008).

Elongation and division of the peptidoglycan sacculus

Rod-shaped bacteria always divide through the same medial plane, while in some cocci such as *S. aureus* the division plane is consecutively rotated by 90°, thus resulting in three different perpendicular division planes (Giesbrecht *et al.*, 1998). During bacterial growth, the stress-bearing peptidoglycan sacculus has to be elongated (in the case of rod-shaped bacteria) and/or divided without loosing its strength. Especially in Gram-negative bacteria, where the peptidoglycan sacculus is mostly single-layered, this is hard to envisage. Several models have been proposed to describe this task. Most models propose the insertion of new glycan strands into the existing peptidoglycan sacculus by the combined action of peptidoglycan synthesizing enzymes (murein synthases) and murein degrading enzymes (murein hydrolases). *E. coli* contains at least 20 murein hydrolases, fulfilling different functions as *N*-acetylmuramyl-L-alanine amidases (Ami), lytic transglycosylases (LT), endopeptidases (D,D-EP and L,D-EP) and carboxypeptidases (D,D-CP and L,D-CP) (Vollmer and Bertsche, 2008; Vollmer *et al.*, 2008b) (Fig. 1.2). All of these enzymes are active within the cytoplasm. In addition, three murein hydrolases, which are specific for 1,6-anhydromuropeptides, exist. These are the N-acetylglucosaminidase NagZ (Yem and Wu, 1976; Cheng *et al.*, 2000; Vötsch and Templin, 2000), and the 1,6-anhydroMur*N*Ac-L-alanine amidases AmpD (Höltje *et al.*, 1994) and AmiD (Uehara and Park, 2007). NagZ and AmpD are located in the cytoplasm, while AmiD is a lipoprotein of the outer membrane (Uehara and Park, 2007). These enzymes degrade the murein turnover products caused by the activity of the lytic transglycosylases and endopeptidases into smaller molecules, which can then enter the murein recycling pathway.

Genome analysis proposed 18 different murein hydrolases for *S. aureus* (Vollmer *et al.*, 2008b), four of which have been characterized. The gene *atl* encodes for two proteins synthesized as one proprotein: *N*-acetylmuramyl-L-alanine amidase (Ami) and *N*-acetylglucosaminidase (GL) (Fig. 1.2). How this proprotein is processed into two mature proteins is still unknown (Oshida *et al.*, 1995). In addition, *S. aureus* contains an endopeptidase (EP) (Giesbrecht *et al.*, 1998) and AAA, a recently discovered autolysin/adhesin (Heilmann *et al.*, 2005).

Burman and Park proposed a model in which some of the cross-links are hydrolysed first, to give room for the insertion of new glycan strands by the formation of cross-links between old and new material. After a while, new material would also be added to recently incorporated glycan strands (Burman and Park, 1984). However, this model does not take the observed peptidoglycan turnover of about 50% of the entire murein content during each cell cycle into account (Goodell and Schwarz, 1985). Ninety per cent of the turnover products are recycled and used either for *de novo* peptidoglycan biosynthesis or for other synthesizing pathways (Goodell, 1985). This model is not in accordance with the 'make-before-break' strategy, which suggests that a safe enlargement can only be accomplished by the primary formation of new covalent bonds followed by the hydrolyses of other bonds (Koch, 1995).

A model which considers both was proposed by Höltje as the 'three-for-one' model (Höltje, 1993) (Fig. 1.4). For cell elongation it suggests the addition of three newly synthesized glycan strands underneath every second glycan strand of the peptidoglycan sacculus, thereby forming short-time trimeric cross-links with the free m-A_2pm. The old 'docking strand' is degraded by murein hydrolases simultaneously with the synthesis of the new triplet. If every second glycan strand acts as a docking strand the surface of the peptidoglycan sacculus would exactly double in length without loosing its integrity. Provided that the inserted triplet has the same length as the old 'docking strand' the cell diameter would remain the same if the glycan strands run perpendicular to the long cell axis. The triplet is thought to be synthesized starting from an already existing free glycan strand made earlier by monofunctional transglycosylases (Höltje, 1998). The newly synthesized murein is characterized by the existence of tetra-tetra cross-bridges resulting from the activity of D,D-carboxypeptidases. During maturation they are converted into tetra-tri cross-links by L,D-carboxypeptidases. This pulse of L,D-carboxypeptidases activity at the end of the elongation phase could function as a signal for the start of cell division (Glauner and Höltje, 1990). The expansion and concomitant insertion of the new peptidoglycan could be explained by the surface stress theory. The raising of the osmotic pressure within the cell would lead to an overall stretching of the existing sacculus and thereby pull the newly synthesized structures into the existing monolayer (Koch and Woeste, 1992) (see also *Control of peptidoglycan biosynthesis by multienzyme complexes*).

Cell division can also be explained by the same model. In this case all three strands of the new triplet would be synthesized simultaneously and the middle strand of the newly inserted triplet would serve as 'docking strand' for the next set and be immediately degraded. For the new strand to serve as a docking strand, the final D-Ala at position five must be cleaved off by a D,D-carboxypeptidase because the cell division-specific transpeptidase PBP3 seems to recognize tetra–tetra cross-links only (de Jonge *et al.*, 1989; Obermann and Höltje, 1994). The murein synthesizing enzyme complexes are pulled inward by the help of the proteins of the divisome (Höltje, 1998).

Pulse-chase experiments have indeed shown that there is new material incorporated into the side wall during elongation. However, while there are sites of homogenous incorporation there are also areas where growth seems to be achieved in patches and arcs. There is no incorporation of peptidoglycan into the existing cell poles, which are therefore called 'inert' (Koch and Woldringh, 1994; de Pedro *et al.*, 1997). During cell division peptidoglycan incorporation takes place primarily at the septation site (de Pedro *et al.*, 2003a). Most recent experiments show an immediate release of over 30% of the newly synthesized peptidoglycan during septal growth. This can be explained by the separation of the new poles of the two daughter cells (Uehara and Park, 2008).

Control of peptidoglycan biosynthesis by multienzyme complexes

During growth and division of the peptidoglycan sacculus the control of the murein synthases and especially the murein hydrolases, which can act as autolysins, is crucial. These two types of enzymes are thought to work in balance with each other, thereby ensuring the safe enlargement and division of the stress-bearing sacculus (Weidel and Pelzer, 1964). If this balance is disturbed, for example by the loss of precursors or the inactivation of the synthetic enzymes, the lytic actions dominate and the cells lyse. The best way to control these enzymes would be a multienzyme complex consisting of a synthetic and a lytic subcomplex. These complexes would enable the cell to spatially and temporarily control peptidoglycan biosynthesis (Höltje, 1996, 1998).

In recent years, evidence for the existence of these multienzyme complexes *in vivo* has increased. It has been shown that the

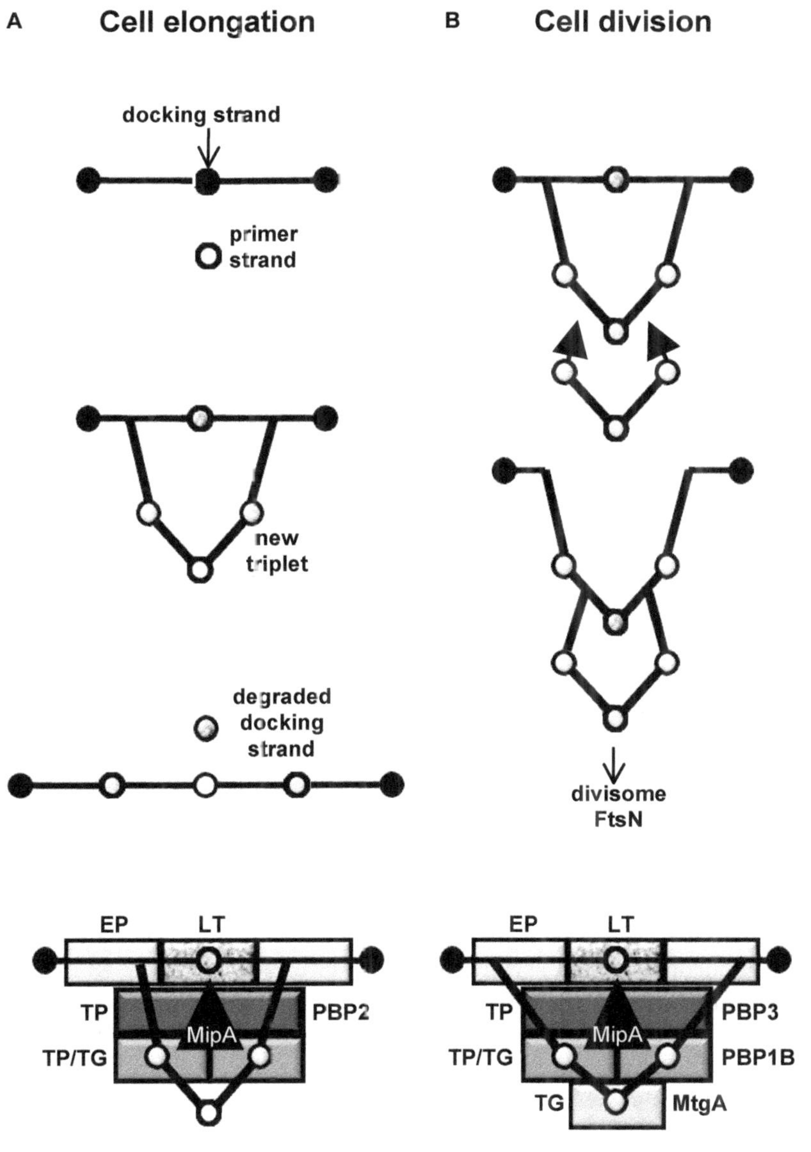

Figure 1.4 'Three-for-one' model and the peptidoglycan synthesizing multienzyme complexes in *Escherichia coli*. The model depicts murein growth following the make-before-break strategy. **A** During cell elongation, murein biosynthesis is thought to use a pre-existing primer strand. Two additional glycan strands are synthesized by the TG/TP activity of the multienzyme complex and cross-linked to both sides of the primer strand. The resulting triplet is attached to either side of the docking strand by the TP activity of PBP2. Pre-existing cross-links and the docking strand are degraded by the murein hydrolases of the multienzyme complex (EP and LT) and the new triplet is pulled into the existing sacculus. **B** During cell division all three glycan strands of the murein triplet are synthesized simultaneously by the TG activities of MtgA and PBP1B and cross-linked by the TP activity of PBP1B. This triplet is cross-linked to the peptides of both sides of the docking strand by the TP activity of PBP3. Again, the docking strand is degraded by murein hydrolases. The middle strand of the first triplet then serves as docking strand for the next triplet and so on. The murein biosynthesis complex is pulled inward by the divisome, another multienzyme complex containing FtsN and other enzymes, resulting in constriction of the sacculus. The glycan strands are shown as black circles, the primer strand as a white circle and the newly synthesized murein triplet as grey circles. The peptide bridges are shown as straight lines. EP, endopeptidase; LT, lytic transglycosylase; TP, transpeptidase; TG, transglycosylase; MipA, MltA-interacting protein A; PBP, penicillin-binding protein.

overproduction of inactive variants of PBP1B of *E. coli* results in cell lysis, while the overproduction of active variants has no effect. This suggests that the inactive variants of the protein are incorporated in a multienzyme complex with a non-functional synthetic subcomplex and a functional lytic subcomplex. This would cause the excision of glycan strands from the existing sacculus without the insertion of new peptidoglycan, or insertion of glycan strands which are not cross-linked. This would lead to the formation of nicks in the cell wall, so that cells would eventually lyse. In mutants carrying deletions in several of the genes encoding for murein hydrolases, lysis was strongly decreased. If such an enzyme complex containing synthases and hydrolases did not exist, these effects would be hard to explain. One could assume a limited number of peptidoglycan synthesis sites per cell, which would be blocked by the inactive PBP1B variants. Active synthases could no longer bind and murein biosynthesis is inhibited. However, if this would lead to lysis is questionable (Meisel *et al.*, 2003).

In rod shaped *E. coli*, two different complexes must exist: one containing the elongation-specific PBP2 and another harbouring the cell division-specific PBP3 for the attachment of the new triplet to the docking strand by a transpeptidase reaction. Both multienzyme complexes must also contain a dimer of bifunctional PBPs for synthesis of the two outer glycan strands and their addition to the middle strand. It was proposed that PBP1A plays this role for the elongation complex, while PBP1B seems to be part of the division complex (Höltje, 1998). Dimerization of PBP1B was confirmed (Zijderveld *et al.*, 1991, 1995). The suggestion that the elongation complex forms a triplet on a free glycan strand is in accordance with the finding that *in vitro* PBP1A preferentially cross-links a new glycan strand to an already existing one rather than cross-linking two newly synthesized strands (Born *et al.*, 2006). On the other hand, PBP1B immediately cross-links newly synthesized glycan strands *in vitro* but hardly uses existing glycan strands as a substrate for the TP reaction (Bertsche *et al.*, 2005). So far it has not been shown whether PBP1B is capable of acting together with the monofunctional transglycosylase MtgA. However, the latter has been shown to synthesize glycan strands at the septation site and to interact with PBP3 as well as with the cell division proteins FtsW and FtsN (Derouaux *et al.*, 2008). An *in vivo* and *in vitro* interaction could also be shown for PBP1B, PBP3 and FtsN (Bertsche *et al.*, 2006; Müller *et al.*, 2007). Both findings suggest that the multienzyme complex for peptidoglycan biosynthesis could be controlled by another multienzyme complex: the divisome. This is supported by the hypothesis that both the TG and the TP activity of PBP1B are positively influenced by the addition of FtsN during *in vitro* peptidoglycan synthesis (Müller *et al.*, 2007).

Interactions between murein synthases and murein hydrolases have also been reported, e.g. the interaction between PBP1B and the membrane bound lytic transglycosylase A (MltA) via a scaffolding protein called MltA interacting protein A (MipA) (Vollmer *et al.*, 1999). Affinity chromatography experiments with immobilized PBP1B and crude membrane extract revealed an additional murein hydrolase as an (in)direct interaction partner: AmiC (Müller *et al.*, 2007), which is also part of the divisome (Heidrich *et al.*, 2001; Bernhardt and de Boer, 2003).

Localization studies with immunofluorescence and GFP fusion proteins revealed the co-localization of PBP1B and PBP3 at the cell division site (Bertsche *et al.*, 2006). However, PBP1B is also distributed diffusely all over the side walls of *E. coli*. As expected for an elongation specific transpeptidase, GFP-PBP2 was shown to localize at the side wall, but was also seen at the septum during the beginning of constriction of the cell walls (den Blaauwen *et al.*, 2003). This suggests a role in the regulation of septation (Vinella *et al.*, 1993).

To date, no data on the other organism discussed in this chapter – *S. aureus* – are available. What has been shown is the localization of the bifunctional PBP2, which is essential for cell division, at the septation site in a ring-like manner (Pinho and Errington, 2005). The septation site is also the region where new peptidoglycan is mainly, if not only, incorporated (Pinho and Errington, 2003), although some additional inside-to-outside growth was proposed because of high rates of peptidoglycan turnover (Giesbrecht *et al.*, 1998). At the same site two murein hydrolases are also localized in a ring-like manner

(Yamada *et al.*, 1996; Sugai *et al.*, 1997). Latest electron microscopy studies suggest an immediate segregation of the new cell wall material into two parts: one for each new cell. This speaks in favour of a coordinated action between murein synthases and hydrolases in Gram-positive cocci (Matias and Beveridge, 2007). Multienzyme complexes containing PBPs have also been reported for other organisms. These include *Haemophilus influenzae* (Alaedini and Day, 1999), *Bacillus subtilis* (Simon and Day, 2000) and *Caulobacter crescentus* (Figge *et al.*, 2004).

Recently, it was discovered that rod-shaped bacteria contain a cytoskeleton, with MreB being the actin homologue (van den Ent *et al.*, 2001), which forms helical filaments coiling around the lateral wall of the cell underneath the inner membrane (Jones *et al.*, 2001). In *Caulobacter crescentus* MreB was shown to interact with PBP2, RodA, MreC, and itself (Figge *et al.*, 2004; Divakaruni *et al.*, 2005). The tubulin homologue FtsZ (Erickson, 1995; Erickson, 1998; Löwe and Amos, 1999) is part of the divisome, but was recently shown to also oscillate in helices through the cytoplasm of *E. coli* cells during early elongation phase (Thanedar and Margolin, 2004).

A proposed model for the transition between cell elongation and cell division is as follows: during cell elongation peptidoglycan synthesizing multienzyme complexes (containing PBP2) are either formed at the MreB filaments and then move away or they are guided along the lateral wall by these filaments. FtsZ stops oscillating and then localizes with other early cell division proteins (Aarsman *et al.*, 2005) at the future cell division site, well before constriction is visible. There, the elongation peptidoglycan synthesizing complex is redirected from MreB to FtsZ, leading to preseptal growth. This is in accordance with the fact that PBP2 is found at midcell (see Table 1.2). The complex has then to be remodelled into the division peptidoglycan synthesizing complex which contains MtgA, PBP1B and PBP3 and is controlled by the divisome. This multiprotein ring inserts new peptidoglycan at the leading edge of constriction, resulting in concentric circles, which get smaller and smaller. After completion of division, the FtsZ-ring complex disassembles and growth is redirected to cell elongation and the MreB filaments. (For a schematic model of the transition between cell elongation and cell division and the involved protein complexes see Vollmer and Bertsche, 2008.)

The influence of stress on the peptidoglycan sacculus

The peptidoglycan sacculus is the stress bearing structure of the bacterial cell. Experiments with isolated sacculi showed that they are quite elastic and can be expanded to 200–300% of their original size (Koch and Woeste, 1992). This high degree of elasticity might be the reason why most cell envelope stress responsive systems are not activated by osmotic shock (Morbach and Kramer, 2002; Jordan *et al.*, 2008). Nevertheless, some changes in peptidoglycan structure have been reported as being associated with high salt concentrations. Studies with *Lactobacillus casei* showed a decrease in oligomeric muropeptides and an increase in free amino groups. Both effects can be explained by a decrease in cross-linkage due to the fact that the genes encoding PBPs were either expressed at lower levels or had decreased activity under these conditions. This was shown by a decrease in binding affinity to penicillin G under high-salt conditions. A loss in activity could be shown for two of the three essential PBPs: PBP1 and PBP2a. In addition, cells grown under high-salt conditions were more susceptible to mutanolysin and antibiotics targeting to the peptidoglycan (Piuri *et al.*, 2005).

As antibiotic stress is the most common threat to the peptidoglycan sacculus, the rest of this chapter will deal with the counter measures developed by different bacteria. The cell wall-specific antibiotics, especially β-lactams, are of great danger to the integrity of the peptidoglycan sacculus. Therefore, several detoxification modules for sensing changes such as an increased turnover rate have evolved. These include the BceRS-/VanRS-like two component system (TCS) and BlaR1/Mec1. Peptidoglycan synthesis is also influenced by an increase in *pbp* gene expression through the action of TCS which is not directly regulated by antibiotics (Jordan *et al.*, 2008).

Cell wall stress stimulons

A very recent study in *E. coli* (Laubacher and Ades, 2008) revealed the importance of the Rcs phosphorelay for the response to murein damage.

This was the only stress response activated by all tested cell wall specific antibiotics at (near) MIC concentrations. Cefsoludin inhibits PBP1A and PBP1B, mecillinam inhibits PBP2, PBP3 is inhibited by aztreonam and ampicillin inhibits all high molecular weight PBPs as well as some low molecular weight PBPs. Antibiotics against other targets had only little effect on the Rcs regulon, indicating that the Rcs phosphorelay senses stress caused by cell wall damage rather than the antibiotics themselves. Rcs is also activated by osmotic stress and polymyxin B, a disorganizer of bacterial membranes. Therefore Rcs is most likely induced by alterations in the turgor pressure caused by inhibition of peptidoglycan synthesis. A second possibility would be the direct sensing of the peptidoglycan status.

Rapid lysis caused by the inhibition of PBP1A and PBP1B through cefsoludin results in a different gene expression pattern than the inactivation of PBP2 or PBP3, which results in slow lysis. The only up-regulated PBP in this study was the non-essential PBP6, encoded by *dacC*, which was slightly induced by cefsoludin/mecillinam. This indicates that the Rcs regulon, which is widely conserved among enteritic pathogens, has not specifically evolved to counteract damages in the peptidoglycan sacculus or to confer antibiotic resistance. Rather, it responds to stress occurring during the normal cell cycle, when the murein sacculus is elongated and divided. However, it also enhances growth in the presence of antibiotics of near MIC concentrations and therefore might allow bacteria to survive longer during antibiotic treatment (Laubacher and Ades, 2008).

A genome-wide transcriptional profiling of the *S. aureus* response to D-cycloserine, oxacillin (a β-lactam) and bacitracin, which inhibits recycling of the lipid carrier, suggests the existence of a 'cell-wall stress stimulon', including genes for cell wall metabolism (*pbpB*, *murZ*, *fmt*, and *vraS*) and stress response (e.g. *msrA*, *htrA*, *psrA*, and *hslO*). The effect of D-cycloserine is less pronounced than the effects of oxacillin and bacitracin, meaning the latter two cause an up-regulation which is about twice as much as for D-cycloserine. The increase in expression of *pbpB*, which encodes for PBP2, the only bifunctional PBP, and *murZ*, which catalyses one of the first steps in Lipid II-formation, might be an attempt to raise the rate of peptidoglycan biosynthesis in order to compensate for the damaged and partially missing cell wall. Fmt-proteins are methicillin resistance proteins while VraS is the histidine kinase sensor of a vancomycin associated two-component system (Utaida *et al.*, 2003).

Resistance against antibiotics

Many Gram-negative bacteria such as *Citrobacter freundii* and *Enterobacter cloacae* contain an inducible *ampC* β-lactamase gene, which is not native in *E. coli*. *AmpC* expression is regulated by AmpR (Lindquist *et al.*, 1993), a LysR type regulator with a DNA and a cofactor binding domain (Normark *et al.*, 1994). In the presence of UDP-MurNAc-pentapeptide, a signal for ongoing peptidoglycan biosynthesis, AmpR serves as a repressor. However, in the presence of β-lactams peptidoglycan turnover increases dramatically, thus resulting in high amounts of 1,6-anhydro-MurNAc-tripeptide (products of the lytic transglycosylases and other murein hydrolases of the recycling pathway) in the cytoplasm. Thereby the co-repressor is displaced from AmpR, resulting in the expression of *ampC* and finally the expression of *blaZ* (Jacobs *et al.*, 1997).

E. coli cells, which do not carry the gene for β-lactamase resistance *ampC*, have nevertheless found a way to strengthen the peptidoglycan sacculus when stressed by cell wall active antibiotics. Examination of the muropeptide composition of cells before treatment with antibiotics and after 75% of the cells had lysed, showed clear changes. Addition of D-cycloserine, which inhibits the formation of D-alanyl-D-alanine, penicillin G, an inhibitor of PBPs, and the TG inhibitor moenomycin revealed an increase in the degree of cross-linkage due to the formation of L,D-cross-links between two m-A_2PM (Fig. 1.2). This seems to be an attempt to replace the D,D-transpeptidase activities of the inhibited PBPs. However, there was also an increase in anhydro-compounds showing ongoing peptidoglycan hydrolysis (Kohlrausch and Höltje, 1991). An overview of the muropeptide composition of *E. coli* treated with penicillin G is given in Table 1.1. In addition to changes in the muropeptide pattern, the amount of glycan chains with only one DS increased four- to fivefold in the presence of moenomycin and penicillin G, respectively (Kohlrausch and Höltje, 1991).

While Gram-negatives sense the presence of β-lactams indirectly by the increase of peptidoglycan turnover products, resistant Gram-positives contain two sensor proteins within the cytoplasmic membrane: BlaR1 and MecR1. Upon binding of β-lactams, BlaR1 and MecR1 are cleaved autocatalytically, thus leading to the activation of a zinc peptidase domain. This domain then cleaves the repressor BlaI or MecI, respectively, and allows the transcription of *blaZ* and *mecA*, thus resulting in the production of β-lactamase and PBP2A respectively (Berger-Bächi and Rohrer, 2002). Transcription of *mecA* might also be influenced by a regulatory signal caused by changes in the transcription of *pbpB*. Decrease of PBP2 results in a decline of PBP2A (Gardete *et al.*, 2006). The stem peptide and the pentaglycine bridge of the staphylococcal peptidoglycan is also associated with methicillin-resistance. Shortening of this pentaglycine bridge by deletion of *femA* or *femB* results in a loss or reduction of resistance (Henze *et al.*, 1993; Stranden *et al.*, 1997).

PBP2A, a penicillin-binding protein with low affinity to β-lactams, was shown to cross-link glycan strands by functional interaction with PBP2. The latter has an inactive TP domain, but retains a fully functional TG domain. This is a first hint that protein–protein interactions between the murein synthases of Gram-positive bacteria may also take place (Pinho *et al.*, 2001).

Another cell wall active antibiotic is vancomycin, which is detected by the TCS VanRS. Vancomycin binds to the terminal D-Ala-D-Ala of the stem peptide, thereby preventing cross-linking of the peptidoglycan. VanRS activates the expression of lactate dehydrogenase (VanH), a D-Ala-D-Lac ligase (VanA) and a D-Ala-D-Lac-peptidase (VanX). This results in the replacement of D-Ala-D-Ala by D-Ala-D-Lac. However, this stem peptide is not recognized by FemX, the protein adding the first glycine of the pentaglycine bridge to the Lys of the stem peptide. This function is fulfilled by VanK (Hong *et al.*, 2005). Stem peptides still containing D-Ala-D-Ala termini are processed by VanY, which cleaves off the last D-Ala, thus resulting in lipid II with a tetrapeptide only. This prevents binding to vancomycin. The combined action of VanX and VanY results in a complete elimination of the target for glycopeptides (Mainardi *et al.*, 2008). Instead of D-Lac, the terminal D-Ala can also be substituted by D-Ser, but this confers only low-level resistance (Courvalin, 2005).

Because more and more pathogens, particularly the resistant *S. aureus* variants MRSA and VRSA, are becoming resistant against β-lactams and vancomycin, new cell wall active antibiotics must be studied. One example is lysostaphin, an endopeptidase encoded on a plasmid of *S. simulans* biovar *staphylolyticus*, which cleaves the pentaglycine bridge of staphylococci (Heath *et al.*, 1989). However, a gene called *lif* (lysostaphin immunity factor) encoded on the same plasmid confers resistance by the incorporation of serine residues instead of glycine in the interpeptide bridge (Strauss *et al.*, 1998) (Fig. 1.3), which could eventually result in other lysostaphin-resistant bacteria.

Conclusion

Peptidoglycan is a polysaccharide which forms a sacculus around the bacterial cell to protect it from internal turgor pressure. The rigid glycan strands are cross-linked by flexible peptide bridges, resulting in an overall structure which can be stretched to two to three times its original size as a reaction to osmotic changes. During bacterial growth the murein sacculus must be elongated and/or divided without losing its strength. In order to achieve a safe insertion of new material, multienzyme complexes containing murein synthases and murein hydrolases have been proposed and evidence for the interaction of these proteins exist.

Peptidoglycan biosynthesis is a target for various antibiotics. Therefore, a large number of resistance mechanisms have evolved. Resistance strategies include changing the peptide structure of peptidoglycan. For example, replacing the terminal D-Ala-D-Ala with D-Ala-D-Lac confers resistance against vancomycin- and penicillin-insensitive L,D-transpeptidases and leads to L,D- instead of D,D-cross-links. Activation of the 'cell-wall stress stimulon' by antibiotics results in overexpression of peptidoglycan biosynthesis-associated genes, suggesting a higher biosynthesis rate in order to cope with damages of the cell wall.

Acknowledgements

The author thanks Volkmar Braun and Evelyn Tichy for critical reading of the manuscript and helpful suggestions. Research in Bertsche's group is financed by the Deutsche Forschungsgesellschaft (SFB 766 'Bacterial Cell Envelope') and by the Strukturfond of the University of Tuebingen.

References

Aarsman, M.E., Piette, A., Fraipont, C., Vinkenvleugel, T.M., Nguyen-Disteche, M., and den Blaauwen, T. (2005). Maturation of the *Escherichia coli* divisome occurs in two steps. Mol. Microbiol. *55*, 1631–1645.

Alaedini, A., and Day, R.A. (1999). Identification of two penicillin-binding multienzyme complexes in *Haemophilus influenzae*. Biochem. Biophys. Res. Commun. *264*, 191–195.

Anderson, J.S., Matsuhashi, M., Haskin, M.A., and Strominger, J.L. (1965). Lipid-phosphoacetylmuramyl-pentapeptide and lipid-phosphodisaccharide-pentapeptide: presumed membrane transport intermediates in cell wall synthesis. Proc. Natl. Acad. Sci. USA *53*, 881–889.

Asoh, S., Matsuzawa, H., Matsuhashi, M., and Ohta, T. (1983). Molecular cloning and characterization of the genes (*pbpA* and *rodA*) responsible for the rod shape of *Escherichia coli* K-12: analysis of gene expression with transposon Tn5 mutagenesis and protein synthesis directed by constructed plasmids. J. Bacteriol. *154*, 10–16.

Barnickel, G., Labischinski, H., Bradaczek, H., and Giesbrecht, P. (1979). Conformational energy calculation on the peptide part of murein. Eur. J. Biochem. *95*, 157–165.

Bera, A., Biswas, R., Herbert, S., and Götz, F. (2006). The presence of peptidoglycan O-acetyltransferase in various staphylococcal species correlates with lysozyme resistance and pathogenicity. Infect. Immun. *74*, 4598–4604.

Bera, A., Herbert, S., Jakob, A., Vollmer, W., and Götz, F. (2005). Why are pathogenic staphylococci so lysozyme resistant? The peptidoglycan O-acetyltransferase OatA is the major determinant for lysozyme resistance of *Staphylococcus aureus*. Mol. Microbiol. *55*, 778–787.

Berger-Bächi, B., Barberis-Maino, L., Strässle, A., and Kayser, F.H. (1989). FemA, a host-mediated factor essential for methicillin resistance in *Staphylococcus aureus*: molecular cloning and characterization. Mol. Gen. Genet. *219*, 263–269.

Berger-Bächi, B., and Rohrer, S. (2002). Factors influencing methicillin resistance in staphylococci. Arch. Microbiol. *178*, 165–171.

Bernhardt, T.G., and de Boer, P.A. (2003). The *Escherichia coli* amidase AmiC is a periplasmic septal ring component exported via the twin-arginine transport pathway. Mol. Microbiol. *48*, 1171–1182.

Bertsche, U., Breukink, E., Kast, T., and Vollmer, W. (2005). *In Vitro* Murein (peptidoglycan) synthesis by dimers of the bifunctional transglycosylase-transpeptidase PBP1B from *Escherichia coli*. J. Biol. Chem. *280*, 38096–38101.

Bertsche, U., Kast, T., Wolf, B., Fraipont, C., Aarsman, M.E., Kannenberg, K., von Rechenberg, M., Nguyen-Disteche, M., den Blaauwen, T., Höltje, J.-V., and Vollmer, W. (2006). Interaction between two murein (peptidoglycan) synthases, PBP3 and PBP1B, in *Escherichia coli*. Mol. Microbiol. *61*, 675–690.

Boneca, I.G., Huang, Z.H., Gage, D.A., and Tomasz, A. (2000). Characterization of *Staphylococcus aureus* cell wall glycan strands, evidence for a new beta-N-acetylglucosaminidase activity. J. Biol. Chem. *275*, 9910–9918.

Born, P., Breukink, E., and Vollmer, W. (2006). *In vitro* synthesis of cross-linked murein and its attachment to sacculi by PBP1A from *Escherichia coli*. J. Biol. Chem. *281*, 26985–26993.

Bouhss, A., Trunkfield, A.E., Bugg, T.D.H., and Mengin-Lecreulx, D. (2008). The biosynthesis of peptidoglycan lipid-linked intermediates. FEMS Microbiol. Rev. *32*, 208–233.

Braun, V. (1975). Covalent lipoprotein from the outer membrane of *Escherichia coli*. Biochim. Biophys. Acta *415*, 335–377.

Braun, V., and Rehn, K. (1969). Chemical characterization, spatial distribution and function of a lipoprotein (murein-lipoprotein) of the *E. coli* cell wall. The specific effect of trypsin on the membrane structure. Eur. J. Biochem *10*, 426–438.

Braun, V., and Wolff, H. (1970). The murein-lipoprotein linkage in the cell wall of *Escherichia coli*. Eur. J. Biochem *14*, 387–391.

Broome-Smith, J.K., Edelman, A., Yousif, S., and Spratt, B.G. (1985). The nucleotide sequences of the *ponA* and *ponB* genes encoding penicillin-binding protein 1A and 1B of *Escherichia coli* K12. Eur. J. Biochem. *147*, 437–446.

Broome-Smith, J.K., Ioannidis, I., Edelman, A., and Spratt, B.G. (1988). Nucleotide sequences of the penicillin-binding protein 5 and 6 genes of *Escherichia coli*. Nucleic Acids Res. *16*, 1617.

Burman, L.G., and Park, J.T. (1984). Molecular model for elongation of the murein sacculus of *Escherichia coli*. Proc. Natl. Acad. Sci. USA *81*, 1844–1848.

Cheng, Q., Li, H., Merdek, K., and Park, J.T. (2000). Molecular characterization of the beta-N-acetylglucosaminidase of *Escherichia coli* and its role in cell wall recycling. J. Bacteriol. *182*, 4836–4840.

Claessen, D., Rink, R., de Jong, W., Siebring, J., de Vreugd, P., Boersma, F.G.H., Dijkhuizen, L., and Wosten, H.A.B. (2003). A novel class of secreted hydrophobic proteins is involved in aerial hyphae formation in *Streptomyces coelicolor* by forming amyloid-like fibrils. Genes Dev. *17*, 1714–1726.

Courvalin, P. (2005). Genetics of glycopeptide resistance in gram-positive pathogens. Int. J. Med. Microbiol. *294*, 479–486.

de Jonge, B.L., Chang, Y.S., Gage, D., and Tomasz, A. (1992). Peptidoglycan composition of a highly methicillin-resistant *Staphylococcus aureus* strain. The role of penicillin binding protein 2A. J. Biol. Chem. *267*, 11248–11254.

de Jonge, B.L., Wientjes, F.B., Jurida, I., Driehuis, F., Wouters, J.T., and Nanninga, N. (1989). Peptidoglycan synthesis during the cell cycle of *Escherichia coli*: composition and mode of insertion. J. Bacteriol. *171*, 5783–5794.

de Pedro, M.A., Quintela, J.C., Höltje, J.-V., and Schwarz, H. (1997). Murein segregation in *Escherichia coli*. J. Bacteriol. *179*, 2823–2834.

de Pedro, M.A., Schwarz, H., and Koch, A.L. (2003a). Patchiness of murein insertion into the sidewall of *Escherichia coli*. Microbiology *149*, 1753–1761.

de Pedro, M.A., Young, K.D., Höltje, J.-V., and Schwarz, H. (2003b). Branching of *Escherichia coli* cells arises from multiple sites of inert peptidoglycan. J. Bacteriol. *185*, 1147–1152.

den Blaauwen, T., Aarsman, M.E., Vischer, N.O., and Nanninga, N. (2003). Penicillin-binding protein PBP2 of *Escherichia coli* localizes preferentially in the lateral wall and at mid-cell in comparison with the old cell pole. Mol. Microbiol. *47*, 539–547.

Derouaux, A., Wolf, B., Fraipont, C., Breukink, E., Nguyen-Disteche, M., and Terrak, M. (2008). The monofunctional glycosyltransferase of *Escherichia coli* localizes to the cell division site and interacts with penicillin-binding protein 3, FtsW, and FtsN. J. Bacteriol. *190*, 1831–1834.

Di Berardino, M., Dijkstra, A., Stuber, D., Keck, W., and Gubler, M. (1996). The monofunctional glycosyltransferase of *Escherichia coli* is a member of a new class of peptidoglycan-synthesising enzymes. FEBS Lett. *392*, 184–188.

Divakaruni, A.V., Loo, R.R., Xie, Y., Loo, J.A., and Gober, J.W. (2005). The cell-shape protein MreC interacts with extracytoplasmic proteins including cell wall assembly complexes in *Caulobacter crescentus*. Proc. Natl. Acad. Sci. USA *102*, 18602–18607.

Dmitriev, B.A., Ehlers, S., and Rietschel, E.T. (1999). Layered murein revisited: a fundamentally new concept of bacterial cell wall structure, biogenesis and function. Med. Microbiol. Immunol. (Berl) *187*, 173–181.

Dmitriev, B.A., Toukach, F.V., Holst, O., Rietschel, E.T., and Ehlers, S. (2004). Tertiary structure of *Staphylococcus aureus* cell wall murein. J. Bacteriol. *186*, 7141–7148.

Dmitriev, B.A., Toukach, F.V., Schaper, K.J., Holst, O., Rietschel, E.T., and Ehlers, S. (2003). Tertiary structure of bacterial murein: the scaffold model. J. Bacteriol. *185*, 3458–3468.

Domanski, T.L., and Bayles, K.W. (1995). Analysis of *Staphylococcus aureus* genes encoding penicillin-binding protein 4 and an ABC-type transporter. Gene *167*, 111–113.

Doublet, P., van Heijenoort, J., and Mengin-Lecreulx, D. (1994). The glutamate racemase activity from *Escherichia coli* is regulated by peptidoglycan precursor UDP-N-acetylmuramoyl-L-alanine. Biochemistry *33*, 5285–5290.

Doublet, P., van Heijenoort, J., and Mengin-Lecreulx, D. (1996). Regulation of the glutamate racemase of *Escherichia coli* investigated by site-directed mutagenesis. Microb. Drug Resist. *2*, 43–49.

Dougherty, T.J., Kennedy, K., Kessler, R.E., and Pucci, M.J. (1996). Direct quantitation of the number of individual penicillin-binding proteins per cell in *Escherichia coli*. J. Bacteriol. *178*, 6110–6115.

Dramsi, S., Magnet, S., Davison, S., and Arthur, M. (2008). Covalent attachment of proteins to peptidoglycan. FEMS Microbiol. Rev. *32*, 307–320.

Elliot, M.A., Karoonuthaisiri, N., Huang, J., Bibb, M.J., Cohen, S.N., Kao, C.M., and Buttner, M.J. (2003). The chaplins: a family of hydrophobic cell-surface proteins involved in aerial mycelium formation in *Streptomyces coelicolor*. Genes Dev. *17*, 1727–1740.

Erickson, H.P. (1995). FtsZ, a prokaryotic homolog of tubulin? Cell *80*, 367–370.

Erickson, H.P. (1998). Atomic structures of tubulin and FtsZ. Trends Cell Biol. *8*, 133–137.

Figge, R.M., Divakaruni, A.V., and Gober, J.W. (2004). MreB, the cell shape-determining bacterial actin homologue, co-ordinates cell wall morphogenesis in *Caulobacter crescentus*. Mol. Microbiol. *51*, 1321–1332.

Fotheringham, I.G., Bledig, S.A., and Taylor, P.P. (1998). Characterization of the genes encoding D-amino acid transaminase and glutamate racemase, two D-glutamate biosynthetic enzymes of *Bacillus sphaericus* ATCC 10208. J. Bacteriol. *180*, 4319–4323.

Fraipont, C., Sapunaric, F., Zervosen, A., Auger, G., Devreese, B., Lioux, T., Blanot, D., Mengin-Lecreulx, D., Herdewijn, P., Van Beeumen, J., *et al.* (2006). Glycosyl transferase activity of the *Escherichia coli* penicillin-binding protein 1b: specificity profile for the substrate. Biochemistry *45*, 4007–4013.

Galperin, M.Y., and Koonin, E.V. (1997). A diverse superfamily of enzymes with ATP-dependent carboxylate-amine/thiol ligase activity. Protein Sci. *6*, 2639–2643.

Gardete, S., de Lencastre, H., and Tomasz, A. (2006). A link in transcription between the native *pbp3* and the acquired *mecA* gene in a strain of *Staphylococcus aureus*. Microbiology *152*, 2549–2558.

Ghuysen, J.M. (1991). Serine beta-lactamases and penicillin-binding proteins. Annu. Rev. Microbiol. *45*, 37–67.

Ghuysen, J.M., Bricas, E., Lache, M., and Leyh-Bouille, M. (1968). Structure of the cell walls of *Micrococcus lysodeikticus*. 3. Isolation of a new peptide dimer, N-alpha-[L-alanyl-gamma-(alpha-D-glutamylglycine)]-L-lysyl-D-alanyl-N-alph a-[L-alanyl-gamma-(alpha-D-glutamylglycine)]-L-lysyl-D-alanine. Biochemistry *7*, 1450–1460.

Giesbrecht, P., Kersten, T., Maidhof, H., and Wecke, J. (1998). Staphylococcal cell wall: morphogenesis and fatal variations in the presence of penicillin. Microbiol. Mol. Biol. Rev. *62*, 1371–1414.

Glauner, B. (1988). Separation and quantification of muropeptides with high-performance liquid chromatography. Anal. Biochem. *172*, 451–464.

Glauner, B., and Höltje, J.-V. (1990). Growth pattern of the murein sacculus of *Escherichia coli*. J. Biol. Chem. *265*, 18988–18996.

Glauner, B., Höltje, J.-V., and Schwarz, U. (1988). The composition of the murein of *Escherichia coli*. J. Biol. Chem. *263*, 10088–10095.

Goffin, C., and Ghuysen, J.M. (1998). Multimodular penicillin-binding proteins: an enigmatic family of orthologs and paralogs. Microbiol. Mol. Biol. Rev. *62*, 1079–1093.

Goffin, C., and Ghuysen, J.M. (2002). Biochemistry and comparative genomics of SxxK superfamily acyltransferases offer a clue to the mycobacterial paradox: presence of penicillin-susceptible target proteins versus lack of efficiency of penicillin as therapeutic agent. Microbiol. Mol. Biol. Rev. *66*, 702–738.

Goodell, E.W. (1985). Recycling of murein by *Escherichia coli*. J. Bacteriol. *163*, 305–310.

Goodell, E.W., and Schwarz, U. (1985). Release of cell wall peptides into culture medium by exponetially growing *Escherichia coli*. J. Bacteriol. *162*, 391–397.

Hackbarth, C.J., Kocagoz, T., Kocagoz, S., and Chambers, H.F. (1995). Point mutations in *Staphylococcus aureus* PBP 2 gene affect penicillin-binding kinetics and are associated with resistance. Antimicrob Agents Chemother. *39*, 103–106.

Halouska, S., Chacon, O., Fenton, R.J., Zinniel, D.K., Barletta, R.G., and Powers, R. (2007). Use of NMR Metabolomics To Analyze the Targets of D-Cycloserine in Mycobacteria: Role of D-Alanine Racemase. J. Proteome Res. *6*, 4608–4614.

Hara, H., Yasuda, S., Horiuchi, K., and Park, J.T. (1997). A promoter for the first nine genes of the *Escherichia coli mra* cluster of cell division and cell envelope biosynthesis genes, including *ftsI* and *ftsW*. J. Bacteriol. *179*, 5802–5811.

Hartman, B.J., and Tomasz, A. (1984). Low-affinity penicillin-binding protein associated with beta-lactam resistance in *Staphylococcus aureus*. J. Bacteriol. *158*, 513–516.

Harz, H., Burgdorf, K., and Höltje, J.-V. (1990). Isolation und separation of the glycan strands from murein of *Escherichia coli* by reversed phase high-performance liquid chromatography. Anal. Biochem. *190*, 120–128.

Heath, H.E., Heath, L.S., Nitterauer, J.D., Rose, K.E., and Sloan, G.L. (1989). Plasmid-encoded lysostaphin endopeptidase resistance of staphylococcus simulans biovar staphylolyticus. Biochem. Biophys. Res. Commun. *160*, 1106.

Heidrich, C., Templin, M.F., Ursinus, A., Merdanovic, M., Berger, J., Schwarz, H., de Pedro, M.A., and Höltje, J.-V. (2001). Involvement of N-acetylmuramyl-L-alanine amidases in cell separation and antibiotic-induced autolysis of *Escherichia coli*. Mol. Microbiol. *41*, 167–178.

Heilmann, C., Hartleib, J., Hussain, M.S., and Peters, G. (2005). The multifunctional *Staphylococcus aureus* autolysin *aaa* mediates adherence to immobilized fibrinogen and fibronectin. Infect. Immun. 73, 4793–4802.

Henriques, A.O., Glaser, P., Piggot, P.J., and Moran, C.P.J. (1998). Control of cell shape and elongation by the *rodA* gene in *Bacillus subtilis*. Mol. Microbiol. *28*, 235–247.

Henze, U., Sidow, T., Wecke, J., Labischinski, H., and Berger-Bachi, B. (1993). Influence of *femB* on methicillin resistance and peptidoglycan metabolism in *Staphylococcus aureus*. J. Bacteriol. *175*, 1612–1620.

Hirashima, A., Wu, H.C., Venkateswaran, P.S., and Inouye, M. (1973). Two forms of a structural lipoprotein in the envelope of *Escherichia coli*. Further characterization of the free form. J. Biol. Chem. *248*, 5654–5659.

Höltje, J.-V. (1993). 'Three for one'- A simple growth mechanism that guarantees a precise copy of the thin, rod-shaped murein sacculus of *Escherichia coli*, In Bacterial Growth and Lysis-Metabolism and Structure of the Bacterial Sacculus, M.A. de Pedro, J.-V. Höltje, and W. Löffelhardt, eds. (Plenum Press, New York), pp. 419–426.

Höltje, J.-V. (1996). A hypothetical holoenzyme involved in the replication of the murein sacculus of *Escherichia coli*. Microbiology *142 (Pt 8)*, 1911–1918.

Höltje, J.-V. (1998). Growth of the stress-bearing and shape-maintaining murein sacculus of *Escherichia coli*. Microbiol. Mol. Biol. Rev. *62*, 181–203.

Höltje, J.-V., Kopp, U., Ursinus, A., and Wiedemann, B. (1994). The negative regulator of beta-lactamase induction AmpD is a N-acetyl-anhydromuramyl-L-alanine amidase. FEMS Microbiol. Lett *122*, 159–164.

Höltje, J.-V., Mirelman, D., Sharon, N., and Schwarz, U. (1975). Novel type of murein transglycosylase in *Escherichia coli*. J. Bacteriol. *124*, 1067–1076.

Hong, H.-J., Hutchings, M.I., Hill, L.M., and Buttner, M.J. (2005). The Role of the Novel Fem Protein VanK in Vancomycin Resistance in *Streptomyces coelicolor*. J. Biol. Chem. *280*, 13055–13061.

Ikeda, M., Sato, T., Wachi, M., Jung, H.K., Ishino, F., Kobayashi, Y., and Matsuhashi, M. (1989). Structural similarity among *Escherichia coli* FtsW and RodA proteins and *Bacillus subtilis* SpoVE protein, which function in cell division, cell elongation, and spore formation, respectively. J. Bacteriol. *171*, 6375–6378.

Ikeda, M., Wachi, M., Jung, H.K., Ishino, F., and Matsuhashi, M. (1991). The *Escherichia coli mraY* gene encoding UDP-N-acetylmuramoyl-pentapeptide: undecaprenyl-phosphate phospho-N-acetylmuramoyl-pentapeptide transferase. J. Bacteriol. *173*, 1021–1026.

Ishino, F., and Matsuhashi, M. (1981). Peptidoglycan synthetic enzyme activities of highly purified penicillin-binding protein 3 in *Escherichia coli*: a septum-forming reaction sequence. Biochem. Biophys. Res. Commun. *101*, 905–911.

Ishino, F., Mitsui, K., Tamaki, S., and Matsuhashi, M. (1980). Dual enzyme activities of cell wall peptidoglycan synthesis, peptidoglycan transglycosylase and penicillin-sensitive transpeptidase, in purified preparations of *Escherichia coli* penicillin-binding protein 1A. Biochem. Biophys. Res. Commun. *97*, 287–293.

Ishino, F., Park, W., Tomioka, S., Tamaki, S., Takase, I., Kunugita, K., Matsuzawa, H., Asoh, S., Ohta, T., Spratt, B.G., and Matsuhashi, M. (1986). Peptidoglycan synthetic activities in membranes of *Escherichia coli* caused by overproduction of penicillin-binding protein 2 and RodA protein. J. Biol. Chem. *261*, 7024–7031.

Ishino, F., Tamaki, S., Spratt, B.G., and Matsuhashi, M. (1982). A mecillinam-sensitive peptidoglycan crosslinking reaction in *Escherichia coli*. Biochem. Biophys. Res. Commun. *109*, 689–696.

Ito, E., and Strominger, J.L. (1962). Enzymatic synthesis of the peptide in bacterial uridine nucleotides. II. Enzymatic synthesis and addition of D-alanyl-D-alanine. J. Biol. Chem. *237*, 2696–2703.

Jacobs, C., Frere, J.M., and Normark, S. (1997). Cytosolic intermediates for cell wall biosynthesis and degradation control inducible beta-lactam resistance in gram-negative bacteria. Cell *88*, 823–832.

Jones, L.J., Carballido-Lopez, R., and Errington, J. (2001). Control of cell shape in bacteria: helical, actin-like filaments in *Bacillus subtilis*. Cell *104*, 913–922.

Jordan, S., Hutchings, M.I., and Mascher, T. (2008). Cell envelope stress response in Gram-positive bacteria. FEMS Microbiol. Rev *32*, 107–146.

Kern, T., Hediger, S., Mueller, P., Giustini, C., Joris, B., Bougault, C., Vollmer, W., and Simorre, J.-P. (2008). Toward the characterization of peptidoglycan structure and protein-peptidoglycan interactions by solid-state NMR spectroscopy. J. Am. Chem. Soc. *130*, 5618–5619.

Koch, A.L. (1983). The surface stress theory of microbial morphogenesis. Adv. Microb. Physiol. *24*, 301–366.

Koch, A.L. (1995). Bacterial Growth and Form (New York: Chapman & Hall).

Koch, A.L. (1998). Orientation of the peptidoglycan chains in the sacculus of *Escherichia coli*. Res. Microbiol. *149*, 689–701.

Koch, A.L. (2000). Simulation of the conformation of the murein fabric: the oligoglycan, penta-muropeptide, and cross-linked nona-muropeptide. Arch Microbiol. *174*, 429–439.

Koch, A.L., and Woeste, S. (1992). Elasticity of the sacculus of *Escherichia coli*. J. Bacteriol. *174*, 4811–4819.

Koch, A.L., and Woldringh, C. (1994). The metabolic inertness of the pole wall of a Gram-negative rod. J. Theor Biol. *171*, 415–425.

Kohlrausch, U., and Höltje, J.-V. (1991). Analysis of murein and murein precursors during antibiotic-induced lysis of *Escherichia coli*. J. Bacteriol. *173*, 3425–3431.

Kunin, C.M., and Rudy, J. (1991). Effect of NaCl-induced osmotic stress on intracellular concentrations of glycine betaine and potassium in *Escherichia coli*, *Enterococcus faecalis*, and staphylococci. J. Lab. Clin. Med. *118*, 217–224.

Labischinski, H., Barnickel, G., Bradaczek, H., and Giesbrecht, P. (1979). On the secondary and tertiary structure of murein. Low and medium-angle X-ray evidence against chitin-based conformations of bacterial peptidoglycan. Eur. J. Biochem. *95*, 147–155.

Labischinski, H., Barnickel, G., Naumann, D., and Keller, P. (1985). Conformational and topological aspects of the three-dimensional architecture of bacterial peptidoglycan. Ann. Inst. Pasteur Microbiol. *136A*, 45–50.

Labischinski, H., Goodell, E.W., Goodell, A., and Hochberg, M.L. (1991). Direct proof of a 'more-than-single-layered' peptidoglycan architecture of *Escherichia coli* W7: a neutron small-angle scattering study. J. Bacteriol. *173*, 751–756.

Lara, B., and Ayala, J.A. (2002). Topological characterization of the essential *Escherichia coli* cell division protein FtsW. FEMS Microbiol. Lett. *216*, 23–32.

Laubacher, M.E., and Ades, S.E. (2008). The Rcs phosphorelay is a cell envelope stress response activated by peptidoglycan stress and contributes to intrinsic antibiotic resistance. J. Bacteriol. *190*, 2065–2074.

Lavollay, M., Arthur, M., Fourgeaud, M., Dubost, L., Marie, A., Veziris, N., Blanot, D., Gutmann, L., and Mainardi, J.-L. (2008). The peptidoglycan of stationary phase *mycobacterium tuberculosis* predominantly contains cross-links generated by L,D-transpeptidation. J. Bacteriol. *190*, 4360–4366.

Lindquist, S., Weston-Hafer, K., Schmidt, H., Pul, C., Korfmann, G., Erickson, J., Sanders, C., Martin, H.H., and Normark, S. (1993). AmpG, a signal transducer in chromosomal beta-lactamase induction. Mol. Microbiol. *9*, 703–715.

Lovering, A.L., de Castro, L.H., Lim, D., and Strynadka, N.C. (2007). Structural insight into the transglycosylation step of bacterial cell-wall biosynthesis. Science *315*, 1402–1405.

Löwe, J., and Amos, L.A. (1999). Tubulin-like protofilaments in Ca2+-induced FtsZ sheets. Embo J. *18*, 2364–2371.

Magnet, S., Bellais, S., Dubost, L., Fourgeaud, M., Mainardi, J.-L., Petit-Frere, S., Marie, A., Mengin-Lecreulx, D., Arthur, M., and Gutmann, L. (2007). Identification of the L,D-Transpeptidases responsible for attachment of the Braun lipoprotein to *Escherichia coli* peptidoglycan. J. Bacteriol. *189*, 3927–3931.

Magnet, S., Dubost, L., Marie, A., Arthur, M., and Gutmann, L. (2008). Identification of the L,D-transpeptidases for peptidoglycan cross-linking in *Escherichia coli*. J. Bacteriol. *190*, 4782–4785.

Mainardi, J.-L., Villet, R., Bugg, T.D., Mayer, C., and Arthur, M. (2008). Evolution of peptidoglycan biosynthesis under the selective pressure of antibiotics in Gram-positive bacteria. FEMS Microbiol. Rev. *32*, 386–408.

Margolin, W. (2000). Themes and variations in prokaryotic cell division. FEMS Microbiol. Rev *24*, 531–548.

Marraffini, L.A., DeDent, A.C., and Schneewind, O. (2006). Sortases and the art of anchoring proteins to the envelopes of Gram-positive bacteria. Microbiol. Mol. Biol. Rev. *70*, 192–221.

Marraffini, L.A., and Schneewind, O. (2005). Anchor structure of staphylococcal surface proteins. V. Anchor structure of the sortase B substrate IsdC. J. Biol. Chem. *280*, 16263–16271.

Marraffini, L.A., and Schneewind, O. (2006). Targeting proteins to the cell wall of sporulating *Bacillus anthracis*. Mol. Microbiol. *62*, 1402–1417.

Matias, V.R., and Beveridge, T.J. (2006). Native cell wall organization shown by cryo-electron microscopy confirms the existence of a periplasmic space in *Staphylococcus aureus*. J. Bacteriol. *188*, 1011–1021.

Matias, V.R.F., and Beveridge, T.J. (2007). Cryo-electron microscopy of cell division in *Staphylococcus aureus* reveals a mid-zone between nascent cross walls. Mol. Microbiol. *64*, 195–206.

Mazmanian, S.K., Liu, G., Ton-That, H., and Schneewind, O. (1999). *Staphylococcus aureus* sortase, an enzyme that anchors surface proteins to the cell wall. Science *285*, 760–763.

Mazmanian, S.K., Ton-That, H., Su, K., and Schneewind, O. (2002). An iron-regulated sortase anchors a class of surface protein during *Staphylococcus aureus* pathogenesis. Proc. Natl. Acad. Sci. USA *99*, 2293–2298.

Meisel, U., Höltje, J.-V., and Vollmer, W. (2003). Overproduction of inactive variants of the murein synthase PBP1B causes lysis in *Escherichia coli*. J. Bacteriol. *185*, 5342–5348.

Meroueh, S.O., Bencze, K.Z., Hesek, D., Lee, M., Fisher, J.F., Stemmler, T.L., and Mobashery, S. (2006). Three-dimensional structure of the bacterial cell wall peptidoglycan. Proc. Natl. Acad. Sci. USA *103*, 4404–4409.

Morbach, S., and Kramer, R. (2002). Body shaping under water stress: osmosensing and osmoregulation of solute transport in bacteria. ChemBioChem 3, 384–397.

Müller, P., Ewers, C., Bertsche, U., Anstett, M., Kallis, T., Breukink, E., Fraipont, C., Terrak, M., Nguyen-Distèche, M., and Vollmer, W. (2007). The essential cell division protein FtsN interacts with the murein (peptidoglycan) synthase PBP1B in *Escherichia coli*. J. Biol. Chem. *282*, 36394–36402.

Murakami, K., Fujimura, T., and Doi, M. (1994). Nucleotide sequence of the structural gene for the penicillin-binding protein 2 of *Staphylococcus aureus* and the presence of a homologous gene in other staphylococci. FEMS Microbiol. Lett. *117*, 131–136.

Nakagawa, J., Tamaki, S., Tomioka, S., and Matsuhashi, M. (1984). Functional biosynthesis of cell wall peptidoglycan by polymorphic bifunctional polypeptides. Penicillin-binding protein 1Bs of *Escherichia coli* with activities of transglycosylase and transpeptidase. J. Biol. Chem. *259*, 13937–13946.

Neuhaus, F.C. (1962). The Enzymatic Synthesis of D-Alanyl-D-alanine. I. Purification and properties of D-alanyl-D-alanine synthetase. J. Biol. Chem. *237*, 778–786.

Normark, S., Bartowski, E., Erickson, J., Jacobs, C., Lindberg, F., Lindquist, S., Weston-Hafer, K., and Wikström, M. (1994). Mechanisms of chromosomal β-lactamase induction in Gram-negative bacteria. In Bacterial Cell Wall, J.M. Ghuysen, and R. Hakenbeck, eds. (Amsterdam: Elsevier), pp. 485–504.

Obermann, W., and Höltje, J.-V. (1994). Alterations of murein structure and of penicillin-binding proteins in minicells from *Escherichia coli*. Microbiology *140*, 79–87.

Oshida, T., Sugai, M., Komatsuzawa, H., Hong, Y.M., Suginaka, H., and Tomasz, A. (1995). A *Staphylococcus aureus* autolysin that has an N-acetylmuramoyl-L-alanine amidase domain and an endo-beta-N-acetylglucosaminidase domain: cloning, sequence analysis, and characterization. Proc. Natl. Acad. Sci. USA *92*, 285–289.

Pereira, S.F.F., Henriques, A.O., Pinho, M.G., de Lencastre, H., and Tomasz, A. (2007). Role of PBP1 in cell division of *Staphylococcus aureus*. J. Bacteriol. *189*, 3525–3531.

Perlstein, D.L., Zhang, Y., Wang, T.S., Kahne, D.E., and Walker, S. (2007). The direction of glycan chain elongation by peptidoglycan glycosyltransferases. J. Am. Chem. Soc. *129*, 12674–12675.

Pinho, M.G., de Lencastre, H., and Tomasz, A. (2000). Cloning, characterization, and inactivation of the gene *pbpC*, encoding penicillin-binding protein 3 of *Staphylococcus aureus*. J. Bacteriol. *182*, 1074–1079.

Pinho, M.G., de Lencastre, H., and Tomasz, A. (2001). An acquired and a native penicillin-binding protein cooperate in building the cell wall of drug-resistant staphylococci. Proc. Natl. Acad. Sci. USA *98*, 10886–10891.

Pinho, M.G., and Errington, J. (2003). Dispersed mode of *Staphylococcus aureus* cell wall synthesis in the absence of the division machinery. Mol. Microbiol. *50*, 871–881.

Pinho, M.G., and Errington, J. (2005). Recruitment of penicillin-binding protein PBP2 to the division site of *Staphylococcus aureus* is dependent on its transpeptidation substrates. Mol. Microbiol. *55*, 799–807.

Pisabarro, A.G., de Pedro, M.A., and Vazquez, D. (1985). Structural modifications in the peptidoglycan of *Escherichia coli* associated with changes in the state of growth of the culture. J. Bacteriol. *161*, 238–242.

Piuri, M., Sanchez-Rivas, C., and Ruzal, S.M. (2005). Cell wall modifications during osmotic stress in *Lactobacillus casei*. J. Appl. Microbiol. *98*, 84–95.

Pucci, M.J., and Dougherty, T.J. (2002). Direct quantitation of the numbers of individual penicillin-binding proteins per cell in *Staphylococcus aureus*. J. Bacteriol. *184*, 588–591.

Pucci, M.J., Thanassi, J.A., Discotto, L.F., Kessler, R.E., and Dougherty, T.J. (1997). Identification and characterization of cell wall-cell division gene clusters in pathogenic gram-positive cocci. J. Bacteriol. *179*, 5632–5635.

Quintela, J.C., Caparros, M., and de Pedro, M.A. (1995). Variability of peptidoglycan structural parameters in gram-negative bacteria. FEMS Microbiol. Lett *125*, 95–100.

Rogers, H.J., Perkins, H.R., and Ward, J.B. (1980). Microbial Cell Walls and Membranes: Chapman & Hall, London).

Salmond, G.P., Lutkenhaus, J.F., and Donachie, W.D. (1980). Identification of new genes in a cell envelope-cell division gene cluster of *Escherichia coli*: cell envelope gene *murG*. J. Bacteriol. *144*, 438–440.

Sauvage, E., Kerff, F., Terrak, M., Ayala, J.A., and Charlier, P. (2008). The penicillin-binding proteins: structure and role in peptidoglycan biosynthesis. FEMS Microbiol. Rev. *32*, 234–258.

Schiffer, G., and Höltje, J.-V. (1999). Cloning and characterization of PBP 1C, a third member of the multimodular class A penicillin-binding proteins of *Escherichia coli*. J. Biol. Chem. *274*, 32031–32039.

Schleifer, K.H., and Kandler, O. (1972). Peptidoglycan types of bacterial cell walls and their taxonomic implications. Bacteriol. Rev *36*, 407–477.

Schneider, T., Senn, M.M., Berger-Bächi, B., Tossi, A., Sahl, H.G., and Wiedemann, I. (2004). *In vitro* assembly of a complete, pentaglycine interpeptide bridge containing cell wall precursor (lipid II-Gly5) of *Staphylococcus aureus*. Mol. Microbiol. *53*, 675–685.

Simon, M.J., and Day, R.A. (2000). Improved Resolution of Hydrophobic Penicillin-binding Proteins and Their Covalently Linked Complexes on a Modified C18 Reversed Phase Column. Analytical Letters *33*, 861–867.

Song, M.D., Wachi, M., Doi, M., Ishino, F., and Matsuhashi, M. (1987). Evolution of an inducible penicillin-target protein in methicillin-resistant *Staphylococcus aureus* by gene fusion. FEBS Lett. *221*, 167–171.

Spratt, B.G. (1975). Distinct penicillin binding proteins involved in the division, elongation, and shape of *Escherichia coli* K12. Proc. Natl. Acad. Sci. USA *72*, 2999–3003.

Stranden, A.M., Ehlert, K., Labischinski, H., and Berger-Bächi, B. (1997). Cell wall monoglycine cross-bridges and methicillin hypersusceptibility in a femAB null mutant of methicillin-resistant *Staphylococcus aureus*. J. Bacteriol. *179*, 9–16.

Strauss, A., Thumm, G., and Götz, F. (1998). Influence of Lif, the lysostaphin immunity factor, on acceptors of surface proteins and cell wall sorting efficiency in *Staphylococcus carnosus*. J. Bacteriol. *180*, 4960–4962.

Struve, W.G., and Neuhaus, F.C. (1965). Evidence for an initial acceptor of UDP-NAc-muramyl-pentapeptide in the synthesis of bacterial mucopeptide. Biochem. Biophys. Res. Commun. *18*, 6–12.

Sugai, M., Yamada, S., Nakashima, S., Komatsuzawa, H., Matsumoto, A., Oshida, T., and Suginaka, H. (1997). Localized perforation of the cell wall by a major autolysin: *atl* gene products and the onset of penicillin-induced lysis of *Staphylococcus aureus*. J. Bacteriol. *179*, 2958–2962.

Tanizawa, K., Masu, Y., Asano, S., Tanaka, H., and Soda, K. (1989). Thermostable D-amino acid aminotransferase from a thermophilic Bacillus species. Purification, characterization, and active site sequence determination. J. Biol. Chem. *264*, 2445–2449.

Terrak, M., Ghosh, T.K., van Heijenoort, J., Van Beeumen, J., Lampilas, M., Aszodi, J., Ayala, J.A., Ghuysen, J.M., and Nguyen-Distèche, M. (1999). The catalytic, glycosyl transferase and acyl transferase modules of the cell wall peptidoglycan-polymerizing penicillin-binding protein 1b of *Escherichia coli*. Mol. Microbiol. *34*, 350–364.

Terrak, M., and Nguyen-Distèche, M. (2006). Kinetic characterization of the monofunctional glycosyltransferase from *Staphylococcus aureus*. J. Bacteriol. *188*, 2528–2532.

Thanedar, S., and Margolin, W. (2004). FtsZ exhibits rapid movement and oscillation waves in helix-like patterns in *Escherichia coli*. Curr. Biol. *14*, 1167–1173.

Thunnissen, A.M., Isaacs, N.W., and Dijkstra, B.W. (1995). The catalytic domain of a bacterial lytic transglycosylase defines a novel class of lysozymes. Proteins *22*, 245–258.

Ton-That, H., and Schneewind, O. (2003). Assembly of pili on the surface of *Corynebacterium diphtheriae*. Mol. Microbiol. *50*, 1429–1438.

Touhami, A., Jericho, M.H., and Beveridge, T.J. (2004). Atomic force microscopy of cell growth and division in *Staphylococcus aureus*. J. Bacteriol. *186*, 3286–3295.

Uehara, T., and Park, J.T. (2007). An anhydro-N-acetyl-muramyl-L-alanine amidase with broad specificity tethered to the outer membrane of *Escherichia coli*. J. Bacteriol. *189*, 5634–5641.

Uehara, T., and Park, J.T. (2008). Growth of *Escherichia coli*: significance of peptidoglycan degradation during elongation and septation. J. Bacteriol. *190*, 3914–3922.

Utaida, S., Dunman, P.M., Macapagal, D., Murphy, E., Projan, S.J., Singh, V.K., Jayaswal, R.K., and Wilkinson, B.J. (2003). Genome-wide transcriptional profiling of the response of *Staphylococcus aureus* to cell-wall-active antibiotics reveals a cell-wall-stress stimulon. Microbiology *149*, 2719–2732.

van den Ent, F., Amos, L.A., and Lowe, J. (2001). Prokaryotic origin of the actin cytoskeleton. Nature *413*, 39.

van Heijenoort, J. (2001). Recent advances in the formation of the bacterial peptidoglycan monomer unit. Nat. Prod. Rep. *18*, 503–519.

van Heijenoort, J. (2007). Lipid intermediates in the biosynthesis of bacterial peptidoglycan. Microbiol. Mol. Biol. Rev. *71*, 620–635.

Vazquez-Laslop, N., Lee, H., Hu, R., and Neyfakh, A.A. (2001). Molecular sieve mechanism of selective release of cytoplasmic proteins by osmotically shocked *Escherichia coli*. J. Bacteriol. *183*, 2399–2404.

Verwer, R.W., Beachey, E.H., Keck, W., Stoub, A.M., and Poldermans, J.E. (1980). Oriented fragmentation of *Escherichia coli* sacculi by sonication. J. Bacteriol. *141*, 327–332.

Verwer, R.W., Nanninga, N., Keck, W., and Schwarz, U. (1978). Arrangement of glycan chains in the sacculus of *Escherichia coli*. J. Bacteriol. *136*, 723–729.

Vinella, D., Joseleau-Petit, D., Thevenet, D., Bouloc, P., and D'Ari, R. (1993). Penicillin-binding protein 2 inactivation in *Escherichia coli* results in cell division inhibition, which is relieved by FtsZ overexpression. J. Bacteriol. *175*, 6704–6710.

Vollmer, W., and Bertsche, U. (2008). Murein (peptidoglycan) structure, architecture and biosynthesis in *Escherichia coli*. Biochim. Biophys. Acta *1778*, 1714–1734.

Vollmer, W., Blanot, D., and de Pedro, M.A. (2008a). Peptidoglycan structure and architecture. FEMS Microbiol. Rev. *32*, 149–167.

Vollmer, W., and Höltje, J.-V. (2004). The architecture of the murein (peptidoglycan) in Gram-negative bacteria: Vertical scaffold or horizontal layer(s)? J. Bacteriol. *186*, 5978–5987.

Vollmer, W., Joris, B., Charlier, P., and Foster, S. (2008b). Bacterial peptidoglycan (murein) hydrolases. FEMS Microbiol. Rev. *32*, 259–286.

Vollmer, W., von Rechenberg, M., and Höltje, J.-V. (1999). Demonstration of molecular interactions between the murein polymerase PBP1B, the lytic transglycosylase MltA, and the scaffolding protein MipA of *Escherichia coli*. J. Biol. Chem. *274*, 6726–6734.

Vötsch, W., and Templin, M.F. (2000). Characterization of a beta -N-acetylglucosaminidase of *Escherichia coli* and elucidation of its role in muropeptide recycling and beta-lactamase induction. J. Biol. Chem. *275*, 39032–39038.

Wang, Q.M., Peery, R.B., Johnson, R.B., Alborn, W.E., Yeh, W.K., and Skatrud, P.L. (2001). Identification and characterization of a monofunctional glycosyltransferase from *Staphylococcus aureus*. J. Bacteriol. *183*, 4779–4785.

Weidel, W., and Pelzer, H. (1964). Bagshaped macromolecules – a new outlook on bacterial cell walls. Advanc. Enzymol. *26*, 193–232.

Weiss, D.S., Pogliano, K., Carson, M., Guzman, L.M., Fraipont, C., Nguyen-Disteche, M., Losick, R., and Beckwith, J. (1997). Localization of the *Escherichia coli* cell division protein FtsI (PBP3) to the division site and cell pole. Mol. Microbiol. *25*, 671–681.

Wientjes, F.B., Woldringh, C.L., and Nanninga, N. (1991). Amount of peptidoglycan in cell walls of gram-negative bacteria. J. Bacteriol. *173*, 7684–7691.

Wilkinson, B.J. (1997). The staphylococci in human disease. In Biology, K.B. Crossley, and G.L. Archer, eds. (New York: Churchill Livingstone), pp. 1–38.

Wood, W.A., and Gunsalus, I.C. (1951). D-alanine formation: a racemase in *Streptococcus faecalis*. J. Biol. Chem. *190*, 403–416.

Wu, D., Hu, T., Zhang, L., Chen, J., Du, J., Ding, J., Jiang, H., and Shen, X. (2008). Residues Asp164 and Glu165 at the substrate entryway function potently in substrate orientation of alanine racemase from *E. coli*: Enzymatic characterization with crystal structure analysis. Protein Sci. *17*, 1066–1076.

Wu, H., and Fives-Taylor, P.M. (2001). Molecular strategies for fimbrial expression and assembly. Crit. Rev. Oral Biol. Med. *12*, 101–115.

Yamada, S., Sugai, M., Komatsuzawa, H., Nakashima, S., Oshida, T., Matsumoto, A., and Suginaka, H. (1996). An autolysin ring associated with cell separation of *Staphylococcus aureus*. J. Bacteriol. *178*, 1565–1571.

Yao, X., Jericho, M., Pink, D., and Beveridge, T. (1999). Thickness and elasticity of gram-negative murein sacculi measured by atomic force microscopy. J. Bacteriol. *181*, 6865–6875.

Yem, D.W., and Wu, H.C. (1976). Purification and properties of beta-N-acetylglucosaminidase from *Escherichia coli*. J. Bacteriol. *125*, 324–331.

Yuan, Y., Barrett, D., Zhang, Y., Kahne, D., Silz, P., and Walker, S. (2007). Crystal structure of a peptidoglycan glycosyltransferase suggests a model for processive glycan chain synthesis. Proc. Natl. Acad. Sci. USA *104*, 5348–5353.

Zijderveld, C.A., Aarsman, M.E., den Blaauwen, T., and Nanninga, N. (1991). Penicillin-binding protein 1B of *Escherichia coli* exists in dimeric forms. J. Bacteriol. *173*, 5740–5746.

Zijderveld, C.A., Aarsman, M.E., and Nanninga, N. (1995). Differences between inner membrane and peptidoglycan-associated PBP1B dimers of *Escherichia coli*. J. Bacteriol. *177*, 1860–1863.

Genetics and Regulation of Bacterial Lipopolysaccharide Synthesis

2

Mikael Skurnik and José Antonio Bengoechea

Abstract

Lipopolysaccharide (LPS) is the major component of the outer leaflet of the outer membrane of Gram-negative bacteria. The LPS molecule is composed of two biosynthetic entities: the lipid A–core and the O-polysaccharide (O-antigen). Most biological effects of LPS are due to the lipid A part, however, there is an increasing body of evidence indicating that O-antigen (O-Ag) plays an important role in effective colonization of host tissues, resistance to complement-mediated killing and in the resistance to cationic antimicrobial peptides that are key elements of the innate immune system. Recently, data have started to accumulate on the intricacies in the genetic regulation of the structural components of this molecule and this is highly relevant to the biological function of the molecule. In this review we discuss the regulation, mainly using yersiniae as model organisms but also discussing other bacteria where relevant.

Polysaccharide biosynthesis in bacteria

Outline of the structure and biosynthesis of LPS

Bacteria carry on their outer membrane a heterogeneous population of LPS molecules. This is clearly seen when isolated LPS is analysed in polyacrylamide gel electrophoresis; a wide range of LPS molecules of different sizes are present. The smallest LPS molecules are composed of lipid A and (inner) core moieties, the slightly larger molecules have lipid A and complete core structure and the largest contain lipid A, a complete core and a variable number of O-Ag repeat units. Therefore, these LPS populations appear as a ladder in the stained gel. If the repeating unit is a single sugar, the O-Ag is called homopolymeric, if it is composed of two or more different sugar residues, the O-Ag is called heteropolymeric.

Polysaccharide biosynthesis precursors

The biosynthesis of O-Ag requires numerous enzymatic activities that take care of the biosynthetic pathways in the cytoplasm leading to individual NDP-activated precursor sugars. Various glycosyltransferases transfer the sugar residues from the NDP-sugars in a linkage and sugar-specific manner to the growing oligosaccharide structure. To complete the polysaccharide structure and translocate it to periplasmic space the activities of the O-unit flippase, O-Ag polymerase and O-chain length determinant (Wzx, Wzy and Wzz, respectively) are required. Based on this enzymatic mode of biosynthesis, LPSs isolated from bacteria carry a heterogeneous population of O-Ag molecules; some do not carry any O-Ag while others are substituted with O-Ag chains of variable lengths (Whitfield *et al.*, 1997).

LPS biosynthesis starts by the activation of sugar-1-phosphate (P), such as glucose-1-P or fructose-1-P by reaction with NTP to form NDP-activated sugars. The activation reactions take place in the cytoplasm and are catalysed by soluble enzymes. The various NDP-sugars are then synthesized in several enzymatic steps involving epimerases, reductases, hydratases, etc. For example, GTP is used to activate mannose and D-rhamnose, CTP, abequose and paratose, UTP, galactose and glucose, and dTTP, L-rhamnose; the reason for the choice of NTP is unclear. Some NDP-sugars are intermediates of general metabolism, however, most used in LPS biosynthesis need to be synthesized specifically for this purpose.

Lipid A–core biosynthesis
Lipid A is assembled on the cytoplasmic face of inner membrane (IM). In several enzymatic steps UDP-GlcNAc and fatty acids (supplied by acyl carrier proteins) are converted into lipid A, which is a polar lipid with a backbone of βGlcN(1→6)-GlcN substituted with 4–7 N- and O-linked saturated fatty acid residues and one or two phosphate groups (Raetz and Whitfield, 2002). The core oligosaccharide, which typically contains 10–15 sugar residues, is built on lipid A sequentially by specific glycosyltransferases. The completed lipid A–core molecules are translocated to the periplasmic face of IM where O-antigen ligase, WaaL, ligates the O-specific polysaccharide (see below) to some of them. From IM the LPS molecules are finally translocated to OM.

O-antigen biosynthesis

The O-polysaccharide can be either homo- or heteropolymer, i.e. a polymer of a single sugar or a polymer of repeats of different sugars (O-units), respectively, and their biosynthesis differs in some respects. Both types of O-polysaccharides have been found in genus *Yersinia.*

Wzy-dependent O-antigen biosynthesis
The heteropolymeric O-units are synthesized on a lipid carrier molecule, undecaprenyl phosphate (Und-P), a 55-carbon isoprenoid carrier lipid. A complete O-unit is flipped to the periplasmic face of the inner membrane where Wzy (the O-Ag polymerase) polymerizes the O-units and Wzz (the O-Ag chain length determinator) somehow controls the length of the O-Ag although in a yet not understood manner (Daniels and Morona, 1999; Franco *et al.*, 1998; Whitfield *et al.*, 1997).

Most steps in the O-antigen biosynthesis take place in the cytoplasmic face of IM. In the *heteropolymeric pathway,* each identical repeat unit is first assembled on a carrier lipid, Und-P. Remarkably, the initiation reaction, transfer of the first sugar-1-P to Und-P, is in many strains accomplished by WecA, a *N*-acetylglucosamine-1-P transferase, the gene of which is located in the enterobacterial common antigen (ECA) gene cluster. In a few cases another dedicated transferase, such as a Gal-1-P transferase, is used for initiation. The O-unit assembly is then continued by dedicated glycosyltransferases that use NDP-activated sugar precursors as sugar sources. The 'flippase' protein, Wzx (Liu *et al.*, 1996), translocates the completed Und-PP-O-unit to the periplasmic side of IM. There, O-antigen polymerase, Wzy, polymerizes the O-units into long chains still carried by Und-P. The chain length determinant protein Wzz regulates the chain length. The polymerized chains are transferred from Und-P to the preformed lipid A–core structure by O-antigen ligase, WaaL.

Wzy-independent O-antigen biosynthesis
In the *homopolymeric pathway,* the O-antigen polymer is synthesized on Und-P and completely elongated to full length in the cytoplasmic side of IM by sequential transfer of sugar residues to the non-reducing end of the growing polysaccharide chain. The completed O-antigen is translocated to the periplasmic side by an ATP driven transporter system, composed of Wzt and Wzm, that belong to the ATP binding cassette (ABC) transporter family (Whitfield, 1995). The O-antigen ligase, WaaL, transfers and ligates also the homopolymeric O-antigen onto lipid A–core.

Enterobacterial common antigen (ECA) and LPS
ECA is a polysaccharide with a trisaccharide repeat unit expressed by all Enterobacteriaceae including *Yersinia* (Kuhn *et al.*, 1988; Radziejewska-Lebrecht *et al.*, 1998; Rick and Silver, 1996). ECA can be in three different structural forms: (i) linked to L-glycerophosphatide, (ii) linked to LPS, or (iii) as a cyclic structure (Rick and Silver, 1996). When linked to LPS, ECA is ligated specifically to those lipid A–core molecules that do not carry O-antigen. ECA biosynthesis on Und-P takes place using similar reactions as described above for heteropolymeric O-antigen, and ligation to LPS is WaaL dependent (Rick and Silver, 1996). LPS-linked ECA is immunogenic and antibodies against ECA are formed during infection.

Genetics of lipopolysaccharide biosynthesis

LPS is a very complex polysaccharide structure with a wide variety of sugar specificities and glycosidic linkages. In bacteria the genes involved in LPS biosynthesis are usually organized in gene clusters comprising one or more operons. The LPS genes encode NDP-sugar biosynthetic enzymes, glycosyltransferases forming the glycosidic bonds between the sugar residues, the O-unit polymerases and the machinery required to translocate the structure to the CM.

Genetic organization of the O-Ag and outer core gene clusters of *Yersinia*

In the genus *Yersinia* the genomic locus of ca. 20 kb in size between the *hemH* and *gsk* genes is occupied by the genes required for the biosynthesis of the heteropolymeric O-Ags (Skurnik, 1999). In strains which express homopolymeric O-Ags, such as *Y. enterocolitica* O:3 and O:9, the *hemH-gsk* locus is occupied by the outer core (OC) gene cluster (Skurnik *et al.*, 1995). In these strains the O-Ag gene cluster is located elsewhere in the genome.

Genetic setup of O-Ag gene clusters of Y. pseudotuberculosis *and* Y. pestis

The genetic set-up of the O-Ag gene clusters of different *Y. pseudotuberculosis* serotypes has been recently elucidated in the laboratories of Peter Reeves and Mikael Skurnik both by gene-specific PCRs and by sequencing (Bogdanovich *et al.*, 2003; Pacinelli *et al.*, 2002; Reeves *et al.*, 2003; Skurnik, 2003; Skurnik and Bengoechea, 2003; Skurnik *et al.*, 2000). Details of the gene clusters is available in the Bacterial Polysaccharide Gene Database (http://www.mmb.usyd.edu.au/BPGD/default.htm). The gene clusters carry biosynthetic gene modules and the O-Ag variability is based on the module make-up. The known structures of the *Y. pseudotuberculosis* O-Ags carry four or five sugar residues, however, some of the sugars (such as glucose, galactose and N-acetylglucosamine) are also present in other oligo- or polysaccharides, thus their biosynthetic genes may not be within the O-antigen gene clusters. Characteristic biosynthetic gene modules are those carrying genes involved in the biosynthesis of CDP-dideoxyhexoses (the *ddhABCD* and *prt*, *tyv*, *ascEF* or *abe* genes), GDP-mannose (the *manBC* genes) and GDP-fucose (the *gmd* and *fcl* genes). Those serotypes that have N-acetylgalactosamine in their structures (O:2b, O:3, O:5a, O:5b and O:4a) carry the *gne* gene encoding for UDP-*N*-acetylglucosamine-4-epimerase that will provide the bacteria with UDP-*N*-acetylgalactosamine. The biosynthetic modules usually also include genes for the specific glycosyltransferases and for the O-unit structure-specific proteins Wzx and Wzy. The gene clusters of bacteria that contain 6-deoxy-*manno*-heptose in their O-antigens have the *wbyB–gmhB* region dedicated for the biosynthesis of GDP-6-deoxy-*manno*-heptose and the necessary glycosyltransferases.

In the *Y. pestis* O-Ag gene cluster that is >95% identical to that of *Y. pseudotuberculosis* serotype O:1b, five genes are inactivated either by frameshift mutations (four genes) or a small deletion (Skurnik *et al.*, 2000). The mutations prevent the biosynthesis of GDP-fucose and CDP-paratose, the GDP-mannose pathway is however, intact. The biological significance of these mutations is not clear at present.

Genetics and biosynthesis of Y. enterocolitica *O-antigens and the OC hexasaccharide of* Y. enterocolitica *serotype O:3*

O-Ag of serotype O:8 The O-Ag polysaccharide of *Y. enterocolitica* serotype O:8 (hereafter *Ye* O:8) contains branched pentasaccharide O-units of N-acetyl-galactosamine (GalNAc), L-fucose (Fuc), D-galactose (Gal), D-mannose (Man) and 6-deoxy-D-gulose (6d-Gul). The repeat units are linked together by a (1→4) glycosidic bond between GalNAc and Man residues (Tomshich *et al.*, 1987). The O-Ag gene cluster contains 18 genes which are sufficient to express the O-Ag in a heterologous host like *Escherichia coli*. The NDP-activated sugar precursors needed for the *Ye* O:8 O-unit biosynthesis are: (i) CDP-6-deoxygulose, synthesized enzymatically from glucose-1-phosphate by the action of DdhA, DdhB, WbcA and WbcB, (ii) GDP-mannose, synthesized from fructose-1-phosphate by ManA, ManB and ManC, (iii) GDP-fucose, synthesized from GDP-mannose by Gmd and Fcl, (iv) UDP-N-acetylgalactosamine, synthesized from UDP-N-acetylglucosamine by Gne (Bengoechea *et al.*,

2002a), and (v) UDP-galactose, synthesized from UDP-glucose by GalE. The glycosyltransferases needed to assemble the O-unit on Und-P are WecA (transfers the N-acetylgalactosamine residue to Und-P), WbcC, WbcG, WbcH and WbcI. The cluster also contains the genes encoding for Wzx and Wzy (Bengoechea *et al.*, 2002a) both required to build up the heteropolymeric O-antigen. The Wzz protein, also present in the cluster, is required to control the O-Ag length which, in *Ye* O:8, is of 7 to 10 repeats (Bengoechea *et al.*, 2002a; Zhang *et al.*, 1997; Zhang *et al.*, 1996). At present we do not have a function for the *wbcF* gene product. Thus, all the genes needed are located in the O-Ag gene cluster except *manA*, *galE* and *wecA*. The enzymatic activities of the last three gene products are used in other cellular processes in addition to the O-Ag biosynthesis.

Promoter cloning experiments revealed that the O-Ag gene cluster contains two promoters, one promoter upstream of *ddhA*, the first gene in the O-Ag cluster, and another one between the *manB* and *gne* genes (designated Pwb_1 and Pwb_2 respectively) (Bengoechea *et al.*, 2002b). These promoters are expressed under a complex regulatory circuit (see below).

O-Ag and OC of serotypes O:3 and O:9 The structures of the LPSs of *Ye* O:3 and O:9 (and presumably also some other serotypes) have some peculiarities rarely seen in other enterobacteria. The serotype O:3 O-Ag is a homopolymer of 1–2 linked 6-deoxy-L-altrose (Hoffman *et al.*, 1980) and that of O:9, a homopolymer of N-formylperosamine (Caroff *et al.*, 1984) that are attached to the inner core region of the LPS. In addition to the O-Ag of these serotypes also the OC hexasaccharide is attached to the inner core thus forming a short branch in the LPS molecule (Skurnik, 2004; Skurnik, 2007). This peculiar structure has made it possible to construct mutants that are missing either the O-antigen, the OC or both, and that has been extensively studied with serotype O:3 (Sirisena and Skurnik, 2003; Skurnik *et al.*, 1995, 1999). Based on PCR-analysis the O:9 OC gene cluster is very closely related to that of O:3. In both serotypes, the OC gene clusters are located between the *hemH* and *gsk* genes while the genomic location of the O-Ag clusters are presently unknown (Skurnik, 2004; Skurnik and Bengoechea, 2003).

Yersinia O-antigen and virulence

To characterize the role of *Y. enterocolitica* O-Ag in virulence two rough mutants were isolated from *Ye* O:8 and *Ye* O:3 by resistance to lysis by bacteriophages and they were used to infect mice orogastrically (Al-Hendy *et al.*, 1992; Zhang *et al.*, 1997). Significantly, both mutants were attenuated with LD_{50} values 50–100 times higher than the wild-type (wt) strains. To pinpoint the phase of infection in which the mutants manifest the attenuation, mice were co-infected with the wt strains and each of the rough strains. Neither of the mutants colonized the Peyer's patches as efficiently as the wt strains and they did not multiply in spleen, liver and mesenteric lymph nodes. The importance of *Ye* O:8 O-Ag as virulence factor has been further corroborated by signature-tagged transposon mutagenesis (STM) experiment as 23% of the attenuated mutants had insertions inactivating genes in the O-antigen gene cluster (Darwin and Miller, 1999). The *Ye* O:8 rough mutant was also attenuated in two other animal models, rabbit and pig, which are relevant in the context of yersiniosis (Najdenski *et al.*, 2006; Najdenski *et al.*, 2003).

We have also studied the relative contribution of OC to *Ye* O:3 virulence. The LD_{50} of the OC mutant was approximately 1000 times higher than the wt in orogastrically infected mice (Skurnik *et al.*, 1999). However, in contrast to the rough mutant, co-infection experiments revealed that the OC mutant did colonize the Peyer's patches as efficiently as the wt but it was much less efficient in colonizing deeper organs and at 5 days post infection it was completely eliminated from Peyer's patches. It is then clear that *Ye* O:3 OC and O-Ag play different roles during infection. It seems that O-Ag is needed during the first hours of infection whereas the OC is required for prolonged survival of the bacteria in Peyer's patches and for invasion of deeper tissues like liver and spleen.

Interestingly, O-Ag also plays a role in virulence of *Y. pseudotuberculosis*. Using signature-tagged transposon mutagenesis (STM), it was found that 15–30% of the isolated mutants do not express O-Ag (Karlyshev *et al.*, 2001; Mecsas *et al.*, 2001). Similar to *Y. enterocolitica* rough strains, these *Y. pseudotuberculosis* mutants were unable to reach the Peyer's patches and, furthermore, they did not invade epithelial cells as efficiently as the wild type (Mecsas *et al.*, 2001).

On the contrary, O-Ag does not play any role in *Y. pestis* virulence since this species does not express O-Ag (Darveau *et al.*, 1983; Hartley *et al.*, 1974; Hitchen *et al.*, 2002; Prior *et al.*, 2001; Skurnik *et al.*, 2000). Interestingly, expression of a heterologous O-Ag in *Y. pestis* did not affect significantly its virulence (Oyston *et al.*, 2003). However, we cannot at present rule out that expression of *Y. pestis* homologous O-Ag might affect the virulence of this species. Nevertheless, the function of the plasminogen-activating protease (Pla) was compromised in a *Y. pestis* strain expressing heterologous O-Ag (Kukkonen *et al.*, 2004). Pla is essential for the development of pneumonic and bubonic forms of plague (Lathem *et al.*, 2005; Sebbane *et al.*, 2006).

We can now only speculate on the specific role of O-Ag in virulence. It could be that O-Ag prevents the access of harmful molecules into the outer membrane as has been demonstrated for many pathogens. For example, it is generally accepted that O-Ag is involved in the resistance to complement-mediated killing and to antimicrobial peptides. However, our initial results indicated that this is not the case in *Y. enterocolitica* (Bengoechea *et al.*, 2004; Biedzka-Sarek *et al.*, 2005; Skurnik *et al.*, 1999, 2007). It is also conceivable that some virulence factors located in the outer membrane require the presence of the O-Ag for its proper expression or functionality and hence the O-Ag role in virulence would be indirect (see below).

Regulation of LPS genes

JUMP start and ops sequences

The gene clusters for *Y. enterocolitica* O:3 outer core and O-antigen, for *Y. enterocolitica* O:8 O-antigen, for *Y. enterocolitica* O:9 outer core and O-antigen and for all *Y. pseudotuberculosis* O-antigens studied this far, carry a characteristic sequence called JUMPstart sequence. The JUMPstart sequence has been identified in all surface polysaccharide biosynthesis gene clusters and also for sex pilus assembly and haemolysin production (Hobbs and Reeves, 1994; Nieto *et al.*, 1996). The sequence is generally located downstream of functional promoters of the gene cluster. Nieto *et al* reported that a short 8-bp sequence within the JUMPstart sequence plays a role in gene regulation, and this element was named as *ops* (operon polarity suppressor). *ops* was shown to function as a *cis*-acting element that together with RfaH, a transcriptional regulator, allows efficient transcription of long operons, especially of the distal genes (Bailey *et al.*, 1996). It is postulated that *ops* is giving the specificity for RfaH for the subset of operons it is regulating. The role of *ops* and RfaH in regulation of *Yersinia* LPS operons has not yet been studied but they are very likely to be involved.

Regulation of O-antigen expression in *Yersinia*

Y. enterocolitica expresses less O-Ag when grown at 37°C. This has been demonstrated by bacteriophage susceptibility tests (Kawaoka *et al.*, 1983) exposure of antigenic determinants to antibodies (Acker *et al.*, 1981) and by chemical analysis of isolated LPS (Wartenberg *et al.*, 1983). By Northern blot analysis we could reveal that O-Ag gene cluster transcription was down-regulated at 37°C (Al-Hendy *et al.*, 1991; Bengoechea *et al.*, 2002b). We constructed reporter strains in which *lucFF* was introduced downstream of each of the promoters identified in the O-Ag gene clusters to study more closely the regulation of O-Ag expression (Bengoechea *et al.*, 2002b; Lahtinen *et al.*, 2003). These *in vitro* studies confirmed that O-Ag expression is down-regulated at 37°C and that there are signals in addition to temperature that modulate O-Ag expression. So far, these signals are the bacterial growth phase, low pH, iron concentration, oxygen tension, ionic strength (Lahtinen *et al.*, 2003) (Bengoechea *et al.*, unpublished). Considering that these signals are likely to be found *in vivo* we speculated whether or not *Y. enterocolitica* expresses O-Ag *in vivo*. Recent experiments have demonstrated that *Ye* O:8 O-Ag expression is higher *in vivo* than *in vitro* at 37°C. It is noteworthy that expression is tissue specific being maximal in Peyer's patches and almost absent in spleen and liver (Bengoechea *et al.*, unpublished). This finding might indicate that O-Ag is indeed expressed during infection at least during colonization of the gut. Giving support to this, a study designed to identify *Y. enterocolitica* O:8 *loci* expressed *in vivo* found that several genes of the O-Ag cluster were expressed in Peyer's patches whereas none was expressed in spleen (Gort and Miller, 2000; Young, 2007).

Identification of loci involved in temperature regulation of O-antigen expression

The data summarized above clearly indicate that O-Ag expression is highly regulated although the molecular basis is largely unknown. Therefore, we started the identification of loci that could be involved in the *Ye* O:8 O-Ag regulation. We mapped a region downstream of the O-Ag gene cluster containing an operon formed by the *rosA* and *rosB* genes (for regulation of O-Ag synthesis) that conferred O-Ag temperature dependent expression (Bengoechea *et al.*, 2002b). These studies also revealed that the RosAB system required the expression of *Ye* O:8 Wzz, the O-Ag chain length determinant, to exert its regulatory effect. Expression of RosAB without Wzz did not down-regulate O-Ag expression at 37°C. RosA is a proton motive force-driven efflux pump involved in resistance to amphipathic compound and RosB is a potassium antiporter that together with RosA are also involved in the resistance to antimicrobial peptides (Bengoechea and Skurnik, 2000). To address the role of RosAB and Wzz in O-Ag regulation in *Yersinia*, we constructed a *rosA-rosB* double mutant (YeO8-RosA$^-$B$^-$) and a *wzz* mutant (YeO8-ΔWzzGB) and analysed O-Ag expression (Bengoechea *et al.*, 2002b). The *rosAB* double mutant but also *rosA* or *rosB* non-polar mutants expressed significantly more O-Ag than *Ye* O:8 when grown at 37°C. Analysis of O-Ag expression by the *Ye* O:8 *wzz* mutant demonstrated that in addition to displaying a random distribution of O-Ag chain lengths this mutant also expressed more O-antigen than when the wt was grown at 37°C (Bengoechea *et al.*, 2002b).

RosAB and Wzz modulate the O-Ag transcription

Taking into consideration that O-Ag expression is transcriptionally regulated; it was quite feasible to postulate that RosAB affects the transcriptional regulation of the O-Ag gene cluster. Indeed, transcriptional analysis using promoter fusions to of each of the two promoters identified in the O-Ag gene cluster, P*wb*$_1$ and P*wb2*, demonstrated that only the activity of the second promoter was higher in the *rosAB* background than in the wt background (Bengoechea *et al.*, 2002b). Thus, in the *rosAB* strain at 37°C the P*wb2* activity is derepressed leading to overexpression of both Gne and Wzz which is linked to an increase in the O-Ag expression. In good agreement, overexpression of *gne* or *wzz* in the wt strain caused an increase in O-Ag expression at 37°C (Bengoechea *et al.*, 2002b).

To study the role of Wzz in the transcriptional regulation of the O-Ag gene cluster we followed a similar approach as before. Results showed that both promoter fusions were up-regulated in the *wzz* mutant background (Bengoechea *et al.*, 2002b). Since Wzz is overexpressed in the *rosAB* mutant we studied the promoter activities in wt overexpressing Wzz. Somewhat unexpectedly both promoter fusions were down-regulated, the effect being more pronounced for the P*wb2* than for the P*wb1* promoter and more evident in bacteria grown at 37°C than in bacteria grown at RT (Bengoechea *et al.*, 2002b). This finding was in sharp contrast to the LPS phenotype of *Ye* O:8 overexpressing Wzz that produced more O-antigen when grown at 37°C (Bengoechea *et al.*, 2002b). Since more O-antigen is produced from a decreased number of transcripts the efficiency of the post-transcriptional events in the O-antigen biosynthesis must be better. At present, we find two non-exclusive explanations for this.

1 Since Wzz, Wzy, Wzx and perhaps also WaaL may form a complex involved in the O-antigen assembly/translocation it is possible that overexpression of Wzz could facilitate the Wzy polymerase function. Thus, changes in the stoichiometry of the complex could modulate the efficiency of the O-antigen polymerization and translocation resulting in increased expression of O-antigen.
2 Overexpression of Wzz could also affect the expression or function of systems that are involved in down-regulation of the O-antigen expression at the post-transcriptional level. Supporting this explanation

the expression of *rosAB* was affected by Wzz (Bengoechea *et al.*, 2002b).

Coordinate regulation of O-antigen and other *Yersinia* virulence factors

One hallmark of pathogenic bacteria is their ability to coordinate the expression of different virulence factors. The outer membrane, and particularly the O-Ag, plays a critical role in the bacterial interaction with the environment and hence it is likely that changes in the O-Ag affect bacterial virulence as discussed in the preceding sections. In this context, we put forward the hypothesis that the expression of other virulence factors could be coordinated with the O-Ag regulation. Initially we focused our attention in those factors located in the outer membrane and analysed whether their expression is affected in *Ye* O:8 rough strain. Results have shown that *inv* expression was one twelfth in the mutant compared with that in the wt (Bengoechea *et al.*, 2004). Of note, Inv is the main factor for penetration of the intestinal epithelium by *Y. enterocolitica* (Pepe and Miller, 1993; Pepe *et al.*, 1995). Significantly, this finding may account for the low colonization of Peyer's patches by the rough mutant. Indeed, invasion of epithelial culture cells by the rough mutant was severely impaired (Bengoechea *et al.*, 2004). Our studies have also shown that the expression of the flagellar master regulatory operon, *flhDC*, was 60-fold higher in the rough mutant than in the wt (Bengoechea *et al.*, 2004). Furthermore, the rough strain was hypermotile and expressed higher amounts of flagellins than the wt. Noteworthy, in *Yersinia* these two phenotypes have been previously linked to *flhDC* overexpression (Young *et al.*, 1999). It has been demonstrated that the *Yersinia* flagellum basal body functions as a type III secretion system and secretes proteins designated as Fops (flagellar outer proteins) (Young and Young, 2002). Moreover, the amount of secreted proteins correlates with *flhDC* expression. One of these proteins is YlpA, a virulence-associated phospholipase A (Schmiel *et al.*, 1998), which, as we expected, was secreted in higher quantities by the rough strain than the wt (Bengoechea *et al.*, 2004). In a recent study, we analysed whether the absence of O-Ag also affected the type III secretion system encoded in the virulence plasmid pYV. Indeed, expression of the secreted proteins, named Yops, was downregulated in the rough mutant (Perez-Gutierrez *et al.*, 2007). Mechanistically, this was as a consequence of the abnormally elevated levels of FlhDC found in the rough mutant. Most probably the global regulator H-NS underlies *flhDC* up-regulation in the rough strain and our data clearly indicated that there is a relationship between O-Ag and H-NS levels (Perez-Gutierrez *et al.*, 2007).

Altogether these findings give experimental support to our hypothesis and serve as evidence that the expression of *Ye* O:8 membrane components might be coordinated by a regulatory network. A tantalizing speculation could be that *Y. enterocolitica* somehow senses the O-Ag status and this acts as a regulatory signal that, so far, affects the expression of outer membrane components.

Variation of the lipid A structure in *Yersinia*

Recent studies have shown that yersiniae also regulate the structure of lipid A. Thus, the number and structure of the lipid A fatty acids and the substitutions of the 1- and 4′-positions in the glucosamine disaccharide can vary. Rebeil and coworkers (Rebeil *et al.*, 2004) elegantly demonstrated that a shift in temperature induces a change in the number and type of acyl groups on the lipid A of the three *Yersinia* species. At 21°C, lipid As are mainly hexa-acylated whereas at 37°C they are tetra-acylated (Rebeil *et al.*, 2004). Likely the temperature-regulation of the lipid A acyltransferases explains the shift in lipid A acylation both in *Y. pestis* (Rebeil *et al.*, 2006) and *Y. enterocolitica* (Pérez-Gutiérrez and Bengoechea, unpublished).

The endotoxic activity of LPS depends on the lipid A acylation pattern such that hexa-acylated lipid A is most endotoxic and tetra-acylated almost non-toxic. Not surprisingly, lipid A from yersiniae grown at 37°C stimulated less TNFα from human monocytes than lipid A from bacteria grown at 21°C (Rebeil *et al.*, 2004). The importance of expressing a less acylated lipid A at the host temperature has been corroborated in *Y. pestis*. A constitutively hexa-acylated *Y. pestis* appears to be severely attenuated (Montminy *et al.*, 2006) thus extending the importance of the LPS regulation to early phases of infection where hexa-acylated LPS functions as a triggering signal for innate immunity.

The substitution of the phosphate at the 1′ end of the glucosamine disaccharide with aminoarabinose is also temperature regulated in *Y. pestis* (Anisimov *et al.*, 2005; Rebeil *et al.*, 2004) and in *Y. enterocolitica* (Reinés and Bengoechea, unpublished). Aminoarabinose content is higher in lipid As from bacteria grown at 21°C than those from bacteria grown at 37°C. Similarly to other Gram-negative pathogens, this lipid A modification is associated with increased resistance to antimicrobial peptides, key weapons of the innate immune system against infections. Thus, yersiniae are more resistant to antimicrobial peptides when grown at 21°C than at 37°C (Anisimov *et al.*, 2005; Bengoechea *et al.*, 1998a,b, 1996). The impact in virulence, if any, of the substitution of lipid A with aminoarabinose is not clearly established although a *Y. pseudotuberculosis* mutant lacking this modification is as virulent as the wt strain (Marceau *et al.*, 2004).

Conclusions

LPS is the major component of the outer leaflet of the outer membrane of Gram-negative bacteria. Recent data from a variety of organisms demonstrate that in addition to the endotoxic activity residing in the lipid A part of LPS, the O-antigen also plays a significant biological role in resistance to elements of the host innate immune system. The structures of LPS components are genetically regulated and this is reflected to the biological function of the molecule.

Acknowledgements

Our work has been supported by grants from the Academy of Finland and the European Commission (contract QLK-99-007800) to M.S. and Fondo de Investigación Sanitaria (Instituto de Salud Carlos III) and Govern Balear (Grant for Competitive groups) to J.A.B.

References

Acker, G., Knapp, W., Wartenberg, K., and Mayer, H. (1981). Localization of enterobacterial common antigen in *Yersinia enterocolitica* by the immunoferritin technique. J. Bacteriol. *147*, 602–611.

Al-Hendy, A., Toivanen, P., and Skurnik, M. (1991). The effect of growth temperature on the biosynthesis of *Yersinia enterocolitica* O:3 lipopolysaccharide: temperature regulates the transcription of the *rfb* but not of the *rfa* region. Microb. Pathog. *10*, 81–86.

Al-Hendy, A., Toivanen, P., and Skurnik, M. (1992). Lipopolysaccharide O side chain of *Yersinia enterocolitica* O:3 is an essential virulence factor in an orally infected murine model. Infect. Immun. *60*, 870–875.

Anisimov, A.P., Dentovskaya, S.V., Titareva, G.M., Bakhteeva, I.V., Shaikhutdinova, R.Z., Balakhonov, S.V., Lindner, B., Kocharova, N.A., Senchenkova, S.N., Holst, O., *et al.* (2005). Intraspecies and temperature-dependent variations in susceptibility of *Yersinia pestis* to the bactericidal action of serum and to polymyxin B. Incection and Immunity *73*, 7324–7331.

Bailey, M. J. A., Hughes, C., and Koronakis, V. (1996). Increased distal gene transcription by the elongation factor RfaH, a specialized homologue of NusG. Mol. Microbiol. *22*, 729–737.

Bengoechea, J.A., Brandenburg, K., Seydel, U., Díaz, R., and Moriyón, I. (1998a). *Yersinia pseudotuberculosis* and *Yersinia pestis* show increased outer membrane permeability to hydrophobic agents which correlates with lipopolysaccharide acyl-chain fluidity. Microbiology *144*, 1517–1526.

Bengoechea, J.A., Díaz, R., and Moriyón, I. (1996). Outer membrane differences between pathogenic and environmental *Yersinia enterocolitica* biogroups probed with hydrophobic permeants and polycationic peptides. Infect. Immun. *64*, 4891–4899.

Bengoechea, J.A., Lindner, B., Seydel, U., Díaz, R., and Moriyón, I. (1998b). *Yersinia pseudotuberculosis* and *Yersinia pestis* are more resistant to bactericidal cationic peptides than *Yersinia enterocolitica*. Microbiology *144*, 1509–1515.

Bengoechea, J.A., Najdenski, H., and Skurnik, M. (2004). Lipopolysaccharide O antigen status of *Yersinia enterocolitica* O:8 is essential for virulence and absence of O antigen affects the expression of other *Yersinia* virulence factors. Mol. Microbiol. *52*, 451–469.

Bengoechea, J.A., Pinta, E., Salminen, T., Oertelt, C., Holst, O., Radziejewska-Lebrecht, J., Piotrowska-Seget, Z., Venho, R., and Skurnik, M. (2002a). Functional characterization of Gne (UDP-N-acetylglucosamine-4-epimerase), Wzz (chain length determinant), and Wzy (O-antigen polymerase) of *Yersinia enterocolitica* serotype O:8. J. Bacteriol. *184*, 4277–4287.

Bengoechea, J.A., and Skurnik, M. (2000). Temperature-regulated efflux pump/potassium antiporter system mediates resistance to cationic antimicrobial peptides in *Yersinia*. Mol. Microbiol. *37*, 67–80.

Bengoechea, J.A., Zhang, L., Toivanen, P., and Skurnik, M. (2002b). Regulatory network of lipopolysaccharide O-antigen biosynthesis in *Yersinia enterocolitica* includes cell envelope-dependent signals. Mol. Microbiol. *44*, 1045–1062.

Biedzka-Sarek, M., Venho, R., and Skurnik, M. (2005). Role of YadA, Ail, and lipopolysaccharide in serum resistance of *Yersinia enterocolitica* serotype O:3. Infect. Immun. *73*, 2232–2244.

Bogdanovich, T.M., Carniel, E., Fukushima, H., and Skurnik, M. (2003). Genetic (sero)typing of *Yersinia pseudotuberculosis*. In The Genus *Yersinia*: entering the functional genomic era, M. Skurnik, K. Granfors, and

J. A. Bengoechea, eds. (New York, Kluwer Academic/Plenum Publishers), pp. 337–340.

Caroff, M., Bundle, D.R., and Perry, M. B. (1984). Structure of the O-chain of the phenol-phase soluble cellular lipopolysaccharide of *Yersinia enterocolitica* serotype O:9. Eur. J. Biochem. *139*, 195–200.

Daniels, C., and Morona, R. (1999). Analysis of *Shigella flexneri* Wzz (Rol) function by mutagenesis and cross-linking: Wzz is able to oligomerize. Mol. Microbiol. *34*, 181–194.

Darveau, R.P., Charnetzky, W.T., Hurlbert, R.E., and Hancock, R. E. W. (1983). Effects of growth temperature, 47-megadalton plasmid, and calcium deficiency on the outer membrane protein porin and lipopolysaccharide composition of *Yersinia pestis* EV76. Infect. Immun. *42*, 1092–1101.

Darwin, A.J., and Miller, V. L. (1999). Identification of *Yersinia enterocolitica* genes affecting survival in an animal host using signature-tagged transposon mutagenesis. Mol. Microbiol. *32*, 51–62.

Franco, A.V., Liu, D., and Reeves, P. R. (1998). The Wzz (Cld) protein in *Escherichia coli*: amino acid sequence variation determines O-antigen chain length specificity. J. Bacteriol. *180*, 2670–2675.

Gort, A.S., and Miller, V. L. (2000). Identification and characterization of *Yersinia enterocolitica* genes induced during systemic infection. Infect. Immun. *68*, 6633–6642.

Hartley, J.L., Adams, G.A., and Tornabene, T. G. (1974). Chemical and physical properties of lipopolysaccharide of *Yersinia pestis*. J. Bacteriol. *118*, 848–854.

Hitchen, P.G., Prior, J.L., Oyston, P.C., Panico, M., Wren, B.W., Titball, R.W., Morris, H.R., and Dell, A. (2002). Structural characterization of lipo-oligosaccharide (LOS) from *Yersinia pestis*: regulation of LOS structure by the PhoPQ system. Mol. Microbiol. *44*, 1637–1650.

Hobbs, M., and Reeves, P. R. (1994). The JUMPstart sequence: A 39 bp element common to several polysaccharide gene clusters. Mol. Microbiol. *12*, 855–856.

Hoffman, J., Lindberg, B., and Brubaker, R. R. (1980). Structural studies of the O-specific side-chains of the lipopolysaccharide from *Yersinia enterocolitica* Ye 128. Carbohydr. Res. *78*, 212–214.

Karlyshev, A.V., Oyston, P.C., Williams, K., Clark, G.C., Titball, R.W., Winzeler, E.A., and Wren, B. W. (2001). Application of high-density array-based signature-tagged mutagenesis to discover novel *Yersinia* virulence-associated genes. Infect. Immun. *69*, 7810–7819.

Kawaoka, Y., Otsuki, K., and Tsubokura, M. (1983). Growth temperature-dependent variation in the bacteriophage-inactivating capacity and antigenicity of *Yersinia enterocolitica* lipopolysaccharide. J. Gen. Microbiol. *129*, 2739–2747.

Kuhn, H.-M., Meier-Dieter, U., and Mayer, H. (1988). ECA, the enterobacterial common antigen. FEMS Microbiol. Rev. *54*, 195–222.

Kukkonen, M., Suomalainen, M., Kyllonen, P., Lahteenmaki, K., Lang, H., Virkola, R., Helander, I.M., Holst, O., and Korhonen, T. K. (2004). Lack of O-antigen is essential for plasminogen activation by Yersinia pestis and Salmonella enterica. Mol. Microbiol. *51*, 215–225.

Lahtinen, P., Brzezinska, A., and Skurnik, M. (2003). Temperature and growth phase regulate the transcription of the O-antigen gene cluster of *Yersinia enterocolitica* O:3. In The Genus *Yersinia*: entering the functional genomic era, M. Skurnik, K. Granfors, and J. A. Bengoechea, eds. (New York, Kluwer Academic/Plenum Publishers), pp. 289–292.

Lathem, W.W., Crosby, S.D., Miller, V.L., and Goldman, W. E. (2005). Progression of primary pneumonic plague: a mouse model of infection, pathology, and bacterial transcriptional activity. Proc. Natl. Acad. Sci. USA *102*, 17786–17791.

Liu, D., Cole, R.A., and Reeves, P. R. (1996). An O-antigen processing function for Wzx (RfbX): a promising candidate for O-unit flippase. J. Bacteriol. *178*, 2102–2107.

Marceau, M., Sebbane, F., Ewann, F., Collyn, F., Lindner, B., Campos, M.A., Bengoechea, J.A., and Simonet, M. (2004). The *pmrF* polymyxin-resistance operon of *Yersinia pseudotuberculosis* is up-regulated by the PhoP-PhoQ two-component system but not by PmrA-PmrB, and is not required for virulence. Microbiology *150*, 3947–3957.

Mecsas, J., Bilis, I., and Falkow, S. (2001). Identification of attenuated *Yersinia pseudotuberculosis* strains and characterization of an orogastric infection in BALB/c mice on day 5 postinfection by signature-tagged mutagenesis. Infect. Immun. *69*, 2779–2787.

Montminy, S.W., Khan, N., McGrath, S., Walkowicz, M.J., Sharp, F., Conlon, J.E., Fukase, K., Kusumoto, S., Sweet, C., Miyake, K., *et al.* (2006). Virulence factors of *Yersinia pestis* are overcome by a strong lipopolysaccharide response. Nature Immunology *7*, 1066–1073.

Najdenski, H., Golkocheva, E., Kussovski, V., Ivanova, E., Manov, V., Iliev, M., Vesselinova, A., Bengoechea, J.A., and Skurnik, M. (2006). Experimental pig yersiniosis to assess attenuation of *Yersinia enterocolitica* O:8 mutant strains. FEMS Immunol. Med. Microbiol. *47*, 425–435.

Najdenski, H., Golkocheva, E., Vesselinova, A., Bengoechea, J.A., and Skurnik, M. (2003) Proper expression of the O-antigen of lipopolysaccharide is essential for the virulence of *Yersinia enterocolitica* O:8 in experimental oral infection of rabbits. FEMS Immunol. Med. Microbiol. *38*, 97–106.

Nieto, J.M., Bailey, M. J. A., Hughes, C., and Koronakis, V. (1996). Suppression of transcription polarity in the *Escherichia coli* haemolysin operon by a short upstream element shared by polysaccharide and DNA transfer determinants. Mol. Microbiol. *19*, 705–713.

Oyston, P.C., Prior, J.L., Kiljunen, S., Skurnik, M., Hill, J., and Titball, R. W. (2003). Expression of heterologous O-antigen in *Yersinia pestis* KIM does not affect virulence by the intravenous route. J. Med. Microbiol. *52*, 289–294.

Pacinelli, E., Wang, L., and Reeves, P. R. (2002). Relationship of *Yersinia pseudotuberculosis* O antigens IA, IIA, and IVB: the IIA gene cluster was derived from that of IVB. Infect. Immun. *70*, 3271–3276.

Pepe, J.C., and Miller, V. L. (1993). *Yersinia enterocolitica* invasin: A primary role in the initiation of infection. Proc. Natl. Acad. Sci. USA *90*, 6473–6477.

Pepe, J.C., Wachtel, M.R., Wagar, E., and Miller, V. L. (1995). Pathogenesis of defined invasion mutants of *Yersinia enterocolitica* in a BALB/c mouse model of infection. Infect. Immun. *63*, 4837–4848.

Perez-Gutierrez, C., Llompart, C.M., Skurnik, M., and Bengoechea, J. A. (2007). Expression of the *Yersinia enterocolitica* pYV-encoded type III secretion system is modulated by lipopolysaccharide O-antigen status. Infect. Immun. *75*, 1512–1516.

Prior, J.L., Parkhill, J., Hitchen, P.G., Mungall, K.L., Stevens, K., Morris, H.R., Reason, A.J., Oyston, P.C., Dell, A., Wren, B.W., and Titball, R. W. (2001). The failure of different strains of *Yersinia pestis* to produce lipopolysaccharide O-antigen under different growth conditions is due to mutations in the O-antigen gene cluster. FEMS Microbiol. Lett. *197*, 229–233.

Radziejewska-Lebrecht, J., Skurnik, M., Shashkov, A.S., Brade, L., Rozalski, A., Bartodziejska, B., and Mayer, H. (1998). Immunochemical studies on R mutants of *Yersinia enterocolitica* O:3. Acta Biochimica Polonica *45*, 1011–1019.

Raetz, C.R., and Whitfield, C. (2002). Lipopolysaccharide endotoxins. Annu. Rev. Biochem. *71*, 635–700.

Rebeil, R., Ernst, R.K., Gowen, B.B., Miller, S.I., and Hinnebusch, B. J. (2004). Variation in lipid A structure in the pathogenic yersiniae. Mol. Microbiol. *52*, 1363–1373.

Rebeil, R., Ernst, R.K., Jarrett, C.O., Adams, K.N., Miller, S.I., and Hinnebusch, B. J. (2006). Characterization of late acyltransferase genes of *Yersinia pestis* and their role in temperature-dependent lipid A variation. J. Bacteriol. *188*, 1381–1388.

Reeves, P.R., Pacinelli, E., and Wang, L. (2003). O antigen Gene Clusters of *Yersinia pseudotuberculosis*. In The Genus *Yersinia*: entering the functional genomic era, M. Skurnik, K. Granfors, and J. A. Bengoechea, eds. (Kluwer Academic/Plenum Publishers), pp. 199–206.

Rick, P.D., and Silver, R.P. (1996). Enterobacterial common antigen and capsular polysaccharides. In *Escherichia coli* and *Salmonella:* cellular and molecular biology, F.C. Neidhardt, R. Curtiss III, J.L. Ingraham, E.C.C. Lin, K.B. Low, B. Magasanik, W. S. Reznikoff, M. Riley, M. Schaechter, and H. E. Umbarger, eds. (Washington, D.C., American Society for Microbiology), pp. 104–118.

Schmiel, D.H., Wagar, E., Karamanou, L., Weeks, D., and Miller, V.L. (1998). Phospholipase A of *Yersinia enterocolitica* contributes to pathogenesis in a mouse model. Infect. Immun. *66*, 3941–3951.

Sebbane, F., Jarrett, C.O., Gardner, D., Long, D., and Hinnebusch, B.J. (2006). Role of the *Yersinia pestis* plasminogen activator in the incidence of distinct septicemic and bubonic forms of flea-borne plague. Proc. Natl. Acad. Sci. USA *103*, 5526–5530.

Sirisena, D.M., and Skurnik, M. (2003). Mutations in the genes for synthesis of the outer core region of the lipopolysaccharide of *Yersinia enterocolitica* O:3. J. Appl. Microbiol. *94*, 686–692.

Skurnik, M. (1999). Molecular genetics of *Yersinia* lipopolysaccharide. In Genetics of Bacterial Polysaccharides, J. Goldberg, ed. (Boca Raton, FL, CRC Press), pp. 23–51.

Skurnik, M. (2003). Molecular genetics, biochemistry and biological role of *Yersinia* lipopolysaccharide. In The Genus *Yersinia*: entering the functional genomic era, M. Skurnik, K. Granfors, and J. A. Bengoechea, eds. (New York, Kluwer Academic/Plenum Publishers), pp. 187–189.

Skurnik, M. (2004). Lipopolysaccharides of *Yersinia*. In *Yersinia:* molecular and cellular biology, E. Carniel, and B. J. Hinnebusch, eds. (Wymondham, Horizon Bioscience), pp. 215–241.

Skurnik, M. (2007). My life with *Yersinia*. Adv. Exp. Med. Biol. *603*, 44–73.

Skurnik, M., and Bengoechea, J.A. (2003). The biosynthesis and biological role of lipopolysaccharide O-antigens of pathogenic *Yersiniae*. Carbohydr. Res. *338*, 2521–2529.

Skurnik, M., Biedzka-Sarek, M., Lubeck, P.S., Blom, T., Bengoechea, J.A., Perez-Gutierrez, C., Ahrens, P., and Hoorfar, J. (2007). Characterization and biological role of the O-polysaccharide gene cluster of *Yersinia enterocolitica* serotype O:9. J. Bacteriol. *189*, 7244–7253.

Skurnik, M., Peippo, A., and Ervelä, E. (2000). Characterization of the O-antigen gene clusters of *Yersinia pseudotuberculosis* and the cryptic O-antigen gene cluster of *Yersinia pestis* shows that the plague bacillus is most closely related to and has evolved from *Y. pseudotuberculosis* serotype O:1b. Mol. Microbiol. *37*, 316–330.

Skurnik, M., Venho, R., Bengoechea, J.-A., and Moriyón, I. (1999). The lipopolysaccharide outer core of *Yersinia enterocolitica* serotype O:3 is required for virulence and plays a role in outer membrane integrity. Mol. Microbiol. *31*, 1443–1462.

Skurnik, M., Venho, R., Toivanen, P., and Al-Hendy, A. (1995). A novel locus of *Yersinia enterocolitica* serotype O:3 involved in lipopolysaccharide outer core biosynthesis. Mol. Microbiol. *17*, 575–594.

Tomshich, S.V., Gorshkova, R.P., and Ovodov, Y.S. (1987). Structural investigation of the lipopolysaccharide of *Y. enterocolitica* serovar O:8. Khimia Prirodnykh Soedinenii, 657–664.

Wartenberg, K., Knapp, W., Ahamed, N.M., Widemann, C., and Mayer, H. (1983). Temperature-dependent changes in the sugar and fatty acid composition of lipopolysaccharide from *Yersinia enterocolitica* strains. Int. J. Med. Microb. *253*, 523–530.

Whitfield, C. (1995). Biosynthesis of lipopolysaccharide O antigens. Trends Microbiol. *3*, 178–185.

Whitfield, C., Amor, P.A., and Koplin, R. (1997). Modulation of the surface architecture of gram-negative bacteria by the action of surface polymer: lipid a-core ligase and by determinants of polymer chain length. Mol. Microbiol. *23*, 629–638.

Young, B.M., and Young, G.M. (2002). YplA is exported by the Ysc, Ysa, and flagellar type III secretion systems of *Yersinia enterocolitica*. J. Bacteriol. *184*, 1324–1334.

Young, G.M. (2007). The Ysa type 3 secretion system of *Yersinia enterocolitica* biovar 1B. Adv. Exp. Med. Biol. *603*, 286–297.

Young, G.M., Smith, M.J., Minnich, S.A., and Miller, V.L. (1999). The *Yersinia enterocolitica* motility master regulatory operon, *flhDC*, is required for flagellin production, swimming motility, and swarming motility. J. Bacteriol. *181*, 2823–2833.

Zhang, L., Radziejewska-Lebrecht, J., Krajewska-Pietrasik, D., Toivanen, P., and Skurnik, M. (1997). Molecular and chemical characterization of the lipopolysaccharide O-antigen and its role in the virulence of *Yersinia enterocolitica* serotype O:8. Mol. Microbiol. *23*, 63–76.

Zhang, L., Toivanen, P., and Skurnik, M. (1996). The gene cluster directing O-antigen biosynthesis in *Yersinia enterocolitica* serotype O:8: Identification of the genes for mannose and galactose biosynthesis and the gene for the O-antigen polymerase. Microbiology *142*, 277–288.

Mycobacterial Cell Wall Arabinogalactan: a Detailed Perspective on Structure, Biosynthesis, Functions and Drug Targeting

3

Suresh Bhamidi, Michael S. Scherman and Michael R. McNeil

Abstract

Mycobacteria have an intricate cell wall core structure consisting of a hydrophobic mycolate layer and a peptidoglycan layer tethered together by a polysaccharide, arabinogalactan. The biosynthetic pathways of all the three major cell wall core components have been studied in great deal for their role as potential drug targets for tuberculosis. This chapter specifically deals with the structural details of the arabinogalactan and its biosynthetic pathways, starting from the precursors to its polymerization, and subsequent attachment to peptidoglycan and the mycolic acids. Further, various aspects of drug targeting are discussed in detail for each of these enzymes. The available enzyme assays and the difficulties associated with others will be discussed. Finally, future aspects of tuberculosis research with respect to the arabinogalactan will be presented.

Introduction

The disease tuberculosis

Mycobacterium tuberculosis is the causative organism of tuberculosis (TB), which is a major cause of illness and death worldwide, mainly in Southeast Asia and Africa. World Health Organization (WHO) reported 9.2 million new cases globally and 1.7 million deaths from TB in the year 2006, of which 0.7 million cases and 0.2 million deaths were among HIV-positive individuals (WHO Report 2008: Global tuberculosis control surveillance, planning, financing). The traditional way of treating TB is using a combination of drugs such as isoniazid (INH), rifampin (RM), pyrazinamide (PZA) and ethambutol (EMB). However, these treatments require longer periods for curing the disease and the treatment time is much longer in case of latent TB. Emergence of MDR (multidrug resistant) and XDR (extensively drug resistant) TB cases in recent years has reawakened TB drug research efforts with an urgent need towards more effective treatment. The recommended approach to TB control is DOTS (directly observed therapy short-course), by which a 95% cure rate has been reported. DOTS also reduces development of drug resistance by ensuring that the full course of treatment is followed (Onyebujoh and Rook, 2004).

Cell envelope of *Mycobacteria*

The *M. tuberculosis* cell envelope differs significantly from the cell envelope structures of other Gram-positive and Gram-negative bacteria. The cell envelope consists of the plasma membrane, the cell wall core, the outer membrane, and the capsule (Fig. 3.1). These components are briefly described herein to put in context the subject of this chapter, the arabinogalactan.

The plasma membrane is asymmetrical and typically consists of a thicker outer leaflet and a thinner inner leaflet. Various components have been reported to be present in the plasma membrane including but not limited to the major phosphatidylinositol mannosides (PIMs) which are preferentially located in the outer leaflet (Brennan, 1988). The phospholipids in mycobacterial cell envelopes are derived from phosphatidic acid. The common ones are phosphatidylglycerol, diphosphatidylglycerol,

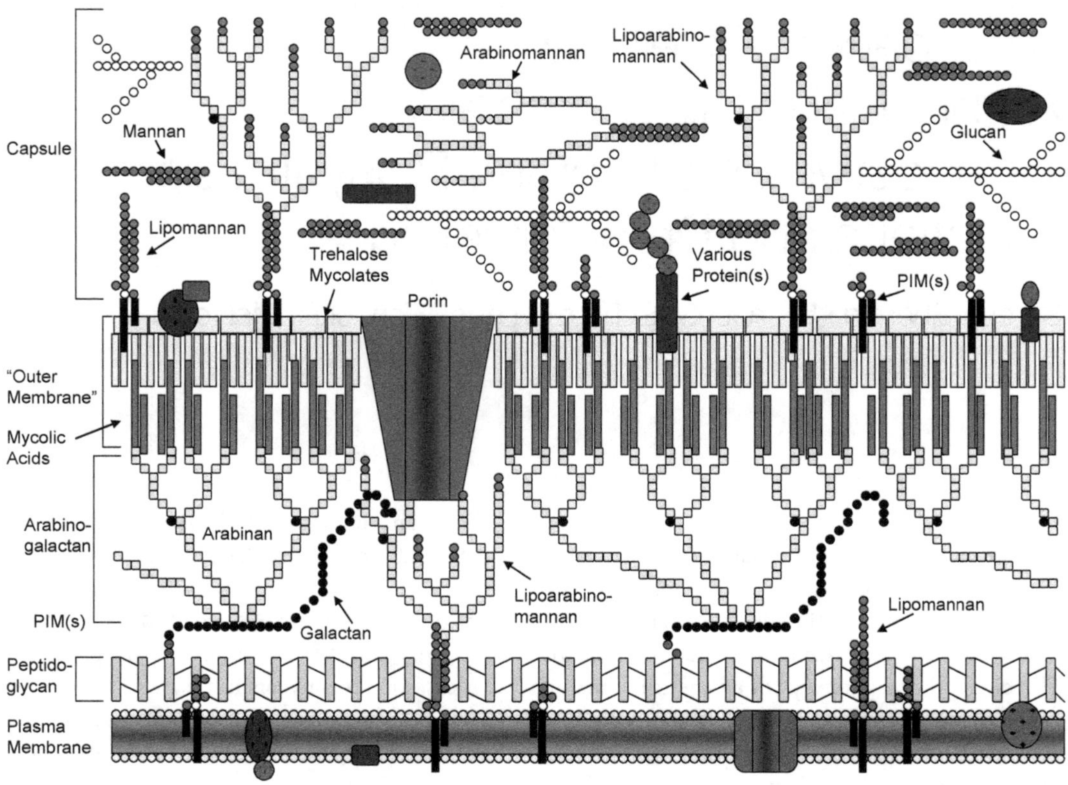

Figure 3.1 Schematic model of cell envelope of *Mycobacterium tuberculosis*. The cell envelope consists of a plasma membrane, a cell wall core, an outer membrane and a capsule. The plasma membrane is shown to have an important polysaccharide lipoarabinomannan (LAM) that is further extended to the periplasmic space and phosphatidylinositol mannosides (PIMs). The cell wall core is made up of hydrophobic mycolic acids, arabinogalactan polysaccharide and a peptidoglycan. The orientation of peptidoglycan is shown here as perpendicular to the plasma membrane based on the recent report by Dmitriev *et al.*, (2003). The space in between the mycolates and PG is termed as 'periplasmic space' analogous to that of Gram-negative bacteria. The outer membrane is recently confirmed to be a 'lipid bilayer', mainly containing mycolic acids, and porins (Hoffman *et al.*, 2008). At the cell surface is the capsule mainly containing polysaccharides such as LAM, glucans, arabinomannan and mannan. The placement of LAM is still not clear, although a recent report suggests its location is only limited to plasma membrane but not on the outer membrane (Alsteens *et al.*, 2008).

phosphatidylethanolamine, and phosphatidylinositol and its mannosides (PIMs). PIMs are restricted to actinomycetes and form the lipid base of lipoarabinomannan (LAM) and lipomannan (LM). It was also reported that the plasma membrane contains other components such as polyterpene-based products (carotenoids that protect from photolytic damage and is also responsible for the characteristic yellow-orange colour in some species such as *M. gordonae* and *M. kansasii*) and the menaquinones that take part in electron transport (Brennan, 1988). Additionally, the plasma membrane has glycosylphosphopolyprenols, associated with cell wall biosynthesis and is discussed later in the chapter.

The mycolic acids (part of the cell wall core discussed below) and various lipids such as trehalose dimycolates form an 'outer membrane' that has been hypothesized to be a 'mycolyl lipid bilayer' (Minnikin, 1982). The outer membrane also contains, phospholipids (PIMs), species-specific lipids (dimycocerosate of phthiocerol) and type-specific lipids (PGL, and lipooligosaccharides) along with trehalose mono and dimycolates (Daffe *et al.*, 1991; Watanabe *et al.*, 1994). A very recent report confirmed the existence of this 'outer membrane' bi-layer by cryo-electron tomography (which preserves the architecture of cells) and using vitreous cryosections (for structure identification in cross-sections of ultra thin specimens) (Hoffmann *et al.*, 2008). According to this study, when wild-type *Corynebacterium* and a non-mycolate-producing strain Δ*pks*13 were compared, the cryo-electron micrographs of vitreous cryosections showed the absence of outer membrane in the Δ*pks*13 mutant. Thus, this study redefined the concept of the mycobacterial outermost layer in *M. smegmatis*, *M. bovis* and *C. glutamicum*. The outer membrane also contains proteins such as porins, which are water-filled channels that allow diffusion of small hydrophilic molecules and ions into the periplasm. In this context, porins from *M. smegmatis* MspA, B, C and D (Faller *et al.*, 2004), and from *M. tuberculosis* and *M. bovis BCG* OmpATb are studied in detail (Alahari *et al.*, 2007; Kartmann *et al.*, 1999).

The substance exterior to the outer membrane is generally termed as the 'capsule' (Fig. 3.1). This cell component has been understood in more detail recently by improved extraction and microscopic techniques (Daffe and Etienne, 1999). Being at the interface between the bacterium and host cells, the capsule is likely to be involved in bacterial pathogenicity (Daffe and Draper, 1998). The capsule largely contains polysaccharides and proteins along with few lipid moieties. The major capsular polysaccharides are glucans, arabinomannan and mannan. Additionally, LAM could be present in this region (Fig. 3.1), although its exact location is still not clear. Capsular proteins are a complex mixture of polypeptides, with a population of secreted and cell envelope associated, or cytoplasmic proteins (Ortalo-Magne *et al.*, 1995). In particular, secreted enzymes such as catalase/peroxidase and superoxide dismutase that are involved in detoxification of reactive oxygen intermediates (ROI) have been reported to be present in this region (Zhang *et al.*, 1991).

The cell wall core consists of three components; namely the mycolic acids, peptidoglycan (PG), and arabinogalactan (AG). These three covalently attached molecules are known as the cell wall core or as the mycolyl arabinogalactan-peptidoglycan (mAGP). The cell wall core is essential for the integrity of the mycobacterial envelope. Therefore, the three components of the cell wall core have been studied in detail in the perspective of drug targeting. Early research demonstrated that isoniazid (INH) inhibits mycolate synthesis (Winder, 1982), ethambutol inhibits arabinogalactan biosynthesis (Takayama and Kilburn, 1989) and cycloserine inhibits the synthesis of peptidoglycan (PG) (Jarlier *et al.*, 1991). Here, the mycolic acids and peptidoglycan are briefly discussed; where as in the next sections, arabinogalactan, the subject of this chapter is discussed.

The mycolic acids are present in the outer membrane and are responsible for the highly hydrophobic nature of the cell wall core. The structure of the mycolic acids was determined in detail primarily by Minnikin (1982). The mycolic acids (Fig. 3.2), contain 76–89 carbon atoms and are in essentially three major forms: α (70%), keto (10–15%) and methoxy (10–15%). The mycolic acids are typically composed of a shorter β-hydroxy chain with a longer α-alkyl side chain. The exact number of carbons varies in different species and is used for the purpose of classification as well (Barry, III *et al.*, 1998). There is also evidence that all the various forms are required for pathogenicity of *M. tuberculosis* (Takayama *et al.*, 2005). The mycolic acids are responsible for some of the resistance of the bacteria towards various drugs and aid in the survival of the bacteria in the macrophages. Biosynthesis of mycolic acids has been studied in detail in the context of drug targeting, and however, is not detailed here.

The peptidoglycan (PG) is the backbone of the cell envelope and is responsible for maintaining the cell shape and size. It has been a topic of debate whether the peptidoglycan is parallel or perpendicular to the plasma membrane. In this context, Dmitriev *et al.* (2003) have proposed in their scaffold model that glycan chains within the murein (peptidoglycan) are perpendicular to the plasma membrane (Dmitriev *et al.*, 2003; Dmitriev *et al.*, 2000) (Fig. 3.1). Early studies in mycobacteria revealed a chemotype IV peptidoglycan similar to that of *E. coli* with both diaminopimelic acid (DAP)-Ala and DAP-DAP cross-linking (Wietzerbin *et al.*, 1974). The glycan consists of alternating units of N-acetylglucosamine (GlcNAc) and muramic acid (Mur). The tetrapeptide side chains

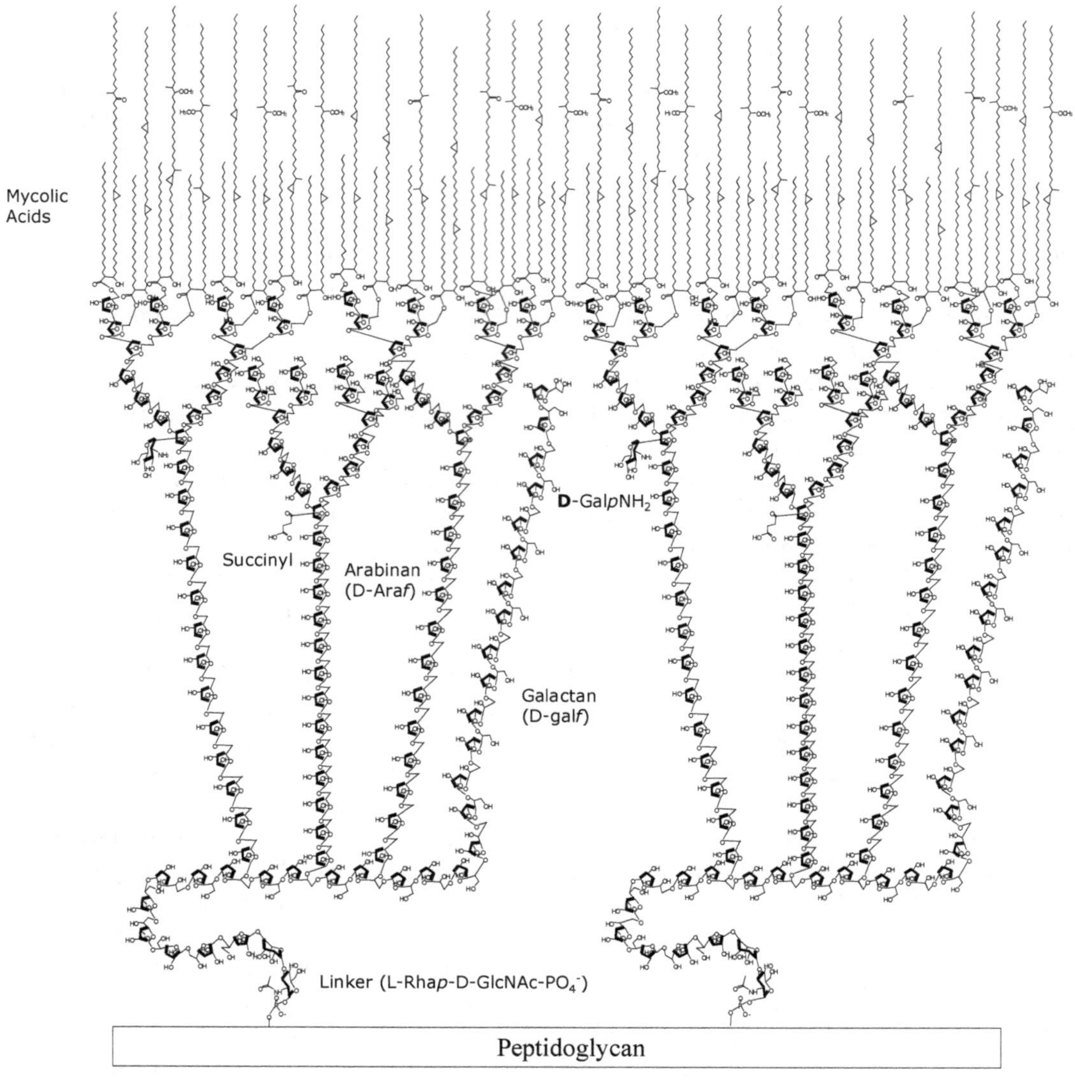

Figure 3.2 The structure of the mycolyl arabinogalactan and its attachment to peptidoglycan. It is shown to include the 31 unit long arabinan of which two-thirds are substituted with mycolyl residues. The non-mycolylated chains are substituted with succinyl groups. Finally, one-third of the chain (whether mycolated or not) are substituted with galactosaminosyl residues. Three arabinan chains are present per galactan core at positions 8, 10, and 12. The absence of mycolylation on succinylated arabinan further suggests that succinylation negatively controls mycolylation. The other significant structural characteristics are presented in the text.

of PG consist of L-alanyl-D-isoglutaminyl-meso-diaminopimelyl-D-alanine (L-Ala-D-Glu-A_2pm-D-Ala), with Glu being further amidated (Lederer *et al.*, 1975). In mycobacterial PG, *N*-acylation of Mur residues with glycolic acid (MurNGly), and the cross-linking between two diaminopimelyl residues (A_2pm) as well as A_2pm and D-Ala is observed (Lederer *et al.*, 1975). However, in *M. leprae* PG, the muramic acid residues are exclusively *N*-acetylated, in contrast to *M. tuberculosis* (Mahapatra *et al.*, 2008). Recent studies on *M. smegmatis* PG showed amidation on the free carboxyl group of the *meso*-diaminopimelic acid residue of the peptide side-chains of lipid II along with the diamidated species (Mahapatra *et al.*, 2005). According to this study, the lipid II intermediates of *M. smegmatis* act as substrates for various enzymes that introduce modifications to the sugar and amino acid residues prior to the synthesis of peptidoglycan.

The space in between the peptidoglycan and mycolates is analogous to the 'periplasmic space' of other Gram-negative bacteria. This space mainly consists of hydrophilic molecules, mostly arabinogalactan (AG). Additionally, lipoarabinomannan (LAM) and lipomannan (LM) may also be present in periplasm, although they are also considered to be anchored in outer membrane. Recently, Alsteens *et al.* (2008) contradicted the presence of LAM in the outer membrane. Using atomic force microscopy combined with chemical force microscopy and immunogold labelling, the cell surface hydrophobicity is measured in the presence and absence of cell wall drugs. The uniform distribution of hydrophobicity and the absence of labelling with anti-LAM antibodies suggested that LAM is not abundant at the cell surface. However, in the presence of drugs such as INH and EMB, the mycolic acids were removed exposing the hydrophilic LAM at the surface (Alsteens *et al.*, 2008). Hence, the localization of LAM and LM, is still not unequivocally established. LAM might be localized in either or in both, anchored in the outer membrane or in the periplasm (Fig. 3.1), but this uncertainty must be kept in mind.

Structure of mycobacterial arabinogalactan

The arabinogalactan is conveniently explained here as three major constituents: the reducing end, the galactan region, and the non-reducing end.

The reducing end

The primary structure of arabinogalactan (AG) is very complex and unlike most bacterial polysaccharides, it is not composed of a simple repeating unit. The detailed structural arrangement of AG is shown in Fig. 3.2. Previous research led to the identification of the linker unit on the reducing end of AG that connects arabinogalactan with peptidoglycan (McNeil *et al.*, 1990; McNeil *et al.*, 1991). Elucidation of this structure was advanced by the isolation and characterization of per-*O*-alkylated oligosaccharide alditols together with GC/MS and NMR analysis. The results demonstrated that the reducing end of AG consists of the terminal sequence →5)-D-Gal*f*-(1→4)-L-Rha*p*-(1→3)-D-GlcNAc. Additionally, a muramyl-6-P was also found within the PG moiety. Together, it was articulated that the massive mycolylarabinogalactan of mycobacteria is attached to the PG by the actinomycete-specific diglycosylphosphoryl bridge, L-Rha*p*-(1→3)-D-GlcNAc-(1→P) (Fig. 3.2).

The galactan component and the middle region

Arabinogalactan contains a galactan chain, with alternating 5-linked β-D-galactofuranosyl (Gal*f*) and 6-linked β-D-Gal*f* residues (Fig. 3.2) (Daffe *et al.*, 1990). Characterization of per-*O*-alkylated oligosaccharide alditols revealed that the arabinan chains are attached to C-5 of some of the 6-linked Gal*f* residues. Daffe *et al.*, (1990) have also proposed three major structural domains for arabinan; one consists of linear 5-linked α-D-Ara*f* residues, a second domain with branched 3,5-linked α-D-Ara*f* residues substituted with 5-linked α-D-Ara*f* units at both branched positions; and the terminal non-reducing end arabinan consists of a 3,5-linked α-D-Ara*f* residue substituted at both branched positions with the disaccharide β-D-Ara*f*-(1→2)-α-D-Ara*f*. The places of attachment of arabinan on the galactan chain was recently identified. The major problem for these kinds of studies is the essentiality of AG in mycobacteria. Thus, an alternative in the form of *Corynebacterium glutamicum* was found to be very useful, where AG is non-essential. Knock-out studies using *C. glutamicum* combined with mass spectrometry revealed that three arabinan chains are attached to the galactan at residues 8, 10, and 12 (Fig. 3.2) (Alderwick *et al.*, 2005). These researchers knocked out the gene of the arabinosyl transferase, *emb*, whose product extends the arabinan after a different arabinosyl transferase, AftA (Alderwick *et al.*, 2006) attaches the first arabinosyl unit. Methylation of the cell wall core of the *emb* knockout strain yielded galactan substituted with three terminal Ara*f* units (attached by the persistent activity of arabinosyl transferase AftA) in the form of a methyl rhamnoside unit at the reducing end. The biosynthetic aspects and corresponding enzymes are discussed later in the chapter.

The non-reducing end

The non-reducing end of arabinogalactan is covalently attached to the mycolic acids of the outer membrane. The hydrophobicity of mycolic acids has been a challenging barrier to drug entry. Additionally, the mycolyl arabinogalactan peptidoglycan (mAGP) is responsible for aspects of disease pathogenesis and much of the antibody response in infections. Hence, it was essential to elucidate the structural details of the non-reducing end as well as the linkage patterns. Esterification of the mycolic acids at C-5 of some arabinofuranosyl residues was reported previously (Azuma *et al.*, 1969; Amar-Nagasch and Vilkas, 1970). More detailed information on the location of mycolyl components and their attachment to the arabinan residues was obtained by McNeil *et al.* (1991). Analysis of methylated mAGP and per-*O*-alkylated oligoglycosyl alditols revealed that the mycolyl substituents were selectively and equally distributed on the 5-hydroxyl functions of terminal- and the penultimate 2-linked Ara*f* residues. Further, the mycolyl residues were clustered in groups of four on the non reducing terminal pentaarabinosyl unit (β-Ara*f*-(1→2)-α-Araf)$_2$_3,5-α-Ara*f* (Fig. 3.2). Thus, the major part (66%) of the pentaarabinosyl units are substituted by mycolic acids, leaving the unsubstituted minor region (33%), that is available for interaction with the immune system. A follow up study by Besra *et al.* (1995) extended the structure of the arabinan at the non-reducing end from the hexa-arabinoside to include the last 17–22 arabinosyl residues, but until very recently no additional progress on the structure of AG has been made.

In recent times, an endogenous arabinase, which releases large oligoarabinosides, has been used effectively for structural studies (Lee *et al.*, 2006; Shi *et al.*, 2006). This enzyme was partially purified from the non-virulent strain *Mycobacterium smegmatis*, and an assay methodology was developed (Dong *et al.*, 2006). Surprisingly, cell wall arabinan was found to be not rapidly turning over, although there is active endogenous arabinase (Dong *et al.*, 2006). Therefore, the actual role of this enzyme is not understood. However, it could be speculated that it is involved in 'slippage' between the peptidoglycan layer and the mycolate layer. Endogenous arabinase solubilizes the arabinan from AG and this opened up possibilities for searches for base-labile groups on AG. This enzyme can do single, double and even triple cleavages of the arabinan molecule and arabinan fragments of various lengths are isolated. Use of this enzyme, combined with the matrix-assisted laser desorption/ionization-time of flight mass spectrometry (MALDI-TOF MS), allowed the sequencing of large oligoarabinosides (Lee *et al.*, 2006). These studies revealed that in *M. tuberculosis* a non-*N*-acetylated galactosamine residue ($GalNH_2$) (Draper *et al.*, 1997) is located on the interior branched arabinosyl residue (Lee *et al.*, 2006).

Another recent application of the endogenous arabinase is the identification of succinyl groups on the arabinan chains towards the non-reducing end (Bhamidi *et al.*, 2008). Endogenous arabinase has been used to solubilize the arabinan region of the cell wall and it was shown using mass spectrometry and NMR (HSQC, gHMBC and 1D-TOCSY) that succinyl esters are present on O-2 of the inner-branched 1,3,5-α-D-arabinofuranosyl residues. The data suggest that approximately one of the three arabinosyl chains attached to the galactan chain contains succinyl groups (Fig. 3.2). It was also shown that although one succinyl group was most common, up to three succinyl groups per released arabinan fragment could be found on oligo-arabinans. However, arabinan fragments substituted with $GalNH_2$ were not succinylated (Fig. 3.2). Importantly, in the case of *M. tuberculosis*, and most likely in all slow growers, both positive charge (protonated $GalNH_2$ as $GalNH_3^+$) and negative charge (succinyl) are present in the middle regions of the arabinan, specifically at O-2 of the inner 3,5-α-D-Ara*f* units. The succinyl residues were shown to be on the non-mycolylated chain. Finally, since there is approximately one $GalNH_2$ per entire AG molecule (Draper *et al.*, 1997) it seems likely that mycolylated arabinan chains are both galactosaminylated and non-substituted. The observation of non-succinylation on the mycolylated arabinan chains suggests a possibility that succinylation negatively controls mycolylation.

This work also redefined the arabinan chain length with regard to the number of Ara*f* units

(Bhamidi *et al.*, 2008). Previous results using mass spectra of arabinans produced by partial acid hydrolysis, showed up to 22 Ara*f* residues from base solubilized AG from *M. smegmatis* (Besra *et al.*, 1995). On the other hand, the details on the exact number of arabinosyl residues that are present before attachment to the galactan are not known. However, with recent structural studies using endogenous arabinase, Bhamidi *et al.* (2008) have further elucidated the structural aspects, proposing a complete primary model of AG. In this study, the released arabinans were purified by a P-2 sizing chromatography column; the former fractions contained $GalNH_2$ rich arabinans and the latter fractions contained succinyl rich arabinans. Initial MALDI-TOF studies on the $GalNH_2$ rich fraction revealed a series of Ara*f* fragments up to 31 residues that are associated with $GalNH_2$ [from the major $Ara_{20}(GalNH_2)_1$ up to Ara_{30}(GalNH2)]. The succinyl rich portion from the P-2 column initially showed only 22 Ara residues, but increased laser power on the MALDI-TOF instrument resulted in 31 residues on par with the $GalNH_2$ rich portion (30 released and one 'remnant' on the galactan). Thus, in summary, the Ara_{17} non-reducing end of the arabinan is attached to the galactan *via* a linear α-1,5-arabinan with up to 14 units (Bhamidi *et al.*, 2008).

AG and LAM-comparison of arabinans

LAM (lipoarabinomannan) is a major glycolipid in the cell envelope (Fig. 3.1) exhibiting a wide spectrum of immunomodulatory functions (Chatterjee and Khoo, 1998). Unlike AG, LAM is non-covalently linked. As mentioned earlier in this chapter, its location however, is still not clear, it may be present on the outer membrane and/or anchored in the plasma membrane via its phosphatidyl-*myo*-inositol (PI) unit (Fig. 3.1). The linkage pattern of Ara*f* units and the hepta-arabinoside non-reducing end are similar to that of AG as shown in Fig. 3.3. One major difference between AG and LAM is that the non-reducing end of the LAM arabinan is substituted with α-mannosyl residues rather than with mycolic acids (Chatterjee *et al.*, 1992). Recently its been shown using endoarabinase from *M. smegmatis* that even the interior regions of LAM arabinan are fundamentally different than that of AG arabinan

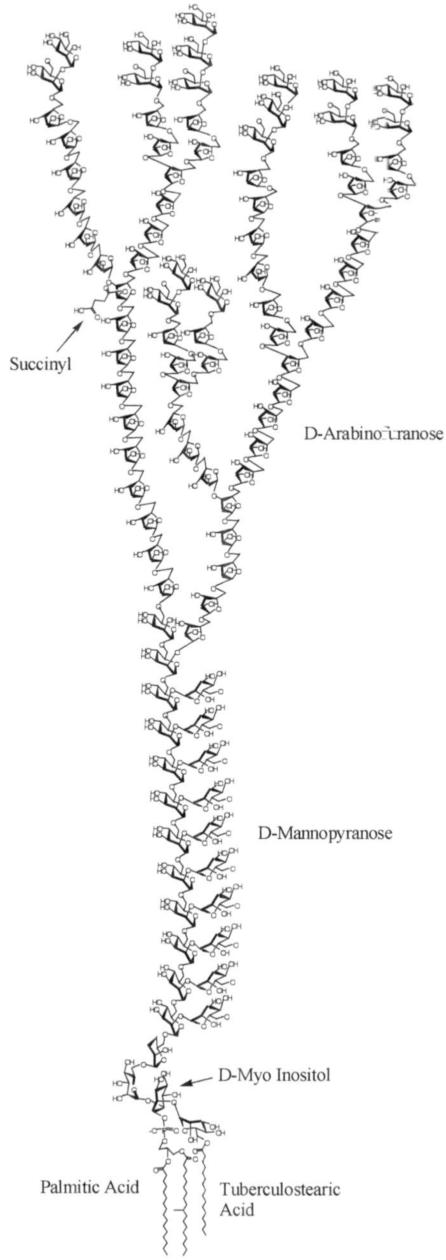

Figure 3.3 Structure of lipoarabinomannan (LAM). The backbone of LAM consists of a linear α-1,6-mannosyl polymer that is linked to a phosphatidylinositol anchor. Most mannosyl residues are substituted on their C-2 with another single mannose. In *M. tuberculosis*, branched arabisonyl side chains (α-D-Ara*f*) are attached to α-mannosyl residues rather than mycolates as in case with AG. Although LAM with two branched arabinan chains is shown here to demonstrate various arabinan structural features, recent biochemical and mass spectral data strongly suggest the existence of a single arabinan chain (Devinder Kaur, personal communication).

(Shi *et al.*, 2006). In LAM the precise Ara_{18} arrangement is not always present, but rather chains of various lengths extend from interior 3,5-Ara*f* residues (Shi *et al.*, 2006). Such differences may well reflect the function of the two molecules. Thus, the number of glycosyl residues separating the PG and mycolate layers would be reasonably constant in the case of AG, in order to form the distinct two-layered cell wall core. Contrastingly, in the case of LAM, the arabinan region extends to be a part of the extracellular environment and the lipid anchor is likely in the mycolic acids layer. Due to this special arrangement, precise arabinan lengths and branching would not be mandatory.

Implications of this model

The model presented in Fig. 3.2 is more of a primary structure (similar to an amino acid sequence) rather than a tertiary model (analogous to a protein structure). The space between the mycolates and the peptidoglycan is analogous to the periplasmic space of Gram-negative bacteria, which is large depending on the conformation(s) adapted by the polysaccharide. This space is hydrophilic because of the arabinogalactan and the negatively (succinyl) and positively (galactosamine) charged groups effect the environment of the 'periplasmic' space. Draper *et al.*, (1997) proposed that the protonated $GalNH_2$ (as $GalNH_3^+$) interacts with anionic substances such as phosphates of glycolipid anchors like those of LAM (Draper *et al.*, 1997). The negatively charged succinates would possibly interact with $GalNH_3^+$ leading to a more rigid and tightened AG structure. It is certain that some aspects of cell wall biosynthesis as well as other biological events occur in this region. The size of the AG polymer and its furanose nature allows considerable flexibility between the mycolate and peptidoglycan layer as has been previously pointed out (Liu *et al.*, 1996) but would, on the other hand, keep the two layers connected within certain spatial limits.

Arabinogalactan biosynthesis

It is an amazing topic of research to know how this very complex polysaccharide, arabinogalactan is synthesized. To start with, the complete genome sequencing of *M. tuberculosis* made a way for better understanding of the enzymatic pathways (Cole *et al.*, 1998). Biosynthesis of AG can be classified into three important steps carried by specific enzymes: (1) precursor formation, (2) polymerization and (3) attachment to the other two cell wall components, mycolates and peptidoglycan.

Precursor formation

Most of the AG precursor enzymes are soluble and all have been expressed and characterized to some extent (Scherman *et al.*, 2003; Ma *et al.*, 1997, 2001; Stern *et al.*, 1999; Weston *et al.*, 1998; Lee *et al.*, 1996). Insertion mutagenesis experiments revealed that all of these enzymes are essential for growth of *M. tuberculosis* (Sassetti *et al.*, 2003). The precursors for AG biosynthesis are discussed here.

Decaprenyl phosphate (DP) is a key carrier molecule for the synthesis of AG (Mikusova *et al.*, 1996), PG, and possibly mycolic acids. DP consists of 50 carbon atoms unlike the more commonly utilized undecaprenyl phosphate (Takayama and Goldman, 1970; Mahapatra *et al.*, 2005). It was also reported that DP is synthesized in higher rates in fast growing *M. smegmatis* compared with that of slow growing *M. tuberculosis* (Crick *et al.*, 2000). DP was shown to be a rate limiting factor for cell wall synthesis in some of the bacteria such as *Staphylococcus aureus* (Higashi *et al.*, 1970) and *Bacillus sps* (Anderson and Page, 1972). The synthesis of DP is done in two major steps: Pre-polymerization [i.e. the formation of isopentyl diphosphate (IPP) and dimethylallyl diphosphate (DMAPP)] and polymerization (events leading to DP). Specifically, IPP synthesis is studied in detail as a potentially rich source of drug targets in bacteria.

IPP biosynthesis in some bacteria differs from that of eukaryotes. In eukaryotes, the formation of IPP is by a condensation reaction of acetyl coenzyme A (CoA) forming 3-hydroxy-3-methylglutaryl-CoA (HMG CoA) and its reduction to mevalonate. Subsequent decarboxylation leads to the formation of the five carbon molecule IPP, which is isomerized to the allylic diphosphate–dimethyl allyl diphosphate (DMAPP) (Rip *et al.*, 1985). The polymerization progresses by the addition of IPP molecules to DMAPP forming geranyl diphosphate (GPP) and farnesyl diphosphate (FPP). FPP then primes the reaction for the synthesis of long chain DP *via* a 1′-4 head to tail condensation (Rilling, 1985).

However, a series of studies (Zhou and White, 1991; Rohmer *et al.*, 1993; Horbach *et al.*, 1993; Sprenger *et al.*, 1997; Duvold *et al.*, 1997; Putra *et al.*, 1998; Lois *et al.*, 1998) proposed a different pathway for IPP/DMAPP biosynthesis in some bacteria including mycobacteria, which is

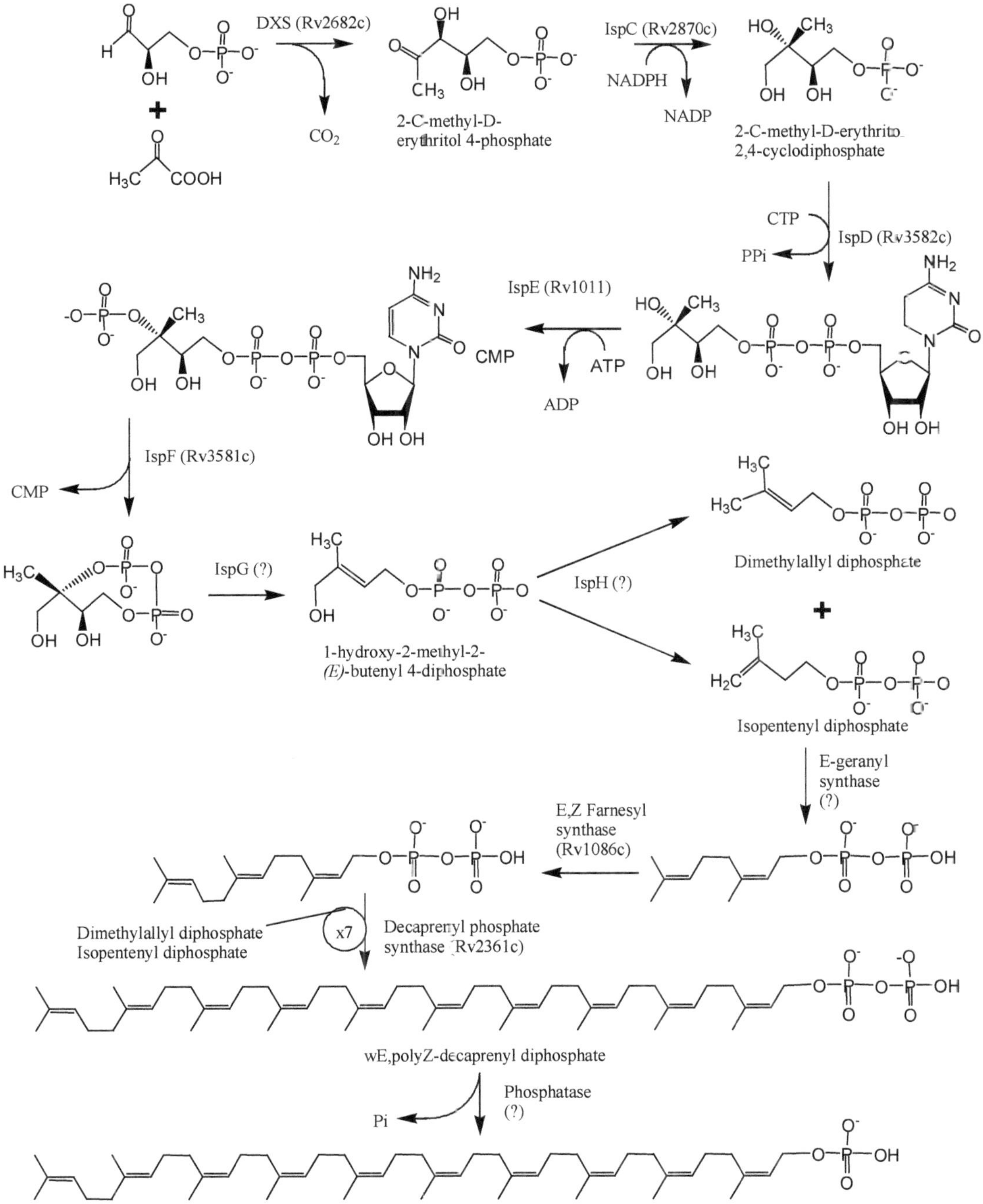

Figure 3.4 The biosynthesis of decaprenyl phosphate (DP). It proceeds mainly in two phases, the formation of intermediates, Isopentinyl diphosphate (IPP) and Dimethylallyl diphosphate (DMAPP) and their subsequent polymerization to form decaprenyl phosphate. The biosynthetic details of each of these enzymes and their targeting are discussed in the text.

known as the MEP pathway, wherein 2-C-methyl-D-erythritol 4-phosphate (MEP) is the first committed precursor. This pathway is shown in Fig. 3.4. In this context, Duvold *et al.*, (1997) have proposed a biosynthetic pathway in which the enzyme DXS (Rv2682c: 1-deoxy-D-xylulose-5-phosphate synthase) carries out a condensation reaction between pyruvate and glyceraldehyde 3-phosphate (GAP) forming 1-deoxyxylulose 5-phosphate (DXP). This reaction has been described as a thiamine diphosphate dependent acylation condensation between C2 and C1 of pyruvate and GAP to yield DXP (Sprenger *et al.*, 1997). DXP is further converted to MEP by 1-deoxy-D-xylulose 5-phosphate reductoisomerase (IspC-Rv2870c). Recently it was demonstrated that IspD (Rv3582c) catalyses the formation of 4-diphosphocytidyl-2-*C*-methyl-D-erythritol (CDP-ME) in the presence of MEP and CTP (Eoh *et al.*, 2007). CDP-ME further is converted to IPP by a series of enzymatic reactions involving, IspE (4-diphosphocytidyl-2-*C*-methyl-D-erythritol kinase), IspF (2-*C*-methyl-D-erythritol 2,4-cyclodiphosphate synthase), IspG (1-hydroxy-2-methyl-2-(*E*)-butenyl 4-diphosphate synthase), and IspH (1-hydroxy-2-methyl-2-(*E*)-butenyl 4-diphosphate reductase) (Herz *et al.*, 2000; Luttgen *et al.*, 2000; Rohdich *et al.*, 1999; Rohdich *et al.*, 2001). Interestingly, the single enzyme, IspH, forms both IPP and DMAPP (Rohdich *et al.*, 2003).

The polymerization of IPP and DMAPP to longer polyprenyl diphosphate molecules happens through sequential addition of IPP to allylic diphosphates mediated by the enzymes known as polyprenyl diphosphate synthases. Thus, IPP is added to DMAPP by an unknown enzyme to form the ten-carbon molecule geranyl diphosphate (GPP). *M. tuberculosis* is unusual in that it has two genes (*Rv1086* and *Rv2361c*) encoding proteins with significant sequence similarity to undecaprenyl diphosphate synthase (UPS). *Rv1086* is a unique prenyldiphosphate synthase (present in cell membrane) as it adds only one isoprene unit to geranyl diphosphate forming the 15 carbon *E,Z*-farnesyl diphosphate (Schulbach *et al.*, 2000; Schulbach *et al.*, 2001). The chain length control of Rv1086 is explained recently by structural studies with bound citronellyl diphosphate and the product mimic E,E-farnesyl diphosphate (Wang *et al.*, 2008). Subsequently, Rv2361c, encoding a decaprenyl diphosphate synthase (present in cytosol) adds seven isoprene units to *E,Z*-farnesyl diphosphate to generate 50 carbon prenyl diphosphate (Schulbach *et al.*, 2000; Kaur *et al.*, 2004). The subsequent biosynthetic step for the formation of decaprenyl phosphate (DP) from decaprenyl diphosphate probably involves de-phosphorylation and the respective enzyme is not yet identified.

Decaprenylphosphoryl D-arabinofuranose (DPA) is the only known precursor of polymerized Ara*f* residues in mycobacteria (Wolucka *et al.*, 1994; Mikusova *et al.*, 2005). It was demonstrated that DPA was formed from 5-phosphoribose diphosphate (pRpp) based on the fact that D-Ara*f* residues are derived from the pentose phosphate shunt (Klutts *et al.*, 2002; Scherman *et al.*, 1995). Mikusova *et al.*, (2005) have elucidated the biosynthetic pathway of DPA formation in mycobacterial arabinan (Fig. 3.5A). First, pRpp is transferred to a decaprenyl phosphate (DP) by an enzyme phosphoribosyl transferase (Rv3806c) to form 5-phosphoryl-decaprenylphosphoryl-β-D-ribofuranose. This is further dephosphorylated to form decaprenylphosphoryl-β-D-ribofuranose, which in turn is acted on by the enzyme decaprenylphosphoryl-D-ribose oxidase (Rv3790 that oxidizes the hydroxyl group at the 2 position of the ribose) to form decaprenylphosphoryl-2-keto-β-D-erythro-pentofuranose. This is finally reduced to DPA by the enzyme, decaprenylphosphoryl-2-keto-D-erythropentose reductase (Rv3791). It was also found that the epimerization is only possible when both proteins (Rv3790 and 3791) are together whereas individually they could not support the activity (Mikusova *et al.*, 2005).

The molecule dTDP-β-L-Rha*p* is an important precursor that is associated in the formation of the linker unit. Glucose-1-phosphate is converted to dTDP-Rha by a series of enzymatic reactions involving *RmlA* (Rv0334), *RmlB* (Rv3464), *RmlC* (Rv3465), and *RmlD* (Rv3266c) encoding an α-D-glucose-1-phosphate thymidylyl transferase (Ma *et al.*, 1997), dTDP-D-glucose 4,6-dehydratase (Ma *et al.*, 2001), dTDP-4-keto-6-deoxy-D-glucose 3,5-epimerase (Stern *et al.*, 1999) and dTDP-rhamnose synthetase (Hoang *et al.*, 1999) respectively (Fig. 3.5B).

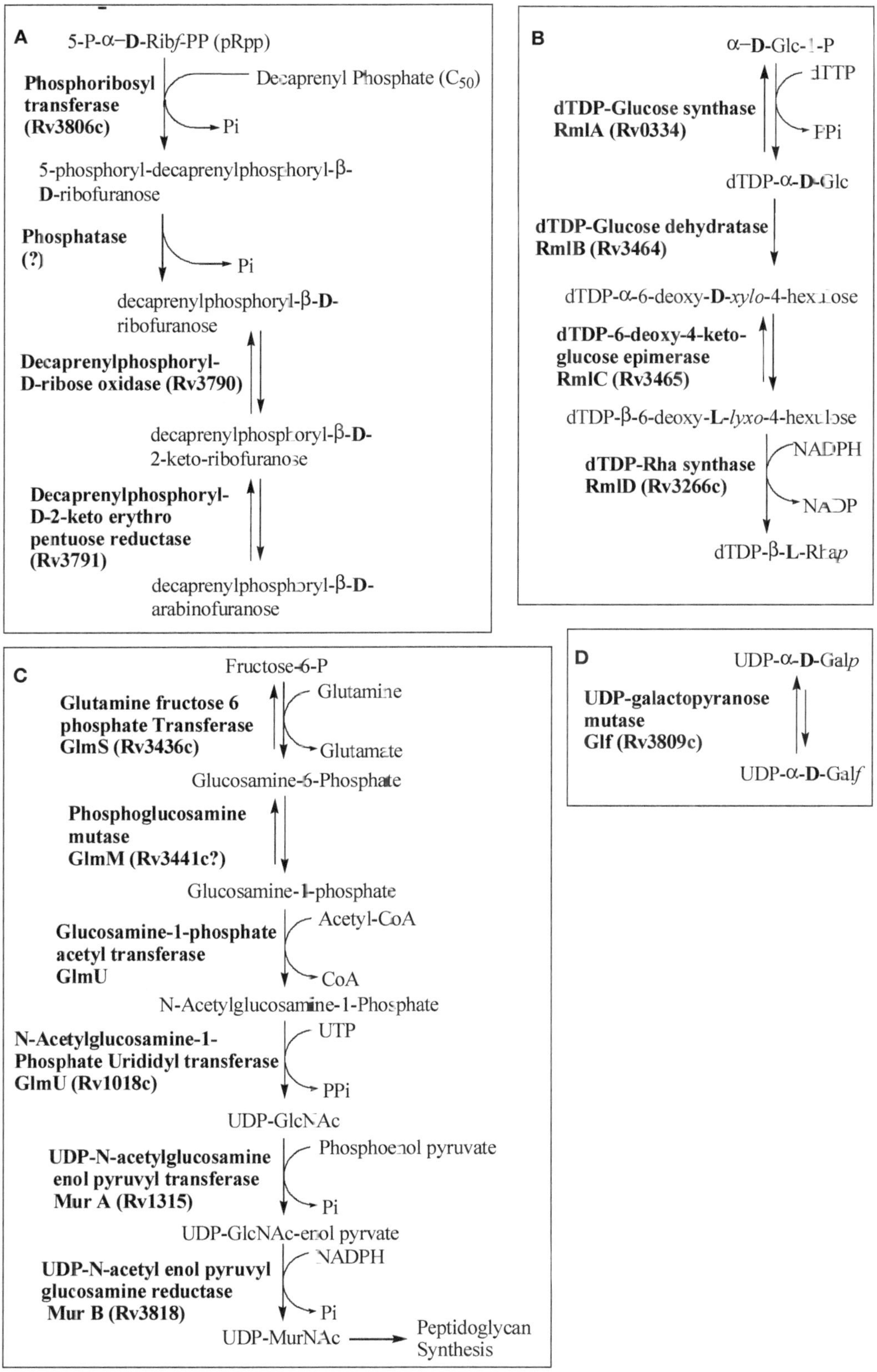

Figure 3.5 The biosynthesis of various sugar donors required for AG biosynthesis. **A** Decaprenylphosphoryl-D-arabinose DPA. **B** dTDP-Rha biosynthesis. **C** UDP-GlcNAc. **D** UDP-Gal*f*.

Basically, all the genes in *E. coli* are similar to that of *M. tuberculosis* for peptidoglycan biosynthesis. GlcNAc is an essential component in both PG as well as the rhamnose-GlcNAc linker of the cell wall in mycobacteria. The biosynthetic pathway of UDP-GlcNAc, an activated nucleotide sugar of GlcNAc was first elucidated in *E. coli* (Mengin-Lecreulx and Vanheijenoort, 1994) (Fig. 3.5C). In the first step, fructose-6-phosphate and glutamine are converted to glucosamine-6-phosphate and glutamate by the enzyme glutamine fructose-6-phosphate transferase, encoded by the *glmS* (Rv3436c) gene. Subsequently, the enzyme phosphoglucosamine mutase (*glmM*), converts glucosamine-6-phosphate ($GlcNH_2$-6-P) to glucosamine-1-phosphate ($GlcNH_2$-1-P). The bifunctional enzyme GlmU (Rv1018c) carries out the next two reactions. The initial reaction of GlmU (glucosamine-1-phosphate acetyltransferase) catalyses the formation of N-acetylglucosamine-1-phosphate (GlcNAc-1-P) from glucosamine-1-phosphate using acetyl-CoA as the acetyl donor. The second activity of GlmU (N-acetylglucosamine-1-phosphate uridylyltransferase) catalyses the formation of UDP-GlcNAc from GlcNAc-1-P and uridine triphosphate (UTP). A homologue of *E. Coli* GlmU in *M. tuberculosis*, Rv1018c has been recently reported and its activity is also proved (Zhang *et al.*, 2008b).

The UDP-Gal*f* is the donor of the gal*f* residues of the galactan in *M. tuberculosis* AG biosynthesis (Weston *et al.*, 1998). The first reaction in the synthesis of UDP-Gal*f* is the conversion of UDP-glucose to UDP-galactopyranose (UDP-Gal*p*) catalysed by UDP-galactopyranose epimerase (Rv3634) (Weston *et al.*, 1998). The *glf*-encoded enzyme, UDP-galactopyranose mutase (Rv3809c), then converts UDP-Gal*p* to UDP-Gal*f* (Fig. 3.5D).

Polymerization of arabinogalactan

The entire arabinogalactan molecule is synthesized on the carrier molecule-decaprenyl phosphate (DP). The biosynthetic pathway of AG polymerization is shown in Fig. 3.6, with special reference to the arabinosyl transferases in Fig. 3.7. It is conveniently discussed here as four phases, formation of the linker unit, the galactan biosynthesis, arabinan biosynthesis and finally the attachment of AG to PG and mycolates. The first sugar, GlcNAc, is attached as α-GlcNAc-1-P (transferred from UDP-GlcNAc to DP, Fig. 3.5A and C) forming a diphosphate bridge to the carrier lipid. The enzyme WbbL (Rv3265c) (Mills *et al.*, 2004) attaches the rhamnosyl residues (from dTDP-Rha, Fig. 3.5B) to the prenyl diphosphoryl GlcNAc Rha-GlcNAc-P-P-decaprenyl, to form the 'linker region' of the AG (Mikusova *et al.*, 1996). Mikusova *et al.*, (2000) have shown the synthesis of a prenyl diphosphate linked oligosaccharide consisting of GlcNAc, Rha, Gal, and Ara in cell-free preparations from *M. smegmatis*. Therefore, it is thought that the formation of the linker region is essentially followed by the addition of the Gal*f* and Ara*f* residues to form the mature lipid-linked AG.

The galactan is synthesized by two bifunctional galactofuranosyl transferases, with the first two residues (from UDP-Gal*f*, Fig. 3.5D) added to the rhamnosyl residue by the galactofuranosyl transferase Rv3782 (Mikusova *et al.*, 2006) while, the remaining 1→5 and 1→6 galactofuranosyl residues are added by the bifunctional galactofuranosyl transferase GlfT (Rv3808c) (Rose *et al.*, 2006; Kremer *et al.*, 2001; Mikusova *et al.*, 2000; Alderwick *et al.*, 2008) (Fig. 3.6). The galactofuranosyl transferase activity for both the enzymes was further proved by *in vitro* assays using glycolipid acceptors, and characterization of the products.

The next step in AG polymerization is the addition of Ara*f* molecules to the galactan chain. Until recently, little information on arabinosyl transferases was available, because of the fact that AG is essential for *M. tuberculosis* survival. In this regard, for better understanding of the arabinosyl transferases, the non-essentiality of AG in a closely related actinomycete *Corynebacterium* was very helpful. Moreover, the glycosidic linkage analysis of *C. glutamicum* AG showed the presence of similar arabinosyl linkage patterns to that of *M. tuberculosis* AG (Besra *et al.*, 1995; Alderwick *et al.*, 2005). Dr Besra's lab at The University of Birmingham, UK, successfully carried out knock-out studies in *Corynebacterium* to identify various arabinosyltransferases.

The step-wise action of various arabinofuranosyl transferases is shown in Fig. 3.7. All evidence suggests that the arabinan is built-up

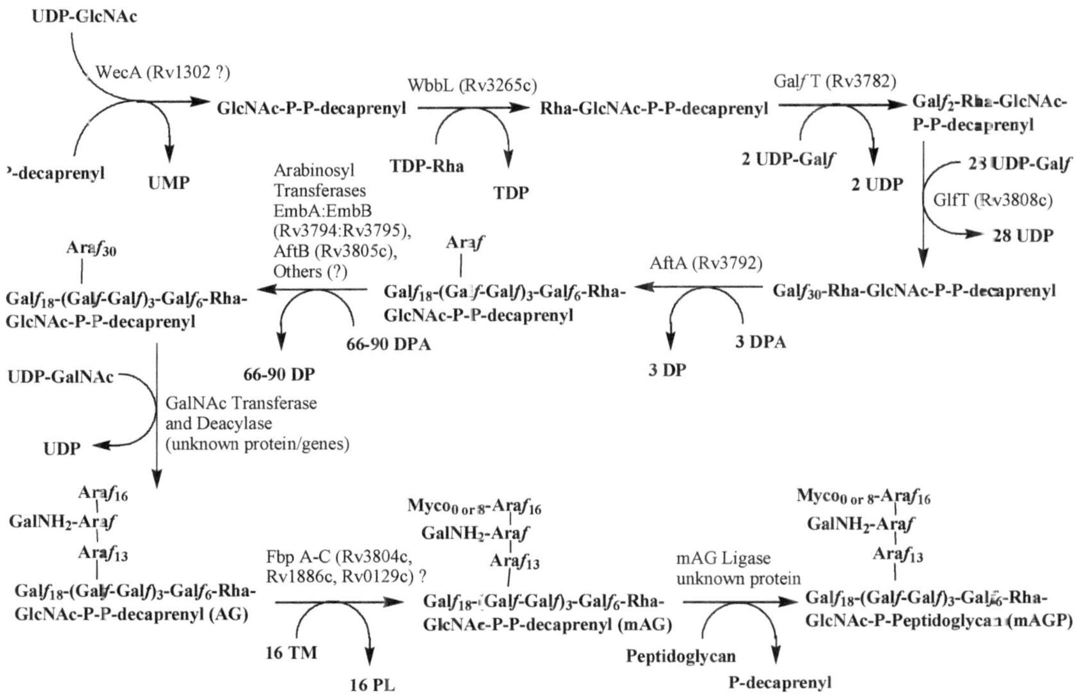

Figure 3.6 The biosynthetic pathway of mycolyl arabinogalactan peptidoglycan complex (mAGP). Various glycosyl transferases that are involved in this process are shown. The formation of the linker unit proceeds by further addition of galactose molecules, mediated by galactosyl transferases. Subsequently, the recently identified arabinosyl transferases add Ara*f* residues to complete the polymerization of AG. The later events of formation of mAGP complex are unknown, however for convenience, it is presented here that mycolic acids are attached to the lipid linked AG molecule. Finally, the mycolyl arabinogalactan is attached to the peptidoglycan by the action of ligase.

on the galactan directly, without any biosynthetic intermediates such as lipid linked arabinans. The Ara*f* molecules from the sugar donor β-D-arabinofuranosyl-1-monophosphoryldecaprenol (DPA) are added to the C-5 hydroxyl group of a β-(1→6) linked Gal*f* in the presynthesized galactan chain, by the enzyme α-1,5 Ara*f* transferase, (AftA, Rv3792) priming the galactan for further expansion by other arabinofuranosyl transferases (Alderwick *et al.*, 2005; Alderwick *et al.*, 2006). As mentioned before, the addition of arabinose on the galactan chain is specifically done at 8, 10 and 12 positions. Interestingly, the transfer of the first Ara*f* to the galactan was shown to be essential for viability in case of *M. smegmatis* by knock-out studies of the gene *MSMEG_6386* (an orthologue of *Rv3792*) (Shi *et al.*, 2008).

The next enzyme adds on the α-1,5 Ara*f* units to the three already existing Ara*f* residues on the galactan core. This α-1,5 arabinosyl transferase enzyme has been identified as 'Emb' by knock-out experiments in *C. glutamicum* that resulted in a slow growing, yet viable mutant. Characterization of the cell wall components of *C. glutamicum* Δ*emb* showed a truncated AG structure possessing only the terminal Ara*f* residues that are provided by AftA (Alderwick *et al.*, 2005). Thus, in the knock-out mutant, the entire α-1,5-linked arabinan is not synthesized showing the α-1,5 transferase capability of Emb. Moreover, *in vitro* studies have shown that ethambutol, known to act against Emb proteins (Belanger *et al.*, 1996) specifically inhibits the α-1,5 activity, using arabinooligosaccharide acceptors (Alderwick *et al.*, 2006; Seidel *et al.*, 2007).

Analogous to '*emb*' of *Corynebacterium*, a three gene cluster '*embCAB*' is present in mycobacteria (Telenti *et al.*, 1997). Individual knock-out mutants of *emb A,B* and *C* were obtained in *M. smegmatis*, and biochemical characterization of the cell wall revealed that, *embA* and *embB* affect AG, while *embC* affects LAM (Escuyer *et* al., 2001; Zhang et al., 2003). The effect of *embC* on LAM is thus consistent with the loss of α-1,5 transferase activity that builds the arabinans on the mannose backbone. Thus, *embC* inhibition of arabinosylation of LM to mature LAM is parallel to the *emb* mediated inhibition of arabinosylation of galactan in corynebacterial AG confers the α-1,5 transferase activity of Emb proteins.

Surprisingly, there are two genes (*emb A, B*) in *M. tuberculosis* and only one (*emb*) in *Corynebacterium* that are involved in AG biosynthesis. Even though AG formation is essential for growth of mycobacteria, either *embA* or *embB* can be knocked out in *M. smegmatis* (Escuyer et al., 2001). In this case, although AG biosynthesis is affected, it still occurs. This result can be rationalized by assuming that the Emb arabinosyl transferases function as dimers. Thus for AG biosynthesis, EmbA and EmbB (which are very similar to each other) preferentially form a heterodimer in mycobacteria, whereas the single protein Emb, for AG biosynthesis in *Corynebacterium* and EmbC for LAM biosynthesis in mycobacteria act as homodimers.

The other enzyme in the polymerization of AG is identified as AftB (Rv3805c). AftB transfers Ara*f* residues from DPA to the arabinan domain to form terminal β-(1→2) linked Ara*f* residues, at the non-reducing end of the AG (Seidel et al., 2007) (Fig. 3.7). These researchers deleted the orthologous gene *NCgl2780* in *C. glutamicum*, which resulted in reduction of cell wall-bound mycolic acids, consistent with a partial loss of mycolylation sites. The linkage analysis of AG in AftB mutant revealed the complete absence of terminal 2-linked Ara*f* residues, supporting the fact that AftB attaches the t-Ara*f* residues to the 2- position of the penultimate Ara*f* units at the non-reducing end of the arabinan chain (Seidel et al., 2007). The β-(1→2) arabinosyltransferase activity of AftB has also been proven with *in vitro* assays in the presence of glycolipid acceptors and subsequent characterization of the products.

Yet another interesting work by Birch et al., (2008) identified a α-(1→3) arabinofuranosyltransferase (AftC), at the non-reducing end of AG (Birch et al., 2008). Knock-out studies using *M. smegmatis* revealed the function of AftC that puts an α-Ara*f* on position 3 of the internal and possibly the external Ara*f* unit at the non-reducing end (Fig. 3.7). In this context, earlier reports noting that the AG produced by either the *embA* or *embB* knockouts had diminished amounts of the β-Ara*f*-(1→2) disaccharide attached to position 3 of the normally 3,5-linked Ara*f* at the non-reducing end, were consistent with a role of EmbA/B in the synthesis of the non-reducing terminal end, perhaps as an α-1,3 arabinosyl transferase. Since the α-1,5 arabinosyl transferase activity of EmbA/B is established, these results are difficult to interpret; possibly the effect on the non-reducing terminal is a secondary effect not directly caused by lack of EmbA/B activity,

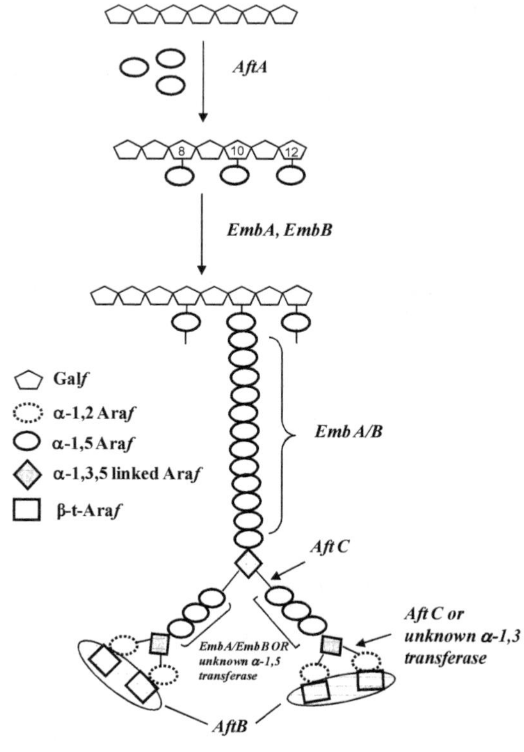

Figure 3.7 Schematic representation of the action of various arabinosyl transferases in the polymerization of arabinogalactan molecule. Briefly, the arabinosyl transferases AftA, Emb B/A. AftC and AftB act in succession, carrying out the addition of the first Ara*f* units to the galactan chain; α-1,5 Ara*f* to the growing chain; α-1,3 Ara*f* at the interior of non-reducing end, and terminal β-1,2-linked Ara*f* at the terminal non-reducing end, respectively to complete the AG polymerization. Research is in progress on identification of other unknown enzymes, especially the branching ones. Similar polymerization steps occur with the other two galactan linked Ara*f* units; however, this is not shown in the figure.

or, less likely, that EmbA/B might have more than one arabinosyl transferase activity. Apart from this, further unknown enzymes (including AftB) that specifically form the non-reducing structure have yet to be identified.

Finally, the order of the later events in the pathway are not known; specifically the transfer of the growing AG molecule to peptidoglycan – it could be before, during, or after arabinosylation, but probably before mycolylation; or after all these events have occurred. There are no enzymes reported to show the attachment of these three molecules, AG, PG and mycolates, except an *in vitro* assay demonstrating the 'ligation' of AG to PG (Yagi *et al.*, 2003). Using cell-free extracts of *M. smegmatis* as an enzyme source; and UDP-[^{14}C]Gal*p* (as the source of galactan) and 5-P-[^{14}C]ribosyl-P-P (donor of Ara*f*) as substrates, Yagi *et al.*, (2003) demonstrated the synthesis of the simpler Pol-P-P-Linker Unit-(Galf)*n* glycolipid intermediates, followed by the Pol-P-P-LU-AG and, finally, ligation of the P-LU-AG to PG. From these data, it could be speculated that mycolylation is not a pre-requisite for ligation of AG to PG although it is not biochemically proven. Finally, it is not known as to when the mycolic acids are attached; they might well be attached after attachment of AG to PG rather than before. Mycolic acid biosynthesis itself is fairy well understood (Barry *et al.*, 2007; Barry, III *et al.*, 1998) but it is out of the scope of this chapter. Nevertheless, the antigen 85 proteins, encoded by three independent genes, *fbpA*, *fbpB* and *fbpC* are studied in the genomes of *M. tuberculosis*, *M. leprae* and *M. avium* for the mycolyltransferase activity. *In vitro* enzymatic assays revealed that all three purified antigens catalyse the transfer of mycolates on trehalose, leading to the formation of trehalose monomycolate (TMM) and trehalose dimycolate (TDM, cord factor) (Belisle *et al.*, 1997) but no transfer from trehalose mycolates to AG has been reported. However, it has been shown by knock-out studies that antigen 85C is involved in the transfer of mycoloyl residues to the cell wall AG in *M. tuberculosis* (Jackson *et al.*, 1999), although there is no biochemical evidence for this. Similarly, the activity of mycoloyl transferases (myt) in the transfer of mycolyl residues to AG is successfully demonstrated in *C. glutamicum* (Kacem *et al.*, 2004).

Targeting the mycobacterial cell wall biosynthetic enzymes

The most important aim of studying the biosynthetic aspects of cell wall enzymes is to design new drugs, which inhibit the enzymes, thereby targeting the organism itself. The AG biosynthetic enzymes, including precursors and polymerizing enzymes, discussed earlier in this chapter will be reviewed here for their targeting studies and for any assay development.

Targeting arabinogalactan precursor molecules

Decaprenyl phosphate synthesis

It is generally considered better to target the initial enzymes of a biosynthetic pathway as this blocks the further steps in a chain of reactions. Decaprenyl phosphate (DP) is an important carrier molecule on which arabinogalactan is formed. Thus, targeting its biosynthetic enzymes in the MEP pathway is highly advantageous. As discussed earlier in the text, the two major events in the MEP pathway are the formation of isopentyl diphosphate (IPP) and dimethylallyl diphosphate (DMAPP) and their subsequent polymerization into DP (Fig. 3.4). Especially, the inhibition of enzymes in the pre-polymerization events shuts down both energy formation as well as cell wall biosynthesis, making them a high priority for targeting. Hence, it is obvious to target the enzyme DXS that catalyses the first step of the MEP pathway through condensation of pyruvate and glyceraldehyde-3-phopshate (GAP). Interestingly, DXS has also been reported as an attractive target for the development of novel antibiotics, antimalarials, and herbicides. DXS has been cloned from various bacteria including *M. tuberculosis* (Sprenger *et al.*, 1997; Kuzuyama *et al.*, 2000) as well as from several plant species (Lange *et al.*, 1998). In addition, DXS from *M. tuberculosis* has been enzymatically characterized (Bailey *et al.*, 2000) and its essentiality has been proven (Sassetti *et al.*, 2003). The crystal structure of DXS, from *E. coli* and *D. radiodurans*, in complex with the coenzyme thiamine pyrophosphate (TPP) has been solved, showing its homology to eukaryotic transketolases at the amino acid level (Xiang *et al.*, 2007), a condition that is tricky in terms of drug development.

The next enzyme to study in DP biosynthesis is IspC, and it carries out the rearrangement and reduction of DXP to form MEP (Baba and Allen, 1978; Dhiman *et al.*, 2005). It has been cloned from a number of bacteria and plants, plus detailed enzyme kinetics were obtained from *E. coli* (Koppisch *et al.*, 2002) and *M. tuberculosis* (Dhiman *et al.*, 2005). Structural studies of this enzyme from several sources provided encouraging information making this enzyme a potential drug target (Mac Sweeney *et al.*, 2005; Reuter *et al.*, 2002; Ricagno *et al.*, 2004; Steinbacher *et al.*, 2003; Yajima *et al.*, 2004; Yajima *et al.*, 2002). Co-crystallization studies on DXR were carried out in the presence of the inhibitor fosmidomycin, magnesium and the essential co-substrate NADPH. Following the formation of a ternary complex, DXR undergoes a remarkable conformational change that is induced by the binding of substrate or inhibitor in the presence of NADPH, which further plays an important role in substrate binding and catalysis (MacSweeney *et al.*, 2005). Unexpectedly, the gene was shown to be non-essential by high density transposon mutagenesis (Sassetti *et al.*, 2003). However, until shown clearly by the characterization of a knock-out mutant, this result needs to be interpreted with caution.

Interestingly, the antibiotic fosmidomycin that inhibits IspC (Kuzuyama *et al.*, 1998; Zeidler *et al.*, 1998) in a number of organisms, is not active against *M. tuberculosis* (Dhiman *et al.*, 2005), even though the MEP pathway appears to be the sole IPP biosynthetic pathway in this organism. Further it was shown that the gene encoding IspC, *Rv2870c* was able to complement an inactivated chromosomal copy of *IspC* in *Salmonella enterica* and the complemented strain was sensitive to fosmidomycin (Dhiman *et al.*, 2005). This led to the conclusion that the resistance to fosmidomycin in *M. tuberculosis* is not due to intrinsic properties of Rv2870c, and the enzyme still appears to be a valid drug target in this pathogen (Dhiman *et al.*, 2005). In contrast with this interpretation, a very recent work showed that the resistance to fosmidomycin is due to the lack of cellular penetration into hydrophobic mycobacterial cell wall based on the fact that the uptake of fosmidomycin, a hydrophilic molecule, is an active process involving a cAMP-dependent glycerol-3-phosphate transporter (GlpT) (Brown and Parish, 2008). Furthermore, in *M. tuberculosis* no homologue for *glpT* was identified, suggesting that the lack of intracellular accumulation of fosmidomycin was not due to efflux, based on the fact that efflux pump inhibitors had no effect on fosmidomycin resistance (Brown and Parish, 2008).

Later on in DP biosynthesis, a series of enzymes, namely Isp D, E and F convert MEP to 2-C-methyl-D-erythritol 2,4-cyclodiphosphate followed by the consecutive action of IspG and IspH finally leading to isopentinyl diphosphate (IPP) and dimethylallyl diphosphate (DMAPP) (Fig. 3.4). The enzymes IspD (Rv3582c) (Eoh *et al.*, 2007), IspE (Rv1011) and IspF (Rv3581c) have been cloned, expressed and characterized to some extent (Dean Crick personal communication). It is generally considered that all of the MEP pathway enzymes after IspC would also be good drug targets. In particular, IspE is a potential drug target vowing to a great deal of information on the development of kinase inhibitors (Miallau *et al.*, 2003; Wada *et al.*, 2003). IspE from *E. coli* displays the α/β fold characteristic of the galactose kinase/homoserine kinase/mevalonate kinase/phosphomevalonate kinase superfamily. However, its binding to adenosine is reported to be in an unusual *syn* orientation of the base with respect to the ribose across the glycosidic bond, suggesting it as an unusual kinase (Miallau *et al.*, 2003; Wada *et al.*, 2003). Although, the proven essentiality and available crystal structure data make the enzymes IspE–H ideal drug targets, the unavailability of required substrates makes them difficult targets to develop HTS assays.

The next enzymes to consider for drug targeting studies are polyprenyl diphosphate synthases, which polymerize IPP and DMAPP to longer polyprenyl diphosphate molecules. The enzymes E,Z Farnesyl synthase (Rv1086c) and Decaprenyldiphosphate synthase (Rv2361c) that catalyse IPP polymerization to DP have been cloned and expressed (Schulbach *et al.*, 2000; Kaur *et al.*, 2004), and is further supplemented with recent X-ray structural work using various substrates and analogues (Zhang *et al.*, 2008a). Efforts to obtain specific inhibitors are in progress, keeping in mind that human dolichol synthase (or *cis*-isoprenyltransferase) may have a similar structure (Shridas *et al.*, 2003). Using high density transposon mutagenesis, Sassetti *et al.*, (2003) have shown that in *M. tuberculosis*, Rv2361c is likely to be essential while Rv1086 is not. In addition, Rv2361c was shown to compensate for the deletion of Rv1086 and also is shown to synthesize *E,Z*-FPP, although with lesser catalytic rates (Kaur *et al.*, 2004). Based on this, it could be stated that Rv1086 is not a good drug target in *M. tuberculosis* as speculated earlier (Schulbach *et al.*, 2001), even though the synthesis of DP is essential.

Arabinogalactan precursor molecules

The biosynthetic pathways of major sugar donors for arabinogalactan biosynthesis, β-D-DPA*f*, dTDP-β-L-Rha*p*, UDP-GlcNAc (which is an important constituent of both peptidoglycan and the linker unit) and UDP-α-D-Gal*f* are explained earlier (Fig. 3.5). The biosynthetic details for most of these enzymes are well understood and HTS assays are available for many. Most of these enzymes were shown to be essential using insertion mutagenesis (Sassetti *et al.*, 2003) with no human analogues/absent in humans, which makes them attractive drug targets.

Decaprenylphosphoryl arabinofuranose (DPA) is the donor of the D-Ara*f* residues of mycobacterial arabinan and its biosynthesis is discussed in detail in the earlier section (Fig. 3.5A). The gene *Rv3806c* has been cloned and expressed in *E. coli* and the conversion of phosphoribose diphosphate and decaprenyl phosphate into decaprenylphosphoryl-5-phosphoribose has also been demonstrated (Huang *et al.*, 2005). An assay for Rv 3806c amenable to HTS based on the detection of released PPi might be possible. The last two essential enzymes of DPA biosynthesis, decaprenylphosphoryl-D-ribose oxidase (Rv3790) and decaprenylphosphoryl-2-D-keto erythropentose reductase (Rv3791), were shown to work together in the conversion of decaprenyl-P-ribose (DPR) to decaprenyl-P-arabinose (DPA) (Mikusova *et al.*, 2005). Although, no robust HTS assays are known for these enzymes, a recent report is very encouraging in that benzothiazinones (BTZ), a class of antitubercular drug, inhibits the activity of Rv3790 *in vitro*. Biochemical evidence for the same was provided by overexpressing the *Rv3790* gene in *M. tuberculosis, M. bovis* BCG and *M. smegmatis* and demonstrating higher resistance levels to BTZ (Manina *et al.*, 2008).

The biosynthetic pathway for dTDP-β-L-Rha*p* is shown in Fig. 3.5B. A microtitre plate-based HTS assay is available for RmlB-D (Ma *et al.*, 2001) supplemented with crystal structures in *M. tuberculosis* as well as in other bacteria (Beis *et al.*, 2005; Blankenfeldt *et al.*, 2002; Allard *et al.*, 2001; Sanders *et al.*, 2001; Allard *et al.*, 2000; Blankenfeldt *et al.*, 2000; Giraud *et al.*, 2000). The HTS assay for Rml B–D is rather simple wherein the reduction of NADPH is monitored at an absorbance of 340nm (Ma *et al.*, 2001). Our lab is currently focused on assaying for RmlC, together with RmlB that is used for pre-incubation to prepare the dTDP-4-keto-6-deoxyglucose substrate and RmlD, in large excess as an 'aid' enzyme. In particular, RmlC has many advantages to be an attractive drug target because it is essential, soluble, crystal structures with substrate analogues are available (Dong *et al.*, 2007), evidence that its inhibition could be lethal, has been presented in view of the fact that inhibition of WbbL later in the same pathway results in cell death (Mills *et al.*, 2004). Further, the details on the structure of the transition state, and conformation changes of the sugar residue are also available (Dong *et al.*, 2007).

GlcNAc is a crucial component in PG as well as the linker unit in the cell wall of mycobacteria (Fig. 3.5C). Recently in our lab, an HTS assay was developed for the transacetylase activity of the enzyme GlmU that transfers an acetate group from acetyl CoA to glucosamine-1-phosphate resulting in the formation of *N*-acetylglucosamine. The released CoA was detected by its reaction with DTNB [5,5′-dithiobis(2-nitrobenzoic acid)] at an absorbance of 412 nm. Optimization studies for pH, temperature, substrate and enzyme stability were done in 96-well microtitre plates followed by detailed kinetic studies. The Z′-factor was calculated to be 0.9, indicating a high-quality and robust assay. Currently, assays are in progress on >200,000 compound from the NIH Small Molecule Repository at Southern Research Institute in a 384-well plate format (Bhamidi *et al.*, unpublished results).

The other important enzyme is Glf (Rv3809c), which is a mutase converting UDP-α-D-gal*p* to UDP-α-D-gal*f* (Fig. 3.5D). For this, an *in vitro* assay based on the enzymatic activity is available (Scherman *et al.*, 2003). This radioactive assay iş based on the fact that tritiated formaldehyde is released only from UDP-galactofuranose but not UDP-galactopyranose, following the action of sodium periodate. One of the limitations in this assay is the requirement of the FAD co-factor in the reduced $FADH_2$ form. An appropriate reducing agent such as dithionite is not compatible with the micro titre plate assay that uses sodium periodate. A more sensitive assay based on the aspect of fluorescence polarization (FP) was developed where in the emission of a fluorescent compound excited with plane-polarized light is measured (Carlson *et al.*, 2006; Soltero-Higgin *et al.*, 2004). The FP technique relies upon changes in the rotational properties of molecules in solution. The readings are based on the differences in polarization depending on whether the fluorescent probe is bound to Glf (tumbling slowly) or released into solution by a competitive inhibitor (tumbling rapidly). Inhibitors of Glf have also been identified using a fluorescence polarization assay (Soltero-Higgin *et al.*, 2004).

Arabinogalactan polymerization biosynthetic enzymes

The AG polymerizing enzymes, commonly known as glycosyl transferases (Figs 3.6 and 3.7) are not as well studied as the enzymes that form the AG precursors. Interestingly, all these enzymes are shown to be essential with no human homologues, which make them attractive drug targets. However, the difficulties associated with these enzymes as ideal drug targets are substrate preparation, non-availability of crystal structures and the fact that these are integral or membrane bound. Specifically, the rhamnosyl transferase, WbbL (Mills *et al.*, 2004), and major galactofuranosyl transferase, GlfT (Rose *et al.*, 2006; Kremer *et al.*, 2001; Mikusova *et al.*, 2000; Alderwick *et al.*, 2008) were studied in detail (Fig. 3.6). Research is in progress towards getting reasonable amounts of membrane associated WbbL (McNeil *et al.*, unpublished), improved expression studies (Grzegorzewicz *et al.*, 2008), and acceptor (GlcNAc-P-P-decaprenyl) synthesis (Imperiali and Zimmerman, 1990). The donor dTDP-Rha can be prepared both enzymatically (Mikusova *et al.*, 1996; Mills *et al.*, 2004) and chemically (Ben Davis personal communication). Microtitre plate assays are highly possible for WbbL based on detection of dTDP, either by release of phosphate, coupling to lactate dehydrogenase (Gosselin *et al.*, 1994), or by coupling to ATP production.

The two galactosyl transferases, Rv3782 (which adds two Gal*f* units to the rhamnosyl linker unit) (Mikusova *et al.*, 2006) and GlfT (Rv3808c-major polymerization enzyme), are also studied for the purpose of drug targeting. GlfT has been produced in reasonable amounts in soluble form (Rose *et al.*, 2006; Alderwick *et al.*, 2008). Furthermore, its dual galactofuranosyl transferase activities were proved using β(1→5) and β(1→6) linked digalactofuranosyl neoglycolipid acceptors (Alderwick *et al.*, 2008). The donor substrate UDP-Gal*f* can be prepared in reasonable yields both chemically (Marlow and Kiessling, 2001) and chemically/enzymatically (Errey *et al.*, 2004). A variety of synthetic acceptors have also been prepared chemically (Rose *et al.*, 2006; Pathak *et al.*, 2001b). A HTS assay based on the detection of UDP appears to be practical.

The most current topic for drug targeting is the relatively new arabinosyl transferases that are found to be essential (Figs 3.6 and 3.7). The two well-understood arabinosyl transferases are AftA and AtfB, responsible for transfer of the first Ara*f* units on to the galactan and the terminal β-(1→5)-linked Ara*f* at the non-reducing end, respectively (Alderwick *et al.*, 2006; Seidel *et al.*, 2007). AftA has been expressed from *E. coli* in soluble form (Alderwick *et al.*, 2006) and an improved chemical synthesis of the donor decaprenylphosphoryl-D-arabinose (DPA) has been recently reported (Liav *et al.*, 2006). The transferase activity of AftA has been further proven by an *in vitro* assay using *M. tuberculosis aftA* overexpressed in *E. coli* as enzyme source and cell wall galactan as an acceptor (Alderwick *et al.*, 2005). Currently, the galactan acceptor for AftA has been prepared from the *aftA* mutant of *C. glutamicum,* however, it is unclear whether this could be used for inhibitor screening. Similarly, an *in vitro* assay has been developed to confirm AftB activity using a *C. glutamicum aftB* knock-out mutant, in the presence of the sugar donor β-D-arabinofuranosyl-1-monophosphoryl-decaprenol and the neoglycolipid acceptor α-D-Araf-(1→5)-α-D-Araf-O-C_8 (Seidel *et al.*, 2007).

More arabinosyl transferase activities were identified using specific acceptors in cell-free assays of *M. smegmatis* (Pathak *et al.*, 2001a; Lee *et al.*, 1997; Khasnobis *et al.*, 2006). Khasnobis *et al.*, (2006) have developed a model assay for arabinosyltransferase activity using chemically synthesized oligo arabinan as an acceptor, cell wall as enzyme source and DPA (prepared either chemically or enzymatically by treating pRpp with membranes) (Khasnobis *et al.*, 2006). Another *in vitro* study by the same group revealed that only α-Araf is specifically transferred from DPA on a newly synthesized branched acceptor (α-D-Araf)$_2$-3,5-α-D-Araf-(1 →5)-α-D-Araf-(1→5)-α-D-Araf (Zhang *et al.*, 2007). The on-going work to screen inhibitors for arabinosyl transferases by the attachment of the acceptor to scintillation proximity beads could be an exciting possibility (Delphi Chatterjee, personal communication).

Once the AG biosynthesis is completed, it is attached to PG and mycolic acids by unknown mechanisms. The sole report in this context is the identification of the ligase that was shown to attach AG to PG using an *in vitro* assay (Yagi *et al.*, 2003). Mycobacterial ligase is reported to have functional similarity with other unidentified Gram positive bacterial ligases that attach teichoic acids (which have a similar linker disaccharide) to PG. Finally, targeting of the AG/PG ligase remains unsolved due to the unavailability of a simple assay procedure and because its encoding protein/gene is unidentified.

Conclusions and possible future directions of AG research and drug targeting

Overall, the structural features and the biosynthetic aspects of arabinogalactan are substantially understood, although, the drug targeting studies still need to be explored. Still, some research is needed to fill the gaps of our understanding of the AG structure. For instance, the role of $GalNH_2$ and succinates in the context of cell wall integrity is not clearly understood and only hypothesized. The $GalNH_2$ is shown to be present only in slow growing bacteria such as *M. tuberculosis*, but not in fast growing ones like *M. smegmatis*. This could further correlate to the virulence and indicate that the $GalNH_2$ pathway in *M. tuberculosisis* is probably not essential. In such case, targeting the $GalNH_2$ biosynthetic pathway would be very helpful. The most likely pathway to the formation of $GalNH_2$ is from UDP-GlcNAc, which is epimerized to UDP-GalNAc, that donates a GalNAc residue to AG (either directly or after transfer to decaprenyl phosphate to make decaprenylphosphoryl-D-GalNAc) to make GalNAc arabinan. Subsequent deacetylation leads to the formation of $GalNH_2$ arabinan. Hence, targeting the epimerization would be interesting, as this will inhibit the formation of $GalNH_2$-arabinan, in turn affecting the virulence. Similarly, mutants of *M. tuberculosis* that do not succinate AG and/or LAM could be useful to find out if succinylation has any role in bacterial virulence.

Most of the available data on structure and biosynthetic aspects are with cells grown under controlled environment *in vitro*. However, the fate of the bacterium within the infected tissue is still not known. A major difficulty to hasten the development of anti-tuberculosis drugs is the inability to screen drug candidates in an *in vitro* system that mimics the *in vivo* environment in which the drug is ultimately used. The environment that *M. tuberculosis* encounters in the human host is rather different from that of a typical *in vitro* culture media. In an infected tissue, *M. tuberculosis* resides in granulomatous lesions at various developmental stages, resolution, or necrosis, thereby adding more difficulty to mimic the *in vivo* growth environment. In general, during normal infections, metabolic alterations occur within pathogens that reflect their intracellular environments. In this regard, there is no experimental evidence defining active metabolic pathways that are essential to sustain *M. tuberculosis* growth *in vivo*. Like other pathogenic bacteria, the *M. tuberculosis* genome encodes regulatory networks that respond to environmental signals. Thus, the bacterium can translate information about the environment into biochemical and transcriptional signals that can be exploited as biomarkers in the development of *in vivo* mimicking culture conditions. Our lab is currently working on this very aspect of cell wall composition, *in vivo* compared with *in vitro*, and for any evidence of changed morphology or structural variations.

Another major challenge in drug targeting efficacy of cell wall targets is the condition of the cells that are actively dividing versus those in dormancy. For example, it was shown that drugs such as isoniazid that are effective against actively growing cells, are much less effective in stationary phase cultures (Herbert *et al.*, 1996; Paramasivan *et al.*, 2005). However, in stationary cells, some cell wall processes, such as release of antigens, including some cell wall components occur (Ulrichs *et al.*, 2005). For a total cure of tuberculosis, it is required to eradicate both replicating as well as non-replicating bacteria. A recent report on the selective killing of non-replicating bacteria is very helpful in this regard and paves a way for further research for drug targeting latent tuberculosis (Bryk *et al.*, 2008). These researchers have found that selective 'rhodanine' based compounds inhibit the enzyme dihydrolipoamide acetyltransferase (DlaT). DlaT is required to cause tuberculosis in guinea pigs and used by the mycobacteria to resist the immuno response stress, caused by reactive nitrogen intermediates (RNI) in the host. An earlier report by the same group demonstrated that *in vitro*, RNI at short exposures or low levels prevents *M. tuberculosis* replication, whereas longer exposures or higher concentrations kill the organism (Nathan and Ehrt, 2004).

A different approach to target dormant bacteria is to induce replication, followed by treatment with anti-tubercular drugs. Research in this direction led to the identification of 'resuscitation-promoting factors' (Rpf), encoded by many different bacteria including mycobacteria. They were initially reported in *Micrococcus luteus* to be capable of resuscitating the non-replicating dormant bacteria (Mukamolova *et al.*, 2002). Specifically in all bacteria (including mycobacteria), during the final stages of cell division, the peptidoglycan layer between two daughter cells is degraded by enzymes such as CwIM and Rv2719c (Deng *et al.*, 2005; Chauhan *et al.*, 2006). Likewise, Rpf proteins that constitute a family of lytic transglycosylase enzymes, are capable of hydrolysing glycosidic bonds in the PG layer (Mukamolova *et al.*, 2006). RpfB and RpfE interact with RipA and all the three proteins form a complex at the dividing septum (Hett *et al.*, 2008; Tufariello *et al.*, 2006). Presently, extensive research is on going to better understand their role in mycobacterial latent infections.

Finally, focusing on arabinogalactan, the data concerning regulation of the AG biosynthetic gene expression and control of the cell wall degradation are still not completely available and these enzymes share little homology with other well-characterized enzymes. Furthermore, AG biosynthesis is shown to be down-regulated under non-replicating conditions (Hampshire *et al.*, 2004). This indicates that targeting the AG polymerization enzymes would not be a good idea in the case of stationary cells. However, it is noted that early pre-polymerization enzymes in the MEP pathway are required for the production of the sugar donor decaprenyl phosphate, on which the entire AG molecule is built (Fig. 3.4). Thus, targeting the early stages of isoprene biosynthesis would shut down mechanisms of energy formation as well as cell wall biosynthesis. Therefore, inhibitors of the arabinogalactan polymerizing enzymes should kill actively growing bacteria and targeting MEP pathway enzymes should kill both replicating, as well as latent mycobacteria.

Acknowledgements

The McNeil laboratory gratefully acknowledges the support of the National Institute of Health (AI33706). The authors also thank Dr Devinder Kaur for helpful discussions and Victoria Jones for editorial contributions.

References

Alahari, A., Saint, N., Campagna, S., Molle, V., Molle, G., and Kremer, L. (2007). The N-terminal domain of OmpATb is required for membrane translocation and pore-forming activity in mycobacteria. J. Bacteriol. *189*, 6351–6358.

Alderwick, L.J., Dover, L.G., Veerapen, N., Gurcha, S.S., Kremer, L., Roper, D.L., Pathak, A.K., Reynolds, R.C., and Besra, G.S. (2008). Expression, purification and characterisation of soluble GlfT and the identification of a novel galactofuranosyltransferase Rv3782 involved in priming GlfT-mediated galactan polymerisation in *Mycobacterium tuberculosis*. Protein Expression and Purification *58*, 332–341.

Alderwick, L.J., Radmacher, E., Seidel, M., Gande, R., Hitchen, P.G., Morris, H.R., Dell, A., Sahm, H., Eggeling, L., and Besra, G.S. (2005). Deletion of Cg-emb in Corynebacterianeae leads to a novel truncated cell wall arabinogalactan, whereas inactivation of CgubiA results in an arabinan-deficient mutant with a cell wall galactan core. J. Biol. Chem. *280*, 32362–32371.

Alderwick, L.J., Seidel, M., Sahm, H., Besra, G.S., and Eggeling, L. (2006). Identification of a novel arabinofuranosyl transferase (AftA) involved in cell wall arabinan biosynthesis in *mycobacterium tuberculosis*. J. Biol. Chem. *281*, 15653–15661.

Allard, S.T.M., Giraud, M.F., Whitfield, C., Graninger, M., Messner, P., and Naismith, J.H. (2001). The crystal structure of dTDP-D-glucose 4,6-dehydratase (RmlB) from *Salmonella enterica* serovar *typhimurium*, the second enzyme in the dTDP-L-rhamnose pathway. J. Mol. Biol. *307*, 283–295.

Allard, S.T.M., Giraud, M.-F., Whitfield, C., Messner, P., and Naismith, J.H. (2000). The purification, crystallization and structural elucidation of dTDP-D-glucose 4,6-dehydratase (RmlB), the second enzyme of the dTDP-L-rhamnose synthesis pathway from *Salmonella enterica* serovar typhimurium. Acta Crystallogr. D. Biol. Crystallogr. *56 (Pt 2)*, 222–225.

Alsteens, D., Verbelen, C., Dague, E., Raze, D., Baulard, A.R., and Dufrene, Y.F. (2008). Organization of the mycobacterial cell wall: a nanoscale view. Pflügers Archiv Eur. J. Physiol. *456*, 117–125.

Amar-Nagasch, C., and Vilkas, E. (1970). Etude des parois de *Mycobacterium tuberculosis*. II. Mise en evidence d'un mycolate d'arabinobiose et d'un glucane dans les parois de *M. tuberculosis* H37Ra. Bull. Soc. Chim. Biol. *52*, 145–151.

Anderson, J.S. and Page, R.L. (1972). Biosynthesis of the polysaccharide of *Micrococcus lysodeikticus* Cell Walls – Part II. J. Biol. Chem. *247*, 2480–2485.

Azuma, I., Yamamura, Y., and Misaki, A. (1969). Isolation and characterization of arabinose mycolate from firmly bound lipids of mycobacteria. J. Bacteriol. *98*, 331–333.

Baba, T. and Allen, C.M. (1978). Substrate-specificity of undecaprenyl pyrophosphate synthetase from *Lactobacillus plantarum*. Biochemistry *17*, 5598–5604.

Bailey, A.M., Mahapatra, S., Brennan, P.J., and Crick, D.C. (2000). Purification and characterization of 1-deoxyxylulose 5-phosphate synthase from *Mycobacterium tuberculosis*. Glycobiology *10*, 46.

Barry, C.E., Crick, D.C., and McNeil, M.R. (2007). Targeting the formation of the cell wall core of *M. tuberculosis*. Infect. Disorders-Drug Targets 7, 182–202.

Barry, C.E., III, Lee, R.E., Mdluli, K., Sampson, A.E., Schroeder, B.G., Slayden, R.A., and Yuan, Y. (1998). Mycolic acids: structure, biosynthesis and physiological functions. Prog. Lipid Res. *37*, 143–179.

Beis, K., Srikannathasan, V., Liu, H., Fullerton, S.W.B., Bamford, V.A., Sanders, D.A.R., Whitfield, C., McNeil, M.R., and Naismith, J.H. (2005). Crystal structures of Mycobacteria tuberculosis and *Klebsiella pneumoniae* UDP-galactopyranose mutase in the oxidised state and *Klebsiella pneumoniae* UDP-galactopyranose mutase in the (Active) reduced state. J. Mol. Biol. *348*, 971–982.

Belanger, A.E., Besra, G.S., Ford, M.E., Mikusova, K., Belisle, J., Brennan, P.J., and Inamine, J.M. (1996). The embAB genes of Mycobacterium avium encode an arabinosyl transferase involved in cell wall arabinan biosynthesis that is the target for the antimycobacterial drug ethambutol. Proc. Natl. Acad. Sci. USA *93*, 11919–11924.

Belisle, J.T., Vissa, V.D., Sievert, T., Takayama, K., Brennan, P.J., and Besra, G.S. (1997). Role of the major antigen of *Mycobacterium tuberculosis* in cell wall biogenesis. Science *276*, 1420–1422.

Besra, G.S., Khoo, K.-H., McNeil, M., Dell, A., Morris, H.R., and Brennan, P.J. (1995). A new interpretation of the structure of the mycolyl-arabinogalactan complex of *Mycobacterium tuberculosis* as revealed through characterization of oligoglycosylalditol fragments by fast-atom bombardment mass spectrometry and ^{1}H nuclear magnetic resonance spectroscopy. Biochemistry *34*, 4257–4266.

Bhamidi, S., Scherman, M.S., Rithner, C.D.P.J.E.C.D., Khoo, K.-H., and McNeil, M.R. (2008). The Identification and location of succinyl residues and the characterization of the interior arabinan region allow for a model of the complete primary structure of *Mycobacterium tuberculsis* Mycolyl Arabinogalactan. J. Biol. Chem. *283*, 12992–13000.

Birch, H.L., Alderwick, L.J., Bhatt, A., Rittmann, D., Krumbach, K., Singh, A., Bai, Y., Lowary, T.L., Eggeling, L., and Besra, G.S. (2008). Biosynthesis of mycobacterial arabinogalactan: identification of a novel alpha(1→3) arabinofuranosyltransferase. Mol. Microbiol. *69*, 1191–1206.

Blankenfeldt, W., Asuncion, M., Lam, J.S., and Naismith, J.H. (2000). The structural basis of the catalytic mechanism and regulation of glucose-1-phosphate thymidylyltransferase (RmlA). EMBO J. *19*, 6652–6663.

Blankenfeldt, W., Kerr, I.D., Giraud, M.F., McMiken, H.J., Leonard, G., Whitfield, C., Messner, P., Graninger, M., and Naismith, J.H. (2002). Variation on a theme of SDR. dTDP-6-deoxy-L-lyxo-4-hexulose reductase (RmlD) shows a new Mg2+-dependent dimerization mode. Structure. (Camb.) *10*, 773–786.

Brennan, P.J. (1988). *Mycobacterium* and other actinomycetes. In Microbial Lipids Vol. I, C.Ratledge and S.G.Wilkinson, eds. (London: Academic Press), pp. 203–298.

Brown, A.C. and Parish, T. (2008). Dxr is essential in *Mycobacterium tuberculosis* and fosmidomycin resistance is due to a lack of uptake. BMC Microbiology *8*.

Bryk, R., Gold, B., Venugopal, A., Singh, J., Samy, R., Pupek, K., Cao, H., Popescu, C., Gurney, M., Hotha, S., Cherian, J., Rhee, K., Ly, L., Converse, P.J., Ehrt, S., Vandal, O., Jiang, X.J., Schneider, J., Lin, G., and Nathan, C. (2008). Selective killing of nonreplicating mycobacteria. Cell Host & Microbe 3, 137–145.

Carlson, E.E., May, J.F., and Kiessling, L.L. (2006). Chemical probes of UDP-galactopyranose mutase. Chemistry & Biology *13*, 825–837.

Chatterjee, D. and Khoo, K.H. (1998). Mycobacterial lipoarabinomannan: an extraordinary lipoheteroglycan with profound physiological effects. Glycobiology *8*, 113–120.

Chatterjee, D., Lowell, K., Rivoire, B., McNeil, M., and Brennan, P.J. (1992). Lipoarabinomannan of *Mycobacterium tuberculosis*: capping with mannosyl residue in some strains. J. Biol. Chem. *267*, 6234–6239.

Chauhan, A., Lofton, H., Maloney, E., Moore, J., Fol, M., Madiraju, M.V.V.S., and Rajagopalan, M. (2006). Interference of *Mycobacterium tuberculosis* cell division by Rv2719c, a cell wall hydrolase. Mol. Microbiol. *62*, 132–147.

Cole, S.T., Brosch, R., Parkhill, J., Garnier, T., Churcher, C., Harris, D., Gordon, S.V., Eiglmeier, K., Gas, S., Barry, C.E., Tekaia, F., Badcock, K., Basham, D., Brown, D., Chillingworth, T., Connor, R., Davies, R., Devlin, K., Feltwell, T., Gentles, S., Hamlin, N., Holroyd, S., Hornby, T., Jagels, K., Krogh, A., McLean, J., Moule, S., Murphy, L., Oliver, K., Osborne, J., Quail, M.A., Rajandream, M.A., Rogers, J., Rutter, S., Seeger, K., Skelton, J., Squares, R., Squares, S., Sulston, J.E., Taylor, K., Whitehead, S., and Barrell, B.G. (1998). Deciphering the biology of *Mycobacterium tuberculosis* from the complete genome sequence. Nature *393*, 537–544.

Crick, D.C., Schulbach, M.C., Zink, E.E., Macchia, M., Barontini, S., Besra, G.S., and Brennan, P.J. (2000). Polyprenyl phosphate biosynthesis in *Mycobacterium tuberculosis* and *Mycobacterium smegmatis*. J. Bacteriol. *182*, 5771–5778.

Daffe, M., Brennan, P.J., and McNeil, M. (1990). Predominant structural features of the cell wall arabinogalactan of *Mycobacterium tuberculosis* as revealed through characterization of oligoglycosyl alditol fragments by gas chromatography/mass spectrometry and by ^{1}H and ^{13}C-NMR analyses. J. Biol. Chem. *265*, 6734–6743.

Daffe, M., Brennan, P.J., and McNeil, M. (1991). Novel type-specific lipo-oligosaccharides from *Mycobacterium tuberculosis*. Biochemistry *30*, 378–388.

Daffe, M. and Draper, P. (1998). The envelope layers of mycobacteria with reference to their pathogenicity. Advan. Microb. Physiol. 131–203.

Daffe, M. and Etienne, G. (1999). The capsule of *Mycobacterium tuberculosis* and its implications for pathogenecity. Tuber. Lung Dis. *79*, 153–169.

Deng, L.L., Humphries, D.E., Arbeit, R.D., Carlton, L.E., Smole, S.C., and Carroll, J.D. (2005). Identification of a novel peptidoglycan hydrolase CwlM in *Mycobacterium tuberculosis*. Biochim. Biophys. Acta *1747*, 57–66.

Dhiman, R.K., Schaeffer, M.L., Bailey, A.M., Testa, C.A., Scherman, H., and Crick, D.C. (2005). 1-Deoxy-D-xylulose 5-phosphate reductoisomerase (IspC) from *Mycobacterium tuberculosis*: towards understanding mycobacterial resistance to fosmidomycin. J. Bacteriol. *187*, 8395–8402.

Dmitriev, B.A., Ehlers, S., Rietschel, E.T., and Brennan, P.J. (2000). Molecular mechanics of the mycobacterial cell wall: from horizontal layers to vertical scaffolds. Int. J. Med. Microbiol. *290*, 251–258.

Dmitriev, B.A., Toukach, F.V., Schaper, K.M., Holst, O., Rietschel, E.T., and Ehlers, S. (2003). Tertiary structure of bacterial murein: The scaffold model. J. Bacteriol. *185*, 3458–3468.

Dong, C., Major, L.L., Srikannathasan, V., Errey, J.C., Giraud, M.F., Lam, J.S., Graninger, M., Messner, P., McNeil, M.R., Field, R.A., Whitfield, C., and Naismith, J.H. (2007). RmlC, a C3′ and C5′ carbohydrate epimerase, appears to operate via an intermediate with an unusual twist boat conformation. J. Mol. Biol. *365*, 146–159.

Dong, X., Bhamidi, S., Scherman, M., Xin, Y., and McNeil, M.R. (2006). Development of a quantitative assay for mycobacterial endogenous arabinase and ensuing studies of arabinase levels and arabinan metabolism in Mycobacterium smegmatis. Appl. Environ. Microbiol. *72*, 2601–2605.

Draper, P., Khoo, K.H., Chatterjee, D., Dell, A., and Morris, H.R. (1997). Galactosamine in walls of slow-growing mycobacteria. Biochem. J. *327 (Pt 2)*, 519–525.

Duvold, T., Bravo, J.M., PaleGrosdemange, C., and Rohmer, M. (1997). Biosynthesis of 2-C-methyl-D-erythritol, a putative C-5 intermediate in the mevalonate independent pathway for isoprenoid biosynthesis. Tetrahedron Lett. *38*, 4769–4772.

Eoh, H., Brown, A.C., Buetow, L., Hunter, W.N., Parish, T., Kaur, D., Brennan, P.J., and Crick, D.C. (2007). Characterization of the *Mycobacterium tuberculosis* 4-diphosphocytidyl-2-C-methyl-D-erythritol synthase: Potential for drug development. J. Bacteriol. *189*, 8922–8927.

Errey, J.C., Mukhopadhyay, B., Kartha, K.P.R., and Field, R.A. (2004). Flexible enzymatic and chemo-enzymatic approaches to a broad range of uridine-diphospho-sugars. Chem. Comm. 2706–2707.

Escuyer, V.E., Lety, M.A., Torrelles, J.B., Khoo, K.H., Tang, J.B., Rithner, C.D., Frehel, C., McNeil, M.R., Brennan, P.J., and Chatterjee, D. (2001). The role of the embA and embB gene products in the biosynthesis of the terminal hexaarabinofuranosyl motif of *Mycobacterium smegmatis* arabinogalactan. J. Biol. Chem. *276*, 48854–48862.

Faller, M., Niederweis, M., and Schulz, G.E. (2004). The structure of a mycobacterial outer-membrane channel. Science *303*, 1189–1192.

Giraud, M.-F., Leonard, G.A., Field, R.A., Berlind, C., and Naismith, J.H. (2000). RmlC, the third enzyme of dTDP-L-rhamnose pathway, is a new class of epimerase. Nature Struct. Biol. *7*, 398–402.

Gosselin, S., Alhussaini, M., Streiff, M.B., Takabayashi, K., and Palcic, M.M. (1994). A continuous spectrophotometric assay for glycosyltransferases. Anal. Biochem. *220*, 92–97.

Grzegorzewicz, A.E., Ma, Y., Jones, V., Crick, C., Liav, A., and McNeil, M.R. (2008). Development of a microtitre plate based assay for lipid linked glycosyl transferase products using the mycobacterial cell wall rhamnosyl transferase WbbL. Microbiol. *154*, 3724-3730.

Hampshire, T., Soneji, S., Bacon, J., James, B.W., Hinds, J., Laing, K., Stabler, R.A., Marsh, P.D., and Butcher, P.D. (2004). Stationary phase gene expression of *Mycobacterium tuberculosis* following a progressive nutrient depletion: a model for persistent organisms? Tuberculosis *84*, 228–238.

Herbert, D., Paramasivan, C.N., Venkatesan, P., Kubendiran, G., Prabhakar, R., and Mitchison, D.A. (1996). Bactericidal action of ofloxacin, sulbactam-ampicillin, rifampin, and isoniazid on logarithmic- and stationary-phase cultures of *Mycobacterium tuberculosis*. Antimicrobial Agents Chemother. *40*, 2296–2299.

Herz, S., Wungsintaweekul, J., Schuhr, C.A., Hecht, S., Luttgen, H., Sagner, S., Fellermeier, M., Eisenreich, W., Zenk, M.H., Bacher, A., and Rohdich, F. (2000). Biosynthesis of terpenoids: YgbB protein converts 4-diphosphocytidyl-2C-methyl-D-erythritol 2-phosphate to 2C-methyl-D-erythritol 2,4-cyclodiphosphate. Proc. Natl. Acad. Sci. USA *97*, 2486–2490.

Hett, E.C., Chao, M.C., Deng, L.L., and Rubin, E.J. (2008). A mycobacterial enzyme essential for cell division synergizes with resuscitation-promoting factor. PLoS Pathog. *4*, e1000001.

Higashi, Y., Strominger, J.L., and Sweeley, C.C. (1970). Biosynthesis of the Peptidoglycan of Bacterial Cell Walls. J. Biol. Chem. *245*, 3697–3702.

Hoang, T.T., Ma, Y.F., Stern, R.J., McNeil, M.R., and Schweizer, H.P. (1999). Construction and use of low-copy number T7 expression vectors for purification of problem proteins: purification of *Mycobacterium tuberculosis* RmlD and *Pseudomonas aeruginosa* LasI and RhlI proteins, and functional analysis of purified RhlI. Gene *237*, 361–371.

Hoffmann, C., Leis, A., Niederweis, M., Plitzko, J.M., and Engelhardt, H. (2008). Disclosure of the mycobacterial outer membrane: Cryo-electron tomography and vitreous sections reveal the lipid bilayer structure. Proc. Natl. Acad. Sci. USA *105*, 3963–3967.

Horbach, S., Sahm, H., and Welle, R. (1993). Isoprenoid biosynthesis in bacteria – 2 different pathways. FEMS Microbiol. Lett. *111*, 135–140.

Huang, H., Scherman, M.S., D'Haeze, W., Vereecke, D., Holsters, M., Crick, D.C., and McNeil, M.R. (2005). Identification and active expression of the M. tuberculosis gene encoding, 5-phospho-alpha -D-ribose-1-diphosphate: decaprenyl-phosphate 5-phosphoribosyltransferase, the first enzyme committed to decaprenylphosphoryl-D-arabinose synthesis. J. Biol. Chem. *280*, 24539–24543.

Imperiali, B. and Zimmerman, J.W. (1990). Synthesis of dolichypyrophosphate-linked oligosaccharides. Tetrahedron Lett. *31*, 6485–6488.

Jackson, M., Raynaud, C., Laneelle, M.A., Gullhot, C., Laurent-Winter, C., Ensergueix, D., Gicquel, B., and Daffe, M. (1999). Inactivation of the antigen 85C gene profoundly affects the mycolate content and alters the permeability of the *Mycobacterium tuberculosis* cell envelope. Mol. Microbiol. *31*, 1573–1587.

Jarlier, V., Gutmann, L., and Nikaido, H. (1991). Interplay of cell wall barrier and beta-lactamase activity determines high resistance to beta-lactam antibiotics in Mycobacterium chelonae. Antimicrob. Agents Chemother. *35*, 1937–1939.

Kacem, R., Sousa-D'Auria, C., Tropis, M., Chami, M., Gounon, P., Leblon, R., Houssin, C., and Daffe, M. (2004). Importance of mycoloyltransferases on the physiology of *Corynebacterium glutamicum*. Microbiology-Sgm *150*, 73–84.

Kartmann, B., Stengler, S., and Niederweis, M. (1999). Porins in the cell wall of *Mycobacterium tuberculosis*. J. Bacteriol. *181*, 6543–6546.

Kaur, D., Brennan, P.J., and Crick, D.C. (2004). Decaprenyl diphosphate synthesis in *Mycobacterium tuberculosis*. J. Bacteriol. *186*, 7564–7570.

Khasnobis, S., Zhang, J., Angala, S.K., Amin, A.G., McNeil, M.R., Crick, D.C., and Chatterjee, D. (2006). Characterization of a specific arabinosyltransferase activity involved in mycobacterial arabinan biosynthesis. Chem. Biol. *13*, 787–795.

Klutts, J.S., Hatanaka, K., Pan, Y.T., and Elbein, A.D. (2002). Biosynthesis of D-arabinose in *Mycobacterium smegmatis*: Specific labeling from D-glucose. Arch. Biochem. Biophys. *398*, 229–239.

Koppisch, A.T., Fox, D.T., Blagg, B.S.J., and Poulter, C.D. (2002). *E. coli* MEP synthase: Steady-state kinetic analysis and substrate binding. Biochemistry *41*, 236–243.

Kremer, L., Dover, L.G., Morehouse, C., Hitchin, P., Everett, M., Morris, H.R., Dell, A., Brennan, P.J., McNeil, M.R., Flaherty, C., Duncan, K., and Besra, G.S. (2001). Galactan biosynthesis in *Mycobacterium tuberculosis*: Identification of a bifunctional UDP-galactofuranosyltransferase. J. Biol. Chem. *276*, 26430–26440.

Kuzuyama, T., Shimizu, T., Takahashi, S., and Seto, H. (1998). Fosmidomycin, a specific inhibitor of 1-deoxy-D-xylulose 5-phosphate reductoisomerase in the nonmevalonate pathway for terpenoid biosynthesis. Tetrahedron Lett. *39*, 7913–7916.

Kuzuyama, T., Takagi, M., Takahashi, S., and Seto, H. (2000). Cloning and characterization of 1-deoxy-D-xylulose 5-phosphate synthase from Streptomyces sp strain CL190, which uses both the mevalonate and nonmevalonate pathways for isopentenyl diphosphate biosynthesis. J. Bacteriol. *182*, 891–897.

Lange, B.M., Wildung, M.R., McCaskill, D., and Croteau, R. (1998). A family of transketolases that directs isoprenoid biosynthesis via a mevalonate-independent pathway. Proc. Natl. Acad. Sci. USA *95*, 2100–2104.

Lederer, E., Adam, A., Ciorbaru, R., Petit, J.-F., and Wietzerbin, J. (1975). Cell walls of mycobacteria and related organisms; chemistry and immunostimulent properties. Mol. Cell Biochem. *7*, 87–104.

Lee, A., Wu, S.W., Scherman, M.S., Torrelles, J.B., Chatterjee, D., McNeil, M.R., and Khoo, K.H. (2006). Sequencing of Oligoarabinosyl units released from mycobacterial arabinogalactan by endogenous arabinanase: identification of distinctive and novel structural motifs. Biochemistry *45*, 15817–15828.

Lee, R., Monsey, D., Weston, A., Duncan, K., Rithner, C., and McNeil, M. (1996). Enzymatic synthesis of UDP-Galactofuranose and assays for UDP-galactopyranosyl mutase based on HPLC and on linked enzymatic production of NADH. Anal. Biochem. *242*, 1–7.

Lee, R.E., Brennan, P.J., and Besra, G.S. (1997). Mycobacterial arabinan biosynthesis: the use of synthetic arabinoside acceptors in the development of an arabinosyl transfer assay. Glycobiology *7*, 1121–1128.

Liav, A., Huang, H.R., Ciepichal, E., Brennan, P.J., and McNeil, M.R. (2006). Stereoselective synthesis of decaprenylphosphoryl beta-D-arabinofuranose. Tetrahedron Lett. *47*, 545–547.

Liu, J., Barry, C.E., Besra, G.S., and Nikaido, H. (1996). Mycolic acid structure determines the fluidity of the mycobacterial cell wall. J. Biol. Chem *271*, 29545–29551.

Lois, L.M., Campos, N., Putra, S.R., Danielsen, K., Rohmer, M., and Boronat, A. (1998). Cloning and characterization of a gene from *Escherichia coli* encoding a transketolase-like enzyme that catalyzes the synthesis of D-1-deoxyxylulose 5-phosphate, a common precursor for isoprenoid, thiamin, and pyridoxol biosynthesis. Proc. Natl. Acad. Sci. USA *95*, 2105–2110.

Luttgen, H., Rohdich, F., Herz, S., Wungsintaweekul, J., Hecht, S., Schuhr, C.A., Fellermeier, M., Sagner, S., Zenk, M.H., Bacher, A., and Eisenreich, W. (2000). Biosynthesis of terpenoids: YchB protein of *Escherichia coli* phosphorylates the 2-hydroxy group of 4-diphosphocytidyl-2C-methyl-D-erythritol. Proc. Natl. Acad. Sci. USA *97*, 1062–1067.

Ma, Y., Mills, J.A., Belisle, J.T., Vissa, V., Howell, M., Bowlin, K., Scherman, M.S., and McNeil, M. (1997). Drug targeting rhamnose biosynthesis in mycobacteria: determination of the pathway for rhamnose biosynthesis in mycobacteria and cloning, sequencing, expressing and determining the genetic organization of the *Mycobacterium tuberculosis* gene encoding a-D-glucose-1-phosphate thymidylyltransferase. Microbiology *143*, 937–945.

Ma, Y., Stern, R.J., Scherman, M.S., Vissa, V., Yan, W., Jones, V.C., Zhang, F., Franzblau, S.G., Lewis, W.H., and McNeil, M.R. (2001). Drug targeting *M. tuberculosis* cell wall synthesis: the genetics of dTDP-rhamnose synthetic enzymes and development of a microtiter plate based screen for inhibitors of the conversion of dTDP-glucose to dTDP-rhamnose. Antimicrob. Agents Chemother. *45*, 1407–1416.

Mac Sweeney, A., Lange, R., Fernandes, R.P.M., Schulz, H., Dale, G.E., Douangamath, A., Proteau, P.J., and Oefner, C. (2005). The crystal structure of *E. coli* 1-deoxy-D-xylulose-5-phosphate reductoisomerase in a ternary complex with the antimalarial compound fosmidomycin and NADPH reveals a tight-binding closed enzyme conformation. J. Mol. Biol. *345*, 115–127.

Mahapatra, S., Crick, D.C., McNeil, M.R., and Brennan, P.J. (2008). Unique structural features of the pepidoglycan of Mycobacterium leprae. J. Bacteriol. *190*, 655–661.

Mahapatra, S., Yagi, T., Belisle, J.T., Espinosa, B.J., Hill, P.J., McNeil, M.R., Brennan, P.J., and Crick, D.C. (2005). Mycobacterial lipid II is composed of a complex mixture of modified muramyl and peptide moieties linked to decaprenyl phosphate. J. Bacteriol. *187*, 2747–2757.

Manina, G, Makarov, V, Pasca, M. R, Ryabova, O, Buroni, S, Rossi, E. D, Milano, A, Moellmann, U, Kordulakova, J, Mikusova, K., Riccardi, G, and Cole, S. T. Identification and characterization of the cellular target for benzothiazinones, a new potential antitubercular drug class. Seventh International Conference on the Pathogenesis of Mycobacterial Infections, Saltsiobaden, Stockholm, Sweden. 2008. 6–26–2008. Ref Type: Conference Proceeding.

Marlow, A.L. and Kiessling, L.L. (2001). Improved chemical synthesis of UDP-galactofuranose. Organic Lett. *3*, 2517–2519.

McNeil, M., Daffe, M., and Brennan, P.J. (1990). Evidence for the nature of the link between the arabinogalactan and peptidoglycan components of mycobacterial cell walls. J. Biol. Chem. *265*, 18200–18206.

McNeil, M., Daffe, M., and Brennan, P.J. (1991). Location of the mycolyl ester substituent in the cell walls of mycobacteria. J. Biol. Chem. *266*, 13217–13223.

Mengin-Lecreulx, D. and Vanheijenoort, J. (1994). Copurification of glucosamine-1-phosphate acetyltransferase and n-acetylglucosamine-1-phosphate uridyltransferase activities of *Escherichia coli* – characterization of the glmu gene-product as a bifunctional enzyme catalyzing 2 subsequent steps in the pathway for Udp-N-acetylglucosamine synthesis. J. Bacteriol. *176*, 5788–5795.

Miallau, L., Alphey, M.S., Kemp, L.E., Leonard, G.A., McSweeney, S.M., Hecht, S., Bacher, A., Eisenreich, W., Rohdich, F., and Hunter, W.N. (2003). Biosynthesis of isoprenoids: Crystal structure of 4-diphosphocytidyl-2C-methyl-D-erythritol kinase. Proc. Natl. Acad. Sci. USA *100*, 9173–9178.

Mikusova, K., Belanova, M., Kordulakova, J., Honda, K., McNeil, M.R., Mahapatra, S., Crick, D.C., and Brennan, P.J. (2006). Identification of a novel galactosyl transferase involved in biosynthesis of the mycobacterial cell wall. J. Bacteriol. *118*, 6592–6598.

Mikusova, K., Huang, H., Yagi, T., Holsters, M., Vereecke, D., D'Haeze, W., Scherman, M.S., Brennan, P.J., McNeil, M.R., and Crick, D.C. (2005). Decaprenylphosphoryl arabinofuranose, the donor of the D-arabinofuranosyl residues of mycobacterial arabinan, is formed via a two-step epimerization of decaprenylphosphoryl ribose. J. Bacteriol. *187*, 8020–8025.

Mikusova, K., Mikus, M., Besra, G., Hancock, I., and Brennan, P.J. (1996). Biosynthesis of the linkage region of the mycobacterial cell wall. J. Biol. Chem. *271*, 7820–7828.

Mikusova, K., Yagi, T., Stern, R., McNeil, M.R., Besra, G.S., Crick, D.C., and Brennan, P.J. (2000). Biosynthesis of the galactan component of the mycobacterial cell wall. J. Biol. Chem. *275*, 33890–33897.

Mills, J.A., Motichka, K., Jucker, M., Wu, H.P., Uhlik, B.C., Stern, R.J., Scherman, M.S., Vissa, V.D., Pan, F., Kundu, M., Ma, Y.F., and McNeil, M. (2004). Inactivation of the mycobacterial rhamnosyltransferase, which is needed for the formation of the arabinogalactan-peptidoglycan linker, leads to irreversible loss of viability. J. Biol. Chem. *279*, 43540–43546.

Minnikin, D.E. (1982). Lipids: complex lipids, their chemistry, biosynthesis and roles. In The Biology of the Mycobacteria, Vol. I, C.Ratledge and J.L.Stanford, eds. (London: Academic Press), pp. 95–184.

Mukamolova, G.V., Murzin, A.G., Salina, E.G. Demina, G.R., Kell, D.B., Kaprelyants, A.S., and Young, M. (2006). Muralytic activity of Micrococcus luteus Rpf and its relationship to physiological activity in promoting bacterial growth and resuscitation. Mol. Microbiol. *59*, 84–98.

Mukamolova, G.V., Turapov, O.A., Kazarian, K., Telkov, M., Kaprelyants, A.S., Kell, D.B., and Young, M. (2002). The rpf gene of Micrococcus luteus encodes an essential secreted growth factor. Mol. Microbiol. *46*, 611–621.

Nathan, C. and Ehrt, S. (2004). Nitric oxide in tuberculosis. In Tuberculosis, W.N.Rom and S.M.Garay, eds. (Philadelphia: Lippincott Williams & Wilkins), pp. 215–235.

Onyebujoh, P. and Rook, G.A.W. (2004). Tuberculosis. Nat. Rev. Microbiol. 2, 930–932.

Ortalo-Magne, A., Dupont, M.A., Lemassu, A., Andersen, A.B., Gounon, P., and Daffe, M. (1995). Molecular composition of the outermost capsular material of the tubercle bacillus. Microbiology *141*, 1609–1620.

Paramasivan, C.N., Sulochana, S., Kubendiran, G., Venkatesan, P., and Mitchison, D.A. (2005). Bactericidal action of gatifloxacin, rifampin, and isoniazid on logarithmic- and stationary-phase cultures of *Mycobacterium tuberculosis*. Antimicrob. Agents Chemother. *49*, 627–631.

Pathak, A.K., Pathak, V., Maddry, J.A., Suling, W.J., Gurcha, S.S., Besra, G.S., and Reynolds, R.C. (2001a). Studies on α(1→5) linked octyl arabinofuranosyl disaccharides for mycobacterial arabinosyl transferase activity. Bioorg. Med. Chem. *9*, 3145–3151.

Pathak, A.K., Pathak, V., Seitz, L., Maddry, J.A., Gurcha, S.S., Besra, G.S., Suling, W.J., and Reynolds, R.C. (2001b). Studies on (β,1→5) and (β,1→6) linked octyl Gal(f) disaccharides as substrates for mycobacterial galactosyltransferase activity. Bioorg. Med. Chem. *9*, 3129–3143.

Putra, S.R., Lois, L.M., Campos, N., Boronat, A., and Rohmer, M. (1998). Incorporation of [2,3-C-13(2)]- and [2,4-C-13(2)]-D-1-deoxyxylulose into ubiquinone of *Escherichia coli* via the mevalonate-independent pathway for isoprenoid biosynthesis. Tetrahedron Letters *39*, 23–26.

Reuter, K., Sanderbrand, S., Jomaa, H., Wiesner, J., Steinbrecher, I., Beck, E., Hintz, M., Klebe, G., and Stubbs, M.T. (2002). Crystal structure of 1-deoxy-D-xylulose-5-phosphate reductoisomerase, a crucial enzyme in the non-mevalonate pathway of isoprenoid biosynthesis. J. Biol. Chem. *277*, 5378–5384.

Ricagno, S., Grolle, S., Bringer-Meyer, S., Sahm, H., Lindqvist, Y., and Schneider, G. (2004). Crystal structure of 1-deoxy-D-xylulose-5-phosphate reductoisomerase from Zymomonas mobilis at 1.9-angstrom resolution. Biochim. Biophys. Acta Proteins and Proteomics *1698*, 37–44.

Rilling, H.C. (1985). The mechanism of the condensation-reactions of cholesterol-biosynthesis. Biochem. Soc. Transactions *13*, 997–1003.

Rip, J.W., Rupar, C.A., Ravi, K., and Carroll, K.K. (1985). Distribution, metabolism and function of dolichol and polyprenols. Progr. Lipid Res. *24*, 269–309.

Rohdich, F., Kis, K., Bacher, A., and Eisenreich, W. (2001). The non-mevalonate pathway of isoprenoids: genes, enzymes and intermediates. Curr. Opin. Chem. Biol. *5*, 535–540.

Rohdich, F., Wungsintaweekul, J., Fellermeier, M., Sagner, S., Herz, S., Kis, K., Eisenreich, W., Bacher, A., and Zenk, M.H. (1999). Cytidine 5′-triphosphate-dependent biosynthesis of isoprenoids: YgbP protein of *Escherichia coli* catalyzes the formation of 4-diphosphocytidyl-2-C-methylerythritol. Proc. Natl. Acad. Sci. USA *96*, 11758–11763.

Rohdich, F., Zepeck, F., Adam, P., Hecht, S., Kaiser, J., Laupitz, R., Grawert, T., Amslinger, S., Eisenreich, W., Bacher, A., and Arigoni, D. (2003). The deoxyxylulose phosphate pathway of isoprenoid biosynthesis: Studies on the mechanisms of the reactions catalyzed by IspG and IspH protein. Proc. Natl. Acad. Sci. USA *100*, 1586–1591.

Rohmer, M., Knani, M., Simonin, P., Sutter, B., and Sahm, H. (1993). Isoprenoid Biosynthesis in Bacteria – A Novel Pathway for the Early Steps Leading to Isopentenyl Diphosphate. Biochemical Journal *295*, 517–524.

Rose, N.L., Completo, G.C., Lin, S.J., McNeil, M., Palcic, M.M., and Lowary, T.L. (2006). Expression, Purification, and Characterization of a Galactofuranosyltransferase Involved in *Mycobacterium tuberculosis* Arabinogalactan Biosynthesis. J. Am. Chem. Soc. *128*, 6721–6729.

Sanders, D.R., Steins, A.G., McMahon, S.A., McNeil, M.R., Whitfield, C., and Naismith, J.H. (2001). UDP-galactopyranose mutase has a novel structure and mechanism. Nat. Struct. Biol. *8*, 858–863.

Sassetti, C.M., Boyd, D.H., and Rubin, E.J. (2003). Genes required for mycobacterial growth defined by high density mutagenesis. Mol. Microbiol. *48*, 77–84.

Scherman, M., Weston, A., Duncan, K., Whittington, A., Upton, R., Deng, L., Comber, R., Friedrich, J.D., and McNeil, M. (1995). Biosynthetic Origin of Mycobacterial Cell-Wall Arabinosyl Residues. J. Bacteriol. *177*, 7125–7130.

Scherman, M.S., Winans, K., Bertozzi, C.R., Stern, R., Jones, V.C., and McNeil, M.R. (2003). Drug targeting *Mycobacterium tuberculosis* cell wall synthesis: development of a microtiter plate-based screen for UDP-galactopyranose mutase and identification of an inhibitor from a uridine-based library . Antimicrob. Agents Chemother. *47*, 378–382.

Schulbach, M.C., Brennan, P.J., and Crick, D.C. (2000). Identification of a short (C-15) chain Z-isoprenyl diphosphate synthase and a homologous long (C-50) chain isoprenyl diphosphate synthase in *Mycobacterium tuberculosis*. J. Biol. Chem. *275*, 22876–22881.

Schulbach, M.C., Mahapatra, S., Macchia, M., Barontini, S., Papi, C., Minutolo, F., Bertini, S., Brennan, P.J., and Crick, D.C. (2001). Purification, enzymatic characterization, and inhibition of the Z-farnesyl diphosphate synthase from *Mycobacterium tuberculosis*. J. Biol. Chem. *276*, 11624–11630.

Seidel, M., Alderwick, L.J., Birch, H.L., Sahm, H., Eggeling, L., and Besra, G.S. (2007). Identification of a novel arabinofuranosyltransferase AftB involved in a terminal step of cell wall arabinan biosynthesis in Corynebacterianeae, such as *Corynebacterium glutamicum* and *Mycobacterium tuberculosis*. J. Biol. Chem. *282*, 14729–14740.

Shi, L., Zhou, R., Liu, Z., Lowary, T.L., Seeberger, P.H., Stocker, B.L., Crick, D.C., Khoo, K.H., and Chatterjee, D. (2008). Transfer of the first arabinofuranose residue to galactan is essential for *Mycobacterium smegmatis* viability. J. Bacteriol. *190*, 5248–5255.

Shi, L.B., Berg, S., Lee, A., Spencer, J.S., Zhang, J., Vissa, V., McNeil, M.R., Khoo, K.H., and Chatterjee, D. (2006). The carboxy terminus of EmbC from *Mycobacterium smegmatis* mediates chain length extension of the arabinan in lipoarabinomannan. J. Biol. Chem. *281*, 19512–19526.

Shridas, P., Rush, J.S., and Waechter, C.J. (2003). Identification and characterization of a cDNA encoding a long-chain cis-isoprenyltranferase involved in dolichyl monophosphate biosynthesis in the ER of brain cells. Biochem. Biophys. Res. Commun. *312*, 1349–1356.

Soltero-Higgin, M., Carlson, E.E., Phillips, J.H., and Kiessling, L.L. (2004). Identification of inhibitors for UDP-galactopyranose mutase. J. Am. Chem. Soc. *126*, 10532–10533.

Sprenger, G.A., Schorken, U., Wiegert, T., Grolle, S., deGraaf, A.A., Taylor, S.V., Begley, T.P., BringerMeyer, S., and Sahm, H. (1997). Identification of a thiamin-dependent synthase in *Escherichia coli* required for the formation of the 1-deoxy-D-xylulose 5-phosphate precursor to isoprenoids, thiamin, and pyridoxol. Proc. Natl. Acad. Sci. USA *94*, 12857–12862.

Steinbacher, S., Kaiser, J., Eisenreich, W., Huber, R., Bacher, A., and Rohdich, F. (2003). Structural basis of fosmidomycin action revealed by the complex with 2-C-methyl-D-erythritol 4-phosphate synthase (IspC) – Implications for the catalytic mechanism and anti-malaria drug development. J. Biol. Chem. *278*, 18401–18407.

Stern, R.J., Lee, T.Y., Lee, T.J., Yan, W., Scherman, M.S., Vissa, V.D., Kim, S.K., Wanner, B.L., and McNeil, M.R. (1999). Conversion of dTDP-4-keto-6-deoxyglucose to free dTDP-4-keto-rhamnose by the *rmlC* gene products of *Escherichia coli* and *Mycobacterium tuberculosis*. Microbiology *145*, 663–671.

Takayama, K. and Goldman, D.S. (1970). Enzymatic synthesis of mannosyl-1-phosphoryldecaprenol by a cell- free system of *Mycobacterium tuberculosis*. J. Biol. Chem. *245*, 6251–6257.

Takayama, K. and Kilburn, J.O. (1989). Inhibition of synthesis of arabinogalactan by ethambutol in *Mycobacterium smegmatis*. Antimicrob. Agents Chemother. *33*, 1493–1499.

Takayama, K., Wang, C., and Besra, G.S. (2005). Pathway to synthesis and processing of mycolic acids in *Mycobacterium tuberculosis*. Clin. Microbiol. Rev. *18*, 81–101.

Telenti, A., Philipp, W.J., Sreevatsan, S., Bernasconi, C., Stockbauer, K.E., Wieles, B., Musser, J.M., and Jacobs, W.R. (1997). The emb operon, a gene cluster of *Mycobacterium tuberculosis* involved in resistance to ethambutol. Nature Med. *3*, 567–570.

Tufariello, J.M., Mi, K.X., Xu, J.Y., Manabe, Y.C., Kesavan, A.K., Drumm, J., Tanaka, K., Jacobs, W.R., and Chan, J. (2006). Deletion of the *Mycobacterium tuberculosis* resuscitation-promoting factor Rv1009 gene results in delayed reactivation from chronic tuberculosis. Infect. Immun. *74*, 2985–2995.

Ulrichs, T., Kosmiadi, G.A., Jorg, S., Pradl, L., Titukhina, M., Mishenko, V., Gushina, N., and Kaufmann, S.H.E. (2005). Differential organization of the local immune response in patients with active cavitary tuberculosis or with nonprogressive tuberculoma. J. Infect. Dis. *192*, 89–97.

Wada, T., Kuzuyama, T., Satoh, S., Kuramitsu, S., Yokoyama, S., Unzai, S., Tame, J.R.H., and Park, S.Y. (2003). Crystal structure of 4-(cytidine 5 '-diphospho)-2-C-methyl-D-erythritol kinase, an enzyme in the non-mevalonate pathway of isoprenoid synthesis. J. Biol. Chem. *278*, 30022–30027.

Wang, W., Dong, C., McNeil M, R., Kaur, D., Mahapatra, S., Crick, D., and and Naismith, J.H. (2008). The structural basis of chain length control in Rv1086. J. Mol. Biol. *381*, 129–140.

Watanabe, M., Yamada, Y., Iguchi, K., and Minnikin, D.E. (1994). Structural Elucidation of new phenolic glycolipids from *Mycobacterium tuberculosis*. Biochim. Biophys. Acta *1210*, 174–180.

Weston, A., Stern, R.J., Lee, R.E., Nassau, P.M., Monsey, D., Martin, S.L., Scherman, M.S., Besra, G.S., Duncan, K., and McNeil, M.R. (1998). Biosynthetic origin of mycobacterial cell wall galactofuranosyl residues. Tuber. Lung Dis. *78*, 123–131.

Wietzerbin, J., Das, B.C., Petit, J.F., Lederer, E., Leyh Bouille, M., and Ghuysen, J.M. (1974). Occurrence of D-alanyl-(D)-meso-diaminopimelic acid and meso-diaminopimelyl-meso-diaminopimelic acid interpeptide linkages in the peptidoglycan of mycobacteria. Biochemistry *13*, 3471–3476.

Winder, F.G. (1982). Mode of action of the antimycobacterial agents and associated aspects of the molecular biology of the mycobacteria. In The Biology of the Mycobacteria Vol 1, C.Ratledge and J.Standford, eds. (London: Academic Press), pp. 354–442.

Wolucka, B.A., McNeil, M.R., de Hoffmann, E., Chojnacki, T., and Brennan, P.J. (1994). Recognition of the lipid intermediate for arabinogalactan/arabinomannan biosynthesis and its relation to the mode of action of ethambutol on mycobacteria. J. Biol. Chem. *269*, 23328–23335.

Xiang, S., Usunow, G., Lange, G., Busch, M., and Tong, L. (2007). Crystal structure of 1-deoxy-D-xylulose 5-phosphate synthase, a crucial enzyme for isoprenoids biosynthesis. J. Biol. Chem. *282*, 2676–2682.

Yagi, T., Mahapatra, S., Mikusova, K., Crick, D.C., and Brennan, P.J. (2003). Polymerization of mycobacterial arabinogalactan and ligation to peptidoglycan. J. Biol. Chem. *278*, 26497–26504.

Yajima, S., Hara, K., Sanders, J.M., Yin, F.L., Ohsawa, K., Wiesner, J., Jomaa, H., and Oldfield, E. (2004). Crystallographic structures of two bisphosphonate : 1-deoxyxylulose-5-phosphate reductoisomerase complexes. J. Am. Chem. Soc. *126*, 10824–10825.

Yajima, S., Nonaka, T., Kuzuyama, T., Seto, H., and Ohsawa, K. (2002). Crystal structure of 1-deoxy-D-xylulose 5-phosphate reductoisomerase complexed with cofactors: Implications of a flexible loop movement upon substrate binding. Journal of Biochemistry *131*, 313–317.

Zeidler, J., Schwender, J., Muller, C., Wiesner, J., Weidemeyer, C., Beck, E., Jomaa, H., and Lichtenthaler, H.K. (1998). Inhibition of the non-mevalonate 1-deoxy-D-xylulose-5-phosphate pathway of plant isoprenoid biosynthesis by fosmidomycin. Zeitschrift fur Naturforschung C-A J. Biosci. *53*, 980–986.

Zhang, J., Khoo, K.H., Wu, S.W., and Chatterjee, D. (2007). Characterization of a distinct arabinofuranosyltransferase in Mycobacterium smegmatis. J. Am. Chem. Soc. *129*, 9650–9662.

Zhang, N., Torrelles, J.B., McNeil, M.R., Escuyer, V.E., Khoo, K.H., Brennan, P.J., and Chatterjee, D. (2003). The Emb proteins of mycobacteria direct arabinosylation of lipoarabinomannan and arabinogalactan via an N-terminal recognition region and a C-terminal synthetic region. Mol. Microbiol. *50*, 69–76.

Zhang, W., Jones, V.C., Scherman, M.S., Mahapatra, S., Crick, D., Bhamidi, S., McNeil M, R., and Ma, Y.F. (2008a). *Mycobacterium tuberculosis* Rv1018c exhibits glucosamine-1-phosphate acetyltransferase and N-acetylglucosamine-1-phosphate uridyltransferase activities and it is required for mycobacterial growth. Int. J. Biochem. Cell Biol. *40*, 2560–2571.

Zhang, W., Jones, V.C., Scherman, M.S., Mahapatra, S., Crick, D., Xin, Y., McNeil, M.R., and Ma, Y. (2008b). Expression, essentiality, and a microtiter plate assay for mycobacterial GlmU, the bifunctional glucosamine-1-phosphate acetyltransferase and N-acetylglucosamine-1-phosphate uridyltransferase. Int. J. Biochem. Cell Biol. *40*, 2560–2571.

Zhang, Y., Lathigra, R., Garbe, T., Catty, D., and Young, D. (1991). Genetic analysis of superoxide dismutase, the 23 kilodalton antigen of *Mycobacterium tuberculosis*. Mol. Microbiol. *5*, 381–391.

Zhou, D. and White, R.H. (1991). Early steps of isoprenoid biosynthesis in *Escherichia coli*. Biochem. J. *273*, 627–634.

Part II

Microbial Polysaccharides in Human Pathogens

Genetics and Regulation of Bacterial Polysaccharide Expression in Human Pathogenic Bacteria

4

David Corbett and Ian S. Roberts

Abstract

The presentation of a polysaccharide capsule on the cell surface is a common feature of many bacteria. The presence of a polysaccharide capsule mediates interactions between the bacterium and its immediate environment. As such, regulation of capsule expression will be vital in the survival of bacteria in new environments, as will moderating capsule expression in response to local changes in the environment as a consequence of bacterial growth. This ability to regulate capsule expression is particularly important for pathogenic bacteria where the local environment will change as a consequence of the host response to infection and where more or less capsule expression will dictate the likely survival of the bacteria in a hostile environment. In this chapter we discuss the regulation of capsule gene expression in a number of pathogenic bacteria and describe the regulation of group 2 capsule gene expression in *Escherichia coli* in detail.

Introduction

Polysaccharides are utilized by prokaryotes for a wide range of functions, including the formation of cell wall structures and cell-surface components such as the serologically important surface-linked capsular K antigens, and the O antigens of lipopolysaccharide (LPS) (Jann and Jann, 1997). LPS is the most abundant lipid in the outer leaf of the Gram-negative bacterial outer membrane (Marolda and Valvano, 1998), protecting cells from potentially hazardous substances. Bacteria are also able to produce extracellular polysaccharides (EPS), such as colanic acid, that are not covalently linked to the cell surface but form an amorphous, highly hydrated gel around the bacterial cell. EPSs are not K antigens, although they are often produced in addition to them. The major components of bacterial capsules are highly hydrated, high-molecular-weight acidic polysaccharides that confer upon bacteria an overall negative charge and hydrophilic properties. These polysaccharides have been studied extensively in *E. coli*, *Neisseria meningitidis* and *Klebsiella pneumoniae* (Jann and Jann, 1997). There are more than 80 different K antigens in *E. coli*, and this variety is achieved through the utilization of various monosaccharide residues, allowing the formation of homo- or hetero-polymers that may be linked in several different conformations along with the formation of branched chains.

Capsules form the outermost layer of many species of bacteria, mediating interactions between them and the immediate environment. Consequently, the capsule has been implicated as an important virulence factor in encapsulated strains. The K antigens of *E. coli* are often associated with particular diseases. For example, the K1 capsule is the most common capsule type found in isolates of *E. coli* causing neonatal meningitis and urinary tract infection. The K5 polysaccharide is associated both with urinary tract infection and sepsis, but not with meningitis. In both cases, these capsules are more often found associated with infection than in the normal intestinal flora of healthy individuals (Kaijser and Jodal, 1984). Perhaps the most important function of the polysaccharide capsule is protection from the host's non-specific immune system, specifically the actions of complement and phagocytosis (Jann and Jann, 1997; Roberts, 1996). The role of polysaccharide capsules in protecting bacteria from the action of complement has been repeatedly demonstrated in a range of bacteria. For example, *Staphylococcus aureus* and *Streptococcus pneumoniae* capsules can reduce C3 binding and inhibit phagocytosis because the capsule prevents interaction between cell wall-bound C3b and phagocyte receptors (Brown *et al.*, 1983; Cunnion *et al.*, 2001; Verbrugh *et al.*, 1982). Although they do not generally protect against opsonization by capsule-specific antibodies, some capsules are poorly immunogenic because they closely resemble polysaccharides found on host cell surfaces. The *E. coli* K1 capsule, identical to the capsule expressed by *N. meningitidis* group B, is a polymer of α-2,8-linked N-acetylneuraminic acid (NeuNAc, also known as sialic acid) (Jann and Jann, 1997; McGuire and Binkley, 1964), which is identical to the composition of polysialic acid expressed on neural cellular adhesion molecules (nCAM) in the human body. Therefore, bacteria expressing these capsules elicit a very weak humoral immune response. In contrast, isolates of *N. meningitidis* group C express a polysialic acid capsule containing α-2,9 linkages, which is substantially different enough for this capsule to elicit an immune response in the host (Jann and Jann, 1997).

Given the substantial energy investment that the biosynthesis and export of a capsular polysaccharide represents to bacteria, and that expression of a capsule is not appropriate for bacteria in all circumstances, it is to be expected that capsule expression be tightly regulated. For example, under certain circumstances expression of a polysaccharide capsule protects bacteria from the effects of serum-mediated killing and may be beneficial for dissemination into the host, whereas under other conditions the capsule may disrupt intimate adhesion of bacteria to host tissues, and inhibit colonization. The following sections review our current understanding of capsule regulation in a range of bacteria, and a detailed discussion of the regulation of expression of the model *E. coli* group 2 K5 capsule.

Escherichia coli K30 and K-12 colanic acid

Colanic acid (M antigen) is a polymer of glucose, galactose, fucose and glucuronic acid modified with acetyl and pyruvyl groups (Gottesman and Stout, 1991). It is expressed at environmental temperatures (below 30°C), and does not play a role in virulence (Lopez-Torres and Stout, 1996) a conclusion supported by the fact that colanic acid interferes with specific adhesion of uropathogenic *E. coli* (Hanna *et al.*, 2003). The principal role of colanic acid appears to be in protection from the external environment, being necessary both for prevention of desiccation and biofilm formation, and its expression is induced in response to osmotic shock (Ophir and Gutnick, 1994; Sledjeski and Gottesman, 1996). Colanic acid can be co-expressed with *E. coli* polysaccharide capsules of groups 2, 3, and 4, but not with those of group 1. Unlike K-antigens, significant proportions of colanic acid are secreted into the growth medium (Whitfield, 2006). However, the regulation of both colanic acid and group 1 K antigens is similar, involving the Rcs phosphorelay, a highly conserved regulatory system amongst pathogenic or commensal enterobacteria. Across diverse species of bacteria, this phosphorelay regulates systems involved in biofilm formation, cell envelope formation, adhesion to mammalian cells, capsule production, cell envelope stress, systemic infection of mice and intrinsic antibiotic resistance (Erickson and Detweiler, 2006; Hinchliffe *et al.*, 2008; Laubacher and Ades, 2008).

The Rcs phosphorelay

The Rcs phosphorelay was discovered upon investigation of the mucoid phenotype observed in capsule-expressing *E. coli* bearing mutations in the ATP-dependent protease Lon (Majdalani and Gottesman, 2005). Since then, the Rcs phosphorelay has been found to be induced by osmotic shock, membrane perturbation, and disruptions to the integrity of the peptidoglycan layer (Laubacher and Ades, 2008). The phosphorelay was originally considered a two-component regulatory system involving RcsC, RcsB and an accessory protein, RcsA (Gottesman and Stout, 1991). Two-component regulatory systems allow bacteria to couple a stimulus, detected by an autophosphorylation-capable sensor kinase protein, to a specific response in the form of altered gene expression. This occurs via a response regulator protein, typically a transcriptional regulator, whose activity, usually DNA binding and transcription modulation, is controlled by the sensor protein through phosphorylation of conserved aspartate residues (Stock *et al.*, 2000). The model has been further refined to include the outer-membrane lipoprotein RcsF that acts as a sensor and transduces the signal to RcsC an inner membrane-bound sensor histidine kinase; RcsD, a membrane-bound phosphotransfer protein; and RcsB, a transcriptional response regulator (Fig. 4.1) (Majdalani and Gottesman, 2005). The accessory protein RcsA is also involved in colanic acid gene expression, and provides a layer of regulation independent of the phosphorelay system (Majdalani and Gottesman, 2005). The phosphorelay involves the transfer of phosphate from His to Asp to His to Asp. Atypically, RcsC contains a response regulator domain in addition to a histidine kinase domain, but it lacks a histidine phosphotransfer domain, a function provided by RcsD (Takeda *et al.*, 2001). Upon activation, phosphate is transferred from the kinase domain to the response regulator domain within RcsC. The phosphate moiety is then transferred to a histidine residue within RcsD, and from there to the response regulator RcsB, resulting in its activation. Once activated dimeric RcsB binds near the –35 promoter region, probably through interaction with RNA polymerase (Majdalani and Gottesman, 2005). In addition, RcsB is also capable of forming heterodimers with RcsA, the levels of which are low due to the action of the

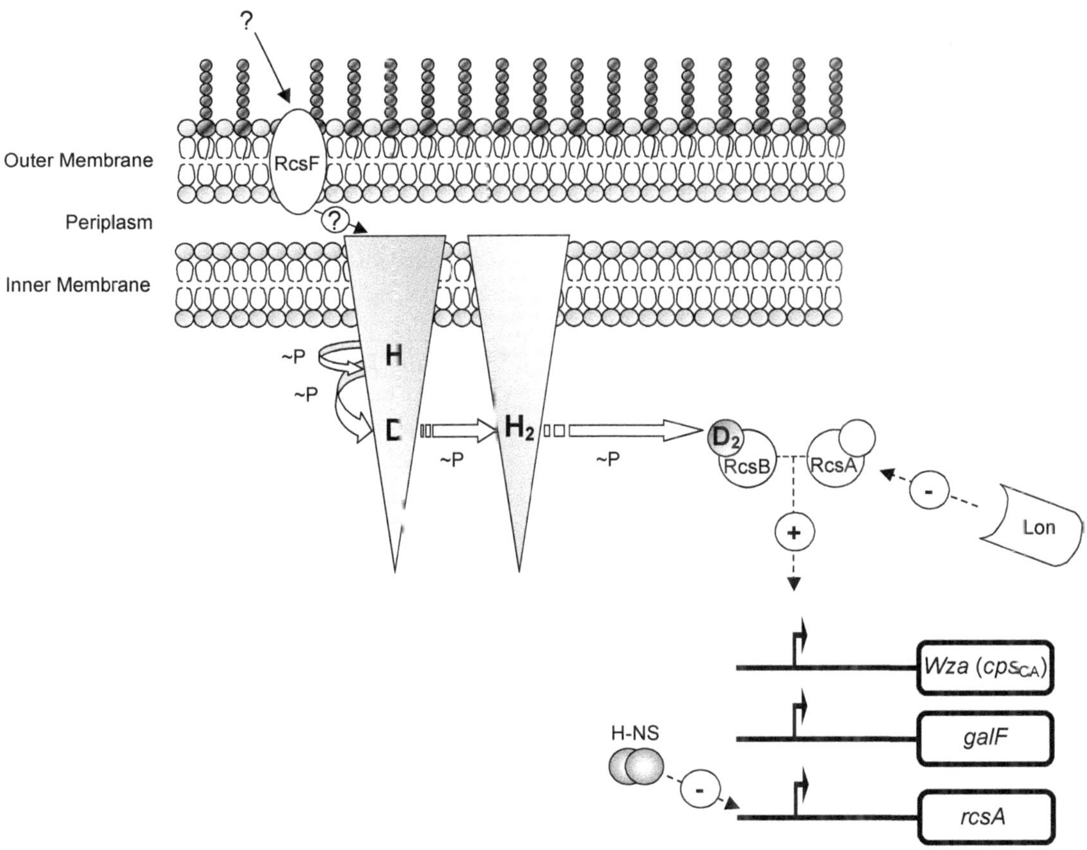

Figure 4.1 Model of the RCS phosphorelay system. The inducing signal for RcsF is unknown, as is that between RcsF and RcsC. Phosphate transfer from RcsC to RcsB is shown via RcsD travelling from His (H) to Asp (D). Genes involved in capsular polysaccharide production whose transcription is specifically activated by RcsAB are shown. H-NS represses transcription from *PrcsA*, whilst the Lon protease is responsible for proteolytic degradation of RcsA.

ATP-dependent protease Lon (Majdalani and Gottesman, 2005). RcsAB heterodimer formation shields RcsA from the action of Lon, and permits subsequent direct activation of RcsA-dependent target promoters, including the colanic acid biosynthetic operon. Mutations in the *lon* gene result in RcsC independent overexpression of colanic acid and present a mucoid phenotype, as the half-life of RcsA is dramatically increased (Gottesman and Stout, 1991). This also suggests that an alternative mechanism for activating RcsB must exist. The binding site is altered to a distantly upstream region containing what has been termed the RcsAB box with a core consensus sequence of TaAGaatatTCctA (Wehland and Bernhard, 2000). Transcription of *rcsA* is also repressed by the global regulator H-NS, which is in turn negatively regulated by the small non-coding RNA, DsrA. DsrA is specifically expressed at low temperature, thus it may also play a role in colanic acid expression at environmental temperatures (Majdalani and Gottesman, 2005).

The Rcs phosphorelay is not involved in wild-type levels of capsule gene expression as grown under laboratory conditions (Jayaratne *et al.*, 1993). Unlike the colanic acid biosynthetic operon of *E. coli* K-12, Group 1 capsular polysaccharides are not temperature-regulated and possess no RcsAB box immediately upstream of the capsule gene cluster. Basal transcription of these genes is unaffected by the Rcs phosphorelay system (Rahn and Whitfield, 2003). However, upstream of *galF*, whose gene product is involved in the synthesis of UDP-glucose, there is an RcsAB box and multi-copy RcsB was found to induce the transcription of *galF*, resulting in an increase in K30 expression and a mucoid phenotype (Rahn and Whitfield, 2003). Thus, Group 1 capsule expression is in part regulated by the availability of activated sugar precursors, which in turn may be increased through activation of the Rcs phosphorelay system (Rahn and Whitfield, 2003).

Both *E. coli* group 1 capsule gene clusters and the colanic acid gene cluster are dependent upon a 39-bp sequence downstream of their promoters known as JUMPStart (Just Upstream of Many Polysaccharide gene Starts) (Hobbs and Reeves, 1994). This sequence is present in the 5′ untranslated region (UTR) of many polysaccharide genes, and within it is a conserved GGCGGTAG motif known as *ops* (operon polarity suppressor) (Nieto *et al.*, 1996). During transcription, RNA polymerase pauses at the *ops* sequence whilst the anti-termination factor RfaH interacts with the exposed, non-coding DNA strand in the melted transcription bubble. Transcription elongation is then accelerated, and the RNA polymerase–RfaH complex is subsequently less sensitive to transcription termination signals (Artsimovitch and Landick, 2002; Bailey *et al.*, 2000). As such, RfaH is often essential for the transcription of large polysaccharide gene clusters (Bailey *et al.*, 1997). This is true for the K30 gene cluster, which contains a stem–loop transcription-terminating structure between the *wzc* and *wbaP* genes. Transcription of the genes downstream of *wzc*, and eventual expression of the K30 polysaccharide requires RfaH (Rahn and Whitfield, 2003).

In summary, the regulation of colanic acid and group 1 capsule gene expression in *E. coli* share similarities: due to the size of the operons responsible for their production, both require the presence of the RfaH protein and *ops* sequences upon which it acts to be present in the mRNA. Their regulation is distinguished by the fact that group 1 capsules are not directly regulated by the Rcs phosphorelay, a regulatory cascade common to the control of capsule gene expression in a number of bacteria.

Klebsiella pneumoniae

One of the most common capsule serotypes in *K. pneumoniae* is the K2 polysaccharide consisting of [→)4-Glc-(1→3)-α-Glc-(1→4)-β-Man-(3←1)-α-GlcA)-(1→]*n*, which is synthesized in a manner similar to *E. coli* group 1 capsules (Wacharotayankun *et al.*, 1993; Whitfield and Roberts, 1999). *K. pneumoniae* is naturally mucoid, a phenotype induced by the large, naturally occurring plasmid pLVPK (Nassif *et al.*, 1989). This plasmid contains the gene *rmpA*, which enhances the virulence of *K. pneumoniae* 1000-fold in mice (Nassif *et al.*, 1989). Overexpression of *rmpA* in *E. coli* induces the expression of colanic acid, but only at temperatures of 30°C or less. However, this cannot explain the mucoid effect seen in *K. pneumoniae* as it does not produce colanic acid (Nassif *et al.*, 1989). When the *K. pneumoniae* K2 ($K2_{Kp}$) capsule gene cluster was cloned into *E. coli*, the K2 capsular polysaccharide was expressed to similar levels as seen in *K. pneumoniae* at 30°C (but not at 37°C), and only in the presence of *rmpA* (Arakawa *et al.*, 1991). Multi-copy *E. coli rcsB* was found to be a positive regulator of $K2_{Kp}$ transcription, resulting in expression of a thick $K2_{Kp}$ capsule and colanic acid simultaneously in *E. coli* (Wacharotayankun *et al.*, 1992). Later studies found that *rmpA* was actually part of a larger locus, renamed *rmpA2*, and that the mucoid phenotype was the result of overproduction of serotype-specific polysaccharide presented as a thick extra capsular polysaccharide (McCallum and Whitfield, 1991; Wacharotayankun *et al.*, 1993). A more recent study identified variance in the *rmpA2* gene in different *K. pneumoniae* isolates, with G-tracts ranging from 9 to 12 nucleotides in length (Lai *et al.*, 2003). The authors found that only the G_{11} variant was able to encode full length RmpA2 and activate capsule gene transcription, and only the full length protein was able to bind to the ORF_{1-2} and ORF_{3-15} promoters (Lai *et al.*, 2003). The relevance of this variation in *rmpA2* in relation to the mucoid phenotype and virulence is unclear, as all of the variants were found in clinical isolates. Additionally, although RmpA2 was able to activate transcription of the capsule locus, constitutive transcription of the capsule genes was low and effectively unaltered in *rmpA2*, *rcsB*, or *rcsBrmpA2* mutants, possibly because *rmpA2* is transcribed at very low levels (Lai *et al.*, 2003). The fact that RmpA2 is not involved with basal capsular polysaccharide is in agreement with earlier studies reporting that strains lacking the virulence plasmid were still encapsulated (Lai *et al.*, 2003; Nassif *et al.*, 1989; Wacharotayankun *et al.*, 1992). Strains unable to express the Lon protease showed the greatest increase in capsule gene transcription, believed to be due to increased stability of RmpA2 in this strain, although it is possible that this is also due to increased RcsA stability (Lai *et al.*, 2003). Nevertheless, overexpression of RmpA2 increases capsule gene transcription more than 40-fold, and so it has been proposed that RmpA2 functions to maintain capsule gene expression at high levels.

In addition to RmpA2, a JUMPStart sequence has been identified upstream of the capsule gene cluster, implicating RfaH in the regulation of *K. pneumoniae* capsule gene expression, although its role has never been experimentally proven (Rahn *et al.*, 1999). Another feature of the *K. pneumoniae* capsule gene cluster similar to *E. coli* group 1 gene clusters is the presence of an RcsAB box upstream of the *galF* gene (Wehland and Bernhard, 2000). Whether or not this box, or indeed up-regulation of *galF* expression play any role in expression of the K2 capsule, remains to be investigated. However, the Rcs phosphorelay is likely to be involved as multi-copy RcsA is able to increase capsule expression in serotype K20, which lacks the large virulence plasmid containing *rmpA2* (McCallum and Whitfield, 1991).

In conclusion, the most important regulator of *K. pneumoniae* capsule overexpression and mucoidy appears to be the transcriptional regulator RmpA, although the Rcs phosphorelay and RfaH may both also play a role.

Salmonella enterica serovar Typhi

The enteric Gram-negative bacterium *S.* Typhi is a facultative intracellular pathogen and the cause of typhoid fever in humans. This *Salmonella* capsule comprises α-1,4-linked 2-deoxy-2-N-acetylgalactosamine, which is variably O-acetylated (Szu *et al.*, 1991), and prevents recognition of Salmonella LPS by the toll-like receptor TLR-4 (Wilson *et al.*, 2008). The loci responsible for capsule production, *viaA* and *viaB*, map to two different regions of the chromosome. The *viaA* locus contains genes homologous to RcsB and RcsC of *E. coli*, whereas *viaB* consists of 10 ORFs transcribed in the same orientation responsible for capsule regulation, synthesis, and transport (Hashimoto *et al.*, 1993; Virlogeux *et al.*, 1995; Waxin *et al.*, 1993).

TviA, encoded by the first gene of the *viaB* locus, positively controls transcription of *tviB* from the *tviA* promoter in concert with RcsB by enabling transcription to proceed past a proposed transcriptional terminator located between *tviA* and *tviB* (Virlogeux *et al.*, 1996). However, there is also a JUMPStart sequence upstream of the *tviB* gene, suggesting that RfaH may also play a role in transcription elongation (Hobbs and Reeves, 1994). TviA is not essential for capsule expression, but inactivation of *rcsB* abolishes it (Virlogeux *et al.*, 1996; Virlogeux *et al.*, 1995). Given the dependency of capsule expression on both proteins, it was anticipated that TviA was functionally homologous to RcsA, with which

RcsB interacts in *E. coli*. However, TviA cannot complement an *E. coli rcsA* mutant, and TviA is not susceptible to degradation mediated by the Lon protease, as is the case for RcsA (Virlogeux *et al.*, 1996). RcsA does not itself play a role in V_i capsule expression (Lee *et al.*, 2006; Virlogeux *et al.*, 1996). TviA may play a role in assisting RcsB-DNA interaction as it is able to restore capsule expression to strains carrying mutations in the DNA-binding domain of RcsB, but it is unable to restore capsule expression to an *rcsB* knockout strain (Virlogeux *et al.*, 1996). An RcsAB box has been identified upstream of the *tviA* gene that was bound by *E. coli* RcsAB *in vitro* (Wehland and Bernhard, 2000). If TviA and RcsB do interact *in vivo*, this could represent a potential target for binding.

One environmental factor influencing V_i capsule expression is osmolarity. High levels of NaCl, such as those found in the human gut, inhibit capsule gene transcription in wild-type cells and result in the secretion of invasion-promoting Sip proteins, which, whilst the bacterium is expressing a capsule, are not secreted but stockpiled in the cell (Arricau *et al.*, 1998; Pickard *et al.*, 1994; Santander *et al.*, 2007). Growth in high osmolarity media (>300 mM NaCl) yields an increase in expression of the alternative sigma factor RpoS, which represses capsule expression. Agglutination assays dependent on V_i capsule masking of O-antigens revealed that $RpoS^-$ cells express capsule below 300 mM NaCl, but fail to express it above this osmolarity. This indicates that another factor must be involved in the repression of capsule gene expression above 300 mM NaCl. Wild-type cells also fail to express capsule above this apparent threshold concentration of NaCl, but below this level produce a mixed population of V_i^+ and V_i^- cells, otherwise known as the VW phenotype (Santander *et al.*, 2007). The repressive effects of RpoS might also account for the fact that capsule expression is maximally expressed during exponential phase, i.e. when RpoS expression is minimal (Arricau *et al.*, 1998; Lee *et al.*, 2006; Santander *et al.*, 2007). The OmpR–EnvZ two-component regulator system has also been proposed to positively regulate V_i antigen synthesis (Pickard *et al.*, 1994), although its precise role has not been defined. A recent study has also implicated the IHF and Fis proteins, both global regulators of gene expression, in the positive control of transcription of the *viaB* locus (Lee *et al.*, 2006).

In summary, regulation of the *S.* Typhi capsule gene cluster is controlled by an Rcs-like system involving the proteins TviA and RcsB, the latter of which is essential, the former of which is not and cannot complement *rcsA* mutants of *E. coli*. The RfaH protein may also enhance transcription of the capsule gene cluster, as may the global regulators IHF, Fis, and the OmpR-EnvZ two-component systems, although their exact roles remain to be elucidated.

Streptococcus pneumoniae

More than 90 capsule serotypes have been identified in *S. pneumoniae*, a Gram-positive organism carried asymptomatically in the nasopharynx of healthy individuals that can cause a range of diseases including pneumonia, meningitis and otitis media. The capsule is able to protect *S. pneumoniae* from phagocytosis but inhibits adherence. With the exception of types 3 and 37, the first four genes (*cpsABCD*) involved in capsule biosynthesis are common to nearly all *S. pneumoniae*, whilst downstream genes are considered serotype-specific (Llull *et al.*, 2001b). Both encapsulated (opaque) and unencapsulated (transparent) colonies can be observed in clinical isolates of this bacterium, but serotypes associated with invasive infection have been shown to have elevated *cpsA* expression (Hathaway *et al.*, 2007). Phase variation from an opaque to a transparent phenotype occurs spontaneously with a frequency of 10^{-3} to 10^{-6} (Weiser *et al.*, 1994). Opaque variants express 1.2 to 5.6-fold more capsule than transparent variants of the same strain which correlates with the virulence of each variant *in vivo* (Kim and Weiser, 1998). Small acapsular colonies of type 3 *S. pneumoniae* isolated from Sorbarod biofilms have been shown to revert with high frequency to an encapsulated phenotype when sub cultured on blood agar (Waite *et al.*, 2001). The cause of this phase variation was found to be the generation of perfect tandem duplications of varying size within the *cap3A* gene, which encodes a UDP-glucose dehydrogenase required for capsule biosynthesis. These duplications introduced either premature stop codons or structural changes into the *cap3A* sequence resulting in loss of function of Cap3A, and duplication-induced phase variation could be reproduced with artificial constructs (Waite *et al.*, 2001). Similar observations were later reported in *cap8E* for the type 8 capsule gene cluster (Waite *et al.*, 2003), and in type 37 pneumococci, tandem duplications occur within the single gene required for capsule expression, *tts* (Llull *et al.*, 2001a; Waite *et al.*, 2003). In all cases, reversion to an encapsulated state occurred via a precise excision of the duplicated sequence, and the frequency of phase variation was found to increase with the size of the duplication (Waite *et al.*, 2003; Waite *et al.*, 2001). An additional explanation for the opaque to transparent variation might be due to loss of CpsD phosphorylation (Weiser *et al.*, 2001) which plays an important role in capsule regulation (see below).

To date, the involvement of only one transcriptional regulator has been demonstrated. RegM, a homologue of the catabolite control protein CcpA from *S. aureus*, is required for full transcription of *cps2A* in sucrose- or glucose-containing media (Giammarinaro and Paton, 2002). Cells lacking a functional *regM* gene display reduced capsule expression and are attenuated for virulence in mice (Giammarinaro and Paton, 2002). Cps2A is itself 44% similar to the transcriptional regulator LytR from *Bacillus subtilis*, and deletion of *cpsIA* in *S. agalactiae* resulted in a two- to threefold reduction in transcription of the downstream gene *cpsIaD*, as well as reduction in surface capsule presentation (Cieslewicz *et al.*, 2001). Capsule expression is reduced in *S. pneumoniae* type 2 and type 19 *cpsA* mutants (Bender *et al.*, 2003; Morona *et al.*, 2004; Morona *et al.*, 2000), but it has not been directly demonstrated that this is the result of reduced transcription of the common genes (Morona *et al.*, 2004). Whatever its role in capsule expression, there is no significant difference in virulence between a *cps2A* mutant and wild-type *S. pneumoniae* (Morona *et al.*, 2004). Despite these findings, transcriptional regulation does not appear to be the major method by which this bacterium regulates capsule expression, as transcription of the *cps2A* gene does not change significantly between different *in vivo* environments including the nasopharynx, lungs and blood, despite concurrent phenotypic variation of capsule expression (LeMessurier *et al.*, 2006). The primary regulatory mechanism appears to be post-translational, centred around the tyrosine phosphorylation of CpsD (LeMessurier *et al.*, 2006).

CpsD is an auto-phosphorylating protein-tyrosine kinase that contains a C-terminal $[YGX]_4$ repeat domain (Morona *et al.*, 2000), and promotes attachment of the capsular polysaccharide to teichoic acid in the bacterial cell wall (Morona *et al.*, 2000; Morona *et al.*, 2006). At least three of these tyrosine residues are required for full CpsD function, whilst two or fewer result in a mucoid phenotype (Morona *et al.*, 2003). CpsC is proposed to interact with unphosphorylated, active CpsD, enabling CpsD to bind ATP and interact with other biosynthetic enzymes to promote capsule biosynthesis. Upon tyrosine phosphorylation of CpsD, interactions between the capsule biosynthetic proteins change such that capsule biosynthesis is reduced but attachment of capsular polysaccharide to the cell wall is increased. Mutation of tyrosine residues to phenylalanine confers a mucoid phenotype on pneumococci owing to the continuous production of capsular polysaccharide but reduction in its attachment to the cell wall (Morona *et al.*, 2000; Morona *et al.*, 2003). CpsC and CpsD are the Gram-positive functional equivalent of the N-terminus and C-terminus, respectively, of Wzc in *E. coli* group 1 capsules. CpsD is dephosphorylated by CpsB, a manganese-dependent phosphotyrosine-protein phosphatase, and the cycle begins again (Morona *et al.*, 2002). Thus, in *cpsB* mutants, regulation of capsule expression through CpsD phosphorylation is lost, as CpsD cannot be cycled, resulting in a reduction of capsule biosynthesis but an increase in the attachment of existing capsular polysaccharide to the cell wall (Morona *et al.*, 2006). Some *cpsB* isolates exhibit a mucoid colony morphology on blood agar that is due to additional mutations located in *cpsC*. These mutations result in a 50% reduction in the amount of capsular polysaccharide attached to the cell wall, despite producing wild-type levels of capsular polysaccharide, thus demonstrating the importance of CpsC in addition to CpsD in cell wall attachment of the polysaccharide (Morona *et al.*, 2006). The importance of the tyrosine phosphorylation cycle is demonstrated by the fact that *S. pneumoniae cpsB* and *cpsD* mutants are avirulent, whilst mutants displaying a mucoid phenotype, are unable to cause systemic disease in mice after intranasal infection (Morona *et al.*, 2004; Morona *et al.*, 2006). Oxygen levels may be an important environmental factor influencing capsule expression as opaque variants of *S. pneumoniae* grown under microaerophilic conditions show increased capsule expression concurrent with an increase in the level of phosphorylated CpsD in the cell (Weiser *et al.*, 2001). Transparent clinical isolates show no response of *cpsA* expression to oxygen tension (Hathaway *et al.*, 2007).

In summary, the *S. pneumoniae* capsule is phase variable due to tandem duplications in the *capA* gene. Additionally, this bacterium controls capsule expression at the post-translational level, through cyclic phosphorylation and dephosphorylation of CpsD by CpsC and CpsB.

Streptococcus pyogenes

Group A streptococci (GAS) express a capsule composed of hyaluronic acid identical to that found in human connective tissue (Dougherty and van de Rijn, 1994). During incubation in human blood, this bacterium up-regulates its hyaluronic capsule, conferring upon it the ability to avoid opsonophagocytosis promoted by antibodies directed at surface exposed antigens (Dinkla *et al.*, 2007). The capsule is a high molecular weight linear glycosaminoglycan composed of β-1,4-linked glucuronic acid disaccharides β-1,3-linked to GlcNAc. The genes responsible for hyaluronic acid synthesis are organized as a single transcriptional unit, *hasABC*. The capsule is not expressed during stationary phase (van de Rijn, 1983), which appears to be the result of transcriptional regulation as no transcription of the *has* operon occurs in either stationary phase cells or unencapsulated isolates (Crater and van de Rijn, 1995). Sequences upstream of –47 relative to the transcription start point are not involved in the regulation of *has* operon transcription (Alberti *et al.*, 1998), and it was suggested that differences in the structure between the –35 and –10 promoter sequences could be the cause of higher levels of *has* operon transcription and capsule expression observed in M18 strains of GAS compared with M3 strains (Bernish and van de Rijn, 1999). However, a later study failed to find a correlation between these sequences and variations in capsule expression (Bernish and van de Rijn, 1999). Both studies found that the relative level of capsule expression varied in different strains, both within and between M-types, by as much as three- to 20-fold, and therefore other as-yet unidentified strain-specific factors independent of the differences in promoter sequence are responsible for the observed differences in capsule expression (Alberti *et al.*, 1998; Bernish and van de Rijn, 1999).

A two-component regulatory system, CovR–CovS (also known as CsrR–CsrS), was later identified and shown to be involved in the repression of capsule expression, in addition to other virulence factors including streptolysin S and exotoxin B (Federle *et al.*, 1999; Heath *et al.*, 1999; Levin and Wessels, 1998) and CovR has been shown to regulate the expression of 15% of all genes in GAS (Graham *et al.*, 2002). The C-terminus of CovS shows homology to other histidine kinase proteins and possesses two theoretical membrane spanning domains indicating that this protein functions as the sensor. Inactivation of *covR* results in a twofold increase in *has* operon transcription, corresponding with a sixfold increase in capsule expression, increased resistance to serum-mediated killing and a 500-fold reduction of the LD_{50} of GAS in a mouse model of infection (Levin and Wessels, 1998). CovR is phosphorylated at an aspartate residue in its N-terminus, causing it to oligomerize (Dalton and Scott, 2004; Gusa *et al.*, 2006; Miller *et al.*, 2001). CovR binds directly to the *has* promoter region (P*has*), and DNA-binding efficiency is increased when CovR is phosphorylated, and in the presence of RNA polymerase (Federle and Scott, 2002; Gao *et al.*, 2005; Gusa *et al.*, 2006; Miller *et al.*, 2001). Phosphorylation is also essential for repression of transcription from P*has in vivo*, and enhances the ability of CovR to repress transcription from P*has* more than sixfold *in vitro* (Dalton and Scott, 2004; Gusa *et al.*, 2006). Phosphorylation of CovR also enhances its ability to repress transcription from P*cov* and P*sagA* (controlling expression of streptolysin S) *in vitro*, although RNA polymerase affords no enhancement at these promoters (Gao *et al.*, 2005; Gusa *et al.*, 2006; Gusa and Scott, 2005). The extent of these enhancements in binding efficiency and ability to repress transcription differ between CovR-regulated promoters, an observation anticipated to be the result of differences in promoter sequence and binding modes as described below (Gusa *et al.*, 2006).

A consensus DNA-binding sequence of ATTARA has been identified for CovR using uracil interference assays (Federle and Scott, 2002). The pair of thymine residues in this sequence is essential for CovR-mediated repression at P*has* (Federle and Scott, 2002). At P*cov*, where CovR directly represses its own transcription (Federle *et al.*, 1999; Gusa and Scott, 2005), this situation is less clear-cut. Here, CovR binds to seven regions in which the consensus motif is present, but in only five of these regions is the thymine pair essential for CovR binding. In addition, at P*cov*, CovR is able to bind regions lacking the consensus, and to regions in which the thymine pair is mutated, indicating that CovR-DNA interaction is not totally dependent on the presence of the previously identified thymine pair (Gusa and Scott, 2005). At P*sagA*, ATTARA sequences are present, but unlike other promoters, the DNA binding sites for CovR are poorly defined, and binding appears to be cooperative (Gao *et al.*, 2005). A model has been proposed to account for these differences, which implies that there are two binding modes. The first is dependent on ATTARA sequences, to which CovR is able to bind without significant contribution from cooperative binding. This would account for the relatively small effect that phosphorylation has on DNA binding at P*has*. The second requires polymerization of phosphorylated CovR molecules bound at multiple phased TTA sites, hence phosphorylation under these circumstances results in a relatively large increase in binding affinity (Gao *et al.*, 2005). These models imply that the extent of phosphorylation of CovR, which could be influenced by environmental signals, will influence differential gene expression by this protein. CovR is itself positively regulated by the GAS-specific RocA protein, and as a result *rocA* mutants show increased *has* operon transcription and capsule expression (Biswas and Scott, 2003).

One potential signal to which the CovS protein could respond is environmental Mg^{2+}. GAS capsule gene expression is repressed by high levels of Mg^{2+}, as are other Cov-regulated genes, including *sagA* and *ska* encoding streptokinase. The *covRS* operon is not itself regulated by Mg^{2+}, suggesting that it is the phosphorylation state of CovS that is changed, resulting in a change in the activation state of CovR (Gryllos *et al.*, 2003). CovS is dispensable for CovR activation, which suggests that an alternative method of activation (phosphorylation) must exist in the cell (Dalton and Scott, 2004). However, it appears that the principal function of CovS is to inactivate CovR rather than to activate it, as inactivation of *covR* restores viability to *covS* mutants under stress (Dalton and Scott, 2004).

Expression of the *S. pneumoniae* capsule genes is therefore primarily controlled at the level of transcription by the CovRS two-component global regulatory system, where CovR represses transcription from P*has*. CovRS activity might respond at the post-translational level to environmental signals such as Mg2+ availability. Changes in the phosphorylation state of CovR modulate its ability to recognize various promoters and therefore to repress transcription.

Neisseria meningitidis

This bacterium is a Gram-negative diplococcus and the cause of septicaemia and infant meningitis. The most common capsule serogroups of *N. meningitidis* are B, C, W-135, and Y, but serogroups B and C, with α-2,8 and α-2,9 linked polysialic acid respectively, cause more than 95% of meningococcal disease in the northern hemisphere (Jennings *et al.*, 1977). The capsule gene cluster is composed of three regions: region A is responsible for the synthesis of NeuNAc and its polymerization to sialic acid; region B contains the genes responsible for translocation of the capsular polysaccharide from the cytoplasm to the periplasm, region C completes the transport process from the periplasm across the outer membrane. Wild-type isolates of serogroup B are subject to phase variation with a frequency of 10^{-4}, one cause of which is the insertion of IS1301 into the *siaA* gene in approximately 20% of acapsular isolates (Hammerschmidt *et al.*, 1996a). SiaA is responsible for the epimerization of N-acetyl glucosamine to N-acetyl mannosamine. This forms the substrate for the generation of N-acetyl neuraminic acid in a condensation reaction with phosphoenolpyruvate mediated by SiaC (Edwards *et al.*, 1994). A second cause of phase variation was found to be the insertion or deletion of a cytidine residue in a run of seven cytidines within the coding region of *siaD*, resulting in a premature stop codon downstream of this $(dC)_7$ box (Hammerschmidt *et al.*, 1996b). SiaD is the α-2,8 polysialyl transferase gene responsible for polysaccharide chain elongation and essential for expression of the polysaccharide capsule (Edwards *et al.*, 1994).

Expression of the meningococcal capsule may also be regulated at the level of transcription. The 5′ UTR of *siaA* harbours a 70-bp *cis*-acting negative regulatory element that represses transcription of the region A *sia* and region C *ctr* genes, which are divergently transcribed from abutting promoters (Tzeng *et al.*, 2001; Von Loewenich *et al.*, 2001). Deletion of the promoter region upstream of position –46 relative to the transcription start point of *siaA* (P*sia*) has no effect on transcription, indicating that no regulatory factors act upstream of the basic promoter and UP elements (Von Loewenich *et al.*, 2001). During intimate adhesion to epithelial cells, transcription of the *sia* genes is reduced (Deghmane *et al.*, 2002). This reduction is brought about directly by the CrgA protein, a LysR-type regulator that is induced upon host cell contact (Deghmane *et al.*, 2000; Deghmane and Taha, 2003), although the CrgA binding site is upstream of the aforementioned UTR region (Deghmane *et al.*, 2002). CrgA is expressed at low levels in the absence of host cells and represses its own transcription (Deghmane and Taha, 2003; Ieva *et al.*, 2005). This is achieved through interaction of the CrgA C-terminal domain with RNA polymerase, an interaction that is essential for CrgA to inhibit transcription, and the two proteins bind to the *crgA* promoter (P*crgA*) synergistically (Deghmane *et al.*, 2004; Deghmane and Taha, 2003). CrgA does not block open complex formation or transcription initiation, rather, it inhibits transcription elongation by RNA polymerase leaving the transcription initiation complex (Deghmane *et al.*, 2004). Comparison of *in vivo* DMS footprinting studies performed on *N. meningitidis* either associated with epithelial cells or grown in isolation have demonstrated that CrgA dissociates from its target sites within P*crgA* upon binding (Deghmane *et al.*, 2004). This results in increased expression of CrgA and subsequent repression of capsule expression (Deghmane and Taha, 2003). It has been proposed that the central region of this protein might respond to a signal molecule induced upon host cell contact, although the signal has yet to be identified (Deghmane and Taha, 2003). Two studies have challenged the involvement of CrgA in meningococcal capsule regulation. The first found no evidence that CrgA affected the expression of a P*sia–lacZ* fusion (Von Loewenich *et al.*, 2001). The second, using primer extension assays performed on RNA extracted from wild-type and *crgA*– meningococci, failed to show any effect of CrgA on transcription from P*sia*, even when the bacteria were grown in the presence of α-methylene-γ-butyrolactone (MBL), which has been shown to regulate transcription from several promoters in a CrgA-dependent manner (Ieva *et al.*, 2005). In light of these results it has been suggested that the previously observed binding of CrgA to P*sia in vitro* may be biologically irrelevant (Ieva *et al.*, 2005). Given that previous data from RT-PCR amplification of RNA from meningococci adhered to epithelial cells demonstrated CrgA-dependent down-regulation of *sia* transcription (Deghmane *et al.*, 2002), it is possible that the response of CrgA to host cell contact is fundamentally different to that observed for MBL and involves additional factors or conditions absent from the more recent study. The difference may also be attributable to the fact that different serogroups of *N. meningitidis* were used in these studies.

The *N. meningitidis* capsule is phase variable due to changes at the genetic level within *siaA* and *siaD*. The capsule genes are also regulated at the transcriptional level by the host epithelial cell contact-induced regulator CrgA, which down regulates capsule expression in response to intimate adhesion with host cells.

E. coli group 2 capsules (K5)

Genetic organization

Group 2 capsule gene clusters are organized into three principal regions. Regions 1 and 3 contain the genes responsible for transport of newly synthesized capsular polysaccharide from the cytoplasm to the bacterial cell surface, and are conserved in all group 2 capsule gene clusters. Region 2 encodes the genes responsible for the synthesis of the polysaccharide and its precursors

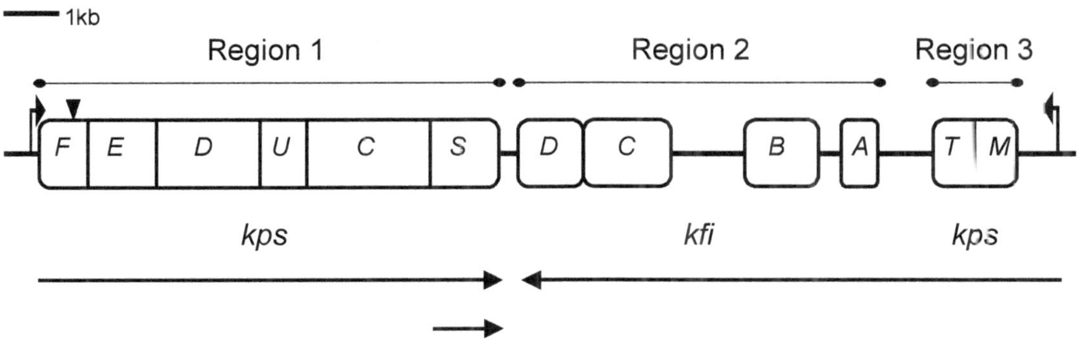

Figure 4.2 Genetic organization of the *E. coli* K5 capsule gene cluster. Bent arrows upstream of regions 1 and 3 indicate transcription start points. Horizontal arrows show the major transcript and processed *kpsS*-specific transcript from region 1, and the major transcript originating from region 3 that proceeds through region 2. The inverted triangle represents the Rho-dependent transcriptional terminator in *kpsF*.

(Fig. 4.2) (Whitfield, 2006). Transcription of the K5 polysaccharide gene cluster is driven by two convergent temperature-regulated promoters located upstream of regions 1 and 3 (Simpson *et al.*, 1996; Stevens *et al.*, 1997), designated PR1 and PR3 respectively. Region 1 is a single transcriptional unit containing six genes, *kpsFEDUCS* (Simpson *et al.*, 1996). PR1, a σ^{70} promoter, is located 225 bp upstream of *kpsF* and there are no alternative sigma factor recognition sites present in PR1 (Simpson *et al.*, 1996). Transcription from PR1 generates an 8-kb polycistronic message that is later processed to yield a separate 1.3 kb *kpsS*-specific transcript, and this processing has been proposed to play a role in the differential expression of this gene (Roberts, 2000). Within the *kpsF* gene there is a Rho-dependent transcriptional terminator. Rho is an RNA-binding protein that exists as a hexamer, requiring a particular DNA sequence – CACA(Y)$_3$ and C(N)$_3$CC, also known as *rut* – approximately 200 bp upstream of the termination site (Ciampi, 2006). It is proposed that the Rho protein tracks behind RNA polymerase as it proceeds through transcription and if RNA polymerase pauses, transcription is terminated when the Rho protein catches up (Ciampi, 2006). These intragenic terminators are believed to play a role in reducing unnecessary transcription during times of physiological stress. Region 3 contains two genes, *kpsMT*, which are involved in transport of the capsular polysaccharide across the cytoplasmic membrane (Bliss and Silver, 1996). PR3 is located 741 bp upstream of the first of two genes in region 3, *kpsM*. Transcription from this promoter is driven through region 2, past a Rho-dependent transcriptional terminator that exists between *kpsT* and *kfiA*, the first gene of region 2.

Region 2 is serotype-specific (Roberts, 1996), and in the K5 gene cluster, this region consists of *kfiA*, *kfiB*, *kfiC* and *kfiD* genes (Petit *et al.*, 1995). A fifth ORF was identified, although expression of the predicted 16 kDa gene product could not be demonstrated (Petit *et al.*, 1995). In contrast, in the K1 capsule gene cluster, region 2 consists of six genes, *neuDBACES* (Whitfield, 2006). Region 2 of the K5 capsule gene cluster has a higher A+T content (66%) than either region 1 or 3, which possess 50% and 57% AT, respectively, values more typical for the *E. coli* chromosome (Petit *et al.*, 1995; Blattner *et al.*, 1997). This suggests that region 2 has evolved through the acquisition of capsule gene DNA sequences from other bacteria. The mechanism by which acquisition and exchange of region 2 occurs is unknown but it may involve recombination events between regions 1 and 3 of incoming and resident capsule DNA (Roberts, 1996). This *en bloc* acquisition of biosynthetic loci is supported by the observation that the C-terminal regions of *kpsS* (region 1) and *kpsT* (region 3) from K1 and K5 which flank, and therefore form junctions with, region 2 demonstrate the greatest divergence in sequence of the conserved regions (Boulnois and Jann, 1989; Pavelka *et al.*, 1994; Petit *et al.*, 1995; Roberts, 1996). Additionally, it has been suggested that *kfiD* might have been acquired from a streptococcal species (Munoz *et al.*,

1998). There are long intergenic regions in the K5 region 2; *kfiA* and *kfiB* are separated by 340 bp, and *kfiB* is separated from *kfiC* by 1293 bp (Petit *et al.*, 1995). Currently no function has been assigned to these intergenic regions although three promoters have been identified upstream of *kfiA*, *kfiB*, and *kfiC*, generating 8-, 6- and 3.5-kb transcripts respectively. However, these promoters are relatively weak (Petit *et al.*, 1995), and on their own are insufficient for levels of the biosynthetic proteins necessary to express a K5 capsule (Stevens *et al.*, 1997). It would appear that the vast majority of region 2 message is produced by read-through transcription from PR3, generating a polycistronic transcript containing both the region 3 and 2 genes (Stevens *et al.*, 1997).

Regulation of transcription

Unlike many of the other capsule types discussed here, the Rcs (or an Rcs-like) phosphorelay system is not involved in expression of the group 2 gene clusters (Keenleyside *et al.*, 1993; Stevens *et al.*, 1994). Group 2 capsule gene clusters are temperature regulated, being expressed at 37°C, but not at temperatures below 25°C. Temperature-dependent expression of the capsule genes is controlled in part at the level of transcription, as no detectable transcription originates from either promoter at 20°C (Cieslewicz and Vimr, 1996; Rowe *et al.*, 2000). Temperature regulation involves the global regulatory proteins H-NS and BipA which play an unusual dual role in the temperature regulation of transcription at PR1: both are required for maximum transcription at 37°C, yet both also contribute to repression of transcription at 20°C, and the transcriptional effects are reflected in the levels of K5 capsule produced by the respective mutants (Rowe *et al.*, 2000). H-NS is a highly abundant protein bearing a functional resemblance to eukaryotic histones (Atlung and Ingmer, 1997; Dorman, 2004) that forms an integral component of the bacterial nucleoid and is involved in the compaction of the chromosome of enteric bacteria (Spassky *et al.*, 1984; Dame *et al.*, 2000), and the silencing of horizontally acquired genetic material (Dorman, 2007; Lucchini *et al.*, 2006; Navarre *et al.*, 2006). The effect of the *bipA* mutant is unlikely to be the result of modulation of H-NS expression as quantitative western blots have shown the level of H-NS to be unchanged in the absence of *bipA* (Grant *et al.*, 2003). BipA is a ribosome-binding GTPase that is able to regulate gene expression at the level of translation (Owens *et al.*, 2004). Thus, the role of this protein in regulating transcription from PR1 is probably indirect, for example by controlling the expression of a transcriptional activator (not SlyA), but the identity of this putative regulator is currently unknown.

Recently a further transcriptional regulator, SlyA, was shown to be involved in activating transcription from PR1 at 37°C (Corbett *et al.*, 2007). SlyA is a dimeric protein essential for virulence of *S.* Typhimurium in mice, resistance to the oxidative burst and survival within murine macrophages (Buchmeier *et al.*, 1997; Libby *et al.*, 1994; Stapleton *et al.*, 2002) and has been characterized as a regulator of a variety of virulence factors in *S.* Typhimurium, the majority of which occur in either the cell membrane, periplasm, or are surface exposed (Ellison and Miller, 2006; Navarre *et al.*, 2005; Spory *et al.*, 2002). SlyA has been suggested to act as an H-NS anti-repressor (Heroven *et al.*, 2004), however the situation at PR1 is more complex. H-NS and SlyA can bind simultaneously to PR1, and H-NS is required for maximal SlyA-mediated activation of transcription at this promoter, explaining the observation that *hns* mutants show reduced levels of capsule gene transcription at 37°C, and suggesting that at the capsule permissive temperature, the two proteins are co-activators of transcription (Corbett *et al.*, 2007). The precise means by which the resulting remodelled nucleoprotein complex is conducive to transcription is unclear: it may directly result in enhanced open complex formation, or may allow the binding of a further, as-yet unidentified transcriptional activator, as seen at the *E. coli csiD* promoter (Germer *et al.*, 2001). This complex interplay between H-NS and SlyA, along with further regulators and RNA polymerase has also been observed at the *E. coli hlyE* and *S.* Typhimurium *pagC* and *ugtL* promoters (Lithgow *et al.*, 2007; Perez *et al.*, 2008). It is also formally possible that SlyA could modulate H-NS activity at the post-translational level, either interacting directly with it, or even mediating the chemical modulation of H-NS (Reusch *et al.*, 2002).

The IHF protein is required for maximum transcription from PR1 at 37°C, binding to a

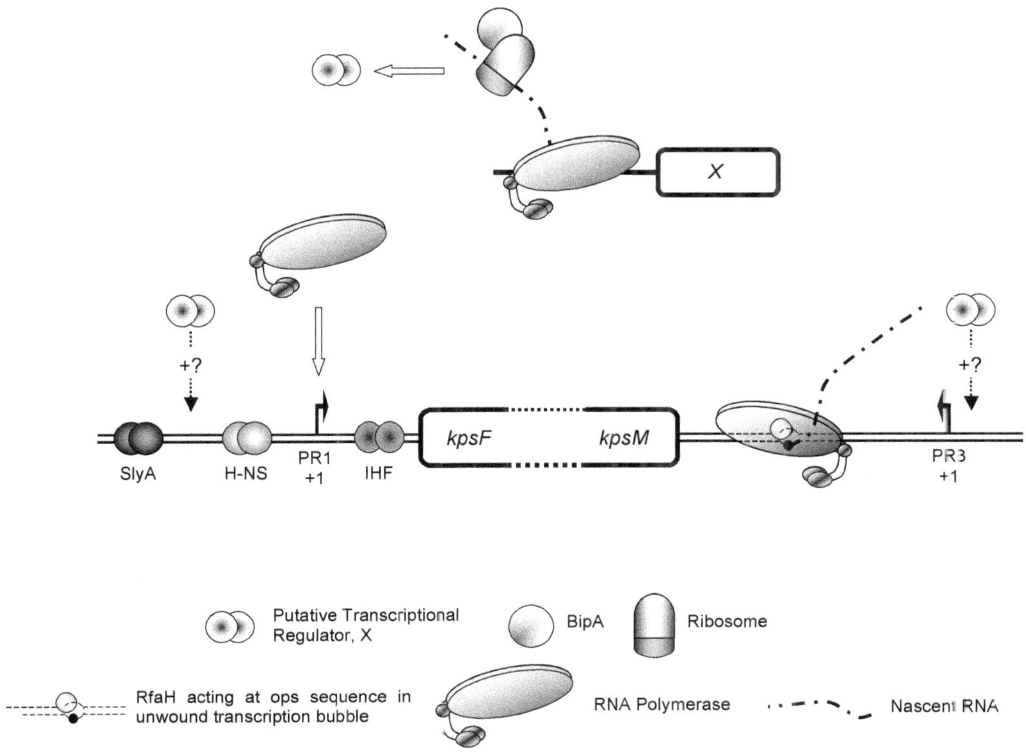

Figure 4.3 Current model of transcriptional regulation of the K5 capsule gene cluster occurring at 37°C at PR1 and PR3. BipA is shown enhancing the translation of an unknown transcriptional regulator which presumably promotes transcription from PR1. SlyA and H-NS are bound to PR1 in a manner which promotes RNA polymerase binding to the promoter. An RNA Polymerase-RfaH complex is shown at PR3 elongating a nascent RNA chain at the *ops* sequence.

single site located 130 bp downstream of the transcription start point at PR1 (Rowe *et al.*, 2000) (Fig. 4.3). This site encompasses the IHF consensus WATCAANNNNTTR (W = A/T, R = A/G). IHF is able to bend DNA to an angle of more than 160° at the W base pair, and, in doing so, can potentially bring other activators or functionally important regions of DNA into contact with RNA polymerase, and thus potentiate transcription (Engelhorn and Geiselmann, 1998). In the absence of H-NS, IHF appears to be largely redundant, suggesting a role for IHF in H-NS anti-repression. Additionally, there appears to be no interplay between the SlyA- and IHF-mediated regulation of this promoter, unlike the situation that exists for H-NS and SlyA (Corbett and Roberts, unpublished).

Transcription of the region 2 genes is driven by PR3, and because the region 2 promoters are too weak to enable the bacterium to express a capsule (Petit *et al.*, 1995), with expression of the K5 capsular polysaccharide being dependent upon RfaH, which acts on the *ops* sequence present in the 5′ UTR of *kpsM* (Fig. 4.3) (Stevens *et al.*, 1994; 1997). Unlike PR3, there is no dependency upon RfaH for transcription elongation at PR1, and no JUMPStart sequence is present upstream of *kpsF* (Simpson *et al.*, 1996; Stevens *et al.*, 1997). A further difference between the regulation of the two promoters is that IHF plays no role in the regulation of PR3. H-NS, SlyA and BipA, however, appear to play broadly similar roles at both promoters (Xue and Roberts, unpublished). The function of the unusually long untranslated region between PR3 and the 5′ end of *kpsM* in the regulation of capsule gene expression is currently unknown, although it is not believed to encode any additional regulators or to affect RNA stability (Xue and Roberts, unpublished).

The expression of group 2 capsules in *E. coli* is therefore subject to complex transcriptional regulation involving the action of several global regulators acting at two convergent promoters, which are apparently differentially regulated. The differences in regulation at the two promoters may reflect the need for subtle adjustments in the stoichiometry of the proteins required for capsule synthesis and export. At present, temperature is the only known environmental signal responsible for regulating transcription of the capsule gene cluster, which is mediated through the H-NS, BipA and SlyA proteins.

Concluding remarks and future issues

The regulation of capsule expression in pathogenic bacteria is complex and reflects the needs of bacteria to balance the protective benefits of capsule expression with the substantial energy investment and potential hindrance to the progression of infection that their production represents. Control of capsule expression is exerted at almost every level in the hierarchy of control of gene expression: examples discussed here range from phase-variation, through globally acting transcription factors to regulation at the post-translational level through the employment of phosphorelays.

With reference to the control of *E. coli* group 2 capsule gene expression, several issues remain to be addressed, foremost amongst which is the identification of the transducer responsible for the temperature-dependent activation of capsule expression. The mechanism by which the translational regulator BipA exerts its control on transcription also remains to be elucidated. Further investigation of both areas will have broader implications for our understanding of the control of virulence gene expression in bacteria. Knowledge of the full repertoire of transcription factors controlling capsule gene transcription will allow us to predict the environmental signals utilized by bacteria to vary capsule expression *in vivo*. It will also enable us to understand how various virulence factors are cross-regulated in order to understand their co-ordinated expression. The exciting discovery of the host cell contact-dependent transcriptional regulator (CrgA) of *Neisseria meningitidis* is an example of how further environmental signals (e.g. the microenvironment of the host cell surface rather than more general nutrient availability) may be integrated into the control of capsule gene expression. Such studies could be facilitated through the use of comparative microarray analyses and ChIP-Chip analysis of bacteria during the course of infection.

References

Alberti, S., Ashbaugh, C.D., and Wessels, M.R. (1998). Structure of the has operon promoter and regulation of hyaluronic acid capsule expression in group A Streptococcus. Mol. Microbiol. *28*, 343–353.

Arakawa, Y., Ohta, M., Wacharotayankun, R., Mori, M., Kido, N., Ito, H., Komatsu, T., Sugiyama, T., and Kato, N. (1991). Biosynthesis of Klebsiella K2 capsular polysaccharide in *Escherichia coli* HB101 requires the functions of rmpA and the chromosomal cps gene cluster of the virulent strain *Klebsiella pneumoniae* Chedid (O1:K2). Infect. Immun. *59*, 2043–2050.

Arricau, N., Hermant, D., Waxin, H., Ecobichon, C., Duffey, P.S., and Popoff, M.Y. (1998). The RcsB-RcsC regulatory system of Salmonella typhi differentially modulates the expression of invasion proteins, flagellin and Vi antigen in response to osmolarity. Mol. Microbiol. *29*, 835–850.

Artsimovitch, I., and Landick, R. (2002). The transcriptional regulator RfaH stimulates RNA chain synthesis after recruitment to elongation complexes by the exposed nontemplate DNA strand. Cell *109*, 193–203.

Atlung, T., and Ingmer, H. (1997). H-NS: a modulator of environmentally regulated gene expression. Mol. Microbiol. *24*, 7–17.

Bailey, M.J., Hughes, C., and Koronakis, V. (1997). RfaH and the ops element, components of a novel system controlling bacterial transcription elongation. Mol. Microbiol. *26*, 845–851.

Bailey, M.J., Hughes, C., and Koronakis, V. (2000). In vitro recruitment of the RfaH regulatory protein into a specialised transcription complex, directed by the nucleic acid ops element. Mol. Gen. Genet. *262*, 1052–1059.

Bender, M.H., Cartee, R.T., and Yother, J. (2003). Positive correlation between tyrosine phosphorylation of CpsD and capsular polysaccharide production in *Streptococcus pneumoniae*. J. Bacteriol. *185*, 6057–6066.

Bernish, B., and van de Rijn, I. (1999). Characterization of a two-component system in *Streptococcus pyogenes* which is involved in regulation of hyaluronic acid production. J. Biol. Chem. *274*, 4786–4793.

Biswas, I., and Scott, J.R. (2003). Identification of rocA, a positive regulator of covR expression in the group A streptococcus. J. Bacteriol. *185*, 3081–3090.

Bliss, J.M., and Silver, R.P. (1996). Coating the surface: a model for expression of capsular polysialic acid in *Escherichia coli* K1. Mol. Microbiol. *21*, 221–231.

Boulnois, G.J., and Jann, K. (1989). Bacterial polysaccharide capsule synthesis, export and evolution of structural diversity. Mol. Microbiol. *3*, 1819–1823.

Brown, E.J., Joiner, K.A., Cole, R.M., and Berger, M. (1983). Localization of complement component 3 on *Streptococcus pneumoniae*: anti-capsular antibody causes complement deposition on the pneumococcal capsule. Infect. Immun. *39*, 403–409.

Buchmeier, N., Bossie, S., Chen, C.Y., Fang, F.C., Guiney, D.G., and Libby, S.J. (1997). SlyA, a transcriptional regulator of *Salmonella typhimurium*, is required for resistance to oxidative stress and is expressed in the intracellular environment of macrophages. Infect. Immun. *65*, 3725–3730.

Ciampi, M.S. (2006). Rho-dependent terminators and transcription termination. Microbiology *152*, 2515–2528.

Cieslewicz, M., and Vimr, E. (1996). Thermoregulation of kpsF, the first region 1 gene in the kps locus for polysialic acid biosynthesis in *Escherichia coli* K1. J. Bacteriol. *178*, 3212–3220.

Cieslewicz, M.J., Kasper, D.L., Wang, Y., and Wessels, M.R. (2001). Functional analysis in type Ia group B streptococcus of a cluster of genes involved in extracellular polysaccharide production by diverse species of streptococci. J. Biol. Chem. *276*, 139–146.

Corbett, D., Bennett, H.J., Askar, H., Green, J., and Roberts, I.S. (2007). SlyA and H-NS regulate transcription of the *Escherichia coli* K5 capsule gene cluster, and expression of slyA in *Escherichia coli* is temperature-dependent, positively autoregulated, and independent of H-NS. J. Biol. Chem. *282*, 33326–33335.

Crater, D.L., and van de Rijn, I. (1995). Hyaluronic acid synthesis operon (has) expression in group A streptococci. J. Biol. Chem. *270*, 18452–18458.

Cunnion, K.M., Lee, J.C., and Frank, M.M. (2001). Capsule production and growth phase influence binding of complement to Staphylococcus aureus. Infect. Immun. *69*, 6796–6803.

Dalton, T.L., and Scott, J.R. (2004). CovS inactivates CovR and is required for growth under conditions of general stress in Streptococcus pyogenes. J. Bacteriol. *186*, 3928–3937.

Deghmane, A.E., Giorgini, D., Larribe, M., Alonso, J.M., and Taha, M.K. (2002). Down-regulation of pili and capsule of *Neisseria meningitidis* upon contact with epithelial cells is mediated by CrgA regulatory protein. Mol. Microbiol. *43*, 1555–1564.

Deghmane, A.E., Giorgini, D., Maigre, L., and Taha, M.K. (2004). Analysis *in vitro* and *in vivo* of the transcriptional regulator CrgA of *Neisseria meningitidis* upon contact with target cells. Mol. Microbiol. *53*, 917–927.

Deghmane, A.E., Petit, S., Topilko, A., Pereira, Y., Giorgini, D., Larribe, M., and Taha, M.K. (2000). Intimate adhesion of *Neisseria meningitidis* to human epithelial cells is under the control of the crgA gene, a novel LysR-type transcriptional regulator. Embo J. *19*, 1068–1078.

Deghmane, A.E., and Taha, M.K. (2003). The *Neisseria meningitidis* adhesion regulatory protein CrgA acts through oligomerization and interaction with RNA polymerase. Mol. Microbiol. *47*, 135–143.

Dinkla, K., Sastalla, I., Godehardt, A.W., Janze, N., Chhatwal, G.S., Rohde, M., and Medina, E. (2007). Up-regulation of capsule enables Streptococcus pyogenes to evade immune recognition by antigen-specific antibodies directed to the G-related alpha2-macroglobulin-binding protein GRAB located on the bacterial surface. Microbes Infect. *9*, 922–931.

Dorman, C.J. (2004). H-NS: a universal regulator for a dynamic genome. Nat. Rev. Microbiol. *2*, 391–400.

Dorman, C.J. (2007). H-NS, the genome sentinel. Nat. Rev. Microbiol. *5*, 157–161.

Dougherty, B.A., and van de Rijn, I. (1994). Molecular characterization of hasA from an operon required for hyaluronic acid synthesis in group A streptococci. J. Biol. Chem. *269*, 169–175.

Ebah, L., and Roberts, I.S. (2001). Regulation of the *Escherichia coli* K5 Capsule Gene Cluster Durin Biofilm Development. In Biofilm Community Interactions: Chance or Necessity?, P. Gilbert, D. Allison, M. Brading, J. Verran, and J. Walker, eds. (Gregynog Hall, Powys, Wales, BioLine), pp. 211–219.

Edwards, U., Muller, A., Hammerschmidt, S., Gerardy-Schahn, R., and Frosch, M. (1994). Molecular analysis of the biosynthesis pathway of the alpha-2,8 polysialic acid capsule by *Neisseria meningitidis* serogroup B. Mol. Microbiol. *14*, 141–149.

Ellison, D.W., and Miller, V.L. (2006). Regulation of virulence by members of the MarR/SlyA family. Curr. Opin. Microbiol. *9*, 153–159.

Engelhorn, M., and Geiselmann, J. (1998). Maximal transcriptional activation by the IHF protein of *Escherichia coli* depends on optimal DNA bending by the activator. Mol. Microbiol. *30*, 431–441.

Erickson, K.D., and Detweiler, C.S. (2006). The Rcs phosphorelay system is specific to enteric pathogens/commensals and activates ydeI, a gene important for persistent *Salmonella* infection of mice. Mol. Microbiol. *62*, 883–894.

Federle, M.J., McIver, K.S., and Scott, J.R. (1999). A response regulator that represses transcription of several virulence operons in the group A streptococcus. J. Bacteriol. *181*, 3649–3657.

Federle, M.J., and Scott, J.R. (2002). Identification of binding sites for the group A streptococcal global regulator CovR. Mol. Microbiol. *43*, 1161–1172.

Gao, J., Gusa, A.A., Scott, J.R., and Churchward, G. (2005). Binding of the global response regulator protein CovR to the sag promoter of Streptococcus pyogenes reveals a new mode of CovR–DNA interaction. J. Biol. Chem. *280*, 38948–38956.

Germer, J., Becker, G., Metzner, M., and Hengge-Aronis, R. (2001). Role of activator site position and a distal UP-element half-site for sigma factor selectivity at a CRP/H-NS-activated sigma(s)-dependent promoter in *Escherichia coli*. Mol. Microbiol. *41*, 705–716.

Giammarinaro, P., and Paton, J.C. (2002). Role of RegM, a homologue of the catabolite repressor protein CcpA, in the virulence of *Streptococcus pneumoniae*. Infect. Immun. *70*, 5454–5461.

Gottesman, S., and Stout, V. (1991). Regulation of capsular polysaccharide synthesis in *Escherichia coli* K12. Mol. Microbiol. *5*, 1599–1606.

Graham, M.R., Smoot, L.M., Migliaccio, C.A., Virtaneva, K., Sturdevant, D.E., Porcella, S.F., Federle, M.J., Adams, G.J., Scott, J.R., and Musser, J.M. (2002). Virulence control in group A Streptococcus by a two-component gene regulatory system: global expression profiling and *in vivo* infection modeling. Proc. Natl. Acad. Sci. USA *99*, 13855–13860.

Grant, A.J., Farris, M., Alefounder, P., Williams, P.H., Woodward, M.J., and O'Connor, C.D. (2003) Co-ordination of pathogenicity island expression by the BipA GTPase in enteropathogenic *Escherichia coli* (EPEC). Mol. Microbiol. *48*, 507–521.

Gryllos, I., Levin, J.C., and Wessels, M.R. (2003). The CsrR/CsrS two-component system of group A

Streptococcus responds to environmental Mg2+. Proc. Natl. Acad. Sci. USA *100*, 4227–4232.

Gusa, A.A., Gao, J., Stringer, V., Churchward, G., and Scott, J.R. (2006). Phosphorylation of the group A Streptococcal CovR response regulator causes dimerization and promoter-specific recruitment by RNA polymerase. J. Bacteriol. *188*, 4620–4626.

Gusa, A.A., and Scott, J.R. (2005). The CovR response regulator of group A streptococcus (GAS) acts directly to repress its own promoter. Mol. Microbiol. *56*, 1195–1207.

Hammerschmidt, S., Hilse, R., van Putten, J.P., Gerardy-Schahn, R., Unkmeir, A., and Frosch, M. (1996a). Modulation of cell surface sialic acid expression in *Neisseria meningitidis* via a transposable genetic element. Embo J. *15*, 192–198.

Hammerschmidt, S., Muller, A., Sillmann, H., Muhlenhoff, M., Borrow, R., Fox, A., van Putten, J., Zollinger, W.D., Gerardy-Schahn, R., and Frosch, M. (1996b). Capsule phase variation in *Neisseria meningitidis* serogroup B by slipped-strand mispairing in the polysialyltransferase gene (siaD): correlation with bacterial invasion and the outbreak of meningococcal disease. Mol. Microbiol. *20*, 1211–1220.

Hanna, A., Berg, M., Stout, V., and Razatos, A. (2003). Role of capsular colanic acid in adhesion of uropathogenic *Escherichia coli*. Appl. Environ. Microbiol. *69*, 4474–4481.

Hashimoto, Y., Li, N., Yokoyama, H., and Ezaki, T. (1993). Complete nucleotide sequence and molecular characterization of ViaB region encoding Vi antigen in Salmonella typhi. J. Bacteriol. *175*, 4456–4465.

Hathaway, L.J., Battig, P., and Muhlemann, K. (2007). In vitro expression of the first capsule gene of *Streptococcus pneumoniae*, cpsA, is associated with serotype-specific colonization prevalence and invasiveness. Microbiology *153*, 2465–2471.

Heath, A., DiRita, V.J., Barg, N.L., and Engleberg, N.C. (1999). A two-component regulatory system, CsrR-CsrS, represses expression of three Streptococcus pyogenes virulence factors, hyaluronic acid capsule, streptolysin, S.,and pyrogenic exotoxin B. Infect. Immun. *67*, 5298–5305.

Heroven, A.K., Nagel, G., Tran, H.J., Parr, S., and Dersch, P. (2004). RovA is autoregulated and antagonizes H-NS-mediated silencing of invasin and rovA expression in Yersinia pseudotuberculosis. Mol. Microbiol. *53*, 871–888.

Hinchliffe, S.J., Howard, S.L., Huang, Y.H., Clarke, D.J., and Wren, B.W. (2008). The importance of the Rcs phosphorelay in the survival and pathogenesis of the enteropathogenic yersiniae. Microbiology *154*, 1117–1131.

Hobbs, M., and Reeves, P.R. (1994). The JUMPstart sequence: a 39 bp element common to several polysaccharide gene clusters. Mol. Microbiol. *12*, 855–856.

Ieva, R., Alaimo, C., Delany, I., Spohn, G., Rappuoli, R., and Scarlato, V. (2005). CrgA is an inducible LysR-type regulator of Neisseria meningitidis, acting both as a repressor and as an activator of gene transcription. J. Bacteriol. *187*, 3421–3430.

Jann, K., and Jann, B. (1997). Capsules of *Escherichia coli*. In *Escherichia coli* Mechanisms of Virulence, M. Sussman, ed. (Cambridge, The Press Syndicate of the University of Cambridge), pp. 113–144.

Jayaratne, P., Keenleyside, W.J., MacLachlan, P.R., Dodgson, C., and Whitfield, C. (1993). Characterization of rcsB and rcsC from *Escherichia coli* O9:K30:H12 and examination of the role of the rcs regulatory system in expression of group I capsular polysaccharides. J. Bacteriol. *175*, 5384–5394.

Jennings, H.J., Bhattacharjee, A.K., Bundle, D.R., Kenny, C.P., Martin, A., and Smith, I.C. (1977). Strucutres of the capsular polysaccharides of *Neisseria meningitidis* as determined by 13C-nuclear magnetic resonance spectroscopy. J. Infect Dis *136* Suppl., S78–83.

Kaijser, B., and Jodal, U. (1984). *Escherichia coli* K5 antigen in relation to various infections and in healthy individuals. J. Clin. Microbiol. *19*, 264–266.

Kim, J.O., and Weiser, J.N. (1998). Association of intrastrain phase variation in quantity of capsular polysaccharide and teichoic acid with the virulence of *Streptococcus pneumoniae*. J. Infect. Dis. *177*, 368–377.

Lai, Y.C., Peng, H.L., and Chang, H.Y. (2003). RmpA2, an activator of capsule biosynthesis in *Klebsiella pneumoniae* CG43, regulates K2 cps gene expression at the transcriptional level. J. Bacteriol. *185*, 788–800.

Laubacher, M.E., and Ades, S.E. (2008). The Rcs phosphorelay is a cell envelope stress response activated by peptidoglycan stress and contributes to intrinsic antibiotic resistance. J. Bacteriol. *190*, 2065–2074.

Lee, F.K., Morris, C., and Hackett, J. (2006). The *Salmonella enterica* serovar Typhi Vi capsule and self-association pili share controls on expression. FEMS Microbiol. Lett *261*, 41–46.

LeMessurier, K.S., Ogunniyi, A.D., and Paton, J.C. (2006). Differential expression of key pneumococcal virulence genes *in vivo*. Microbiology *152*, 305–311.

Levin, J.C., and Wessels, M.R. (1998). Identification of csrR/csrS, a genetic locus that regulates hyaluronic acid capsule synthesis in group A Streptococcus. Mol. Microbiol. *30*, 209–219.

Libby, S.J., Goebel, W., Ludwig, A., Buchmeier, N., Bowe, F., Fang, F.C., Guiney, D.G., Songer, J.G., and Heffron, F. (1994). A cytolysin encoded by Salmonella is required for survival within macrophages. Proc. Natl. Acad. Sci. USA *91*, 489–493.

Lithgow, J.K., Haider, F., Roberts, I.S., and Green, J. (2007). Alternate SlyA and H-NS nucleoprotein complexes control hlyE expression in *Escherichia coli* K-12. Mol. Microbiol. *66*, 685–698.

Llull, D., Garcia, E., and Lopez, R. (2001a). Tts, a processive beta-glucosyltransferase of *Streptococcus pneumoniae*, directs the synthesis of the branched type 37 capsular polysaccharide in Pneumococcus and other gram-positive species. J. Biol. Chem. *276*, 21053-21061.

Llull, D., Lopez, R., and Garcia, E. (2001b). Genetic bases and medical relevance of capsular polysaccharide biosynthesis in pathogenic streptococci. Curr. Mol. Med *1*, 475–491.

Lopez-Torres, A.J., and Stout, V. (1996). Role of colanic acid polysaccharide in serum resistance *in vivo* and in adherence. Curr. Microbiol. 33, 383–389.

Lucchini, S., Rowley, G., Goldberg, M.D., Hurd, D., Harrison, M., and Hinton, J.C. (2006). H-NS mediates the silencing of laterally acquired genes in bacteria. PLoS Pathog. *2*, e81.

Majdalani, N., and Gottesman, S. (2005). The Rcs phosphorelay: a complex signal transduction system. Annu. Rev. Microbiol. *59*, 379–405.

Marolda, C.L., and Valvano, M.A. (1998). Promoter region of the *Escherichia coli* O7-specific lipopolysaccharide gene cluster: structural and functional characterization of an upstream untranslated mRNA sequence. J. Bacteriol. *180*, 3070–3079.

McCallum, K.L., and Whitfield, C. (1991). The rcsA gene of *Klebsiella pneumoniae* O1:K20 is involved in expression of the serotype-specific K (capsular) antigen. Infect. Immun. *59*, 494–502.

McGuire, E.J., and Binkley, S.B. (1964). The structure and chemistry of colominic acid. Biochemistry *3*, 247–251.

Miller, A.A., Engleberg, N.C., and DiRita, V.J. (2001). Repression of virulence genes by phosphorylation-dependent oligomerization of CsrR at target promoters in S. pyogenes. Mol. Microbiol. *40*, 976–990.

Morona, J.K., Miller, D.C., Morona, R., and Paton, J.C. (2004). The effect that mutations in the conserved capsular polysaccharide biosynthesis genes cpsA, cpsB, and cpsD have on virulence of *Streptococcus pneumoniae*. J. Infect. Dis. *189*, 1905–1913.

Morona, J.K., Morona, R., Miller, D.C., and Paton, J.C. (2002). *Streptococcus pneumoniae* capsule biosynthesis protein CpsB is a novel manganese-dependent phosphotyrosine-protein phosphatase. J. Bacteriol. *184*, 577–583.

Morona, J.K., Morona, R., Miller, D.C., and Paton, J.C. (2003). Mutational analysis of the carboxy-terminal (YGX)4 repeat domain of CpsD, an autophosphorylating tyrosine kinase required for capsule biosynthesis in *Streptococcus pneumoniae*. J. Bacteriol. *185*, 3009–3019.

Morona, J.K., Morona, R., and Paton, J.C. (2006). Attachment of capsular polysaccharide to the cell wall of *Streptococcus pneumoniae* type 2 is required for invasive disease. Proc. Natl. Acad. Sci. USA *103*, 8505–8510.

Morona, J.K., Paton, J.C., Miller, D.C., and Morona, R. (2000). Tyrosine phosphorylation of CpsD negatively regulates capsular polysaccharide biosynthesis in *Streptococcus pneumoniae*. Mol. Microbiol. *35*, 1431–1442.

Munoz, R., Garcia, E., and Lopez, R. (1998). Evidence for horizontal transfer from Streptococcus to *Escherichia coli* of the kfiD gene encoding the K5-specific UDP-glucose dehydrogenase. J. Mol. Evol *46*, 432–436.

Nassif, X., Fournier, J.M., Arondel, J., and Sansonetti, P.J. (1989). Mucoid phenotype of *Klebsiella pneumoniae* is a plasmid-encoded virulence factor. Infect. Immun. *57*, 546–552.

Navarre, W.W., Halsey, T.A., Walthers, D., Frye, J., McClelland, M., Potter, J.L., Kenney, L.J., Gunn, J.S., Fang, F.C., and Libby, S.J. (2005). Co-regulation of Salmonella enterica genes required for virulence and resistance to antimicrobial peptides by SlyA and PhoP/PhoQ. Mol. Microbiol. *56*, 492–508.

Navarre, W.W., Porwollik, S., Wang, Y., McClelland, M., Rosen, H., Libby, S.J., and Fang, F.C. (2006). Selective silencing of foreign DNA with low GC content by the H-NS protein in Salmonella. Science *313*, 236–238.

Nieto, J.M., Bailey, M.J., Hughes, C., and Koronakis, V. (1996). Suppression of transcription polarity in the *Escherichia coli* haemolysin operon by a short upstream element shared by polysaccharide and DNA transfer determinants. Mol. Microbiol. *19*, 705–713.

Ophir, T., and Gutnick, D.L. (1994). A role for exopolysaccharides in the protection of microorganisms from desiccation. Appl. Environ. Microbiol. *60*, 740–745.

Owens, R.M., Pritchard, G., Skipp, P., Hodey, M., Connell, S.R., Nierhaus, K.H., and O'Connor, C.D. (2004). A dedicated translation factor controls the synthesis of the global regulator Fis. EMBO J. *23*, 3375–3385.

Pavelka, M.S., Jr., Hayes, S.F., and Silver, R.P. (1994). Characterization of KpsT, the ATP-binding component of the ABC-transporter involved with the export of capsular polysialic acid in *Escherichia coli* K1. J. Biol. Chem. *269*, 20149–20158.

Perez, J.C., Latifi, T., and Groisman, E.A. (2008). Overcoming H-NS-mediated transcriptional silencing of horizontally acquired genes by the PhoP and SlyA proteins in *Salmonella enterica*. J. Biol. Chem. *283*, 10773–10783.

Petit, C., Rigg, G.P., Pazzani, C., Smith, A., Sieberth, V., Stevens, M., Boulnois, G., Jann, K., and Roberts, I.S. (1995). Region 2 of the *Escherichia coli* K5 capsule gene cluster encoding proteins for the biosynthesis of the K5 polysaccharide. Mol. Microbiol. *17*, 611–620.

Pickard, D., Li, J., Roberts, M., Maskell, D., Hone, D., Levine, M., Dougan, G., and Chatfield, S. (1994). Characterization of defined ompR mutants of Salmonella typhi: ompR is involved in the regulation of Vi polysaccharide expression. Infect. Immun. *62*, 3984–3993.

Rahn, A., Drummelsmith, J., and Whitfield, C. (1999). Conserved organization in the cps gene clusters for expression of *Escherichia coli* group 1 K antigens: relationship to the colanic acid biosynthesis locus and the cps genes from *Klebsiella pneumoniae*. J. Bacteriol. *181*, 2307–2313.

Rahn, A., and Whitfield, C. (2003). Transcriptional organization and regulation of the *Escherichia coli* K30 group 1 capsule biosynthesis (cps) gene cluster. Mol. Microbiol. *47*, 1045–1060.

Reusch, R.N., Shabalin, O., Crumbaugh, A., Wagner, R., Schroder, O., and Wurm, R. (2002). Posttranslational modification of *E. coli* histone-like protein H-NS and bovine histones by short-chain poly-(R)-3-hydroxybutyrate (cPHB). FEBS Lett. *527*, 319–322.

Roberts, I.S. (1996). The biochemistry and genetics of capsular polysaccharide production in bacteria. Annu. Rev. Microbiol. *50*, 285–315.

Roberts, I.S. (2000). Transcriptional organisation and regulation of *E. coli* group 2 capsule expression. Adv. Exp. Med. Biol. *485*, 95–101.

Rowe, S., Hodson, N., Griffiths, G., and Roberts, I.S. (2000). Regulation of the *Escherichia coli* K5 capsule gene cluster: evidence for the roles of H-NS, BipA, and integration host factor in regulation of group 2 capsule gene clusters in pathogenic *E. coli*. J. Bacteriol. *182*, 2741–2745.

Santander, J., Wanda, S.Y., Nickerson, C.A., and Curtiss, R., 3rd (2007). Role of RpoS in fine-tuning the synthesis of Vi capsular polysaccharide in *Salmonella enterica* serotype Typhi. Infect. Immun. 75, 1382–1392.

Schwan, W.R., Beck, M.T., Hultgren, S.J., Pinkner, J., Woolever, N.L., and Larson, T. (2005). Down-regulation of the kps region 1 capsular assembly operon following attachment of *Escherichia coli* type 1 fimbriae to D-mannose receptors. Infect. Immun. 73, 1226–1231.

Simpson, D.A., Hammarton, T.C., and Roberts, I.S. (1996). Transcriptional organization and regulation of expression of region 1 of the *Escherichia coli* K5 capsule gene cluster. J. Bacteriol. *178*, 6466–6474.

Sledjeski, D.D., and Gottesman, S. (1996). Osmotic shock induction of capsule synthesis in *Escherichia coli* K-12. J. Bacteriol. *178*, 1204–1206.

Spory, A., Bosserhoff, A., von Rhein, C., Goebel, W., and Ludwig, A. (2002). Differential regulation of multiple proteins of *Escherichia coli* and Salmonella enterica serovar Typhimurium by the transcriptional regulator SlyA. J. Bacteriol. *184*, 3549–3559.

Stapleton, M.R., Norte, V.A., Read, R.C., and Green, J. (2002). Interaction of the *Salmonella typhimurium* transcription and virulence factor SlyA with target DNA and identification of members of the SlyA regulon. J. Biol. Chem. *277*, 17630–17637.

Stevens, M.P., Clarke, B.R., and Roberts, I.S. (1997). Regulation of the *Escherichia coli* K5 capsule gene cluster by transcription antitermination. Mol. Microbiol. *24*, 1001–1012.

Stock, A.M., Robinson, V.L., and Goudreau, P.N. (2000). Two-component signal transduction. Annu. Rev. Biochem. *69*, 183–215.

Szu, S.C., Li, X.R., Stone, A.L., and Robbins, J.B. (1991). Relation between structure and immunologic properties of the Vi capsular polysaccharide. Infect. Immun. *59*, 4555–4561.

Takeda, S., Fujisawa, Y., Matsubara, M., Aiba, H., and Mizuno, T. (2001). A novel feature of the multistep phosphorelay in *Escherichia coli*: a revised model of the RcsC -> YojN -> RcsB signalling pathway implicated in capsular synthesis and swarming behaviour. Mol. Microbiol. *40*, 440–450.

Tzeng, Y.L., Swartley, J.S., Miller, Y.K., Nisbet, R.E., Liu, L.J., Ahn, J.H., and Stephens, D.S. (2001). Transcriptional regulation of divergent capsule biosynthesis and transport operon promoters in serogroup B Neisseria meningitidis. Infect. Immun. *69*, 2502–2511.

van de Rijn, I. (1983). Streptococcal hyaluronic acid: proposed mechanisms of degradation and loss of synthesis during stationary phase. J. Bacteriol. *156*, 1059–1065.

Verbrugh, H.A., Peterson, P.K., Nguyen, B.Y., Sisson, S.P., and Kim, Y. (1982). Opsonization of encapsulated *Staphylococcus aureus*: the role of specific antibody and complement. J. Immunol. *129*, 1681–1687.

Virlogeux, I., Waxin, H., Ecobichon, C., Lee, J.O., and Popoff, M.Y. (1996). Characterization of the rcsA and rcsB genes from Salmonella typhi: rcsB through tviA is involved in regulation of Vi antigen synthesis. J. Bacteriol. *178*, 1691–1698.

Virlogeux, I., Waxin, H., Ecobichon, C., and Popoff, M.Y. (1995). Role of the viaB locus in synthesis, transport and expression of Salmonella typhi Vi antigen. Microbiology *141 (Pt 12)*, 3039–3047.

Von Loewenich, F.D., Wintermeyer, E., Dumig, M., and Frosch, M. (2001). Analysis of transcriptional control mechanisms of capsule expression in *Neisseria meningitidis*. Int. J. Med Microbiol. *291*, 361–369.

Wacharotayankun, R., Arakawa, Y., Ohta, M., Hasegawa, T., Mori, M., Horii, T., and Kato, N. (1992). Involvement of rcsB in *Klebsiella* K2 capsule synthesis in *Escherichia coli* K-12. J. Bacteriol. *174*, 1063–1067.

Wacharotayankun, R., Arakawa, Y., Ohta, M., Tanaka, K., Akashi, T., Mori, M., and Kato, N. (1993). Enhancement of extracapsular polysaccharide synthesis in *Klebsiella pneumoniae* by RmpA2, which shows homology to NtrC and FixJ. Infect. Immun. *61*, 3164–3174.

Waite, R.D., Penfold, D.W., Struthers, J.K., and Dowson, C.G. (2003). Spontaneous sequence duplications within capsule genes cap8E and tts control phase variation in *Streptococcus pneumoniae* serotypes 8 and 37. Microbiology *149*, 497–504.

Waite, R.D., Struthers, J.K., and Dowson, C.G. (2001). Spontaneous sequence duplication within an open reading frame of the pneumococcal type 3 capsule locus causes high-frequency phase variation. Mol. Microbiol. *42*, 1223–1232.

Waxin, H., Virlogeux, I., Kolyva, S., and Popoff, M.Y. (1993). Identification of six open reading frames in the *Salmonella enterica* subsp. *enterica* ser. Typhi viaB locus involved in Vi antigen production. Res. Microbiol. *144*, 363–371.

Wehland, M., and Bernhard, F. (2000). The RcsAB box. Characterization of a new operator essential for the regulation of exopolysaccharide biosynthesis in enteric bacteria. J. Biol. Chem. *275*, 7013–7020.

Weiser, J.N., Austrian, R., Sreenivasan, P.K., and Masure, H.R. (1994). Phase variation in pneumococcal opacity: relationship between colonial morphology and nasopharyngeal colonization. Infect. Immun. *62*, 2582–2589.

Weiser, J.N., Bae, D., Epino, H., Gordon, S.B., Kapoor, M., Zenewicz, L.A., and Shchepetov, M. (2001). Changes in availability of oxygen accentuate differences in capsular polysaccharide expression by phenotypic variants and clinical isolates of *Streptococcus pneumoniae*. Infect. Immun. *69*, 5430–5439.

Whitfield, C. (2006). Biosynthesis and assembly of capsular polysaccharides in *Escherichia coli*. Annu Rev Biochem 75, 39–68.

Whitfield, C., and Roberts, I.S. (1999). Structure, assembly and regulation of expression of capsules in *Escherichia coli*. Mol. Microbiol. *31*, 1307–1319.

Wilson, R.P., Raffatellu, M., Chessa, D., Winter, S.E., Tukel, C., and Baumler, A.J. (2008). The Vi-capsule prevents Toll-like receptor 4 recognition of Salmonella. Cell Microbiol. *10*, 876–890.

Therapies Directed at *Pseudomonas aeruginosa* Polysaccharides

5

Joanna B. Goldberg

Abstract

Given the critical importance of *Pseudomonas aeruginosa* as an opportunistic pathogen, it is necessary to consider novel targets for therapeutic development. This is especially true as this bacterium is naturally resistant to many antimicrobials and with the over use of antibiotics has become resistant to those it was once sensitive. Thus, there is a real need for new drugs and approaches to combat the myriad of diseases caused by this pathogen. Polysaccharides would appear to be suitable candidates in this regard. They are generally the outermost molecules on the surface of *P. aeruginosa*. Some have roles in adherence and/or biofilm formation. Other polysaccharides are considered virulence factors and are immunodominant antigens. Thus, this group of molecules would seem ideal for the development of therapeutic interventions. Unfortunately, the identification of reagents to specifically inhibit polysaccharides has lagged far behind the obvious need for such drugs.

Introduction

Pseudomonas aeruginosa, while generally considered an innocuous soil-inhabiting bacterium, is also an important opportunistic pathogen. Healthy people are typically able to resist infections with *P. aeruginosa,* but the population that is particularly susceptible is the immunocompromised, including those undergoing chemotherapy and others with immunodeficiencies. These individuals can rapidly succumb to acute respiratory infections and septicaemia. In addition, *P. aeruginosa* is one of the most important pathogens responsible for acute infections following burns, leading to rapid and lethal dissemination. *P. aeruginosa* is also the most prominent cause of bacterial keratitis following eye injury or in contact lens wearers; in this case the infection stays localized to the eye. This Gram-negative bacterium also is the major cause of morbidity and mortality in individuals with the inherited disease, cystic fibrosis (CF) (Strausbaugh and Davis, 2007). The *P. aeruginosa* infections in CF patients are exclusive to the lungs and are both chronic and recurrent.

Here we focus on approaches directed at inhibiting polysaccharide production and expression, and biofilm formation, and thereby the virulence of this bacterium. The topic of *P. aeruginosa* lipopolysaccharide biosynthesis, expression, and as a virulence factor and vaccine candidate has been extensively reviewed recently (Rocchetta *et al.*, 1999; Knirel *et al.*, 2006; Pier, 2007; Kintz and Goldberg, 2008) and will not be covered here.

Alginate

The phenotype of strains of *P. aeruginosa* from acute infections resembles those from the environment. The most apparent difference between these isolates and those found associated with chronic CF lung infections is that the latter produce copious amounts of the exopolysaccharide alginate (Linker and Jones, 1966; Evans and Linker, 1973), which gives these strains a mucoid phenotype (Fig. 5.1). This conversion likely reflects adaptation that allows bacteria to survive in the hostile CF lung environment (Speert *et al.*, 1990).

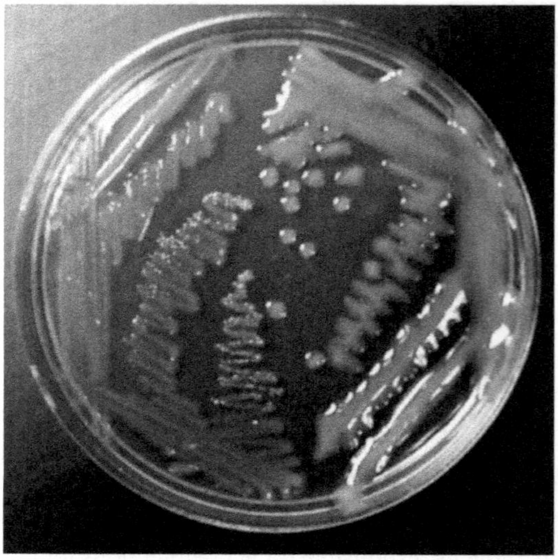

Figure 5.1 *Pseudomonas aeruginosa* growing on rich agar media. The strain on the left hand side of the plate shows the nonmucoid phenotype that is typical of those found in the environment in acute infections, and in early lung infections in cystic fibrosis (CF) patients. The strain on the right hand side of the plate shows the mucoid phenotype due to the over production of the exopolysaccharide alginate. This phenotype is generally unique to strains isolated from chronic lung infections in CF patients.

Alginate is a high-molecular-weight linear polymer of β-1,4 linked β-D-mannuronic acid and its C5 epimer α-L-guluronic acid. In the case of *P. aeruginosa*, the sugars are non-repeating and arranged in either polymannuronic blocks or heteropolymeric mannuronic and guluronic blocks, giving these polysaccharides a flexible structure distinct from polyguluronic blocks, which are rigid and are never found in *P. aeruginosa*. The lack of polyguluronic blocks makes *P. aeruginosa* alginate distinct from that isolated from the bacteria *Azotobacter vinelandii*, which uses alginate in both its vegetative state and for the process of encystment (Sadoff, 1975; Page *et al.*, 2001).

The thick *P. aeruginosa* alginate coat, as is typical for bacterial capsules, confers resistance to phagocytosis (Schwarzmann and Boring, 1971; Baltimore and Mitchell, 1980). It has also be noted that it provides resistance to reactive oxygen species and other immune effectors such as IFN-γ (Leid *et al.*, 2005), as well as a hydrophobic and anionic barrier to protect against dehydration and cationic peptides (Dunne, 2002; Harrison *et al.*, 2005). Alginate has been reported to reduce the chemotaxis of polymorphonuclear leucocytes (PMN) into the CF lungs and by itself to inhibit activation of the complement system (Pedersen *et al.*, 1990). Compared with non-mucoid strains, mucoid strains are inherently more resistant to phagocytosis by PMN and macrophages (Meshulam *et al.*, 1984; Cabral *et al.*, 1987; Krieg *et al.*, 1988). *P. aeruginosa* alginate has also been shown to provide protection from cationic quaternary ammonium compounds, which are biocides (Campanac *et al.*, 2002) and to bind to positively charged antibiotics such the aminoglycosides and inhibit their activity (Nichols *et al.*, 1988).

Using an enzyme that degrades alginate as well as specific monoclonal antibodies, the alginate coat has been shown to have a role in adherence to dacron fibre microcolumns (Mai *et al.*, 1993). However, using gentamicin exclusion assays, others have observed that mucoid *P. aeruginosa* strains associate with and invade CF respiratory epithelial cells less efficiently than non-mucoid strains (Massengale *et al.*, 2000), suggesting that the interaction between alginate on the bacteria and host cells is more complex than simple surface interactions. In fact, the situation is even more complicated *in vivo*, where non-mucoid *P. aeruginosa* strains actually express detectable alginate one hour after infection in the mouse lung environment (Pier *et al.*, 2004).

The biosynthesis and regulation of *P. aeruginosa* alginate production has been studied for many years and has been reviewed recently (Jain and Ohman, 2004; Ramsey and Wozniak, 2005; Muhammadi and Ahmed, 2007). The genes and their functions have been based in large part by the analysis of mutants defective in the production of this polysaccharide. Many of the earliest studies were performed on mucoid *P. aeruginosa* strains that were isolated from chronic CF lung infections (Darzins and Chakrabarty, 1984; Goldberg and Ohman, 1984). Further analysis performed by numerous groups has helped define the pathway for the biosynthesis of alginate (Fig. 5.2).

Interestingly, the alginate genes are not unique to CF isolates and have been identified and localized to the chromosome of the non-mucoid laboratory strain PAO1. Most of the genes

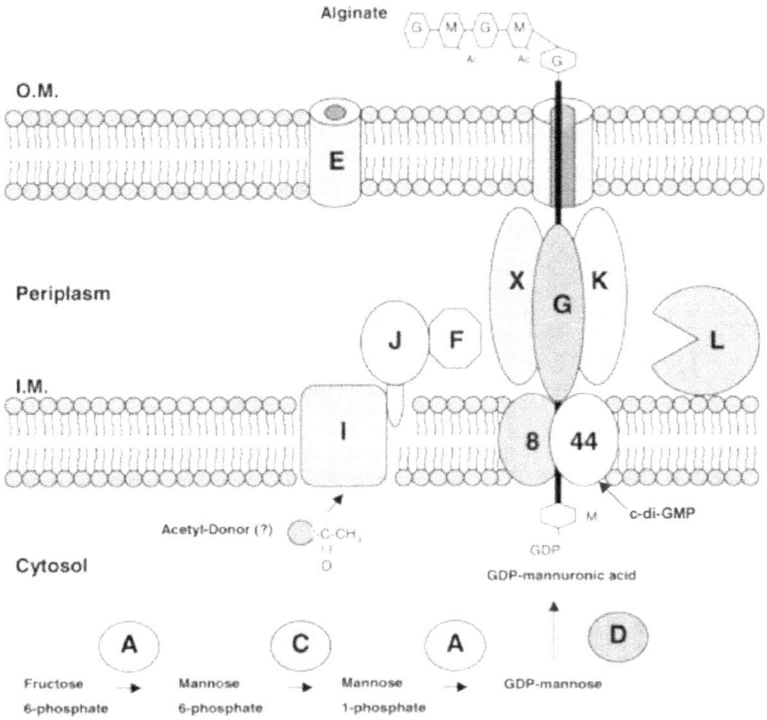

Figure 5.2 Model of alginate synthesis in *P. aeruginosa* modified from Ramsey and Wozniak (Ramsey and Wozniak, 2005). Shown are steps involved in the synthesis of the nucleotide sugar GDP-mannuronic acid and the modification and export through the periplasm via the alginate secretion scaffold. The letters and numbers indicate the alginate biosynthetic proteins. O.M. and I.M. are outer membrane and inner membrane, respectively. G and M are the saccharides guluronic acid and mannuronic acid, respectively. Ac represents the addition of acetyl groups to particular mannuronic acid residues. The interaction of Alg44 with the second messenger cyclic di-GMP (Merighi *et al.*, 2007) is also shown.

encoding the proteins for the synthesis of alginate are clustered in a 12-gene operon (PA3540-PA3551), referred to as the *algD* operon (Fig. 5.3a). These include the enzymes in the pathway for the synthesis of the precursors as well as the proteins that form a complex for the polymerization, epimerization, acetylation, and export of the alginate polymer (Jain and Ohman, 2004; Muhammadi and Ahmed, 2007). The exception is the *algC* gene, which is located to a distinct region of the genome (PA5322).

Biosynthesis

For the generation of the nucleotide sugar precursor GDP-mannuronic acid, carbon enters the alginate system via the Entner–Doudoroff pathway (Banerjee *et al.*, 1985). Fructose-6-phosphate is converted to mannose-6-phosphate through the action of phosphomannose isomerase, encoded by the *algA* gene (Darzins *et al.*, 1985; Gill *et al.*, 1986).

Mannose-6-phosphate is converted to mannose-1-phosphate by phosphomannomutase encoded by the *algC* gene (Zielinski *et al.*, 1991). As mentioned, this gene is located outside the *algD* operon. AlgC is a bifunctional enzyme with both phosphomannomutase and phosphoglucomutase. The phosphomannomutase activities is involved in alginate biosynthesis, while its phosphoglucomutase activity responsible for the synthesis of the lipopolysaccharide core (Coyne *et al.*, 1994).

The AlgA protein is also a bifunctional enzyme that has an additional GDP-mannose pyrophosphorylase activity, which converts mannose-1-phosphate to GDP-mannose (Shinabarger *et al.*, 1991). In the final reaction, GDP-mannose is oxidized to GDP-mannuronic

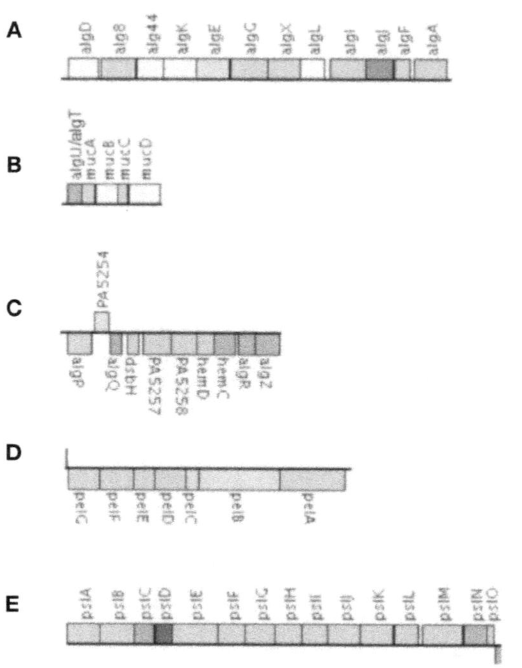

Figure 5.3 Genetic loci involved in *P. aeruginosa* polysaccharide production. Diagrams were obtained from www.pseudomonas.com. Genes above the line are transcribed from left to right, those below are transcribed right to left. **A** The *algD* operon (PA3540-PA3551). **B** The alginate regulatory operon (PA0762–PA0766). **C** The *algR* regulatory region (PA5253-PA5261). **D** The *pel* operon (PA3058-PA3064). **E** The *psl* operon (PA2231–PA2245).

by GDP-mannose dehydrogenase encoded by the *algD* gene (Shinabarger *et al.*, 1991), representing the committed step in alginate biosynthesis (May *et al.*, 1991) (Fig. 5.2).

Secretion and polymerization of mannuronic acid residues across the inner membrane is due to the action of the alginate gene product Alg8, which has a membrane-spanning as well as a glycosyltransferase domain, making this the likely alginate polymerase. Alg8 acts along with Alg44, which appears to be a membrane anchor protein (Oglesby *et al.*, 2008). Interestingly Alg44 has a PilZ domain and has been shown to bind to the ubiquitous second messenger cyclic di-GMP (D'Argenio and Miller, 2004; Jenal, 2004) and that this binding is essential for alginate production (Merighi *et al.*, 2007). This finding suggests how this protein regulates the polymerization and/or transport of the polysaccharide (Fig. 5.2).

Poly-mannuronic acid is the initial polymer of alginate that is made. It undergoes further modifications by the periplasmic protein AlgG, which is a C-5 epimerase that converts some of the D-mannuronic acids to L-guluronic acids (Franklin *et al.*, 1994) and also has a role in secretion (Jain *et al.*, 2003). Interestingly, the regulation of epimerization and therefore the ratio the amount of mannuronic acid residues to guluronic residues is not well understood, but has an obvious impact on the final alginate structure. The AlgIJF proteins *O*-acetylate the mannuronic acids at the C-2 and/or C-3 position (Franklin and Ohman, 2002). The acetylation of alginate also impacts a number of physical properties of the polysaccharide. In addition, there is a role for alginate acetylation in protection against the opsonic killing in the context of CF. *P. aeruginosa* with non-acetylated alginate is readily killed by normal human serum, while a wild-type strain with normal levels of acetylated alginate is resistant to killing by complement in the absence of antibodies (Pier *et al.*, 2001).

The roles of the *algK*, *algG*, and *algX* gene products have been determined by mutant studies. It has been suggested that together these form a periplasmic scaffold to bring the polymers to the alginate-specific outer membrane porin AlgE, while protecting it from the action of the alginate lyase AlgL (Jain *et al.*, 2003) (Fig. 5.2).

Regulation

The regulation of alginate production is complex. The alginate genes are normally silent and highly regulated, and under hierarchical control. The *algD* promoter controls the expression of the alginate biosynthetic operon. The key regulator is the alternative sigma factor AlgT (also known as AlgU or σ22) (Flynn and Ohman, 1988; Deretic *et al.*, 1994; Yu *et al.*, 1995) which induces the expression of both itself (DeVries and Ohman, 1994; Schurr *et al.*, 1995) and *algD* at the consensus σ22 promoter sites (Hershberger *et al.*, 1995). The *algT* gene is in an operon with four other genes (PA0762-PA0766), encoding MucA, MucB, MucC, and MucD (Fig. 5.3b). MucA is an inner membrane protein that appears to be a major regulator; it is an anti-sigma factor that binds to and sequesters AlgT and thereby inhibits its ability to initiate transcription. MucB is also a negative regulator that binds to MucA and helps to keep transcription by AlgT low. Interesting, mutations in *mucA* are those most typically found to be responsible for the conversion to the mucoid phenotype observed in CF isolates. One common mutation is *mucA22*, which is a deletion of a single base within a stretch of G residues that results in a truncated form of MucA (Boucher *et al.*, 1997). Recently a number of single nucleotide polymorphisms were identified within the *mucABD* locus in clinical and environmental strains of *P. aeruginosa* (Bragonzi *et al.*, 2006). Even within this varied group of strains, it was found that the *mucA22* allele was prevalent. It has been suggested that the deletion of a G residue in a homopolymeric tract is due to exposure of *P. aeruginosa* to reactive oxygen species in the context of biofilms in the CF lung environment (Mathee *et al.*, 1999). However, it is interesting to note that conversion to the non-mucoid phenotype in these mucoid strains is not due to the restoration of the wild-type *mucA* allele, but rather is due to spontaneous mutations in the *algT* gene (DeVries and Ohman, 1994), suggesting an advantage to this mechanism of alginate suppression. The roles of MucC and MucD are less clear, however MucD has homology to *Escherichia coli* HtrA (DegP), which is a periplasmic serine protease that appears to have a role in degradation of abnormal proteins and resistance to heat and oxidative stress (Boucher *et al.*, 1996).

AlgT also induces the expression of the key regulatory genes, *algR*, *algB*, and *armZ*. The *algR* gene is part of the *algR* regulatory region (Fig. 5.3c) and encodes a two-component regulatory protein that binds to two segments far upstream of the *algD* promoter and another site just upstream of the transcriptional start site (Mohr *et al.*, 1990; Kato and Chakrabarty, 1991; Mohr *et al.*, 1991; Mohr *et al.*, 1992). AlgR also binds upstream and downstream of the *algC* transcriptional start site (Fujiwara *et al.*, 1993). Just upstream of the *algR* gene is *algZ* (also called *fimS*), which encodes the sensor component that interacts with AlgR (Whitchurch *et al.*, 1996; Yu *et al.*, 1997). Interestingly, AlgZ is not required for alginate production, suggesting that alginate production controlled by AlgR is phosphorylation independent (Ma *et al.*, 1998). Also, within this region is *algQ* (*algR2*) a positive activator of *algD* transcription (Kato *et al.*, 1989; Konyecsni and Deretic, 1990).

The *algB* gene (PA5483) encodes a two component regulatory protein (Wozniak and Ohman, 1991; Goldberg and Dahnke, 1992). AlgB binds upstream of the *algD* promoter (Leech *et al.*, 2008). Immediately downstream of *algB* is *kinB*, which appears to be the cognate environmental sensor of AlgB (Ma *et al.*, 1997). While AlgB appears to be an NtrC-type regulator, it is distinct in that the alternative sigma factor RpoN, which controls other NtrC-type two component regulators, has no effect on alginate production (Mohr *et al.*, 1990; Totten *et al.*, 1990). A microarray analysis showed many genes require both *algB* and *kinB* for maximal expression, but interestingly mutation in *kinB* does not affect the mucoid phenotype, suggesting that these genes have a more global impact on gene expression that simply on alginate regulation (Leech *et al.*, 2008).

AmrZ (PA3385) is another activator of *algD* transcription (Baynham *et al.*, 2006). This protein (which was previously also known as *algZ*) (Baynham and Wozniak, 1996; Baynham *et al.*, 1999) binds upstream of the *algD* transcriptional start site. It is under the control of AlgT and is a homologous ribbon–helix–helix family of DNA-binding proteins.

The global regulator integrative host factor (IHF) has been shown to play a role in the expression of *algD* (Wozniak and Ohman, 1993; Delic-Attree *et al.*, 1995). Similarly AlgP (Fig. 5.3c) (also known as AlgR3) is a histone-like protein that is required for *algD* expression (Kato *et al.*, 1990; Konyecsni and Deretic, 1990).

More recently a number of proteases have been shown to be directly or indirectly involved in the regulation of alginate biosynthesis (Qiu *et al.*, 2007; Qiu *et al.*, 2008). In some cases, these have been shown to have an effect on degradation of the MucA protein.

Conditions affecting alginate expression

The conversion of non-mucoid *P. aeruginosa* to a mucoid form occurs virtually exclusively during chronic infection in the lungs of patients with CF suggesting that this niche induces alginate production (Doggett *et al.*, 1964). However, the exact conditions responsible for this are not completely understood. Treatment of non-mucoid *P. aeruginosa* PAO1 with activated PMNs results in conversion to the mucoid phenotype (Mathee *et al.*, 1999). *In vitro* the environmental conditions inducing alginate production, including high osmolarity and dehydration, appear to vary depending on the *P. aeruginosa* strain (Berry *et al.*, 1989; DeVault *et al.*, 1990, 1991; Terry *et al.*, 1991; Zielinski *et al.*, 1992). In fact, it has been observed the subinhibitory concentrations of some antibiotics actually increased alginate gene transcription (Bagge *et al.*, 2004). Recent elegant studies of Wood *et al.* revealed that cell wall-perturbing antibiotics, in particular, induced gene expression of alginate genes regulated by *algT* (σ 22) (Wood *et al.*, 2006), begging the question of the efficacy of using these agents for treatment of *P. aeruginosa* infection. In addition, when non-mucoid *P. aeruginosa* are grown under anaerobic conditions there is an increase in alginate gene transcription and polysaccharide expression both *in vitro* and *in vivo* suggesting that this state may contribute to early persistence in the CF lung environment (Bragonzi *et al.*, 2005). Understanding the regulation of alginate production will be critical in the development of new approaches to inhibit production.

Specifically targeting alginate

As alginate is such an important component of the outer surface in the context of chronic *P. aeruginosa* lung infections in CF, it makes an ideal target for therapeutic interventions. The same group that first identified alginate on the surface of *P. aeruginosa* (Linker and Jones, 1966; Evans and Linker, 1973) recognized enzymes that could degrade this polysaccharide (Linker and Evans, 1984). As mentioned, an alginase has been shown to decrease adherence of *P. aeruginosa* to an abiotic surface (Mai *et al.*, 1993). Others have found that an aminoglycoside diffusion barrier generated by the addition of alginate could be disrupted with a preparation of *P. aeruginosa* alginate lyase (Hatch and Schiller, 1998). Most importantly, treatment of bacterial infections caused by mucoid strains of *P. aeruginosa* with alginate lyase has been shown to improve the efficacy of antibiotics (Bayer *et al.*, 1992). Altogether these finding suggested a role for alginate lyase as an adjunct to antibiotic treatment. However, the immunogenicity of administration this enzyme to the lungs of CF patients would need to be addressed.

With respect to directly inhibiting alginate synthesis, the crystal structure of two alginate enzymes has set the stage for the development of drugs to specifically inhibit these targets. The structure of the *algD* gene product indicates that it is a dimer with sites for binding to the cofactor NAD(H) and the product GDP-mannuronic acid. Each of the polypeptide chains contributes to the active site of this enzyme (Snook *et al.*, 2003). Further studies of the enzymatic activity of AlgD have indicated that penicillic acid is an irreversible inactivator of GDP-mannose dehydrogenase suggesting the importance of cysteine residues in the active site of this protein (Kimmel and Tipton, 2005).

The enzyme phosphomannomutatase/phosphoglucomutase has also been crystallized. The enzyme has four relatively compact domains that form a 'heart' shape. Each of these domains contributes to the formation of a large active site in the protein (Regni *et al.*, 2002). Site directed mutagenesis has revealed the importance of the 'hinge regions' linking the domains of this protein that form a lid over the active site of this protein (Schramm *et al.*, 2008). While molecules that inhibit this enzyme have been described (Liu *et al.*, 2004; Regni *et al.*, 2006), none of these has been tested to see whether they specifically inhibit *P. aeruginosa* alginate production.

Owing to its extracellular location alginate would appear to represent a major antigenic target for *P. aeruginosa*. However, problems have been found when trying to use alginate as an immunogen for vaccine development. As is typical for many purified polysaccharides, alginate is poorly immunogenic. During chronic infection patients with CF actually make antibodies to alginate, which are obviously ineffective at clearing the infection. Studies have shown that these antibodies are non-opsonic and failed to mediate killing of *P. aeruginosa in vitro* (Pier *et al.*, 1987). It has been suggested that if opsonic antibodies to alginate could be generated, they would be a valuable therapeutic agent. Opsonic, but not non-opsonic, antibodies to alginate protect animals against chronic endobronchial infection (Pier *et al.*, 1990). However, when purified alginate has injected into healthy volunteers only a small proportion produced an increased level of opsonic antibodies to alginate (Pier, 2005).

Conjugation of purified alginate to protein carriers has improved the immunogenicity in animals (Cryz *et al.*, 1991; Theilacker *et al.*, 2003), but studies in humans have not been reported. There are also problems associated with identifying the epitopes within alginate that will elicit a protective antibody response. While antibodies to specific epitopes can mediate killing of mucoid strains *in vitro* (Pier *et al.*, 1983), antibodies to alginate produced during chronic lung infection do not mediate killing (Meluleni *et al.*, 1995). It has been suggested that the random arrangement of the two monosaccharides in alginate as well as the variation in the amount and placement of the acetyl groups makes it difficult to define the appropriate epitopes.

Fully human monoclonal antibodies to alginate have been obtained that are opsonic and protective against both mucoid and non-mucoid strains in a mouse model of acute pneumonia. The target of these antibodies was identified as blocks of unacetylated mannuronic acid residues (Pier *et al.*, 2004). One of these fully humanized IgG1 monoclonal antibodies (F429) to *P. aeruginosa* alginate has been recently used in a completely different model: infection after eye injury (Zaidi and Pier, 2008). Topical application of this antibody showed protection both prior to as well as subsequent to injury and infection, suggesting its potential use for *P. aeruginosa*-induced ulcerative keratitis. While alginate overproducing strains are not found in eye infections, the presence of this antigen on the bacterial cell surface apparently makes it a target for these antibodies. Additionally or alternatively, this antibody may also block adherence to the corneal epithelial cell surface and/or invasion into host tissues. This antibody would appear to have real potential either alone or as an adjunct to antibiotic therapy to treat this type of *P. aeruginosa* infection.

Polysaccharides and biofilm formation

Bacterial biofilms have been defined as surface-attached communities of cells encased within a self-produced extracellular matrix. Biofilm synthesis proceeds via a number of well-programmed and organized steps, including initial surface attachment, the formation of these three-dimensional structures, and finally, the development of a mature biofilm (Costerton *et al.*, 1999; O'Toole and Kolter, 1998a; Klausen *et al.*, 2003).

The overproduction of alginate has generally been considered a hallmark of *P. aeruginosa* biofilms. Many years ago 'microcolonies' or clusters of mucoid strains were detected in the lungs of CF patients (Lam *et al.*, 1980; Baltimore *et al.*, 1989). It is now appreciated that this 'microcolony' formation is a step in biofilm development. Alginate has been recognized as a critical component in these structures as degradation by alginate lyase has aided in the effectiveness of antibiotics to kill biofilm grown *P. aeruginosa* (Alkawash *et al.*, 2006) and prevented microcolony formation (Tielen *et al.*, 2005).

With the advent of an easy high-throughput assay that monitors biofilm production (O'Toole and Kolter, 1998b), many genes involved in the expression of this phenotype have been identified in *P. aeruginosa*. Together with more sophisticated techniques, the pathway for biofilm development has been defined and characterized.

Bacterial biofilms appear to be highly resistant to killing by phagocytes (Meluleni *et al.*, 1995) and neutrophils (Jesaitis *et al.*, 2003). Compared with those grown under planktonic conditions, biofilm-grown *P. aeruginosa* show reduced activation of complement compared with planktonic bacteria (Jensen *et al.*, 1993). Altogether the persistence of bacteria in these biofilms results in the production of 'frustrated phagocytes' that damage host cells (Costerton *et al.*, 1999; Davies, 2003; Jesaitis *et al.*, 2003).

With respect to resistance to antimicrobial agents, it has been found that a mucoid strain had a more highly organized biofilm structure that was more resistant to tobramycin than that formed by an isogenic non-mucoid strain (Hentzer *et al.*, 2001). It has been suggested that bacteria within biofilms are more antibiotic resistant due to the protective shielding afforded by the biofilm matrix (Gilbert *et al.*, 1990; Costerton *et al.*, 1999; Gilbert *et al.*, 2002; Worlitzsch *et al.*, 2002). However, the exact mechanism responsible for this increased resistance is not clear and may be different under different conditions (Yoon *et al.*, 2002; Tart and Wozniak, 2008).

Other polysaccharides

Alginate overproduction is considered pathognomonic with lung infections in CF. Also, alginate expression was one of the first obvious components defining *P. aeruginosa* biofilms. However, with the observations made over the last 5 years that even non-mucoid, alginate non-producing strains could make biofilms, it became clear that the definition of biofilms being composed exclusively of alginate was overly simplistic (Wozniak *et al.*, 2003). Even non-mucoid strains from CF patients appeared to be capable of forming biofilms (Lee *et al.*, 2005), indicating that other polysaccharides are involved in biofilm formation in *P. aeruginosa*. Also, strains deficient in lectins showed biofilm defects (Tielker *et al.*, 2005; Diggle *et al.*, 2006). It should be noted that both bacterial and host DNA and proteins can also contribute and enhance the biofilm phenotype (Whitchurch *et al.*, 2002; Matsukawa and Greenberg, 2004; Walker *et al.*, 2005). Altogether these observations lead to the search for additional components of biofilms.

The *pel* locus was identified as being involved in the formation of a pellicle in *P. aeruginosa* PA14, a strain that was fist characterized for its virulence in both mammalian and plant systems (Rahme *et al.*, 1995). Friedman and Kolter screened a transposon library of PA14 for those mutants unable to make a pellicle, and identified a seven-gene cluster. The homologous genes were identified and highly conserved to genes PA3058-PA3064 on the PAO1 chromosome (Fig. 5.3d). For the most part, the genes are similar to those encoding enzymes involved in polysaccharide synthesis (Friedman and Kolter, 2004a,b). A similar screen of a non-piliated derivative of strain PAK also identified the *pel* locus (Vasseur *et al.*, 2005). Currently the structure of Pel is only reported as a glucose-rich matrix, which is distinct from cellulose (Friedman and Kolter, 2004a).

A second polysaccharide locus was recognized as essential for biofilm formation in PAO1 and ZK2870 (Friedman and Kolter, 2004b; Jackson *et al.*, 2004; Overhage *et al.*, 2005; Ma *et al.*, 2006). This locus, *psl*, encodes a 15-gene operon (PA2231-PA2245) (Fig. 5.3e), which encodes enzymes involved in polysaccharide synthesis. In particular, this polysaccharide appears to have a role in adhesin and maintaining the biofilm structure (Ma *et al.*, 2006). Psl is a mannose-rich and galactose-rich polysaccharide (Ma *et al.*, 2007).

While the determination of the exact structure and biosynthetic pathways of both Pel and Psl polysaccharides remains an active area of many laboratories, it has been suggested that these distinct polysaccharides (and possibly other matrix components) together contribute to biofilm formation and persistence prior to (or concurrent with) the conversion to the alginate producing chronically infecting form of *P. aeruginosa* (Ramsey and Wozniak, 2005). Supporting this is the findings that *pel* and *psl* genes are commonly regulated. In particular, both are regulated by cyclic di-GMP (D'Argenio and Miller, 2004; Goodman *et al.*, 2004; Hickman *et al.*, 2005; Ventre *et al.*, 2006), which itself is regulated by the activities of diguanylate cyclase (GGDEF) and phosphodiesterase (EAL) proteins.

Compounds to disrupt or disperse biofilms or prevent their formation

As the biofilm mode of growth of *P. aeruginosa* is so critical for initiating infections, particularly in the lung, it has been suggested as a rational target for therapeutic development; antimicrobials alone have been shown to be relatively ineffective at killing *P. aeruginosa* in a biofilm. *P. aeruginosa* strains from CF patients grown *in vitro* as adherent monolayers and as biofilms were less susceptible to bactericidal killing by individual antibiotics compared with those grown planktonically and biofilm-grown bacteria were more resistant to combinations of antibiotics (Aaron *et al.*, 2002). It has been speculated that the biofilm matrix may be responsible for this recalcitrance to antibiotic treatments. Incubation of *P. aeruginosa* with subinhibitory concentrations of the macrolide antibiotic, clarithromycin, altered the structure and architecture of the *P. aeruginosa* biofilm, but did not inhibit the formation of the biofilm (Wozniak and Keyser, 2004). In other studies, azithromycin inhibited biofilm formation but did not affect previously formed mature biofilms (Gillis and Iglewski, 2004). Further studies showed that azithromycin inhibited the products involved in biofilm formation in the laboratory strains PAO1 but this was not consistent among the clinical isolates of *P. aeruginosa* that were tested (Wagner *et al.*, 2005). Interestingly, phenotypic variants of *P. aeruginosa* with an enhanced ability to form biofilms that are more resistant to antimicrobials may be selected for in the CF lung by the therapy itself (Drenkard and Ausubel, 2002). In other cases it has been found that subinhibitory concentrations of aminoglycoside antibiotics actually induced biofilm formation in *P. aeruginosa*. The tobramycin-induced biofilm formation was inhibited by exogenous GTP, which is known to inhibit EAL protein activity suggesting the regulation by cyclic di-GMP (Hoffman *et al.*, 2005).

Since biofilm organisms are generally considered more resistant to innate immune effector and antimicrobial treatment than the same bacteria under planktonic conditions, it is reasonable to assume that if a method was developed to inhibit biofilm formation, *P. aeruginosa* infections could be easier for the host to clear and physicians to treat. Towards this end, a number of different approaches based on the physiology surrounding biofilm formation and maintenance have been attempted.

Initiation of a biofilm depends on the ability of *P. aeruginosa* to attach to a surface, and conditions that disrupt this would prevent the establishment of a biofilm. In this regard, the chelation of iron with lactoferrin, a ubiquitous and abundant component of human secretions and an innate immune effector, overstimulates twitching motility and thwarts surface contacts and thus biofilm formation (Singh *et al.*, 2002).

Biofilm dissolution and cell detachment has been observed in *P. aeruginosa* and may represent a progression in natural history of biofilm development, whereby environmental cues signal some bacteria leave the safety and security of the biofilm and return to their planktonic form to find new niches. A better understanding of this process of may provide hints as to how to induce *P. aeruginosa* to degrade its own biofilm and thus make this pathogen more susceptible to conventional treatment therapies.

The proteome of *P. aeruginosa* undergoing dispersion resembles that of planktonically growing cells rather than those of biofilm grown cells (Sauer *et al.*, 2002). Increased carbon substrate concentration can trigger immediate large-scale release of cells from the biofilm (Sauer *et al.*, 2004). The extent of detachment seems to vary depending on the carbon sources. This effect was inhibited by protein phosphatase inhibitors, suggesting that dephosphorylation is critical for this detachment process.

A number of different distinct molecules have been found to induce biofilm dispersion. Barraud *et al.* (2006) demonstrated that sublethal concentrations nitric oxide (NO) caused *P. aeruginosa* biofilm dissolution and an NO reductase mutant had increased biofilm dispersion. NO is a biological signalling molecule in eukaryotic systems regulating diverse processes including apoptosis, and cell proliferation and differentiation. This suggests commonalities in the role of NO in the development of eukaryotes and biofilm formation. These findings further indicate that a thorough understanding of its effects on both host and pathogen must be considered before NO (or any inhibitor) can be considered as a therapeutic agent.

Interestingly, the chelator EDTA also induced dispersal of biofilms. In combination with gentamicin, EDTA treatment resulted in complete killing of *P. aeruginosa* in biofilms. Divalent cations magnesium, calcium, and iron protected against this EDTA-induced biofilm dispersion (Banin *et al.*, 2006). This finding fits well with the observed effects of lactoferrin, which also binds iron, and inhibits biofilm formation (Singh *et al.*, 2002).

Because iron is so important for bacterial growth, molecules that disrupt iron metabolism could be used as antimicrobial agents. Gallium is chemically similar to iron and can substitute for iron in many biological process. It was found that gallium not only inhibits *P. aeruginosa* growth, but also biofilm formation, killing both biofilm and planktonic cells. and may provide a novel therapy for *P. aeruginosa* infections (Kaneko *et al.*, 2007).

The biosurfactant, rhamnolipid, has been shown to induce the detachment of *P. aeruginosa* from biofilms (Boles *et al.*, 2005). Expression of this molecule by *P. aeruginosa* is likely part of the natural process of detachment that provides the bacteria the ability to leave a particular location when the conditions become unfavourable.

A search for *P. aeruginosa* gene products that could trigger biofilm dispersion identified a protein, BdlA. A *bdlA* mutant was dispersion-deficient and a *bdlA* biofilm was more resistant to the effects of hydrogen peroxide. This biofilm also had increased adherence and increased levels of cyclic di-GMP, suggesting a link between environmental sensing via this second message and biofilm disruption (Morgan *et al.*, 2006).

Quorum sensing inhibitors as antagonists of biofilms

In vitro biofilm formation is dependent on the quorum-sensing system (Davies et al., 1998), a mechanism of communication used by a bacterium to sense and response to itself or other species within its immediate environment (Davies, 2003). The quorum-sensing molecules (including both acyl homoserine lactones and AI-2) have been found to be present in sputum of CF patients (Singh *et al.*, 2000; Duan and Surette, 2007) suggesting their importance in the CF lung environment. Due to its effect on gene expression (De Kievit and Iglewski, 1999; Parsek and Greenberg, 1999; Schuster et al., 2003; Wagner et al., 2003), quorum-sensing signalling may come into play at multiple stages of biofilm formation from initiation, promoting motility and/or attachment to surfaces, biofilm structure, controlling the size of the biofilm, inducing detachment and dispersion (Irie and Parsek, 2008). For *P. aeruginosa*, strains mutated for the production of quorum-sensing molecules have shown increased sensitivity to SDS, tobramycin, and H_2O_2 (Davies *et al.*, 1998; Bjarnsholt *et al.*, 2005). While difficult to separate from its role in biofilm formation, quorum sensing also has a role in virulence (Parsek and Greenberg, 2000).

Quorum sensing is important at distinct steps in biofilm development, leading to the identification of new targets for *P. aeruginosa* biofilm inhibition. Furanone derivatives have been synthesized and shown to inhibit both *P. aeruginosa* quorum sensing itself and biofilm formation (Kim *et al.*, 2008). Others have developed high-throughput assays to screen compound libraries that inhibit quorum sensing. Interestingly, this screen identified two compounds that resembled the acyl homoserine lactone quorum-sensing molecules, and had a general effect in quorum sensing regulation, suggesting their mechanism of action (Muh *et al.*, 2006b). This same group recognized inhibitors that are structurally unrelated to quorum sensing molecules (Muh *et al.*, 2006a). It is highly anticipated that these quorum-sensing inhibitors will have a general effect on bacterial pathogenesis, but particularly biofilm formation and therefore could be used alone or as an adjunct to antimicrobial therapies to promote clearance and reduction in *P. aeruginosa* infections (Rasmussen and Givskov, 2006).

Conclusions and perspectives

The emergence of the mucoid phenotype of *P. aeruginosa* is coincident with a poor outcome in CF patients (Pedersen *et al.*, 1992; Strausbaugh and Davis, 2007). Overproduction of the polysaccharide alginate is responsible for this phenotype. Alginate expression is normally silent and multiple layers of regulation control alginate production. While specific environmental conditions found in the CF lung induce the conversion to the mucoid phenotype, little is known about what these conditions are or how to control them. Reagents to inhibit this conversion or interfere for alginate regulatory networks have not been reported.

While the alginate biosynthetic pathway is relatively well understood and crystal structures have been determined for some enzymes, specific reagents to inhibit alginate production are still out of reach. Enzymes that degrade alginate have been recognized and they may have a role as an adjunct to antibiotic therapy. And antibodies to alginate appear efficacious in various clinical settings.

It is now appreciated that *P. aeruginosa* biofilms are not simply composed of the alginate, but also other polysaccharides including Pel and Psl. While the genes required for Pel and Psl expression are known, the structures of these polysaccharides have not been reported. Once determined, this will help define the functions of the enzymes involved in the synthesis of these polysaccharides; these enzymes would represent rational targets for drug development to inhibit biofilm formation. Also the polysaccharides themselves may be good candidates for vaccine development.

While the steps in biofilm formation are becoming understood, the physiology of how *P. aeruginosa* grow in the biofilm or what promotes their formation is still a topic of great interest and must be considered as new antimicrobials are developed (Hassett *et al.*, 2002). Inhibiting various steps in the natural history of biofilms progression or inducing their dispersion may allow the host innate immune system or antibiotics to be more effective. The control of biofilm formation by quorum sensing molecules also represents a reasonable avenue of investigation to inhibit this mode of growth and thereby the virulence of this important opportunistic pathogen.

Acknowledgement

This work is supported by a grant from the National Institutes of Health (R01 AI068112).

References

Aaron, S.D., Ferris, W., Ramotar, K., Vandemheen, K., Chan, F., and Saginur, R. (2002). Single and combination antibiotic susceptibilities of planktonic, adherent, and biofilm-grown *Pseudomonas aeruginosa* isolates cultured from sputa of adults with cystic fibrosis. J. Clin. Microbiol. *40*, 4172–4179.

Alkawash, M.A., Soothill, J.S., and Schiller, N.L. (2006). Alginate lyase enhances antibiotic killing of mucoid *Pseudomonas aeruginosa* in biofilms. APMIS *114*, 131–138.

Bagge, N., Schuster, M., Hentzer, M., Ciofu, O., Givskov, M., Greenberg, E.P., and Hoiby, N. (2004). *Pseudomonas aeruginosa* biofilms exposed to imipenem exhibit changes in global gene expression and beta-lactamase and alginate production. Antimicrob. Agents Chemother. *48*, 1175–1187.

Baltimore, R.S., Christie, C.D., and Smith, G.J. (1989). Immunohistopathologic localization of *Pseudomonas aeruginosa* in lungs from patients with cystic fibrosis. Implications for the pathogenesis of progressive lung deterioration. Am. Rev. Resp. Dis. *140*, 1650–1661.

Baltimore, R.S., and Mitchell, M. (1980). Immunologic investigations of mucoid strains of *Pseudomonas aeruginosa*: comparison of susceptibility to opsonic antibody in mucoid and nonmucoid strains. J. Infect. Dis. *141*, 238–247.

Banerjee, P.C., Vanags, R.I., Chakrabarty, A.M., and Maitra, P.K. (1985). Fructose 1,6-bisphosphate aldolase activity is essential for synthesis of alginate from glucose by *Pseudomonas aeruginosa*. J. Bacteriol. *161*, 458–460.

Banin, E., Brady, K.M., and Greenberg, E.P. (2006). Chelator-induced dispersal and killing of *Pseudomonas aeruginosa* cells in a biofilm. Appl. Environ. Microbiol. *72*, 2064–2069.

Barraud, N., Hassett, D.J., Hwang, S.H., Rice, S.A., Kjelleberg, S., and Webb, J.S. (2006). Involvement of nitric oxide in biofilm dispersal of *Pseudomonas aeruginosa*. J. Bacteriol. *188*, 7344–7353.

Bayer, A.S., Park, S., Ramos, M.C., Nast, C.C., Eftekhar, F., and Schiller, N.L. (1992). Effects of alginase on the natural history and antibiotic therapy of experimental endocarditis caused by mucoid *Pseudomonas aeruginosa*. Infect. Immun. *60*, 3979–3985.

Baynham, P.J., Brown, A.L., Hall, L.L., and Wozniak, D.J. (1999). *Pseudomonas aeruginosa* AlgZ, a ribbon-helix-helix DNA-binding protein, is essential for alginate synthesis and *algD* transcriptional activation. Mol. Microbiol. *33*, 1069–1080.

Baynham, P.J., Ramsey, D.M., Gvozdyev, B.V., Cordonnier, E.M., and Wozniak, D.J. (2006). The *Pseudomonas aeruginosa* ribbon-helix-helix DNA-binding protein AlgZ (AmrZ) controls twitching motility and biogenesis of type IV pili. J. Bacteriol. *188*, 132–140.

Baynham, P.J., and Wozniak, D.J. (1996). Identification and characterization of AlgZ, an AlgT-dependent DNA-binding protein required for *Pseudomonas aeruginosa algD* transcription. Mol. Microbiol. *22*, 97–108.

Berry, A., DeVault, J.D., and Chakrabarty, A.M. (1989). High osmolarity is a signal for enhanced *algD* transcription in mucoid and nonmucoid *Pseudomonas aeruginosa* strains. J. Bacteriol. *171*, 2312–2317.

Bjarnsholt, T., Jensen, P.O., Burmolle, M., Hentzer, M., Haagensen, J.A., Hougen, H.P., Calum, H., Madsen, K.G., Moser, C., Molin, S., Høiby, N., and Givskov, M. (2005). *Pseudomonas aeruginosa* tolerance to tobramycin, hydrogen peroxide and polymorphonuclear leukocytes is quorum-sensing dependent. Microbiology *151*, 373–383.

Boles, B.R., Thoendel, M., and Singh, P.K. (2005). Rhamnolipids mediate detachment of *Pseudomonas aeruginosa* from biofilms. Mol. Microbiol. *57*, 1210–1223.

Boucher, J.C., Martinez-Salazar, J., Schurr, M.J., Mudd, M.H., Yu, H., and Deretic, V. (1996). Two distinct loci affecting conversion to mucoidy in *Pseudomonas aeruginosa* in cystic fibrosis encode homologs of the serine protease HtrA. J. Bacteriol. *178*, 511–523.

Boucher, J.C., Yu, H., Mudd, M.H., and Deretic, V. (1997). Mucoid *Pseudomonas aeruginosa* in cystic fibrosis: characterization of *muc* mutations in clinical isolates and analysis of clearance in a mouse model of respiratory infection. Infect. Immun. *65*, 3838–3846.

Bragonzi, A., Wiehlmann, L., Klockgether, J., Cramer, N., Worlitzsch, D., Doring, G., and Tummler, B. (2006). Sequence diversity of the *mucABD* locus in *Pseudomonas aeruginosa* isolates from patients with cystic fibrosis. Microbiology *152*, 3261–3269.

Bragonzi, A., Worlitzsch, D., Pier, G.B., Timpert, P., Ulrich, M., Hentzer, M., Andersen, J.B., Givskov, M., Conese, M., and Doring, G. (2005). Nonmucoid *Pseudomonas aeruginosa* expresses alginate in the lungs of patients with cystic fibrosis and in a mouse model. J. Infect. Dis. *192*, 410–419.

Cabral, D.A., Loh, B.A., and Speert, D.P. (1987). Mucoid *Pseudomonas aeruginosa* resists nonopsonic phagocytosis by human neutrophils and macrophages. Pediatr. Res. *22*, 429–431.

Campanac, C., Pineau, L., Payard, A., Baziard-Mouysset, G., and Roques, C. (2002). Interactions between biocide cationic agents and bacterial biofilms. Antimicrob. Agents Chemother. *46*, 1469–1474.

Costerton, J.W., Stewart, P.S., and Greenberg, E.P. (1999). Bacterial biofilms: a common cause of persistent infections. Science *284*, 1318–1322.

Coyne, M.J., Jr., Russell, K.S., Coyle, C.L., and Goldberg, J.B. (1994). The *Pseudomonas aeruginosa algC* gene encodes phosphoglucomutase, required for the synthesis of a complete lipopolysaccharide core. J. Bacteriol. *176*, 3500–3507.

Cryz, S.J., Jr., Furer, E., and Que, J.U. (1991). Synthesis and characterization of a *Pseudomonas aeruginosa* alginate-toxin A conjugate vaccine. Infect. Immun. *59*, 45–50.

D'Argenio, D.A., and Miller, S.I. (2004). Cyclic di-GMP as a bacterial second messenger. Microbiology *150*, 2497–2502.

Darzins, A., and Chakrabarty, A.M. (1984). Cloning of genes controlling alginate biosynthesis from a mucoid cystic fibrosis isolate of *Pseudomonas aeruginosa*. J. Bacteriol. *159*, 9–18.

Darzins, A., Nixon, L.L., Vanags, R.I., and Chakrabarty, A.M. (1985). Cloning of *Escherichia coli* and *Pseudomonas aeruginosa* phosphomannose isomerase genes and their expression in alginate-negative mutants of *Pseudomonas aeruginosa*. J. Bacteriol. *161*, 249–257.

Davies, D. (2003). Understanding biofilm resistance to antibacterial agents. Nature Rev. *2*, 114–122.

Davies, D.G., Parsek, M.R., Pearson, J.P., Iglewski, B.H., Costerton, J.W., and Greenberg, E.P. (1998). The involvement of cell-to-cell signals in the development of a bacterial biofilm. Science *280*, 295–298.

De Kievit, T.R., and Iglewski, B.H. (1999). Quorum sensing, gene expression, and *Pseudomonas* biofilms. Methods Enzymol. *310*, 117–128.

Delic-Attree, I., Toussaint, B., and Vignais, P.M. (1995). Cloning and sequence analyses of the genes coding for the integration host factor (IHF) and HU proteins of *Pseudomonas aeruginosa*. Gene *154*, 61–64.

Deretic, V., Schurr, M.J., Boucher, J.C., and Martin, D.W. (1994). Conversion of *Pseudomonas aeruginosa* to mucoidy in cystic fibrosis: environmental stress and regulation of bacterial virulence by alternative sigma factors. J. Bacteriol. *176*, 2773–2780.

DeVault, J.D., Hendrickson, W., Kato, J., and Chakrabarty, A.M. (1991). Environmentally regulated *algD* promoter is responsive to the cAMP receptor protein in *Escherichia coli*. Mol. Microbiol. *5*, 2503–2509.

DeVault, J.D., Kimbara, K., and Chakrabarty, A.M. (1990). Pulmonary dehydration and infection in cystic fibrosis: evidence that ethanol activates alginate gene expression and induction of mucoidy in *Pseudomonas aeruginosa*. Mol. Microbiol. *4*, 737–745.

DeVries, C.A., and Ohman, D.E. (1994). Mucoid-to-nonmucoid conversion in alginate-producing *Pseudomonas aeruginosa* often results from spontaneous mutations in *algT*, encoding a putative alternate sigma factor, and shows evidence for autoregulation. J. Bacteriol. *176*, 6677–6687.

Diggle, S.P., Stacey, R.E., Dodd, C., Camara, M., Williams, P., and Winzer, K. (2006). The galactophilic lectin, LecA, contributes to biofilm development in *Pseudomonas aeruginosa*. Env. Microbiol. *8*, 1095–1104.

Doggett, R.G., Harrison, G.M., and Wallis, E.S. (1964). Comparison of some properties of *Pseudomonas aeruginosa* isolated from infections in persons with and without cystic fibrosis. J. Bacteriol. *87*, 427–431.

Drenkard, E., and Ausubel, F.M. (2002). *Pseudomonas* biofilm formation and antibiotic resistance are linked to phenotypic variation. Nature *416*, 740–743.

Duan, K., and Surette, M.G. (2007). Environmental regulation of *Pseudomonas aeruginosa* PAO1 Las and Rhl quorum-sensing systems. J. Bacteriol. *189*, 4827–4836.

Dunne, W.M., Jr. (2002). Bacterial adhesion: seen any good biofilms lately? Clin. Microbiol. Rev. *15*, 155–166.

Evans, L.R., and Linker, A. (1973). Production and characterization of the slime polysaccharide of *Pseudomonas aeruginosa*. J. Bacteriol. *116*, 915–924.

Flynn, J.L., and Ohman, D.E. (1988). Cloning of genes from mucoid *Pseudomonas aeruginosa* which control spontaneous conversion to the alginate production phenotype. J. Bacteriol. *170*, 1452–1460.

Franklin, M.J., Chitnis, C.E., Gacesa, P., Sonesson, A., White, D.C., and Ohman, D.E. (1994). *Pseudomonas aeruginosa* AlgG is a polymer level alginate C5-mannuronan epimerase. J. Bacteriol. *176*, 1821–1830.

Franklin, M.J., and Ohman, D.E. (2002). Mutant analysis and cellular localization of the AlgI, AlgJ, and AlgF proteins required for O acetylation of alginate in *Pseudomonas aeruginosa*. J. Bacteriol. *184*, 3000–3007.

Friedman, L., and Kolter, R. (2004a). Genes involved in matrix formation in *Pseudomonas aeruginosa* PA14 biofilms. Mol. Microbiol. *51*, 675–690.

Friedman, L., and Kolter, R. (2004b). Two genetic loci produce distinct carbohydrate-rich structural components of the *Pseudomonas aeruginosa* biofilm matrix. J. Bacteriol. *186*, 4457–4465.

Fujiwara, S., Zielinski, N.A., and Chakrabarty, A.M. (1993). Enhancer-like activity of A1gR1-binding site in alginate gene activation: positional, orientational, and sequence specificity. J. Bacteriol. *175*, 5452–5459.

Gilbert, P., Collier, P.J., and Brown, M.R. (1990). Influence of growth rate on susceptibility to antimicrobial agents: biofilms, cell cycle, dormancy, and stringent response. Antimicrob. Agents Chemother. *34*, 1865–1868.

Gilbert, P., Maira-Litran, T., McBain, A.J., Rickard, A.H., and Whyte, F.W. (2002). The physiology and collective recalcitrance of microbial biofilm communities. Advanc. Microbial Physiol. *46*, 202–256.

Gill, J.F., Deretic, V., and Chakrabarty, A.M. (1986). Overproduction and assay of *Pseudomonas aeruginosa* phosphomannose isomerase. J. Bacteriol. *167*, 611–615.

Gillis, R.J., and Iglewski, B.H. (2004). Azithromycin retards *Pseudomonas aeruginosa* biofilm formation. J. Clin. Microbiol. *42*, 5842–5845.

Goldberg, J.B., and Dahnke, T. (1992). *Pseudomonas aeruginosa* AlgB, which modulates the expression of alginate, is a member of the NtrC subclass of prokaryotic regulators. Mol. Microbiol. *6*, 59–66.

Goldberg, J.B., and Ohman, D.E. (1984). Cloning and expression in *Pseudomonas aeruginosa* of a gene involved in the production of alginate. J. Bacteriol. *158*, 1115–1121.

Goodman, A.L., Kulasekara, B., Rietsch, A., Boyd, D., Smith, R.S., and Lory, S. (2004). A signaling network reciprocally regulates genes associated with acute infection and chronic persistence in *Pseudomonas aeruginosa*. Dev. Cell *7*, 745–754.

Harrison, J.J., Turner, R.J., and Ceri, H. (2005). Persister cells, the biofilm matrix and tolerance to metal cations in biofilm and planktonic *Pseudomonas aeruginosa*. Env. Microbiol. *7*, 981–994.

Hassett, D.J., Cuppoletti, J., Trapnell, B., Lymar, S.V., Rowe, J.J., Yoon, S.S., Hilliard, G.M., Parvatiyar, K., Kamani, M.C., Wozniak, D.J., Hwang, S.H., McDermott, T.R., and Ochsner, U.A. (2002). Anaerobic metabolism and quorum sensing by *Pseudomonas aeruginosa* biofilms in chronically infected cystic fibrosis airways: rethinking antibiotic treatment strategies and drug targets. Advanc. Drug Delivery Rev. *54*, 1425–1443.

Hatch, R.A., and Schiller, N.L. (1998). Alginate lyase promotes diffusion of aminoglycosides through the extracellular polysaccharide of mucoid *Pseudomonas aeruginosa*. Antimicrob. Agents Chemother. *42*, 974–977.

Hentzer, M., Teitzel, G.M., Balzer, G.J., Heydorn, A., Molin, S., Givskov, M., and Parsek, M.R. (2001). Alginate overproduction affects *Pseudomonas aeruginosa* biofilm structure and function. J. Bacteriol. *183*, 5395–5401.

Hershberger, C.D., Ye, R.W., Parsek, M.R., Xie, Z.D., and Chakrabarty, A.M. (1995). The *algT* (*algU*) gene of *Pseudomonas aeruginosa*, a key regulator involved in alginate biosynthesis, encodes an alternative sigma factor (sigma E). Proc. Natl. Acad. Sci. USA *92*, 7941–7945.

Hickman, J.W., Tifrea, D.F., and Harwood, C.S. (2005). A chemosensory system that regulates biofilm formation through modulation of cyclic diguanylate levels. Proc. Natl. Acad. Sci. USA *102*, 14422–14427.

Hoffman, L.R., D'Argenio, D.A., MacCoss, M.J., Zhang, Z., Jones, R.A., and Miller, S.I. (2005). Aminoglycoside antibiotics induce bacterial biofilm formation. Nature *436*, 1171–1175.

Irie, Y., and Parsek, M.R. (2008). Quorum sensing and microbial biofilms. In Bacterial Biofilms, Current Topics in Microbiology and Immunology, T. Romeo, ed. (Berlin Heidelberg, Srpinger-Verlag), pp. 67–84

Jackson, K.D., Starkey, M., Kremer, S., Parsek, M.R., and Wozniak, D.J. (2004). Identification of *psl*, a locus encoding a potential exopolysaccharide that is essential for *Pseudomonas aeruginosa* PAO1 biofilm formation. J. Bacteriol. *186*, 4466–4475.

Jain, S., Franklin, M.J., Ertesvag, H., Valla, S., and Ohman, D.E. (2003). The dual roles of AlgG in C-5-epimerization and secretion of alginate polymers in *Pseudomonas aeruginosa*. Mol. Microbiol. *47*, 1123–1133.

Jain, S., and Ohman, D.E. (2004). Alginate biosynthesis. In *Pseudomonas*, Vol. 3, J.-L. Ramos, ed. (New York, Kluwer Academic/Plenum Publishers), pp. 53–81.

Jenal, U. (2004). Cyclic di-guanosine-monophosphate comes of age: a novel secondary messenger involved in modulating cell surface structures in bacteria? Curr. Opin. Microbiol. *7*, 185–191.

Jensen, E.T., Kharazmi, A., Garred, P., Kronborg, G., Fomsgaard, A., Mollnes, T.E., and Hoiby, N. (1993). Complement activation by *Pseudomonas aeruginosa* biofilms. Microb. Pathogen. *15*, 377–388.

Jesaitis, A.J., Franklin, M.J., Berglund, D., Sasaki, M., Lord, C.I., Bleazard, J.B., Duffy, J.E., Beyenal, H., and Lewandowski, Z. (2003). Compromised host defense on *Pseudomonas aeruginosa* biofilms: characterization of neutrophil and biofilm interactions. J. Immunol. *171*, 4329–4339.

Kaneko, Y., Thoendel, M., Olakanmi, O., Britigan, B.E., and Singh, P.K. (2007). The transition metal gallium disrupts *Pseudomonas aeruginosa* iron metabolism and has antimicrobial and antibiofilm activity. Journal of Clinical Investigations *117*, 877–888.

Kato, J., and Chakrabarty, A.M. (1991). Purification of the regulatory protein AlgR1 and its binding in the far upstream region of the *algD* promoter in *Pseudomonas aeruginosa*. Proc. Natl. Acad. Sci. USA *88*, 1760–1764.

Kato, J., Chu, L., Kitano, K., DeVault, J.D., Kimbara, K., Chakrabarty, A.M., and Misra, T.K. (1989). Nucleotide sequence of a regulatory region controlling alginate synthesis in *Pseudomonas aeruginosa*: characterization of the *algR2* gene. Gene *84*, 31–38.

Kato, J., Misra, T.K., and Chakrabarty, A.M. (1990). AlgR3, a protein resembling eukaryotic histone H1, regulates alginate synthesis in *Pseudomonas aeruginosa*. Proc. Natl. Acad. Sci. USA *87*, 2887–2891.

Kim, C., Kim, J., Park, H.Y., Park, H.J., Lee, J.H., Kim, C.K., and Yoon, J. (2008). Furanone derivatives as quorum-sensing antagonists of *Pseudomonas aeruginosa*. Appl. Microbiol. Biotechnol. *80*, 37–47.

Kimmel, J.L., and Tipton, P.A. (2005). Inactivation of GDP-mannose dehydrogenase from *Pseudomonas aeruginosa* by penicillic acid identifies a critical active site loop. Archives of Biochemistry and Biophysics *441*, 132–140.

Kintz, E., and Goldberg, J.B. (2008). Regulation of lipopolysaccharide O antigen expression in *Pseudomonas aeruginosa*. Future Microbiol. *3*, 191–203.

Klausen, M., Heydorn, A., Ragas, P., Lambertsen, L., Aaes-Jorgensen, A., Molin, S., and Tolker-Nielsen, T. (2003). Biofilm formation by *Pseudomonas aeruginosa* wild type, flagella and type IV pili mutants. Mol. Microbiol. *48*, 1511–1524.

Knirel, Y.A., Bystrova, O.V., Kocharova, N.A., Zahringer, U., and Pier, G.B. (2006). Conserved and variable structural features in the lipopolysaccharide of *Pseudomonas aeruginosa*. J. Endotoxin Res. *12*, 324–336.

Konyecsni, W.M., and Deretic, V. (1990). DNA sequence and expression analysis of *algP* and *algQ*, components of the multigene system transcriptionally regulating mucoidy in *Pseudomonas aeruginosa*: *algP* contains multiple direct repeats. J. Bacteriol. *172*, 2511–2520.

Krieg, D.P., Helmke, R.J., German, V.F., and Mangos, J.A. (1988). Resistance of mucoid *Pseudomonas aeruginosa* to nonopsonic phagocytosis by alveolar macrophages *in vitro*. Infect. Immun. *56*, 3173–3179.

Lam, J., Chan, R., Lam, K., and Costerton, J.W. (1980). Production of mucoid microcolonies by *Pseudomonas aeruginosa* within infected lungs in cystic fibrosis. Infect. Immun. *28*, 546–556.

Lee, B., Haagensen, J.A., Ciofu, O., Andersen, J.B., Hoiby, N., and Molin, S. (2005). Heterogeneity of biofilms formed by nonmucoid *Pseudomonas aeruginosa* isolates from patients with cystic fibrosis. J. Clin. Microbiol. *43*, 5247–5255.

Leech, A.J., Sprinkle, A., Wood, L., Wozniak, D.J., and Ohman, D.E. (2008). The NtrC family regulator AlgB, which controls alginate biosynthesis in mucoid

Pseudomonas aeruginosa, binds directly to the *algD* promoter. J. Bacteriol. *190*, 581–589.

Leid, J.G., Willson, C.J., Shirtliff, M.E., Hassett, D.J., Parsek, M.R., and Jeffers, A.K. (2005). The exopolysaccharide alginate protects *Pseudomonas aeruginosa* biofilm bacteria from IFN-gamma-mediated macrophage killing. J. Immunol. *175*, 7512–7518.

Linker, A., and Evans, L.R. (1984). Isolation and characterization of an alginase from mucoid strains of *Pseudomonas aeruginosa*. J. Bacteriol. *159*, 958–964.

Linker, A., and Jones, R.S. (1966). A new polysaccharide resembling alginic acid isolated from Pseudomonads. J. Biol. Chem. *241*, 3845–3851.

Liu, H.Y., Wang, Z., Regni, C., Zou, X., and Tipton, P.A. (2004). Detailed kinetic studies of an aggregating inhibitor; inhibition of phosphomannomutase/phosphoglucomutase by disperse blue 56. Biochemistry *43*, 8662–8669.

Ma, L., Jackson, K.D., Landry, R.M., Parsek, M.R., and Wozniak, D.J. (2006). Analysis of *Pseudomonas aeruginosa* conditional *psl* variants reveals roles for the *psl* polysaccharide in adhesion and maintaining biofilm structure postattachment. J. Bacteriol. *188*, 8213–8221.

Ma, L., Lu, H., Sprinkle, A., Parsek, M.R., and Wozniak, D.J. (2007). *Pseudomonas aeruginosa* Psl is a galactose- and mannose-rich exopolysaccharide. J. Bacteriol. *189*, 8353–8356.

Ma, S., Selvaraj, U., Ohman, D.E., Quarless, R., Hassett, D.J., and Wozniak, D.J. (1998). Phosphorylation-independent activity of the response regulators AlgB and AlgR in promoting alginate biosynthesis in mucoid *Pseudomonas aeruginosa*. J. Bacteriol. *180*, 956–968.

Ma, S., Wozniak, D.J., and Ohman, D.E. (1997). Identification of the histidine protein kinase KinB in *Pseudomonas aeruginosa* and its phosphorylation of the alginate regulator *algB*. J. Biol. Chem. *272*, 17952–17960.

Mai, G.T., McCormack, J.G., Seow, W.K., Pier, G.B., Jackson, L.A., and Thong, Y.H. (1993). Inhibition of adherence of mucoid *Pseudomonas aeruginosa* by alginase, specific monoclonal antibodies, and antibiotics. Infect. Immun. *61*, 4338–4343.

Massengale, A.R., Quinn, F.J., Williams, A., Gallagher, S., and Aronoff, S.C. (2000). The effect of alginate on the invasion of cystic fibrosis respiratory epithelial cells by clinical isolates of *Pseudomonas aeruginosa*. Exp. Lung Res. *26*, 163–178.

Mathee, K., Ciofu, O., Sternberg, C., Lindum, P.W., Campbell, J.I., Jensen, P., Johnsen, A.H., Givskov, M., Ohman, D.E., Molin, S., Høiby, N., and Kharazmi, A. (1999). Mucoid conversion of *Pseudomonas aeruginosa* by hydrogen peroxide: a mechanism for virulence activation in the cystic fibrosis lung. Microbiology *145*, 1349–1357.

Matsukawa, M., and Greenberg, E.P. (2004). Putative exopolysaccharide synthesis genes influence *Pseudomonas aeruginosa* biofilm development. J. Bacteriol. *186*, 4449–4456.

May, T.B., Shinabarger, D., Maharaj, R., Kato, J., Chu, L., DeVault, J.D., Roychoudhury, S., Zielinski, N.A., Berry, A., Rothmel, R.K. (1991). Alginate synthesis by *Pseudomonas aeruginosa*: a key pathogenic factor in chronic pulmonary infections of cystic fibrosis patients. Clin. Microbiol. Rev. *4*, 191–206.

Meluleni, G.J., Grout, M., Evans, D.J., and Pier, G.B. (1995). Mucoid *Pseudomonas aeruginosa* growing in a biofilm *in vitro* are killed by opsonic antibodies to the mucoid exopolysaccharide capsule but not by antibodies produced during chronic lung infection in cystic fibrosis patients. J. Immunol. *155*, 2029–2038.

Merighi, M., Lee, V.T., Hyodo, M., Hayakawa, Y., and Lory, S. (2007). The second messenger bis-(3′-5′)-cyclic-GMP and its PilZ domain-containing receptor Alg44 are required for alginate biosynthesis in *Pseudomonas aeruginosa*. Mol. Microbiol. *65*, 876–895.

Meshulam, T., Obedeanu, N., Merzbach, D., and Sobel, J.D. (1984). Phagocytosis of mucoid and nonmucoid strains of *Pseudomonas aeruginosa*. Clin. Immunol. Immunopathol. *32*, 151–165.

Mohr, C.D., Hibler, N.S., and Deretic, V. (1991). AlgR, a response regulator controlling mucoidy in *Pseudomonas aeruginosa*, binds to the FUS sites of the *algD* promoter located unusually far upstream from the mRNA start site. J. Bacteriol. *173*, 5136–5143.

Mohr, C.D., Leveau, J.H., Krieg, D.P., Hibler, N.S., and Deretic, V. (1992). AlgR-binding sites within the *algD* promoter make up a set of inverted repeats separated by a large intervening segment of DNA. J. Bacteriol. *174*, 6624–6633.

Mohr, C.D., Martin, D.W., Konyecsni, W.M., Govan, J.R., Lory, S., and Deretic, V. (1990). Role of the far-upstream sites of the *algD* promoter and the *algR* and *rpoN* genes in environmental modulation of mucoidy in *Pseudomonas aeruginosa*. J. Bacteriol. *172*, 6576–6580.

Morgan, R., Kohn, S., Hwang, S.H., Hassett, D.J., and Sauer, K. (2006). BdlA, a chemotaxis regulator essential for biofilm dispersion in *Pseudomonas aeruginosa*. J. Bacteriol. *188*, 7335–7343.

Muh, U., Hare, B.J., Duerkop, B.A., Schuster, M., Hanzelka, B.L., Heim, R., Olson, E.R., and Greenberg, E.P. (2006a). A structurally unrelated mimic of a *Pseudomonas aeruginosa* acyl-homoserine lactone quorum-sensing signal. Proc. Natl. Acad. Sci. USA *103*, 16948–16952.

Muh, U., Schuster, M., Heim, R., Singh, A., Olson, E.R., and Greenberg, E.P. (2006b). Novel *Pseudomonas aeruginosa* quorum-sensing inhibitors identified in an ultra-high-throughput screen. Antimicrob. Agents Chemother. *50*, 3674–3679.

Muhammadi, and Ahmed, N. (2007). Genetics of bacterial alginate: alginate genes distribution, organization and biosynthesis in bacteria. Curr. Genomics *8*, 191–202.

Nichols, W.W., Dorrington, S.M., Slack, M.P., and Walmsley, H.L. (1988). Inhibition of tobramycin diffusion by binding to alginate. Antimicrob. Agents Chemother. *32*, 518–523.

O'Toole, G.A., and Kolter, R. (1998a). Flagellar and twitching motility are necessary for *Pseudomonas aeruginosa* biofilm development. Mol. Microbiol. *30*, 295–304.

O'Toole, G.A., and Kolter, R. (1998b). Initiation of biofilm formation in *Pseudomonas fluorescens* WCS365 proceeds via multiple, convergent signalling pathways: a genetic analysis. Mol. Microbiol. *28*, 449–461.

Oglesby, L.L., Jain, S., and Ohman, D.E. (2008). Membrane topology and roles of *Pseudomonas aeruginosa* Alg8 and Alg44 in alginate polymerization. Microbiology *154*, 1605–1615.

Overhage, J., Schemionek, M., Webb, J.S., and Rehm, B.H. (2005). Expression of the *psl* operon in *Pseudomonas aeruginosa* PAO1 biofilms: PslA performs an essential function in biofilm formation. Appl. Environ. Microbiol. *71*, 4407–4413.

Page, W.J., Tindale, A., Chandra, M., and Kwon, E. (2001). Alginate formation in *Azotobacter vinelandii* UWD during stationary phase and the turnover of poly-beta-hydroxybutyrate. Microbiology *147*, 483–490.

Parsek, M.R., and Greenberg, E.P. (1999). Quorum sensing signals in development of *Pseudomonas aeruginosa* biofilms. Methods in Enzymology *310*, 43–55.

Parsek, M.R., and Greenberg, E.P. (2000). Acyl-homoserine lactone quorum sensing in gram-negative bacteria: a signaling mechanism involved in associations with higher organisms. Proc. Natl. Acad. Sci. USA *97*, 8789–8793.

Pedersen, S.S., Hoiby, N., Espersen, F., and Koch, C. (1992). Role of alginate in infection with mucoid *Pseudomonas aeruginosa* in cystic fibrosis. Thorax *47*, 6–13.

Pedersen, S.S., Kharazmi, A., Espersen, F., and Hoiby, N. (1990). *Pseudomonas aeruginosa* alginate in cystic fibrosis sputum and the inflammatory response. Infect. Immun. *58*, 3363–3368.

Pier, G. (2005). Application of vaccine technology to prevention of *Pseudomonas aeruginosa* infections. Expert Rev. Vaccines *4*, 645–656.

Pier, G.B. (2007). *Pseudomonas aeruginosa* lipopolysaccharide: a major virulence factor, initiator of inflammation and target for effective immunity. Int. J. Med. Microbiol. *297*, 277–295.

Pier, G.B., Boyer, D., Preston, M., Coleman, F.T., Llosa, N., Mueschenborn-Koglin, S., Theilacker, C., Goldenberg, H., Uchin, J., Priebe, G.P., Grout, M., Posner, M., and Cavacini, L. (2004). Human monoclonal antibodies to *Pseudomonas aeruginosa* alginate that protect against infection by both mucoid and nonmucoid strains. J. Immunol. *173*, 5671–5678.

Pier, G.B., Coleman, F., Grout, M., Franklin, M., and Ohman, D.E. (2001). Role of alginate O acetylation in resistance of mucoid *Pseudomonas aeruginosa* to opsonic phagocytosis. Infect. Immun. *69*, 1895–1901.

Pier, G.B., Matthews, W.J., Jr., and Eardley, D.D. (1983). Immunochemical characterization of the mucoid exopolysaccharide of *Pseudomonas aeruginosa*. J. Infect. Dis. *147*, 494–503.

Pier, G.B., Saunders, J.M., Ames, P., Edwards, M.S., Auerbach, H., Goldfarb, J., Speert, D.P., and Hurwitch, S. (1987). Opsonophagocytic killing antibody to *Pseudomonas aeruginosa* mucoid exopolysaccharide in older noncolonized patients with cystic fibrosis. N. Engl. J. Med. *317*, 793–798.

Pier, G.B., Small, G.J., and Warren, H.B. (1990). Protection against mucoid *Pseudomonas aeruginosa* in rodent models of endobronchial infections. Science *249*, 537–540.

Qiu, D., Eisinger, V.M., Head, N.E., Pier, G.B., and Yu, H.D. (2008). ClpXP proteases positively regulate alginate overexpression and mucoid conversion in *Pseudomonas aeruginosa*. Microbiology *154*, 2119–2130.

Qiu, D., Eisinger, V.M., Rowen, D.W., and Yu, H.D. (2007). Regulated proteolysis controls mucoid conversion in *Pseudomonas aeruginosa*. Proc. Natl. Acad. Sci. USA *104*, 8107–8112.

Rahme, L.G., Stevens, E.J., Wolfort, S.F., Shao, J., Tompkins, R.G., and Ausubel, F.M. (1995). Common virulence factors for bacterial pathogenicity in plants and animals. Science *268*, 1899–1902.

Ramsey, D.M., and Wozniak, D.J. (2005). Understanding the control of *Pseudomonas aeruginosa* alginate synthesis and the prospects for management of chronic infections in cystic fibrosis. Mol. Microbiol. *56*, 309–322.

Rasmussen, T.B., and Givskov, M. (2006). Quorum-sensing inhibitors as anti-pathogenic drugs. Int. J. Med. Microbiol. *296*, 149–161.

Regni, C., Shackelford, G.S., and Beamer, L.J. (2006). Complexes of the enzyme phosphomannomutase/phosphoglucomutase with a slow substrate and an inhibitor. Acta Crystallogr. Section F Structural Biol. Crystalliz. Commun. *62*, 722–726.

Regni, C., Tipton, P.A., and Beamer, L.J. (2002). Crystal structure of PMM/PGM: an enzyme in the biosynthetic pathway of *P. aeruginosa* virulence factors. Structure *10*, 269–279.

Rocchetta, H.L., Burrows, L.L., and Lam, J.S. (1999). Genetics of O-antigen biosynthesis in *Pseudomonas aeruginosa*. Microbiol. Mol. Biol. Rev. *63*, 523–553.

Sadoff, H.L. (1975). Encystment and germination in *Azotobacter vinelandii*. Bacteriol. Rev. *39*, 516–539.

Sauer, K., Camper, A.K., Ehrlich, G.D., Costerton, J.W., and Davies, D.G. (2002). *Pseudomonas aeruginosa* displays multiple phenotypes during development as a biofilm. J. Bacteriol. *184*, 1140–1154.

Sauer, K., Cullen, M.C., Rickard, A.H., Zeef, L.A., Davies, D.G., and Gilbert, P. (2004). Characterization of nutrient-induced dispersion in *Pseudomonas aeruginosa* PAO1 biofilm. J. Bacteriol. *186*, 7312–7325.

Schramm, A.M., Mehra-Chaudhary, R., Furdui, C.M., and Beamer, L.J. (2008). Backbone flexibility, conformational change, and catalysis in a phosphohexomutase from *Pseudomonas aeruginosa*. Biochemistry *47*, 9154–9162.

Schurr, M.J., Yu, H., Boucher, J.C., Hibler, N.S., and Deretic, V. (1995). Multiple promoters and induction by heat shock of the gene encoding the alternative sigma factor AlgU (sigma E) which controls mucoidy in cystic fibrosis isolates of *Pseudomonas aeruginosa*. J. Bacteriol. *177*, 5670–5679.

Schuster, M., Lostroh, C.P., Ogi, T., and Greenberg, E.P. (2003). Identification, timing, and signal specificity of *Pseudomonas aeruginosa* quorum-controlled genes: a transcriptome analysis. J. Bacteriol. *185*, 2066–2079.

Schwarzmann, S., and Boring, J.R. (1971). Antiphagocytic effect of slime from a mucoid strain of *Pseudomonas aeruginosa*. Infect. Immun. *3*, 762–767.

Shinabarger, D., Berry, A., May, T.B., Rothmel, R., Fialho, A., and Chakrabarty, A.M. (1991). Purification and characterization of phosphomannose isomerase-guanosine diphospho-D-mannose pyrophosphorylase. A bifunctional enzyme in the alginate biosynthetic pathway of *Pseudomonas aeruginosa*. J. Biol. Chem. *266*, 2080–2088.

Singh, P.K., Parsek, M.R., Greenberg, E.P., and Welsh, M.J. (2002). A component of innate immunity prevents bacterial biofilm development. Nature *417*, 552–555.

Singh, P.K., Schaefer, A.L., Parsek, M.R., Moninger, T.O., Welsh, M.J., and Greenberg, E.P. (2000). Quorum-sensing signals indicate that cystic fibrosis lungs are infected with bacterial biofilms. Nature *407*, 762–764.

Snook, C.F., Tipton, P.A., and Beamer, L.J. (2003). Crystal structure of GDP-mannose dehydrogenase: a key enzyme of alginate biosynthesis in *P. aeruginosa*. Biochemistry *42*, 4658–4668.

Speert, D.P., Farmer, S.W., Campbell, M.E., Musser, J.M., Selander, R.K., and Kuo, S. (1990). Conversion of *Pseudomonas aeruginosa* to the phenotype characteristic of strains from patients with cystic fibrosis. J. Clin. Microbiol. *28*, 188–194.

Strausbaugh, S.D., and Davis, P.B. (2007). Cystic fibrosis: a review of epidemiology and pathobiology. Clin. Chest Med. *28*, 279–288.

Tart, A. and Wozniak, D.J. (2008). Shifting paradigms in *Pseudomonas aeruginosa* biofilm research. In Bacterial Biofilms, Current Topics in Microbiology and Immunology, T. Romeo, ed. (Berlin: Srpinger-Verlag), pp. 67–84.

Terry, J.M., Pina, S.E., and Mattingly, S.J. (1991). Environmental conditions which influence mucoid conversion *Pseudomonas aeruginosa* PAO1. Infect. Immun. *59*, 471–477.

Theilacker, C., Coleman, F.T., Mueschenborn, S., Llosa, N., Grout, M., and Pier, G.B. (2003). Construction and characterization of a *Pseudomonas aeruginosa* mucoid exopolysaccharide-alginate conjugate vaccine. Infect. Immun. *71*, 3875–3884.

Tielen, P., Strathmann, M., Jaeger, K.E., Flemming, H.C., and Wingender, J. (2005). Alginate acetylation influences initial surface colonization by mucoid *Pseudomonas aeruginosa*. Microbiol. Res. *160*, 165–176.

Tielker, D., Hacker, S., Loris, R., Strathmann, M., Wingender, J., Wilhelm, S., Rosenau, F., and Jaeger, K.E. (2005). *Pseudomonas aeruginosa* lectin LecB is located in the outer membrane and is involved in biofilm formation. Microbiology *151*, 1313–1323.

Totten, P.A., Lara, J.C., and Lory, S. (1990). The *rpoN* gene product of *Pseudomonas aeruginosa* is required for expression of diverse genes, including the flagellin gene. J. Bacteriol. *172*, 389–396.

Vasseur, P., Vallet-Gely, I., Soscia, C., Genin, S., and Filloux, A. (2005). The *pel* genes of the *Pseudomonas aeruginosa* PAK strain are involved at early and late stages of biofilm formation. Microbiology *151*, 985–997.

Ventre, I., Goodman, A.L., Vallet-Gely, I., Vasseur, P., Soscia, C., Molin, S., Bleves, S., Lazdunski, A., Lory, S., and Filloux, A. (2006). Multiple sensors control reciprocal expression of *Pseudomonas aeruginosa* regulatory RNA and virulence genes. Proc. Natl. Acad. Sci. USA *103*, 171–176.

Wagner, T., Soong, G., Sokol, S., Saiman, L., and Prince, A. (2005). Effects of azithromycin on clinical isolates of *Pseudomonas aeruginosa* from cystic fibrosis patients. Chest *128*, 912–919.

Wagner, V.E., Bushnell, D., Passador, L., Brooks, A.I., and Iglewski, B.H. (2003). Microarray analysis of *Pseudomonas aeruginosa* quorum-sensing regulons: effects of growth phase and environment. J. Bacteriol. *185*, 2080–2095.

Walker, T.S., Tomlin, K.L., Worthen, G.S., Poch, K.R., Lieber, J.G., Saavedra, M.T., Fessler, M.B., Malcolm, K.C., Vasil, M.L., and Nick, J.A. (2005). Enhanced *Pseudomonas aeruginosa* biofilm development mediated by human neutrophils. Infect. Immun. *73*, 3693–3701.

Whitchurch, C.B., Alm, R.A., and Mattick, J.S. (1996). The alginate regulator AlgR and an associated sensor FimS are required for twitching motility in *Pseudomonas aeruginosa*. Proc. Natl. Acad. Sci. USA *93*, 9839–9843.

Whitchurch, C.B., Tolker-Nielsen, T., Ragas, P.C., and Mattick, J.S. (2002). Extracellular DNA required for bacterial biofilm formation. Science *295*, 1487.

Wood, L.F., Leech, A.J., and Ohman, D.E. (2006). Cell wall-inhibitory antibiotics activate the alginate biosynthesis operon in *Pseudomonas aeruginosa*: Roles of sigma (AlgT) and the AlgW and Prc proteases. Mol. Microbiol. *62*, 412–426.

Worlitzsch, D., Tarran, R., Ulrich, M., Schwab, U., Cekici, A., Meyer, K.C., Birrer, P., Bellon, G., Berger, J., Weiss, T., Botzenhart, K., Yankaskas, J.R., Randell, S., Boucher, R.C., and Döring, G. (2002). Effects of reduced mucus oxygen concentration in airway *Pseudomonas* infections of cystic fibrosis patients. J. Clin. Invest. *109*, 317–325.

Wozniak, D.J., and Keyser, R. (2004). Effects of subinhibitory concentrations of macrolide antibiotics on *Pseudomonas aeruginosa*. Chest *125*, 62S-69S; quiz 69S.

Wozniak, D.J., and Ohman, D.E. (1991). *Pseudomonas aeruginosa* AlgB, a two-component response regulator of the NtrC family, is required for *algD* transcription. J. Bacteriol. *173*, 1406–1413.

Wozniak, D.J., and Ohman, D.E. (1993). Involvement of the alginate *algT* gene and integration host factor in the regulation of the *Pseudomonas aeruginosa algB* gene. J. Bacteriol. *175*, 4145–4153.

Wozniak, D.J., Wyckoff, T.J., Starkey, M., Keyser, R., Azadi, P., O'Toole, G.A., and Parsek, M.R. (2003). Alginate is not a significant component of the extracellular polysaccharide matrix of PA14 and PAO1 *Pseudomonas aeruginosa* biofilms. Proc. Natl. Acad. Sci. USA *100*, 7907–7912.

Yoon, S.S., Hennigan, R.F., Hilliard, G.M., Ochsner, U.A., Parvatiyar, K., Kamani, M.C., Allen, H.L., DeKievit, T.R., Gardner, P.R., Schwab, U., Rowe, J.J., Iglewski, B.H., McDermott, T.R., Mason, R.P., Wozniak, D.J., Hancock, R.E., Parsek, M.R., Noah, T.L., Boucher, R.C., and Hassett, D.J. (2002). *Pseudomonas aeruginosa* anaerobic respiration in biofilms: relationships to cystic fibrosis pathogenesis. Dev. Cell *3*, 593–603.

Yu, H., Mudd, M., Boucher, J.C., Schurr, M.J., and Deretic, V. (1997). Identification of the *algZ* gene upstream of the response regulator *algR* and its participation in control of alginate production in *Pseudomonas aeruginosa*. J. Bacteriol. *179*, 187–193.

Yu, H., Schurr, M.J., and Deretic, V. (1995). Functional equivalence of *Escherichia coli* sigma E and *Pseudomonas aeruginosa* AlgU: *E. coli rpoE* restores mucoidy and reduces sensitivity to reactive oxygen intermediates in *algU* mutants of *P. aeruginosa*. J. Bacteriol. *177*, 3259–3268.

Zaidi, T., and Pier, G.B. (2008). Prophylactic and therapeutic efficacy in murine keratitis infection of a fully human IgG1 monoclonal antibody to *Pseudomonas aeruginosa* alginate. Infect. Immun. *76*, 4720–4725.

Zielinski, N.A., Chakrabarty, A.M., and Berry, A. (1991). Characterization and regulation of the *Pseudomonas aeruginosa algC* gene encoding phosphomannomutase. J. Biol. Chem. *266*, 9754–9763.

Zielinski, N.A., Maharaj, R., Roychoudhury, S., Danganan, C.E., Hendrickson, W., and Chakrabarty, A.M. (1992). Alginate synthesis in *Pseudomonas aeruginosa*: environmental regulation of the *algC* promoter. J. Bacteriol. *174*, 7680–7688.

Immune Responses to Microbial Polysaccharides

6

Darran J. Wigelsworth, Erin B. Troy and Dennis L. Kasper

Abstract

The immune system is capable of recognizing almost any biological polymer (including proteins and glycolipids) and presenting it to T-cells via major histocompatibility proteins. These adaptive immune responses have long been considered the territory of antigenic proteins, whereas carbohydrates are considered T-cell-independent antigens and are not recognized by the complete adaptive machinery. However, recent work reveals a larger role for sugars in immune recognition. The zwitterionic polysaccharide PSA from the commensal organism *Bacteroides fragilis* has recently been added to the list of antigens presented to T-cells. This molecule has been shown not only to be presented on MHCII but also to play a role in immunomodulation and suppression of inflammatory diseases such as colitis. Bacteria have developed polysaccharide capsules to take advantage of the fact that polysaccharides are generally not recognized by the human immune system. However, the recent finding that sugars play a role in immune recognition makes them more attractive targets in the search for antigenic epitopes. This chapter will discuss in detail the immune response to foreign antigens, the function of zwitterionic polysaccharides in immune recognition, and finally the use of carbohydrates in glycoconjugated vaccines, undoubtedly one of the biggest scientific breakthroughs of the past century.

Principles of innate and adaptive immunity

The macrophages and neutrophils of the innate immune system provide a first line of defence against many common microorganisms and are essential for the control of many infections. However, these cells cannot always eliminate infectious microorganisms. The lymphocytes of the adaptive immune system have evolved to provide a more versatile means of defence and to offer greater protection from re-infection with the same pathogen. There is generally a delay of 4–7 days before an effective adaptive immune response can be generated, and during this interval the function of the innate immune system is critical.

Until recently, immunological doctrine held that bacterial protein antigens – but not polysaccharide antigens – can be processed and presented to the lymphocytes of the adaptive immune system. This dogma has now been overturned: not only can polysaccharides activate the adaptive immune system, but they do so in a manner similar to that documented for proteins. It is important to understand the basic principles underlying each type of immune response.

The innate immune response

The innate immune system functions to clear an infectious agent within minutes to hours. Only if a pathogen is able to breach innate defences does an adaptive response develop. Innate immunity relies on germline-encoded receptors that recognize specific patterns on microorganisms and that can effectively distinguish between self and non-self. The first line of innate defence against outside microorganisms is the epithelial layer of the skin. Other components of innate immunity include the commensal bacteria of the gut, which prevent colonization by pathogenic bacteria. If a microorganism crosses an epithelial barrier, it is immediately recognized by mononuclear phagocytes or macrophages that reside within these tissues.

Two major families of phagocytes exist. *Macrophages,* which mature continuously from monocytes that leave the circulation to migrate into tissues throughout the body, are found in the gastrointestinal tract, the lungs, the blood cells of the liver (where they are known as Kupffer cells), and throughout the spleen. *Neutrophils* or *polymorphonuclear leukocytes* (PMNs) are short-lived cells that abound within the blood but that, except in inflammation, are only rarely found within tissues. Both phagocyte families play a crucial role in recognizing, ingesting, and destroying pathogens. Macrophages and neutrophils recognize pathogens by means of their cell surface receptors, including macrophage mannose receptors (only on macrophages); scavenger receptors, which bind negatively charged ligands such as lipoteichoic acid; and CD14 and the Toll-like receptors (TLRs), which can recognize lipopolysaccharide (LPS) and pure polysaccharides as discussed later in this chapter.

Receptors of the innate immune system

Macrophage mannose receptor (MMR)

This C-type (calcium-dependent) lectin cell-surface receptor directly recognizes pathogens' surfaces and binds to sugar molecules found on the surface of many bacteria. A multihead molecule with several carbohydrate recognition domains (CRDs), MMR is a transmembrane cell-surface receptor and as such can function as a phagocytic receptor. The CRD binds to mannose and fructose on bacteria that are a specific distance apart. Any deviation of spacing between the mannose or fructose will result in loss of recognition by the receptor.

Dectin 1

Dendritic cell-associated lectin 1 (dectin 1, also known as C-type lectin domain family 7, member A, or CLEC7A) is a membrane-expressed human receptor. Dectin 1 functions as a pattern recognition receptor for β-1,3-linked and β-1,6-linked glucans from a variety of sources.

DC-SIGN

Also known as CD209, DC-SIGN (dendritic cell-specific ICAM-3 grabbing non-integrin) is another C-type lectin bound to macrophages and dendritic cells (DCs). Like MMR, DC-SIGN recognizes mannose-type carbohydrates found on viruses, bacteria, and fungi. Binding of DC-SIGN to pathogens activates endocytosis and phagocytic removal of the invader.

SIGN-R1

Another C-type lectin related to DC-SIGN, SIGN-R1 (SIGN-related 1) is an uptake receptor expressed at high levels in spleen and lymph node macrophages. SIGNR1 is the principal receptor for the capsular polysaccharides of *Streptococcus pneumoniae* and other bacterial glucans (Geijtenbeek *et al.*, 2002; Kang *et al.*, 2003) (see Chapter 1).

Scavenger receptors

Scavenger receptors, which recognize various anionic polymers and acetylated low-density lipoproteins, exist in at least six distinct molecular forms. Some scavenger receptors recognize structures that are shielded by sialic acid on normal host cells; these receptors can be involved in the removal of senescent red blood cells that have lost their own sialic acid. Other targets of scavenger receptors are still being characterized. In addition to triggering phagocytosis, some scavenger receptors are involved in inducing factors that are related to innate immunity (e.g. chemokines and cytokines) and that induce macrophages and DCs to express co-stimulatory molecules involved in activating the adaptive immune system. Among the most important and well-characterized of the scavenger receptors related to polysaccharides are the Toll-like receptors.

Toll-like receptors

Most mammalian species have an estimated 10–15 types of TLRs. Ten TLRs have been identified in humans, and equivalent forms of many of these TLRs have been found in other mammalian species (Du *et al.*, 2000; Chuang and Ulevitch, 2000; Tabeta *et al.*, 2004). One of the best-characterized TLR responses is that of TLR4 to bacterial LPS.

Bacterial LPS and TLR4

LPS is a cell wall component of Gram-negative bacteria and a potent stimulator of the immune system, including TLR4. A dramatic result of human exposure to LPS, septic shock, is caused by the overwhelming secretion of cytokines (especially tumour necrosis factor α [TNF-α]) in response to uncontrolled systemic bacterial infection. It is interesting that mice, which lack TLR4 are resistant to septic shock but are highly sensitive to LPS-bearing bacteria such as *Salmonella*, which is a natural mouse pathogen.

LPS activates TLR4 through the action of two accessory molecules. In plasma, LPS is initially bound by the soluble LPS-binding protein, which then loads its attached LPS onto the glycosylphosphatidylinositol (GPI)-tethered membrane protein CD14. The LPS–CD14 complex then interacts with membrane-bound TLR4, which consequently sends a signal to the nucleus of the antigen presenting cell and thereby activates the transcription factor nuclear factor κB (NFκB), which in turn activates genes involved in defence against microorganisms. Several of these genes expressed code for the co-stimulatory molecules CD80 and CD86 – cell surface proteins produced by macrophages and DCs.

The complement system

The complement system is made up of many distinct plasma proteins reacting with one another to opsonize pathogens and induce a series of inflammatory responses that help fight infection. Early in an infection, the complement system can be activated on the surface of a pathogen through antigen–antibody immune complexes that bind to the bacterial cell to activate C1q. In each of the multiple complement pathways (Fig. 6.1), a sequence of reactions generates C3 convertase, a protease. This C3 convertase cleaves C3 into C3a, a mediator of inflammation, and C3b, which binds to pathogens and acts as an opsonin, targeting them for phagocytosis. In addition, later complement components combine to form a membrane-attack complex (MAC) that causes bacteriolysis.

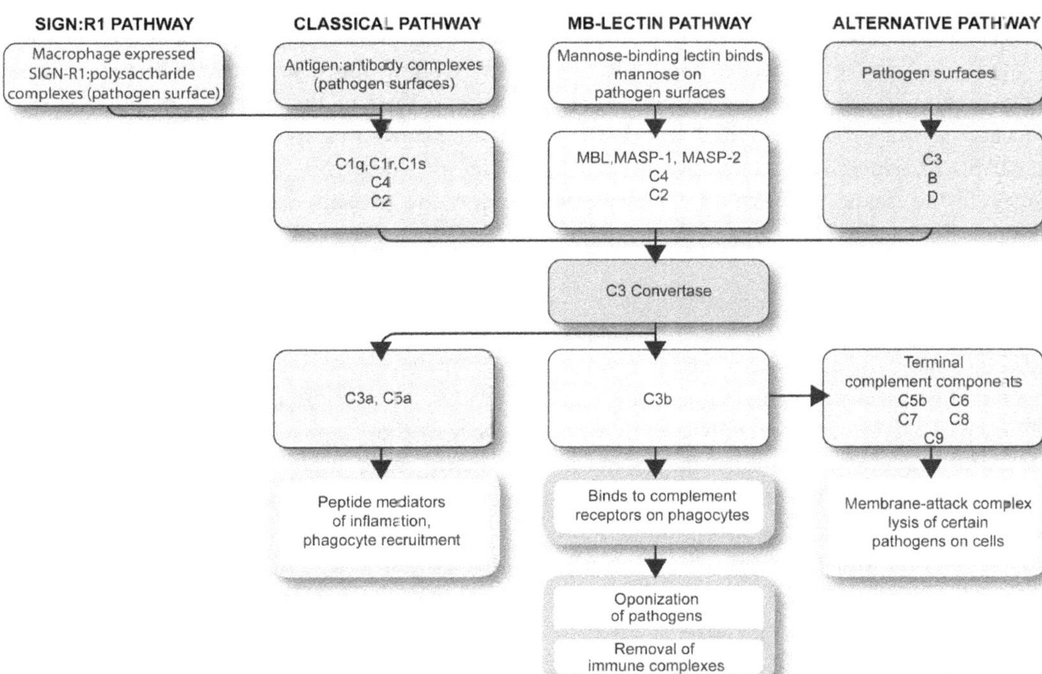

Figure 6.1 Overview of the complement pathways. Initial events surrounding each pathway results in the formation of a common component known as C3 convertase. At this point all pathways converge and generate the effector functions of complement. C3 convertase cleaves C3 into C3a and C3b. C3b functions to bind to bacterial surfaces promoting opsonization. C3a and terminal components of complement are involved in mediating inflammatory responses. Figure adapted from Janeway *et al.* (2005).

The classic pathway

The C1 complex comprises a single C1q molecule bound to two molecules each of the zymogens C1r and C1s. C1q has six globular heads; the binding of more than one of these heads to a pathogen's surface results in a conformational change in the C1 complex. This change leads to the activation of autocatalytic enzymatic activity in C1r. The active C1r cleaves its associated C1s to generate a serine protease. The protease then acts on the next two complement components, C4 and C2, generating two large fragments, C4b and C2b, which together form the C3 convertase.

The mannose-binding lectin pathway

This pathway is activated by mannose-binding lectin (MBL), a protein similar to C1q. As its name suggests, MBL binds to mannose residues on the surface of bacterial cells. As this binding takes place only when the residues are arranged in a specific configuration, random binding and activation of this destructive pathway does not occur. MBL does not bind to vertebrate cells, as sialic acid residues cover mannose. The expression of MBL can be up-regulated by the liver during the acute phase of an infection.

MBL is a two- to six-headed molecule that forms a complex with two zymogens, MBL-associated serine proteases 1 and 2 (MASP-1 and MASP-2), which have homology to C1r and C1s. As in the classic pathway, a C3 convertase is formed from C2b bound to C4b in the MBL pathway.

The alternative pathway

Unlike the classic and mannose-binding pathways, the alternative pathway of complement does not rely on a pathogen-binding protein for activation. Instead, this pathway depends on the spontaneous hydrolysis of C3, which yields the distinct C3 convertase designated C3bBb through the following mechanism. The spontaneous hydrolysis of C3 leads to C3(H_2O), which binds plasma protein factor B. This binding allows factor D to cleave factor B to Ba and Bb. Bb remains bound to C3(H_2O) to form C3(H_2O)Bb, a plasma C3 convertase that cleaves C3 to C3a and C3b; C3b attaches to the surface of pathogens and can then be bound by factor B and cleaved by factor D to form the full complex, C3bBb.

SIGN-R1 complement pathway for polysaccharides

The recently discovered complement pathway for polysaccharides (Kang *et al.*, 2006) is in some ways similar to the classic pathway but uses a unique lectin known as SIGN-R1. This lectin captures microbial polysaccharides in the spleen and acts together with complement to provide resistance to blood-borne infections, especially those due to *S. pneumoniae*. Unlike the classic pathway, which relies on antigen-antibody immune complexes bound to the bacterium and activating C1q, this pathway depends on the direct binding of macrophage expressed SIGN-R1 to the bacterial surface. C1q then binds SIGN-R1, and the same sequence of events seen in the classic pathway occurs.

Convergence of the complement pathways

The major role of the complement pathways is to form the C3 convertase, which, as mentioned previously, cleaves C3 into C3a and C3b. The deposition of large amounts of C3b on the surface of pathogenic cells triggers phagocytosis. C3b receptors on leukocytes are called CR1 (complement recognition receptor 1) and the binding of the CR1 to the C3b on the pathogens triggers phagocytosis in the presence of other complement factors. In response to some Gram-negative bacteria (e.g. *Neisseria meningitidis*), several additional proteins are activated after C3b, and a MAC consisting of C7, C8, and C9 is assembled on the surface of the organism. The MAC creates a pore-forming channel, and the organism subsequently lyses and dies.

Innate immune responses

Once macrophages recognize a pathogen through pattern recognition receptors or complement, their primary function is to remove the pathogen through phagocytosis, initiation of an inflammatory response, and activation of the adaptive immune system.

Phagocytosis

Phagocytosis is an active process in which the bound pathogen is first surrounded by the phagocytic membrane and then internalized in a membrane-bound vesicle known as a phagosome or endocytic vacuole. The phagosome undergoes an oxidative burst, releasing toxic oxygen radicals, and then becomes acidified, killing most pathogens. In addition to being phagocytic, macrophages and neutrophils possess membrane-bound granules called lysosomes, which contain enzymes, proteins, and peptides that can mediate antimicrobial responses. The phagosome can fuse with a lysosome to form a phagolysosome; the lysosomal contents are then released and destroy the pathogen.

In addition to the lysosomal contents, macrophages and neutrophils produce a variety of toxic products that aid in the killing of pathogens and the breakdown and presentation of bacterial polysaccharides, as will be described later. The most important of these products are nitric oxide (NO), superoxide anion (O_2^-), and hydrogen peroxide (H_2O_2). NO is produced by a high-output form of nitric oxide synthase ($iNOS_2$). Superoxide is generated by a multicomponent, membrane-associated NADPH oxidase (in a process known as the respiratory burst because it is accompanied by an increase in oxygen consumption) and is converted by the enzyme superoxide dismutase into H_2O_2. Further chemical and enzymatic reactions produce a range of toxic chemicals from H_2O_2, including the hydroxyl radical (OH) and the hypochlorite (OCl^-) and hypobromite (OBr^-) ions.

Another important effect of bacterial interaction with macrophages and other phagocytes is the release of cytokines and chemokines, which are responsible for establishing inflammation. Cytokines induce the expression of co-stimulatory molecules (e.g. on DCs) that activate the adaptive immune response.

Inflammatory responses

Inflammation plays three major roles in immunity: (1) the delivery of effector molecules to the site of infection; (2) the formation of a physical barrier to prevent microorganisms from entering the bloodstream; and (3) the repair of injured tissue. Inflammatory responses manifest as pain, redness, heat, and swelling. The first change seen in blood vessels is an increase in diameter, which contributes to heat, redness, and reduced blood-flow velocity. A second change is the expression of adhesion molecules on endothelial cells, with consequent binding and extravasation (movement into infected tissue sites) of circulating leukocytes. These changes are activated by cytokines and chemokines. A third change involves leakage in blood vessels and the movement of leukocytes and other cells to infected sites, which causes causing more swelling and pain as well as the accumulation of plasma proteins (including complement) that aid in the host's defence. Together, inflammatory changes facilitate efficient phagocytosis and activation of the adaptive immune response.

Cytokines and chemokines

Cytokines and chemokines are small proteins (~25 kDa) released by a variety of cells (e.g. macrophages) in response to an activating signal resulting from pattern recognition of pathogens. Bound to specific receptors, cytokines can exert autocrine effects on the same cell that secreted them, paracrine effects on adjacent cells, or rare endocrine effects on distant targets reached through the bloodstream. There are three families of cytokines: (1) the haematopoietin family, which includes growth hormones and the structurally diverse interleukins; (2) the cxc and c-c chemokine families; and (3) the tumour necrosis factor (TNF) family, which encompasses important mediators of innate and adaptive immunity, especially for LPS.

Chemokines are chemoattractant cytokines (all related by amino acid structure) that bind to integral membrane-associated receptors and recruit leukocytes, neutrophils, and other effector cells from the blood to the site of infection. CC chemokines have two adjacent cysteines near the amino terminus, whereas CXC chemokines have two N-terminal cysteines separated by a single amino acid. CC chemokines bind to CC receptors and CXC chemokines to CXC receptors.

The adaptive immune response

In an adaptive immune response, antigen is recognized by two distinct sets of highly variable receptor molecules: immunoglobulins, which serve as antigen receptors on B-cells, and antigen-specific receptors on T-cells. T-cells recognize only antigens that are presented on cell surface major histocompatibility complex (MHC) glycoproteins. Like protein antigens, zwitterionic polysaccharides are presented on MHC class II (MHCII) molecules.

Antigen presentation on MHCII molecules

Unlike MHCI molecules, which deliver antigens originating in the cytosol (e.g. intracellular virus antigens) to the surface, MHCII molecules deliver antigens of extracellular origin that have been endocytosed and degraded in acidified endocytic vesicles. As mentioned earlier, immunological doctrine until recently held that only proteins and peptides are presented on MHCII and, in contrast, that polysaccharides cannot be presented and therefore cannot activate T-cells. This viewpoint was contradicted by the discovery and characterization of ZPSs such as polysaccharide A (PSA) from the intestinal anaerobic commensal *Bacteroides fragilis*. Once ZPSs are cleaved through the action of oxygen and nitrogen species, they are loaded onto MHCII. In this process, endocytic vesicles containing degraded polysaccharides combine with lysosomes to form endolysosomes. The endolysosome combines with an MHCII molecule-carrying vesicle that originated in the endoplasmic reticulum (ER), forming the MHCII compartment (MIIC) where loading of the depolymerized polysaccharide onto MHCII occurs.

Generation of MHCII

The generation of MHCII begins in the ER. At this early stage, a peptide called the invariant chain (li) is bound to the MHCII glycoprotein. The li forms a trimer that binds to the αβ heterodimer of MHCII; part of the li binds within the binding groove (peptide-binding cleft), thereby blocking the binding of other peptides. This chain also functions to deliver MHCII to the compartment (MIIC) where antigen loading can occur. The li:MHCII is retained in the MIIC for 2–4 hours. During this period, acid proteases such as cathepsin S act on the li (Fig. 6.2). Initial cleavage truncates the li; after further cleavage, a single peptide called CLIP (*cl*ass II–associated *i*nvariant *p*eptide) remains bound to the MHCII binding groove. In order to bind antigens, this peptide must be displaced, and this function is carried out by a specialized molecule.

This molecule was identified in human B-cell lines with a defective antigen presentation pathway. In this setting, MHCII was still presented, but with CLIP bound rather than antigen. The defective molecule was found to be an MHCII-like molecule called HLA-DM (H-2M in mice). HLA-DM, which is found within the MIIC, binds and stabilizes 'empty' MHCII, catalyses the release of CLIP, and promotes binding of antigen to the newly emptied binding groove. HLA-DM can continually bind and rebind different antigens,

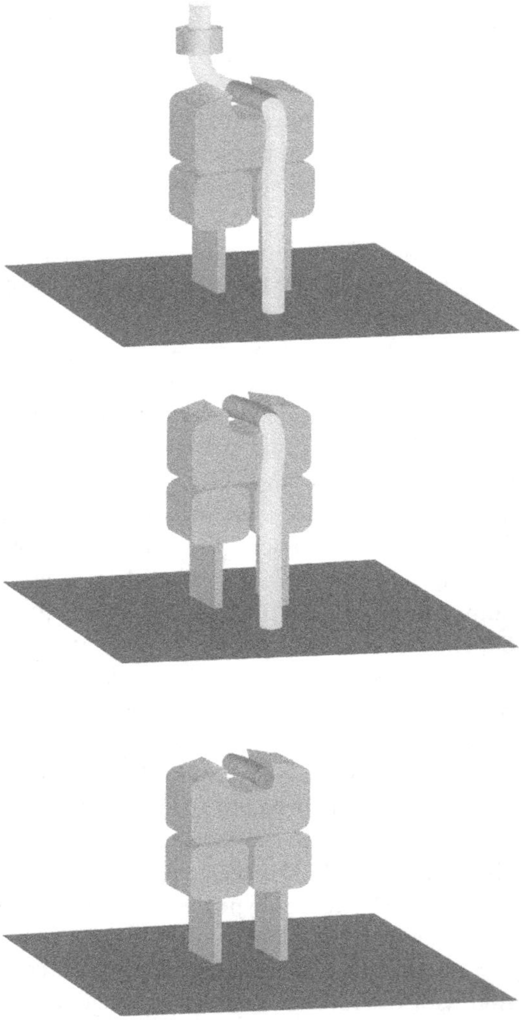

Figure 6.2 Cleavage of the invariant chain leaves CLIP bound to MHCII. Within the endoplasmic reticulum the invariant chain binds to the MHCII such that the CLIP peptide is within the binding groove. Cleavage in acidified vesicles with cathepsin S results in CLIP remaining bound to the MHCII groove. The release of CLIP and exchange of presented peptides occurs through the action of HLA-DM. Figure adapted from Janeway *et al.* (2005).

ensuring the presence of peptides with the strongest binding affinity; this function is important, as the MHCII–antigen complex may have to survive on the surface of APCs for days before encountering $CD4^+$ T-cells.

T-cell-mediated immunity

T-cell activation

To become an effector cell, a naïve T-cell must not only recognize antigen in the presence of MHCII but also be activated by co-stimulatory molecules on professional APCs. Only macrophages, DCs and B-cells are capable of expressing both MHCII-antigen complexes and co-stimulatory molecules.

Unlike innate immune responses, adaptive immune responses are not initiated at the site of infection but rather in lymphoid tissues. The innate response delivers APCs (including macrophages and DCs) to the lymphoid system; the APCs accumulate in lymphoid tissues, where T-cells also reside. Naïve T-cells then sample MHCII-antigen complexes within these tissues. Since only 1 in ~10^6 T-cells can recognize a specific antigen, activation of the adaptive response may take a few days; at that point, the T-cell ceases to migrate and generates effector T-cells.

For naïve T-cells to become active, three events must occur: initial weak binding to bring the T-cell and the APC together, recognition of the MHCII–antigen complex by the T-cell receptor (TCR), and co-stimulation by accessory molecules.

Initial binding of T-cell and APC

T-cells migrate from the circulation to specific lymph nodes through the action of the integrin lymphocyte function–associated molecule-1 (LFA-1). Through activation of APC receptors (including TLRs), professional APCs (e.g., DCs) migrate to make contact with the T-cells. (For example, if DCs come into contact with LPS, TLR4 is activated, NF-κB and subsequently TNF-α are released, and the DCs migrate through tissues to contact T-cells.) The T-cells (through LFA-1) and the APCs [through intercellular adhesion molecules (ICAMs) 1, 2, and 3] bind.

As described earlier, T-cell recognition of a foreign antigen expressed by MHCII relies on the TCR. However, activation of the T-cell by its TCR requires a signalling event initiated by an accessory receptor on the T-cell – i.e. either $CD4^+$ or $CD8^+$. With MHCII, $CD4^+$ molecules are required; with MHCI, a $CD8^+$ co-signal is necessary. $CD4^+$ alerts the T-cell to the detection by the TCR of foreign antigen bound via MHCII.

Clonal expansion of naïve T-cells

Once T-cells have been activated by contact with foreign antigens, other co-stimulatory receptors are required for their clonal expansion. The B7 co-stimulatory molecules CD80 (B7.1) and CD86 (B7.2) are the best characterized. These two molecules, which are structurally and functionally similar, are present on professional APCs and bind to a TCR known as CD28. This interaction is absolutely required for clonal expansion of T-cells. Once this signal is generated, the T-cell expresses more proteins that help sustain its expansion and differentiation; for example, CD40 ligand (CD154) binds to the CD40 receptor on the APC. To control the expansion response, a T-cell can also express CTLA-4, a receptor that is similar to CD28 but that binds B7 molecules with 20-fold greater avidity.

Upon activation, T-cells are driven to expand by IL-2, which is produced by the activated T-cell itself. Expression of the IL-2 receptor enhances the response of the same T-cell to IL-2. The production of IL-2 is driven in part by the transcription factor nuclear factor of activated T-cells (NFAT), which is produced after antigen recognition by the T-cell. Without IL-2, T-cells will not proliferate in response to infection; indeed, IL-2 can be used as an effective target for suppression of immune responses in settings such as transplant rejection. After 4 or 5 days of rapid expansion induced by IL-2, activated T-cells differentiate into effector T-cells (helper T-cells) whose exposure to specific antigen does not require mediation by co-stimulation. Helper T-cells play an important role in humoral immunity and in the activation of B-cells to produce antibodies.

The humoral immune response

Pathogens (e.g. viruses) can often enter cells without being phagocytosed; there they reside and even replicate, but they generally must pass from cell to cell via the extracellular space. The function of the humoral immune response is to destroy pathogens in the extracellular space; to this end, antibodies are produced by B-cells that have differentiated into antibody-secreting cells. This process generally requires helper T-cells (usually the T_H2 type, although a T_H1 subset can also carry out this function). By binding to pathogens antibodies contribute to immunity in three ways.

- They prevent the pathogen from entering a cell (and thereby protecting itself from extracellular recognition) in a process known as neutralization.
- They facilitate phagocytosis through opsonization, which is achieved in either of two ways: antibodies may be recognized by F_C receptors on phagocytic cells, or they may activate complement.

B-cell activation

B-cells have several immunological functions. B-cells can internalize and present antigen on MHCII. Signals from T-cells, upon recognition of the B-cell MHCII-antigen complex, stimulate the B-cells' differentiation into antibody-secreting cells.

Antibody responses to antigens are classified as either thymus-dependent (TD) responses, which require helper T-cells, or thymus-independent (TI) responses, which are elicited directly by polysaccharides on the surface of pathogens.

Thymus-dependent antigens

As described earlier, T-cells respond to specific antigens by proliferating into helper T-cells. This event must occur before these T-cells can activate B-cells to produce antibodies. A T-cell facilitates B-cell proliferation only when the B-cell displays an antigen that the T-cell recognizes. This type of T-cell-mediated response is extremely important in the case of polysaccharide-based vaccines, such as that to *Haemophilus influenzae* type b.

When a helper T-cell recognizes antigen bound to MHC on a B-cell, several T-cell molecules that activate that B-cell are synthesized. One such molecule, CD40 ligand, binds to CD40 on the B-cell. The B-cell responds by expressing B7 molecules, which sustain T-cell growth and differentiation. Both CD40 ligand and IL-4 (a cytokine that is also produced by T-cells, specifically T_H2 cells) drive the clonal expansion, or proliferation, of B-cells that precedes antibody production. IL-5 and IL-6, both secreted by helper T-cells, allow B-cells to differentiate into antibody-secreting plasma cells. As will be discussed later, these antibodies are important molecules in effecting antigen binding.

Thymus-independent antigens

Although antibody responses are in many instances dependent on helper T-cells, mice and humans with T-cell deficiencies can still generate antibody responses, especially those to bacterial polysaccharides. The responsible antigens in these latter cases are known as TI antigens and are of two classes.

TI-1 antigens are also known as B-cell mitogens because, at high concentrations, they stimulate B-cell proliferation and differentiation. In a process known as polyclonal activation, this TI-1 stimulation occurs regardless of the B-cell's antigenic specificity. A classic example of a TI-1 antigen is LPS, which binds to LPS-binding protein and CD14 and then associates with TLR4 on B-cells. An important point is that LPS acts as a mitogen or polyclonal activator only at a concentration at least 100 times that needed to stimulate DCs; a lower concentration stimulates only B-cells with a specific receptor for LPS.

TI-2 antigens consist of molecules (e.g. capsular polysaccharides) that have highly repetitive structures. Unlike TI-1 antigens, TI-2 antigens have no intrinsic ability to activate B-cells. Instead, it is thought that activation of B-cells by these antigens results from B-cell receptor (BCR) cross-linking due to the long and repetitive structure. If there are too few cross-links, no activation occurs; paradoxically, too much activation can result in inactivation of a B-cell. The antigen concentration is therefore critical.

Why is it important for the immune system to recognize TI-2 antigens? Bacteria have evolved a number of mechanisms for evading immune responses, among them the production of capsular polysaccharides that enable the bacteria to avoid phagocytosis and thereby prevent the presentation of bacterial antigens to T-cells. In short, through capsular polysaccharides, both the innate and adaptive immune responses are bypassed. TI-2-responding B-cells, however, can rapidly produce IgG and IgM antibodies that bind to TI-2 capsular polysaccharides and aid in their uptake by professional APCs.

Structure of an antibody

Antibodies are the secreted form of the BCR. Each antibody (Fig. 6.3) is composed of three fairly equal-sized structures connected by a flexible hinge region. The antibody contains two heavy chains (50 kDa each) and two light chains (25 kDa each). At the tip of each arm is a variable region that permits antigen recognition. The major difference between an antibody and a BCR is that the C terminus of the BCR has a hydrophobic membrane-anchoring region, whereas the antibody has a hydrophilic region allowing secretion. The specificity of the antigen-binding site is determined by two variable domains, V_H and V_L (H and L are the *h*eavy and *l*ight chains). The effector action of the antibody is determined by the isotype of its heavy chain constant region, of which there are five isotypes IgA, IgD, IgE, IgG, and IgM. All naïve B-cells express IgM and IgD; however, the most abundant isotype in plasma is IgG. In a process known as isotype switching, a B-cell switches to production of a different antibody isotype in response to a positive stimulus. Numerous signals released from helper T-cells allow isotype switching to occur. T_H2 cells release IL-4, which induces a switch to IgG1 and IgE; transforming growth factor β (TGF-β) induces IgG2b and IgA. T_H1 cells, in contrast, are poor inducers of isotype switching but contribute to the process through the release of interferon (IFN) γ, which promotes the production of IgG2a and IgG3.

Effector functions of antibodies

Listed below are some of the main effects of antibodies that aid the immune system in protecting the host from infection by bacteria and viruses.

Blocking bacterial and viral adherence to host cells

Bacteria contain surface proteins, knows as adhesins, that facilitate their binding to the host cell surface. For example, *Neisseria gonorrhoeae*, the causative agent of gonorrhoea, binds to epithelial

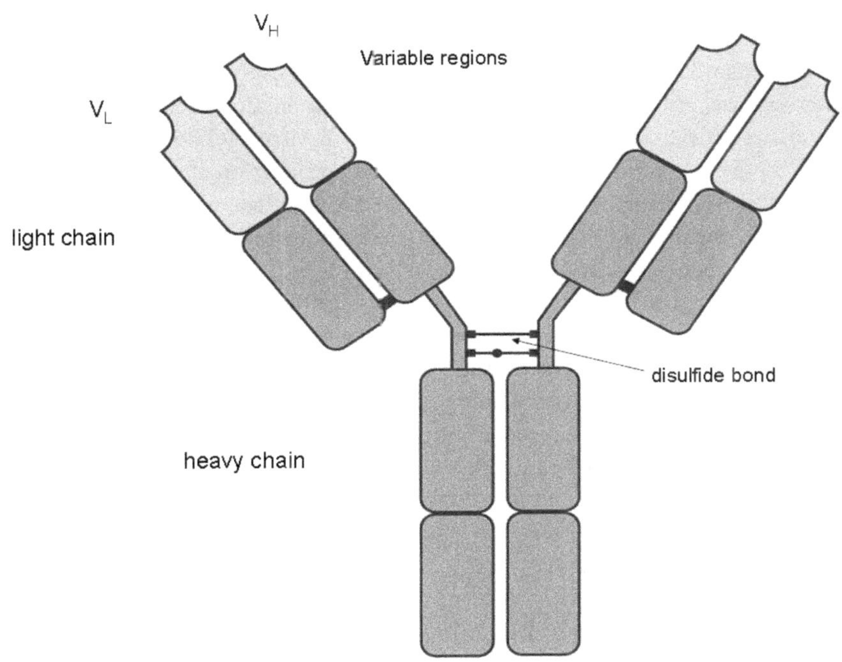

Figure 6.3 A simplified representation of an antibody molecule. Three globular regions form a Y shaped antibody structure, with a flexible hinge region. Variable regions at the N terminus of the antibody allow interactions with specific antigens. The heavy chain (denoted with the letter H) determines the class of antibody (IgM, IgD, IgA, IgE and IgG). Figure adapted from Janeway *et al*. (2005).

cells in the urinary tract via pilin. IgA antibodies can bind to pilin and prevent infection. IgA and IgG can have a similar effect against viruses. Haemagglutinin on influenza viruses binds to sialic acid residues on carbohydrates on host cells in the respiratory tract. Antibodies can bind to haemagglutinin and prevent infection.

Activating complement: the classic pathway

In the classic pathway, complement is activated when antibodies bind to C1q (which, along with C1r and C1s, makes up C1, the first component of complement). Because of structural binding requirements, it is generally IgG or IgM that binds to C1q. What is interesting is that these antibodies cannot activate complement in solution but must be bound to multiple sites on a pathogen surface. The binding of C1q to antibody activates enzymatic activity in C1r, triggering the complement cascade.

Destroying pathogens

Antibodies can effectively neutralize pathogens during infection, but these pathogens still need to be removed and destroyed. Effector cells express specific F_C receptors that allow very tight binding between effector cells and pathogens coated with F_C antibodies. This binding promotes phagocytosis by certain cells with F_C receptors, including macrophages, natural killer cells, and DCs. Phagocytes are activated by IgG antibodies. The binding of the F_C receptors on phagocytes to the IgG bound to bacteria and viruses results in phagocytosis, destruction of the pathogen, and a consequent immune response to degraded antigens. It is therefore critical for F_C receptors to distinguish antibody bound to pathogen from free soluble antigen; this distinction is made through binding avidity. F_C receptors bind much more avidly to antibodies that are multimerized (like those on pathogenic surfaces) than to individual immunoglobulin monomers.

T-cell activation by zwitterionic polysaccharides

The immune system has evolved to permit T-cells to recognize a huge array of biological antigens (including peptides, superantigens, and glycolipids) through their presentation by MHC proteins such as MHCI, MHCII, and CD1 (Watts and Powis, 1999). Carbohydrates, in contrast, have been considered to be exclusively T-cell-independent antigens. However, it has recently been demonstrated that sugars play an important role in immune recognition. Unexpected additions to the list of T-cell antigens are the zwitterionic polysaccharide A (PSA) from *B. fragilis* (Tzianabos *et al.*, 2000) and the type 1 polysaccharide of *S. pneumoniae*. PSA is the first polysaccharide shown to activate T-cells via presentation on MHCII proteins.

ZPSs as T-cell antigens

Carbohydrate-containing structures are abundant on the surface of Gram-negative and Gram-positive bacteria and include LPSs, teichoic and lipoteichoic acids, peptidoglycans, and glycoproteins. These carbohydrates can consist of several hundred repeating units; each repeating unit is made up of one to eight sugars that are usually linked by glycosidic bonds. Differences in sugar composition, ring forms, linkage positions, and other factors contribute to the variety of epitopes and the huge number of structures that can interact with the immune system.

Several decades ago, studies in mice showed that carbohydrates are T-cell-independent antigens. Purified polysaccharides induced specific IgM responses with no detectable IgG. This failure to induce immunoglobulin class switching from IgM to IgG and a failure to increase antibody production after re-challenge are hallmarks of T-cell-independent antigens. Only conjugation of carbohydrates to protein antigens has allowed carbohydrate-specific responses and engagement of T-cell responses. A series of recent investigations, however, identified a new type of polysaccharide, the ZPS, whose immunological activity departs from the traditional view of carbohydrates as exclusively T-cell-dependent antigens.

Thirty years ago, investigations into the cause of intra-abdominal abscesses in patients recovering from surgery highlighted the role of commensal bacteria in abscess induction. The most prominent bacterium recovered from these abscesses was the Gram-negative obligate anaerobe *B. fragilis*, a gut commensal in all mammals. Investigators found that administration of *B. fragilis* PSA alone protects animals from abscess formation. This observation raised intriguing questions about a possible immunomodulatory role for commensal organisms.

PSA from the surface of *B. fragilis* is a ZPS that functions as an immunomodulatory molecule, specifically activating T-cells. The immunodominant constituent of *B. fragilis* capsular polysaccharides, PSA, possesses both a positive and a negative charge within its repeating unit (the hallmark of ZPSs). A distinctive feature of *B. fragilis* is the large proportion of its genome devoted to carbohydrate metabolism, including the degradation of dietary polysaccharides and the production of surface capsular polysaccharides (Coyne and Comstock, 2008; Wexler, 2007; Xu *et al.*, 2003). Thus far, eight different capsular polysaccharides, each synthesized from a distinct genomic locus, have been identified within a single strain of *B. fragilis*. Each polysaccharide synthesis locus contains 11–21 genes under the control of a single promoter. Seven of the eight loci are regulated by the reversible inversion of the DNA containing the promoter, which results in an ON-OFF expression phenotype known as phase variation. The loci are regulated independently, and a bacterium may simultaneously express any combination of polysaccharides; thus there are 256 possible distinct surface profiles (Coyne *et al.*, 2003; Krinos *et al.*, 2001). It has been proposed that this great potential variation results in a bacterial population best fit to survive in the intestinal milieu. Multiple polysaccharides have been shown to enhance intestinal colonization by *B. fragilis* (Liu *et al.*, 2008).

Initial evidence that PSA can indeed activate T-cells was provided by the observation that T-cells from mice immunized with purified PSA were associated with elevated levels of the T-cell-activation cytokine IL-2. Subsequent *in vitro* studies showed that PSA induces the proliferation of CD4$^+$ T-cells cultured in the presence of APCs. Depletion of CD4$^+$ T-cells results in the loss of PSA-induced proliferation. Only specific T-cell subsets (i.e. CD4$^+$ but not CD8$^+$) are activated. The major challenge to the rule that polysaccharides are T-cell-independent antigens was the demonstrated requirement for APCs for T-cell activation. These early investigations marked the beginning of a line of research into the ability of commensal organisms to elicit T-cell protective mechanisms.

Structure of ZPSs

B. fragilis is unique among both *Bacteroides* and non-*Bacteroides* species in that it has no fewer than eight distinct polysaccharides on its surface. High-resolution nuclear magnetic resonance (NMR) showed that two of these polysaccharides, PSA and PSB, have unique structures and contain positively and negatively charged motifs in each repeating unit. As most bacterial polysaccharides are either negatively charged or neutral, it was proposed that the additional positive charge enables these ZPSs to activate T-cells. Chemical neutralization of either charged group abrogated the polysaccharides' ability to protect against abscess formation and proved the importance of the zwitterionic motif. In addition to PSA and PSB of *B. fragilis*, ZPSs have been identified in different bacterial species, including *S. pneumoniae* (type 1 capsular polysaccharide, CP1) and *Staphylococcus aureus* (types 5 and 8). All of the known ZPSs have immunomodulatory properties that are dependent on the presence of both positively and negatively charged groups within the repeating unit structure (Tzianabos *et al.*, 2000; Tzianabos *et al.*, 2001; Tzianabos *et al.*, 1993). Of all known ZPSs, PSA is the best characterized.

Structure of PSA

PSA (Fig. 6.4) consists of several hundred repeating units of a tetrasaccharide and has a molecular weight of ~110 kDa. The three-dimensional structure of the highly related PSA2 molecule (clinical isolate 638R) shows a right-handed helix with two repeating units per turn and a pitch of 20 Å. The molecule is covered with positive and negative charges that alternate along the sides of the helical backbone and are exposed on the outermost surface. This conformation offers a plausible explanation for the interactions of ZPSs with other molecules. A proposed 'groove-binding' model (Wang *et al.*, 2000) provided an explanation for PSA's T-cell stimulatory ability. On the basis of computer modelling studies, the authors proposed that PSA forms a complex with MHC by docking onto the α-helices that make up the lateral boundaries of the peptide groove of MHC molecules. Therefore, they proposed that PSA – a pure polysaccharide – is presented on MHC molecules to T-cells. This mechanism had previously been reserved for protein antigens.

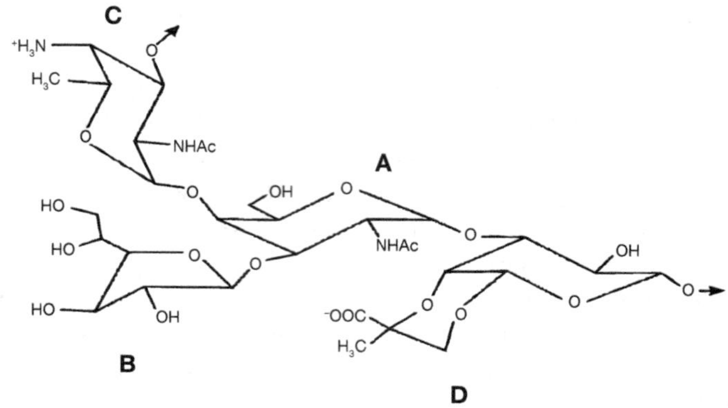

Figure 6.4 Chemical structure of PSA A2. The repeating unit of PSA consists of 2-acetamido-2-deoxy-α-D-galactopyranose (residue **A**) β-D-galactofuranose (residue **B**) 2-acetamido-4-amino-2,4,6-tri-deoxy-α-D-galactopyranose (residue **C**) and β-D-galactopyranose with a pyruvic acid acetal substituent (residue **D**). Zwitterionic motifs are highlighted.

Presentation of ZPSs on MHC molecules

Although ZPSs do initiate T-cell responses, how they do so is not yet clear. MHC molecules comprise of MHC class I, which are present on most tissues and constitutively sample peptides produced through the normal turnover of cells and MHCII molecules that are expressed by a population of professional APCs, including DCs, macrophages, and B-cells. It has long been accepted that MHC-like molecules (CD1) can present non-protein antigens to T-cell-like cells (natural killer cells). Until recently, however, the possibility that pure polysaccharides could be displayed on MHC molecules was considered unlikely.

In vitro studies of the potential involvement of MHC molecules in PSA presentation showed that all three types of professional APCs were required for PSA activation of T-cells. This activity also required physical contact between the APC and the T-cell. Therefore, it was hypothesized that PSA is presented directly to the TCR by MHCII molecules. Consistent with this hypothesis, B-cells lacking MHCII did not activate T-cells in the presence of PSA, and antibody-mediated neutralization of MHC haplotypes identified the involvement of HLA-DR. Immunoprecipitation showed direct physical attachment of PSA to HLA-DR, a result confirming the role of MHCII in PSA presentation.

Processing of PSA

The oxidative burst

Among the earliest events induced in bacterial phagocytosis by APCs is the oxidative burst, which is characterized by the production of strong oxidizing molecules stemming primarily from the formation of superoxide and NO through two separate pathways (Foreman *et al.*, 2003). First, superoxide anion is synthesized from oxygen through the enzymatic action of NADPH oxidase (Pollock *et al.*, 1995). The downstream products that are then made include superoxide dismutase (SOD), which converts superoxide into hydrogen peroxide, and myeloperoxidase (MPO), which converts hydrogen peroxide into hypochlorous acid; these products are collectively known as reactive oxygen species (ROS). In the second arm of the burst, represented by the reactive nitrogen species (RNS), the central player is the free radical NO formed from L-arginine by the enzyme inducible nitric oxide synthase (iNOS) (Laubach *et al.*, 1995).

Degradation and presentation of PSA within APC endosomes

Once within the APC endosome/lysosome, PSA is degraded to a smaller molecule (~10–15 kDa) before being presented to the CD4$^+$ T-cell. The surprising dependence of this degradation on iNOS indicates that carbohydrates are degraded by a different mechanism than protein antigens. Specifically, PSA appears to be degraded predominantly by RNS through deamination (Duan *et al.*, 2008).

Once PSA is degraded, can it be presented on MHC molecules in a similar manner to protein antigens? Several lines of evidence suggest that this is indeed the case. First, similar to protein antigens, PSA traffics through the endocytic pathway after internalization by APCs, co-localizing with the endosomal markers lysosomal-associated membrane protein 1 (LAMP1) and HLA-DM in a process that requires the polymerization of actin and microtubules and the acidification of endosomes (Cobb *et al.*, 2004). Second, confocal microscopy reveals the presence of PSA at the immunological synapse between APCs and T-cells – that is, at the site of engagement between MHCII molecules and the αβ-TCR. This observation constitutes strong evidence that a tertiary complex of PSA, MHC, and the TCR mediates the activities of PSA. Further, in MHCII-deficient cells, PSA remains in the intracellular compartment and cannot be trafficked to the cell surface (Cobb *et al.*, 2004). Third, PSA binds to MHCII in a manner that requires the peptide-exchange factor HLA-DM.

PSA and the immune response

TLR2 is activated by several microbial products, including peptidoglycans and lipoproteins from both Gram-positive and Gram-negative bacteria (Takeda *et al.*, 2003). TLR2 is also the receptor that recognizes PSA. PSA-activated signalling through TLR2 is necessary to initiate the innate and adaptive immune responses to the carbohydrate. PSA stimulation of TLR2 on the APC results in activation of transcription factor NF-κB, with subsequent production of NO, the proinflammatory cytokine TNF-α, and other immunologically important molecules such as IL-12 (Wang *et al.*, 2006). NO is necessary for efficient chemical degradation of PSA in the MIIC (Cobb and Kasper, 2005). TLR2 also plays a critical role in the up-regulation of MHCII, CD86 CC and chemokine receptor 7 (CCR7) on the APC. CCR7 is associated with DC homing to secondary lymphoid organs (Forster *et al.*, 1999).
From these results, it is apparent that TLR2 signalling is required for optimal DC activation and initiation of the adaptive immune response to PSA. Consequently, bone marrow DCs (BMDCs) lacking TLR2 ($TLR2^{-/-}$) have a decreased ability to activate $CD4^+$ T-cells in response to PSA. $TLR2^{-/-}$ BMDCs produce significantly less IL-12 than do wild-type BMDCs. The decrease in IL-12 reduces but does not eliminate IFN-γ production by T-cells; the residual IFN-γ production in the absence of TLR2 may be the result of inefficient antigen processing and presentation. Other immune pathways may also be involved in PSA-mediated production of IFN-γ. Blocking of the acidification of the MIIC with chloroquine completely eliminates production of the cytokine, a result indicating that TLR2 signalling on DCs is necessary for the differentiation of T_H1 cells and the establishment of a proper T_H1/T_H2 balance (Wang *et al.*, 2006).

PSA and the inflammatory response

Much research has focused on the response of the immune system to polysaccharides such as PSA. It is important to understand that the response of the immune system does not have to be proinflammatory. Some molecules, including PSA, can actually suppress the immune response. This phenomenon is not entirely surprising, given that *B. fragilis* is a gut commensal that coexists with the mammalian host and provides immunological benefits. A critical aspect of commensalism is the capacity of the host's immune system to tolerate the presence of a microorganism without mounting an attack against it.

Prevention of inflammatory disease

Inflammatory bowel disease (IBD) is associated with uncontrolled inflammatory $CD4^+$ T-cell responses to bacteria in the gastrointestinal tract (Bouma and Strober, 2003; Cong *et al.*, 1998). Inflammatory responses resulting in IBD are frequently directed against bacteria such as *Helicobacter* and *Clostridium* – i.e. species that abound in the gut microflora of healthy individuals (Pamer, 2007; Balish und Warner, 2002); symptoms of IBD are alleviated if bacterial numbers in the intestine are decreased by antibiotic treatment (Videla *et al.*, 1994; Hoentjen *et al.*, 2003). Recent research has demonstrated that PSA produced by *B. fragilis* in the gut may be capable of preventing IBD induced by bacterial agents.

What changes are seen with the presence of *B. fragilis*? Lymphocyte populations in the spleens of germ-free (GF) mice include a reduced proportion of $CD4^+$ T-cells. However, colonization of these mice with *B. fragilis* restores the $CD4^+$ population to levels found in conventional colonized mice. Characterization of the $CD4^+$ T-cell population in the spleen has revealed that GF mice have a smaller proportion of $CD4^+CD45Rb^{low}$ T-cells than conventionally colonized mice. The latter cells are thought to be 'antigen experienced' and have anti-inflammatory properties that can provide protection in animal models of IBD (Izcue *et al.*, 2006). Once again, however, colonization of GF mice with *B. fragilis* increases the proportion of $CD4^+CD45Rb^{low}$ cells in the spleen to conventional levels. This effect is PSA dependent: colonization with a *B. fragilis* mutant not expressing PSA (ΔPSA) has no effect.

Protection against experimental colitis

PSA protects mice in an experimental model of colitis. Mice lacking recombination-activating genes ($Rag^{-/-}$) have no T or B-cells. If $CD4^+CD45Rb^{high}$ T-cells are transferred into $Rag^{-/-}$ mice and the mice are colonized with *Helicobacter hepaticus*, the animals develop colitis within 2 months (Maloy *et al.*, 2003). If, at the time of colonization with *H. hepaticus*, the mice are co-colonized with a PSA-producing strain of *B. fragilis*, they are protected from disease and the associated pro-inflammatory cytokine responses. However, if the mice are co-colonized with a mutant deficient in PSA production rather than with the wild-type strain, they develop full-blown colitis.

Thus, the presence of a single polysaccharide produced by a single organism is sufficient to prevent colitis induced by colonization with a commensal bacterium that has pathogenic potential. PSA appears to dampen the pro-inflammatory response leading to colitis by repressing a particular set of T-cells: T_H17 cells. Classically, $CD4^+$ helper cells were thought to differentiate into only two lineages, T_H1 and T_H2. However, recent research has shown that $CD4^+$ cells can also differentiate into T_H17 cells, which are characterized by the production of IL-17 and not by that of IFN-γ or IL-4 (Chen and O'Shea, 2008). T_H17 cells are key mediators of inflammation, as IL-17 is a potent chemoattractant for neutrophils (Matsuzaki and Umemura, 2007). Increased levels of IL-17 have been found in inflamed mucosa and sera from patients with IBD (Fujino *et al.*, 2003). Protection against colitis mediated by *B. fragilis* PSA was found to be the result of modulation of cytokine production. Levels of cytokines critical in the T_H17 lineage (IL-1β, IL-23, and TNF-α) are elevated in the colons of $Rag^{-/-}$ mice colonized with *H. hepaticus*. Colonization with *B. fragilis* in conjunction with *H. hepaticus* reduces levels of these pro-inflammatory cytokines to those found in SPF $Rag^{-/-}$ mice.

Therapeutic activity of PSA against IBD

PSA has been given therapeutically to decrease pro-inflammatory cytokine production in an experimental model of colonic irritation (Elson *et al.*, 1996). The cytokines induced by rectal trinitrobenzene sulfonic acid (TNBS) mimic those found in the *H. hepaticus* colitis model. $CD4^+$ T-cells from the spleens and mesenteric lymph nodes of TNBS-treated animals produce elevated levels of IL-23, TNF-α, and IL-17. Therapeutic administration of PSA to TNBS-treated animals significantly reduces levels of these proinflammatory cytokines. In addition, PSA decreases the expression of RORγt, the transcription factor that directs the T_H17 differentiation pathway (Ivanov *et al.*, 2006).

PSA suppression of the inflammatory immune response is dependent on IL-10, one of the most potent anti-inflammatory cytokines in the immune system. Previous research has shown that IL-10 plays a critical role in protection against inflammation in several animal models, including the *H. hepaticus* model of colitis (Xavier and Podolsky, 2005; Jankovic *et al.*, 2002).

Conclusions

B. fragilis represents <1% of the intestinal flora but exerts a powerful effect on the host immune system. PSA, the major capsular polysaccharide of *B. fragilis*, not only has the unique potential to be presented on MHCII molecules and to stimulate T-cell-dependent mechanisms but also can suppress the inflammatory response (as seen in IBD). Clearly, the impact of polysaccharides on the immune system warrants further research, particularly with regard to vaccine development.

Polysaccharides and vaccines

Polysaccharides make up a large proportion of the surface molecules of bacteria and thus are attractive targets for vaccinology. Unfortunately, their use as vaccines is complicated by several issues. For example, polysaccharides are generally T-cell-independent antigens and as such can be poorly immunogenic. Moreover, many of the polysaccharides on the bacterial surface closely resemble human polysaccharides. In the 1920s, substantial progress toward solving the first problem was made discovering that conjugation of polysaccharides to protein carriers rendered the carbohydrates more immunogenic (Avery *et al.*, 1929). However, the remarkable success of antibiotic treatment coincided with Avery's research, and this approach remained on the sidelines until relatively recently, when the limitations of antibiotics were fully comprehended and the need for alternatives became pressing.

Although protein vaccines, particularly toxoids, have proved successful, protective vaccines against important bacterial pathogens carrying polysaccharide capsules (e.g. *N. meningitidis*, *H. influenzae*, and *S. pneumoniae*) have been hampered by many drawbacks including low immunogenicity in young children, poor induction of memory response, and the provision of only temporary protection (Lesinski and Westerink, 2001a). The polysaccharide molecules used in these vaccines have repeated carbohydrate domains and can activate B-cells directly without T-cell help (i.e. they are TI-2 antigens). Although these types of polysaccharides are characterized as poor inducers of immunological memory with exceptions (Obukhanych, 2006), they do, in fact, rapidly induce IgM responses and can induce IgG in adults. In developing vaccines that target encapsulated bacteria, the differences between the immunological contributions of proteins and polysaccharides must be understood and taken into account so that immune responses can be optimized.

Conjugate vaccines

In an attempt to capitalize on the advantages of both types of antigen, glycoconjugate vaccines have been produced wherein the polysaccharide molecule is conjugated to a protein carrier (Fig. 6.5) (Siber, 1994). In these preparations, the recruitment of T-cell help for polysaccharide antigens relies on the covalent linkage of the

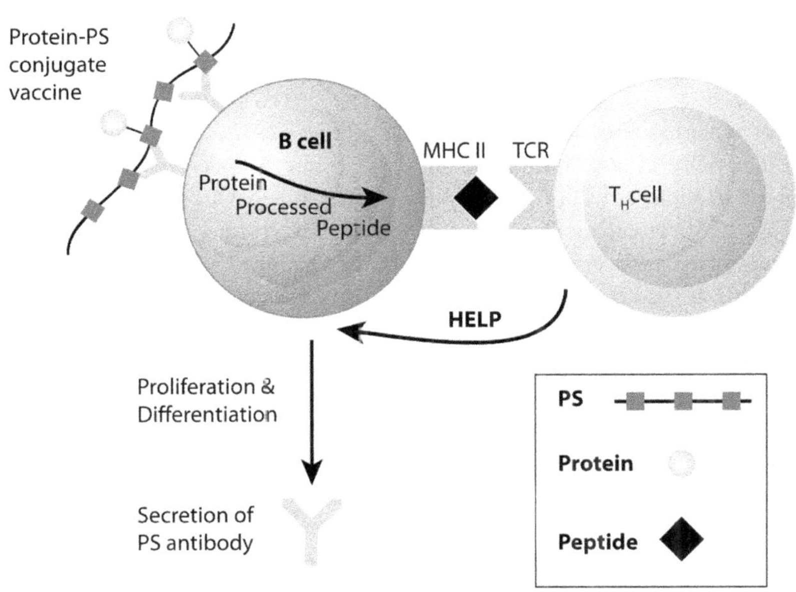

Figure 6.5 Recruitment of T-cell help for polysaccharide conjugate vaccines. Conjugated polysaccharide (PS) vaccines bind to B-cells specific for the PS and are internalized. The protein is digested and presented on MHCII molecules and recognized by T-cells, which in turn promote the proliferation of the PS specific B-cells into antibody producing cells. Figure adapted from Siber (1994).

polysaccharides with protein carriers. In this model, B-cells with surface immunoglobulins specific for the polysaccharide bind the covalent complex, internalize it, and process the protein antigen into peptides that are displayed on MHCII to helper T-cells. The activated helper T-cells stimulate B-cell proliferation and differentiation into antibody-secreting cells.

One reason for the great success of protein-polysaccharide conjugate vaccines is that polysaccharides alone do not – for poorly understood reasons – generate good responses in children <2 years of age. Since polysaccharides are attractive targets, other ways to generate protection in this age group are required.

Polysaccharide immunogenicity in young children

Young children are prone to infection with encapsulated bacteria. It is known that children <2 years are deficient in IgG2 and IgG4 (Zegers *et al.*, 1980); although IgG1 and IgG3 reach normal levels by the age of 2 or 3 years, levels of IgG2 and IgG4 rise so slowly that they may not become optimal until 13–17 years of age (Garty *et al.*, 1994). Since IgG2 is the most effective immunoglobulin against some carbohydrates (Barrett and Ayoub, 1986), the susceptibility of young children to infection with encapsulated bacteria may be due to this specific immunodeficiency along with the child's inability to mount a full humoral response to polysaccharides.

Complete tolerance to some bacterial polysaccharides is another cause of poor responses, not just in children but in adults as well. For example, Lancefield group A *Streptococcus* (*Streptococcus pyogenes*) is encapsulated and causes invasive and mucosal disease in the nasopharynx of children (Finn, 2004). The problem in mounting a defence is that the major polysaccharide of *S. pyogenes* is hyaluronic acid, which is also a significant component of human connective tissues (Wessels, 1999). Likewise, the poor immunogenicity of the capsule of *N. meningitidis* group B may be explained by structural similarities between the capsular polysaccharide and polysialic acid groups found in mammalian tissues, including the nervous system (Finne *et al.*, 1987). Covalent conjugation of polysaccharide to protein carriers renders these antigens more immunogenic; the humoral immune system is stimulated via B-cells, and T-cell help is recruited via the protein carrier.

Some non-conjugated TI-2 vaccines based on capsular polysaccharide have been developed and used against bacteria such as *S. pneumoniae, N. meningitidis,* and *H. influenzae* type b. These vaccines have tended to confer protection against multiple bacterial serotypes. For example, the first pneumococcal vaccine provided protection against 14 serogroups; it was replaced in 1983 by a vaccine protective against 23 serogroups (Levine *et al.*, 2006). One of the first meningococcal vaccine was tetravalent, covering serogroups A, C, Y, and W-135 (Danzig, 2004). Unfortunately, these vaccines were not immunogenic in infants <2 years of age. The meningococcal polysaccharides, when given after 12 months of age, provided effective antibody titres in only 20% of 2- to 6-year-olds (Pollabauer *et al.*, 2005), whereas adults developed higher titres lasting for 5–10 years. To overcome these disadvantages, conjugate vaccines against all three of these encapsulated bacterial species were developed.

Glycoconjugate vaccines

Without doubt, conjugate vaccines are one of the greatest success stories in science. Some of the most successful conjugate vaccines target debilitating diseases associated with high mortality rates.

Pneumococcal vaccines

S. pneumoniae, a Gram-positive bacterium classified into >90 serotypes in 46 serogroups (Henrichsen, 1995), continues to be a major cause of respiratory tract infection, invasive disease, and very high mortality rates in both young children and adults worldwide (McGee, 2007). For many years, pneumococcal infections were treated with penicillin-type antibiotics, but in the 1960s resistance to these agents emerged. New non-β-lactam antibiotics were subsequently used, but the number of antibiotic-resistant strains continued to grow. In fact, >80% of pneumococcal infections worldwide are now caused by just five resistant serotypes: 6B, 9V, 14, 19F, and 23F (Clarke, 2006; Clarke *et al.*, 2006). The readiness with which *S. pneumoniae* shares its DNA – and therefore its resistance and virulence factors – with other members of its family increases the transmissibility of this organism within the population. In 1983, a 23-valent polysaccharide vaccine (PPV) became available and was recommended for persons >2 years of age. Although this vaccine effectively reduced rates of invasive pneumococcal disease among adults (including the elderly), several problems remained:

1 The vaccine was ineffective in infants <2 years of age.
2 Carriage of the organism and transmission of the disease did not decrease.
3 The vaccine was less effective in the setting of immunocompromise. Another strategy was required, and in 2000 Wyeth introduced their seven-valent conjugate vaccine.

Conjugate pneumococcal vaccine

The introduction of the conjugate vaccine proved to be a turning point in the control of pneumococcal diseases. In this vaccine, called PCV7 (Prevnar), pneumococcal polysaccharide of seven serotypes (4, 6B, 9V, 14, 18C, 19F, and 23F) was conjugated to diphtheria proteins; the targeted serotypes included the five serotypes mentioned previously that account for 80% of resistant infections in the United States, Canada, Europe, and Africa and for 65% of invasive disease cases in Asia and Latin America (Whitney and Klugman, 2004).

Effect of PCV7 on disease and transmission

After the introduction of PCV7, the incidence of invasive pneumococcal disease decreased dramatically in the USA (McGee, 2007). Most important, a reduced incidence among children <2 years of age demonstrated the superiority of conjugate vaccines over other strategies. An impact on incidence has also been seen in other countries that have adopted this vaccine and used it according to various strategies (Aristegui *et al.*, 2007; Dubos *et al.*, 2007). Furthermore, this conjugate vaccine has reduced rates of nasopharyngeal colonization with the vaccine-targeted serotypes, although some increases in non-vaccine-targeted serotypes has also been noted (Frazao *et al.*, 2005; Huang *et al.*, 2005). Substantial reductions in rates of invasive pneumococcal disease among older children and adults suggest that this vaccine interferes with the transmissibility of pneumococci in the community (Kyaw *et al.*, 2006; Hammitt *et al.*, 2006). Another potential benefit of increased use of effective vaccines like PCV7 is a reduction in the use of antimicrobial agents that may promote resistant serotypes.

Group B streptococcal vaccines

For more than 25 years, the group B *Streptococcus* (GBS) has been the leading cause of death among newborns in the USA (Walsh and Hutchins, 1989). Manifestations of GBS bacteraemia and meningitis can develop as soon as 1 day after birth (Schuchat, 1998). The majority of isolates responsible in the United States are of serotypes Ia, Ib, II, III, and V; worldwide, other serotypes (e.g. VI and VIII) are prevalent (Schuchat, 1998). Early-onset GBS disease is acquired during delivery. Thus, to protect neonates, a vaccine eliciting IgG responses that could cross the placenta was required. An initial polysaccharide vaccine administered to women at 31 weeks of pregnancy (Baker *et al.*, 1988) evoked IgG antibodies that did indeed cross the placenta, but the level of immunogenicity was low.

GBS conjugate vaccine

GBS conjugate vaccines using tetanus toxoid have now been developed against all nine serotypes, including types Ia and Ib (Brigtsen *et al.*, 2002), type II (Baker *et al.*, 2000), type III (Paoletti *et al.*, 2000), types IV and VII (Paoletti and Kasper, 2002), and type V (Palazzi *et al.*, 2004). To effectively target GBS serotypes, a multivalent conjugate vaccine is probably required.

Initial testing showed that several monovalent (type II or type III) and bivalent (types Ia and Ib) vaccines were highly successful at protecting newborns. In a study of mice, maternal IgG antibodies were transferred from pregnant mothers to offspring after vaccination with type III-TT (tetanus toxoid) vaccine, and all mouse pups were protected when challenged with GBS (Paoletti *et al.*, 2000). Moreover, in phase 1 and 2 clinical trials, healthy human adults have safely received conjugate vaccines prepared with capsular polysaccharides of GBS types Ia, Ib, II, III, and V. These vaccines have elicited specific antibody that opsonizes GBS for *in vitro* killing by human peripheral blood leukocytes in the presence of complement (Paoletti and Kasper, 2003). These trials have also shown that GBS conjugate vaccines are safe, well tolerated, and immunogenic in healthy adults (Paoletti and Madoff, 2002).

The initial goal in developing these conjugate vaccines was to induce IgG switching in pregnant women, who would then produce antibody that could cross the placenta and protect neonates. This was no easy task and represented a novel approach to protection against a widespread disease. The success attained illustrates how effective a conjugate vaccine can be at protecting neonates.

H. influenzae *type b (Hib)*

Early vaccines using polyribosyl ribitol phosphate (PRP), a capsular polysaccharide purified by detergent extraction from Hib cultures, protected older children but had limited immunogenicity in children <2 years of age (Castagnola and Fioredda, 2003). Subsequent Hib conjugate vaccines, however, have been very successful (Steiniger *et al.*, 2006). At least four different carrier proteins have been used, as have saccharides of different structures and conformations (Timens *et al.*, 1989). The different Hib vaccines have displayed different immunogenicities. One vaccine (PRP-OMP, Merck) incorporates a type b saccharide with meningococcal outer-membrane protein, requires two priming doses (according to the U.S. protocol), and elicits good T-cell-dependent antibody responses in infancy, perhaps in part because the carrier has adjuvant properties (Cobb *et al.*, 2004).

Hib vaccines are generally given with various diphtheria, tetanus, and acellular pertussis vaccines (e.g. DTaP). Although infants appeared to have lower IgG titres (Soriano-Gabarro *et al.*, 2002) after three doses of the combined vaccine good immunological memory was evoked with good responses in booster programs (Pollabauer *et al.*, 2005). Through Hib vaccination, rates of Hib infection, especially among young children, have been reduced to such an extent that these infections are now considered a rarity.

Neisseria meningitidis

In children, *N. meningitidis* is a leading cause of meningitis and septicaemia, with persistent fatality rates of 3–10% (Pollard and Levin, 2000). Serogroups B and C cause the most cases in developed countries (Pace and Pollard, 2007); concern regarding serogroup Y is increasing in the United States (Bilukha and Rosenstein, 2005). Serogroup W-135 is a minor cause of disease in developed countries but a more important cause in developing countries. Up to 10% of healthy adults carry *N. meningitidis* in the nasopharynx and represent an ongoing reservoir for transmission (Pollard and Levin, 2000; Cartwright *et al.*, 1987).

Meningococcal polysaccharide vaccines

The capsules of meningococcal serogroups B, C, Y, and W-135 are made up of polysialic acid or sialic acid linked to glucose or galactose; the capsule of serogroup A is composed of N-acetylmannosamine-1-phosphate (Jennings *et al.*, 1977). Monovalent (A or C), bivalent (A/C), and tetravalent (ACYW-135) polysaccharide vaccines successfully stimulate B-cell responses with antibodies specific to the polysaccharide(s) contained in the vaccines. Unfortunately, these antibody responses are short-lived and vary with the age of the recipient. Serogroup A (MenA) vaccine induces responses after 3 months of age, whereas serogroup C (MenC) vaccine is poorly immunogenic in children <2 years of age (De Wals *et al.*, 2005; Lepow *et al.*, 1977; Reingold *et al.*, 1985). The increased antibody response of infants after Men A boosting suggests that this response may not be strictly T-cell independent (Gold *et al.*, 1975). In addition to evoking inadequate responses in some key age groups, meningococcal polysaccharide vaccines do not generate memory responses in B-cells.

Meningococcal conjugate vaccines

Conjugation of meningococcal capsular polysaccharides/oligosaccharides to a carrier protein results in a T-cell-dependent immune response, with the production of high-affinity antibodies and the formation of memory B-cells specific for the carbohydrate antigens (Lesinski and Westerink, 2001b). The introduction of this meningococcal conjugate vaccine overcame all the problems encountered by the polysaccharide vaccine alone.

MenC vaccine, one of the first conjugated meningitis vaccines developed during the 1990s, proved reasonably successful in the United

Kingdom, with lower rates of disease, evidence of herd immunity (Maiden and Stuart, 2002), and high short-term vaccine effectiveness (Ramsay *et al.*, 2003). Some B-cell memory was noted in children, but antibody titres decreased significantly within a couple of years if the vaccine was given during the first 2 years of life (Andrews *et al.*, 2003; Snape *et al.*, 2005). In contrast, antibodies persisted for up to 4 years with vaccination after the age of three (Trotter *et al.*, 2004). This vaccine was highly successful in reducing the transmission of serogroup C infection and disease.

The Meningitis Vaccine Project is developing a monovalent MenA vaccine with the specific aim of creating a low-cost product effective in the meningitis belt of Africa. Africa has a high incidence of meningococcal A/C disease and therefore is a good candidate for these types of low cost vaccines. In fact, MenA/C vaccines have been shown to be safe and immunogenic in clinical trials of young adults (Zhang *et al.*, 2001; Anderson *et al.*, 1994; Lakshman *et al.*, 2002) as well as infants (Chippaux *et al.*, 2004; Twumasi *et al.*, 1995; Fairley *et al.*, 1996).

A tetravalent vaccine has been licensed for use in 11- to 55-year-olds since January 2005 (Bilukha and Rostenstein, 2005). Although published data are limited so far, this vaccine appears to evoke immunological memory and higher antibody titres than previous vaccines.

Conclusions

The introduction of conjugate vaccines represents an enormous breakthrough in science and vaccinology. Hib disease is now rare in young children. Both young children and adults are being protected against deadly pneumococcal and GBS disease; in particular, neonates can be protected against serious GBS infections. Conjugate vaccines have reduced the transmissibility of these infections and have targeted bacterial strains that were becoming resistant to many antibiotics.

Acknowledgements

The authors would like to thank Tom DiCesare (MIT) for the construction of the figures and Julie McCoy (Brigham Women's Hospital) for the editing of this chapter.

References

Anderson, E.L., Bowers, T., Mink, C.M., Kennedy, D.J., Belshe, R.B., Harakeh, H., Pais, L., Holder, P., and Carlone, G.M. (1994). Safety and immunogenicity of meningococcal A and C polysaccharide conjugate vaccine in adults. Infect. Immun. *62*, 3391–3395.

Andrews, N., Borrow, R., and Miller, E. (2003). Validation of serological correlate of protection for meningococcal C conjugate vaccine by using efficacy estimates from postlicensure surveillance in England. Clin. Diagn Lab. Immunol. *10*, 780–786.

Aristegui, J., Bernaola, E., Pocheville, I., García, C., Arranz, L., Durán, G., Péres, L., Bastida, M., Canduela, C., Herranz Aguirre, M., Garrote, E., Fletcher, M.A., and Péres, C. (2007). Reduction in pediatric invasive pneumococcal disease in the Basque Country and Navarre, Spain, after introduction of the heptavalent pneumococcal conjugate vaccine. Eur. J. Clin. Microbiol. Infect Dis. *26*, 303–310.

Avery, O.T, Goebel, W.F. (1929) Chemo-immunological studies on conjugated carbohydrate-proteins. II Immunological specificity of synthetic sugar-protein antigens. J. Exp Med. *50*, 533–550.

Baker, C.J., Rench, M.A., Edwards, M.S., Carpenter, R.J., Hays, B.M., and Kasper, D.L. (1988). Immunization of pregnant women with a polysaccharide vaccine of group B streptococcus. N. Engl J. Med. *319*, 1180–1185.

Baker, C.J., Paoletti, L.C., Rench, M.A., Guttormsen, H.K., Carey, V.J., Hickman, M.E., and Kasper, D.L. (2000). Use of capsular polysaccharide-tetanus toxoid conjugate vaccine for type II group B Streptococcus in healthy women. J. Infect. Dis. *182*, 1129–1138.

Balish, E., and Warner, T. (2002). Enterococcus faecalis induces inflammatory bowel disease in interleukin-10 knockout mice. Am. J. Pathol. *160*, 2253–2257.

Barrett, D.J. and Ayoub, E.M. (1986). IgG2 subclass restriction of antibody to pneumococcal polysaccharides. Clin. Exp. Immunol. *63*, 127–134.

Bilukha, O.O., Rosenstein, N. (2005). Prevention and control of meningococcal disease. Recommendations of the Advisory Committee on Immunization Practices (ACIP). MMWR Recomm. Rep. *54*, 1–21.

Bouma, G., and Strober, W. (2003). The immunological and genetic basis of inflammatory bowel disease. Nat. Rev. Immunol. *3*, 521–533.

Brigtsen, A.K., Kasper, D.L., Baker, C.J., Jennings, H J., and Guttormsen, H.K. (2002). Induction of cross-reactive antibodies by immunization of healthy adults with types Ia and Ib group B streptococcal polysaccharide-tetanus toxoid conjugate vaccines. J. Infect. Dis. *185*, 1277–1284.

Candy, D., E. Graham Davies and Ross, E. (2001). Clinical Paediatrics and Child Health. London: Saunders (WB) Co. Ltd. .

Cartwright, K.A., Stuart, J.M., Jones, D.M., and Noah, N.D. (1987). The Stonehouse survey: nasopharyngeal carriage of meningococci and *Neisseria lactamica*. Epidemiol. Infect. *99*, 591–601.

Castagnola, E. and Fioredda, F. (2003). Prevention of life-threatening infections due to encapsulated bacteria in children with hyposplenia or asplenia: a brief review of current recommendations for practical purposes. Eur J. Haematol. *71*, 319–326.

Chen, Z., and O'Shea, J.J. (2008). Th17 cells: a new fate for differentiating helper T-cells. Immunol. Res. *41*, 87–102.

Chippaux, J.P., Garba, A., Ethevenaux, C., Campagne, G., de Chabalier, F., Djibo, S., Nicolas, P., Ali, H., Charrondiére, M., Ryall, R., Bybel, M., and Schuchat, A. (2004). Immunogenicity, safety, and memory of different schedules of *Neisseria meningitidis* A/C-diphtheria toxoid conjugate vaccine in infants in Niger. Vaccine. *22*, 3303–3311.

Chuang, T.H., and Ulevitch, R.J. (2000). Cloning and characterization of a sub-family of human toll-like receptors: hTLR7, hTLR8 and hTLR9. Eur Cytokine Netw. *11*, 372–378.

Clarke, S.C. (2006). Control of pneumococcal disease in the United Kingdom – the start of a new era. J. Med Microbiol. *55*, 975–980.

Clarke, S.C., Jefferies, J.M., Smith, A.J., McMenamin, J., Mitchell, T.J., and Edwards, G.F. (2006). Pneumococci causing invasive disease in children prior to the introduction of pneumococcal conjugate vaccine in Scotland. J. Med Microbiol. *55*, 1079–1084.

Cobb, B.A., Wand, Q., Tzianabos, A.O., and Kasper, D.L. (2004). Polysaccharide processing and presentation by the MHCII pathway. Cell, *117*, 677–687.

Cobb, B.A., and Kasper, D.L. (2005). Zwitterionic capsular polysaccharides: the new MHCII-dependent antigens. Cell Microbiol. *7*, 1398–1403.

Cong, Y., Brandwein, S.L., McCabe, R.P., Lazenby, A., Birkenmeier, E.Hl, Sundberg, J.P., and Elson, C.O. (1998). CD4+ T-cells reactive to enteric bacterial antigens in spontaneously colitic C3H/HeJBir mice: increased T helper cell type 1 response and ability to transfer disease. J. Exp. Med. *187*, 855–864.

Coyne, M.J., Weinacht, K.G., Krinos, C.M., and Comstock, L.E. (2003). Mpi recombinase globally modulates the surface architecture of a human commensal bacterium. Proc. Natl. Acad. Sci. USA, *100*, 10446–10451.

Coyne, M.J., and Comstock, L.E. (2008). Niche-specific features of the intestinal bacteroidales. J. Bacteriol. *190*, 736–742.

Danzig, L. (2004). Meningococcal vaccines. Pediatr Infect Dis J. *23*, 285–292.

De Wals, P., Deceuninck, G., De Serres, G., Boivin, J.F., Duval, B., Remis, R., and Massé, R. (2005). Effectiveness of serogroup C meningococcal polysaccharide vaccine: results from a case-control study in Quebec. Clin. Infect Dis. *40*, 1116–1122.

Du, X., Poltorak, A.; Wei, Y., and Beutler, B. (2000). Three novel mammalian toll-like receptors: gene structure, expression, and evolution. Eur. Cytokine Netw. *11*, 362–371.

Duan, J., Avci, F.Y., and Kasper, D.L. (2008). Microbial carbohydrate depolymerization by antigen-presenting cells: deamination prior to presentation by the MHCII pathway. Proc Natl Acad Sci. USA, *105*, 5183–5188.

Dubos, F., Marechal, I., Husson, M.O., Courouble, C., Aurel, M., and Martinot, A. (2007). Decline in pneumococcal meningitis after the introduction of the heptavalent-pneumococcal conjugate vaccine in northern France. Arch Dis Child. *92*, 1009–1012.

Elson, C.O., Beagley, K.W., Sharmanov, A.T., Fujihashi, K., Kiyono, H., Tennyson, G.S., Cong, Y., Black, C.A., Ridwan, B.W., and McGhee, J.R. (1996). Hapten-induced model of murine inflammatory bowel disease: mucosa immune responses and protection by tolerance. J. Immunol. *157*, 2174–2185.

Fairley, C.K., Begg, N., Borrow, R., Fox, A.J., Jones, D.M., and Cartwright, K. (1996). Conjugate meningococcal serogroup A and C vaccine: reactogenicity and immunogenicity in United Kingdom infants. J. Infect Dis. *174*, 1360–1363.

Finn, A. (2004). Bacterial polysaccharide-protein conjugate vaccines. Br Med Bull. *70*, 1–14.

Finne, J., Bitter-Suermann, D., Goridis, C., and Finne U. (1987). An IgG monoclonal antibody to group B meningococci cross-reacts with developmentally regulated polysialic acid units of glycoproteins in neural and extraneural tissues. J. Immunol. *138*, 4402–4407.

Foreman, J., Demidchik, V., Bothwell, J.H., Mylona, P., Miedema, H., Torres, M.A., Linstead, P., Costa, S., Brownlee, C., Jones, J.D., Davies, J.M., and Dolan, L. (2003). Reactive oxygen species produced by NADPH oxidase regulate plant cell growth. Nature, *422*, 442–446.

Forster, R., Schubel, A., Breitfeld, D., Kremmer, E., Renner-Müller, I., Wolf, E., and Lipp, M. (1999). CCR7 coordinates the primary immune response by establishing functional microenvironments in secondary lymphoid organs. Cell, *99*, 23–33.

Frazão, N., Brito-Avô, A., Simas, C., Saldanha, J., Mato, R., Nunes, S., Sousa, N.G., Carriço, J.A., Almeida, J.S., Santos-Sanches, I., and de Lencastre, H. (2005) Effect of the seven-valent conjugate pneumococcal vaccine on carriage and drug resistance of *Streptococcus pneumoniae* in healthy children attending day-care centers in Lisbon. Pediatr Infect Dis J. *24*, 243–252.

Fujino, S., Andoh, A., Bamba, S., Ogawa, A., Hata, K., Araki, Y., Bamba, T., and Fujiyama, Y. (2003). Increased expression of interleukin 17 in inflammatory bowel disease. Gut *52*, 65–70.

Garty, B.Z., Ludomirsky, A., Danon, Y.L., Peter, J.B., and Douglas, S.D. (1994). Placental transfer of immunoglobulin G subclasses. Clin. Diagn. Lab. Immunol. *1*, 667–669.

Geijtenbeek, T.B., Groot, P.C., Nolte, M.A., van Vliet, S.J., Gangaram-Panday, S.T., van Duijnhoven, G.C., Kraal, G., van Oosterhout, A.J., and van Kooyk, Y. (2002). Marginal zone macrophages express a murine homologue of DC-SIGN that captures blood-borne antigens *in vivo*. Blood, *100*, 2908–2916.

Gold, R., Lepow, M.L., Goldschneider, I., Draper, T.L., and Gotschlich, E.C. (1975). Clinical evaluation of group A and group C meningococcal polysaccharide vaccines in infants. J. Clin. Invest. *56*, 1536–1547.

Hammitt, L.L., Bruden, D.L., Butler, J.C., Barrett, H.C., Hurlburt, D.A., Reasonover, A., and Hennessy, T.W. (2006). Indirect effect of conjugate vaccine on adult carriage of *Streptococcus pneumoniae*: an explanation of trends in invasive pneumococcal disease. J. Infect. Dis. *193*, 1487–1494.

Henrichsen, J. (1995). Six newly recognized types of *Streptococcus pneumoniae*. J. Clin. Microbiol. *33*, 2759–2762.

Hoentjen, F., Harmsen, H.J., Braat, H., Torrice, C.D., Mann, B.A., Sartor, R.B., and Dieleman, L.A. (2003). Antibiotics with a selective aerobic or anaerobic spectrum have different therapeutic activities in various regions of the colon in interleukin 10 gene deficient mice. Gut, *52*, 1721–1727.

Huang, S.S., Platt, R., Rifas-Shiman, S.L., Pelton, S.I., Goldmann, D., and Finkelstein, J.A. (2005) Post-PCV7 changes in colonizing pneumococcal serotypes in 16 Massachusetts communities, 2001 and 2004. Pediatrics. *116*, 408–413.

Ivanov, I.I., McKenzie, B.S., Zhou, L., Tadokoro, C.E., Lepelley, A., Lafaille, J.J., Cua, D.J., and Littman, D.R. (2006). The orphan nuclear receptor RORgammat directs the differentiation program of proinflammatory IL-17+ T helper cells. Cell, *126)*, 1121–1133.

Izcue, A., Coombes, J.L., and Powrie, F. (2006). Regulatory T-cells suppress systemic and mucosal immune activation to control intestinal inflammation. Immunol. Rev. *212*, 256–271.

Janeway, C.A, Walport, M., Travers, P., and Shlomchik, M. J. (2005). Immunobiology: the Immune System in Health and Disease, 6th edn: Garland Science Publishing.

Jankovic, D., Kullberg, M.C., Hieny, S., Caspar, P., Collazo, C.M., and Sher, A. (2002). In the absence of IL-12, CD4(+) T-cell responses to intracellular pathogens fail to default to a Th2 pattern and are host protective in an IL-10(–/–) setting. Immunity, *16*, 429–439.

Jennings, H.J., Bhattacharjee, A.K., Bundle, D.R., Kenny, C.P., Martin, A., and Smith, I.C. (1977). Strucutres of the capsular polysaccharides of *Neisseria meningitidis* as determined by 13C-nuclear magnetic resonance spectroscopy. J. Infect Dis. *136*, 78–83.

Kang, Y.S., Yamazaki, S., Iyoda, T., Pack, M., Bruening, S.A., Kim, J.Y., Takahara, K., Inaba, K., Steinman, R.M., and Park, C.G. (2003). SIGN-R1, a novel C-type lectin expressed by marginal zone macrophages in spleen, mediates uptake of the polysaccharide dextran. Int. Immunol. *15*, 177–186.

Kang, Y.S.; Do, Y.Y., Lee, H.K., Park, S.H., Cheong, C., Lynch, R.M., Loeffler, J.M., Steinmann, R.M., and Park, C.G. (2006). A dominant complement fixation pathway for pneumococcal polysaccharides initiated by SIGN-R1 interacting with C1q. Cell. *125*, 47–58.

Krinos, C.M., Coyne, M.J., Weinacht, K.G., Tzianabos, A.O., Kasper, D.L., and Comstock, L.E. (2001). Extensive surface diversity of a commensal microorganism by multiple DNA inversions. Nature, *414*, 555–558.

Kyaw, M.H., Lynfield, R., Schaffner, W., Craig, A.S., Hadler, J., Reingold, A., Thomas, A.R., Harrison, L.H., Bennett, N.M., Farley, M.M., Facklam, R.R., Jorgensen, J.H., Besser, J., Zell, E.R., Schuchat, A., and Whitney, C.G. (2006). Effect of introduction of the pneumococcal conjugate vaccine on drug-resistant *Streptococcus pneumoniae*. N. Engl. J. Med. *354*, 1455–1463.

Lakshman, R., Burkinshaw, R., Choo, S., and Finn, A. (2002). Prior meningococcal A/C polysaccharide vaccine does not reduce immune responses to conjugate vaccine in young adults. Vaccine *20*, 3778–3782.

Laubach, V.E., Shesely, E.G. Smithies, O., and Sherman, P.A. (1995). Mice lacking inducible nitric oxide synthase are not resistant to lipopolysaccharide-induced death. Proc. Natl. Acad. Sci. USA *92*, 10688–10692.

Lepow, M.L., Goldschneider, I., Gold, R., Randolph, M., and Gotschlich, E.C. (1977). Persistence of antibody following immunization of children with groups A and C meningococcal polysaccharide vaccines. Pediatrics *60*, 673–680.

Lesinski, G.B., and Westerink, M.A. (2001a). Vaccines against polysaccharide antigens. Curr. Drug Targets Infect. Disord. *1*, 325–34.

Lesinski, G.B., and Westerink, M.A. (2001b). Novel vaccine strategies to T-independent antigens. J. Microbiol. Methods *47*, 135–149.

Levine, O.S., O`Brien, K.L., Knoll, M., Adegbola, R.A., Black, S., Cherian, T., Dagan, R., Goldblatt, D., Grange, A., Greenwood, B., Hennessy, T., Klugman, K.P., Madhi, S.A., Mulholland, K., Nohynek, H., Santosham, M., Saha, S.K., Scott, J.A., Sow, S., Whitney, C.G., and Cutts, F. (2006). Pneumococcal vaccination in developing countries. Lancet *367*, 1880–1882.

Liu, C.H., Lee, S.M., Vanlare, J.M., Kasper, D.L., and Mazmanian, S.K. (2008). Regulation of surface architecture by symbiotic bacteria mediates host colonization. Proc. Natl. Acad. Sci. USA *105*, 3951–3956.

Maiden, M.C., and Stuart, J.M. (2002). Carriage of serogroup C meningococci 1 year after meningococcal C conjugate polysaccharide vaccination. Lancet. *359*, 1829–1831.

Maloy, K.J., Salaun, L., Cahill, R., Dougan, G., Saunders, N.J., and Powrie, F. (2003). CD4+CD25+ T(R) cells suppress innate immune pathology through cytokine-dependent mechanisms. J. Exp Med. *197*, 111–119.

Matsuzaki, G., and Umemura, M. (2007). Interleukin-17 as an effector molecule of innate and acquired immunity against infections. Microbiol. Immunol. *51*, 1139–1147.

McGee, L. (2007). The coming of age of niche vaccines? Effect of vaccines on resistance profiles in *Streptococcus pneumoniae*. Curr. Opin. Microbiol. *10*, 473–478.

Obukhanych, T.V., and Nussenzweig, M.C. (2006). T-independent type II immune responses generate memory B-cells. J. Exp. Med. *203*, 305–310.

Pace, D. and Pollard, A.J. (2007). Meningococcal, A., C, Y and W-135 polysaccharide-protein conjugate vaccines. Arch. Dis. Child. *92*, 909–915.

Palazzi, D.L., Rench, M.A., Edwards, M.S., and Baker, C.J. (2004). Use of type V group B streptococcal conjugate vaccine in adults 65–85 years old. J. Infect Dis. *190*, 558–564.

Pamer, E.G. (2007). Immune responses to commensal and environmental microbes. Nat Immunol. *8*, 1173–1178.

Paoletti, L.C., Pinel, J., Kennedy, R.C., and Kasper, D.L. (2000). Maternal antibody transfer in baboons and mice vaccinated with a group B streptococcal polysaccharide conjugate. J. Infect Dis. *181*, 653–658.

Paoletti, L.C. and Kasper, D.L. (2002). Conjugate vaccines against group B Streptococcus types IV and VII. J. Infect. Dis. *186*, 123–126.

Paoletti, L.C. and Madoff, L.C. (2002). Vaccines to prevent neonatal GBS infection. Semin. Neonatol. 7, 315–323.

Paoletti, L.C. and Kasper, D.L. (2003). Glycoconjugate vaccines to prevent group B streptococcal infections. Expert Opin. Biol. Ther. *3*, 975–984.

Pollabauer, E.M., Petermann, R., and Ehrlich, H.J. (2005). Group C meningococcal polysaccharide-tetanus toxoid conjugate vaccine: a meta-analysis of immunogenicity, safety and posology. Hum. Vaccin. *1*, 131–139.

Pollard, A.J., and Levin, M. (2000). Vaccines for prevention of meningococcal disease. Pediatr. Infect. Dis. J. *19*, 333–344.

Pollock, J.D., Williams, D.A., Gifford, M.A., Li, L.L., Du, X., Fisherman, J., Orkin, S.H., Doerschuk, C.M., and Dinauer, M.C. (1995). Mouse model of X-linked chronic granulomatous disease, an inherited defect in phagocyte superoxide production. Nat. Genet, *9*, 202–209.

Ramsay, M.E., Andrews, N.J., Trotter, C.L., Kaczmarski, E.B., and Miller, E. (2003). Herd immunity from meningococcal serogroup C conjugate vaccination in England: database analysis. Bmj. *326*, 365–366.

Reingold, A.L., Broome, C.V., Hightower, A.W., Ajello, G.W., Bolan, G.A., Adamsbaum, C., Jones, E.E., Phillips, C., Tiendrebeogo, H., and Yada, A. (1985). Age-specific differences in duration of clinical protection after vaccination with meningococcal polysaccharide A vaccine. Lancet. *2*, 114–118.

Schuchat, A. (1998). Epidemiology of group B streptococcal disease in the United States: shifting paradigms. Clin. Microbiol. Rev. *11*, 497–513.

Siber, G.R. (1994). Pneumococcal disease: prospects for a new generation of vaccines. Science *265*, 1385–1387.

Snape, M.D., Kelly, D.F., Green, B., Moxon, E.R., Borrow, R., and Pollard, A.J. (2005). Lack of serum bactericidal activity in preschool children two years after a single dose of serogroup C meningococcal polysaccharide-protein conjugate vaccine. Pediatr Infect Dis J. *24*, 128–131.

Soriano-Gabarro, M., Stuart, J.M., and Rosenstein, N.E. (2002). Vaccines for the prevention of meningococcal disease in children. Semin Pediatr Infect Dis. *13*, 182–189.

Steiniger, B., Timphus, E.M., and Barth, P.J. (2006). The splenic marginal zone in humans and rodents: an enigmatic compartment and its inhabitants. Histochem Cell Biol. *126*, 641–648.

Tabeta, K., Georgel, P., Janssen, E., Du, X., Hoebe, K., Crozat, K., Mudd, S., Shamel, L., Sovath, S., Goode, J., Alexopoulou, L., Flavell, R.A., and Beutler, B. (2004). Toll-like receptors 9 and 3 as essential components of innate immune defense against mouse cytomegalovirus infection. Proc. Natl. Acad. Sci. USA *101*, 3516–3521.

Takeda, K., Kaisho, T., and Akira, S. (2003). Toll-like receptors. Annu. Rev. Immunol. *21*, 335–376.

Timens, W., Boes, A., Rozeboom-Uiterwijk, T., and Poppema, S. (1989). Immaturity of the human splenic marginal zone in infancy. Possible contribution to the deficient infant immune response. J. Immunol. *143*, 3200–3206.

Trotter, C.L., Andrews, N.J., Kaczmarski, E.B., Miller, E., and Ramsay, M.E. (2004). Effectiveness of meningococcal serogroup C conjugate vaccine 4 years after introduction. Lancet *364*, 365–367.

Twumasi, P.A., Jr., Kumah, S., Leach, A., O'Dempsey, T.J., Ceesay, S.J., Todd, J., Broome, C.V., Carlone, G.M., Pais, L.B., and Holder, P.K. (1995). A trial of a group A plus group C meningococcal polysaccharide-protein conjugate vaccine in African infants. J. Infect. Dis. *171*, 632–638.

Tzianabos, A.O., Onderdonk, A.B., Rosner, B., Cisneros, R.L., and Kasper, D.L. (1993). Structural features of polysaccharides that induce intra-abdominal abscesses. Science *262*, 416–419.

Tzianabos, A.O., Finberg, R.W., Wang, Y., Chan, M., Oderdonk, A.B., Jennings, H.J., and Kasper, D.L. (2000).T-cells activated by zwitterionic molecules prevent abscesses induced by pathogenic bacteria. J. Biol. Chem. *275*, 6733–6740.

Tzianabos, A.O., Wang, J.Y., and Lee, J.C. (2001). Structural rationale for the modulation of abscess formation by *Staphylococcus aureus* capsular polysaccharides. Proc. Natl. Acad. Sci. USA *98*, 9365–9370.

Videla, S., Vilaseca, J., Guarner, F., Salas, A., Treserra, F., Crespo, E., Antonlín, M., and Malagelada, J.R. (1994). Role of intestinal microflora in chronic inflammation and ulceration of the rat colon. Gut *35*, 1090–1097.

Walsh, J.A., and Hutchins, S. (1989). Group B streptococcal disease: its importance in the developing world and prospect for prevention with vaccines. Pediatr Infect Dis J. *8*, 271–277.

Wang, Y., Kalka-Moll, W.M., Roehrl, M.H., and Kasper, D.L. (2000). Structural basis of the abscess-modulating polysaccharide A2 from *Bacteroides fragilis*. Proc. Natl. Acad. Sci. USA *97*, 13478–13483.

Wang, Q., McLoughlin, R.M., Cobb, B.A., Charrel-Dennis, M., Zahleski, K.J., Golenbock, D., Tzianabos, A.O., and Kasper, D.L. (2006). A bacterial carbohydrate links innate and adaptive responses through Toll-like receptor 2. J. Exp. Med. *203*, 2853–2863.

Watts, C., and Powis, S. (1999). Pathways of antigen processing and presentation. Rev. Immunogenet. *1*, 60–74.

Wessels, M.R. (1999). Regulation of virulence factor expression in group A streptococcus. Trends Microbiol. *7*, 428–430.

Wexler, H.M. (2007). Bacteroides: the good, the bad, and the nitty-gritty. Clin. Microbiol. Rev. *20*, 593–621.

Whitney, C.G., and Klugman, K.P. (2004). Vaccines as tools against resistance: the example of pneumococcal conjugate vaccine. Semin. Pediatr. Infect. Dis. *15*, 86–93.

Xavier, R., and Podolsky, D.K. (2005). Commensal flora: wolf in sheep's clothing. Gastroenterology *128*, 1122–1126.

Xu, J., Bjursell, M.K., Himrod, Jl, Deng, S., Carmichael, L.K., Chiang, H.C., Hooper, L.V., and Gordon, J.I. (2003). A genomic view of the human-*Bacteroides thetaiotaomicron* symbiosis. Scienc *299*, 2074–2076.

Zegers, B.J., van der Giessen, M., Reerink-Brongers, E.E., and Stoop, J.W. (1980). The serum IgG subclass levels in healthy infants of 13–62 weeks of age. Clin. Chim Acta, *101*, 265–269.

Zhang, Q., Lakshman, R., Burkinshaw, R., Choo, S., Everard, J., Akhtar, S., and Finn, A. (2001). Primary and booster mucosal immune responses to meningococcal group A and C conjugate and polysaccharide vaccines administered to university students in the United Kingdom. Infect. Immun. *69*, 4337–4341.

Part III

Polysaccharides in Pathogens of the Oral Cavity

Polysaccharides of Gram-negative Periodontopathic Bacteria

7

Ryoma Nakao

Abstract

Periodontitis is caused by periodontal biofilms, complex communities of microorganisms, and is one of the most common oral diseases of adults. In this chapter, recent reports concerning lipopolysaccharides, capsular polysaccharides, and glycoproteins of periodontopathic bacteria, mainly *Porphyromonas gingivalis*, are reviewed in the context of the their particular roles in attachment and biofilm formation. Understanding these roles of the bacterial polysaccharides provides helpful insights in bacteria–bacteria and bacteria–host interactions, which are crucial for the progress of periodontitis.

Introduction

The discovery of the 'DNA–mRNA–protein' central dogma has led to remarkable innovations in the life sciences. The concept has been predominant and indispensable for the development of medicine, dentistry, and pharmacy since the mid-twentieth century. However, there are unknown phenomena and there is an amazing diversity in bacterial cells that can not be explained using just this central dogma. In the post-genomics and post-proteomics era, polysaccharide biochemistry, i.e. glycomics has been used to explain the diversity of organisms since polysaccharides confer remarkable variations in the cell structure of bacterial organisms. This potential of polysaccharides is also associated with the complexity in cellular immunology and evolution of organisms. The importance of polysaccharides has been recognized not only in eukaryotes but also in prokaryotes. For instance, bacterial polysaccharides or glycoconjugates, e.g. lipopolysaccharides (LPS), capsular polysaccharides (CPS), and glycoproteins, often face the external environment, suggesting their possible roles in the interaction between bacteria, bacteria and hosts, or bacteria and inorganic materials.

Periodontitis is known as one of the most common oral diseases in adults and is defined as an inflammation-based infectious disease in supporting structures of teeth characterized by progressive destruction of the periodontal ligament and alveolar bone. Progression and severity of the disease depend on both local factors at the periodontal tissues (oral bacterial infection, malocclusion, etc.) and the systemic factors associated with immunity or hormones (i.e. diabetes, pregnancy, and others). It is well-known that destruction of periodontal tissues is a consequence of immuno-pathogenesis owing to the infection with periodontopathic bacteria. Data suggested that some periodontopathic bacteria are also associated with systemic diseases and complications including cardiovascular diseases (Beck *et al.*, 1996; Demmer and Desvarieux, 2006; Genco *et al.*, 2002; Gibson *et al.*, 2004; Sherlock *et al.*, 2006) and pre-term births (Goldenberg *et al.*, 2008; Offenbacher *et al.*, 1996).

At present, more than 700 different bacterial species have been identified in the oral cavity and more than 400 of these species are detected in the periodontal pockets (Paster *et al.*, 2001, 2006), where complex communities of microorganisms develop oral biofilms. In the periodontal pockets, oral streptococci such as *Streptococcus gordonii* and *Streptococcus sanguinis* initially colonize the tooth surface and form the physical foundation of periodontal biofilms. Thereafter, various secondary colonizers such as Gram-negative anaerobes increase in numbers in the bacterial communities via intra- and inter-species interactions and result in the formation of mature biofilms. Many reports concerning the interactions among oral bacteria have been published for decades and were recently reviewed (Kolenbrander *et al.*, 2006). For examples, *Porphyromonas gingivalis*, a primary component in forming subgingival plaque biofilms of periodontitis, can adhere to *S. gordonii* using the minor fimbrial antigen, MfaI, and polypeptide, SspB, respectively (Brooks *et al.*, 1997; Chung *et al.*, 2000). *Fusobacterium nucleatum*, a long rod with tapering ends, coaggregates with *P. gingivalis* (Kolenbrander *et al.*, 1985) and with streptococci to form so-called 'corncobs' (Lancy *et al.*, 1983). *P. gingivalis*, *Tannerella forsythensis*, and *Treponema denticola* are clinically associated with severe forms of periodontitis and are referred to as the so-called 'red complex' (Socransky *et al.*, 1998). A cooperative synergistic effect of mixed cultures with *P. gingivalis* and *T. denticola* in biofilm formation is observed *in vitro* (Hashimoto *et al.*, 2003b; Onagawa *et al.*, 1994). *Aggregatibacter actinomycetemcomitans*, a facultatively anaerobic rod, plays a primary etiological role in localized aggressive periodontitis (LAP) (Meyer and Fives-Taylor, 1997; Zambon, 1985). *Prevotella intermedia*, a black-pigmented anaerobic rod, is dominant in periodontal pockets of patients with gingivitis (Slots and Listgarten, 1988).

Nowadays, scientific discussion concentrates on bacterial polysaccharides, because the bacterial components such as LPS and CPS are essential not only for their survival, but also as the primary component of the extracellular polymeric substances that affect the development of oral biofilms. Furthermore, recent reports suggested that protein glycosylation by bacteria is associated with host-bacteria interactions (Comstock and Kasper, 2006; Varki, 2006). Although only some glycoproteins were studied in Gram-negative bacteria so far, more reports will be forthcoming because of the demand to understand the bacteria–bacteria and bacteria–host interactions.

In the following three sections of this chapter, bacterial polysaccharides of periodontal pathogens, especially *Porphyromonas gingivalis*, a primary agent in periodontitis, are discussed. The first section describes the biological roles of LPS, especially one of its components, O-polysaccharides. In the second section, the role of CPS in the context of the relation of CPS structure to dynamics on the bacterial surface is discussed. In the third section, recent advances in protein glycosylation of Gram-negative periodontopathic anaerobes are reviewed in detail.

Lipopolysaccharides of periodontopathic bacteria

General features of the bacterial LPS

Only Gram-negative bacteria produce LPS, a major integral component of the outer membrane. LPS consists of three major domains: the lipid A portion of the molecule anchored in the outer membrane, the core polysaccharide, and the O-polysaccharides extending from the cell to the external environment. LPS strongly induces inflammatory responses and most of theses activities of LPS reside in the lipid A portion of the molecule, although the activities and structures of lipid A vary among Gram-negative bacteria. The lipid A signalling receptor in mammalian cells is the membrane protein Toll-like receptor 4 (TLR4) (Hoshino *et al.*, 1999; Poltorak *et al.*, 1998). The O-polysaccharide is usually highly antigenic and therefore elicits a strong immunogenic response. While O-serotyping is used to classify bacterial strains, the wide variety of O-polysaccharides enables bacteria to evade the host immune system. For instance, *E. coli* is known to synthesize at least 173 different O-polysaccharides. Additionally, previous evidence indicated that O-polysaccharide plays an important role in effective colonization of host tissue (Bouchet *et al.*, 2003; Camprubi *et al.*, 1993a,b; Harvill *et al.*, 2000; Nesper *et al.*, 2001; Nevola *et al.*, 1985, 1987) as well as in resistance to some bactericidal effects (Boman, 1995; Nicolas and Mor, 1995; Thjotta and Waaler, 1932). The physical location of the O-polysaccharide places it at the interface between the bacteria and the outer environment. Hence, a critical density of the long O-antigen chain is likely to prevent access of detrimental molecules into the outer membrane where long O-antigens were found to have an effect on the bacterial attachment and biofilm formation.

Structure of the LPS of *P. gingivalis*

The biological properties of *P. gingivalis* LPS are different from those of *E. coli* LPS, mainly in the relative lack of endotoxicity of the *P. gingivalis* LPS as reviewed by Dixon and Darveau (2005). The difference in the biological properties of LPS is likely due to the difference in the chemical structure of lipid A (Darveau *et al.*, 2004; Kumada *et al.*, 1995; Ogawa, 1993). The biological activity via the TLR2 receptor found to be induced by LPS preparations from *P. gingivalis* (Martin *et al.*, 2001) was recently found to be elicited not by lipid A of *P. gingivalis* but by a lipoprotein encoded by gene, PG1828 (Asai *et al.*, 2005; Hashimoto *et al.*, 2004).

Chemical characterization of the O-polysaccharides from several strains of *P. gingivalis* were performed in the strains W50, W83, ATCC 33277 (Bramanti *et al.*, 1989), and ATCC 53977 (Schifferle *et al.*, 1989). Using the chemical analysis of the monosaccharide composition of the LPS from *P. gingivalis* ATCC 53977, rhamnose, glucose, and galactose were found to be the dominant sugars with lesser amounts of mannose, glucosamine, and galactosamine. For the composition of the O-repeat of *P. gingivalis* W50, two groups independently reported divergent results. While Farquharson *et al.* (2000) reported that the W50 LPS contains O-polysaccharides with regular tetrasaccharide repeating units of -4-α-L-Rhap, -6-α-D-Glcp, -3-α-D-Galp, and GlcNAc, Paramonov *et al.* (2001) reported the O-polysaccharide of W50 to be composed of tetrasaccharide repeats of -6-α-D-Glc*p*-(1–4)-α-L-Rha*p*-(1–3)-β-D-GalNAc-(1–3)-α-D-Gal*p*-1-. These divergent results suggested that there may be several groups of LPS in the same strain. Later, another kind of LPS, the so-called cell surface anionic polysaccharide (APS), was additionally identified in *P. gingivalis* W50 (Slaney *et al.*, 2006). The APS was shown to be a phosphorylated mannan whose backbone is composed of α-1,6-linked mannose residues and the side-chains contain α-1,2-linked mannose oligosaccharides of different lengths (Paramonov *et al.*, 2005). Further studies reported that APS expression on the bacterial surface is the principal mechanism for serum resistance (Slaney *et al.*, 2006) and that APS contains a lipid A portion. Consequently, so the authors renamed APS with the branched mannan repeating units and with conventional O-polysaccharide (Paramonov *et al.*, 2001) as A-LPS and O-LPS, respectively (Rangarajan *et al.*, 2008).

The proportion of the different kinds of *P. gingivalis* LPS might be controlled by a yet-unknown regulatory mechanism since expression of glycoconjugates in some bacterial species were found to be phase variable (Appelmelk *et al.*, 1998; Bouchet *et al.*, 2003; Tong *et al.*, 2000), a phenomenon, which will be described in the last section of this chapter.

The role of O-polysaccharide in the *P. gingivalis* pathobiology

Recently, the relationship between O-polysaccharide and some of the biological activities of *P. gingivalis* was speculated using a *galE* mutant of strain ATCC 33277 (Nakao *et al.*, 2006). The *galE* gene encodes UDP-galactose 4-epimerase, GalE, that catalyses the interconversion of UDP-glucose to UDP-galactose. GalE is not restricted to bacteria but is also universally conserved among mammals including humans. The molecular weight of O-polysaccharide ladder bands of the *P. gingivalis galE* mutant was found to be smaller than that of the wild type. A GalE-positive revertant of the mutant restored the synthesis of the intact O-polysaccharide. GalE seemed to be necessary to provide galactose to the LPS of this strain as galactose can not be assembled into the repeating subunit of the O-polysaccharide in the mutant. The importance of galactose in O-polysaccharide synthesis was also supported by an earlier report in context of the sugar composition of LPS in the *P. gingivalis* strain 33277 (Bramanti *et al.*, 1989). In addition, O-polysaccharide truncations were observed during LPS analysis of *galE* mutants of other bacterial strains (Kwon *et al.*, 1998; Lee *et al.*, 1995) with the exception of *Vibrio cholerae*, whose O-polysaccharide was found to not contain galactose units (Nesper *et al.*, 2001).

Cells of the *galE* mutant of *P. gingivalis* exhibited an aggregative phenotype compared with the wild type. This phenotype may be due to a failure of complete cellular segregation resulting from a compositional alteration of the cell surface that contains a variety of glycoconjugates such as LPS and peptidoglycan. Alternatively, the shortened O-polysaccharide of the *galE* mutant may have made the cell surface stickier because of the higher hydrophobicity associated with a decrease in the glycan content at the cell's surface. Another possible explanation could be that cell aggregation might depend on the steric hindrance associated with the length of O-polysaccharide. When the length of the O-polysaccharide chain becomes too short, another essential component of the outer membrane involved in attachment or biofilm formation may have easily been expose to the environment.

It was furthermore examined whether the *galE* mutant had an altered sensitivity to antibiotics. The *galE* mutant was found to be significantly more susceptible to benzylpenicillin, oxacillin, cefotaxime, imipenem, and vancomycin compared with the wild type. The increased sensitivity to these antibiotics may also dependent on the shortened O-polysaccharide, which was found to be consistent with results of a previous study on truncated O-polysaccharide mutants of *V. cholerae* (Nesper *et al.*, 2001) and *E. coli* (Iguchi *et al.*, 2007).

LPS of other periodontal pathogens

There is only limited information concerning LPS of other periodontal pathogens aside of *P. gingivalis*. The lipid A structure of *Prevotella intermedia* was determined (Hashimoto *et al.*, 2003a) and it was demonstrated to resemble that of *P. gingivalis*. The activity of *P. intermedia* lipid A was weaker than that of *E. coli* lipid A. The immuno-dominant antigen of *Actinobacillus actinomycetemcomitans* was found to be the LPS O-polysaccharide (Lakio *et al.*, 2003; Page *et al.*, 1991; Sims *et al.*, 1991) by which strains of this anaerobe organism were accordingly classified into six distinct O-serotypes termed a to f. O-polysaccharide structures from all serotypes of *A. actinomycetemcomitans* have been characterized (Kaplan *et al.*, 2001; Perry *et al.*, 1996a; Perry *et al.*, 1996b). *A. actinomycetemcomitans* serotype b is clinically associated with LAP (Zambon *et al.*, 1983).

Capsular polysaccharides of periodontopathic bacteria

General features of bacterial CPS

Generally, the bacterial capsule is composed of polysaccharides on the surface and hence is called capsular polysaccharide (CPS) or K-antigen. Similar to the LPS O-antigen, CPS is also highly antigenic and elicits a strong immune response. The diversity of CPS is also used for serotyping of many encapsulated bacteria. One of the well-known critical roles of CPS is evasion from the host immune system (Bulmer and Sans, 1968; Wood and Smith, 1949). CPS-negative (K^-) strains are generally recognized and inactivated by the host innate immunity, i.e. they are easily killed by serum activity and phagocytosis. In contrast, bacterial opsonization via the complement system and antibodies is necessary in order to kill CPS-positive (K^+) strains. In addition, certain glycan structures inhibit activation of complement system in absence of specific antibodies and therefore allow the encapsulated microbes to survive in the blood stream (Jarvis and Vedros, 1987). Further, bacterial CPS mimics host molecules to evade host immunity, i.e. conduct molecular mimicry. The hyaluronic acid capsule of group A streptococci is a clear example of a mammalian structure mimicked by a bacterial polysaccharide (Cywes and Wessels, 2001). Another example of molecular mimicry is shown for the CPS of *Neisseria meningitidis*, which causes meningitis in humans (Finne *et al.*, 1983). A glycopeptide isolated from human fetal brain tissue reacted specifically to meningococcus group B antiserum. The interaction was inhibited by CPS from meningococcus group B but not by CPS of the groups A or C. This result might explain the difficulty for developing an efficient vaccine against group B meningococci in the context of immunological tolerance.

Role of CPS in bacterial surface attachment and biofilm formation.

Differential synthesis of CPS may affect bacterial survival by alternately exposing or masking hydrophobic surface structures. In general, bacterial CPS synthesis inversely correlated with the potential to attach to surfaces and to form biofilms. For example, the CPS of *Vibrio vulnificus*, an opportunistic pathogen that causes severe infections associated with raw oyster consumption, inhibited attachment and biofilm formation (Joseph and Wright, 2004). Encapsulated *Klebsiella pneumoniae* was found to be less adherent than acapsular mutants (Favre-Bonte *et al.*, 1999). In *N. meningitidis* the biofilm-forming clinical isolates exhibit high hydrophobicity and lack CPS (Yi *et al.*, 2004). A K^- strain of *N. meningitidis*, *ctrA* mutant, promoted biofilm formation in comparison to the parental strain (Yi *et al.*, 2004). Expression of genes required for CPS synthesis in *N. meningitidis* was down-regulated by contact with the host cell and this CPS down-regulation promoted a more intimate adhesion to the host cell (Deghmane *et al.*, 2002), further suggesting that CPS synthesis is detrimental for bacterial surface attachment activities.

CPS biosynthesis in *Bacteroides fragilis*

Bacteroides fragilis is an important human intestine symbiont and was found to modulate human intestine surface structures (Krinos *et al.*, 2001). *B. fragilis* produces multiple CPSs, whose synthesis is controlled by a reversible ON/OFF system regulated by a DNA invertase encoded by the *mpi* gene (Coyne *et al.*, 2003). In addition, *B. fragilis* synthesizes multiple fucosidases that cleave the terminal fucose residues from host-derived glycans. The released fucose can then be internalized by the bacteria through the action of a fucose permease. The exogenous fucose is converted to GDP-fucose in the bacterial cell and utilized as building blocks of surface glycan-related molecules, CPS, or bacterial glycoproteins (Coyne *et al.*, 2005). Interestingly, *B. fragilis* is phylogenetically close related to *P. gingivalis* although the genome sizes and the numbers of CPS loci differ between the two species. The genome of *B. fragilis* YCH46 is approximately 5.3 Mb and possesses nine CPS biogenesis loci (Kuwahara *et al.*, 2004), whereas the genome of *P. gingivalis* is approximately 2.3 Mb and possesses only four CPS biosynthetic gene loci (Nelson *et al.*, 2003).

The CPS of *P. gingivalis*

Bacteria in periodontal pockets withstand various stress factors of the oral environment including harsh physical conditions and host immunity. Such physical stress conditions are variable temperatures, pH, oxygen level, mechanical forces, and the constant flow of saliva and gingival crevice fluid. The host immunity represents another important stress factor and is subdivided into highly sophisticated innate and acquired immune systems in the oral cavity. In such a severe environment, some periodontal bacteria possess capsules. In *P. gingivalis*, at least six polysaccharide capsular serotypes (K1 to K6) in addition to the K^- strain have been reported (Laine *et al.*, 1996; van Winkelhoff *et al.*, 1993). Sugar analysis showed that CPS of strain ATCC 53977 (K1) consists of glucose, glucosamine, galactosamine, and galactosaminuronic acid (Schifferle *et al.*, 1989). Farquason *et al.* (2000) reported that CPS of strain W50 (K1) consists of mannuronic acid, glucuronic acid, galactose, and 2-acetamido-2-deoxy-D-glucose. In a clinical study, significantly elevated levels of serum IgG to K1 were reported in patients with chronic periodontitis (Sims *et al.*, 2001). Immunization with purified K1 of *P. gingivalis* strain A7436 prevented the *P. gingivalis*-elicited alveolar bone loss in a mouse model (Gonzalez *et al.*, 2003).

Sequence analysis of the genome of *P. gingivalis* stain W83 suggested four potential loci resembling those involved in capsular biosynthesis in other bacteria (Nelson *et al.*, 2003). The genetic locus designated PG0106-PG0120 was suggested to be a possible CPS locus using a mean of the global comparative genome analysis between W83 (K1) and ATCC 33277 (K^-) (Chen *et al.*, 2004). The sequence from the corresponding CPS locus of strain 381 (K^-) significantly differed from that of W83 (K1) (Aduse-Opoku *et al.*, 2006). Another sequence region from PG0109 to PG0110 in strain W83 was replaced with three unique open reading frames in strain 381 (K^-). The region from PG0112 to PG0114 from strain W83 was found to be deleted in the K^- strain 381 (Genbank accession no. AJ969093). It is tempting to speculate that these genomic differences may be linked to divergent CPS expression. Furthermore, comparative analysis of some mutants in genes of the PG0106-PG0120 locus showed that deletion of region PG0109-PG0118 and region PG0116-PG0120 led to no CPS synthesis although another mutation in locus, PG0117, did not affect the CPS synthesis.

Recently, Davey and Duncan (2006) expanded the role of CPS in oral bacterial ecology. They focused on the relationship between CPS synthesis, cell surface hydrophobicity, and biofilm formation of *P. gingivalis* using a CPS-deficient mutant (ΔPG0106), which originated from the K1 strain W83 and lost the ability to form K1 CPS. As a result, ΔPG0106 had a higher cell surface hydrophobicity and became more aggregative, followed by enhanced biofilm formation compared with the parental strain (Davey and Duncan, 2006). In general, the K^- strains such as 381 and 33277 produce biofilms stronger than K^+ strains such as W50 and W83. Other reports described the involvement of CPS in surface attachment and biofilm formation (Katz *et al.*, 1996; Laine and van Winkelhoff, 1998). All K^+ *P. gingivalis* strains tested in those studies showed a higher invasiveness after subcutaneous inoculation, with recovery of the microorganism from the blood, spleen, lung and kidney, compared to K^- strains, while the K^- strains induced localized abscesses. These results suggested that bacterial CPS might be more important for invasion steps following the prime attachment while these CPS are counterproductive for initial steps in biofilm formation.

Several other genes related to the polysaccharide synthesis were examined in a CPS-negative background using strain ATCC 33277 (K^-) (Nakao *et al.*, 2006). The *P. gingivalis galE* mutant with the shorten O-polysaccharide became more aggregative and had a 4.5-fold greater intensity than that of the parental strain. Similar data were shown in studies of asialylated LPS of *H. influenzae* in a CPS negative background using the chinchilla model for middle ear infections (Bouchet *et al.*, 2003). These data suggested that CPS can mask the hydrophobicity of the bacterial surface and enhance a disseminative phenotype. In the CPS-negative background, surface attachment and ability to form biofilms by *P. gingivalis* may depend on the composition of the LPS O-polysaccharide, which in turn also influences cell surface hydrophobicity.

Glycoproteins of periodontopathic bacteria

Glycoproteins in prokaryotes

DNA, RNA, and proteins are linear or circular polymers and composed of a limited and distinct number of building blocks, either four nucleotides or 20 amino acids, whereas polysaccharides, the fourth large class of biological macromolecules providing biological properties, can exhibit various ways to link together their monosaccharide building blocks. Owing to this characteristic feature of polysaccharides, i.e. the topological diversity and advantages, glycan modification of proteins provides a nearly unlimited variety, which is necessary for the fine tuning of cellular functions and the evolution of organisms.

In eukaryotes, glycosylated proteins are ubiquitous intracellularly and on the exterior leaflet of the plasma membrane. Protein glycosylation was previously considered to be restricted to eukaryotes, however, now it is widely accepted that prokaryotes also possess glycoproteins. The S-layer glycoprotein of *Halobacterium halobium* (*salinarum*) was the first example of a prokaryotic glycoprotein (Mescher *et al.*, 1974). After that, there have been annually increasing numbers of reports of both N-linked and O-linked protein glycosylation pathways in bacteria. The consensus amino acyl sequence of bacterial N-glycosylation sites was defined as a D/E-X-N-Y-S/T (X, Y≠P), which seems more specific than the tripeptide consensus sequence in eukaryotes, N-X-S/T (X≠P) (Kowarik *et al.*, 2006). Additionally, O-glycan attaches to S/T residues without a specific consensus sequence. Glycoproteins increasingly have been found in some Gram-negative bacteria since the 1990s.

In this section, an outline of glycoproteins from medically important Gram-negative pathogens will be given with special emphasis on glycoproteins of Gram-negative periodontopathic bacteria.

Glycoproteins of medically important Gram-negative bacteria

In Gram-negative bacteria, most glycoproteins are located at the outer membrane of the bacteria facing the outer environment and were found to be associated with virulence of the respective pathogens. The current understanding of protein glycosylation in bacterial pathogens was previously reviewed in depth (Comstock and Kasper, 2006; Schmidt *et al.*, 2003; Szymanski and Wren, 2005).

In *Mycobacterium tuberculosis*, three surface antigens (of 45/47, 38 and 19 kDa) have been described as glycoproteins. The 38-kDa antigen was found as a glycoprotein of *M. tuberculosis* and clinically associated with the infection (Espitia *et al.*, 1989; Espitia and Mancilla, 1989). The 19-kDa lipoprotein was associated with the host immune response and its glycan modification has been found to be linked to threonine residues at the N-terminus (Herrmann *et al.*, 1996). The 45/47-kDa Apa antigen complex of *M. tuberculosis* was identified as a glycoprotein, on which the mannose residues were essential for antigenicity of the molecule in a T-cell-dependent immune response *in vitro* and *in vivo* (Romain *et al.*, 1999).

N. meningitidis is a commensal of the human nasopharynx and is also known as the etiological agent of cerebrospinal meningitidis. *N. gonorrhoea* causes gonorrhoea, which remains one of the major sexually transmitted diseases. Neisserial type IV pili allow efficient colonization of the human mucosa (Swanson, 1973). Although 15 proteins are involved in the pilus biogenesis in *N. meningitidis*, the main component of the pilus, PilE, undergoes several post-translational modifications including glycosylation with O-linked glycan (Parge *et al.*, 1995). Genes involved in this type of glycosylation were identified in the *pgl* locus (Jennings *et al.*, 1998; Kahler *et al.*, 2001; Power *et al.*, 2000; Power *et al.*, 2003; Power *et al.*, 2006). However, the biological role of these modifications remained yet unclear.

Campylobacter jejuni is a human gut mucosal pathogen and is one of the main causes of bacterial gastroenteritis. Two distinct glycosylation systems, the O-linked and N-linked glycosylation, have been identified and characterized in *C. jejuni* and have been reviewed recently (Szymanski *et al.*, 2003; Szymanski and Wren, 2005). A flagellar O-linked glycosylation locus has been found adjacent to the flagellum biosynthetic structural genes, *flaA* and *flaB*, in the genome of *C. jejuni* (Parkhill *et al.*, 2000). The carbohydrate structure added to flagellin proteins has been determined as pseudaminic acid (Pse) or its derivatives, which are uncommon carbohydrate residues identified in glycoproteins of Gram-negative bacteria. A second protein glycosylation locus, which was named *pgl* for protein glycosylation, mediates the N-linked glycosylation of multiple proteins in *C. jejuni* (Szymanski *et al.*, 1999). The *pglB* gene in this locus encodes a key enzyme, PglB, which probably acts as an oligosaccharide transferase mediating the N-linkage of the sugar to the peptide backbone (Young *et al.*, 2002). Disruption of N-linked glycosylation in *C. jejuni* resulted in a reduced ability to adhere to and invade host cells *in vitro*, and also a reduced ability to colonized the intestine of chicken and mice (Hendrixson and DiRita, 2004; Jones *et al.*, 2004; Szymanski *et al.*, 2002). However, these results are difficult to interpret due to possible pleiotropic effects caused by changes in the glycosylation status of several proteins.

Opportunistic *Pseudomonas aeruginosa* is an infamous pathogen causing lethal infections in patients with compromised host defences. This microorganism forms biofilms which are responsible for the chronic colonization of the lungs of patient with cystic fibrosis. The expression of *P. aeruginosa* pili is important in cell adhesion (Hahn, 1997) and is responsible for twitching motility (Wall and Kaiser, 1999), a form of cell movement involved in the establishment of multi-cellular surface aggregates and spreading of this microorganism inside the lung. The pili were found to be glycosylated with a trisaccharide, which is linked to a serine residue (Castric *et al.*, 2001). The *P. aeruginosa* flagellin is also glycosylated (Brimer and Montie, 1998) and a genomic cluster of 14 genes responsible for flagellin glycosylation has been identified in the genome of A-type strains harbouring the flagellum biosynthetic genetic locus (Arora *et al.*, 2001).

Helicobacter pylori flaA and *flaB* are two distinct flagellin structural genes. Both FlaA and FlaB proteins are post-translationally modified as glycoproteins by mechanisms similar to those in other Gram-negative bacteria, such as *C. jejuni* and *P. aeruginosa*.

In *E. coli*, three glycoproteins, AIDA-I, TibA, and Ag43, have been characterized which are autotransporter modified by a heptose residue adhesion (Benz and Schmidt, 2001; Lindenthal and Elsinghorst, 1999; Sherlock *et al.*, 2006). In all cases of these glycoproteins, their glycoprotein status might be responsible for attachment to mammalian cells (Benz and Schmidt, 2001; Lindenthal and Elsinghorst, 2001; Sherlock *et al.*, 2006).

In *Borrelia burgdorferi*, the causative agent of Lyme diseases, the outer surface proteins, OspA and OspB, and the flagellin were initially reported as glycoproteins (Ge *et al.*, 1998; Sambri *et al.*, 1992). Recently, both indirect staining and mass spectrometry analysis demonstrated that these proteins are in fact not glycosylated (Sterba *et al.*, 2008). The authors therefore proposed that the previously reported glycoproteins on the surface of *B. burgdorferi* might have been generated in specific culture media. However, this does not rule out the possibility that the organism may glycosylate these or other proteins under the different culture conditions, i.e. *in situ*.

Glycoproteins of Gram-negative periodontopathic bacteria

Currently, three glycoproteins have been characterized in Gram-negative periodontopathic bacteria. One of them is the S-layer protein of *Tannerella (Bacteroides) forsythensis* (Lee *et al.*, 2006). The other two reported glycosylated proteins were gingipains (Curtis *et al.*, 1999) and the outer membrane protein (OMP) OMP85 homologue of *P. gingivalis* (Nakao *et al.*, 2008). All three glycoproteins are *bona fide* virulence factors or associated with pathogenic properties of the organism. Both Gram-negative bacteria mentioned above were regarded as the etiological agents of periodontitis and are members of the so-called 'red complex' microbial consortia strongly associated with the progression of periodontal disease (Socransky *et al.*, 1998).

While a variety of S-layers have been observed in Gram-positive bacteria, only a few Gram-negative bacteria possess this type of surface layer. The S-layer of *T. forsythensis* was identified (Kerosuo, 1988) and characterized, showing that the S-layer is associated with haemagglutination, adhesion, invasion to epithelial cells, and abscess formation (Sabet *et al.*, 2003). The S-layer is composed of two distinct glycoproteins, termed the 200- and 210-kDa proteins encoded by *tfsA* and *tfsB*, respectively (Lee *et al.*, 2006).

Gingipains are specific cysteine proteases and the best-characterized virulence factors in *P. gingivalis* (Sheets *et al.*, 2008). The *rgpA* gene encodes the Arg-X-specific gingipain, RgpA, the *rgpB* gene encodes the Arg-X-specific gingipain, RgpB, and the *kgp* gene encodes the Lys-X-specific gingipain, Kgp. The biological roles of these enzymes were demonstrated to be (i) degradation of exogenous proteins, (ii) supply of the digested peptides for bacterial growth, (iii) destruction of cell-cell junctions in periodontal tissues, and (iv) providing protein components for host immunity. Gingipains were the first to be demonstrated to be post-translationally modified with glycan additions (Curtis *et al.*, 1999). Several genes in the *vim* locus and the *por* locus were found to be involved in glycosylation or regulation of gingipains (Chen *et al.*, 2000; Shoji *et al.*, 2002; Vanterpool *et al.*, 2004, 2005a; Vanterpool *et al.*, 2005b). However, currently there is no detailed information on the enzymology of gingipain glycosylation.

OMP85 is responsible for assembly of outer membrane proteins and is a highly conserved outer membrane protein in all tested Gram-negative bacteria. The OMP85 of *N. meningitidis*, *E. coli*, and *P. aeruginosa* were essential for their survival (Tashiro *et al.*, 2008; Voulhoux *et al.*, 2003; Werner and Misra, 2005). The OMP85 homologue of *P. gingivalis* was recently demonstrated to be a glycoprotein and the same study also provided first direct evidence of OMP85 glycosylation in Gram-negative bacteria (Nakao *et al.*, 2008). Further, an outer-loop peptide antibody specific for OMP85 of *P. gingivalis* inhibited biofilm formation in the deglycosylated *galE* mutant indicating that OMP85 glycosylation may be involved in biofilm formation of *P. gingivalis*. Similar results were obtained for the OMP85 homologue, Opr86, of *P. aeruginosa* (Tashiro *et al.*, 2008). Additionally, OMP85 homologues from some bacterial species were found to be immmunogenic and the immunization of animals by recombinant proteins of each homologue conferred protection against the bacterial challenge (Cameron *et al.*, 2000; Loosmore *et al.*, 1997). These data suggested that OMP85 is a critical and potential candidate as a protective antigen against biofilm formation in Gram-negative pathogens.

Summary and future directions

Sequences of genes and proteins have been quickly determined on large scale by systematic experimental methods through genome and proteome projects. In contrast, in the past glycan structure determination had to face many challenges and difficulties because glycan syntheses are not directly template driven and the topology easily varies according to the branching and anomeric configurations. However, protein and lipid glycosylation, the most common post-translational and membrane-associated modifications *in vivo*, respectively, have become one of the most appealing fields of research in the post-genomic and post-proteomic era since it has been postulated that more than half of all proteins in nature will be found to be glycosylated (Apweiler *et al.*, 1999). Glycans, which frequently attach to extracellular proteins and lipids, are essential for biological activities in eukaryotes (Lowe, 2001; Ohtsubo and Marth, 2006; Rudd *et al.*, 2001).

Likewise, understanding the biological role of the two distinct bacterial glycoconjugates, LPS and CPS, is important, especially in the field of medical microbiology. Most of the glycoproteins in Gram-negative bacteria also have been shown to play critical roles in virulence, which exist at the bacterial surface or are secreted to the external environment. In periodontopathic Gram-negative bacteria, only three glycoproteins have been reported, which are virulence factors or associated with their pathogenic properties.

At present, there are some critical issues which need to be addressed in the light of colonization/biofilm formation by oral bacteria and their immune evasion systems. Until now, data strongly suggest that the bacterial polysaccharides hold the key to solving the questions. For instance, do commensals and pathogens express distinct polysaccharides? Is the glycosylation status strictly associated with the property of colonization/biofilm formation? What is the key structure of bacterial polysaccharides for direct interaction to the host and induction of the host immune system? Can we manage virulence by controlling the expression of bacterial polysaccharides in future? Recently, the whole genome sequence of *P. gingivalis* strain W83 (K1) and ATCC 33277 (K$^-$) was determined (Nelson *et al.*, 2003; Naito *et al.*, 2008). However, the genetic information and manipulative genetic tools have not yet become mature enough for periodontopathic bacteria compared with the other Gram-negative bacteria and the precise roles of the glycosylation system in periodontopathic bacteria have yet to be resolved. New and detailed answers to the above-mentioned questions may in turn give us a better understanding of the molecular and cellular interactions between the host and periodontopathic bacteria.

Acknowledgements

The author thanks Dr Hidenobu Senpuku and Dr Haruo Watanabe for their help and insights on the manuscript. He is also grateful to his family, friends, and colleagues in Japan and Sweden for their encouragement.

References

Aduse-Opoku, J., Slaney, J.M., Hashim, A., Gallagher, A., Gallagher, R.P., Rangarajan, M., Boutaga, K., Laine, M.L., Van Winkelhoff, A.J., and Curtis, M.A. (2006). Identification and characterization of the capsular polysaccharide (K-antigen) locus of *Porphyromonas gingivalis*. Infect. Immun. *74*, 449–460.

Appelmelk, B.J., Shiberu, B., Trinks, C., Tapsi, N., Zheng, P.Y., Verboom, T., Maaskant, J., Hokke, C.H., Schiphorst, W.E., Blanchard, D., *et al.* (1998). Phase variation in *Helicobacter pylori* lipopolysaccharide. Infect. Immun. *66*, 70–76.

Apweiler, R., Hermjakob, H., and Sharon, N. (1999). On the frequency of protein glycosylation, as deduced from analysis of the SWISS-PROT database. Biochim. Biophys. Acta *1473*, 4–8.

Arora, S.K., Bangera, M., Lory, S., and Ramphal, R. (2001). A genomic island in *Pseudomonas aeruginosa* carries the determinants of flagellin glycosylation. Proc. Natl. Acad. Sci. USA *98*, 9342–9347.

Asai, Y., Hashimoto, M., Fletcher, H.M., Miyake, K., Akira, S., and Ogawa, T. (2005). Lipopolysaccharide preparation extracted from *Porphyromonas gingivalis* lipoprotein-deficient mutant shows a marked decrease in toll-like receptor 2-mediated signaling. Infect. Immun. *73*, 2157–2163.

Beck, J., Garcia, R., Heiss, G., Vokonas, P.S., and Offenbacher, S. (1996). Periodontal disease and cardiovascular disease. J. Periodontol *67*, 1123–1137.

Benz, I., and Schmidt, M.A. (2001). Glycosylation with heptose residues mediated by the aah gene product is essential for adherence of the AIDA-I adhesin Mol. Microbiol. *40*, 1403–1413.

Boman, H.G. (1995). Peptide antibiotics and their role in innate immunity. Annu Rev Immunol. *13*, 61–92.

Bouchet, V., Hood, D.W., Li, J., Brisson, J.R., Randle, G.A., Martin, A., Li, Z., Goldstein, R., Schweda, E.K., and

Pelton, S.I. (2003). Host-derived sialic acid is incorporated into *Haemophilus influenzae* lipopolysaccharide and is a major virulence factor in experimental otitis media. Proc. Natl. Acad. Sci. USA *100*, 8898–8903.

Bramanti, T.E., Wong, G.G., Weintraub, S.T., and Holt, S.C. (1989). Chemical characterization and biologic properties of lipopolysaccharide from *Bacteroides gingivalis* strains W50, W83, and ATCC 33277. Oral Microbiol. Immunol. *4*, 183–192.

Brimer, C.D., and Montie, T.C. (1998). Cloning and comparison of *fliC* genes and identification of glycosylation in the flagellin of *Pseudomonas aeruginosa* a-type strains. J. Bacteriol. *180*, 3209–3217.

Brooks, W., Demuth, D.R., Gil, S., and Lamont, R.J. (1997). Identification of a *Streptococcus gordonii* SspB domain that mediates adhesion to *Porphyromonas gingivalis*. Infect. Immun. *65*, 3753–3758.

Bulmer, G.S., and Sans, M.D. (1968). *Cryptococcus neoformans*. 3. Inhibition of phagocytosis. J. Bacteriol. *95*, 5–8.

Cameron, C.E., Lukehart, S.A., Castro, C., Molini, B., Godornes, C., and Van Voorhis, W.C. (2000). Opsonic potential, protective capacity, and sequence conservation of the *Treponema pallidum* subspecies pallidum Tp92. J. Infect Dis *181*, 1401–1413.

Camprubi, S., Merino, S., Benedi, V.J., and Tomas, J.M. (1993a). The role of the O-antigen lipopolysaccharide and capsule on an experimental *Klebsiella pneumoniae* infection of the rat urinary tract. FEMS Microbiol. Lett. *111*, 9–13.

Camprubi, S., Merino, S., Guillot, J.F., and Tomas, J.M. (1993b). The role of the O-antigen lipopolysaccharide on the colonization *in vivo* of the germfree chicken gut by *Klebsiella pneumoniae*. Microb. Pathog. *14*, 433–440.

Castric, P., Cassels, F.J., and Carlson, R.W. (2001). Structural characterization of the *Pseudomonas aeruginosa* 1244 pilin glycan. J. Biol. Chem. *276*, 26479–26485.

Chen, T., Dong, H., Yong, R., and Duncan, M.J. (2000). Pleiotropic pigmentation mutants of *Porphyromonas gingivalis*. Microb. Pathog. *28*, 235–247.

Chen, T., Hosogi, Y., Nishikawa, K., Abbey, K., Fleischmann, R.D., Walling, J., and Duncan, M.J. (2004). Comparative whole-genome analysis of virulent and avirulent strains of *Porphyromonas gingivalis*. J. Bacteriol. *186*, 5473–5479.

Chung, W.O., Demuth, D.R., and Lamont, R.J. (2000). Identification of a *Porphyromonas gingivalis* receptor for the *Streptococcus gordonii* SspB protein. Infect. Immun. *68*, 6758–6762.

Comstock, L.E., and Kasper, D.L. (2006). Bacterial glycans: key mediators of diverse host immune responses. Cell *126*, 847–850.

Coyne, M.J., Reinap, B., Lee, M.M., and Comstock, L.E. (2005). Human symbionts use a host-like pathway for surface fucosylation. Science *307*, 1778–1781.

Coyne, M.J., Weinacht, K.G., Krinos, C.M., and Comstock, L.E. (2003). Mpi recombinase globally modulates the surface architecture of a human commensal bacterium. Proc. Natl. Acad. Sci. USA *100*, 10446–10451.

Curtis, M.A., Thickett, A., Slaney, J.M., Rangarajan, M., Aduse-Opoku, J., Shepherd, P., Paramonov, N., and Hounsell, E.F. (1999). Variable carbohydrate modifications to the catalytic chains of the RgpA and RgpB proteases of *Porphyromonas gingivalis* W50. Infect. Immun. *67*, 3816–3823.

Cywes, C., and Wessels, M.R. (2001). Group A *Streptococcus* tissue invasion by CD44-mediated cell signalling. Nature *414*, 648–652.

Darveau, R.P., Pham, T.T., Lemley, K., Reife, R.A., Bainbridge, B.W., Coats, S.R., Howald, W.N., Way, S.S., and Hajjar, A.M. (2004). *Porphyromonas gingivalis* lipopolysaccharide contains multiple lipid A species that functionally interact with both toll-like receptors 2 and 4. Infect. Immun. *72*, 5041–5051.

Davey, M.E., and Duncan, M.J. (2006). Enhanced biofilm formation and loss of capsule synthesis: deletion of a putative glycosyltransferase in *Porphyromonas gingivalis*. J. Bacteriol. *188*, 5510–5523.

Deghmane, A.E., Giorgini, D., Larribe, M., Alonso, J.M., and Taha, M.K. (2002). Down-regulation of pili and capsule of *Neisseria meningitidis* upon contact with epithelial cells is mediated by CrgA regulatory protein. Mol. Microbiol. *43*, 1555–1564.

Demmer, R.T., and Desvarieux, M. (2006). Periodontal infections and cardiovascular disease: the heart of the matter. J. Am Dent. Assoc *137* Suppl, 14S-20S; quiz 38S.

Dixon, D.R., and Darveau, R.P. (2005). Lipopolysaccharide heterogeneity: innate host responses to bacterial modification of lipid a structure. J. Dent. Res. *84*, 584–595.

Espitia, C., Cervera, I., Gonzalez, R., and Mancilla, R. (1989). A 38-kD *Mycobacterium tuberculosis* antigen associated with infection. Its isolation and serologic evaluation. Clin. Exp Immunol. *77*, 373–377.

Espitia, C., and Mancilla, R. (1989). Identification, isolation and partial characterization of *Mycobacterium tuberculosis* glycoprotein antigens. Clin. Exp. Immunol. *77*, 378–383.

Farquharson, S.I., Germaine, G.R., and Gray, G.R. (2000). Isolation and characterization of the cell-surface polysaccharides of *Porphyromonas gingivalis* ATCC 53978. Oral Microbiol. Immunol. *15*, 151–157.

Favre-Bonte, S., Joly, B., and Forestier, C. (1999). Consequences of reduction of *Klebsiella pneumoniae* capsule expression on interactions of this bacterium with epithelial cells. Infect. Immun. *67*, 554–561.

Finne, J., Leinonen, M., and Makela, P.H. (1983). Antigenic similarities between brain components and bacteria causing meningitis. Implications for vaccine development and pathogenesis. Lancet *2*, 355–357.

Ge, Y., Li, C., Corum, L., Slaughter, C.A., and Charon, N.W. (1998). Structure and expression of the FlaA periplasmic flagellar protein of *Borrelia burgdorferi*. J. Bacteriol. *180*, 2418–2425.

Genco, R., Offenbacher, S., and Beck, J. (2002). Periodontal disease and cardiovascular disease: epidemiology and possible mechanisms. J. Am Dent. Assoc *133 Suppl*, 14S-22S.

Gibson, F.C., 3rd, Hong, C., Chou, H.H., Yumoto, H., Chen, J., Lien, E., Wong, J., and Genco, C.A. (2004). Innate immune recognition of invasive bacteria accelerates atherosclerosis in apolipoprotein E-deficient mice. Circulation *109*, 2801–2806.

Goldenberg, R.L., Culhane, J.F., Iams, J.D., and Romero, R. (2008). Epidemiology and causes of pre-term birth. Lancet *371*, 75–84.

Gonzalez, D., Tzianabos, A.O., Genco, C.A., and Gibson, F.C., 3rd (2003). Immunization with *Porphyromonas gingivalis* capsular polysaccharide prevents *P. gingivalis*-elicited oral bone loss in a murine model. Infect. Immun. *71*, 2283–2287.

Hahn, H.P. (1997). The type-4 pilus is the major virulence-associated adhesin of *Pseudomonas aeruginosa*- a review. Gene *192*, 99–108.

Harvill, E.T., Preston, A., Cotter, P.A., Allen, A.G., Maskell, D.J., and Miller, J.F. (2000). Multiple roles for *Bordetella* lipopolysaccharide molecules during respiratory tract infection. Infect. Immun. *68*, 6720–6728.

Hashimoto, M., Asai, Y., and Ogawa, T. (2004). Separation and structural analysis of lipoprotein in a lipopolysaccharide preparation from *Porphyromonas gingivalis*. Int. Immunol. *16*, 1431–1437.

Hashimoto, M., Asai, Y., Tamai, R., Jinno, T., Umatani, K., and Ogawa, T. (2003a). Chemical structure and immunobiological activity of lipid A from *Prevotella intermedia* ATCC 25611 lipopolysaccharide. FEBS Lett. *543*, 98–102.

Hashimoto, M., Ogawa, S., Asai, Y., Takai, Y., and Ogawa, T. (2003b). Binding of *Porphyromonas gingivalis* fimbriae to *Treponema denticola* dentilisin. FEMS Microbiol. Lett *226*, 267–271.

Hendrixson, D.R., and DiRita, V.J. (2004). Identification of *Campylobacter jejuni* genes involved in commensal colonization of the chick gastrointestinal tract. Mol. Microbiol. *52*, 471–484.

Herrmann, J.L., O'Gaora, P., Gallagher, A., Thole, J.E., and Young, D.B. (1996). Bacterial glycoproteins: a link between glycosylation and proteolytic cleavage of a 19 kDa antigen from *Mycobacterium tuberculosis*. EMBO J. *15*, 3547–3554.

Hoshino, K., Takeuchi, O., Kawai, T., Sanjo, H., Ogawa, T., Takeda, Y., Takeda, K., and Akira, S. (1999). Cutting edge: Toll-like receptor 4 (TLR4)-deficient mice are hyporesponsive to lipopolysaccharide: evidence for TLR4 as the Lps gene product. J. Immunol. *162*, 3749–3752.

Iguchi, A., Iyoda, S., Watanabe, H., and Osawa, R. (2007). O side chain deficiency enhances sensitivity of *Escherichia coli* to Shiga toxin 2-converting bacteriophages. Curr Microbiol. *54*, 14–19.

Jarvis, G.A., and Vedros, N.A. (1987). Sialic acid of group B *Neisseria meningitidis* regulates alternative complement pathway activation. Infect. Immun. *55*, 174–180.

Jennings, M.P., Virji, M., Evans, D., Foster, V., Srikhanta, Y.N., Steeghs, L., van der Ley, P., and Moxon, E.R. (1998). Identification of a novel gene involved in pilin glycosylation in *Neisseria meningitidis*. Mol. Microbiol. *29*, 975–984.

Jones, M.A., Marston, K.L., Woodall, C.A., Maskell, D.J., Linton, D., Karlyshev, A.V., Dorrell, N., Wren, B.W., and Barrow, P.A. (2004). Adaptation of *Campylobacter jejuni* NCTC11168 to high-level colonization of the avian gastrointestinal tract. Infect. Immun. *72*, 3769–3776.

Joseph, L.A., and Wright, A.C. (2004). Expression of *Vibrio vulnificus* capsular polysaccharide inhibits biofilm formation. J. Bacteriol. *186*, 889–893.

Kahler, C.M., Martin, L.E., Tzeng, Y.L., Miller, Y.K., Sharkey, K., Stephens, D.S., and Davies, J.K. (2001). Polymorphisms in pilin glycosylation locus of *Neisseria meningitidis* expressing class II pili. Infect. Immun. *69*, 3597–3604.

Kaplan, J.B., Perry, M.B., MacLean, L.L., Furgang, D., Wilson, M.E., and Fine, D.H. (2001). Structural and genetic analyses of O polysaccharide from *Actinobacillus actinomycetemcomitans* serotype f. Infect. Immun. *69*, 5375–5384.

Katz, J., Ward, D.C., and Michalek, S.M. (1996). Effect of host responses on the pathogenicity of strains of *Porphyromonas gingivalis*. Oral Microbiol. Immunol. *11*, 309–318.

Kerosuo, E. (1988). Ultrastructure of the cell envelope of *Bacteroides forsythus* strain ATCC 43037T. Oral Microbiol. Immunol. 3, 134–137.

Kolenbrander, P.E., Andersen, R.N., and Holdeman, L.V. (1985). Coaggregation of oral *Bacteroides* species with other bacteria: central role in coaggregation bridges and competitions. Infect. Immun. *48*, 741–746.

Kolenbrander, P.E., Palmer, R.J., Jr., Rickard, A.H., Jakubovics, N.S., Chalmers, N.I., and Diaz, P.I. (2006). Bacterial interactions and successions during plaque development. Periodontol 2000 *42*, 47–79.

Kowarik, M., Young, N.M., Numao, S., Schulz, B.L., Hug, I., Callewaert, N., Mills, D.C., Watson, D.C., Hernandez, M., Kelly, J.F. (2006). Definition of the bacterial N-glycosylation site consensus sequence. EMBO J. *25*, 1957–1966.

Krinos, C.M., Coyne, M.J., Weinacht, K.G., Tzianabos, A.O., Kasper, D.L., and Comstock, L.E. (2001). Extensive surface diversity of a commensal microorganism by multiple DNA inversions. Nature *414*, 555–558.

Kumada, H., Haishima, Y., Umemoto, T., and Tanamoto, K. (1995). Structural study on the free lipid A isolated from lipopolysaccharide of *Porphyromonas gingivalis*. J. Bacteriol. *177*, 2098–2106.

Kuwahara, T., Yamashita, A., Hirakawa, H., Nakayama, H., Toh, H., Okada, N., Kuhara, S., Hattori, M., Hayashi, T., and Ohnishi, Y. (2004). Genomic analysis of *Bacteroides fragilis* reveals extensive DNA inversions regulating cell surface adaptation. Proc. Natl. Acad. Sci. USA *101*, 14919–14924.

Kwon, D.H., Woo, J.S., Perng, C.L., Go, M.F., Graham, D.Y., and El-Zaatari, F.A. (1998). The effect of galE gene inactivation on lipopolysaccharide profile of *Helicobacter pylori*. Curr Microbiol. *37*, 144–148.

Laine, M.L., Appelmelk, B.J., and van Winkelhoff, A.J. (1996). Novel polysaccharide capsular serotypes in *Porphyromonas gingivalis*. J. Periodontal Res. *31*, 278–284.

Laine, M.L., and van Winkelhoff, A.J. (1998). Virulence of six capsular serotypes of *Porphyromonas gingivalis* in a mouse model. Oral Microbiol. Immunol. *13*, 322–325.

Lakio, L., Paju, S., Alfthan, G., Tiirola, T., Asikainen, S., and Pussinen, P.J. (2003). *Actinobacillus actinomycetemcomitans* serotype D-specific antigen contains the

O antigen of lipopolysaccharide. Infect. Immun. *71*, 5005–5011.

Lancy, P., Jr., Dirienzo, J.M., Appelbaum, B., Rosan, B., and Holt, S.C. (1983). Corncob formation between *Fusobacterium nucleatum* and *Streptococcus sanguis*. Infect. Immun. *40*, 303–309.

Lee, F.K., Stephens, D.S., Gibson, B.W., Engstrom, J.J., Zhou, D., and Apicella, M.A. (1995). Microheterogeneity of *Neisseria* lipooligosaccharide: analysis of a UDP-glucose 4-epimerase mutant of *Neisseria meningitidis* NMB. Infect. Immun. *63*, 2508–2515.

Lee, S.W., Sabet, M., Um, H.S., Yang, J., Kim, H.C., and Zhu, W. (2006). Identification and characterization of the genes encoding a unique surface (S-) layer of *Tannerella forsythia*. Gene *371*, 102–111.

Lindenthal, C., and Elsinghorst, E.A. (1999). Identification of a glycoprotein produced by enterotoxigenic *Escherichia coli*. Infect. Immun. *67*, 4084–4091.

Lindenthal, C., and Elsinghorst, E.A. (2001). Enterotoxigenic *Escherichia coli* TibA glycoprotein adheres to human intestine epithelial cells. Infect. Immun. *69*, 52–57.

Loosmore, S.M., Yang, Y.P., Coleman, D.C., Shortreed, J.M., England, D.M., and Klein, M.H. (1997). Outer membrane protein D15 is conserved among *Haemophilus influenzae* species and may represent a universal protective antigen against invasive disease. Infect. Immun. *65*, 3701–3707.

Lowe, J.B. (2001). Glycosylation, immunity, and autoimmunity. Cell *104*, 809–812.

Martin, M., Katz, J., Vogel, S.N., and Michalek, S.M. (2001). Differential induction of endotoxin tolerance by lipopolysaccharides derived from *Porphyromonas gingivalis* and *Escherichia coli*. J. Immunol. *167*, 5278–5285.

Mescher, M.F., Strominger, J.L., and Watson, S.W. (1974). Protein and carbohydrate composition of the cell envelope of *Halobacterium salinarium*. J. Bacteriol. *120*, 945–954.

Meyer, D.H., and Fives-Taylor, P.M. (1997). The role of *Actinobacillus actinomycetemcomitans* in the pathogenesis of periodontal disease. Trends Microbiol. *5*, 224–228.

Naito, M., Hirakawa, H., Yamashita, A., Ohara, N., Shoji, M., Yukitake, H., Nakayama, K., Toh, H., Yoshimura, F., Kuhara, S. (2008). Determination of the genome sequence of *Porphyromonas gingivalis* strain ATCC 33277 and genomic comparison with strain W83 revealed extensive genome rearrangements in *P. gingivalis*. DNA Res.

Nakao, R., Senpuku, H., and Watanabe, H. (2006). *Porphyromonas gingivalis* galE is involved in lipopolysaccharide O-antigen synthesis and biofilm formation. Infect. Immun. *74*, 6145–6153.

Nakao, R., Tashiro, Y., Nomura, N., Kosono, S., Ochiai, K., Yonezawa, H., Watanabe, H., and Senpuku, H. (2008). Glycosylation of the OMP85 homolog of *Porphyromonas gingivalis* and its involvement in biofilm formation. Biochem. Biophys. Res. Commun. *365*, 784–789.

Nelson, K.E., Fleischmann, R.D., DeBoy, R.T., Paulsen, I.T., Fouts, D.E., Eisen, J.A., Daugherty, S.C., Dodson, R.J., Durkin, A.S., Gwinn, M., *et al.* (2003). Complete genome sequence of the oral pathogenic bacterium *Porphyromonas gingivalis* strain W83. J. Bacteriol. *185*, 5591–5601.

Nesper, J., Lauriano, C.M., Klose, K.E., Kapfhammer, D., Kraiss, A., and Reidl, J. (2001). Characterization of *Vibrio cholerae* O1 El tor galU and galE mutants: influence on lipopolysaccharide structure, colonization, and biofilm formation. Infect. Immun. *69*, 435–445.

Nevola, J.J., Laux, D.C., and Cohen, P.S. (1987). In vivo colonization of the mouse large intestine and *in vitro* penetration of intestinal mucus by an avirulent smooth strain of *Salmonella typhimurium* and its lipopolysaccharide-deficient mutant. Infect. Immun. *55*, 2884–2890.

Nevola, J.J., Stocker, B.A., Laux, D.C., and Cohen, P.S. (1985). Colonization of the mouse intestine by an avirulent *Salmonella typhimurium* strain and its lipopolysaccharide-defective mutants. Infect. Immun. *50*, 152–159.

Nicolas, P., and Mor, A. (1995). Peptides as weapons against microorganisms in the chemical defense system of vertebrates. Annu. Rev. Microbiol. *49*, 277–304.

Offenbacher, S., Katz, V., Fertik, G., Collins, J., Boyd, D., Maynor, G., McKaig, R., and Beck, J. (1996). Periodontal infection as a possible risk factor for preterm low birth weight. J. Periodontol. *67*, 1103–1113.

Ogawa, T. (1993). Chemical structure of lipid A from *Porphyromonas* (*Bacteroides*) *gingivalis* lipopolysaccharide. FEBS Lett. *332*, 197–201.

Ohtsubo, K., and Marth, J.D. (2006). Glycosylation in cellular mechanisms of health and disease. Cell *126*, 855–867.

Onagawa, M., Ishihara, K., and Okuda, K. (1994). Coaggregation between *Porphyromonas gingivalis* and *Treponema denticola*. Bull. Tokyo Dent. Coll. *35*, 171–181.

Page, R.C., Sims, T.J., Engel, L.D., Moncla, B.J., Bainbridge, B., Stray, J., and Darveau, R.P. (1991). The immunodominant outer membrane antigen of *Actinobacillus actinomycetemcomitans* is located in the serotype-specific high-molecular-mass carbohydrate moiety of lipopolysaccharide. Infect. Immun. *59*, 3451–3462.

Paramonov, N., Bailey, D., Rangarajan, M., Hashim, A., Kelly, G., Curtis, M.A., and Hounsell, E.F. (2001). Structural analysis of the polysaccharide from the lipopolysaccharide of *Porphyromonas gingivalis* strain W50. Eur. J. Biochem. *268*, 4698–4707.

Paramonov, N., Rangarajan, M., Hashim, A., Gallagher, A., Aduse-Opoku, J., Slaney, J.M., Hounsell, E., and Curtis, M.A. (2005). Structural analysis of a novel anionic polysaccharide from *Porphyromonas gingivalis* strain W50 related to Arg-gingipain glycans. Mol. Microbiol. *58*, 847–863.

Parge, H.E., Forest, K.T., Hickey, M.J., Christensen, D.A., Getzoff, E.D., and Tainer, J.A. (1995). Structure of the fibre-forming protein pilin at 2.6 A resolution. Nature *378*, 32–38.

Parkhill, J., Wren, B.W., Mungall, K., Ketley, J.M., Churcher, C., Basham, D., Chillingworth, T., Davies, R.M., Feltwell, T., Holroyd, S., *et al.* (2000). The genome sequence of the food-borne pathogen

Campylobacter jejuni reveals hypervariable sequences. Nature 403, 665–668.

Paster, B.J., Boches, S.K., Galvin, J.L., Ericson, R.E., Lau, C.N., Levanos, V.A., Sahasrabudhe, A., and Dewhirst, F.E. (2001). Bacterial diversity in human subgingival plaque. J. Bacteriol. 183, 3770–3783.

Paster, B.J., Olsen, I., Aas, J.A., and Dewhirst, F.E. (2006). The breadth of bacterial diversity in the human periodontal pocket and other oral sites. Periodontology 2000 42, 80–87.

Perry, M.B., MacLean, L.L., Gmur, R., and Wilson, M.E. (1996a). Characterization of the O-polysaccharide structure of lipopolysaccharide from *Actinobacillus actinomycetemcomitans* serotype b. Infect. Immun 64, 1215–1219.

Perry, M.B., MacLean, L.M., Brisson, J.R., and Wilson, M.E. (1996b). Structures of the antigenic O-polysaccharides of lipopolysaccharides produced by *Actinobacillus actinomycetemcomitans* serotypes a, c, d and e. Eur J. Biochem 242, 682–688.

Poltorak, A., He, X., Smirnova, I., Liu, M.Y., Van Huffel, C., Du, X., Birdwell, D., Alejos, E., Silva, M., and Galanos, C. (1998). Defective LPS signaling in C3H/HeJ and C57BL/10ScCr mice: mutations in Tlr4 gene. Science 282, 2085–2088.

Power, P.M., Roddam, L.F., Dieckelmann, M., Srikhanta, Y.N., Tan, Y.C., Berrington, A.W., and Jennings, M.P. (2000). Genetic characterization of pilin glycosylation in *Neisseria meningitidis*. Microbiology 146 (Pt 4), 967–979.

Power, P.M., Roddam, L.F., Rutter, K., Fitzpatrick, S.Z., Srikhanta, Y.N., and Jennings, M.P. (2003). Genetic characterization of pilin glycosylation and phase variation in *Neisseria meningitidis*. Mol. Microbiol. 49, 833–847.

Power, P.M., Seib, K.L., and Jennings, M.P. (2006). Pilin glycosylation in *Neisseria meningitidis* occurs by a similar pathway to wzy-dependent O-antigen biosynthesis in *Escherichia coli*. Biochem. Biophys. Res. Commun. 347, 904–908.

Rangarajan, M., Aduse-Opoku, J., Paramonov, N., Hashim, A., Bostanci, N., Fraser, O.P., Tarelli, E., and Curtis, M.A. (2008). Identification of a Second Lipopolysaccharide in *Porphyromonas gingivalis* W50. J. Bacteriol..

Romain, F., Horn, C., Pescher, P., Namane, A., Riviere, M., Puzo, G., Barzu, O., and Marchal, G. (1999). Deglycosylation of the 45/47-kilodalton antigen complex of *Mycobacterium tuberculosis* decreases its capacity to elicit *in vivo* or *in vitro* cellular immune responses. Infect. Immun. 67, 5567–5572.

Rudd, P.M., Elliott, T., Cresswell, P., Wilson, I.A., and Dwek, R.A. (2001). Glycosylation and the immune system. Science 291, 2370–2376.

Sabet, M., Lee, S.W., Nauman, R.K., Sims, T., and Um, H.S. (2003). The surface (S-) layer is a virulence factor of *Bacteroides forsythus*. Microbiology 149, 3617–3627.

Sambri, V., Stefanelli, C., and Cevenini, R. (1992). Detection of glycoproteins in *Borrelia burgdorferi*. Arch. Microbiol. 157, 205–208.

Schifferle, R.E., Reddy, M.S., Zambon, J.J., Genco, R.J., and Levine, M.J. (1989). Characterization of a polysaccharide antigen from *Bacteroides gingivalis*. J. Immunol. 143, 3035–3042.

Schmidt, M.A., Riley, L.W., and Benz, I. (2003). Sweet new world: glycoproteins in bacterial pathogens. Trends Microbiol. 11, 554–561.

Sheets, S.M., Robles-Price, A.G., McKenzie, R.M., Casiano, C.A., and Fletcher, H.M. (2008). Gingipain-dependent interactions with the host are important for survival of *Porphyromonas gingivalis*. Front. Biosci. 13, 3215–3238.

Sherlock, O., Dobrindt, U., Jensen, J.B., Munk Vejborg, R., and Klemm, P. (2006). Glycosylation of the self-recognizing *Escherichia coli* Ag43 autotransporter protein. J. Bacteriol. 188, 1798–1807.

Shoji, M., Ratnayake, D.B., Shi, Y., Kadowaki, T., Yamamoto, K., Yoshimura, F., Akamine, A., Curtis, M.A., and Nakayama, K. (2002). Construction and characterization of a nonpigmented mutant of *Porphyromonas gingivalis*: cell surface polysaccharide as an anchorage for gingipains. Microbiology 148, 1183–1191.

Sims, T.J., Moncla, B.J., Darveau, R.P., and Page, R.C. (1991). Antigens of *Actinobacillus actinomycetemcomitans* recognized by patients with juvenile periodontitis and periodontally normal subjects. Infect. Immun. 59, 913–924.

Sims, T.J., Schifferle, R.E., Ali, R.W., Skaug, N., and Page, R.C. (2001). Immunoglobulin G response of periodontitis patients to *Porphyromonas gingivalis* capsular carbohydrate and lipopolysaccharide antigens. Oral Microbiol. Immunol. 16, 193–201.

Slaney, J.M., Gallagher, A., Aduse-Opoku, J., Pell, K., and Curtis, M.A. (2006). Mechanisms of resistance of *Porphyromonas gingivalis* to killing by serum complement. Infect. Immun. 74, 5352–5361.

Slots, J., and Listgarten, M.A. (1988). *Bacteroides gingivalis*, *Bacteroides intermedius* and *Actinobacillus actinomycetemcomitans* in human periodontal diseases J. Clin. Periodontol 15, 85–93.

Socransky, S.S., Haffajee, A.D., Cugini, M.A., Smith, C., and Kent, R.L., Jr. (1998). Microbial complexes in subgingival plaque. J. Clin. Periodontol. 25, 134–144.

Sterba, J., Vancova, M., Rudenko, N., Golovchenko, M., Tremblay, T.L., Kelly, J.F., MacKenzie, C.R., Logan, S.M., and Grubhoffer, L. (2008). Flagellin and outer surface proteins from *Borrelia burgdorferi* are not glycosylated. J. Bacteriol. 190, 2619–2623.

Swanson, J. (1973). Studies on gonococcus infection. IV. Pili: their role in attachment of gonococci to tissue culture cells. J. Exp. Med. 137, 571–589.

Szymanski, C.M., Burr, D.H., and Guerry, P. (2002). *Campylobacter* protein glycosylation affects host cell interactions. Infect. Immun. 70, 2242–2244.

Szymanski, C.M., Logan, S.M., Linton, D., and Wren, B.W. (2003). *Campylobacter* – A tale of two protein glycosylation systems. Trends Microbiol. 11, 233–238.

Szymanski, C.M., and Wren, B.W. (2005). Protein glycosylation in bacterial mucosal pathogens. Nat. Rev. Microbiol. 3, 225–237.

Szymanski, C.M., Yao, R., Ewing, C.P., Trust, T.J., and Guerry, P. (1999). Evidence for a system of general protein glycosylation in *Campylobacter jejuni*. Mol. Microbiol. 32, 1022–1030.

Tashiro, Y., Nomura, N., Nakao, R., Senpuku, H., Kariyama, R., Kumon, H., Kosono, S., Watanabe, H., Nakajima, T., and Uchiyama, H. (2008). Opr86 is essential for viability and is a potential candidate for a protective antigen against biofilm formation by *Pseudomonas aeruginosa*. J. Bacteriol. *190*, 3969–3978.

Thjotta, T., and Waaler, E. (1932). Dissociation and sensitiveness to normal serum in dysentery bacilli of type III. J. Bacteriol. *24*, 301–316.

Tong, H.H., Blue, L.E., James, M.A., Chen, Y.P., and DeMaria, T.F. (2000). Evaluation of phase variation of nontypeable *Haemophilus influenzae* lipooligosaccharide during nasopharyngeal colonization and development of otitis media in the chinchilla model. Infect. Immun. *68*, 4593–4597.

van Winkelhoff, A.J., Appelmelk, B.J., Kippuw, N., and de Graaff, J. (1993). K-antigens in *Porphyromonas gingivalis* are associated with virulence. Oral Microbiol. Immunol. *8*, 259–265.

Vanterpool, E., Roy, F., and Fletcher, H.M. (2004). The vimE gene downstream of vimA is independently expressed and is involved in modulating proteolytic activity in *Porphyromonas gingivalis* W83. Infect. Immun. *72*, 5555–5564.

Vanterpool, E., Roy, F., and Fletcher, H.M. (2005a). Inactivation of vimF, a putative glycosyltransferase gene downstream of vimE, alters glycosylation and activation of the gingipains in *Porphyromonas gingivalis* W83. Infect. Immun. *73*, 3971–3982.

Vanterpool, E., Roy, F., Sandberg, L., and Fletcher, H.M. (2005b). Altered gingipain maturation in vimA- and vimE-defective isogenic mutants of *Porphyromonas gingivalis*. Infect. Immun. *73*, 1357–1366.

Varki, A. (2006). Nothing in glycobiology makes sense, except in the light of evolution. Cell *126*, 841–845.

Voulhoux, R., Bos, M.P., Geurtsen, J., Mols, M., and Tommassen, J. (2003). Role of a highly conserved bacterial protein in outer membrane protein assembly. Science *299*, 262–265.

Wall, D., and Kaiser, D. (1999). Type IV pili and cell motility. Mol. Microbiol. *32*, 1–10.

Werner, J., and Misra, R. (2005). YaeT (OMP85) affects the assembly of lipid-dependent and lipid-independent outer membrane proteins of *Escherichia coli*. Mol. Microbiol. *57*, 1450–1459.

Wood, W.B., Jr., and Smith, M.R. (1949). The inhibition of surface phagocytosis by the capsular slime layer of pneumococcus type III. J. Exp. Med. *90*, 85–96.

Yi, K., Rasmussen, A.W., Gudlavalleti, S.K., Stephens, D.S., and Stojiljkovic, I. (2004). Biofilm formation by *Neisseria meningitidis*. Infect. Immun. *72*, 6132–6138.

Young, N.M., Brisson, J.R., Kelly, J., Watson, D.C., Tessier, L., Lanthier, P.H., Jarrell, H.C., Cadotte, N., St Michael, F., Aberg, E. (2002). Structure of the N-linked glycan present on multiple glycoproteins in the Gram-negative bacterium, *Campylobacter jejuni*. J. Biol. Chem. *277*, 42530–42539.

Zambon, J.J. (1985). Actinobacillus actinomycetemcomitans in human periodontal disease. J. Clin. Periodontol *12*, 1–20.

Zambon, J.J., Slots, J., and Genco, R.J. (1983). Serology of oral *Actinobacillus actinomycetemcomitans* and serotype distribution in human periodontal disease. Infect. Immun. *41*, 19–27.

Bacterial Polysaccharides in Dental Plaque

8

Roy R. B. Russell

Abstract

Dental plaque is a complex biofilm containing several hundred different species of bacteria. While these are normally harmless commensals, shifts in the population structure can lead to the plaque-related diseases such as dental caries and periodontal disease. Surface polysaccharides are important in coaggregation reactions that bind particular species together and aid colonization and metabolic interaction while sucrose-derived glucans and fructans have a significant effect on plaque properties. Fructans serve as an extracellular energy store while glucans contribute to adhesion, modify the permeability of plaque and alter ion-binding capacity and thus have a powerful influence on the creation of conditions in plaque that can lead to dental caries.

Introduction: dental plaque as a biofilm

The biofilm concept that has become firmly established in the last quarter century recognizes that the vast majority of bacteria live on surfaces, in intimate association with other bacteria. This has influenced our understanding of microbial growth on surfaces as diverse as a rock in a mountain stream, a surgically inserted catheter, an oil pipeline or a sewage farm. However, dental plaque is probably the biofilm with which we are all most familiar – you can feel its roughness when you run your tongue over your teeth in the morning and are likely to spend several minutes removing it by brushing (and maybe flossing) a couple of times a day. The fact that that we need to resort to this regular cleaning illustrates the rapid rate of re-growth of plaque once it has been removed, and our inability to remove it completely. The rate of re-growth of plaque varies enormously between individuals but no-one is ever completely free of plaque because the human oral cavity is such a splendid environment for bacteria to colonize. The oral cavity can be compared with other complex ecosystems such as a jungle or tropical coral reef in that it is warm, moist, rich in nutrients, and supporting an abundant variety of species. Not only is there a constant supply of metabolizable carbon and nitrogen available from the catabolism of host macromolecules, principally salivary mucins, but there is the regular bonus of a massive overprovision (for the bacteria) of nutrients at regular mealtimes. Furthermore, some individuals supplement these regular meals by periodic consumption of sugar-rich snacks throughout the day that can disturb the steady state of the bacterial population and have a major influence on the amount and properties of plaque formed.

Plaque-related diseases

The plaque-related diseases of greatest prevalence are dental caries (tooth decay) and periodontal (gum) disease. The first involves acid dissolution of the hard enamel and dentine surfaces of teeth while the second involves the breakdown of supporting ligaments, epithelial tissue and alveolar bone. Despite these two very different pathogenic processes, both have a microbial aetiology and are consequences of adverse population shifts in dental plaque. Supragingival plaque (i.e. visible above the gum margin) is exposed to sugars in the diet so the metabolism of carbohydrate and synthesis of extracellular polysaccharides are of major importance in its composition. Fermentation of dietary carbohydrates (including sucrose) results in acid production in plaque and the resultant pH fall, to pH 4 or below, acts as a selective force that results in increase in the acid-tolerant species that can survive these conditions. Usually, removal of the source of carbohydrate and the buffering capacity of saliva return the plaque pH to near neutrality but high numbers of acidogenic, aciduric species can result in the pH close to the tooth surface being maintained below the critical pH (about pH 5.5) at which net dissolution of enamel occurs, leading to caries.

The microbiology of dental caries has long been a subject of investigation (Table 8.1) and for the last 50 years attention has focused on the streptococci, though more attention is beginning to be paid to other groups of bacteria (Aas *et al.*, 2008; Beighton, 2005; Russell, 2008c). While a number of species of streptococci contribute to acid production, the greatest attention as been paid to the mutans streptococci (*S. mutans* and *S. sobrinus*) because of their implication in the development of dental caries. The mutans streptococci are not considered to be pioneer species in plaque formation and indeed only represent a very minor proportion of the total number of bacteria present, generally less than 1%. Such a level poses no problems so long as the plaque population is maintained in a state of equilibrium, but the relative proportion of these highly aciduric species can be altered drastically by shifts in the environmental conditions. Every exposure to fermentable carbohydrate leads to a transient drop in plaque pH, so successive exposures ratchet up the proportion of mutans streptococci. This is why your dentist advises you not to eat sweets, and certainly not to keep eating sugary snacks all through the day.

Subgingival plaque generally forms below the gum margin when oral hygiene is poor. The thicker, undisturbed plaque shows increased numbers of anaerobic, mostly Gram-negative bacteria that can cause direct tissue damage by proteolysis and also induce the inflammatory tissue responses that manifest as gingivitis or more advanced forms of periodontal disease. The

Table 8.1 Significant developments in caries microbiology

1881	Underwood & Milles	First recognition of bacteria in caries lesions
1890	Miller	Chemicoparasitic theory of caries
1924	Clarke	Description of *Streptococcus mutans*
1960	Fitzgerald & Keyes	Streptococci cause caries in gnotobiotic rats
1965	Carlsson, Gibbons, Guggenheim, Krasse	Importance of sucrose-derived polymers in plaque formation and caries
1968	Carlsson, Guggenheim	Revised taxonomy of oral streptococci
1970	Gibbons	Demonstration of selective adhesion of plaque bacteria
1970	Guggenheim	Characterization of α1,6- and α1,3-linked glucans made by streptococci
1977	Coykendall	*S. mutans* and *S. sobrinus* formally described as separate species of mutans streptococci
1975	Loesche	Epidemiological evidence for major role of mutans streptococci in caries
1980	Marsh, Hamilton and others	Development of Ecological Plaque Hypothesis
1985	Gilpin, Russell & Morrissey	First cloning of a glucosyltransferase gene
1991	Van Houte	Implication of 'low pH, non-mutans' streptococci
1991–1993	Macrina, Bowen, Kuramitsu	Experimental evidence of importance of glucosyltransferases using defined mutants
2002	Ajdic	Genome sequence of *S. mutans*
2008	Paster	Molecular identification methods implicate an increased range of bacteria in caries

basic process of plaque formation is the same as for supragingival plaque but, because the site is anatomically remote from exposure to dietary sugars, carbohydrate metabolism is not a major determinant of bacterial population composition so it will not be considered further.

Dental plaque formation

Current models of dental plaque formation envisage a succession of stages, each of them involving a different range of bacterial species but all stages being characterized by a set of specific molecular interactions. In the initial stage, 'pioneer' species of bacteria colonize the tooth surface by binding to components of the salivary pellicle coating enamel that is made up of selectively adsorbed proteins and glycoproteins from the saliva. Predominant among these initial colonizers are *Streptococcus sanguinis*, *Streptococcus oralis*, *Streptococcus mitis* and the pleomorphic Gram-positive rod *Actinomyces viscosus*. The types and proportions of bacteria present as initial colonizers differs from individual to individual (Diaz *et al.*, 2006) and will help determine the subsequent properties of the plaque. The great variety of types of bacteria, and the variation seen between different individuals, is illustrated by electron microscopy sections through plaque shown in Fig. 8.1 (Nyvad, 1993; Nyvad and Fejerskov, 1989). Some 700 different taxa of bacteria are thought to be capable of flourishing in the oral cavity, though any individual probably only has a few hundred different species at a time (Aas *et al.*, 2005). Plaque first develops on a cleaned tooth surface both by growth of the pioneer species to form microcolonies and by the further accretion of more bacterial cells of the same or other species. This accretion may be mediated by bridging salivary macromolecules which are bound by more than one bacterial cell or can be direct bacterium–bacterium binding in pairs or networks due to specific receptor-ligand binding (Fig. 8.2). The end-result of these processes is a complex structure of many different types of bacteria embedded in an extracellular matrix (Rickard *et al.*, 2003).

Because of its relative accessibility and ease of study, as well as its relationship to disease, dental plaque is one of the best-studied biofilms and there is now a substantial amount of information available on the wide of range of adhesive, nutritional or signalling interactions between different bacteria that result in dynamic changes but overall homeostasis (Kolenbrander *et al.*, 2006; Kuramitsu *et al.*, 2007). Within an individual, stresses due to variation in diet and other environmental factors determine the relative proportion of species in the 'climax community' that may show changes during life but is essentially characteristic. It has even been suggested that a person's oral microflora may be sufficiently distinctive to be used for forensic identification purposes (Rahimi *et al.*, 2005). The established plaque community of 'normal oral flora' may be regarded as consisting of harmless commensals associated with health and this resident microflora resists invasion of incomers. Invasion by pathogens is rarely observed so plaque-related diseases are thus endogenous infections: the shift from health to disease is a consequence of

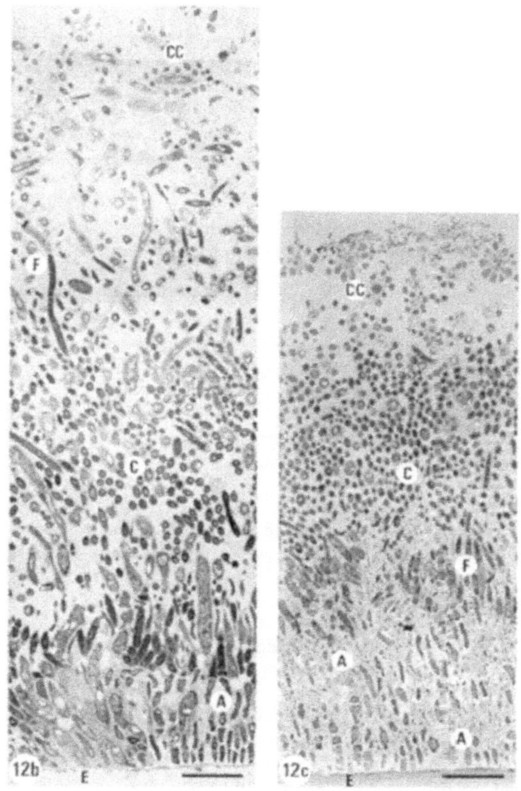

Figure 8.1 Transmission electron microscope image of sections through 14-day-old plaque from 2 individual subjects, illustrating the variety of bacteria present. Enamel surface (**E**) at the bottom. Modified from Nyvad (1993) by permission of S. Karger A.G., Basel.

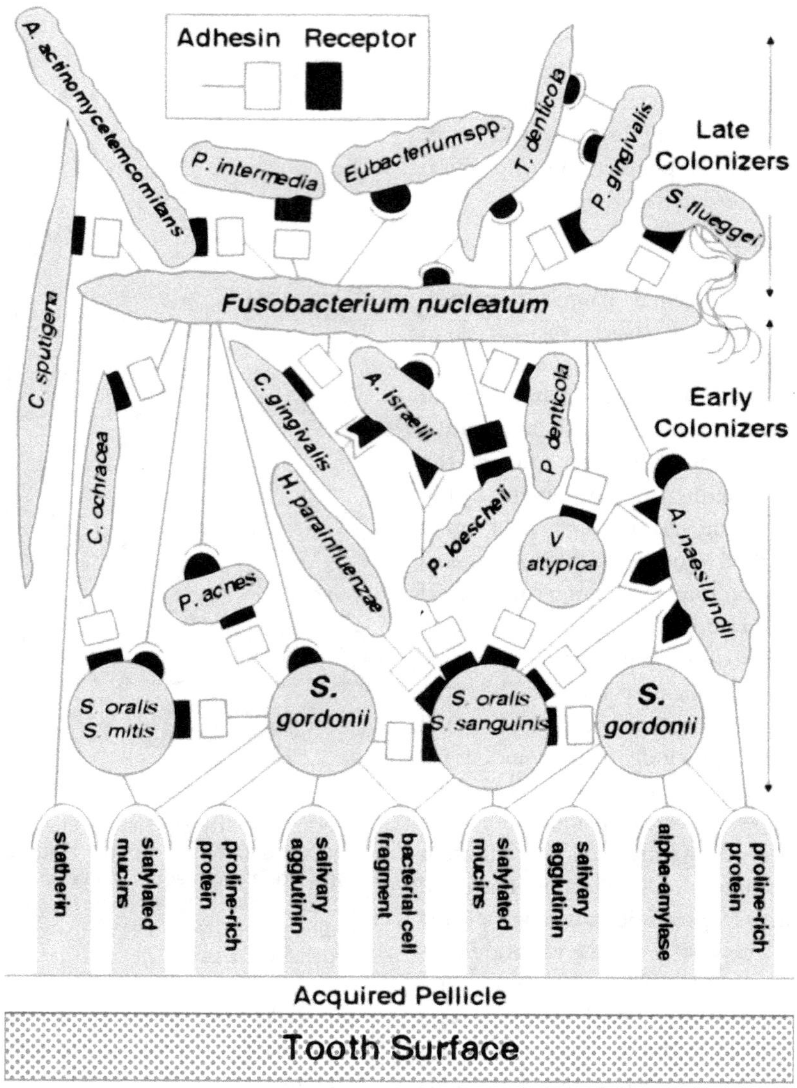

Figure 8.2 Diagrammatic representation of binding to host macromolecule at the tooth surface and coaggregation interactions between species of oral bacteria contributing to plaque formation. Figure kindly provided by P.E. Kolenbrander, redrawn from Kolenbrander and London (1993).

a change in proportions of the species already present in the mouth. Our current understanding of events in plaque therefore relies upon an understanding of factors that can change the relative proportions of different species and employs the same way of thinking used in describing other environments and has led to the development of the 'ecological plaque hypothesis' (Marsh, 2003, 2006; Takahashi and Nyvad, 2008). The microbiology of dental caries is not the concern of this chapter, but the topic has recently been reviewed (Beighton, 2005; Russell, 2008a; Takahashi and Nyvad, 2008).

Contribution of heteropolysaccharides to plaque formation

The colonization of dental plaque by inter-bacterial binding has the obvious advantage of physical stabilization of the biofilm, giving protection against the bacteria being washed away by saliva or displaced by the movements associated with mastication. In addition, there may be metabolic advantages for a particular bacterium in being closely associated with other species (Bradshaw *et al.*, 1998; Jakubovics *et al.*, 2008). The process of adhesion by which cells of genetically distinct species bind to each other is referred to as coaggregation. Coaggregation was first described with dental plaque organisms, but has since been recognized as a phenomenon widespread in biofilms from a diversity of settings (Rickard *et al.*, 2003). Binding between cells of the two participants involves an adhesin on one interacting with a receptor molecule on the other. There may be more than one adhesin-receptor pair, with different specificities, functioning so that complex networks can be built up (Fig. 8.2). The adhesins that have been characterized are known to be proteins whereas the receptors are complex polysaccharides. In one of the best-studied systems, the coaggregation between *Actinomyces naeslundii* and various streptococci has been shown to involve streptococcal receptors that are made up of phosphodiester-linked hexa- or heptasaccharide repeating units of GalNAc→3Gal and Galβ1→3GalNAc (Palmer *et al.*, 2003; Yoshida *et al.*, 2006). Recent advances in sequencing of the genes responsible for synthesis and assembly of the receptor polysaccharides should greatly accelerate our understanding of the molecular basis of polysaccharide structure and define the basis of coaggregation specificity (Yoshida *et al.*, 2005; Yoshida *et al.*, 2008).

Along with these processes of initial plaque formation involving bacterial heteropolysaccharides, plaque build-up is facilitated by the production of bacterial extracellular homopolysaccharide polymers. The polymers of particular interest are those synthesized from sucrose by various species of streptococci.

Sucrose-derived homopolysaccharides in dental plaque

While any fermentable carbohydrate may be utilized by plaque bacteria to generate the acids which attack enamel substance, sucrose is recognized as being particularly important in the caries process because not only can it be fermented but it also serves as substrate for extracellular enzymes of plaque bacteria which synthesize sucrose-derived polymers. These polymers are of central importance in adhesive interactions in plaque, where they mediate attachment of bacteria to the tooth surface and to other bacteria. Not only do they physically stabilize the plaque biofilm (Hamada and Slade, 1980; Kuramitsu, 1993), they also serve as reserve energy stores aiding the survival of plaque bacteria (Burne, 1993), modulate the permeability of plaque and hence the level of acid at the enamel surface (Dibdin and Shellis, 1988), increase resistance to inhibitors (Kreth *et al.*, 2008) and influence the local balance between mineralization and demineralization (Paes Leme *et al.*, 2006).

S. mutans is remarkably efficient in its utilization of sucrose. Not only does it benefit from the monosaccharides released during synthesis of the fructan and glucans but it also exploits the degradation products of these polymers that are taken up by the cell and metabolized through glycolysis (Fig. 8.3). Sucrose thus makes a significant contribution to acid production in the same way as free glucose or fructose (Colby and Russell, 1997; Russell, 2008c). This chapter will concern itself only with the extracellular metabolism of sucrose.

Fructans

Fructans are homopolymers of fructose and those found in plaque are thought to be produced by species of *Streptococcus* and *Actinomyces.* The enzymes responsible are variously referred to as fructosyltransferase, fructansucrase or levansucrase and will be discussed further below. *Streptococcus salivarius* and *Actinomyces viscosus* synthesize a levan-type fructan with predominantly β-2,6-linked units while other streptococci such as *S. mutans* synthesize an inulin-type polymer with predominantly β-2,1-linkages. Consumption of sucrose leads to a rapid accumulation of fructans in plaque, but in the absence of continued exposure to sucrose, the fructans decline to a low level (Burne, 1993; Gold *et al.*, 1974; Hotz *et al.*, 1972). Both types of fructan are broken down and metabolized by a wide range of bacteria, both those which have synthesized it in the first place and others gaining the benefit of the efforts of their neighbours. Fructans are thus thought to act as an extracellular energy store, providing an additional source of fermentable carbohydrate when exogenous sources are lacking. This would result in prolonging the exposure of the tooth sur-

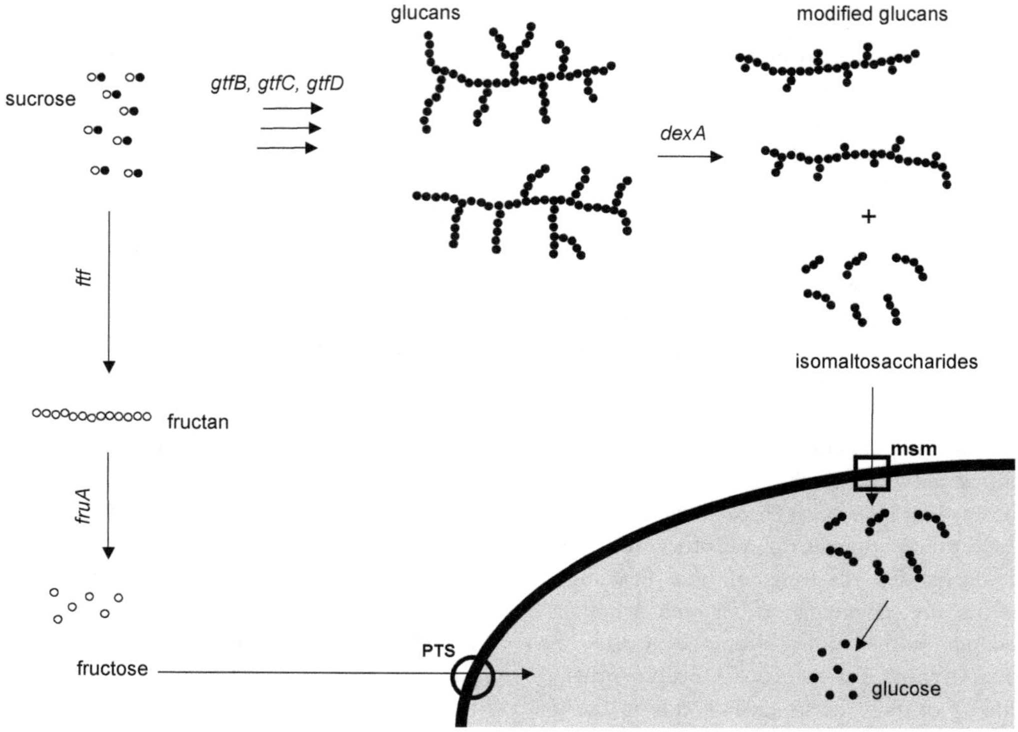

Figure 8.3 The pathways for metabolism of sucrose by *Streptococcus mutans*. •, glucose; ○, fructose; *gtf*, glucosyltransferase; *dexA*, dextranase; *ftf*, fructosyltransferase; *fruA*, fructanase; *msm*, multiple sugar metabolism genes. Modified from Colby and Russell (1997) by permission of Blackwell Publishing.

face to acids produced by streptococci, therefore contributing to maintaining a low pH for dental caries to occur. Fructans are not thought to be involved in the adhesion of streptococci, though they may contribute to the physical stabilization of other species in plaque (Rozen *et al.*, 2001).

Glucans

The glucose half of sucrose is polymerized by enzymes referred to as glucosyltransferases (GTF), glucansucrases or according to the nature of the polymer formed (dextransucrase, mutansucrase, etc.). Streptococcal GTF are capable of synthesizing either α1,6- or α1,3- linkages between the glucose units and the relative proportions of these two linkages and the degree of branching determines the ultimate properties of the glucan. A glucan which is predominantly a linear α1,6-linked chain is referred to as a dextran and is soluble; in contrast, mutan is a water-insoluble glucan with a high proportion of α1,3-linkages. Both types of glucan are believed to be important in dental plaque, with dextran mediating bacterial aggregation and serving as a storage polymer while mutan has been shown to be the major contributor to adherence to hard surfaces (Colby and Russell, 1997; Kuramitsu, 2003). These binding interactions are mediated by GTF themselves as well as by non-enzymatic glucan-binding proteins (Banas and Vickerman, 2003; Lynch *et al.*, 2007; Matsumoto-Nakano *et al.*, 2007). Experimental manipulation of the relative proportions of α1,6- and α1,3-linked glucans by use of mutants defective in particular GTF, or by treatment with degradative enzymes, has helped clarify their contribution to adhesive plaque formation (Cross *et al.*, 2006; Cross *et al.*, 2007; Hayacibara *et al.*, 2004).

In addition to their adhesive function in binding bacteria to surfaces and bacteria to other bacteria, the synthesis of glucan has a major influence on the structure of plaque and the physical spacing of its constituent bacteria. This can be observed in the laboratory (Reese and Guggenheim,

2007; Van Houte *et al.*, 1989) and also clinically, when a patient with high sucrose-consumption is likely to have more bulky, gelatinous plaque on their teeth. More importantly, plaque with a high content of glucan has altered porosity, increasing the ease with which fermentable sugars in the diet can diffuse in to reach bacteria near the tooth surface and enhance their acid production (Dibdin and Shellis, 1988; Hata and Mayanagi, 2003; Van Houte *et al.*, 1989). A further effect of the glucans is on the ion-binding properties of plaque. Of particular importance is calcium, because the local concentration of calcium determines the equilibrium between net mineral loss and remineralization of the enamel surface (Paes Leme *et al.*, 2006; Rose *et al.*, 1993; Rose and Turner, 1998).

GTFs thus contribute in a variety of ways to plaque formation and the dental caries process and there is now a substantial body of experimental evidence to support their importance as virulence determinants (Colby *et al.*, 1999; Kuramitsu, 2003). This has made GTF and glucan-binding proteins attractive targets for the design of specific inhibitors or as targets for experimental caries vaccines (Koga *et al.*, 2002; Russell, 2008b). Just as fructans are broken down in dental plaque, the dextrans are also subject to degradation by dextranases produced by a wide variety of species (Igarashi *et al.*, 1998; Johnson, 1990), followed by utilization as metabolites. By trimming off stretches of α1,6 chains, dextranase also substantially modify the properties of the glucan (Colby *et al.*, 1995; Hayacibara *et al.*, 2004). The final composition of the plaque glucans, and hence the properties of plaque formed in the presence of sucrose, is thus affected by a number of factors: the species of bacteria present, the relative proportions of different GTF produced by those species, the level of expression of the *gtf* genes, the extent to which the product of one GTF may be modified by the chain-extension activities of others, and the activity of the dextranases.

Most attention has focused on the extracellular polysaccharides made by streptococci in plaque because of their association with dental caries. However, another glucan of interest is the amylose-like α1,4-linked glucan produced by species of *Neisseria* (MacKenzie *et al.*, 1978; Parker and Creamer, 1971; Riou *et al.*, 1986; Ruby *et al.*, 1982) and the little-studied *Lautropia mirabilis* (Gerner-Smidt *et al.*, 1994). Although the amylosucrase enzyme that makes the glucan from sucrose is not secreted, the glucan accumulates in large amounts outside the cell. Its contribution to the overall plaque economy is not known, but it is included here because of the substantial amount that is now known about the amylosucrase, discussed below.

Enzymes involved in synthesis and degradation of sucrose-derived homopolymers

Fructosyltransferases (FTF)

Fructan with an inulin-type structure has so far only been observed in the oral species *S. mutans* and *Streptococcus salivarius* strains as well as the Lactic Acid Bacteria *Leuconostoc citreum* and *Lactobacillus reuteri* that are taxonomically quite closely related to the streptococci. Levan production has also been reported for species of *Streptococcus, Lactobacillus* and *Weissella* but is much more widely distributed, being found in many Gram-positive and Gram-negative bacteria (van Hijum *et al.*, 2006).

S. mutans has a single *ftf* gene (Shiroza and Kuramitsu, 1988). Inactivation of *ftf* resulted in a strain which was less pathogenic in a rat model suggesting FTF contributes to pathogenicity (Schroeder *et al.*, 1989) though this was not confirmed by other researchers using a different sucrose diet for their animals (Burne *et al.*, 1996).

Little is yet known about the detailed structure and mechanisms of FTF from other oral species. All FTF, however, belong to family GH68 as defined on the CAZY carbohydrate-active enzyme website: http://afmb.cnrs-mrs.fr/CAZY) (Coutinho and Henrissat, 1999) and share related catalytic domains, though they otherwise vary extensively in size. The FTF from *Bacillus subtilis* has been the subject of extensive investigation (van Hijum *et al.*, 2006) and it is probable that the streptococcal enzymes will share the same mechanism.

Fructanase

Fructanases have been identified in species of *Streptococcus* and *Actinomyces*, the two main genera in plaque that are also known to produce fructans. It would therefore appear that these enzymes have evolved to recycle the fructan that is synthesized in periods of sucrose-excess though within plaque, other bacteria will also benefit from its availability. There is limited detailed information on specificity, but it seems that the enzymes are capable of attacking both β2,6- and β2,1-linked fructans by releasing terminal fructose (Burne *et al.*, 1987; Igarashi *et al.*, 1992; Norman and Gifford, 1996; Takahashi *et al.*, 1985). The genetics of fructanase in *S. mutans* has been explored and there is a single fructanase (*fruA*) that is subject to a complex regulatory circuit (Zeng *et al.*, 2006). There have been no reports of an endofructanase activity in plaque that would generate fructooligosaccharides (FOS). Several oral streptococci can, however, ferment FOS (Hartemink *et al.*, 1995).

Glucosyltransferases (GTF)

The term GTF has been the one preferred by workers with oral bacteria, though this does carry the risk of confusion with the nucleotide sugar-dependent intracellular glycosyltransferases (Lairson *et al.*, 2008). It is important to note that the extracellular GTF have no access to nucleotides and the energy required for the synthesis reaction comes entirely from the scission of the glycosidic bond in sucrose. The alternative term glucansucrase has found favour in recent years, the specificity of the enzyme being indicated by a name derived from the main glucan produced, hence dextransucrase, mutansucrase, alternansucrase, reuteransucrase, amylosucrase.

Early biochemical studies of streptococcal GTF were greatly complicated by the fact that each species produces a number of different GTF and obtaining pure preparations of each was difficult. Cloning and expression of individual *gtf* genes in *Escherichia coli* was thus a valuable advance, though not all *gtf* (or other glucansucrase genes) can readily be expressed at a high level, or even maintained on plasmids, for reasons that are not understood. The first *gtf* genes to be cloned were from *Streptococcus downei*, using phage λ as cloning vector and screening on sucrose-containing medium (Gilpin *et al.*, 1985). GTF activity could be detected by the formation of glucans by GTF released from the bacteria lysed by recombinant phage carrying the *gtf* genes, with watery glucan from *gtfS* and hard, insoluble glucan from *gtfI*. The same screening method was later applied to clone *gtf* from *S. mutans*, *S. salivarius* and *S. gordonii* as well as the amylosucrase from *N. polysaccharaea*. Since not all *gtf* can be detected by expression of activity, a variety of techniques of Southern blotting and a reverse genetics approach using PCR with primers based on conserved sequences have been used to clone related enzymes from *Leuconostoc* and *Lactobacillus* (van Hijum *et al.*, 2006). More recently, genome projects have revealed the presence of homologous genes in other bacteria, such as *Oenococcus*. Currently, 84 sequences are listed on the CAZY database of Carbohydrate Active Enzymes (http://afmb.cnrs-mrs.fr/CAZY/) in Family 70 (Coutinho and Henrissat, 1999).

Cloning and expression of individual GTF provided firm confirmation that a single enzyme was capable of forming more than one type of glycoside link. Of the oral streptococci, *S. mutans* produces three distinct GTF while *S. sobrinus*, the closely related *S. downei* and *S. salivarius* each produce four. *S. gordonii*, *S. sanguinis* and *S. oralis*, in contrast, make only a single GTF while *S. mitis* has none, even though it is closely related and occupies the same environmental niche. Each GTF is encoded by a separate gene and each enzyme has distinctive properties, varying in the proportion of α1,6- and α1,3-linkages and hence the degree of branching it introduces into the glucan, and the total length of the glucan chain produced. The different proportions and relative activity of the various GTF determines the eventual properties of the glucan produced. This can be clearly seen by gene-knockout experiments, where inactivation of *gtfS* and/or *gtfI* has a dramatic effect on the appearance of colonies of *S. downei* grown on sucrose (Colby *et al.*, 1999; Gilmore *et al.*, 1993). The experiments also showed that the short linear α1–6-linked glucan produced by GTFS (Russell *et al.*, 1990) was essential for the subsequent formation of the branched glucan made by GTFI (Gilmore *et al.*, 1993). Within dental plaque, the situation will be much more complex because the overall properties of the glucan represents the combined action of a dozen or more different GTF from different streptococci and on their interactions, since one GTF may modify the product of another. Furthermore, glucans will be modified by the action of dextranase (*dexA*, Fig. 8.3), which is produced by *S. mutans* and various other plaque bacteria and degrades the α1,6-linked chains (Colby *et al.*, 1995).

Variation in GTF activity also affects virulence, as shown by experiments with *S. mutans*. The *S. mutans* GTF are encoded by *gtfB*, *gtfC* and *gtfD*. The enzyme encoded by *gtfB* synthesizes mainly α1–3-linked insoluble glucans (GTF-I). The *gtfC* enzyme synthesizes a mixture of α1–3-linked insoluble and α1–6-linked soluble glucans (GTF-SI) and that of *gtfD* synthesizes α1–6-linked soluble glucan (GTF-S). Inactivation of either *gtfB*, *gtfC* or *gtfD* or all three *gtf* genes has been shown to result in a reduction in smooth surface caries in an animal model (Fujiwara *et al.*, 1996; Munro *et al.*, 1991; Tsumori and Kuramitsu, 1997; Yamashita *et al.*, 1993) though it is important to note that, as with the fructan experiments, different results may be obtained with different animal diets (Yamashita *et al.*, 1993) and there may be differential effects at different tooth sites (Van Der Ploeg and Guggenheim, 2004).

A longstanding question about GTF relates to the mechanisms of chain elongation and the determinants of type of glycosidic linkage introduced to the growing glucan chain. As nucleotide sequences of GTF became available, multiple alignments of the deduced amino acid sequences revealed conserved and unique features which may explain which amino acid residues or domains are responsible for these properties. The basic organization of GTF could be discerned as soon as more than one streptococcal GTF sequence became available (Russell, 1990) but the gene sequences from 15 species of streptococci, and several from species of *Leuconostoc* and *Lactobacillus* are now available and have allowed refinement of the model (Monchois *et al.*, 1999b; van Hijum *et al.*, 2006). All GTF are of high-molecular weight and composed of distinct domains, illustrated in Fig. 8.3. Family 70 glycosylhydrolases are part of the α-amylase super-family and the central catalytic core of all GTF is predicted to have alternating α-helixes and β-sheets, in an arrangement the same as found in the $(\beta/\alpha)_8$ barrel found in amylase. However, in comparison to amylase, the order of the sheets and helices was predicted to be circularly permuted so that helix 3 occurs first, as represented in Fig. 8.4 (MacGregor *et al.*, 1996). The similarity to amylases also allowed prediction of the amino acids contributing to the active site, and these were confirmed by mutagenesis (Devulapalle *et al.*, 1997) while further mutagenic alteration of individual amino acids confirmed the circular permutation (Monchois *et al.*, 2000; Swistowska *et al.*, 2007). By combining information from site-directed mutagenesis with modern methods for characterizing short oligosaccharides, the reaction mechanism of glucansucrases has now been elucidated (Moulis *et al.*, 2006). Catalysis starts by hydrolysis of sucrose, so glucose and fructose are the only carbohydrates detected in the first minutes of reaction. Sucrose immediately starts to act as an acceptor, with glycosylation leading to formation of oligosaccharides (short chains of glucose with a terminal non-reducing fructosyl residue) as proposed earlier (Cheetham *et al.*, 1991). Elongation then proceeds by addition of glucosyl residues onto the non-reducing end of previously formed acceptors. Note that in a complex plaque, these acceptors may be ones formed by the initial enzyme, or chains formed by other GTF made by the same or other bacteria. This processive synthesis of a lengthening chain is aided by the C-terminal domain of the enzyme (Monchois *et al.*, 1999a; Moulis *et al.*, 2006; Wittrock *et al.*, 2008). This terminal one-third of GTF consists of a series of related but non-identi-

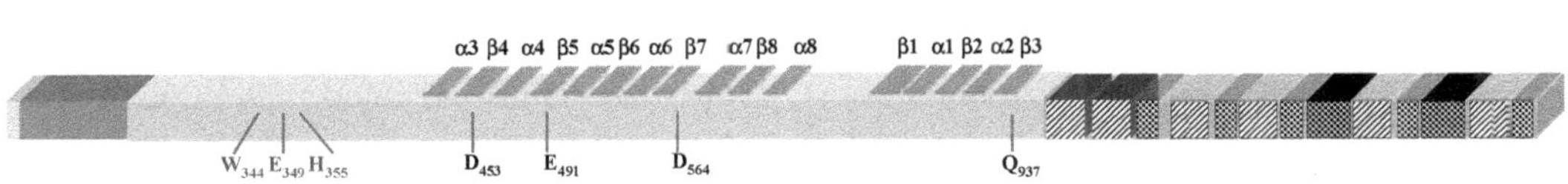

Figure 8.4 Schematic structure of *S. downei* GTF-I showing the N-terminal signal peptide, variable region, catalytic core and C-terminal glucan-binding domain containing amino acid repeats. The predicted locations of the α helix and β strands corresponding to the components of the amylase family $(\beta/\alpha)_8$ barrel and residues contributing to the active site are indicated underneath. Modified from Monchois *et al.* (2000) by permission of Blackwell Publishing.

cal amino acid repeats (Fig. 8.4). The C-terminal region was first shown to be involved in binding exogenous glucans (Ferretti *et al.*, 1987) so it is important both in synthesis of glucan and in the adhesive interactions that help bind bacteria together. Glucan binding domains characteristically have a pattern of repeated tyrosine (Y) and glycine (G) residues and it has been proposed that the basic building unit is a YG repeat of 19–21 amino acids (Giffard and Jacques, 1994; Kingston *et al.*, 2002). A variety of other distinct repeats have been identified (Vickerman *et al.*, 1997), the only one of which that is found in all GTF is 33 amino acids long, named the A repeat (Shah *et al.*, 2004). Comparable repeats have been found in other glucansucrases (Janecek *et al.*, 2000; van Hijum *et al.*, 2006), though the precise way by which they bind glucan is not understood.

The nature of the linkage formed (or rather, the balance between the proportions of different linkages) can be modified by mutation of specific amino acid residues (Funane *et al.*, 2005; Hellmuth *et al.*, 2008) though, again, our understanding is incomplete. There is also evidence that adsorption of GTF to a solid surface can modify the nature of its product glucan, perhaps either by altering enzyme conformation or altering the local concentration of acceptor molecules (Kopec *et al.*, 1997). Complete elucidation of GTF mechanism awaits the availability of a full 3D structure. Successful overexpression and crystallization of a GTF for X-ray crystallography has proved remarkably difficult, though progress has recently been reported (Pijning *et al.*, 2008). The first glucansucrase to be crystallized was amylosucrase (which is not circularly permuted and lacks a C-terminal extension) and much has been learnt from its study that may help understand other enzymes (Albenne *et al.*, 2004, 2007).

While oral microbiologists are clearly interested to understand the processes in plaque, and possibly use the information to develop GTF inhibitors (Wright *et al.*, 2002), the main drive to understand mechanisms comes from the commercial desirability of controlling the molecular size and branching of oligosaccharides or glycosylation of novel acceptors. Historically, dextran has been the main industrial product but many new applications for glucansucrases are envisaged, either as industrial-scale enzymes (Richard *et al.*, 2005) or by generation of transgenic plants expressing bacterial enzymes (Firouzabadi *et al.*, 2007).

Dextranase and mutanase

As already mentioned, dextranase is produced by a variety of plaque bacteria. Rather surprisingly, none produce a mutanase though a number of environmental bacteria and fungi do so (Fuglsang *et al.*, 2000; Shimotsuura *et al.*, 2008; Wiater *et al.*, 2008). The streptococcal dextranases have been best studied, cloned and sequenced though details of their mechanism is still lacking (Igarashi *et al.*, 1995; Wanda and Curtiss, 1994). The *S. mutans* and *S. sobrinus* enzymes are endodextranases, cleaving within the dextran chain to release oligosaccharides. These are transported by the Multiple Sugar Metabolism (*msm*) uptake system (Russell *et al.*, 1992) and further metabolized by a dextran glucosidase to enter glycolysis (Fig. 8.3) (Whiting *et al.*, 1993). Dextran-derived oligosaccharides can thus contribute to bacterial growth and acid production, in the same way as free fructose from the breakdown of fructan, even though the uptake pathways differ. It is an interesting observation relating to bacterial evolution that the homologue of the *msm* system in *Lactobacillus acidophilus* transports fructooligosaccharides, rather the glucooligosaccharides transported by *S. mutans* (Barrangou *et al.*, 2003).

Knockout of the *dexA* dextranase gene in *S. mutans* abolishes the ability to 'trim' α-1,6-linked side chains and shows a striking effect on the nature of the glucan accumulated by colonies grown on sucrose (Colby *et al.*, 1995). Both dextranase or mutanase can thus modify the structure of glucans and have been shown to reduce plaque *in vivo* (Guggenheim *et al.*, 1980; Inoue *et al.*, 1990; Poulsen *et al.*, 1978). Both have also been used in experimental toothpastes or mouthwashes but problems with enzyme stability, interaction with other constituents and adverse reactions have prevented their commercial development. They have, however, found application in denture cleaners (Budtz-Jorgensen, 2007).

Conclusion

Extracellular polysaccharides play a central role in the initiation and development of the complex dental plaque biofilm community. Exposure to sucrose leads to production of glucans and fructans that influence the balance between species and contribute to creation of the conditions in plaque that can lead to dental caries. A substantial amount of information is now available on the major species linked to caries, though the mechanisms of several important enzymes remain to be elucidated and much remains to be discovered concerning regulation. Advances in molecular approaches to complex communities have led to a greater appreciation of the importance of understanding the way in which different species interact with each other but dental plaque provides an excellent model system for the exploration of phenomena relevant to human disease, at the same time giving valuable insights into the biology of extracellular polysaccharides that are relevant to other environmental systems and industrial applications.

References

Aas, J.A., Griffen, A.L., Dardis, S.R., Lee, A.M., Olsen, I., Dewhirst, F.E., Leys, E.J., and Paster, B.J. (2008). Bacteria of dental caries in primary and permanent teeth in children and young adults. J. Clin. Microbiol. *46*, 1407–1417.

Aas, J.A., Paster, B.J., Stokes, L.N., Olsen, I., and Dewhirst, F.E. (2005). Defining the normal bacterial flora of the oral cavity. J. Clin. Microbiol. *43*, 5721–5732.

Albenne, C., Skov, L.K., Mirza, O., Gajhede, M., Feller, G., D'Amico, S., Andre, G., Potocki-Veronese, G., van der Veen, B.A., Monsan, P. (2004). Molecular basis of the amylose-like polymer formation catalyzed by *Neisseria polysaccharea* amylosucrase. J. Biol. Chem. *279*, 726–734.

Albenne, C., Skov, L.K., Tran, V., Gajhede, M., Monsan, P., Remaud-Simeon, M., and Andre-Leroux, G. (2007). Towards the molecular understanding of glycogen elongation by amylosucrase. Proteins *66*, 118–126.

Banas, J.A., and Vickerman, M.M. (2003). Glucan-binding proteins of the oral streptococci. Crit. Rev. Oral Biol. Med. *14*, 89–99.

Barrangou, R., Altermann, E., Hutkins, R., Cano, R., and Klaenhammer, T.R. (2003). Functional and comparative genomic analyses of an operon involved in fructooligosaccharide utilization by *Lactobacillus acidophilus*. Proc. Natl. Acad. Sci. USA *100*, 8957–8962.

Beighton, D. (2005). The complex oral microflora of high-risk individuals and groups and its role in the caries process. Community Dent. Oral Epidemiol. *33*, 248–255.

Bradshaw, D.J., Marsh, P.D., Watson, G.K., and Allison, C. (1998). Role of *Fusobacterium nucleatum* and coaggregation in anaerobe survival in planktonic and biofilm oral microbial communities during aeration. Infect. Immun. *66*, 4729–4732.

Budtz-Jorgensen, E. (2007). A 3-months' study of enzymes as denture cleansers. J. Oral Rehab. *5*, 35–39.

Burne, R.A. (1993). Oral Ecological Disasters: The Role of Short-Term Extracellular Storage Polysaccharides. In Cariology for the Nineties, W.H. Bowen, and L.A. Tabak, eds. (Rochester, NY: University of Rochester Press), pp. 351–364.

Burne, R.A., Chen, Y.Y., Wexler, D.L., Kuramitsu, H., and Bowen, W.H. (1996). Cariogenicity of *Streptococcus mutans* strains with defects in fructan metabolism assessed in a program-fed specific-pathogen-free rat model. J. Dent. Res. *75*, 1572–1577.

Burne, R.A., Schilling, K., Bowen, W.H., and Yasbin, R.E. (1987). Expression, purification, and characterization of an exo-beta-D-fructosidase of *Streptococcus mutans*. J. Bacteriol. *169*, 4507–4517.

Cheetham, N.W.H., Slodki, M.E., and Walker, G.J. (1991). Structure of the linear, low-molecular-weight dextran synthesized by a D-glucosyltransferase (Gtf-S3) of *Streptococcus sobrinus*. Carbohydrate Polymers *16*, 341–353.

Colby, S.M., McLaughlin, R.E., Ferretti, J.J., and Russell, R.R.B. (1999). Effect of inactivation of *gtf* genes on adherence of *Streptococcus downei*. Oral Microbiol. Immunol. *14*, 27–32.

Colby, S.M., and Russell, R.R.B. (1997). Sugar metabolism by mutans streptococci. Soc Appl. Bacteriol. Symp Ser *26*, 80S-88S.

Colby, S.M., Whiting, G.C., Tao, L., and Russell, R.R.B. (1995). Insertional inactivation of the *Streptococcus mutans dexA* (dextranase) gene results in altered adherence and dextran catabolism. Microbiology *141* 2929–2936.

Coutinho, P.M., and Henrissat, B. (1999). Carbohydrate-active enzymes: an integrated database approach. In Recent advances in carbohydrate bioengineering, H.J. Gilbert, G.J. Davies, B. Henrissat, and B. Svennson, eds. (Cambridge: Royal Society of Chemistry), pp. 3–12.

Cross, S.E., Kreth, J., Zhu, L., Qi, F.X., Pelling, A.E., Shi, W.Y., and Gimzewski, J.K. (2006). Atomic force microscopy study of the structure-function relationships of the biofilm-forming bacterium *Streptococcus mutans*. Nanotechnology *17*, S1-S7.

Cross, S.E., Kreth, J., Zhu, L., Sullivan, R., Shi, W., Qi, F., and Gimzewski, J.K. (2007). Nanomechanical properties of glucans and associated cell-surface adhesion of *Streptococcus mutans* probed by atomic force microscopy under *in situ* conditions. Microbiology *153*, 3124–3132.

Devulapalle, K.S., Goodman, S.D., Gao, Q., Hemsley, A., and Mooser, G. (1997). Knowledge-based model of a glucosyltransferase from the oral bacterial group of mutans streptococci. Protein Sci. *6*, 2489–2493.

Diaz, P.I., Chalmers, N.I., Rickard, A.H., Kong, C., Milburn, C.L., Palmer, R.J., Jr., and Kolenbrander, P.E. (2006). Molecular characterization of subject-specific oral microflora during initial colonization of enamel. Appl. Environ. Microbiol. *72*, 2837–2848.

Dibdin, G.H., and Shellis, R.P. (1988). Physical and biochemical studies of *Streptococcus mutans* sediments suggest new factors linking the cariogenicity of plaque with its extracellular polysaccharide content. J. Dent. Res. *67*, 890–895.

Ferretti, J.J., Gilpin, M.L., and Russell, R.R.B. (1987). Nucleotide sequence of a glucosyltransferase gene from *Streptococcus sobrinus* MFe28. J. Bacteriol. *169*, 4271–4278.

Firouzabadi, F.N., Kok-Jacon, G.A., Vincken, J.P., Ji, Q., Suurs, L., and Visser, R.G.F. (2007). Fusion proteins comprising the catalytic domain of mutansucrase and a starch-binding domain can alter the morphology of amylose-free potato starch granules during biosynthesis. Transgenic Res. *16*, 645–656.

Fuglsang, C.C., Berka, R.M., Wahleithner, J.A., Kauppinen, S., Shuster, J.R., Rasmussen, G., Halkier, T., Dalboge, H., and Henrissat, B. (2000). Biochemical analysis of recombinant fungal mutanases. A new family of alpha1,3-glucanases with novel carbohydrate-binding domains. J. Biol. Chem. *275*, 2009–2018.

Fujiwara, T., Tamesada, M., Bian, Z., Kawabata, S., Kimura, S., and Hamada, S. (1996). Deletion and reintroduction of glucosyltransferase genes of *Streptococcus mutans* and role of their gene products in sucrose dependent cellular adherence. Microb. Pathog. *20*, 225–233.

Funane, K., Ishii, T., Ono, H., and Kobayashi, M. (2005). Changes in linkage pattern of glucan products induced by substitution of Lys residues in the dextransucrase. FEBS Lett. *579*, 4739–4745.

Gerner-Smidt, P., Keiser-Nielsen, H., Dorsch, M., Stackebrandt, E., Ursing, J., Blom, J., Christensen, A.C., Christensen, J.J., Frederiksen, W., and Hoffmann, S. (1994). *Lautropia mirabilis* gen. nov., sp. nov., a gram-negative motile coccus with unusual morphology isolated from the human mouth. Microbiology *140* 1787–1797.

Giffard, P.M., and Jacques, N.A. (1994). Definition of a fundamental repeating unit in streptococcal glucosyltransferase glucan-binding regions and related sequences. J. Dent. Res. 73, 1133–1141.

Gilmore, K.S., Russell, R.R.B., and Ferretti, J.J. (1993). Expression of GtfS is essential for normal insoluble glucan synthesis by *Streptococcus downei*. Infect. Immun. *61*, 1246–1250.

Gilpin, M.L., Russell, R.R.B., and Morrissey, P. (1985). Cloning and expression of two *Streptococcus mutans* glucosyltransferases in *Escherichia coli* K-12. Infect. Immun. *49*, 414–416.

Gold, W., Preston, F.B., Lache, M.C., and Blechman, H. (1974). Production of levan and dextran in plaque *in vivo*. J. Dent. Res. *53*, 442–446.

Guggenheim, B., Regolati, B., Schmid, R., and Muhlemann, H.R. (1980). Effects of the topical application of mutanase on rat caries. Caries Res. *14*, 128–135.

Hamada, S., and Slade, H.D. (1980). Biology, immunology, and cariogenicity of *Streptococcus mutans*. Microbiol. Rev *44*, 331–384.

Hartemink, R., Quataert, M.C., van Laere, K.M., Nout, M.J., and Rombouts, F.M. (1995). Degradation and fermentation of fructo-oligosaccharides by oral streptococci. J. Appl. Bacteriol. *79*, 551–557.

Hata, S., and Mayanagi, H. (2003). Acid diffusion through extracellular polysaccharides produced by various mutants of *Streptococcus mutans*. Arch. Oral Biol. *48*, 431–438.

Hayacibara, M.F., Koo, F., Smith, A.M.V., Kopec, L.K., Scott-Anne, K., Cury, J.A., and Bowen, W.H. (2004). The influence of mutanase and dextranase on the production and structure of glucans synthesized by streptococcal glucosyltransferases. Carbohydr. Res. *339*, 2127–2137.

Hellmuth, H., Wittrock, S., Kralj, S., Dijkhuizen, L., Hofer, B., and Seibel, J. (2008). Engineering the glucansucrase GTFR enzyme reaction and glycosidic bond specificity: Toward tailor-made polymer and oligosaccharide products. Biochemistry *47*, 6678–6684.

Hotz, P., Guggenheim, B., and Schmid, R. (1972). Carbohydrates in pooled dental plaque. Caries Res.*6*, 103–121.

Igarashi, T., Yamamoto, A., and Goto, N. (1992). Characterization of an exo-beta-D-fructosidase from *Streptococcus mutans* Ingbritt. Microbiol. Immunol. *36*, 643–647.

Igarashi, T., Yamamoto, A., and Goto, N. (1995). Sequence analysis of the *Streptococcus mutans* Ingbritt *dexA* gene encoding extracellular dextranase. Microbiol. Immunol. *39*, 853–860.

Igarashi, T., Yamamoto, A., and Goto, N. (1998). Detection of dextranase-producing gram-negative oral bacteria. Oral Microbiol. Immunol. *13*, 382–386.

Inoue, M., Yakushiji, T., Mizuno, J., Yamamoto, Y., and Tanii, S. (1990). Inhibition of dental plaque formation by mouthwash containing an endo-alpha-1, 3 glucanase. Clin. Prev. Dent. *12*, 10–14.

Jakubovics, N.S., Gill, S.R., Iobst, S.E., Vickerman, M.M., and Kolenbrander, P.E. (2008). Regulation of gene expression in a mixed-genus community: stabilized arginine biosynthesis in *Streptococcus gordonii* by coaggregation with *Actinomyces naeslundii*. J. Bacteriol. *190*, 3646–3657.

Janecek, S., Svensson, B., and Russell, R.R. (2000). Location of repeat elements in glucansucrases of *Leuconostoc* and *Streptococcus* species. FEMS Microbiol. Lett *192*, 53–57.

Johnson, I.H. (1990). Glucanase-producing organisms in human dental plaques. Microbios *61*, 89–98.

Kingston, K.B., Allen, D.M., and Jacques, N.A. (2002). Role of the C-terminal YG repeats of the primer-dependent streptococcal glucosyltransferase, GtfJ, in binding to dextran and mutan. Microbiology *148*, 549–558.

Koga, T., Oho, T., Shimazaki, Y., and Nakano, Y. (2002). Immunization against dental caries. Vaccine *20*, 2027–2044.

Kolenbrander, P.E., and London, J. (1993). Adhere Today, Here Tomorrow – Oral Bacterial Adherence. J. Bacteriol. *175*, 3247–3252.

Kolenbrander, P.E., Palmer, R.J., Jr., Rickard, A.H., Jakubovics, N.S., Chalmers, N.I., and Diaz, P.I. (2006). Bacterial interactions and successions during plaque development. Periodontol 2000 *42*, 47–79.

Kopec, L.K., Vacca-Smith, A.M., and Bowen, W.H. (1997). Structural aspects of glucans formed in solution and on the surface of hydroxyapatite. Glycobiology *7*, 929–934.

Kreth, J., Zhu, L., Merritt, J., Shi, W., and Qi, F. (2008). Role of sucrose in the fitness of *Streptococcus mutans*. Oral Microbiol. Immunol. *23*, 213–219.

Kuramitsu, H.K. (1993). Virulence factors of mutans streptococci: role of molecular genetics. Crit. Rev. Oral Biol. Med. *4*, 159–176.

Kuramitsu, H.K. (2003). Molecular genetic analysis of the virulence of oral bacterial pathogens: an historical perspective. Crit. Rev. Oral Biol. Med. *14*, 331–344.

Kuramitsu, H.K., He, X., Lux, R., Anderson, M.H., and Shi, W. (2007). Interspecies interactions within oral microbial communities. Microbiol. Mol. Biol. Rev. *71*, 653–670.

Lairson, L.L., Henrissat, B., Davies, G.J., and Withers, S.G. (2008). Glycosyltransferases: structures, functions, and mechanisms. Annu Rev Biochem *77*, 521–555.

Lynch, D.J., Fountain, T.L., Mazurkiewicz, J.E., and Banas, J.A. (2007). Glucan-binding proteins are essential for shaping *Streptococcus mutans* biofilm architecture. FEMS Microbiol. Lett *268*, 158–165.

MacGregor, E.A., Jespersen, H.M., and Svensson, B. (1996). A circularly permuted alpha-amylase-type alpha/beta-barrel structure in glucan-synthesizing glucosyltransferases. FEBS Lett. *378*, 263–266.

MacKenzie, C.R., McDonald, I.J., and Johnson, K.G. (1978). Glycogen metabolism in the genus *Neisseria*: synthesis from sucrose by amylosucrase. Can J. Microbiol. *24*, 357–362.

Marsh, P.D. (2003). Are dental diseases examples of ecological catastrophes? Microbiology-(UK) *149*, 279–294.

Marsh, P.D. (2006). Dental plaque as a biofilm and a microbial community – implications for health and disease. BMC Oral Health *6* (Suppl. 1), S14.

Matsumoto-Nakano, M., Fujita, K., and Ooshima, T. (2007). Comparison of glucan-binding proteins in cariogenicity of *Streptococcus mutans*. Oral Microbiol. Immunol. *22*, 30–35.

Monchois, V., Arguello-Morales, M., and Russell, R.R.B. (1999a). Isolation of an active catalytic core of *Streptococcus downei* MFe28 GTF-I glucosyltransferase. J. Bacteriol. *181*, 2290–2292.

Monchois, V., Vignon, M., Escalier, P.C., Svensson, B., and Russell, R.R.B. (2000). Involvement of Gln937 of *Streptococcus downei* GTF-I glucansucrase in transition-state stabilization. Eur J. Biochem *267*, 4127–4136.

Monchois, V., Willemot, R.M., and Monsan, P. (1999b). Glucansucrases: mechanism of action and structure-function relationships. FEMS Microbiol. Rev. *23*, 131–151.

Moulis, C., Joucla, G., Harrison, D., Fabre, E., Potocki-Veronese, G., Monsan, P., and Remaud-Simeon, M. (2006). Understanding the polymerization mechanism of glycoside-hydrolase family 70 glucansucrases. J. Biol. Chem. *281*, 31254–31267.

Munro, C., Michalek, S.M., and Macrina, F.L. (1991). Cariogenicity of *Streptococcus mutans* V403 glucosyltransferase and fructosyltransferase mutants constructed by allelic exchange. Infect. Immun. *59*, 2316–2323.

Norman, J.M., and Giffard, P.M. (1996). Biochemical studies on LevJ, a fructanase from *Actinomyces naeslundii* T14V. Arch. Oral. Biol. *41*, 565–570.

Nyvad, B. (1993). Microbial colonization of human tooth surfaces. APMIS Suppl. *32*, 1–45.

Nyvad, B., and Fejerskov, O. (1989). Structure of dental plaque and the plaque-enamel interface in human experimental caries. Caries Res. *23*, 151–158.

Paes Leme, A.F., Koo, H., Bellato, C.M., Bedi, G., and Cury, J.A. (2006). The role of sucrose in cariogenic dental biofilm formation – new insight. J. Dent. Res. *85*, 878–887.

Palmer, R.J., Gordon, S.M., Cisar, J.O., and Kolenbrander, P.E. (2003). Coaggregation-mediated interactions of streptococci and actinomyces detected in initial human dental plaque. J. Bacteriol. *185*, 3400–3409.

Parker, R.B., and Creamer, H.R. (1971). Contribution of plaque polysaccharides to growth of cariogenic micro-organisms. Arch. Oral Biol. *16*, 855–862.

Pijning, T., Vujicic-Zagar, A., Kralj, S., Eeuwema, W., Dijkhuizen, L., and Dijkstra, B.W. (2008). Biochemical and crystallographic characterization of a glucansucrase from *Lactobacillus reuteri* 180. Biocatal. Biotransformation *26*, 12–17.

Poulsen, S., Larson, R.H., and Senning, R.S. (1978). Effect of dextranase on plaque formation and caries development in the rat. Scand. J. Dent. Res. *86*, 231–236.

Rahimi, M., Heng, N.C., Kieser, J.A., and Tompkins, G.R. (2005). Genotypic comparison of bacteria recovered from human bite marks and teeth using arbitrarily primed PCR. J. Appl. Microbiol. *99*, 1265–1270.

Reese, S., and Guggenheim, B. (2007). A novel TEM contrasting technique for extracellular polysaccharides in *in vitro* biofilms. Microsc .Res. Tech *70*, 816–822.

Richard, G., Fabre, E., Moulis, C., Joucla, G., Albenne, C., Potocki-Veronese, G., Monsan, P., and Remaud-Simeon, M. (2005). Glucansucrases: Versatile catalysts for the synthesis of functional food. J. Biotechnol. *118*, S135-S135.

Rickard, A.H., Gilbert, P., High, N.J., Kolenbrander, P.E., and Handley, P.S. (2003). Bacterial coaggregation: an integral process in the development of multi-species biofilms. Trends Microbiol. *11*, 94–100.

Riou, J.Y., Guibourdenche, M., Perry, M.B., MacLean, L.L., and Griffith, D.W. (1986). Structure of the exocellular D-glucan produced by *Neisseria polysaccharea*. Can. J. Microbiol. *32*, 909–911.

Rose, R.K., Dibdin, G.H., and Shellis, R.P. (1993). A quantitative study of calcium-binding and aggregation in selected oral bacteria. J. Dent. Res. *72*, 78–84.

Rose, R.K., and Turner, S.J. (1998). Extracellular volume in streptococcal model biofilms: effects of pH, calcium and fluoride. Biochim. Biophys. Acta *1379*, 185–190.

Rozen, R., Bachrach, G., Bronshteyn, M., Gedalia, I., and Steinberg, D. (2001). The role of fructans on dental biofilm formation by *Streptococcus sobrinus*, *Streptococcus mutans*, *Streptococcus gordonii* and *Actinomyces viscosus*. FEMS Microbiol. Lett *195*, 205–210.

Ruby, J.D., Shirey, R.E., Gerencser, V.F., and Stelzig, D.A. (1982). Extracellular iodophilic polysaccharide synthesized by *Neisseria* in human dental plaque. J. Dent. Res. *61*, 627–631.

Russell, R.R.B. (1990). Molecular genetics of glucan metabolism in oral streptococci. Arch. Oral Biol. *35*, S53-S58.

Russell, R.R.B. (2008a). How has genomics altered our view of caries microbiology? Caries Res.42, 319–327.

Russell, R.R.B. (2008b). Might caries control involve immunisation and gene therapy? In Dental Caries: The Disease and its Clinical Management, O. Fejerskov, E. Kidd, B. Nyvad, and V. Baelum, eds. (Copenhagen: Blackwell Munsgaard).

Russell, R.R.B. (2008c). The Molecular Biology of Cariogenic Bacteria. In Molecular Oral Microbiology, A.H. Rogers, ed. (Norfolk, UK, Caister Academic Press), pp. 109–133.

Russell, R.R.B., Aduse-Opoku, J., Sutcliffe, I.C., Tao, L., and Ferretti, J.J. (1992). A binding protein-dependent transport system in *Streptococcus mutans* responsible for multiple sugar metabolism. J. Biol. Chem. *267*, 4631–4637.

Russell, R.R.B., Gilpin, M.L., Hanada, N., Yamashita, Y., Shibata, Y., and Takehara, T. (1990). Characterization of the product of the *gtfS* gene of *Streptococcus downei*, a primer-independent enzyme synthesizing oligo-isomaltosaccharides. J. Gen. Microbiol. *136 (Pt 8)*, 1631–1637.

Schroeder, V.A., Michalek, S.M., and Macrina, F.L. (1989). Biochemical characterization and evaluation of virulence of a fructosyltransferase-deficient mutant of *Streptococcus mutans* V403. Infect. Immun. *57*, 3560–3569.

Shah, D.S., Joucla, G., Remaud-Simeon, M., and Russell, R.R.B. (2004). Conserved repeat motifs and glucan binding by glucansucrases of oral streptococci and *Leuconostoc mesenteroides*. J. Bacteriol. *186*, 8301–8308.

Shimotsuura, I., Kigawa, H., Ohdera, M., Kuramitsu, H.K., and Nakashima, S. (2008). Biochemical and molecular characterization of a novel type of mutanase from *Paenibacillus* sp. strain RM1: identification of its mutan-binding domain, essential for degradation of *Streptococcus mutans* biofilms. Appl. Environ. Microbiol. *74*, 2759–2765.

Shiroza, T., and Kuramitsu, H.K. (1988). Sequence analysis of the *Streptococcus mutans* fructosyltransferase gene and flanking regions. J. Bacteriol. *170*, 810–816.

Swistowska, A.M., Gronert, S., Wittrock, S., Collisi, W., Hecht, H.J., and Hofer, B. (2007). Identification of structural determinants for substrate binding and turnover by glucosyltransferase R supports the permutation hypothesis. FEBS Lett. *581*, 4036–4042.

Takahashi, N., Mizuno, F., and Takamori, K. (1985). Purification and preliminary characterization of exo-beta-D-fructosidase in *Streptococcus salivarius* KTA-19. Infect. Immun. *47*, 271–276.

Takahashi, N., and Nyvad, B. (2008). Caries ecology revisited: microbial dynamics and the caries process. Caries Res. *42*, 409–418.

Tsumori, H., and Kuramitsu, H. (1997). The role of the *Streptococcus mutans* glucosyltransferases in the sucrose-dependent attachment to smooth surfaces: essential role of the GtfC enzyme. Oral Microbiol. Immunol. *12*, 274–280.

Van Der Ploeg, J.R., and Guggenheim, B. (2004). Deletion of *gtfC* of *Streptococcus mutans* has no influence on the composition of a mixed-species *in vitro* biofilm model of supragingival plaque. Eur. J. Oral Sci. *112*, 433–438.

van Hijum, S.A., Kralj, S., Ozimek, L.K., Dijkhuizen, L., and van Geel-Schutten, I.G. (2006). Structure-function relationships of glucansucrase and fructansucrase enzymes from lactic acid bacteria. Microbiol. Mol. Biol. Rev. *70*, 157–176.

Van Houte, J., Russo, J., and Prostak, K.S. (1989). Increased pH-lowering ability of *Streptococcus mutans* cell masses associated with extracellular glucan-rich matrix material and the mechanisms involved. J. Dent. Res. *68*, 451–459.

Vickerman, M.M., Sulavik, M.C., Nowak, J.D., Gardner, N.M., Jones, G.W., and Clewell, D.B. (1997). Nucleotide sequence analysis of the *Streptococcus gordonii* glucosyltransferase gene, *gtfG*. DNA Seq. *7*, 83–95.

Wanda, S.Y., and Curtiss, R., 3rd (1994). Purification and characterization of *Streptococcus sobrinus* dextranase produced in recombinant *Escherichia coli* and sequence analysis of the dextranase gene. J. Bacteriol. *176*, 3839–3850.

Whiting, G.C., Sutcliffe, I.C., and Russell, R.R.B. (1993). Metabolism of polysaccharides by the *Streptococcus mutants dexB* gene product. J. Gen. Microbiol. *139*, 2019–2026.

Wiater, A., Szczodrak, J., and Pleszczynska, M.G. (2008). Mutanase induction in *Trichoderma harzianum* by cell wall of *Laetiporus sulphureus* and its application for mutan removal from oral biofilms. J. Microbiol. Biotechnol. *18*, 1335–1341.

Wittrock, S., Swistowska, A.M., Collisi, W., Hofmann, B., Hecht, H.J., and Hofer, B. (2008). Re- or displacement of invariant residues in the C-terminal half of the catalytic domain strongly affects catalysis by glucosyltransferase R. FEBS Lett. *582*, 491–496.

Wright, W.G., Thelwell, C., Svensson, B., and Russell, R.R.B. (2002). Inhibition of catalytic and glucan-binding activities of a streptococcal GTF forming insoluble glucans. Caries Res. *36*, 353–359.

Yamashita, Y., Bowen, W.H., Burne, R.A., and Kuramitsu, H.K. (1993). Role of the *Streptococcus mutans gtf* genes in caries induction in the specific-pathogen-free rat model. Infect. Immun. *61*, 3811–3817.

Yoshida, Y., Ganguly, S., Bush, C.A., and Cisar, J.O. (2005). Carbohydrate engineering of the recognition motifs in streptococcal co-aggregation receptor polysaccharides. Mol. Microbiol. *58*, 244–256.

Yoshida, Y., Palmer, R.J., Yang, J., Kolenbrander, P.E., and Cisar, J.O. (2006). Streptococcal receptor polysaccharides: recognition molecules for oral biofilm formation. BMC Oral Health *6* Suppl. 1, S12.

Yoshida, Y., Yang, J., Peaker, P.E., Kato, H., Bush, C.A., and Cisar, J.O. (2008). Molecular and antigenic characterization of a *Streptococcus oralis* coaggregation receptor polysaccharide by carbohydrate engineering in *Streptococcus gordonii*. J. Biol. Chem. *283*, 12654–12664.

Zeng, L., Wen, Z.T., and Burne, R.A. (2006). A novel signal transduction system and feedback loop regulate fructan hydrolase gene expression in *Streptococcus mutans*. Mol. Microbiol. *62*, 187–200.

Part IV

Polysaccharides and Biofilm Formation

Composition and Functional Role of Polysaccharides and Extracellular Polymeric Substances in Gram-positive Biofilm Infections

9

Christian Theilacker and Johannes Hübner

Abstract

Biofilms seem to be the default mode of growth for most if not all bacterial species and this phenomenon has profound consequences in numerous clinical settings. For Gram-positive bacteria several mechanisms of biofilm formation involving carbohydrate-containing structures have been identified. Poly-N-acetylglucosamine (PIA/PNAG) has been first described in staphylococci but recently has been shown to be present also in a large number of different bacterial species. Mutants in the gene locus responsible for the synthesis of this molecule lead to a biofilm-negative phenotype. Teichoic acids are polyanionic molecules found in the cell walls of Gram-positive bacteria with low GC content. Teichoic acids are either attached covalently to the peptidoglycan (wall teichoic acids) or inserted into the cell membrane (lipoteichoic acids) and both types have been associated with biofilm formation and adhesion to eukaryotic cells. Recent evidence indicates that extracellular DNA may also be involved in biofilm formation and primary attachment of many Gram-positive organisms and this fact may explain previous observations that autolysin-defective mutants are also impaired in biofilm formation. A more detailed understanding of the molecules involved in biofilm formation is needed to device novel treatment and preventive approaches for Gram-positive infections.

Introduction

It is one of the harsh realities in the life of a microbe that the real world is very different from the cosy environment it encounters in the nutrient-rich broth cultures used by laboratories. In the human host, planktonic bacteria may be confronted with rapid, continual and extreme changes in the environmental conditions. Formation of multicellular communities offers bacteria a shelter to withstand these hostile conditions. In fact, biofilm probably represent the default mode of growth of bacteria (Jefferson, 2004). It therefore comes as no surprise that biofilms are critical in a wide variety of bacterial diseases, ranging from foreign body infections to otitis media (Costerton *et al.*, 1999). Accumulating evidence has shown that this habitat is a highly successful means to escape host defence mechanisms. Bacteria encased in a biofilm may be 10 to 1000 times more resistant to the concentrations of antibiotics needed to kill planktonic bacteria (Lewis, 2001; Sandoe *et al.*, 2006). They are also more resistant to opsonophagocytosis (Cerca *et al.*, 2006) and are protected against the bactericidal activity of antimicrobial peptides (Vuong *et al.*, 2004b).

Bacterial biofilms are composed of bacterial cells, an extracellular matrix and, depending of the type of biofilm examined, often form a system of microcolonies and channels (Hall-Stoodley *et al.*, 2004). Interestingly, bacterial cells do not usually make up more than 2–15% of the total biofilm mass (Frølund *et al.*, 1996). Instead, much of the biofilm matrix – perhaps up to 97% – is actually water (Zhang *et al.*, 1998). Contrary to common belief, the extracellular matrix does not only contain polysaccharides but also a highly variable mixture of proteins, extracellular DNA, glycolipids and glycoproteins. Through hydrophobic interaction, cross-linking by multivalent cations and entanglement by these biopolymers the matrix network provides mechanical support for the spatial arrangement of the bacteria in a biofilm.

Carbohydrates constitute an important component of biofilms. In a genome-wide screening, genes involving polysaccharide synthesis and carbohydrate metabolism are found on a regular basis (Branda *et al.*, 2004; Jefferson, 2004). Because of the problems encountered in purification and analysis of the complex mixture of macromolecules in biofilms our knowledge of the composition and function of extracellular matrix polymers is still incomplete. Currently, the only approach to detect matrix exopolysaccharides *in situ* is the use of fluorescently labelled lectins (Jefferson and Cerca, 2006; Neu *et al.*, 2001). If we think of biofilms we tend to envisage bacteria interacting with plastic surfaces, and methods to screen for microbial factors involved in biofilm formation are largely based on this concept. In the human host, however, plasma proteins like fibrinogen and fibronectin are sequentially absorbed to the polymeric surface of foreign devices and these proteins play an important role in facilitating bacterial binding (Green *et al.*, 1999; Vaudaux *et al.*, 1989). In addition, biofilm-infections are also encountered in the absence of foreign bodies making an adhesive interaction of the multicellular bacterial assemblages with host cells or matrix molecules necessary (Donlan and Costerton, 2002). A wide variety of polysaccharides and biopolymers have been described in biofilms and excellent reviews on matrix polysaccharides have been published (Branda *et al.*, 2004; Sutherland, 2001a; Sutherland, 2001b). In this chapter we will therefore only give an overview about extracellular matrix polysaccharides in human pathogens and will focus on the carbohydrates and polymeric substances in medically important Gram-positive infections.

In Gram-negative pathogens, a number of polysaccharides of the extracellular matrix have been characterized in detail. Alginate, a polyanionic, partially O-acetylated random polymer of D-mannuronic acid and L-guluronic acid, is a major component of the extracellular matrix of mucoid *P. aeruginosa* (Ryder *et al.*, 2007). This antigen is also expressed *in vivo* in non-mucoid strains and has been used to develop vaccine against the chronic biofilm infection in the cystic fibrosis airways and against ulcerative keratitis (Pier *et al.*, 2004; Theilacker *et al.*, 2003; Zaidi and Pier, 2008). Other polysaccharides, however, also contribute to biofilm formation in non-mucoid strains of *P. aeruginosa* and *P. syringae* (Laue *et al.*, 2006; Ma *et al.*, 2006; Ryder *et al.*, 2007). Cellulose, a polysaccharide produced by a number of bacteria including *Salmonella typhimurium* and *Escherichia coli*, has been shown to be a crucial component of the extracellular matrix in these organisms (Solano *et al.*, 2002; Zogaj *et al.*, 2001). Coexpression of cellulose and fimbriae leads to the formation of a highly hydrophobic network with tightly packed cells in a rigid matrix (Zogaj *et al.*, 2001).

Carbohydrate-mediated coaggregation and biofilm formation of oral streptococci play an important role in the microbial colonization of the tooth surface. Highly defined carbohydrate motifs of the cell wall polysaccharides of oral streptococci account for the lectin-mediated coaggregation with other streptococci and *Actinomyces naeslundii* (Yoshida *et al.*, 2006; Yoshida *et al.*, 2008). Polysaccharide intracellular adhesion, also called poly-N-acetylglucosamine (PIA/PNAG) has been shown to be involved in biofilm formation in *S. aureus* and *S. epidermidis* (Gotz, 2002) as well as in some Gram-negative pathogens (Cerca *et al.*, 2007b) and will be reviewed below.

Not only exopolymers, however, are involved in the anchoring of bacteria in biofilms. Numerous studies have demonstrated, that accessory cell wall carbohydrate polymers like lipoteichoic acid and wall teichoic acid are also critical for biofilm formation and these molecules will be discussed more closely in this chapter.

Clinically relevant Gram-positive bacteria and biofilm infections

In recent years, there has been a shift in pathogens associated with nosocomial infections. While there was a predominance of Gram-negative species during the 80s, in the past 20 years there has been a continual rise in Gram-positive nosocomial infections (Karchmer, 2000; Martin *et al.*, 2003). An important factor contributing to the change in epidemiology is the increased use of foreign implant materials in patients and subsequent biofilm infections (Donlan and Costerton, 2002; Fabretti and Huebner, 2005).

The most important Gram-positive nosocomial pathogens are *S. aureus*, *S. epidermidis* and enterococci (*E. faecalis* and *E. faecium*). Together, they account for about 30% of all severe sepsis cases (Bochud *et al.*, 2001) and nosocomial bloodstream infections (Wisplinghoff *et al.*, 2003). Staphylococci and enterococci are well known for their ability to form biofilm, although the underlying mechanisms are not always well understood (O'Toole *et al.*, 2000; Fabretti and Huebner, 2005).

Biofilm infections due to Gram-positive pathogens have been observed in almost every possible setting involving foreign materials used in patients (Box 9.1) (Donlan and Costerton, 2002; Fabretti and Huebner, 2005). These foreign body biofilm infections are often exceedingly difficult to treat and frequently relapse, making it necessary to explant the infected device from the host to achieve long-term cure (Fowler *et al.*, 2003). However, in native valve endocarditis, osteomyelitis, otitis media, peridontitis and the chronic airway infection in cystic fibrosis, biofilms are also found in the absence of plastic surfaces (Donlan and Costerton, 2002).

The development of biofilm infections usually involves several steps (Costerton *et al.*, 1999; Hall-Stoodley *et al.*, 2004; Stoodley *et al.*, 2002):

Box 9.1 Foreign-body materials associated with biofilm infections in humans.

- Prosthetic heart valves
- Ventricular assist devices
- Central-venous catheters
- Implanted central venous catheters (Broviac-Hickman, Port-A-Cath)
- Vascular grafts (dacron patches, etc.)
- Urinary catheters
- Contact lenses
- Intraocular lenses
- Intrauterine devices
- Penile implants
- Dental implants
- Sutures
- Endotracheal tubes
- Biliary stents
- Artificial joints
- Central nervous liquor shunts

1. Seeding to a foreign body. For the initial adherence usually only few bacteria are necessary and several factors and/or mechanisms for adhesion have been described. Initial colonization may occur during surgery or very shortly after insertion of the indwelling device and build-up of a mature biofilm can take a long time.
2. Accumulation and maturation. Bacteria form microcolonies which are considered to be the basic building blocks and structural units of mature biofilms (Fig. 9.1) (Donlan and Costerton, 2002). This is also the phase in which specific intercellular adhesion mechanisms come into play and the bacteria start to elaborate extracellular polymers. The bacterial cells form matrix-enclosed 'towers and mushrooms' and interspersed water channels (Donlan and Costerton, 2002).
3. Detachment/dispersal. For dissemination of the infection and spread to new body compartments, a fraction of the bacteria anchored in a biofilm breaks off if their tensile strength is exceeded. This detachment poses the risk for infective emboli with seeding and

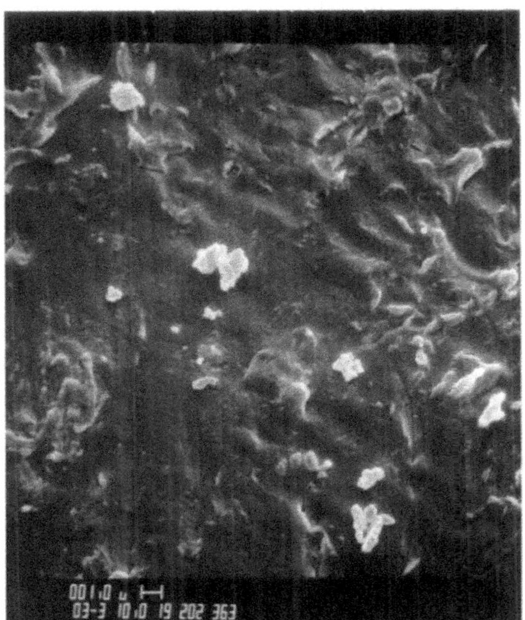

Figure 9.1 Transmission electron-microscopy of microcolonies of an *E. faecalis* biofilm on a plastic catheter.

secondary infections in the next capillary bed (Donlan and Costerton, 2002).

PIA/PNAG

Staphylococcus aureus and *Staphylococcus epidermidis* have been highly successful in adapting to the biofilm mode of growth. In the past, coagulase-negative staphylococci have been labelled 'apathogenic staphylococci'. However, it has become clear that these bacteria exhibit a specific affinity for plastic and other polymers, which, in turn, leads to infection of all kinds of foreign bodies inserted into patients (Box 9.1). Upon removal of the foreign devices, one often sees an amorphous layer, termed 'slime' in the scientific literature. This slime material is loosely bound to the bacterial cells and represents a bacterial biofilm. Polysaccharides, DNA and proteins have been implicated in the formation of biofilm by staphylococci (Gotz, 2002).

The biofilm observed in foreign-body infection is most prominent if *S. epidermidis* is involved. Very early studies identified bacterial glycocomolecules as a major component of 'slime' through agglutination with concanavalin A (Ludwicka *et al.*, 1984). The components of this material are difficult to determine and in some early studies medium-derived sugar compounds were falsely identified (Gotz, 2002).

Tojo and colleagues purified a polysaccharide containing glucosamine from a strong biofilm-producing *S. epidermidis* strain RP-62A (ATCC 35984). While their material was not pure enough to allow determination of the structure, they used this material to raise antibodies that blocked adhesion of the bacteria to catheters *in vitro* (Tojo *et al.*, 1988) and *in vivo* (Kojima *et al.*, 1990; Takeda *et al.*, 1991). Since the first step of biofilm formation seemed to have been blocked, the antigen was named **p**oly**s**accharide/**a**dhesin (PS/A).

In 1994, Mack and coworkers identified a polysaccharide antigen as being essential for cell-to-cell adhesion and also responsible for the accumulation of *S. epidermidis*, the second step in the formation of a mature biofilm. Because of this functional difference to PS/A, this antigen was considered to be a distinct polysaccharide and the researchers therefore named the antigen, for which no structural information was available at the time, polysaccharide intercellular adhesin (PIA) (Mack *et al.*, 1994).

Mack and his group were finally able to solve the structure of PIA in 1996 by NMR spectroscopy as a polysaccharide composed of poly-1→6-β-D-*N*-acetylglucosamine (Fig. 9.2) (Mack *et al.*, 1996). Later, McKenney and colleagues confirmed that PIA and PS/A were in fact identical (McKenney *et al.*, 1998). Previous studies have been done on *S. epidermidis*; McKenney and coworkers provided evidence that the same polysaccharide is also expressed in *S. aureus*, although only elaborated *in vivo* in human and animal infections and named the antigen according to its structure poly-N-acetylglucosamine (PNAG) (McKenney *et al.*, 1999). This antigen was initially named PNSG (poly N-succinyl β-1,6-glucosamine) (McKenney *et al.*, 1999) although it became soon clear that the succinyl groups were an artefact (Maira-Litran *et al.*, 2002). Various animal models have shown the importance of this antigen in virulence and its potential as vaccine target in protection against infections caused by *S. aureus* and *S. epidermidis* (Kropec *et al.*, 2005; McKenney *et al.*, 1999). Surprisingly, genetic homologues of the *ica*-gene have also been found

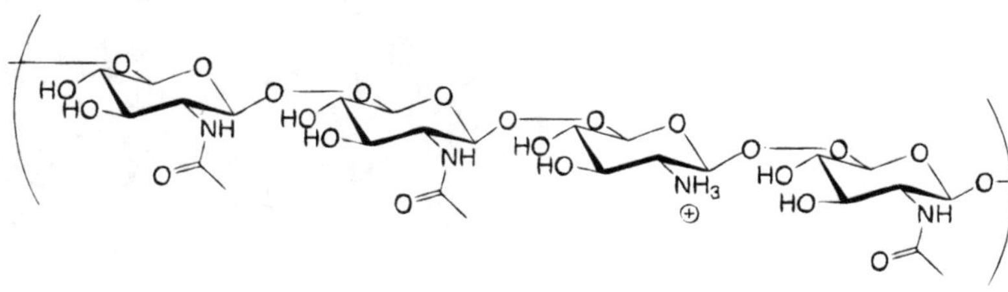

Figure 9.2 Structure of PIA/PNAG. The exopolysaccharide is a linear homoglycan composed of β-(1,6)-N-acetlyglucosamine (Mack *et al.*, 1996). Approximatly 20% of the N-acetyl-residues are removed from the nascent polymer by the protein IcaB giving it an overall positive charge (Vuong *et al.*, 2004a).

in a number of other important Gram-negative pathogens, including *E. coli*, *Yersinia* spp. and *Bordetella* spp., as well as *Actinobacillus actinomycetemcomitans* and *Actinobacillus pleuropneumoniae* (Cerca *et al.*, 2007b).

The PIA/PNAG polymer is loosely attached to the bacterial surface by ionic interaction and it has to be at least partially deacetylated and therefore positively charged to be associated with the bacterial surface. Inactivation of the deacetylase of PIA/PNAG, IcaB, results in the inability of bacteria to retain the polymer on the cell surface and, in consequence, *icaB*-deficient *S. aureus* and *S. epidermidis* loses its ability to form biofilm (Fig. 9.2) (Vuong *et al.*, 2004a). It has been speculated that the polymer is synthesized in its fully acetylated form and subsequently deacetylated by icaB in a second step leading to positively charged N-glucosamine residues (Vuong *et al.*, 2004a). IcaB is found mainly in the culture supernatant and shows sequence similarity with the *Rhizobium* NodB protein, a chitooligosaccharide deacetylase (John *et al.*, 1993). An *icaB* mutant in *S. aureus* expressed significantly less surface-associated PNAG and was also highly susceptible to antibody-independent killing (Cerca *et al.*, 2007a).

Cell wall-associated carbohydrate polymers

As outlined above, biofilm formation on abiotic surfaces is believed to be a two-step process, which is initiated by adhesion of planktonic bacteria to a substrate followed by accumulation of multiple cell layers via aggregation of bacteria (O'Toole *et al.*, 2000). In *S. aureus*, it has been shown that initial adhesion is independent of the expression PIA/PNAG (Gross *et al.*, 2001; Heilmann *et al.*, 1996; Heilmann *et al.*, 1997). Plastic surfaces such as polystyrene, the material most frequently used in the laboratory for initial assessment of biofilm formation, are most often negatively charged polymers. Bacterial adhesion to plastic surfaces is mediated by a complex interplay between Van der Waals forces and repulsive or attractive electrostatic and acid-base forces (Bos *et al.*, 1999). Modulation of cell surface hydrophobicity and electrostatic charge will therefore affect attachment to plastic surfaces. Even if bacteria and surfaces are charged alike, van der Waals forces may predominate over electrostatic repulsion and lead to adhesion (van Loosdrecht *et al.*, 1987a,b).

Structure of teichoic acids

Numerous studies demonstrate that polyanionic carbohydrate cell wall polymers are important determinants to bacterial electrostatic charge and hydrophobicity and play a pivotal role in bacterial attachment to abiotic surfaces and eukaryotic cells. Of the various cell-wall associated carbohydrate polymers, teichoic acid represents the most abundant polyanion in the Gram-positive bacterial cell envelope (Fischer, 1994). While the species of the high G+C branch of Gram-positive bacteria often contain lipoglycans without phosphate in the backbone, those from the low G+C branch typically produce teichoic acids. Due to the polyol-phosphate backbone of teichoic acid, this polymer contains many negatively charged phosphodiester bonds (Neuhaus and Baddiley, 2003; Weidenmaier and Peschel, 2008). Esterification of alanine to the teichoic acid backbone introduces positive charges into the molecule, making it a zwitterionic cell wall glycopolymer.

Two distinct teichoic acids are found in the Gram-positive cell wall. Wall teichoic acid (WTA) is covalently bound to peptidoglycan (PG), whereas lipoteichoic acid (LTA) is inserted into the cell membrane via a glycolipid anchor. Peptidoglycan-anchored WTA is connected to the cell wall by phosphodiester bonds to N-acetylmuramic acid of PG via a linkage unit, which is most frequently composed of an N-acetylglucosamine-N-acetylmannosamine disaccharide (Fig. 9.3) (Araki and Ito, 1989). Owing to its glycosyl substituents, D-alanyl esters, and repeating units WTA is a diverse glycopolymer. Most frequently, the repeating unit is composed of polyribitolphosphate or polyglycerolphosphate. However, complex repeating units consisting of trioses or hexoses or glycosylpolyolphosphate monomers, are also found, such as in *S. pneumoniae* or *E. faecalis* (Fischer *et al.*, 1993; Kaczynski and Theilacker, personal communication). WTA may constitute up to 30–60% of the Gram-positive cell wall (Neuhaus and Baddiley, 2003). Glycosyl substituents in WTA, such in *S. aureus*, frequently consist of glucosyl- and *N*-acetylglucosamine residues (Fig. 9.3) (Neuhaus and Baddiley, 2003). A

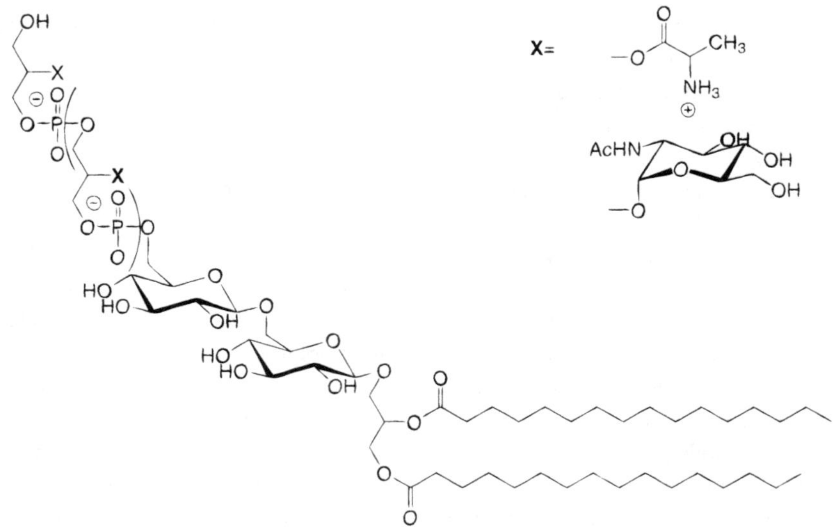

Figure 9.3 Structure of LTA from S. aureus. The average length of the poly-glycerol phosphate repeating unit is 45–50 (Morath *et al.*, 2001). On average, 70% of the C-2 hydroxyls of glycerol are substituted with D-alanine and 15% are decorated by N-acetylglucosamine, giving the polymer an overall polyanionic character (Morath *et al.*, 2001).

non-stochiometric substitution with D-alanine is the major determinant of the positive net charge of WTA. The structure of LTA is less complex. With few exceptions, e.g. the LTA from *S. pneumoniae* and *Lactococcus garvieae,* the repeating unit of the polymer is composed of polyglycerol phosphate (Fig. 9.4) (Fischer, 1990). The major differences between LTA from different bacterial species are the substituents of C-2 of glycerol and the structure of the glycolipid anchor (Fischer, 1990). Decoration of C-2 includes glycosylation and D-alanine esters (Fig. 9.3). In WTA and LTA the alditol substitution is non-stochiometric, which causes considerable microheterogeneity of teichoic acids (Leopold and Fischer, 1991).

Lipoteichoic acid

The first evidence that LTA is involved in the colonization and attachment to host epithelial cells came from work done in streptococci more than three decades ago (Beachey and Ofek, 1976; Hasty *et al.*, 1992). In a series of experiments using LTA as an inhibitor and measuring interaction of LTA to various host tissues its role in the initial adhesion to epithelial cells was delineated (Beachey and Ofek, 1976; Hasty *et al.*, 1992; Nealon and Mattingly, 1984; Simpson and Beachey, 1983) and fibronectin was identified as a binding ligand for LTA on host cells (Courtney *et al.*, 1986; Simpson and Beachey, 1983). In *S. pyogenes,* LTA together with the M-protein is a component of a multimodal adhesion mechanism to host tissues (Hasty *et al.*, 1992). The role of LTA in attachment of streptococci is also corroborated by *in vivo* colonization experiments. Intranasal instillation of LTA in mice blocks colonization and prevents death after intranasal challenge infection with group A streptococci (Dale *et al.*, 1994) Topical vaginal instillation of LTA also reduces neonatal colonization and bacteraemia by group B streptococci (Cox *et al.*, 1993).

Our understanding of the role of LTA in biofilm formation and adhesion was considerably promoted by the availability of mutants in the LTA biosynthesis. Since mutants completely devoid of LTA are lethal (see below) most evidence for the function of LTA was collected in mutants defective in D-alanine esterification. The *dlt* operon, including *dltABCD,* encodes the biosynthetic pathway for D-alanylation of LTA (Neuhaus and Baddiley, 2003). DltA is a cytoplasmic D-alanine-D-alanyl-carrier protein ligase (also designated Dcl) that catalyses the D-alanylation of the D-alanyl carrier protein DltC (or Dcp) (Neuhaus and Baddiley, 2003). D-Alanine esterified to LTA is the precursor for D-alanylation of WTA (Haas *et al.*, 1984). Hence, mutations of the *dlt* operon affect both LTA and WTA alanine substitution. Mutations of

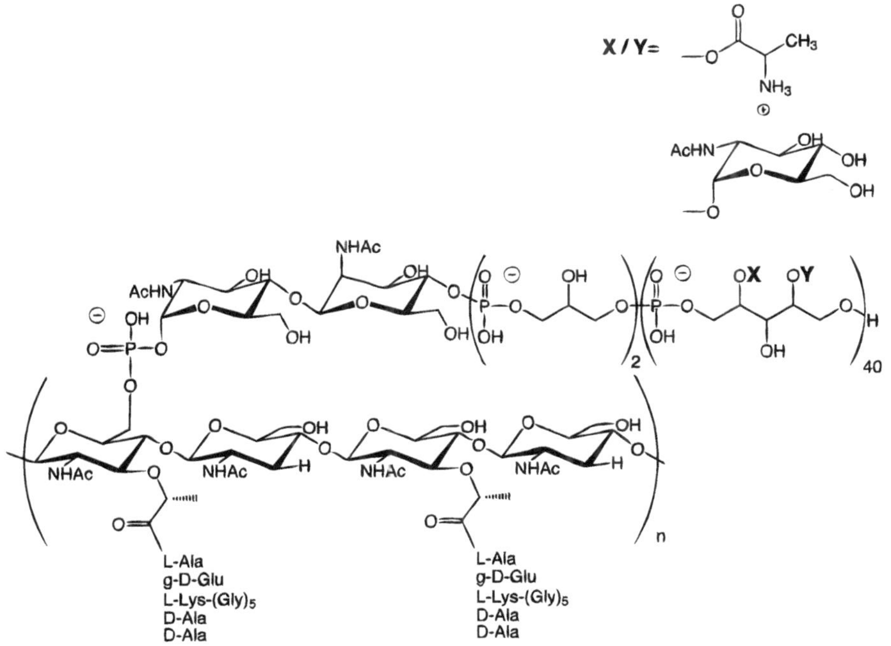

Figure 9.4 Chemical structure of WTA from S. aureus. A linkage unit consisting of ManNAc-β-(1,4)-GlcNAc-$(GroP)_2$ is attached by a phosphodiester to the C-6 hydroxyl of N-acetyl muramic acid sugars of peptidoglycan. In *S. aureus*, the substitution of C-2 and C-3 hydroxyls may be D-alanine or N-acetylglucosamine (Brown *et al.*, 2008).

the D-alanine carrier protein ligase (*dltA*) result in a complete loss of alanine esters in teichoic acids and *dltA*-mutants have been generated in a number of bacteria, including *S. aureus* (Gross *et al.*, 2001), group A streptococci (Kristian *et al.*, 2005), group B streptococci (Henneke *et al.*, 2005), *S. pneumoniae* (Kovacs *et al.*, 2006), *S. gordonii* (Clemans *et al.*, 1999), *E. faecalis* (Fabretti *et al.*, 2006), *Lactococcus reuteri* (Walter *et al.*, 2007) and *L. monocytogenes* (Abachin *et al.*, 2002).

The esterification of D-alanine to teichoic acid has pleiotropic effects including effects on regulation of the activity of autolytic enzymes (Peschel *et al.*, 2000; Wecke *et al.*, 1996), binding of Mg^{2+}, and Ca^{2+} to the cell envelope (Heptinstall *et al.*, 1970; Lambert *et al.*, 1975; Rose and Hogg, 1995), resistance to many antimicrobial cationic peptides/proteins (Fabretti *et al.*, 2006; Koprivnjak *et al.*, 2002; Peschel *et al.*, 1999), virulence (Collins *et al.*, 2002), and increased the proinflammatory activity of LTA (Deininger *et al.*, 2003). In addition, D-alanylation of teichoic acid also promotes coaggregation, biofilm formation and adhesion of bacteria on artificial surfaces (Abachin *et al.*, 2002; Clemans *et al.*, 1999; Fabretti *et al.*, 2006; Gross *et al.*, 2001). The critical role of D-alanylation was first shown in *S. aureus* by Gross and coworkers (Gross *et al.*, 2001). Despite expression of wild-type levels of PIA/PNAG, their *dtlA* mutant was highly compromised in biofilm formation on plastic and glass surfaces (Gross *et al.*, 2001). Even after short contact with plastic surfaces the bacterial density of the *dltA* mutant was reduced, indicating that – in contrast to PIA/PNAG – teichoic acid alanylation is involved in the initial attachment of the bacteria to the substratum. Since D-alanine residues are the major determinant of the net charge of teichoic acid, the defect in attachment is likely to be due to increased ionic repulsion of the bacterial cell wall and abiotic surfaces (Figs 9.3 and 9.4).

A *dltA* mutant in *E. faecalis,* another prominent pathogen causing biofilm infections (Fabretti and Huebner, 2005), was significantly reduced in its capacity to produce biofilm, confirming the observations made in *S. aureus* (Fabretti *et al.*, 2006). In the study by Fabretti and coworkers, secondary effects of the deletion *dtlA* on the structure of LTA were excluded by NMR-analysis. The *dltA* mutant in *E. faecalis* was also impaired in its ability to adhere to HEp-2 cells, an epithelial cell line (Fabretti *et al.*, 2006). In *Lactobacillus reuteri* it was shown that alanine-substitution of LTA is also required for the colonization of the upper gastrointestinal tract of rodents, a process that displays typical characteristics of a biofilm infection (Walter *et al.*, 2007). Decreased adhesion to various eukaryotic cell lines leading to a severe impairment in causing invasive infection has also been described for a $DltA^-$ mutant in *L. monocytogenes* and in group A streptococci (Abachin *et al.*, 2002; Kristian *et al.*, 2005).

Taken together, these results indicate that net charge of teichoic acid is critical not only for biofilm formation on plastic surfaces but also plays a role in biofilms in the gastrointestinal tract and adhesion to target cells of facultative intracellular microorganisms. In oral streptococci, D-alanylation of LTA is also important for bacterial coaggregation. This process is mediated by interaction of LTA with a 100-kDa adhesin and is dependant on the presence of divalent cations (Clemans *et al.*, 1999). Baddiley proposed in this regard that LTA lacking alanine complexes divalent cations, thus preventing interaction with anionic substrates from other bacteria (Baddiley, 2000). Interestingly, the transcription of the *dlt*-operon is regulated by cations such as Mg^{2+} and Ca^{2+}, which repress dlt expression (Koprivnjak *et al.*, 2006). As a result, the equilibrium between alanine esters and environmental Mg^{2+} and Ca^{2+} is highly regulated, ensuring optimal concentration of both components to mediate coaggregation and adherence.

Direct confirmation of the role of LTA in biofilm formation by mutants devoid of this glycoconjugate is more difficult to obtain since LTA is essential for viability (Grundling and Schneewind, 2007b; Jerga *et al.*, 2007). Depletion of the enzyme LtaS, which polymerizes polyglycerolphosphate backbone, causes growth arrest in *S. aureus* (Grundling and Schneewind, 2007b). Mutations in the diacylglycerol kinases, DagK, an other enzyme involved in LTA synthesis, are also lethal (Jerga *et al.*, 2007). Indirect evidence, however, comes from mutants of genes involved in the synthesis of membrane glycolipids, which represent also the substrates for glycolipid anchor of LTA (Fedtke *et al.*, 2007; Grundling and Schneewind, 2007a; Kiriukhin *et al.*, 2001; Lazarevic *et al.*, 2005). In a genomic screening for genes involved in biofilm synthesis in *Bacillus subtilis,* a gene with sequence similarity to an α-phosphoglucomutase, which catalyses the synthesis of UDP-glucose was found (Branda *et al.*, 2004). UDP-glucose is a substrate for the synthesis of glycolipids, the glycosylation of the major wall teichoic acid and the synthesis of the minor wall teichoic acid in *B. subtilis* (Lazarevic *et al.*, 2005; Pooley *et al.*, 1987). Mutants which cannot produce this important precursor for teichoic acid synthesis are unable to produce biofilms although the exact mechanism for this phenotype remains to be determined (Lazarevic *et al.*, 2005). Mutations of the glycosyltransferase YpfP, an enzyme that adds consecutively two glucose residues onto diacylglyerol (DAG), have a variable effect on the biofilm phenotype depending on Gram-positive pathogen and strain background used (Fedtke *et al.*, 2007; Kiriukhin *et al.*, 2001; Lazarevic *et al.*, 2005). YpfP mutants can no longer produce glycolipids and the glycolipid anchor of LTA is replaced by DAG (Fedtke *et al.*, 2007; Grundling and Schneewind, 2007a). In *Bacillus subtilis,* the absence of YpfP has no effect on the LTA content in the cell wall and on the ability to form biofilm (Lazarevic *et al.*, 2005). In *S. aureus* RN4220, a poor biofilm producer, the inactivation of *ypfP* increases LTA turnover and its concentration in the cell wall (Kiriukhin *et al.*, 2001). In *S. aureus* SA113, however, deletion of *ypfP* results in reduced amounts of cellular LTA, increased hydrophobicity and a highly impaired ability to form biofilms on plastic and glass surfaces (Fedtke *et al.*, 2007). The authors of the latter study concluded that the amount of LTA retained in the cell wall correlates directly with the ability to form biofilm.

Interestingly, a different enzyme is responsible for the synthesis of diglucosyldiacylglycerol in *S. agalaticae* and *E. faecalis* (Doran *et al.*, 2005) Theilacker *et al.*, 2009). In these bacteria, a glycosyltransferase utilizes monoglycosyldiacylglycerol as a substrate to form kojibiosediacylglyerol (Edman *et al.*, 2003). Inactivation of IagA, the corresponding glycosyltransferase in group B streptococci, results in the inability to invade the blood–brain barrier (Doran *et al.*, 2005). In *E. faecalis,* a gene with high similarity to *iagA* is also required for the synthesis of diglucosyldiacylglycerol and its inactivation results in a biofilm-negative phenotype (Theilacker *et al.*, 2009). In contrast to *S. aureus* SA 113, the reduced ability of the diglucosyldiacylglycerol-deficient *E. faecalis* mutant to form biofilm is independent of the cellular LTA content. The results by our group may indicate that an other, yet undefined mechanism is involved in the defects in adhesion of these mutants. In this context it should be kept in mind that extracellular LTA has been also suggested to play a role in biofilm detachment although its function in this process has so far been incompletely defined (Neu, 1996). Increased turnover and shedding of LTA into the biofilm matrix could therefore cause detachment of the resident cells and biofilm dispersal (Neu, 1996).

Wall teichoic acid

Recently, also mutants in the biosynthetic pathway of wall teichoic acid became available, (Vergara-Irigaray *et al.*, 2008; Weidenmaier *et al.*, 2004). TagO is an enzyme that synthesizes the transfer of N-acetylglucosamine to bactoprenol, the first step in WTA synthesis (Brown *et al.*, 2008). *tagO* inactivation in *S. aureus* causes a complete loss of WTA without altered viability of bacterial cells (Weidenmaier *et al.*, 2004). The *S. aureus tagO* deletion mutant adhered significantly less to human nasal epithelial cells and was rapidly eliminated from the nares of cotton rats (Weidenmaier *et al.*, 2004). These experiments suggest that WTA is critical for nasal colonization, which has been shown to be the reservoir for systemic *S. aureus* infections (von Eiff *et al.*, 2001). Using the same mutant, Weidenmaier and coworkers also showed a reduced adherence to human umbilical vein endothelial cells and reduced bacterial counts in a model of rat endocarditis (Weidenmaier *et al.*, 2005). To date, the eukaryotic receptor which binds WTA is still unknown. In addition, *tagO* mutants do not elaborate biofilms on polystyrene or glass, indicating that WTA also seems to be involved in the primary attachment and biofilm formation (Vergara-Irigaray *et al.*, 2008). The data on the role of WTA in adhesion to matrix proteins, on the contrary, are somewhat conflicting (Hussain *et al.*, 2001; Weidenmaier *et al.*, 2005). Adherence of the *S. aureus tagO* deletion mutant of Weidenmaier and coworkers to fibrinogen or fibronectin was not impaired (Weidenmaier *et al.*, 2005). On the contrary, WTA from *S. epidermidis* binds to fibronectin and preincubation with purified WTA promotes the binding of *S. epidermidis* to immobilized fibronectin, suggesting a bridging mechanism for WTA (Hussain *et al.*, 2001). Further studies using WTA-deficient mutants will be needed to clarify these findings.

Interestingly, a teichoic acid which is structurally identical to WTA is found abundantly in the biofilm matrix of both *S. aureus* and *S. epidermidis*, including strains isolated from foreign body-associated biofilm infections in humans (Hussain *et al.*, 1992; Sadovskaya *et al.*, 2006; Sadovskaya *et al.*, 2005; Sadovskaya *et al.*, 2004; Vinogradov *et al.*, 2006). In contrast to PIA/PNAG, which may not be expressed by some biofilm- positive strains, extracellular teichoic acid is found on all staphylococcal biofilms (Hussain *et al.*, 1992; Sadovskaya *et al.*, 2006; Sadovskaya *et al.*, 2005). Little is known about the function of extracellular teichoic acid in staphylococcal biofilms. Experiments using dispersin B, an enzyme which specifically cleaves PIA/PNAG, suggest that teichoic acids may be dispensable for biofilm formation (Chaignon *et al.*, 2007). Considering, however, that bacteria commit considerable energy into its biosynthesis it seems unlikely that extracellular teichoic acid does not have any function in staphylococcal biofilms. Further research is needed to clarify the role of this polysaccharide in the biofilm matrix of staphylococci.

Deoxyribonucleic acid (DNA)

It was first noted in *P. aeruginosa* that the abundant extracellular DNA found in biofilm matrix is not simply a 'waste product' of cellular turnover within multicellular communities, but is in fact an important factor in biofilm accumulation and maturation (Allesen-Holm *et al.*, 2006; Wicken and Baddiley, 1963). Whitchurch demonstrated elegantly the importance of extracellular DNA by using DNase I to inhibit *P. aeruginosa* biofilm formation and destabilize an established biofilm (Whitchurch *et al.*, 2002).

More recently, it has been also shown that treatment of wild-type *S. aureus* biofilm with DNase resulted in its destabilization (Rice *et al.*, 2007). Inactivation of *cidA*, a gene inducing murein hydrolysis and cell lysis, leads to a fivefold decrease in extracellular DNA and pronounced impairment of the ability to form biofilm (Bayles, 2007; Rice *et al.*, 2007). Biofilms formed by *cidA* mutants are also less adherent, and in strong contrast to wild-type biofilms are insensitive to treatment with DNAse I. Nevertheless, only 1% of bacteria undergo cell lysis in wild-type *S. aureus* biofilms, indicating that lysis of only a fraction of the population still provides enough genomic DNA for biofilm stability.

These recent findings also shed a different light on role of autolysins in biofilms formation. More than a decade ago, Heilmann *et al.* demonstrated that the autolysin AltE is involved in primary attachment of *S. epidermidis* (Heilmann *et al.*, 1997). In this initial report, a direct physical interaction between AltE and abiotic surfaces and matrix proteins was suggested. More recent work, however, demonstrates that early biofilms of *S. epidermidis* are also sensitive to DNAseI and biofilms of *altE* mutants are highly depleted in extracellular DNA. Therefore, the reduced release of bacterial genomic DNA as a result of the impaired cell lysis, seems to be responsible for the biofilm-negative phenotype of this mutant (Qin *et al.*, 2007). This hypothesis is confirmed by inactivation of a gene encoding an Atl homologue in *S. mutans* (Ahn and Burne, 2006; Shibata *et al.*, 2005). Autolysins have also been associated with biofilm formation in other bacteria. GelE, a serine protease of *E. faecalis*, has been shown to be involved in biofilm formation and primary attachment to plastic surfaces (Hancock and Perego, 2004; Mohamed *et al.*, 2004). Re-evaluation of *gelE* mutants shows that this protease is involved in the regulation of autolysis and release of high molecular weight extracellular DNA (Thomas *et al.*, 2008). It is tempting to speculate that other autolysins involved in primary attachment like Atn of *E. faecalis* and LytB and LytC of *S. pneumoniae* might also mediate biofilm formation by supplying biofilm matrix with extracellular DNA (Mohamed *et al.*, 2004; Moscoso *et al.*, 2006). In summary, extracellular DNA seems to be an important mechanism in biofilm formation and primary attachment in many Gram-positve organisms.

Conclusions

Biofilm infections caused by staphlyococci and enterococci are a major health care concern. Biofilms efficiently protect bacteria from the hostile environment in the human host. Highly charged polymers in biofilm matrix such as the exopolysaccharide PIA/PNAG and extracellular DNA as well as the cell wall polyanions LTA and WTA are involved in the formation of biofilms by these Gram-positive bacteria. The concentrations needed to kill biofilm-encased bacteria are rarely achieved by conventional antibiotic therapy alone. Hence, new therapeutics are in great demand for biofilm infections. Improved artificial surfaces of implant devices that will interact less with the matrix or cell wall carbohydrates could be a promising strategy. Enzymes that degrade biofilm polymers may be an other approach. Dispersin B, an enzyme which specifically cleaves PIA/PNAG, is an example of such a biofilm-specific antimicrobial (Kaplan *et al.*, 2004). Antibiotics interfering with the biosynthesis and the regulation biofilm-associated polymers may represent another means to tackle biofilm infections. With LtaS, DltA, TagO or CidA first targets for biofilm-specific antibiotics have been identifed. Nevertheless, more work will be needed to understand the complex interactions between various biopolymers in the biofilm matrix. Only then we shall be able to develop more specific ways to overcome bacterial resistance in biofilm infections.

Key points

- In recent years, biofilm infections by Gram-positive pathogens such as staphylococci and enterocci are found more frequently in nosocomial infections. This development reflects the increasing use of foreign bodies in health care.
- PIA/PNAG is the major polysaccharide antigen found in the slime of staphylococcal biofilms. It is a β-1,6-linked *N*-acetylglucosamine that embeds bacterial cells, protects against the host immune system, and mediates biofilm accumulation on plastic surfaces.
- Lipoteichoic acid is a polyanionic glycoconjugate anchored by its glycolipid moiety in the bacterial cell membrane. Alanine esterification of LTA is critical for biofilm formation on plastic surfaces as well on host epithelial cells.
- Wall teichoic acid is another polyanionic cell wall associated carbohydrate involved in biofilm formation and colonization of the nares

by staphylococci.

- Extracellular teichoic acid is also an abundant component of the matrix of staphylococcal biofilms although its function remains to be determined.
- Biofilm matrix is rich in DNA and recent evidence demonstrates that the release of genomic bacterial DNA is highly regulated and essential for biofilm formation.

References

Abachin, E., Poyart, C., Pellegrini, E., Milohanic, E., Fiedler, F., Berche, P., and Trieu-Cuot, P. (2002). Formation of D-alanyl-lipoteichoic acid is required for adhesion and virulence of *Listeria monocytogenes*. Mol. Microbiol. *43*, 11–14.

Ahn, S.J., and Burne, R.A. (2006). The atlA operon of *Streptococcus mutans*: role in autolysin maturation and cell surface biogenesis. J. Bacteriol. *188*, 6877–6888.

Allesen-Holm, M., Barken, K.B., Yang, L., Klausen, M., Webb, J.S., Kjelleberg, S., Molin, S., Givskov, M., and Tolker-Nielsen, T. (2006). A characterization of DNA release in *Pseudomonas aeruginosa* cultures and biofilms. Mol. Microbiol. *59*, 1114–1128.

Araki, Y., and Ito, E. (1989). Linkage units in cell walls of gram-positive bacteria. Crit Rev Microbiol. *17*, 121–135.

Baddiley, J. (2000). Teichoic acids in bacterial coaggregation. Microbiology *146 (Pt 6)*, 1257–1258.

Bayles, K.W. (2007). The biological role of death and lysis in biofilm development. Nat. Rev. Microbiol. *5*, 721–726.

Beachey, E.H., and Ofek, I. (1976). Epithelial cell binding of group A streptococci by lipoteichoic acid on fimbriae denuded of M protein. J. Exp. Med. *143*, 759–771.

Bochud, P.Y., Glauser, M.P., and Calandra, T. (2001). Antibiotics in sepsis. Intensive Care Med. 27 Suppl. 1, S33–48.

Bos, R., van der Mei, H.C., and Busscher, H.J. (1999). Physico-chemistry of initial microbial adhesive interactions – its mechanisms and methods for study. FEMS Microbiol. Rev 23, 179–230.

Branda, S.S., Gonzalez-Pastor, J.E., Dervyn, E., Ehrlich, S.D., Losick, R., and Kolter, R. (2004). Genes involved in formation of structured multicellular communities by *Bacillus subtilis*. J. Bacteriol. *186*, 3970–3979.

Brown, S., Zhang, Y.H., and Walker, S. (2008). A revised pathway proposed for *Staphylococcus aureus* wall teichoic acid biosynthesis based on *in vitro* reconstitution of the intracellular steps. Chem. Biol. *15*, 12–21.

Cerca, N., Jefferson, K.K., Maira-Litran, T., Pier, D.B., Kelly-Quintos, C., Goldmann, D.A., Azeredo, J., and Pier, G.B. (2007a). Molecular basis for preferential protective efficacy of antibodies directed to the poorly acetylated form of staphylococcal poly-N-acetyl-beta-(1–6)-glucosamine. Infect. Immun. *75*, 3406–3413.

Cerca, N., Jefferson, K.K., Oliveira, R., Pier, G.B., and Azeredo, J. (2006). Comparative antibody-mediated phagocytosis of *Staphylococcus epidermidis* cells grown in a biofilm or in the planktonic state. Infect. Immun. *74*, 4849–4855.

Cerca, N., Maira-Litran, T., Jefferson, K.K., Grout, M., Goldmann, D.A., and Pier, G.B. (2007b). Protection against *Escherichia coli* infection by antibody to the *Staphylococcus aureus* poly-N-acetylglucosamine surface polysaccharide. Proc. Natl. Acad. Sci. USA *104*, 7528–7533.

Chaignon, P., Sadovskaya, I., Ragunah, C., Ramasubbu, N., Kaplan, J.B., and Jabbouri, S. (2007). Susceptibility of staphylococcal biofilms to enzymatic treatments depends on their chemical composition. Appl. Microbiol. Biotechnol. *75*, 125–132.

Clemans, D.L., Kolenbrander, P.E., Debabov, D.V., Zhang, Q., Lunsford, R.D., Sakone, H., Whittaker, C.J., Heaton, M.P., and Neuhaus, F.C. (1999). Insertional inactivation of genes responsible for the D-alanylation of lipoteichoic acid in *Streptococcus gordonii* DL1 (Challis) affects intrageneric coaggregations. Infect. Immun. *67*, 2464–2474.

Collins, L.V., Kristian, S.A., Weidenmaier, C., Faigle, M., Van Kessel, K.P., Van Strijp, J.A., Gotz, F., Neumeister, B., and Peschel, A. (2002). *Staphylococcus aureus* strains lacking D-alanine modifications of teichoic acids are highly susceptible to human neutrophil killing and are virulence attenuated in mice. J. Infect Dis *186*, 214–219.

Costerton, J.W., Stewart, P.S., and Greenberg, E.P. (1999). Bacterial biofilms: a common cause of persistent infections. Science *284*, 1318–1322.

Courtney, H.S., Ofek, I., Simpson, W.A., Hasty, D.L., and Beachey, E.H. (1986). Binding of *Streptococcus pyogenes* to soluble and insoluble fibronectin. Infect. Immun. *53*, 454–459.

Cox, F., Taylor, L., Eskew, E.K., and Mattingly, S.J. (1993). Prevention of group B streptococcal colonization and bacteremia in neonatal mice with topical vaginal inhibitors. J. Infect Dis *167*, 1118–1122.

Dale, J.B., Baird, R.W., Courtney, H.S., Hasty, D.L., and Bronze, M.S. (1994). Passive protection of mice against group A streptococcal pharyngeal infection by lipoteichoic acid. J. Infect Dis *169*, 319–323.

Deininger, S., Stadelmaier, A., von Aulock, S., Morath, S., Schmidt, R.R., and Hartung, T. (2003). Definition of structural prerequisites for lipoteichoic acid-inducible cytokine induction by synthetic derivatives. J. Immunol. *170*, 4134–4138.

Donlan, R.M., and Costerton, J.W. (2002). Biofilms: survival mechanisms of clinically relevant microorganisms. Clin. Microbiol. Rev *15*, 167–193.

Doran, K.S., Engelson, E.J., Khosravi, A., Maisey, H.C., Fedtke, I., Equils, O., Michelsen, K.S., Arditi, M., Peschel, A., and Nizet, V. (2005). Blood–brain barrier invasion by group B Streptococcus depends upon proper cell-surface anchoring of lipoteichoic acid. J. Clin. Invest. *115*, 2499–2507.

Edman, M., Berg, S., Storm, P., Wikstrom, M., Vikstrom, S., Ohman, A., and Wieslander, A. (2003). Structural features of glycosyltransferases synthesizing major bilayer and nonbilayer-prone membrane lipids in *Acholeplasma laidlawii* and *Streptococcus pneumoniae*. J. Biol. Chem. *278*, 8420–8428.

Fabretti, F., and Huebner, J. (2005). Implant infections due to enterococci: role of capsular polysaccharides and biofilm. Int. J. Artif. Organs *28*, 1079–1090.

Fabretti, F., Theilacker, C., Baldassarri, L., Kaczynski, Z., Kropec, A., Holst, O., and Huebner, J. (2006). Alanine esters of enterococcal lipoteichoic acid play a role in biofilm formation and resistance to antimicrobial peptides. Infect. Immun. *74*, 4164–4171.

Fedtke, I., Mader, D., Kohler, T., Moll, H., Nicholson, G., Biswas, R., Henseler, K., Gotz, F., Zahringer, U., and Peschel, A. (2007). A *Staphylococcus aureus ypfP* mutant with strongly reduced lipoteichoic acid (LTA) content: LTA governs bacterial surface properties and autolysin activity. Mol. Microbiol. *65*, 1078–1091.

Fischer, W. (1990). Bacterial phosphoglycolipids and lipoteichoic acids. In Handbook of Lipid Research, D.J. Hanahan, ed. (New York, Plenum Press), pp. 123–234.

Fischer, W. (1994). Lipoteichoic acid and lipids in the membrane of *Staphylococcus aureus*. Med Microbiol. Immunol. *183*, 61–76.

Fischer, W., Behr, T., Hartmann, R., Peter-Katalinic, J., and Egge, H. (1993). Teichoic acid and lipoteichoic acid of *Streptococcus pneumoniae* possess identical chain structures. A reinvestigation of teichoid acid (C polysaccharide). Eur. J. Biochem. *215*, 851–857.

Fowler, V.G., Jr., Olsen, M.K., Corey, G.R., Woods, C.W., Cabell, C.H., Reller, L.B., Cheng, A.C., Dudley, T., and Oddone, E.Z. (2003). Clinical identifiers of complicated *Staphylococcus aureus* bacteremia. Arch. Intern. Med. *163*, 2066–2072.

Frølund, B., Palmgren, R., Keiding, K., and Halkjær Nielsen, P. (1996). Extraction of extracellular polymers from activated sludge using a cation exchange resin. Water Res. *30*, 1749.

Gotz, F. (2002). *Staphylococcus* and biofilms. Mol. Microbiol. *43*, 1367–1378.

Green, R.J., Davies, M.C., Roberts, C.J., and Tendler, S.J. (1999). Competitive protein adsorption as observed by surface plasmon resonance. Biomaterials *20*, 385–391.

Gross, M., Cramton, S.E., Gotz, F., and Peschel, A. (2001). Key role of teichoic acid net charge in *Staphylococcus aureus* colonization of artificial surfaces. Infect. Immun. *69*, 3423–3426.

Grundling, A., and Schneewind, O. (2007a). Genes required for glycolipid synthesis and lipoteichoic acid anchoring in *Staphylococcus aureus*. J. Bacteriol. *189*, 2521–2530.

Grundling, A., and Schneewind, O. (2007b). Synthesis of glycerol phosphate lipoteichoic acid in *Staphylococcus aureus*. Proc. Natl. Acad. Sci. USA *104*, 8478–8483.

Haas, R., Koch, H.U., and Fischer, W. (1984). Alanyl turnover from lipoteichoic acid to teichoic acid in *Staphylococcus aureus*. FEMS Microbiol. Lett. *21*, 27–31.

Hall-Stoodley, L., Costerton, J.W., and Stoodley, P. (2004). Bacterial biofilms: from the natural environment to infectious diseases. Nat. Rev. Microbiol. 2, 95–108.

Hancock, L.E., and Perego, M. (2004). The *Enterococcus faecalis fsr* two-component system controls biofilm development through production of gelatinase. J. Bacteriol. *186*, 5629–5639.

Hasty, D.L., Ofek, I., Courtney, H.S., and Doyle, R.J. (1992). Multiple adhesins of streptococci. Infect. Immun. *60*, 2147–2152.

Heilmann, C., Gerke, C., Perdreau-Remington, F., and Gotz, F. (1996). Characterization of Tn917 insertion mutants of *Staphylococcus epidermidis* affected in biofilm formation. Infect. Immun. *64*, 277–282.

Heilmann, C., Hussain, M., Peters, G., and Gotz, F. (1997). Evidence for autolysin-mediated primary attachment of *Staphylococcus epidermidis* to a polystyrene surface. Mol. Microbiol. *24*, 1013–1024.

Henneke, P., Morath, S., Uematsu, S., Weichert, S., Pfitzenmaier, M., Takeuchi, O., Muller, A., Poyart, C., Akira, S., Berner, R., *et al.* (2005). Role of lipoteichoic acid in the phagocyte response to group B streptococcus. J. Immunol. *174*, 6449–6455.

Heptinstall, S., Archibald, A.R., and Baddiley, J. (1970). Teichoic acids and membrane function in bacteria. Nature *225*, 519–521.

Hussain, M., Hastings, J.G., and White, P.J. (1992). Comparison of cell-wall teichoic acid with high-molecular-weight extracellular slime material from *Staphylococcus epidermidis*. J. Med Microbiol. *37*, 368–375.

Hussain, M., Heilmann, C., Peters, G., and Herrmann, M. (2001). Teichoic acid enhances adhesion of *Staphylococcus epidermidis* to immobilized fibronectin. Microb. Pathog. *31*, 261–270.

Jefferson, K.K. (2004). What drives bacteria to produce a biofilm? FEMS Microbiol. Lett. *236*, 163–173.

Jefferson, K.K., and Cerca, N. (2006). Bacterial-bacterial cell interactions in biofilms: detection of polysaccharide intercellular adhesins by blotting and confocal microscopy. Methods Mol. Biol. *341*, 119–126.

Jerga, A., Lu, Y.J., Schujman, G.E., de Mendoza, D., and Rock, C.O. (2007). Identification of a soluble diacylglycerol kinase required for lipoteichoic acid production in *Bacillus subtilis*. J. Biol. Chem. *282*, 21738–21745.

John, M., Rohrig, H., Schmidt, J., Wieneke, U., and Schell, J. (1993). Rhizobium NodB protein involved in nodulation signal synthesis is a chitooligosaccharide deacetylase. Proc. Natl. Acad. Sci. USA *90*, 625–629.

Kaplan, J.B., Velliyagounder, K., Ragunath, C., Rohde, H., Mack, D., Knobloch, J.K., and Ramasubbu, N. (2004). Genes involved in the synthesis and degradation of matrix polysaccharide in *Actinobacillus actinomycetemcomitans* and *Actinobacillus pleuropneumoniae* biofilms. J. Bacteriol. *186*, 8213–8220.

Karchmer, A.W. (2000). Nosocomial bloodstream infections: organisms, risk factors, and implications. Clin. Infect Dis *31 Suppl. 4*, S139–143.

Kiriukhin, M.Y., Debabov, D.V., Shinabarger, D.L., and Neuhaus, F.C. (2001). Biosynthesis of the glycolipid anchor in lipoteichoic acid of *Staphylococcus aureus* RN4220: role of YpfP, the diglucosyldiacylglycerol synthase. J. Bacteriol. *183*, 3506–3514.

Kojima, Y., Tojo, M., Goldmann, D.A., Tosteson, T.D., and Pier, G.B. (1990). Antibody to the capsular polysaccharide/adhesin protects rabbits against catheter-related bacteremia due to coagulase-negative staphylococci. J. Infect Dis *162*, 435–441.

Koprivnjak, T., Mlakar, V., Swanson, L., Fournier, B., Peschel, A., and Weiss, J.P. (2006). Cation-induced transcriptional regulation of the *dlt* operon of *Staphylococcus aureus*. J. Bacteriol. *188*, 3622–3630

Koprivnjak, T., Peschel, A., Gelb, M.H., Liang, N.S., and Weiss, J.P. (2002). Role of charge properties of bacterial envelope in bactericidal action of human group IIA phospholipase A2 against *Staphylococcus aureus*. J. Biol. Chem. *277*, 47636–47644.

Kovacs, M., Halfmann, A., Fedtke, I., Heintz, M., Peschel, A., Vollmer, W., Hakenbeck, R., and Bruckner, R. (2006). A functional *dlt* operon, encoding proteins required for incorporation of D-alanine in teichoic acids in gram-positive bacteria, confers resistance to cationic antimicrobial peptides in *Streptococcus pneumoniae*. J. Bacteriol. *188*, 5797–5805.

Kristian, S.A., Datta, V., Weidenmaier, C., Kansal, R., Fedtke, I., Peschel, A., Gallo, R.L., and Nizet, V. (2005). D-alanylation of teichoic acids promotes group a streptococcus antimicrobial peptide resistance, neutrophil survival, and epithelial cell invasion. J. Bacteriol. *187*, 6719–6725.

Kropec, A., Maira-Litran, T., Jefferson, K.K., Grout, M., Cramton, S.E., Gotz, F., Goldmann, D.A., and Pier, G.B. (2005). Poly-N-acetylglucosamine production in *Staphylococcus aureus* is essential for virulence in murine models of systemic infection. Infect. Immun. *73*, 6868–6876.

Lambert, P.A., Hancock, I.C., and Baddiley, J. (1975). Influence of alanyl ester residues on the binding of magnesium ions to teichoic acids. Biochem J. *151*, 671–676.

Laue, H., Schenk, A., Li, H., Lambertsen, L., Neu, T.R., Molin, S., and Ullrich, M.S. (2006). Contribution of alginate and levan production to biofilm formation by *Pseudomonas syringae*. Microbiology *152*, 2909–2918.

Lazarevic, V., Soldo, B., Medico, N., Pooley, H., Bron, S., and Karamata, D. (2005). *Bacillus subtilis* alpha-phosphoglucomutase is required for normal cell morphology and biofilm formation. Appl. Environ. Microbiol. *71*, 39–45.

Leopold, K., and Fischer, W. (1991). Separation of the poly(glycerophosphate) lipoteichoic acids of Enterococcus faecalis Kiel *27738*, *Enterococcus hirae* ATCC 9790 and *Leuconostoc mesenteroides* DSM 20343 into molecular species by affinity chromatography on concanavalin A. Eur. J. Biochem. *196*, 475–482.

Lewis, K. (2001). Riddle of biofilm resistance. Antimicrob. Agents Chemother. *45*, 999–1007.

Ludwicka, A., Uhlenbruck, G., Peters, G., Seng, P.N., Gray, E.D., Jeljaszewicz, J., and Pulverer, G. (1984). Investigation on extracellular slime substance produced by *Staphylococcus epidermidis*. Zentralbl. Bakteriol. Mikrobiol. Hyg. [A] *258*, 256–267.

Ma, L., Jackson, K.D., Landry, R.M., Parsek, M.R., and Wozniak, D.J. (2006). Analysis of *Pseudomonas aeruginosa* conditional psl variants reveals roles for the psl polysaccharide in adhesion and maintaining biofilm structure postattachment. J. Bacteriol. *188*, 8213–8221.

Mack, D., Fischer, W., Krokotsch, A., Leopold, K., Hartmann, R., Egge, H., and Laufs, R. (1996). The intercellular adhesin involved in biofilm accumulation of *Staphylococcus epidermidis* is a linear beta-1,6-linked glucosaminoglycan: purification and structural analysis. J. Bacteriol. *178*, 175–183.

Mack, D., Nedelmann, M., Krokotsch, A., Schwarzkopf, A., Heesemann, J., and Laufs, R. (1994). Characterization of transposon mutants of biofilm-producing *Staphylococcus epidermidis* impaired in the accumulative phase of biofilm production: genetic identification of a hexosamine-containing polysaccharide intercellular adhesin. Infect. Immun. *62*, 3244–3253.

Maira-Litran, T., Kropec, A., Abeygunawardana, C., Joyce, J., Mark, G., 3rd, Goldmann, D.A., and Pier, G.B. (2002). Immunochemical properties of the staphylococcal poly-N-acetylglucosamine surface polysaccharide. Infect. Immun. *70*, 4433–4440.

Martin, G.S., Mannino, D.M., Eaton, S., and Moss, M. (2003). The epidemiology of sepsis in the United States from 1979 through 2000. N. Engl. J. Med. *348*, 1546–1554.

McKenney, D., Hubner, J., Muller, E., Wang, Y., Goldmann, D.A., and Pier, G.B. (1998). The ica locus of *Staphylococcus epidermidis* encodes production of the capsular polysaccharide/adhesin. Infect. Immun. *66*, 4711–4720.

McKenney, D., Pouliot, K.L., Wang, Y., Murthy, V., Ulrich, M., Doring, G., Lee, J.C., Goldmann, D.A., and Pier, G.B. (1999). Broadly protective vaccine for *Staphylococcus aureus* based on an *in vivo*-expressed antigen. Science *284*, 1523–1527.

Mohamed, J.A., Huang, W., Nallapareddy, S.R., Teng, F., and Murray, B.E. (2004). Influence of origin of isolates, especially endocarditis isolates, and various genes on biofilm formation by *Enterococcus faecalis*. Infect. Immun. *72*, 3658–3663.

Morath, S., Geyer, A., and Hartung, T. (2001). Structure-function relationship of cytokine induction by lipoteichoic acid from *Staphylococcus aureus*. J. Exp. Med. *193*, 393–397.

Moscoso, M., Garcia, E., and Lopez, R. (2006). Biofilm formation by *Streptococcus pneumoniae*: role of choline, extracellular DNA, and capsular polysaccharide in microbial accretion. J. Bacteriol. *188*, 7785–7795.

Nealon, T.J., and Mattingly, S.J. (1984). Role of cellular lipoteichoic acids in mediating adherence of serotype III strains of group B streptococci to human embryonic, fetal, and adult epithelial cells. Infect. Immun. *43*, 523–530.

Neu, T., Swerhone, G.D., and Lawrence, J.R. (2001). Assessment of lectin-binding analysis for *in situ* detection of glycoconjugates in biofilm systems. Microbiology *147*, 299–313.

Neu, T.R. (1996). Significance of bacterial surface-active compounds in interaction of bacteria with interfaces. Microbiol. Rev *60*, 151–166.

Neuhaus, F.C., and Baddiley, J. (2003). A continuum of anionic charge: structures and functions of D-alanyl-teichoic acids in gram-positive bacteria. Microbiol. Mol. Biol. Rev. *67*, 686–723.

O'Toole, G., Kaplan, H.B., and Kolter, R. (2000). Biofilm formation as microbial development. Annu. Rev. Microbiol. *54*, 49–79.

Peschel, A., Otto, M., Jack, R.W., Kalbacher, H., Jung, G., and Gotz, F. (1999). Inactivation of the dlt operon in *Staphylococcus aureus* confers sensitivity to defensins, protegrins, and other antimicrobial peptides. J. Biol. Chem. *274*, 8405–8410.

Peschel, A., Vuong, C., Otto, M., and Gotz, F. (2000). The D-alanine residues of *Staphylococcus aureus* teichoic acids alter the susceptibility to vancomycin and the activity of autolytic enzymes. Antimicrob Agents Chemother *44*, 2845–2847.

Pier, G.B., Boyer, D., Preston, M., Coleman, F.T., Llosa, N., Mueschenborn-Koglin, S., Theilacker, C., Goldenberg, H., Uchin, J., Priebe, G.P., *et al.* (2004). Human monoclonal antibodies to *Pseudomonas aeruginosa* alginate that protect against infection by both mucoid and nonmucoid strains. J. Immunol. *173*, 5671–5678.

Pooley, H.M., Paschoud, D., and Karamata, D. (1987). The *gtaB* marker in *Bacillus subtilis* 168 is associated with a deficiency in UDP glucose pyrophosphorylase. J. Gen. Microbiol. *133*, 3481–3493.

Qin, Z., Ou, Y., Yang, L., Zhu, Y., Tolker-Nielsen, T., Molin, S., and Qu, D. (2007). Role of autolysin-mediated DNA release in biofilm formation of *Staphylococcus epidermidis*. Microbiology *153*, 2083–2092.

Rice, K.C., Mann, E.E., Endres, J.L., Weiss, E.C., Cassat, J.E., Smeltzer, M.S., and Bayles, K.W. (2007). The cidA murein hydrolase regulator contributes to DNA release and biofilm development in *Staphylococcus aureus*. Proc. Natl. Acad. Sci. USA *104*, 8113–8118.

Rose, R.K., and Hogg, S.D. (1995). Competitive binding of calcium and magnesium to streptococcal lipoteichoic acid. Biochim. Biophys. Acta *1245*, 94–98.

Ryder, C., Byrd, M., and Wozniak, D.J. (2007). Role of polysaccharides in *Pseudomonas aeruginosa* biofilm development. Curr. Opin. Microbiol. *10*, 644–648.

Sadovskaya, I., Chaignon, P., Kogan, G., Chokr, A., Vinogradov, E., and Jabbouri, S. (2006). Carbohydrate-containing components of biofilms produced *in vitro* by some staphylococcal strains related to orthopaedic prosthesis infections. FEMS Immunol. Med. Microbiol. *47*, 75–82.

Sadovskaya, I., Vinogradov, E., Flahaut, S., Kogan, G., and Jabbouri, S. (2005). Extracellular carbohydrate-containing polymers of a model biofilm-producing strain, *Staphylococcus epidermidis* RP62A. Infect. Immun. *73*, 3007–3017.

Sadovskaya, I., Vinogradov, E., Li, J., and Jabbouri, S. (2004). Structural elucidation of the extracellular and cell-wall teichoic acids of *Staphylococcus epidermidis* RP62A, a reference biofilm-positive strain. Carbohydr. Res. *339*, 1467–1473.

Sandoe, J.A., Wysome, J., West, A.P., Heritage, J., and Wilcox, M.H. (2006). Measurement of ampicillin, vancomycin, linezolid and gentamicin activity against enterococcal biofilms. J. Antimicrob. Chemother. *57*, 767–770.

Shibata, Y., Kawada, M., Nakano, Y., Toyoshima, K., and Yamashita, Y. (2005). Identification and characterization of an autolysin-encoding gene of *Streptococcus mutans*. Infect. Immun. *73*, 3512–3520.

Simpson, W.A., and Beachey, E.H. (1983). Adherence of group A streptococci to fibronectin on oral epithelial cells. Infect. Immun. *39*, 275–279.

Solano, C., Garcia, B., Valle, J., Berasain, C., Ghigo, J.M., Gamazo, C., and Lasa, I. (2002). Genetic analysis of *Salmonella enteritidis* biofilm formation: critical role of cellulose. Mol. Microbiol. *43*, 793–808.

Stoodley, P., Sauer, K., Davies, D.G., and Costerton, J.W. (2002). Biofilms as complex differentiated communities. Annu. Rev. Microbiol. *56*, 187–209.

Sutherland, I. (2001a). Biofilm exopolysaccharides: a strong and sticky framework. Microbiology *147*, 3–9.

Sutherland, I.W. (2001b). The biofilm matrix – an immobilized but dynamic microbial environment. Trends Microbiol. *9*, 222–227.

Takeda, S., Pier, G.B., Kojima, Y., Tojo, M., Muller, E., Tosteson, T., and Goldmann, D.A. (1991). Protection against endocarditis due to *Staphylococcus epidermidis* by immunization with capsular polysaccharide/adhesin. Circulation *84*, 2539–2546.

Theilacker, C., Coleman, F.T., Mueschenborn, S., Llosa, N., Grout, M., and Pier, G.B. (2003). Construction and characterization of a *Pseudomonas aeruginosa* mucoid exopolysaccharide-alginate conjugate vaccine. Infect. Immun. *71*, 3875–3884.

Theilacker, C., Sanchez-Carballo, P., Toma, I., Fabretti, F., Sava, I., Kropec, A., Holst, O., and Huebner, J. (2009). Glycolipids are involved in biofilm accumulation and prolonged bacteremia in *Enterococcus faecalis*. Mol. Microbiol. *71*, 1055–1069.

Thomas, V.C., Thurlow, L.R., Boyle, D., and Hancock, L.E. (2008). Regulation of autolysis-dependent eDNA release by *Enterococcus faecalis* extracellular proteases influences biofilm development. J. Bacteriol. *190*, 5690–5698.

Tojo, M., Yamashita, N., Goldmann, D.A., and Pier, G.B. (1988). Isolation and characterization of a capsular polysaccharide adhesin from *Staphylococcus epidermidis*. J. Infect. Dis. *157*, 713–722.

van Loosdrecht, M.C., Lyklema, J., Norde, W., Schraa, G., and Zehnder, A.J. (1987a). Electrophoretic mobility and hydrophobicity as a measured to predict the initial steps of bacterial adhesion. Appl. Environ. Microbiol. *53*, 1898–1901.

van Loosdrecht, M.C., Lyklema, J., Norde, W., Schraa, G., and Zehnder, A.J. (1987b). The role of bacterial cell wall hydrophobicity in adhesion. Appl. Environ. Microbiol. *53*, 1893–1897.

Vaudaux, P., Pittet, D., Haeberli, A., Huggler, E., Nydegger, U.E., Lew, D.P., and Waldvogel, F.A. (1989). Host factors selectively increase staphylococcal adherence on inserted catheters: a role for fibronectin and fibrinogen or fibrin. J. Infect. Dis. *160*, 865–875.

Vergara-Irigaray, M., Maira-Litran, T., Merino, N., Pier, G.B., Penades, J.R., and Lasa, I. (2008). Wall teichoic acids are dispensable for anchoring the PNAG exopolysaccharide to the *Staphylococcus aureus* cell surface. Microbiology *154*, 865–877.

Vinogradov, E., Sadovskaya, I., Li, J., and Jabbouri, S. (2006). Structural elucidation of the extracellular and cell-wall teichoic acids of *Staphylococcus aureus* MN8m, a biofilm forming strain. Carbohydr. Res. *341*, 738–743.

von Eiff, C., Becker, K., Machka, K., Stammer, H., and Peters, G. (2001). Nasal carriage as a source of

Staphylococcus aureus bacteremia. Study Group. N. Engl. J. Med. *344*, 11–16.

Vuong, C., Kocianova, S., Voyich, J.M., Yao, Y., Fischer, E.R., DeLeo, F.R., and Otto, M. (2004a). A crucial role for exopolysaccharide modification in bacterial biofilm formation, immune evasion, and virulence. J. Biol. Chem. *279*, 54881–54886.

Vuong, C., Voyich, J.M., Fischer, E.R., Braughton, K.R., Whitney, A.R., DeLeo, F.R., and Otto, M. (2004b). Polysaccharide intercellular adhesin (PIA) protects *Staphylococcus epidermidis* against major components of the human innate immune system. Cell. Microbiol. *6*, 269–275.

Walter, J., Loach, D.M., Alqumber, M., Rockel, C., Hermann, C., Pfitzenmaier, M., and Tannock, G.W. (2007). D-alanyl ester depletion of teichoic acids in Lactobacillus reuteri 100–23 results in impaired colonization of the mouse gastrointestinal tract. Environ. Microbiol. *9*, 1750–1760.

Wecke, J., Perego, M., and Fischer, W. (1996). D-alanine deprivation of *Bacillus subtilis* teichoic acids is without effect on cell growth and morphology but affects the autolytic activity. Microb. Drug Resist. *2*, 123–129.

Weidenmaier, C., Kokai-Kun, J.F., Kristian, S.A., Chanturiya, T., Kalbacher, H., Gross, M., Nicholson, G., Neumeister, B., Mond, J.J., and Peschel, A. (2004). Role of teichoic acids in *Staphylococcus aureus* nasal colonization, a major risk factor in nosocomial infections. Nat. Med. *10*, 243–245.

Weidenmaier, C., and Peschel, A. (2008). Teichoic acids and related cell-wall glycopolymers in Gram-positive physiology and host interactions. Nat. Rev. Microbiol. *6*, 276–287.

Weidenmaier, C., Peschel, A., Xiong, Y.Q., Kristian, S.A., Dietz, K., Yeaman, M.R., and Bayer, A.S. (2005). Lack of wall teichoic acids in *Staphylococcus aureus* leads to reduced interactions with endothelial cells and to attenuated virulence in a rabbit model of endocarditis. J. Infect. Dis. *191*, 1771–1777.

Whitchurch, C.B., Tolker-Nielsen, T., Ragas, P.C., and Mattick, J.S. (2002). Extracellular DNA required for bacterial biofilm formation. Science *295*, 1487.

Wicken, A.J., and Baddiley, J. (1963). Structure of intracellular teichoic acids from group D streptococci. Biochem. J. *87*, 54–62.

Wisplinghoff, H., Seifert, H., Wenzel, R.P., and Edmond, M.B. (2003). Current trends in the epidemiology of nosocomial bloodstream infections in patients with hematological malignancies and solid neoplasms in hospitals in the United States. Clin. Infect. Dis. *36*, 1103–1110.

Yoshida, Y., Ganguly, S., Bush, C.A., and Cisar, J.O. (2006). Molecular basis of L-rhamnose branch formation in streptococcal coaggregation receptor polysaccharides. J. Bacteriol. *188*, 4125–4130.

Yoshida, Y., Yang, J., Peaker, P.E., Kato, H., Bush, C.A., and Cisar, J.O. (2008). Molecular and antigenic characterization of a *Streptococcus oralis* coaggregation receptor polysaccharide by carbohydrate engineering in *Streptococcus gordonii*. J. Biol. Chem. *283*, 12654–12664.

Zaidi, T., and Pier, G.B. (2008). Prophylactic and therapeutic efficacy in murine keratitis infection of a fully human IgG1 monoclonal antibody to *Pseudomonas aeruginosa* alginate. Infect. Immun. *76*, 472–04725.

Zhang, X., Bishop, P.L., and Kupferle, M.J. (1998). Measurement of polysaccharides and proteins in biofilm extracellular polymers. Water Sci. Technol. *37*, 345.

Zogaj, X., Nimtz, M., Rohde, M., Bokranz, W., and Romling, U. (2001). The multicellular morphotypes of *Salmonella typhimurium* and *Escherichia coli* produce cellulose as the second component of the extracellular matrix. Mol. Microbiol. *39*, 1452–1463.

Poly-*N*-acetyl-glucosamine as a Mediator of Bacterial Biofilm Formation

10

Kimberly K. Jefferson

Abstract

A number of evolutionarily disparate bacterial species synthesize and secrete a β-(1,6)-linked polymer of *N*-acetylglucosamine. This polysaccharide is known as polysaccharide intercellular adhesin (PIA), PNAG, or PGA depending on the species that produces it, but all are biochemically similar. The conservation of this polysaccharide is a tribute to its success as a virulence factor and it plays dual roles in this capacity. It is an essential scaffold in bacterial biofilms and also plays a role in immune evasion by interfering with opsonophagocytosis. It is loosely associated with the bacterial cell surface and is therefore not considered a capsular polysaccharide. In addition, PIA/PNAG/PGA is more structurally simplistic than most capsular polysaccharides. Nonetheless, the role of this polysaccharide in virulence in such species as *Staphylococcus aureus*, *S. epidermidis*, *Escherichia coli* and *Actinobacillus actinomycetemcomitans* is clear, and advances in our understanding of its synthesis and structure may lead to its use as a therapeutic or vaccine target.

Introduction

If it ain't broke, don't fix it.

Bert Lance, 1977

Genetic changes that ensure fitness gains and increased survival for an organism within its particular niche will result in the propagation of that change, and pathogenic bacterial species are constantly evolving and developing new ways to survive in the face of host immune defences and antimicrobial chemotherapeutics. However, if a virulence factor 'ain't broke' and is particularly effective, then it will tend to be conserved even amongst evolutionarily disparate bacterial species.

It has become increasingly recognized over the past two decades, that the biofilm mode of growth, a physiological state in which bacteria adhere to a solid surface and surround themselves with an exo-polymeric matrix, plays a key role in many types of infections. Many, if not all, commensal and pathogenic bacterial species exhibit some level of biofilm-forming capacity. Biofilm formation plays a particularly important role in medical device-related infections, and *Staphylococcus epidermidis*, the leading cause of catheter and artificial joint infections, has a strong biofilm-forming capacity. The major intercellular adhesin and exo-polymeric matrix component in *S. epidermidis* biofilms was originally referred to as slime associated antigen (SAA) (Ammendolia *et al.*, 1999; Konig *et al.*, 1999). SAA was purified and found to be composed mainly of N-acetylglucosamine (Baldassarri *et al.*, 1996). The chemical structure of SAA was later characterized by HPLC as a β-(1–6)-linked polymer of N-acetylglucosamine of which approximately 10% of the residues were deacetylated (Fig. 10.1) and was referred to as polysaccharide intercellular adhesin or PIA (Mack *et al.*, 1996). The polysaccharide is also synthesized by *S. aureus* but is referred to as poly-*N*-acetylglucosamine or PNAG in this species. PIA/PNAG has also been shown to play a key role in immune evasion (Vuong *et al.*, 2004a). The success and persistence of this

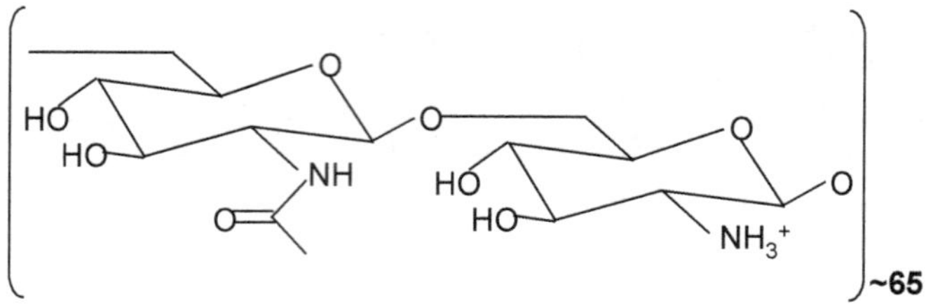

Figure 10.1 Structure of β-(1–6)-linked poly-N-acetylglucosamine. In the staphylococci, approximately 10% of the residues are deacetylated and consequently positively charged.

polysaccharide as a virulence factor has recently come to light as it has been shown to be produced by a number of evolutionarily disparate bacterial pathogens including *Escherichia coli*, *Actinobacillus actinomycetemcomitans*, and *A. pleuropneumoniae*, and it is suspected of being produced by *Yersinia pestis*, *Y. enterocolitica*, *Bordetella pertussis*, *B. parapertussis*, and *Pseudomonas fluorescens* (Hinnebusch and Erickson, 2008; Kaplan *et al.*, 2004b; Wang *et al.*, 2004). In Gram-negative organisms, the polysaccharide is referred to as poly-*N*-acetylglucosamine or PGA, but it should be noted that PGA can also denote poly-glutamic acid. PIA/PNAG/PGA plays multiple roles in virulence and these functions have justly awarded the polysaccharide considerable scientific interest in the past few years.

Role of PIA/PNAG in virulence

Biofilm formation

A biofilm is a community of microorganisms that adhere to one another and to a solid surface and surround themselves in an extracellular biopolymeric matrix. The biofilm phenotype confers a number of advantages upon the constituent bacteria. First, the metabolic potential of the biofilm population is greater than an equivalent planktonic population. The physical proximity of bacterial cells maximizes the activity of extracellular enzymes such as proteases. Furthermore, stratification of chemical conditions throughout the biofilm leads to physiological heterogeneity and a 'division of labour' between subpopulations of bacteria. In other words, bacteria located within the surface of the biofilm, which are frequently exposed to relatively high levels of nutrients and oxygen have a very different metabolic profile compared with bacteria within the deepest layers of the biofilm, which are exposed to lower pH, lower oxygen tension, and fewer nutrients. The bacteria at the bottom of the biofilm therefore need to be geared up to use different nutrients and the metabolic byproducts of their neighbours. The metabolic potential is also increased through the use of nutrients from dead bacterial cells by live cells in the vicinity.

The second advantage that the biofilm phenotype confers upon bacteria is an increase in resistance to all sorts of potentially harmful chemical and physical factors. Biofilms are more resistant to UV light, shear forces, pH changes, disinfectants, and nutrient depletion (Jefferson, 2004; Rupp *et al.*, 2005). They can also withstand concentrations of chemotherapeutic agents 10- to 1000-fold higher that the concentrations needed to kill or inhibit genetically identical planktonic populations of bacteria (Fux *et al.*, 2005; Monroe, 2007).

The formation of a biofilm takes place in stages. Bacteria must first adhere to a solid surface. In the human host, initial adherence of bacteria to host tissues or synthetic medical devices is often mediated by proteins on the bacterial cell surface that interact with extracellular matrix materials. Once the bacteria adhere, they begin to accumulate, through both cell division and adherence of new cells from the surrounding milieu. The accumulation phase of biofilm development is mediated by intercellular adhesins, which are often exo-polysaccharides. For the staphylococci, *E. coli*, and actinobacilli, the major intercellular adhesin that plays a role in this process is PIA/PNAG/PGA (Gotz, 2002; Wang *et al.*, 2004). The charge properties of this polysaccharide and its loose association with the bacterial cell surface allow it to mediate cell-to-cell interactions and to act as the scaffolding for the biofilm. The extracellular matrix material in biofilms frequently makes up a greater part of the total biofilm volume than bacterial cells themselves. Degradation of extracellular matrix generally leads to dispersal of the bacterial cells and dissolution of the biofilm. PIA/PNAG/PGA is sensitive to sodium metaperiodate and consequently, treatment with this chemical dissolves biofilms that utilize this polymer (Chaignon *et al.*, 2007). In addition, certain species synthesize enzymes that degrade the polysaccharide matrix of their own biofilms as a means of dispersing cells to colonize new sites. *A. actinomycetemcomitans* produces the enzyme dispersin B, which degrades PIA/PNAG/PGA and at high enough concentrations, leads to the deconstruction of biofilms (Kaplan *et al.*, 2004a,b).

Infections involving biofilms tend to be indolent, because the bacteria are growing slowly; difficult to treat, because bacteria within biofilms are resistant to antibiotics; and relapsing. Frequently, biofilm-related infections cannot be resolved with chemotherapeutic agents and removal of the infected device or bodily tissue is necessary. In the case of catheter-related infections, this is a straightforward task, but in the case of endocarditis or an artificial joint infection, this is a grave outcome.

Immune evasion

To further complicate therapeutic intervention into biofilm-mediated infections, biofilms are not only highly resistant to antibiotics, but to host immune defences as well. Neutrophils can migrate into biofilms but appear to be unable to ingest the bacteria (Leid *et al.*, 2002). In the staphylococci, PIA/PNAG contributes to resistance against neutrophil phagocytosis (Vuong *et al.*, 2004a,b). The polysaccharide is not generally considered a capsule because it is more loosely associated with the bacterial cell surface than a true capsular polysaccharide. Its anti-phagocytic properties probably stem from properties similar to that of a capsule but with important distinctions, as will be discussed below.

Approximately 10% of the residues in mature PIA/PNAG are de-acetylated and this property results in an overall positive charge. De-acetylation is necessary for adherence of PIA/PNAG to the bacterial surface, for intercellular adhesion, and for biofilm formation (Vuong *et al.*, 2004a). Interestingly, antibodies generated against chemically de-acetylated polysaccharide are more effective against antibody-mediated opsonophagocytosis than antibodies against native PIA/PNAG. This effect is likely due to a tighter association between the deacetylated residues and the cell surface. Antibodies against deacetylated polysaccharide may bind to cell-bound PIA/PNAG whereas antibodies against the acetylated polysaccharide may bind to residues that are dangling away from the bacterial surface (Cerca *et al.*, 2007a). This suggests that *in vivo* the polysaccharide elicits an ineffective immune response against the acetylated residues rather than an effective one. However, even antibodies against de-acetylated PIA/PNAG are not effective against biofilm-embedded bacteria (Cerca *et al.*, 2006). It has been postulated that due to the large quantity of PIA/PNAG within the biofilm, both cell associated and intercalated between cells, the antibodies may adhere to non-cell-associated polysaccharide where they will not be effective in eliciting opsonophagocytosis (Cerca *et al.*, 2006). This means that PIA/PNAG likely inhibits neutrophil killing by a different mechanism than capsular polysaccharides inasmuch as it may thwart anti-polysaccharide antibodies away from the cell surface as a result of its loose association rather than inhibiting engulfment directly.

Regulation of PIA/PNAG synthesis

The staphylococci

Unlike *S. aureus*, which boasts an impressive arsenal of virulence factors, coagulase negative staphylococci, including *S. epidermidis*, possess few. The major virulence factor in *S. epidermidis* is generally considered its ability to form a biofilm. *S. epidermidis* frequently forms a biofilm on inserted medical devices such as intravenous catheters and is consequently the most common cause of catheter-related bloodstream infections. *S. aureus* can form a biofilm on native and replacement heart valves, leading to endocarditis, as well as on prosthetic joints, devitalized bone, and other human tissues and synthetic medical devices. Biofilm formation in the staphylococci is usually dependent upon the production of PIA/PNAG, although under certain growth conditions, other biomolecules such as protein adhesins and extracellular DNA can hold the biofilm together.

In the staphylococci, PIA/PNAG is synthesized by four intercellular adhesin proteins: IcaA, IcaB, IcaC, and IcaD. IcaA is a membrane-spanning glucosyltransferase for which IcaD, appears to augment activity. IcaC, which is also a transmembrane protein, is thought to link short polymer chains together to form the long (approximately 65 residue) mature polysaccharide. Finally, IcaB is a secreted protein that de-acetylates approximately 10% of the residues (Gerke *et al.*, 1998; Vuong *et al.*, 2004a).

The intercellular adhesin genes, *icaADBC*, are co-transcribed. The genes were originally identified in *S. epidermidis* by screening a transposon insertion library for biofilm-negative mutants (Heilmann *et al.*, 1996) and were later found by homology in *S. aureus* (Cramton *et al.*, 1999).

Environmental factors that regulate PIA/PNAG synthesis

Expression of *icaADBC* is regulated externally, by environmental conditions, and internally by the transcription factors that respond to those conditions. It is not completely clear why the bacteria switch to the biofilm mode of growth; some studies suggest that nutrient deprivation induces the switch, others suggest the contrary, that nutrient replete conditions promote biofilm formation, and some studies suggest that conditions that induce a stress response result in the development of a biofilm (Jefferson, 2004).

In support of a role for nutrient availability in PIA/PNAG regulation, it has been well established that glucose augments biofilm formation, and yet, Dobinsky *et al* found that glucose actually results in decreased *icaADBC* expression (Dobinsky *et al.*, 2003). They speculated that the positive effect of glucose on PIA/PNAG production was independent of *icaADBC* expression so this may represent a post-transcriptional or post-translational regulatory mechanism, or glucose may act on another unidentified factor required for production of the polysaccharide.

Iron availability has also been shown to play a role in biofilm formation, but it is not yet clear whether iron depletion induces PIA/PNAG synthesis or whether it acts through another biofilm mediator (Johnson *et al.*, 2005, 2008). Iron deplete conditions could serve as a signal to the bacteria that they are *in vivo*, where the levels of free iron are extremely low.

A small number of clonal strains of *S. aureus* predominate cases of community-acquired infections and these strains have in common the presence of the arginine catabolite mobile element (ACME). ACME appears to have been recently acquired from *S. epidermidis* and is not present in most *S. aureus* lineages (Goering *et al.*, 2007). The role of ACME in virulence in unknown, but arginine appears to be required for biofilm formation and deletion of the arginine deiminase gene results in a decrease in PIA/PNAG elaboration (Zhu *et al.*, 2007).

Supporting the concept that biofilm formation in the staphylococci is induced by the stress response, high osmolarity and ethanol stimulate the expression of PIA/PNAG (Knobloch *et al.*, 2001). Subinhibitory doses of certain antibiotics can also turn up *icaADBC* expression further supporting the hypothesis that PIA/PNAG is induced by stress (Rachid *et al.*, 2000b). External stressors such as ethanol and high salt can repress tricarboxylic acid (TCA) cycle activity and it has been suggested that TCA stress mediates the induction of PIA/PNAG production. In support of this, the TCA cycle inhibitor fluorocitrate stimulates PIA/PNAG synthesis (Vuong *et al.*, 2005).

SigB

Some evidence suggests that stress-induced polysaccharide synthesis acts through the alternative sigma factor SigB; however, as the *icaADBC* promoter does not contain a SigB consensus binding site, it is likely that SigB-mediated effects are indirect. Screening of a library of *S. epidermidis* transposon mutants revealed that a mutation in the gene for the positive regulator of SigB, *rsbU*, resulted in a defect in biofilm formation (Knobloch *et al.*, 2001). Inactivation of *sigB* also leads to the loss of NaCl-induced biofilm formation in *S. aureus* (Rachid *et al.*, 2000a). There is evidence to suggest that repression of SigB activity may be natural regulatory mechanism to restrict PIA/PNAG production as well. In *S. epidermidis*, insertional inactivation of *rsbU* by the insertion sequence IS256 occurs spontaneously and results in reduced PIA/PNAG synthesis and biofilm formation (Conlon *et al.*, 2004).

SarA

The staphylococcal accessory regulator A, SarA, is a DNA-binding protein that influences the expression of secreted proteins and virulence factors (Chien *et al.*, 1999). The *sarA* locus contains three separate promoters that drive the expression of three overlapping transcripts, referred to as *sarA*, *sarB*, and *sarC* (Bayer *et al.*, 1996; Manna *et al.*, 1998). Transcription of *sarA* and *sarB* occurs during exponential growth whereas transcription of *sarC* occurs during post-exponential growth and is dependent upon SigB but it is not yet clear what role the alternative transcripts play (Bischoff *et al.*, 2001). SarA plays a positive role in transcription of *icaADBC* and is necessary for PIA/PNAG production in *S. aureus* (Beenken *et al.*, 2003; Valle *et al.*, 2003) and in *S. epidermidis* (Tormo *et al.*, 2005). It is possible that the positive role of SigB in *icaADBC* expression is related to it role in SarA transcription.

IcaR

Upstream from the *icaADBC* genes and transcribed divergently, is the TetR family transcriptional regulator IcaR. IcaR is a DNA-binding protein and a negative regulator of *icaADBC* transcription that appears to mediate its effect through binding to a site immediately upstream from the *icaA* start codon (Conlon *et al.*, 2002a; Conlon *et al.*, 2002b; Jefferson *et al.*, 2004). The role of SigB on ethanol-induced PIA/PNAG production is mediated through inhibition of IcaR (Knobloch *et al.*, 2004).

Spx

In *Bacillus subtilis*, Spx is involved in adaptation to changes in cellular redox status. It acts as a global regulator by interacting with the RNA polymerase complex (RNApol). Through this interaction Spx can act as a transcriptional activator or it can decrease transcriptional activity of RNApol by preventing its interaction with other, specific transcriptional activators (Nakano *et al.*, 2003; Zuber, 2004). Deletion of the *spx* homologue in *S. aureus* results in defects in adaptation to a variety of stress-inducing conditions (Pamp *et al.*, 2006). Spx is also a positive regulator of *icaR* and consequently a negative regulator of *icaADBC*. Thus, deletion of *spx* results in the suppression of *icaR* transcript levels and an increase in *icaADBC* transcript resulting in elevated PIA/PNAG production (Pamp *et al.*, 2006).

TcaR

The teicoplanin associated locus (Tca) was given its name after it was picked up in a mutagenesis screen for increased resistance to teicoplanin (Brandenberger *et al.*, 2000). The locus is part of the cell wall stress stimulon but its function has not yet been well characterized. The regulator of this locus, TcaR, was identified in a DNA affinity chromatography assay designed to isolate factors that bound to the genetic region upstream from the *ica* locus (Jefferson *et al.*, 2004). TcaR has a weak negative regulatory effect on expression of the *icaADBC* gene transcript and deletion of *tcaR* results in increased *icaADBC* expression.

CodY

CodY was originally identified in *B. subtilis* as a regulator that responds to nutrient-limited conditions and metabolic stress. A CodY homologue was identified in a transposition screen for *S. aureus* mutants with altered biofilm-forming ability (Tu Quoc *et al.*, 2007). Targeted deletion of CodY resulted in an increase in *icaADBC* transcript levels revealing the negative role for this regulator in PIA/PNAG expression although it is not yet known whether this is a direct or an indirect effect (Majerczyk *et al.*, 2008).

SrrB

Conditions of low oxygen induce biofilm formation in the staphylococci (Cramton *et al.*, 2001). The staphylococcal respiratory response regulator SrrAB mediates this effect (Ulrich *et al.*, 2007). SrrAB is a two-component system that is induced in the absence of oxygen and influences the expression of a number of virulence factors (Yarwood *et al.*, 2001). Phosphorylated SrrA binds to the *icaADBC* promoter and positively regulates its expression; consequently, deletion of *srrAB* prevents PIA/PNAG production (Ulrich *et al.*, 2007).

CcpA

As mentioned above, glucose induces PIA/PNAG production as does TCA cycle stress. This may be related to the activity of the catabolite control protein A (CcpA). CcpA regulates a number of genes in response to glucose so that glucose is utilized preferentially over other carbon sources (Jankovic and Bruckner, 2007). CcpA has been shown to positively regulate *icaADBC* expression, PIA/PNAG synthesis, and biofilm formation (Seidl *et al.*, 2008).

Quorum sensing

Bacteria constitutively secrete small chemical messengers referred to as quorum-sensing factors that activate signalling cascades only when the concentration of the factor reaches a threshold level. This will happen when the bacterial population reaches a certain density or when the bacteria are in a viscous environment that limits diffusion of the quorum-sensing factor (Redfield, 2002). The staphylococci are equipped with two quorum-sensing systems; the *agr* system and the *luxS* system. While the *agr* system is specific to the genus *Staphylococcus*, the *luxS* system is common to a number of bacterial species. The quorum-sensing factor in the *agr* system is an octapeptide, AIP, encoded within *agrD*. The *agrB* gene product is involved in post-translational modification and export of AIP. AgrC is a receptor for AIP that becomes auto-phosphorylated in response to AIP binding and transfers its phosphate group to AgrA, which then regulates the transcription of RNAIII. The RNAIII transcript, in conjunction with endoribonuclease III, directly regulates expression of certain virulence factors post-transcriptionally. The quorum-sensing factor in the LuxS system is AI-2. There are LuxS homologues found in both Gram-negative and Gram-positive bacteria, including *S. aureus* and *S. epidermidis*.

Biofilms represent a bacterial quorum and indeed, quorum-sensing appears to be active within biofilms (Yarwood *et al.*, 2004). However, both the *agr* system and the LuxS system negatively regulate the biofilm phenotype and mutations that disrupt the systems augment biofilm formation (Kong *et al.*, 2006; Vuong *et al.*, 2000, 2003; Xu *et al.*, 2006). Population critical mass, limited physical space, limited nutrients, and increased toxic by-products, which are all associated with high population density and lead to activation of quorum-sensing systems could serve as a feedback mechanism for preventing the biofilm from becoming too large or as a mechanism for dispersal. The *agr* system appears to exert its effect on surface-associated proteinaceous adhesins rather than on PIA/PNAG, whereas the LuxS system negatively regulates *icaADBC* and PIA/PNAG synthesis (Xu *et al.*, 2006; Vuong *et al.*, 2003; Vuong *et al.*, 2000).

c-di-GMP

Recently, 3′,5′-cyclic diguanylic acid (c-di-GMP) has been recognized as a bacterial second messenger with an important role in the regulation of certain virulence properties such as biofilm formation. In *S. aureus*, addition of chemically synthesized c-di-GMP inhibited biofilm formation *in vitro* and bacterial colonization in a mouse model of mastitis (Brouillette *et al.*, 2005; Karaolis *et al.*, 2005). The mechanism through which c-di-GMP inhibits biofilm formation has not yet been characterized but experiments showing that it also inhibits intercellular adhesion suggest that it exerts its activity through the inhibition of PIA/PNAG synthesis (Karaolis *et al.*, 2005). GGDEF domain-containing diguanylate cyclases are responsible for c-di-GMP synthesis. An effort to identify staphylococcal proteins that might play a role in c-di-GMP-mediated inhibition of biofilm formation utilized an *in silico* screen to find proteins with GGDEF domains (Holland *et al.*, 2008). Only one conserved domain protein, GdpS, was discovered. Interestingly, mutagenesis of *gdpS* reduced *icaADBC* expression and biofilm formation but this effect was not related to the GGDEF domain or to c-di-GMP synthesis, and diguanylate cyclase activity could not be demonstrated by GdpS *in vitro*. The authors suggested that the staphylococci have only a vestigial system for diguanylate cyclase activity, thus, the effects of c-di-GMP on biofilm formation may or may not be physiologically relevant (Holland *et al.*, 2008).

Phase variation in PIA/PNAG production

While most isolates of *S. aureus* contain the *ica* locus and produce at least low levels of PIA/PNAG, many isolates, especially commensal skin isolates, fail to harbour or express *icaADBC*. A recent study suggests that PIA/PNAG elaboration, while necessary for catheter-related infections, may actually impede skin colonization (Rogers *et al.*, 2008). Thus, while *S. aureus* has a complex regulatory network in play to fine-tune the synthesis of PIA/PNAG, *S. epidermidis* has designed ways to completely shut down expression of the *ica* locus. The insertion sequence IS256 is a transposable genetic element that is frequently found in multiple copies on the *S. epidermidis* chromosome. IS256 was found to insertionally inactivate either the *icaA* or *icaC* genes, effectively shutting down PIA/PNAG production (Ziebuhr *et al.*, 1999). This inactivation was reversible by clean excision of the element. The mechanism for IS256 transposition remains unclear and while there are clearly preferred insertion sites, there is no apparent sequence-specificity to insertion.

Escherichia coli

A genetic locus encoding proteins capable of synthesizing β-1,6-linked poly-N-acetylglucosamine was detected in *E. coli* through a search for genes homologous to the *icaADBC* genes of *S. aureus* and *S. epidermidis* (Wang *et al.*, 2004). The genes were designated *pgaA, B, C,* and *D*, and the polysaccharide is referred to as PGA in *E. coli*. PgaC is homologous to IcaA and possesses glucosyltransferase activity and along with PgaD, it synthesizes the polysaccharide (Fig. 10.2) (Itoh *et al.*, 2008). PgaA appears to form a pore in the outer membrane and is necessary for PGA export. PgaB is homologous to IcaB, deacetylates PGA, and is required for its export (Itoh *et al.*, 2008).

Similar to the staphylococci, PGA plays a key role in biofilm formation in *E. coli*. A role for PGA in immune evasion has not yet been described, but PGA is recognized by staphylococcal PIA/PNAG-specific antiserum, and the antiserum protects against *E. coli* in a mouse bacteraemia model (Cerca *et al.*, 2007b).

Environmental factors that affect PGA

Similar to the staphylococci, expression of *pgaA-BCD* and production of the polysaccharide is induced by stressful stimuli such as high osmolarity and ethanol (Cerca and Jefferson, 2008). Activity of the promoter is highest during post-exponential growth in contrast to the staphylococci, in which expression of *icaADBC* is greater during exponential growth (Cerca and Jefferson, 2008). This may be related to the fact that in *E. coli*, the LuxS quorum-sensing system plays a positive role in biofilm formation, rather than a negative role; although a role for PGA in the effect of quorum-sensing on biofilm formation has not yet been proven (Gonzalez Barrios *et al.*, 2006).

CsrA

Glucose is the preferred carbon source in *E. coli* and will induce carbon catabolite repression of genes involved in the utilization of alternative carbon sources. One regulator of this effect is CsrA. CsrA is an RNA-binding protein that regulates gene expression post-transcriptionally at the levels of transcript stability and translational initiation. The non-coding RNAs CsrB and CsrC counteract CsrA by serving as a sham substrate for its RNA-binding activity (Wang *et al.*, 2005). CsrA represses *pgaABCD* expression and PGA synthesis by binding to the *pgaABCD* transcript and destabilizing it as well as preventing

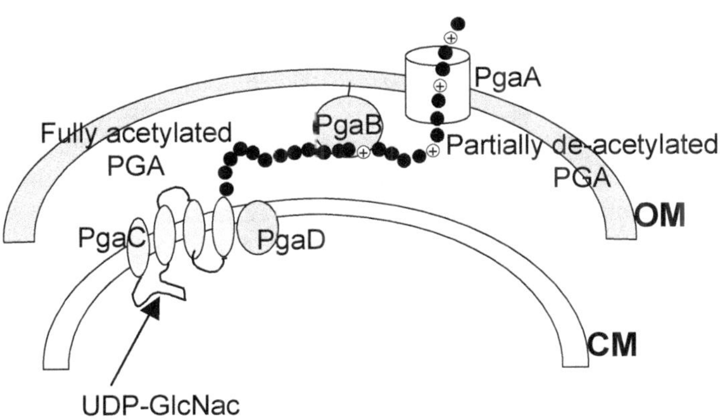

Figure 10.2 Synthesis of PGA in *E. coli*. PgaC is a glucosyltransferase with 4 or 5 transmembrane domains and is associated with the cytoplasmic membrane. PgaD appears to function as a chaperone and increases the glucosyltransferase activity of PgaC. PgaB is a deacetylase; deacetylation is necessary for export by the β-barrel porin protein PgaA.

translation (Wang *et al.*, 2005). Mutation of CsrA resulted in an impressive 3000-fold increase in the expression of a reporter fused to the *pgaABCD* promoter, suggesting that expression of the locus is constitutive in the absence of this repressor (Cerca and Jefferson, 2008).

NhaR

NhaR belongs to the LysR family of transcriptional regulators and regulates NhaA, a proton antiporter that regulates Na^+/H^+ transport under conditions of NaCl stress and pH extremes. Similar to the staphylococci, NaCl induces *pgaABCD* and PGA production, and NhaR has been shown to mediate this effect (Cerca and Jefferson, 2008; Goller *et al.*, 2006). NhaR is therefore a positive regulator of expression and deletion of *nhaR* results in a decrease in PGA synthesis and biofilm formation (Goller *et al.*, 2006).

Actinobacillus and *Aggregatibacter*

Shortly after the discovery of the function of the *pga* locus in *E. coli*, a homologous locus was found in the periodontal pathogen *Actinobacillus* (*Aggregatibacter*) *actinomycetemcomitans* and the respiratory pathogen *Actinobacillus pleuropneumoniae* (Kaplan *et al.*, 2004b). The molecule synthesized by the products of this locus was shown to be biochemically and immunogenically indistinguishable from staphylococcal PIA/PNAG and *E. coli* PGA (Kaplan *et al.*, 2004b).

In the actinobacilli, as in the staphylococci and *E. coli*, PGA plays a key role in intercellular adhesion and biofilm formation. In a screen for *A. actinomycetemcomitans* mutants that failed to disperse, or produce off-shoot colonies, a gene encoding a enzyme with glycohydrolase activity specific for β-(1–6)-linked N-acetylglucosamine was discovered (Kaplan *et al.*, 2004b). The enzyme was named dispersin B because of its ability to disperse the biofilms. Not unexpectedly, because of the biochemical similarity of the polysaccharides, dispersin B is also capable of degrading staphylococcal PIA/PNAG and subsequently dispersing their biofilms (Izano *et al.*, 2007; Kaplan *et al.*, 2004a; Manuel *et al.*, 2007). Owing to its enzymatic specificity and biofilm-degrading capacity, dispersin B has been proposed as a potential therapeutic agent. The interest in downstream applications of dispersin B has lead to its structural characterization by X-ray crystallography (Ramasubbu *et al.*, 2005). Crystallographic analysis revealed a $(\beta/\alpha)_8$ barrel structure characteristic of other β-hexosaminidases and suggested that dispersin B cleaves the β-(1,6)-linked *N*-acetylglucosamine polymer by a mechanism similar to that used by other hexosaminidases which cleave β-(1,4)-linked N-acetylglucosamine.

In addition to its role as an intercellular adhesin and scaffolding matrix for the biofilm, PGA plays an important role in immune evasion for the actinobacilli. A strain of *A. actinomycetemcomitans* was more resistant to killing by an immortalized macrophage cell line than its isogenic PGA mutant (Venketaraman *et al.*, 2008). The PGA-mediated resistance to killing did not depend upon the formation of a biofilm suggesting that the polysaccharide may perform a function similar to a capsular polysaccharide.

Very little is know about the regulation of PGA synthesis in actinobacilli but, similar to *S. epidermidis*, the actinobacilli have a mechanism for insertionally inactivating pga expression. The insertion sequence designated IS1070 was found in *A. pleuropnemoniae* where it has been found to inactivate the *pgaC* gene resulting in PGA-negative, biofilm-negative variants (Liu *et al.*, 2008).

Conclusions

PIA/PNAG plays an important role in virulence and biofilm formation in a number of pathogenic bacterial species. There are still a number of gaps in our knowledge about the synthesis and regulation of this polysaccharide. In particular, in the staphylococci, UDP-*N*-acetylglucosamine can serve as a substrate *in vitro* for the IcaA glucosyltransferase enzyme from *Staphylococcus*, but it is not known if this is the substrate within bacterial cells. A question stemming from this is whether or not PIA/PNAG synthesis could shuttle *N*-acetylglucosamine from the cell wall synthesis pathway. This could partially explain why these organisms tightly regulate the expression of genetic loci responsible for polysaccharide synthesis. Another area that remains to be investigated in the staphylococci is the function of the IcaC protein. IcaC is thought to be involved in synthesizing long polymers from shorter polymers and possibly in export of mature β-(1,6)-linked *N*-acetylglucosamine but this remains somewhat speculative.

There are gaps in our understanding of PGA synthesis and regulation in *E. coli* as well. While the proteins encoded within the *pga* locus have been functionally characterized, the regulatory processes that control expression of PGA are still in the beginning stages of being teased out.

In *Actinobacillus*, the PIA/PNAG/PGA-degrading enzyme dispersin B has received a lot of attention for its potential as a therapeutic agent to disrupt biofilms and potentially to increase the sensitivity of these species to phagocytosis by neutrophils. However, regulation of the *pga* locus in this species has not yet been well characterized in these species.

There are a number of pathogenic bacteria that are suspected of producing the polysaccharide, based on the presence of a locus with significant homology to *ica* and *pga*, or on reactivity with PNAG-specific antibodies, or on degradation of biofilms by the dispersin B enzyme. These species include *Yersinia pestis*, *Y. enterocolitica*, *Bordetella pertussis*, *B. parapertussis*, and *Pseudomonas fluorescens* (Hinnebusch and Erickson, 2008; Kaplan *et al.*, 2004b; Wang *et al.*, 2004). An understanding of the capacity for these organisms to produce PIA/PNAG/PGA is likely just around the corner.

In the future, advances in our understanding of polysaccharide synthesis will improve, and this could lead to novel therapeutic modalities that target this virulence factor. In addition, the conservation of PIA/PNAG amongst a variety of bacterial species holds the promise of a vaccine that would target multiple pathogens and efforts to develop an effective anti-PNAG vaccine are currently under way (Cerca *et al.*, 2007b; Kelly-Quintos *et al.*, 2006).

References

Ammendolia, M.G., Di Rosa, R., Montanaro, L., Arciola, C.R., and Baldassarri, L. (1999). Slime production and expression of the slime-associated antigen by staphylococcal clinical isolates. J. Clin. Microbiol. *37*, 3235–3238.

Baldassarri, L., Donnelli, G., Gelosia, A., Voglino, M.C., Simpson, A.W., and Christensen, G. D. (1996). Purification and characterization of the staphylococcal slime-associated antigen and its occurrence among *Staphylococcus epidermis* clinical isolates. Infect. Immun. *64*, 3410–3415.

Bayer, M.G., Heinrichs, J.H., and Cheung, A. L. (1996). The molecular architecture of the *sar* locus in *Staphylococcus aureus*. J. Bacteriol. *178*, 4563–4570.

Beenken, K.E., Blevins, J.S., and Smeltzer, M. S. (2003). Mutation of *sarA* in *Staphylococcus aureus* limits biofilm formation. Infect. Immun. *71*, 4206–4211.

Bischoff, M., Entenza, J.M., and Giachino, P. (2001). Influence of a functional *sigB* operon on the global regulators *sar* and *agr* in *Staphylococcus aureus*. J. Bacteriol. *183*, 5171–5179.

Brandenberger, M., Tschierske, M., Giachino, P., Wada, A., and Berger-Bächi, B. (2000). Inactivation of a novel three-cistronic operon *tcaR-tcaA-tcaB* increases teicoplanin resistance in *Staphylococcus aureus*. Biochim. Biophys. Acta *1523*, 135–139.

Brouillette, E., Hyodo, M., Hayakawa, Y., Karaolis, D.K., and Malouin, F. (2005). 3′,5′-cyclic diguanylic acid reduces the virulence of biofilm-forming *Staphylococcus aureus* strains in a mouse model of mastitis infection. Antimicrob. Agents Chemother. *49*, 3109–3113.

Cerca, N., and Jefferson, K. K. (2008). Effect of growth conditions on poly-N-acetylglucosamine expression and biofilm formation in *Escherichia coli*. FEMS Microbiol. Lett *283*, 36–41.

Cerca, N., Jefferson, K.K., Maira-Litrán, T., Pier, D.B., Kelly-Quintos, C., Goldmann, D.A., Azeredo, J., and Pier, G. B. (2007a). Molecular basis for preferential protective efficacy of antibodies directed to the poorly acetylated form of staphylococcal poly-N-acetyl-beta-(1–6)-glucosamine. Infect. Immun. *75*, 3406–3413.

Cerca, N., Jefferson, K.K., Oliveira, R., Pier, G.B., and Azeredo, J. (2006). Comparative antibody-mediated phagocytosis of *Staphylococcus epidermidis* cells grown in a biofilm or in the planktonic state. Infect. Immun. *74*, 4849–4855.

Cerca, N., Maira-Litrán, T., Jefferson, K.K., Grout, M., Goldmann, D.A., and Pier, G. B. (2007b). Protection against *Escherichia coli* infection by antibody to the *Staphylococcus aureus* poly-N-acetylglucosamine surface polysaccharide. Proc. Natl. Acad. Sci. USA *104*, 7528–7533.

Chaignon, P., Sadovskaya, I., Ragunah, C., Ramasubbu, N., Kaplan, J.B., and Jabbouri, S. (2007). Susceptibility of staphylococcal biofilms to enzymatic treatments depends on their chemical composition. Appl. Microbiol. Biotechnol. 75, 125–132.

Chien, Y., Manna, A.C., Projan, S.J., and Cheung, A. L. (1999). SarA, a global regulator of virulence determinants in *Staphylococcus aureus*, binds to a conserved motif essential for *sar*-dependent gene regulation. J. Biol. Chem. *274*, 37169–37176.

Conlon, K.M., Humphreys, H., and O'Gara, J. P. (2002a). *icaR* encodes a transcriptional repressor involved in environmental regulation of *ica* operon expression and biofilm formation in *Staphylococcus epidermidis*. J. Bacteriol. *184*, 4400–4408.

Conlon, K.M., Humphreys, H., and O'Gara, J. P. (2002b). Regulation of *icaR* gene expression in *Staphylococcus epidermidis*. FEMS Microbiol. Lett *216*, 171–177.

Conlon, K.M., Humphreys, H., and O'Gara, J. P. (2004). Inactivations of *rsbU* and *sarA* by IS256 represent novel mechanisms of biofilm phenotypic variation in *Staphylococcus epidermidis*. J. Bacteriol. *186*, 6208–6219.

Cramton, S.E., Gerke, C., Schnell, N.F., Nichols, W.W., and Götz, F. (1999). The intercellular adhesion (*ica*) locus is present in *Staphylococcus aureus* and is required for biofilm formation. Infect. Immun. *67*, 5427–5433.

Cramton, S.E., Ulrich, M., Götz, F., and Doring, G. (2001). Anaerobic conditions induce expression of polysaccharide intercellular adhesin in *Staphylococcus aureus* and *Staphylococcus epidermidis*. Infect. Immun. *69*, 4079–4085.

Dobinsky, S., Kiel, K., Rohde, H., Bartscht, K., Knobloch, J.K., Horstkotte, M.A., and Mack, D. (2003). Glucose-related dissociation between *icaADBC* transcription and biofilm expression by *Staphylococcus epidermidis*: evidence for an additional factor required for polysaccharide intercellular adhesin synthesis. J. Bacteriol. *185*, 2879–2886.

Fux, C.A., Costerton, J.W., Stewart, P.S., and Stoodley, P. (2005). Survival strategies of infectious biofilms. Trends Microbiol. *13*, 34–40.

Gerke, C., Kraft, A., Sussmuth, R., Schweitzer, O., and Götz, F. (1998). Characterization of the N-acetylglucosaminyltransferase activity involved in the biosynthesis of the *Staphylococcus epidermidis* polysaccharide intercellular adhesin. J. Biol. Chem. *273*, 18586–18593.

Goering, R.V., McDougal, L.K., Fosheim, G.E., Bonnstetter, K.K., Wolter, D.J., and Tenover, F. C. (2007). Epidemiologic distribution of the arginine catabolic mobile element among selected methicillin-resistant and methicillin-susceptible *Staphylococcus aureus* isolates. J. Clin. Microbiol. *45*, 1981–1984.

Goller, C., Wang, X., Itoh, Y., and Romeo, T. (2006). The cation-responsive protein NhaR of *Escherichia coli* activates *pgaABCD* transcription, required for production of the biofilm adhesin poly-beta-1,6-N-acetyl-D-glucosamine. J. Bacteriol. *188*, 8022–8032.

Gonzalez Barrios, A.F., Zuo, R., Hashimoto, Y., Yang, L., Bentley, W.E., and Wood, T. K. (2006). Autoinducer 2 controls biofilm formation in *Escherichia coli* through a novel motility quorum-sensing regulator (MqsR, B3022). J. Bacteriol. *188*, 305–316.

Götz, F. (2002). *Staphylococcus* and biofilms. Mol. Microbiol. *43*, 1367–1378.

Heilmann, C., Schweitzer, O., Gerke, C., Vanittanakom, N., Mack, D., and Götz, F. (1996). Molecular basis of intercellular adhesion in the biofilm-forming *Staphylococcus epidermidis*. Mol. Microbiol. *20*, 1083–1091.

Hinnebusch, B.J., and Erickson, D. L. (2008). *Yersinia pestis* biofilm in the flea vector and its role in the transmission of plague. Curr. Top. Microbiol. Immunol. *322*, 229–248.

Holland, L.M., O'Donnell, S.T., Ryjenkov, D.A., Gomelsky, L., Slater, S.R., Fey, P.D., Gomelsky, M., and O'Gara, J. P. (2008). A staphylococcal GGDEF domain protein regulates biofilm formation independently of cyclic dimeric GMP. J. Bacteriol. *190*, 5178–5189.

Itoh, Y., Rice, J.D., Goller, C., Pannuri, A., Taylor, J., Meisner, J., Beveridge, T.J., Preston, J.F., 3rd, and Romeo, T. (2008). Roles of *pgaABCD* genes in synthesis, modification, and export of the *Escherichia coli* biofilm adhesin poly-beta-1,6-N-acetyl-D-glucosamine. J. Bacteriol. *190*, 3670–3680.

Izano, E.A., Wang, H., Ragunath, C., Ramasubbu, N., and Kaplan, J. B. (2007). Detachment and killing of *Aggregatibacter actinomycetemcomitans* biofilms by dispersin B and SDS. J. Dent. Res. *86*, 618–622.

Jankovic, I., and Brückner, R. (2007). Carbon catabolite repression of sucrose utilization in *Staphylococcus xylosus*: catabolite control protein CcpA ensures glucose preference and autoregulatory limitation of sucrose utilization. J. Mol. Microbiol. Biotechnol. *12*, 114–120.

Jefferson, K. K. (2004). What drives bacteria to produce a biofilm? FEMS Microbiol. Lett. *236*, 163–173.

Jefferson, K.K., Pier, D.B., Goldmann, D.A., and Pier, G. B. (2004). The teicoplanin-associated locus regulator (TcaR) and the intercellular adhesin locus regulator (IcaR) are transcriptional inhibitors of the *ica* locus in *Staphylococcus aureus*. J. Bacteriol. *186*, 2449–2456.

Johnson, M., Cockayne, A., and Morrissey, J. A. (2008). Iron-regulated biofilm formation in *Staphylococcus aureus* Newman requires *ica* and the secreted protein Emp. Infect. Immun. *76*, 1756–1765.

Johnson, M., Cockayne, A., Williams, P.H., and Morrissey, J. A. (2005). Iron-responsive regulation of biofilm formation in *Staphylococcus aureus* involves fur-dependent and fur-independent mechanisms. J. Bacteriol. *187*, 8211–8215.

Kaplan, J.B., Ragunath, C., Velliyagounder, K., Fine, D.H., and Ramasubbu, N. (2004a). Enzymatic detachment of *Staphylococcus epidermidis* biofilms. Antimicrob. Agents Chemother. *48*, 2633–2636.

Kaplan, J.B., Velliyagounder, K., Ragunath, C., Rohde, H., Mack, D., Knobloch, J.K., and Ramasubbu, N. (2004b). Genes involved in the synthesis and degradation of matrix polysaccharide in *Actinobacillus actinomycetemcomitans* and *Actinobacillus pleuropneumoniae* biofilms. J. Bacteriol. *186*, 8213–8220.

Karaolis, D.K., Rashid, M.H., Chythanya, R., Luo, W., Hyodo, M., and Hayakawa, Y. (2005). c-di-GMP (3′-5′-cyclic diguanylic acid) inhibits *Staphylococcus aureus* cell-cell interactions and biofilm formation. Antimicrob. Agents Chemother. *49*, 1029–1038.

Kelly-Quintos, C., Cavacini, L.A., Posner, M.R., Goldmann, D., and Pier, G. B. (2006). Characterization of the opsonic and protective activity against *Staphylococcus aureus* of fully human monoclonal antibodies specific for the bacterial surface polysaccharide poly-N-acetylglucosamine. Infect. Immun. *74*, 2742–2750.

Knobloch, J.K., Bartscht, K., Sabottke, A., Rohde, H., Feucht, H.H., and Mack, D. (2001). Biofilm formation by *Staphylococcus epidermidis* depends on functional RsbU, an activator of the *sigB* operon: differential activation mechanisms due to ethanol and salt stress. J. Bacteriol. *183*, 2624–2633.

Knobloch, J.K., Jager, S., Horstkotte, M.A., Rohde, H., and Mack, D. (2004). RsbU-dependent regulation of *Staphylococcus epidermidis* biofilm formation is mediated via the alternative sigma factor sigmaB by repression of the negative regulator gene *icaR*. Infect. Immun. *72*, 3838–3848.

Kong, K.F., Vuong, C., and Otto, M. (2006). *Staphylococcus* quorum sensing in biofilm formation and infection. Int. J. Med. Microbiol. *296*, 133–139.

Konig, D.P., Perdreau-Remington, F., Rutt, J., Hilgers, R.D., and Schierholz, J. M. (1999). Adherence to and accumulation of *S. epidermidis* on different biomaterials due to extracellular slime production. In vitro comparison of a slime-producing strain (Rp 62 A) and its isogenic slime negative mutant (M7). Zentralbl. Bakteriol. *289*, 355–364.

Leid, J.G., Shirtliff, M.E., Costerton, J.W., and Stoodley, A. P. (2002). Human leukocytes adhere to, penetrate, and respond to *Staphylococcus aureus* biofilms. Infect. Immun. *70*, 6339–6345.

Liu, J., Tan, C., Li, J., Chen, H., Xu, P., He, Q., Bei, W., and Chen, H. (2008). Characterization of ISApl1, an insertion element identified from *Actinobacillus pleuropneumoniae* field isolate in China. Vet. Microbiol. *132*, 348–354.

Mack, D., Fischer, W., Krokotsch, A., Leopold, K., Hartmann, R., Egge, H., and Laufs, R. (1996). The intercellular adhesin involved in biofilm accumulation of *Staphylococcus epidermidis* is a linear beta-1,6-linked glucosaminoglycan: purification and structural analysis. J. Bacteriol. *178*, 175–183.

Majerczyk, C.D., Sadykov, M.R., Luong, T.T., Lee, C., Somerville, G.A., and Sonenshein, A. L. (2008). *Staphylococcus aureus* CodY negatively regulates virulence gene expression. J. Bacteriol. *190*, 2257–2265.

Manna, A.C., Bayer, M.G., and Cheung, A. L. (1998). Transcriptional analysis of different promoters in the *sar* locus in *Staphylococcus aureus*. J. Bacteriol. *180*, 3828–3836.

Manuel, S.G., Ragunath, C., Sait, H.B., Izano, E.A., Kaplan, J.B., and Ramasubbu, N. (2007). Role of active-site residues of dispersin B, a biofilm-releasing beta-hexosaminidase from a periodontal pathogen, in substrate hydrolysis. FEBS J. *274*, 5987–5999.

Monroe, D. (2007). Looking for chinks in the armor of bacterial biofilms. PLoS Biol. *5*, e307.

Nakano, S., Nakano, M.M., Zhang, Y., Leelakriangsak, M., and Zuber, P. (2003). A regulatory protein that interferes with activator-stimulated transcription in bacteria. Proc. Natl. Acad. Sci. USA *100*, 4233–4238.

Pamp, S.J., Frees, D., Engelmann, S., Hecker, M., and Ingmer, H. (2006). Spx is a global effector impacting stress tolerance and biofilm formation in *Staphylococcus aureus*. J. Bacteriol. *188*, 4861–4870.

Rachid, S., Ohlsen, K., Wallner, U., Hacker, J., Hecker, M., and Ziebuhr, W. (2000a). Alternative transcription factor sigma(B) is involved in regulation of biofilm expression in a *Staphylococcus aureus* mucosal isolate. J. Bacteriol. *182*, 6824–6826.

Rachid, S., Ohlsen, K., Witte, W., Hacker, J., and Ziebuhr, W. (2000b). Effect of subinhibitory antibiotic concentrations on polysaccharide intercellular adhesin expression in biofilm-forming *Staphylococcus epidermidis*. Antimicrob. Agents Chemother. *44*, 3357–3363.

Ramasubbu, N., Thomas, L.M., Ragunath, C., and Kaplan, J. B. (2005). Structural analysis of dispersin, B, a biofilm-releasing glycoside hydrolase from the periodontopathogen *Actinobacillus actinomycetemcomitans*. J. Mol. Biol. *349*, 475–486.

Redfield, R. J. (2002). Is quorum sensing a side effect of diffusion sensing? Trends Microbiol. *10*, 365–370.

Rogers, K.L., Rupp, M.E., and Fey, P. D. (2008). The presence of *icaADBC* is detrimental to the colonization of human skin by *Staphylococcus epidermidis*. Appl. Environ Microbiol. *74*, 6155–6157.

Rupp, C.J., Fux, C.A., and Stoodley, P. (2005). Viscoelasticity of *Staphylococcus aureus* biofilms in response to fluid shear allows resistance to detachment and facilitates rolling migration. Appl. Environ. Microbiol. *71*, 2175–2178.

Seidl, K., Goerke, C., Wolz, C., Mack, D., Berger-Bächi, B., and Bischoff, M. (2008). *Staphylococcus aureus* CcpA affects biofilm formation. Infect. Immun. *76*, 2044–2050.

Tormo, M.A., Marti, M., Valle, J., Manna, A.C., Cheung, A.L., Lasa, I., and Penadés, J.R. (2005). SarA is an essential positive regulator of *Staphylococcus epidermidis* biofilm development. J. Bacteriol. *187*, 2348–2356.

Tu Quoc, P.H., Genevaux, P., Pajunen, M., Savilahti, H., Georgopoulos, C., Schrenzel, J., and Kelley, W. L. (2007). Isolation and characterization of biofilm formation-defective mutants of *Staphylococcus aureus*. Infect. Immun. *75*, 1079–1088.

Ulrich, M., Bastian, M., Cramton, S.E., Ziegler, K., Pragman, A.A., Bragonzi, A., Memmi, G., Wolz, C., Schlievert, P.M., Cheung, A., and Döring, G. (2007). The staphylococcal respiratory response regulator SrrAB induces ica gene transcription and polysaccharide intercellular adhesin expression, protecting *Staphylococcus aureus* from neutrophil killing under anaerobic growth conditions. Mol. Microbiol. *65*, 1276–1287.

Valle, J., Toledo-Arana, A., Berasain, C., Ghigo, J.M., Amorena, B., Penadés, J.R., and Lasa, I. (2003). SarA and not sigmaB is essential for biofilm development by *Staphylococcus aureus*. Mol. Microbiol. *48*, 1075–1087.

Venketaraman, V., Lin, A.K., Le, A., Kachlany, S.C., Connell, N.D., and Kaplan, J. B. (2008). Both leukotoxin and poly-N-acetylglucosamine surface polysaccharide protect *Aggregatibacter actinomycetemcomitans* cells from macrophage killing. Microb. Pathog. *45*, 173–180.

Vuong, C., Gerke, C., Somerville, G.A., Fischer, E.R., and Otto, M. (2003). Quorum-sensing control of biofilm factors in *Staphylococcus epidermidis*. J. Infect. Dis. *188*, 706–718.

Vuong, C., Kidder, J.B., Jacobson, E.R., Otto, M., Proctor, R.A., and Somerville, G.A. (2005). *Staphylococcus epidermidis* polysaccharide intercellular adhesin production significantly increases during tricarboxylic acid cycle stress. J. Bacteriol. *187*, 2967–2973.

Vuong, C., Kocianova, S., Voyich, J.M., Yao, Y., Fischer, E.R., DeLeo, F.R., and Otto, M. (2004a). A crucial role for exopolysaccharide modification in bacterial biofilm formation, immune evasion, and virulence. J. Biol. Chem. *279*, 54881–54886.

Vuong, C., Saenz, H.L., Götz, F., and Otto, M. (2000). Impact of the *agr* quorum-sensing system on adherence to polystyrene in *Staphylococcus aureus*. J Infect Dis *182*, 1688–1693.

Vuong, C., Voyich, J.M., Fischer, E.R., Braughton, K.R., Whitney, A.R., DeLeo, F.R., and Otto, M. (2004b).

Polysaccharide intercellular adhesin (PIA) protects *Staphylococcus epidermidis* against major components of the human innate immune system. Cell Microbiol. *6*, 269–275.

Wang, X., Dubey, A.K., Suzuki, K., Baker, C.S., Babitzke, P., and Romeo, T. (2005). CsrA post-transcriptionally represses *pgaABCD*, responsible for synthesis of a biofilm polysaccharide adhesin of *Escherichia coli*. Mol. Microbiol. *56*, 1648–1663.

Wang, X., Preston, J.F., 3rd, and Romeo, T. (2004). The *pgaABCD* locus of *Escherichia coli* promotes the synthesis of a polysaccharide adhesin required for biofilm formation. J. Bacteriol. *186*, 2724–2734.

Xu, L., Li, H., Vuong, C., Vadyvaloo, V., Wang, J., Yao, Y., Otto, M., and Gao, Q. (2006). Role of the luxS quorum-sensing system in biofilm formation and virulence of *Staphylococcus epidermidis*. Infect. Immun. *74*, 488–496.

Yarwood, J.M., Bartels, D.J., Volper, E.M., and Greenberg, E. P. (2004). Quorum sensing in *Staphylococcus aureus* biofilms. J. Bacteriol. *186*, 1838–1850.

Yarwood, J.M., McCormick, J.K., and Schlievert, P. M. (2001). Identification of a novel two-component regulatory system that acts in global regulation of virulence factors of *Staphylococcus aureus*. J. Bacteriol. *183*, 1113–1123.

Zhu, Y., Weiss, E.C., Otto, M., Fey, P.D., Smeltzer, M.S., and Somerville, G. A. (2007). *Staphylococcus aureus* biofilm metabolism and the influence of arginine on polysaccharide intercellular adhesin synthesis, biofilm formation, and pathogenesis. Infect. Immun. *75*, 4219–4226.

Ziebuhr, W., Krimmer, V., Rachid, S., Lossner, I., Götz, F., and Hacker, J. (1999). A novel mechanism of phase variation of virulence in *Staphylococcus epidermidis*: evidence for control of the polysaccharide intercellular adhesin synthesis by alternating insertion and excision of the insertion sequence element IS256. Mol. Microbiol. *32*, 345–356.

Zuber, P. (2004). Spx-RNA polymerase interaction and global transcriptional control during oxidative stress. J. Bacteriol. *186*, 1911–1918.

Part V

Polysaccharides in Plant-associated Bacteria

11 Surface Polysaccharides as Fitness Factors of Rhizospheric Nitrogen-fixing Bacteria

Elizaveta Krol and Anke Becker

Abstract

Nitrogen-fixing plant-associated bacteria include rhizospheric bacteria, endophytic diazotrophs and rhizobia, the symbiotic root nodule bacteria. Surface polysaccharides produced by bacteria contribute to their survival in the rhizosphere by improvement of the soil structure, enhancement of cell aggregation, and protection against harmful substances. Attachment of bacteria to plant roots and soil particles, which is important for colonization of rhizosphere and roots and for infection of the plant, can be mediated by exopolysaccharides. Symbiotic interactions between root nodule bacteria and leguminous plants include infection of root tissues by bacteria and formation of nitrogen-fixing nodules, where bacteria differentiate into intracellular bacteroids. During early stages of the symbiotic interaction, exopolysaccharides, capsular polysaccharides or lipopolysaccharides can be involved in recognition by the plant and suppression of the immune response. At the stage of intracellular infection, lipopolysaccharides play an important role in maintenance of the bacteroids. Biosynthesis of surface polysaccharides is tightly regulated during both free-living growth and symbiotic development. Two-component regulatory systems, general stress response mechanisms, quorum sensing and symbiotic regulators constitute a flexible network for fine-tuning of surface polysaccharide production. In this chapter, we will focus on the structure of surface polysaccharides, produced by nodule bacteria, mechanisms of genetic regulation of their biosynthesis and their functions as fitness factors for free-living survival and symbiosis with a plant.

Introduction

Bacteria, interacting with plants, are ubiquitous in the nature, and can establish symbiotic, mutualistic, commensalistic or pathogenic interactions with their hosts. The greatest diversity of plant-interacting microorganisms can be found in the rhizosphere, which is defined as soil, immediately adjacent to the roots. Growing plant roots secrete significant amounts of nutrients that can be used by soil-borne bacteria that colonize the rhizosphere. In turn, bacteria can protect plants from pathogens, modify soil structure, improve plant nutrition by increasing solubility of the mineral compounds and even promote plant growth by secretion of plant hormones.

Biological fixation of atmospheric nitrogen by plant-interacting bacteria contributes greatly to plant nitrogen nutrition. The best studied rhizospheric surface-colonizing nitrogen fixers are bacteria of the *Azospirillum* genus. Azospirilla are associated with grasses and promote plant growth by both production of the plant hormone indole-3-acetic acid and nitrogen fixation (extensively reviewed by Steenhoudt and Vanderleiden, 2000). Endophytic diazotrophs are able to invade the root cortex and xylem and fix molecular nitrogen inside the plant. Such bacteria are associated with a wide variety of plants, including important cultured plants such as rice, maize, sugar cane, and coffee. Well-characterized examples include members of *Acetobacteriaceae*, *Azoarcus* and *Azotobacter* species (reviewed in Steenhoudt and Vanderleyden, 2000; Hurek *et al.*, 2003; Saravanan *et al.*, 2008). However, the most efficient form of biological nitrogen fixation and the best studied today is the symbiosis between leguminous plants and root nodule bacteria, collectively called rhizobia. Symbiotic nitrogen fixers include bacteria from the genera *Sinorhizobium, Rhizobium, Mesorhizobium, Bradyrhizobium, Azorhizobium* and *Burkholderia.*

Rhizobia are able to invade the roots of their legume hosts and elicit the formation of nitrogen-fixing nodules. Inside the nodules, bacteria are present in both vegetative, extracellular forms, and symbiotic, intracellular forms, called bacteroids. Bacteroids are organelle-like structures, whose metabolism is almost completely dedicated to nitrogen fixation. Establishment of the symbiotic interaction is a complex process that includes reciprocal recognition of the partners, cell differentiation and exchange of nutrients (for recent reviews, see Gage, 2004; Stacey *et al.*, 2006; Cooper, 2007; Jones *et al.*, 2007). The signalling begins, as rhizobia percept flavonoids, secreted by plant roots, and respond by synthesis and release of lipochitin-oligosaccharide Nod factors. Production of the Nod factor by a small rhizobial colony, attached to the root hair cell, induces curling of the root hair. Bacteria, entrapped in the root hair curl, induce progressive ingrowth of the root hair cell membrane towards the root cortex. This constitutes an infection thread that leads the bacteria to the nodule primordium formed by root cortical cells. Reaching the root cortex, bacteria become engulfed by endocytosis and begin symbiotic differentiation to nitrogen-fixing bacteroids. Two different types of nodules can be formed, depending on the fate of the infection threads and bacteroid division status. Determinate nodules grow further by multiplication of the infected cells and bacteroids inside of them. In indeterminate nodules, bacteroids are not able to divide, therefore these nodules grow due to continuous infection of new cell layers by the bacteria, enclosed in infection threads.

Rhizobial surface polysaccharides include exopolysaccharides (EPS), secreted into the growth medium; periplasmic glucans, which are also partially secreted to the medium, capsular polysaccharides (KPS), attached to the outer membrane; and lipopolysaccharides (LPS), the major structural component of the outer membrane. LPS are essential components of the cell surface and their hydrophobic properties are important for the barrier function of the outer membrane. Mutations in genes controlling the lipid A biosynthesis are lethal, and mutants with altered LPS structure are more sensitive to detergents, oxidative stress and antibiotics. EPS and periplasmic glucans are non-essential, but they contribute greatly to the resistance to osmotic, oxidative, and ionic stresses, protect bacteria from desiccation and, due to their anionic properties, are able to bind cations, thereby improving the supply of oligoelements. Moreover, all kinds of surface polysaccharides are involved in the establishment of nodule symbiosis by rhizobia.

Chemical structures of the surface polysaccharides, produced by symbiotic nitrogen-fixing bacteria and genetic control of their biosynthesis

Exopolysaccharides

EPS produced by rhizobia are hexose homo- or heteropolymers. Typically, they consist of repeating oligosaccharide units, which are mostly branched. Glucose is a principal component of the rhizobial EPS, with exception of EPS produced by *Bradyrhizobium japonicum, Azorhizobium caulinodans* and *Rhizobium* sp. isolated from *Vigna mungo* (L.) Hepper (EPS structures are summarized in Table 11.1). Sugars are often modified with pyruvyl and acetyl substituents, which confer anionic properties to the EPS. Only two homopolymeric EPSs were reported in rhizobia, the cellulose produced by *Rhizobium leguminosarum* and an *Azorhizobium caulinodans* EPS galactan.

In the EPS preparations of *Sinorhizobium meliloti, R. leguminosarum, Rhizobium* sp. NGR234 and *B. japonicum*, two different fractions can be detected, a high molecular weight (HMW) (about 10^3 kDa), and a low molecular weight (LMW) fraction (5–8 kDa) (Gonzalez *et al.*, 1996; 1998; Wang *et al.*, 1999; Djordjevic *et al.* 1986; Louch *et al.* 2001). The relative amounts of the LMW and HMW fractions can vary dependent on the medium salinity (Becker and Pühler, 1998).

In *S. meliloti* reference strain Rm1021 and the closely related strain Rm2011, succinoglycan (EPSI) and galactoglucan (EPSII) biosynthesis

Table 11.1 Structural features of rhizobial EPSs

Bacterium	EPS	Sugar composition	Ratio	Modifications	Structure known	Reference
S. meliloti	Succinoglycan (EPSI)	Glc/Gal	7:1	O-Ac, O-Succ, Pyr	Yes	Reinhold *et al.* (1994)
S. meliloti	Galactoglucan (EPSII)	Glc/Gal	1:1	O-Ac, Pyr	Yes	Her *et al.* (1990)
R. leguminosarum	EPS	Glc/GlcA/Gal	5:2:1	O-Ac, Pyr	Yes	O'Neill *et al.*, (1991)
R. leguminosarum bv. *viciae* 248	EPS	Glc/GlcA/Gal	5:3:1	O-Ac, Pyr	Yes	Canter-Cremers *et al.* (1991)
R. leguminosarum bv. *trifolii* 4s	EPS	Glc/GlcA	5:2	O-Ac, Pyr	Yes	Amemura *et al.* (1983)
R. leguminosarum RBL5523	Cellulose	Glc	No	No	Yes	Drozański *et al.* (1981)
R. leguminosarum RBL5523	Polar glucomannan	Glc/Man	Unknown	Unknown	No	Laus *et al.* (2006)
B. japonicum	EPS	Rha/Gal/GlcA	3:1:1	O-Me	Yes	An *et al.* (1995)
Rhizobium sp. NGR234	EPS	Glc/Gal/GlcA	5:2:2	O-Ac, Pyr	Yes	Djordjevic *et al.* (1986)
Rhizobium sp. KYGT207	EPS	Glc/Gal/ManA	2:1:1	O-Ac, Pyr	Yes	Kaci *et al.* (2005)
Rhizobium sp.	EPS	Rib/Man	Unknown	Unknown	No	Mandal *et al.* (2007)
A. caulinodans	EPS	Gal	No	Pyr	Yes	D'Haezle *et al.* (2004)

Abbreviations: Glc, glucose; Gal, galactose; Man, mannose; Rib, ribose; Rha, rhamnose; ManA, mannuronic acid; GlcA, glucuronic acid; O-Ac, O-acetyl modification; O-Succ, O-succinyl modification; Pyr, pyruvyl modification; O-Me, O-methyl modification.

is controlled by *exo/exs* and *exp* gene clusters, correspondingly (Becker *et al.*, 2002). In *R. leguminosarum*, *pss* genes control biosynthesis of a symbiotically active acidic exopolysaccharide (Sadykov *et al.*, 1998). The *exo* gene cluster is responsible for EPS biosynthesis in *B. japonicum* (Becker *et al.*, 1998). The biosynthesis of *S. meliloti* EPSI, which included the assembly of the repeating unit on an undecaprenol diphosphate, polymerization of the repeating units and export of the mature polysaccharide by different Exo proteins was completely resolved (Glucksmann *et al.*, 1993a; 1993b). A symbiotically active LMW EPSI can be produced either by direct export of the low-polymerized polysaccharide (Becker and Pühler, 1998) or by endoglucanase cleavage of the nascent HMW polysaccharide by ExoK and ExsH (York and Walker, 1998b; Staehelin *et al.*, 2006). Despite a much more simple structure of EPSII, its biosynthesis remains cryptic, because no mutants affecting the intermediate steps of the biosynthetic pathway could be isolated.

Cyclic glucans

Cyclic glucans are produced by rhizobia and a restricted number of other α-proteobacteria. They are mainly present in the periplasmic space but can be also secreted into the growth medium. In most species, they are β-1,2-linked, with an exception of *S. fredii* and *B. japonicum*, in which they are β-1,2- and β-1,6-β-1,3-linked, consist of approximately 17–25 glucose residues and are substituted with *sn*-1-phosphoglycerol (Breedveld and Miller, 1994). Interestingly, the whole biosynthesis of the polysaccharide from UDP-glucose is carried out by a single synthase protein, encoded by *S. meliloti ndvB* and its homologues in other species (Castro *et al.*, 1996). Recently, it was shown that in *Brucella abortus*, a pathogenic bacterium that is very closely related to rhizobia, also the size determination of cyclic glucan is performed by the NdvB homologue. The molecular mechanism involved a glycoside phosphorylase activity of the C-terminal region of the synthase (Ciocchini *et al.*, 2007). In *S. meliloti*, the *ndvA* gene product was suggested to control export of the glucan (Breedveld and Miller, 1994).

Capsular polysaccharides

Rhizobial capsular polysaccharides (Kdo-reach polysaccharides, KPS) are structural analogues of the group II K antigens found in *Escherichia coli.* They are attached to the outer membrane through a lipid anchor and are typically built of dimeric repeating units of a neutral hexose or uronic acid and a 3-keto-D-manno-2-octulosonic acid (Kdo) or a related sugar joined by glycosidic linkages (Table 11.2). However, some bacteria produce KPS that are devoid of Kdo or hexose or include amidic linkages (Gil-Serrano *et al.*, 1999; Rodriguez-Carvajal *et al.*, 2005; Fraysse *et al.*, 2005). The length of the KPS polysaccharide chain is variable, and its visualization on polyacrylamide gels with the Alcian blue staining procedure reveals a typical KPS ladder, which represents Kdo-rich polysaccharides with a different degree of polymerization. A membrane-attached low-molecular weight KPS that consists of poly-Kdo was initially identified in KPS preparations of *S. meliloti* Rm1021 by Fraysse *et al.* (2005). Recently, Chataigné *et al.* (2008) demonstrated that membrane-anchored Kdo homopolymers are

Table 11.2 Complete and partial structures of rhizobial KPS

Strain	Known KPS structure	Reference
S. meliloti 1021	[-β-(2→7)-Kdo-]*n*	Fraysse *et al.* (2005)
S. meliloti AK631	[-β-GlcA→Pse5N(β-OH-But)7NAc-]*n*	Forsberg and Reuhs (1997)
	Poly-Kdo	Chataigné *et al.* (2008)
S. meliloti NGR247	[-α-Glc→α-NeuNAc-]	Reuhs *et al.*, 1998)
S. meliloti NGR185	[-β-GlcNAc→β-Kdo-]*n*	Reuhs *et al.* (1998)
S. fredii USDA208	[-α-Gal→β-Kdo-]*n*	Reuhs *et al.* (1998)
S. fredii USDA201	[-α-Gal→β-Kdo→α-2-*O*-MeHex→β-Kdo]*n*	Reuhs *et al.* (1998)
S. fredii USDA205	[→3)-α-D-Gal*p*-(1→5)-α-D-Kdo*p*-(2→]*n*	Reuhs *et al.* (1993)
	[-2-*O*-MeMan*p*→β-Kdo-]*n*	
S. fredii USDA257	[→3)-β-D-Man*p*-(1→5)-β-D-Kdo*p*-(2→]*n*	Reuhs *et al.* (1993)
	[→3)-β-D-2-*O*-MeMan*p*-(1→5)-β-D-Kdo*p*-(2→]*n*	
S. fredii HH303	[Rha, GalA]*n*	Reuhs *et al.* (1998)
S. fredii HH103	[-7(3OH But)-Pse-]*n*	Reuhs *et al.* (1998)
	Poly-Kdo	Chataigné *et al.* (2008)
S. fredii HWG35	[→6)-2,4-di-*O*-Me-α-D-Gal*p*-(1→4)-β-D-Glc*p*A-(1→]*n*	Rodriguez-Carvajal *et al.* (2005)
(S. fredii) B33	[-α-4-*O*-MeGlc→α-3-*O*-MeGlcA-]*n*	Rodriguez-Carvajal *et al.* (2001)
Rhizobium sp. NGR234	[→4)-β-D-GlcA-(1→4)-β-Pse5Nac7NAc-(2→]*n*	Le Quéré *et al.* (2006)
	Poly-Kdo	Chataigné *et al.* (2008)

Abbreviations: Kdo, 3-deoxy-D-manno-2-octulosonic acid; Pse, 5,7-diamino-3,5,7,9-tetradeoxynonulosonic acid; Neu, 5-amino-3,5-dideoxynonulosonic acid; Hex, hexose; Glc, glucose; Gal, galactose; But, butyryl group, Man, mannose; Rib, ribose; Rha, rhamnose; ManA, mannuronic acid; GlcA, glucuronic acid; Ac, acetyl group; Me, methyl group.

consensus sinorhizobial capsular polysaccharides present in *S. meliloti* Rm41, *S. fredii* HH103 and *Rhizobium* sp. NGR234 as well.

In *S. meliloti* Rm41, KPS biosynthesis is controlled by three *rkp* gene clusters, which include genes required for synthesis of a specific lipid carrier, subunits of the polysaccharide chains and polysaccharide polymerization and transport (Epple *et al.*, 1993; Petrovics *et al.*, 1993; Kiss *et al.*, 1997). Determination of the polysaccharide chain length involves the *rkpZ* gene (Kiss *et al.*, 2001). A *rkp* gene cluster was recently also identified in *Rhizobium* sp. NGR234 genome (Le Quéré *et al.*, 2006).

Although the structure of *M. loti* KPS was not resolved yet, it was recently demonstrated that KPS of strain NZP-2235 are sulphated (Towsend II *et al.*, 2006). So far, sulphated polysaccharides were found in only few bacterial genera, although they are ubiquitous in eukaryotes. The sulphation of *M. loti* KPS was dependent on the *nodPQ* genes, required for biosynthesis of phosphoadenosine phosphosulphate (PAPS), and the sulfotransferase KpsS (Towsend II *et al.*, 2008). PAPS is also required for sulphation of the lipochitin-oligosaccharide Nod factor, a symbiotic signal. Interestingly, KpsS lacked homology to known sulphotransferases but was highly similar to proteins with unknown function in diverse bacteria that interact with eukaryotes. Taken together, the instances mentioned here indicate that polysaccharide sulphation could have a specific function during bacterial–plant interaction.

Lipopolysaccharides

The rhizobial LPS has a typical structure and is composed of the lipid A, core polysaccharide and the O-antigen chain. The rhizobial lipid A generally consists of glucosamine or 2,3-di-aminoglucose disaccharide, 6′-linked to a core polysaccharide through a Kdo, and carry four to six α-hydroxy fatty acids. Disaccharide sugars can be oxidized or phosphorylated. All rhizobia, with a possible exception of *Azorhizobium caulinodans*, contain a very-long-chain fatty acid modification (VLCFA) in their lipid A molecules (Bhat *et al.*, 1991). The *S. meliloti* VLCFA were determined as 27-OHC28:0, 27-O(βOmeC4:0) C28:0, and as an additional minor component 29-OHC30:0 (Ferguson *et al.*, 2004). The lipid A and core polysaccharide structures were resolved for *S. meliloti*, *R. etli* CE3, *R. leguminosarum* 3855, *R. leguminosarum* bv. *phaseoli*, *Mesorhizobium huakuii*, *S. fredii*, and *Rhizobium* sp. NGR234 LPS (Becker *et al.*, 2005).

In *S. meliloti*, *M. loti*, and *Azospirillum brasiliense* LPS are sulphated (Cedergren *et al.*, 1995, Cronan *et al.*, 2004; Vanbleu *et al.*, 2005; Townsend II *et al.*, 2006). Alike KPS sulphation, LPS sulphation was metabolically coupled to Nod factor sulphation via PAPS.

Structures of O-antigen polysaccharides were also determined for *R. etli* (Forsberg *et al.*, 2000), *Rhizobium* sp. NGR234 (Reuhs *et al.*, 2005), *M. loti* (Turska-Szewczuk *et al.*, 2007) and *R. leguminosarum* 3841 (Forsberg and Carlson, 2008). *R. etli* and *R. leguminosarum* 3841 O-antigen polysaccharides displayed some unusual features such as uniform size, low molecular mass, and a backbone containing uncommon and/or previously unreported amino sugars. Extensive O-acetylation and O-methylation of side chain residues were characteristic for the O-antigen polysaccharides from all analysed species.

In contrast to most other rhizobia, some *Sinorhizobium* spp. are either lacking the O-antigen in their LPS or have a conserved O-antigen structure throughout the genus. Sinorhizobia could be classified into two major LPS core serogroups, which did not correspond to species (Reuhs *et al.*, 1998). Since the antigenic properties of these strains are apparently not dependent on LPS, the strain-specific KPS were therefore suggested to function as antigenic determinants.

Surface polysaccharides and rhizospheric root colonization

Modification of the soil structure

During the free-living growth in the soil, bacteria need to cope with abiotic stresses, compete with others for nutrients and protect themselves from plant antimicrobials. Desiccation is a very common abiotic stress that affects both bacteria and plants. Inoculation of plants with EPS-producing rhizobacteria, such as *Pantoea agglomerans* and *Paenibacillus polymyxa* (Amellal *et al.*, 1998; Bezzate *et al.*, 2000), *Rhizobium* sp. YAS34 (Alami *et al.*, 2000) and *Rhizobium* sp. KYGT207 (Kaci *et al.*, 2005) increased root-adhered soil volume per root dry mass, soil macropore volume or its water stability. Thus, EPS production by bacteria in the rhizosphere can enhance water and nutrient supply of the plant and therefore improve the environmental conditions for root colonization by bacteria.

Resistance to stresses in the soil

In arid soils, increased salinity is a common factor that negatively affects root colonization by bacteria. Rhizobial salt stress responses include accumulation of osmoprotectant molecules (for *S. meliloti*, recently reviewed by Vriezen *et al.*, 2007). However, long-term salt tolerance can include cell aggregation and increased production of EPS in *S. meliloti* and *Azospirillum brasiliense* (Rüberg *et al.*, 2003; Chowdhury *et al.*, 2007).

Surface polysaccharides contribute to the protection of rhizobia against plant antimicrobials. *S. meliloti* resistance to the antimicrobial cationic peptide protamine was decreased in mutants affected in EPS and cyclic β-glucan production (Nogales *et al.*, 2006). Competitive maize rhizosphere and root colonization by *Rhizobium tropici* CIAT899 was affected in mutants lacking the O-antigen and part of the lipid A core, which correlated with increased susceptibility to maize antimicrobial 6-methoxy-2-benzoxazolinone (Ormeño-Orrillo *et al.*, 2008).

Attachment, aggregation and biofilm formation

Attachment of the bacteria to the plant roots is important for both rhizospheric root colonization and the initiation of symbiosis. Two steps in the attachment are distinguished (Rodríguez-Navarro *et al.*, 2007). The primary attachment involves either plant lectins that bind the bacterial surface structures (polysaccharides or proteins), or the bacterial protein rhicadhesin, its plant receptor and calcium ions. The primary attachment is weak and reversible. Second, irreversible attachment to the plant surface is mediated by agglutination of the cellulose fibrils, produced by the bacteria.

In the attachment of *R. leguminosarum* bv. *viciae* to the roots of its symbiotic partner *Vicia sativa*, both stages involve exopolysaccharides. The recently discovered *R. leguminosarum* RBL5523 surface polysaccharide glucomannan has a high affinity to the plant lectins and is localized on the cell pole, which is attached to the root surface (Laus *et al.*, 2006). This polysaccharide was specifically required for the attachment to root hairs and biofilm formation on the root hair surface. Glucomannan was not required for root hair infection and formation of the nodules, but it was critical for competitive nodulation, suggesting an important role of the attachment in competitiveness (Williams *et al.*, 2008).

Attachment of rhizobia to roots and soil can involve different surface structures. For example, in *R. leguminosarum*, cellulose fibrils were involved in the second step of attachment and biofilm formation on the root hairs, though they were not essential for nodulation or biofilm formation on inert surfaces (Laus *et al.*, 2005a). Interestingly, EPS produced by *R. leguminosarum* and *A. brasilense* are essential for the attachment of these bacteria to the arbuscular mycorrhizal structures (Bianciotto *et al.*, 2001). Since arbuscular mycorrhizal fungi are common symbionts of the legume plants, this attachment may facilitate colonization of new rhizospheres by the bacteria.

During the free-living growth in the soil, bacterial survival is often linked to the formation of aggregates. A novel mechanism of the *A. brasilense* cell aggregation that involves EPS and an endogenously produced lectin, was recently proposed by Mora *et al.* (2008). In the stationary phase of growth, *A. brasilense* produces an arabinose-rich EPS, which exhibits a great cell aggregation activity (Bahat-Samet *et al.*, 2004). Recently, a lectin produced by this bacterium was discovered, which can specifically recognize and bind EPS (Mora *et al.*, 2008). A function of this lectin was proposed in the adhesion between the cells through EPS bridges.

Plant-associated bacteria can form biofilms on inert surfaces such as sand or glass, on soil and on plant roots (Danhorn and Fugua, 2007; Rudrappa *et al.*, 2008). In *R. leguminosarum* bv. *viciae*, acidic EPSs are indispensable for biofilm formation on the roots. Biofilm formation by *S. meliloti* and *R. leguminosarum* was recently documented (Rinaudi *et al.*, 2006; Fujishige *et al.*, 2006; Wells *et al.*, 2007). Fujishige *et al.* (2006) reported that *S. meliloti* exopolysaccharide-deficient mutants as well as exopolysaccharide overproducers exhibited reduced biofilm phenotypes that correlated with their nodulation abilities. However, another group has shown increase in attachment and biofilm formation capabilities of a succinoglycan overproducing strain (Wells *et al.*, 2007). Thus, the exact role of *S. meliloti* EPS in biofilm formation remains to be elucidated.

EPS functions in the development of nitrogen-fixing symbioses

In the course of nodule development, attachment to the root hair surface is followed by root hair curling and formation of an infection thread. To reach the cortical cell layers, where differentiation to the intracellular bacteroids can occur, bacteria need to survive inside of the infection threads and promote their growth. Depending on the further development, infection can lead to the formation of either determinate or indeterminate nodules. In the spherical determinate nodules, bacteria are released from the infection threads into plant cells during the first few days of nodule development; the nodule grows further due to multiplication of the infected cells and bacteroids inside of them. In the elongated indeterminate nodules, infection threads keep growing during the whole nodule life time and bacteria are continuously released into the new layers of plant cells. Bacterial exopolysaccharides are crucial for the symbiotic process in the indeterminate nodules and are dispensable in the determinate ones. The intriguing fact is that depending on the nodule type, even the same bacterium can require its EPS for infection that results in formation of indeterminate nodules (*M. loti/Leucaena leucocephala*), but not require their EPS if determinate nodules are formed *(M. loti/Lotus pedunculatus)* (Hotter and Scott, 1991). The difference in infection thread morphology, namely much thinner threads in the indeterminate nodules, is considered as one of the reasons for these opposite EPS requirements. A narrow infection thread means a closer physical contact between the symbiotic partners, which may explain the particular importance of EPS as a component of the infection thread matrix. Thus, the whole symbiotic infection process is governed by the plant partner, those structural or functional requirements for the bacterial surface must be fulfilled by a rhizobial partner.

Involvement of EPS in the initiation of symbiosis was documented for *S. meliloti* (Niehaus and Becker, 1998), *Rhizobium* sp. NGR234 (Staehelin *et al.*, 2006), *R. trifolii* (Djordjevic *et al.*, 1987); *R. leguminosarum* (Laus *et al.*, 2004), *M. loti* (Hotter and Scott, 1997) and the recently characterized rhizobium *Mesorhizobium tianshanense* (Wang *et al.*, 2007). Exopolysaccharide-deficient mutants of these species are unable to develop nitrogen-fixing nodules on the host plant roots due to defects at the early stages of nodule formation. If infection threads are initiated, they abort very early and bacteria even are unable to reach the cortical cell layer. As a consequence, only nodule-like structures, called pseudonodules, can be formed, which do not contain nitrogen-fixing bacteroids.

Requirement of EPS for symbiotic infection by rhizobia was first demonstrated as succinoglycan-deficient mutants of *S. meliloti* Rm1021 failed to infect *Medicago sativa* roots (Chakravorty *et al.*, 1982; Leigh *et al.*, 1985). The *S. meliloti–Medicago* symbiosis was extensively studied for more than two decades and presently constitutes a model for symbiotic plant–bacterial interactions. Therefore, in the next subchapters, we will particularly focus on the symbiotic function of *S. meliloti* EPS.

The reference strain, *S. meliloti* Rm1021, does not produce symbiotically active EPSII, due to the natural inactivation of the regulatory gene *expR* (Pellock *et al.*, 2002). Galactoglucan, produced by other *S. meliloti* strains, can substitute for EPSI in the invasion, although it was reported to be less efficient (Gonzalez *et al.*, 1996). Using a constitutively expressed green fluorescent protein to visualize bacteria, Pellock and collaborators (2000) analysed the morphology of the infection threads, promoted by *S. meliloti* strains that produced solely EPSI or EPSII. They observed that EPSII-promoted infection threads aborted in about 50% of the root hairs before reaching the cortex and were wider than the EPSI-promoted ones. In comparison, the EPSI-promoted infection threads successfully reached the root cortex in 98% of all cases. Thus, at least one of the EPS in its symbiotically active form is required for a successful infection thread progression in the *S. meliloti–Medicago* symbiosis.

Role of EPSs in signalling

LMW fractions of succinoglycan or galactoglucan were suggested to act as signals for the plant at the stage of infection thread progression. The infection defects of EPS-deficient mutants could be suppressed by addition of picomolar amounts of the LMW fractions of either EPSI or EPSII, purified from *S. meliloti* EPS preparations, to the inoculated roots (Gonzalez *et al.*, 1996; Wang *et al.*, 1999). Succinyl and acetyl modifications of the succinoglycan polysaccharide chains contributed to its symbiotic function (Cheng and Walker, 1998b). Acetyl modifications enhanced the efficiency of infection thread initiation and elongation but were not absolutely required for nodule formation. In contrast, non-succinylated succinoglycan was only present in the HMW form and was symbiotically inactive. In *S. meliloti*, the endoglucanases ExoK and ExsH contribute to LMW succinoglycan production by cleavage of the nascent HMW polymer during its export (York *et al.*, 1998a). Proper acetylation and succinylation of the succinoglycan were required for an effective cleavage by endoglucanases (York *et al.*, 1998b). Thus, the succinyl and acetyl modifications can be important for both production of LMW succinoglycan and its structural recognition by the plant.

Since EPSs probably act as signal molecules, they could also function in the determination of the host specificity, which is still a matter of discussion. The early studies of the *R. leguminosarum* EPS provided contradictory data about its involvement in the host range determination (Philip-Hollingsworth *et al.*, 1989; O'Neill *et al.*, 1991). Recently, new studies again provided controversial results. In coinoculation experiments with different EPS-producing rhizobia, only the identical or similar EPS produced by *R. leguminosarum* bv. *trifolii* and *R. tropici* could restore infection by an EPS-deficient mutant of *R. leguminosarum* bv. *trifolii* (van Workum *et al.*, 1998; Laus *et al.*, 2005b). This may support the hypothesis that the EPS structure could determine the host specificity of nodulation. However, introduction of the *R. leguminosarum* bv. *viciae* symbiotic plasmid into the heterologous strains, which failed to restore infection in the coinoculation experiments, allowed them to nodulate vetch. This suggests that there are no structural requirements for the host range determination by EPS (Laus *et al.*, 2005b). Recently, studying the strain-ecotype specificity of the *S. meliloti/Medicago truncatula* symbiosis, Simsek and collegues could observe a correlation between the host specificity and the degree of succinoglycan oligosaccharide substitution with succinate and malate (Simsek *et al.*, 2007). Taken together, the instances outlined above indicate a need for a more detailed understanding of the role of EPS in host range determination.

Role of EPS in protection against the plant defence response.

The second important function in infection thread progression assigned to EPS is modulation or evasion of plant defence responses. It is known that infection threads induced by EPS-deficient bacteria exhibit symptoms of plant defence responses such as thickening of the cell walls of the cortical cells and their incrustation with autofluorescing phenolic compounds (Parniske *et al.*, 1994). Addition of *S. meliloti* LMW EPSI to alfalfa cell cultures suppressed the alkalinization induced by yeast elicitor (Niehaus *et al.*, 1996). This effect could not be observed when HMW EPS or heterologous EPS were added. This showed that LMW EPSI is recognized by the plant defence system and can specifically quench plant defence reactions in the infection thread. Moreover, the EPS layer around the bacterial cell can protect it against harmful substances such as reactive oxygen species, antibiotics and cationic peptides acting as a diffusion barrier. *Azorhizobium caulinodans* cells, producing EPS were less susceptible to hydrogen peroxide, generated during the invasion of *Sesbania rostrata* (D'Haeze *et al.*, 2004).

The results of a recent transcriptomic study of *Medicago truncatula* roots inoculated with *S. meliloti* have supported both a signalling function of succinoglycan and its involvement in the suppression of plant defence (Jones *et al.*, 2008). The microarray analysis was carried out to compare the transcriptional profiles of curled root hairs that were formed three days after inoculation with a wild-type or an EPSI-deficient mutant. At this early stage of symbiotic development, the infection threads are still not starting to form and the curled root hairs containing either wild-type bacteria or EPSI-deficient bacteria are morphologically indistinguishable. However, expression of plant defence genes was significantly increased and expression of protein turnover genes was decreased in the absence of EPSI. Therefore, EPSI can act as a signal for the inhibition of plant defence already at the stage of the infection thread initiation.

Shutdown at the nitrogen fixation stage

In contrast to the situation in the infection threads, where EPS production is critical for the successful symbiotic development, in the nitrogen fixation zone of the nodules formed by *S. meliloti* on *Medicago* roots, EPS biosynthesis genes (*exo* and *exp*) are likely to be repressed. The early study by Reuber *et al.* (1991) provided evidences about a repression of *exo* genes in the nitrogen-fixing zone of the nodule. Using a transcriptomic approach, we analysed global gene expression patterns in bacteria and bacteroids, isolated from 21-day-old nodules formed on *Medicago sativa* by *S. meliloti* Rm1021 (Becker *et al.*, 2004). An increase in transcription of *exo* genes in comparison to the free-living cells was observed, which correlated with elevated transcription of *nod* genes and catalase genes, which are characteristic for bacteria in the infection threads. These nodules had a mature nitrogen fixation zone but still contained a significant amount of growing infection threads. In a similar microarray study, performed by Barnett and colleagues (2004), 33- to 35-day-old nodules formed by Rm1021 on *Medicago truncatula* were used for comparison of symbiotic and free-living bacterial expression profiles. In that study, expression of *exo* genes was not detected, with exception of *exoX*, a posttranscriptional inhibitor of EPS synthesis that showed increased expression in bacteroids. This correlated with the absence of significant expression of *nod* genes in the bacteroid preparation. In the older nodules, the infection zone is much less prominent and most part of the nodule is occupied by infected cells that contain nitrogen-fixing bacteroids. Although these two studies were performed with two different host plants, they point to a repression of *exo* gene expression, which correlates with intracellular bacteroid differentiation and nitrogen fixation.

S. meliloti may produce a new symbiotically active polysaccharide

Although the EPS content of *S. meliloti* was studied extensively, there are recent genetic evidences for the existence of a new undiscovered polysaccharide that is synthesized from succinoglycan precursors (Beck *et al.*, 2008). Biosynthesis of this new polysaccharide apparently requires the MsbA2 protein, homologous to the *E. coli* transporter of rough LPS (Griffits *et al.*, 2008). *S. meliloti msbA2* mutants displayed invasion defects, characteristic for EPSI-deficient mutants such as thickened infection threads and enhanced plant defence response (Beck *et al.*, 2008). It was proposed that the *msbA2* mutant phenotype may be caused by an inhibitory accumulation of this new polysaccharide.

Role of cyclic glucans in symbiosis

Periplasmic cyclic glucans were reported to be essential for the symbiosis between legumes and rhizobia since a long time. *S. meliloti*, *S. fredii* and *M. loti* mutants defective in the production of these polysaccharides are not able to promote infection thread progression and only induce the formation of bumps, the nodule-like structures (Dickstein *et al.*, 1988; Bhagwat *et al.*, 1992; D'Antuono *et al.*, 2005). Although a *B. japonicum* cyclic glucan-deficient mutant was able to invade the plant tissue, the bacteroids were not properly maintained resulting in the absence of nitrogen fixation (Wiedemann *et al.*, 2004). The symbiotic defects may be caused by the impaired motility, altered surface protein content and failures in hypoosmotic adaptation, reported for the cyclic glucan mutants (Breedveld and Miller, 1994). Cyclic glucans contribute to the bacterial resistance to inducible plant defence metabolites such as phytoalexin and H_2O_2 (Mithöfer *et al.*, 2001). A periplasmic location of the polysaccharide is likely to be important, since a complementation of the symbiotic defect could not be achieved by adding purified β-glucans (Dylan *et al.*, 1990). Conformation of cyclic glucans in respect to their β-linkage can affect their function in some symbioses. A mutant of *S. fredii* that missed the β-1,2 cyclic glucan but still produced β-1,3 and β-1,6 glucans, was able to form an effective symbiosis with two soybean cultivars, McGall and Peking, but was ineffective on the soybean cv. Williams (de Iannino *et al.*, 1996).

In addition to their passive role in the maintenance of bacterial fitness during the symbiotic interaction, cyclic glucans may play an active role in the symbiotic development as suppressors of the immune response. The *B. japonicum* cyclic glucan suppressed an immune response to the pathogen elicitor β-heptaglucoside, presumably through a competition for the soybean receptor (Mithöfer *et al.*, 1996; Bhagwat *et al.*, 1999). Moreover, in *M. loti–Lotus japonicus* symbiosis, cyclic glucans are required for the normal expression of plant genes, associated with the development of a fully functional nodule (D'Antuono *et al.*, 2008).

Symbiotic role of KPS

Promotion of infection thread development

In contrast to the situation in the indeterminate nodules, where EPS is a primary factor of successful invasion, in the determinate ones KPS can be a principal player. *S. fredii* HH103 produces KPS that is a homopolymer of a pseudaminic acid derivative (Gil-Serrano *et al.*, 1999). It was recently demonstrated that this KPS was required for the effective formation of nodules on soybean and pigeon pea, and that EPS can not substitute for the KPS function (Parada *et al.*, 2006). A dramatic reduction of the plant dry weight and formation of pseudonodules and very small nodules was observed when KPS-deficient *S. fredii* were used for inoculation.

In indeterminate nodules, formed by *S. meliloti* Rm41 on *Medicago sativa,* KPS can substitute for EPS in the promotion of infection thread progression (Reuhs *et al.*, 1995). Similarly to EPSII, KPS produced by Rm41 promoted infection thread progression until the cortical cell layers with 50% efficiency and the infection threads were broader than those promoted by EPSI (Pellock *et al.*, 2000). The *rkpZ* mutant of Rm41 produces KPS with higher molecular weight that lacked a LMW fraction, which correlates with its symbiotic deficiency in the absence of EPS (Reuhs *et al.*, 1995). Therefore, LMW KPS is considered as its symbiotically active fraction. *S. meliloti* strains Rm1021 and Rm2011 produce KPS that consists of poly-Kdo and can not replace EPS in the infection threads (Fraysse *et al.*, 2005). However, the amount produced and the molecular weight of the poly-Kdo KPS is increased in Rm1021 upon introduction of the Rm41 *rkpZ* gene, moreover, this KPS could partially replace EPS in the infection (Sharypova *et al.*, 2006).

Although *S. meliloti* KPS is less effective than EPSI, it functions in promotion of infection thread progression and allows development of nitrogen-fixing nodules. The structural difference between EPS and KPS is dramatic. The only similarities between these polysaccharides are the molecular weight and anionic properties of their LMW forms. It remains to be investigated, if these two polysaccharides could be recognized by the same plant sensory system, or if the plant possesses different sensors that are coupled to the same effectors, which would enable infection thread progression and suppression of plant defence.

Contribution of KPS to the symbiotic efficiency and host range specificity

In indeterminate nodules, KPS can also contribute to effectiveness of the symbiotic interaction. A KPS-deficient mutant of *Rhizobium* sp. NGR234 is able to induce nitrogen-fixing nodules on the roots of *Vigna unguiculata* and other host plants, but the symbiosis is less effective that is reflected by a lower number of nodules and decreased shoot dry mass of the plants (Le Quéré *et al.*, 2006). In *M. loti*, sulphation of KPS and LPS affected both formation of the nodules on *Lotus japonicum* and competitiveness of the bacteria inside the nodule. A mutant of *M. loti* that produced non-sulphated KPS and LPS, formed fewer nodules per plant, but was able to outcompete the wild-type bacteria inside the nodule and has a higher growth rate (Townsend II *et al.*, 2006).

In *S. meliloti* and *S. fredii*, KPS and not LPS determine the antigenic variation of strains (Reuhs *et al.*, 1998), therefore, a possible role of KPS in determining the host specificity of symbiosis was suggested. A correlation between KPS structure and host range specificity was observed for some strains of *S. fredii* (Rodriguez-Carvajal *et al.*, 2005); however, the available structural information was considered insufficient to allow reliable classification.

Role of LPS in the symbiosis

An intact structure of O-antigen is often required for successful symbiosis and is especially important for symbiotic interactions that result in determinant nodules. Studies with rhizobial strains that have defined structural LPS defects have indicated that intact LPS produced at normal levels is required for an efficient symbiosis with the host plant. Truncation or absence of the O-chain or only partial substitution of the core polysaccharide with an O-chain result in defects in endocytosis or bacteroid development in *R. leguminosarum, M. loti* and *R. etli* (Fraysse *et al.*, 2002; Forsberg *et al.*, 2003; Reuhs *et al.*, 2005; Turska-Szewczuk *et al.*, 2007). Recently, an alteration in O-acetylation or O-methylation of the O-antigen produced by *M. loti* mutant 2213.1 (Turska-Szewczuk *et al.*, 2008b) and shifts in the molecular weight distribution of the smooth LPS in the *msbA2* mutant of *S. meliloti* (Beck *et al.*, 2008) were shown to be associated with the symbiotic defects of these strains.

Structural changes of LPS structure during symbiotic differentiation

The role of LPSs in the symbiotic process is complex. First, it ensures membrane integrity and a proper hydrophobicity of the surface in the changing environment that bacteria encounter during symbiosis. The LPS structure and composition changes as a response to differing levels of oxygen, pH, ion composition, carbon source and plant-derived compounds as well as in the symbiotic state (Kannenberg and Brewin, 1989; Bhat and Carlson, 1992; Tao *et al.*, 1992; Reuhs *et al.*, 1994; Kannenberg *et al.*, 1998; Noel and Duelli, 2000; Forsberg and Carlson, 2008). An increase of LPS hydrophobicity is characteristic for bacteroids (Kannenberg and Carlson, 2001; Forsberg and Carlson, 2008). Recent studies showed that in the *Rhizobium* sp. NGR234, an additional LPS fraction appeared in cells treated with apigenin, a plant flavonoid (Fraysse *et al.*, 2002). The LPS structure was shifted from a structurally complex, highly branched anionic rough LPS lacking O-antigen, to an endogenously methylated, hydrophobic rhamnan O-antigen, attached to a modified core-lipid A carrier (Reuhs *et al.*, 2005). This inducible LPS fraction was not produced by a symbiotically ineffective *fixF* mutant and it was detected in the bacteroids isolated from *V. unguiculata* infected with NGR234. Coordination of rhamnan synthesis with rhizobial release from infection threads was considered as part of the symbiotic interaction (Broughton *et al.*, 2006). Interestingly, this rhamnan O-antigen was very similar to the A-band O antigen of *Pseudomonas aeruginosa*, which is associated with a longer duration of the chronic cystic fibrosis lung infection. Another kind of symbiotic modification of the LPS structure that can change the surface hydrophobicity, is an alteration of O-chain methylation LPS from bacteroids and anthocyanin-treated *R. etli* cells was modified by addition of a single methyl group to the 2-position of a fucosyl residue in one of the five O-chain trisaccharide repeat units (Noel *et al.*, 2004). This modification was already sufficient to change the antigenic properties of the strain since the induced cultures lost the reactivity to a particular antibody to which the non-induced cultures readily reacted. In *R. etli* bacteroids, acylation of the lipid A changes. It contains no β-hydrohymyristic acid acyl residue and consists exclusively of tetra-acylated species, compared with the presence of both tetra- and penta-acylated lipid A from laboratory cultures (D'Haeze *et al.*, 2007). Changes in lipid A acylation are characteristic for intracellular pathogens. As an example, addition of palmitate was observed in *Pseudomonas aeruginosa* isolated from infants with cystic fibrosis that developed a chronic airway infection (Trent *et al.*, 2001). Therefore, it can be involved in the adaptation of the bacterial surface to intracellular conditions.

Interestingly, the presence of rhamnose in the O-antigen seems to correlate with the symbiotic activity of LPS. In addition to NGR234 LPS, described above, rhamnose was present in the O-antigen of a symbiotically enhanced *M. loti* strain (Turska-Szewczuk *et al.*, 2008a), although it was absent from the O-antigen of the parental strain.

LPS as a signal molecule

Azorhizobium caulinodans is a nitrogen-fixing symbiont of the tropical water-tolerant plant *Sesbania rostrata*. During invasion by crack entry, bacteria proliferate in the epidermal fissures at the lateral root base or at the adventitious root primordia on the stems, which leads to nodule formation (Goormachtig *et al.*, 2004). A particular feature of this symbiosis is the use of bacterial LPS as a signal molecules during early stages of nodule formation. Mathis and colleagues (2005) have studied the *oac2* mutant of *A. caulinodans* that produces a reduced amount of exopolysaccharides and a modified LPS with reduced rhamnose content and higher electrophoretic mobility. This mutant displayed symbiotic defects, typical for EPS-deficient rhizobia, namely thickened infection threads and absence of the bacteroids. Co-inoculation of this mutant with a strain unable to produce Nod factors, as well as addition of LPS purified from the *A. caulinodans* wild-type cells allowed formation of nitrogen-fixing nodules. It remains to be investigated, whether the signal function of the rhamnose-containing LPS is a unique feature of the *A. caulinodans-S. rostrata* symbiosis or if it could be extended to other symbiotic interactions as well.

Role of LPS in suppression of plant defence

A particular feature of the rhizobial LPS is a VLCFA modification of the LPS core. Recent studies suggest a special role of this fatty acid modification in the survival inside of eukaryotic cells. VLCFA is important for proper *R. leguminosarum* and *S. meliloti* bacteroid differentiation (Vedam *et al.*, 2004; Ferguson *et al.*, 2005; Vedam *et al.*, 2006). Mutants lacking this modification in their lipid A were able to induce nitrogen-fixing nodules, but were delayed in the onset of nodulation and had aberrant large bacteroids or several bacteroids enclosed by a single symbiosome membrane. VLCFA modification of the lipid A was also essential for the intracellular survival of the mammalian pathogen *Brucella abortus*, which is taxonomically very close to rhizobia (Ferguson *et al.*, 2004). Recent studies provided evidences for a possible role of *Arabidopsis* VLCFA in regulation of the hypersensitive response (Raffaele *et al.*, 2008). It is an interesting question, if bacterial VLCFA could modulate the plant immune response by mimicking the internal plant signals.

In plant–pathogenic interactions, bacterial LPS are strong inducers of host defence reactions. In contrast, LPS produced by symbionts can act as suppressors of the host plant immune response. LPS of *S. meliloti* induced an oxidative burst in cell cultures of the non-host plant *Nicotiana tabacum*, but was able to suppress the defence reaction of host plant cell cultures, when this reaction was induced with invertase (Albus *et al.*, 2001; Scheidle *et al.*, 2005). The defence reaction included hydrogen peroxide production and alkalinization of the medium, and it could be suppressed by addition of entire LPS, lipid A or even lipid A lacking the VLCFA modification in a concentration-dependent manner (Scheidle *et al.*, 2005). Moreover, addition of *S. meliloti* LPS lowered expression levels of plant defence-associated genes in invertase-induced *Medicago truncatula* root cell cultures (Tellsröm *et al.*, 2007).

To close this section, it is important to note that surface polysaccharides are at the first site of the contact between a bacterium and a plant and therefore are important for plant recognition. When EPSs are produced, they constitute the outer layer of the surface, and may act in the infection threads of indeterminate nodules as signals, a diffusion barrier for harmful substances and probably also in masking the underlying structures, which could provoke an immune response. During intracellular infection, a correct structure of LPS, which can be different from that of free-living bacteria, is necessary for development and maintenance of bacteroids.

Regulation of surface polysaccharide biosynthesis

To ensure an optimal composition of the bacterial surface during free-living growth, aggregation and interaction with a plant, polysaccharide biosynthesis must be dynamically regulated. Availability of the sugar precursors for polysaccharide biosynthesis, which could be used by different metabolic pathways, can serve as a limiting factor for polysaccharide production. However, the most common mode of regulation is transcriptional and, to less extent, post-transcriptional regulation of polysaccharide biosynthesis gene expression. Our understanding of molecular mechanisms for the environmental and symbiotic regulation of rhizobial surface polysaccharide biosynthesis has been greatly improved in the last years (Fig. 11.1).

ExoR–ExoS–ChvI regulatory cascade

ExoS and *ChvI* constitute a two-component regulatory system that is able to activate succinoglycan

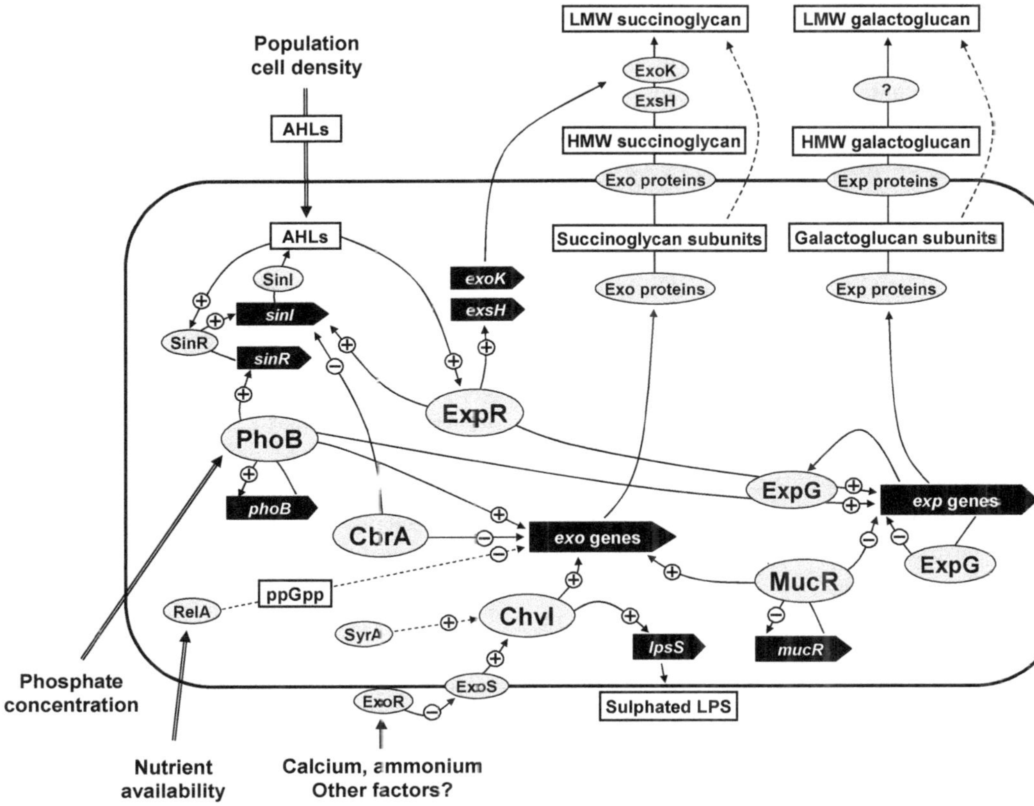

Figure 11.1 Regulation of exopolysaccharide biosynthesis in *S. meliloti*.

production in *S. meliloti* (Cheng and Walker, 1998a). The transcriptional regulator ChvI is activated by the sensor histidine kinase ExoS (Cheng and Walker, 1998a), which is, in turn, apparently negatively controlled by the periplasmic protein ExoR (Reed *et al.*, 1991; Wells *et al.*, 2007). Since *exoS* and *chvI* knock-out mutations are lethal, it is not clear whether ChvI is essential for transcription of *exo* genes or only modulates their expression levels. In *S. meliloti*, a negative control of EPSI production by ExoR is essential for the symbiosis since its *exoR* mutant greatly overproduces succinoglycan and induces only white non-fixing nodules on the plant roots (Doherty *et al.*, 1988). Genetic studies suggested a role of ExoR in the sensing of ammonium and calcium, which inhibits ExoS-ChvI-activated EPSI production and LPS sulphation via repression of *exo* and *lpsS* gene transcription (Ozga *et al.*, 1994; Keating *et al.*, 2007b). Strikingly similar to ExoR, the small symbiotic regulatory protein SyrA (Mulligan and Long, 1989) exerts its activatory action on *exo* and *lpsS* genes through the ExoS–ChvI regulatory pair (Keating *et al.*, 2007a). In *R. leguminosarum* bv. *viciae*, a highly similar *exoR* homologue was identified (Reeve *et al.*, 1997). However, an *exoR* mutant of *R. leguminosarum* bv. *viciae* was able to induce nitrogen-fixing nodules despite EPS overproduction.

CbrA

In the stationary phase of growth, expression of *exo* genes, succinoglycan production and accumulation of its LMW forms are increased in the *cbrA* mutant of *S. meliloti* (Gibson *et al.*, 2006). CbrA is a putative two-component histidine kinase whose action is growth phase dependent. Like the succinoglycan-overproducing *exoR* mutant, the *cbrA* mutant is severely impaired in symbiosis and induces only white non-fixing nodules and pseudonodules on alfalfa. Transcriptomic studies revealed that, in addition to repression of *exo* gene expression, *cbrA* increases transcription of genes related to motility and chemotaxis as well as to envelope functions and lowers transcription of the AHL synthase gene *sinI* and the putative polysaccharide biosynthesis gene *pssF* (Gibson *et al.*, 2007). Therefore, in the *cbrA* mutant the overproduction of EPS correlated also with other alterations in cell physiology. It remains to be investigated, how the activity or abundance of CbrA itself is controlled in *S. meliloti*.

MucR and RosR

In *S. meliloti* Rm2011, both *exo* and *exp* genes are subject to regulation by MucR, a zinc finger protein. *mucR* is necessary for production of HMW EPSI, since only LMW succinoglycan is synthesized if this gene is mutated (Keller *et al.*, 1995). Although MucR is a DNA-binding protein, it only slightly activates *exo* gene transcription and was suggested to post-transcriptionally act on *exo* genes (Bertram-Drogatz *et al.*, 1998). MucR is a principal repressor of *exp* gene transcription in *S. meliloti* Rm2011 (Keller *et al.*, 1995; Rüberg *et al.*, 1999). In the absence of a functional *mucR* gene, Rm2011 constitutively produces EPSII (Keller *et al.*, 1995). However, this EPSII lacks a symbiotically active LMW fraction (González *et al.*, 1996). Binding of the MucR protein to the promoter regions of *exp* genes was recently demonstrated in our laboratory (Bahlawane *et al.*, 2008). MucR was suggested to interact with the internal *exp* regulator ExpG. ExpG regulates *exp* gene expression both positively (in the presence of activated PhoB or ExpR, see below) and negatively (in the absence of MucR) (Bahlawane *et al.*, 2008; McIntosh *et al.*, 2008), which points to possible protein–protein interactions between these regulators. Interestingly, in the *S. meliloti* strain EF1, MucR is an essential activator of *exp* gene expression, although it differs from the corresponding Rm2011 protein in only one amino acid (Martín *et al.*, 2000; Lloret *et al.*, 2002). In *R. leguminosarum* bv. *trifolii*, RosR, a MucR homologue, is a positive regulator of acidic polysaccharide biosynthesis and the *rosR* mutant is symbiotically deficient, in contrast to symbiotically effective *mucR* mutants of *S. meliloti* Rm2011 (Janczarek and Skorupska, 2007). In *R. etli*, *rosR* represses exopolysaccharide biosynthesis genes. The *R. etli rosR* mutant is symbiotically effective but is strongly outcompeted by the wild type in the co-inoculation assays (Bittinger *et al.*, 2000). Although much information on the regulation by MucR and RosR is available, the physiological signals that these proteins could transmit are still unknown.

ExpR

Polysaccharide biosynthesis is often positively regulated by quorum sensing systems that link it to cell aggregation and biofilm formation. Quorum sensing is a mechanism to monitor and control population cell density, which is based on production and perception of small signal molecules (González *et al.*, 2006; Turovskiy *et al.*, 2007; Sanchez-Contreras *et al.*, 2007). In rhizobia, the core elements in quorum sensing regulation are AHL synthases and LuxR-type transcription factors that change their regulatory activity upon binding acyl-homoserine lactons (AHLs) (González and Marketon, 2003). In the reference *S. meliloti* strain Rm1021 and its close relatives, the galactoglucan biosynthesis genes are silent unless the strains are grown in the condition of phosphate starvation. This absence of *exp* gene expression under standard growth conditions is due to the natural inactivation of the *expR* gene by insertion of an IS element (Pellock *et al.*, 2002). In the spontaneous revertant strain Rm8530 that contains a functional *expR* gene, *exp* genes are constitutively expressed and abundant EPSII is produced. ExpR is a central regulator of quorum sensing that responds to AHLs. In *S. meliloti*, production of the AHLs is controlled by the *sinRI*-encoded system, where SinI is an AHL synthase and SinR a LuxR-type transcriptional activator (Marketon *et al.*, 2002). Mutants in the *sinRI* genes are able to induce nitrogen-fixing nodules although their amount per plant is reduced in comparison with a *sinRI* positive strain (Gao *et al.*, 2005). A transcriptomic study of *expR*- and *sinRI*-dependent gene expression revealed that, in the presence of *expR*, the Sin system regulated a multitude of *S. meliloti* genes, including genes that participate in LMW succinoglycan production, motility and chemotaxis and other cellular processes (Hoang *et al.*, 2004).

Recently, transcriptional activation of the endoglucanase genes *exoK* and *exsH* by ExpR was demonstrated (Glenn *et al.*, 2007). Since the majority of the symbiotically active LMW EPSI fraction is generated by the action of these two enzymes (York and Walker, 1997), activation of their expression by the quorum sensing system can greatly contribute to the efficiency of infection. Moreover, *expR* is the only known factor that promotes biosynthesis of LMW galactoglucan (González *et al.*, 1996). Therefore, it can not be ruled out that ExoK and ExsH are also involved in the endoglucanase-mediated cleavage of galactoglucan.

In the course of a study conducted in our laboratory, binding of purified ExpR to the DNA regions upstream of *sinI, expA1* and *expE1*. This binding was enhanced in presence of oxo-C_{14}HL. A consensus sequence of these binding sites was defined (McIntosh *et al.*, 2008), which allowed identification of two additional ExpR-binding sites in the promoter regions of *exoI* and *exsH* genes that are involved in succinoglycan biosynthesis. For the positive regulation of the galactoglucan biosynthesis genes by ExpR, the internal *exp* gene regulator ExpG was required. The relative location of the previously identified ExpG- (Bartels *et al.*, 2003; Baumgarth *et al.*, 2005), MucR- (Bahlawane *et al.*, 2008) and ExpR-binding sites (McIntosh *et al.*, 2008) suggests that protein–protein interactions and probably some competition for the DNA target sites are involved in the complex regulation of *exp* gene expression and EPSII production.

EPS and motility

Opposite regulation of EPS biosynthesis and motility in *S. meliloti* seems to be a conserved feature. Motility and/or motility gene expression are negatively regulated by the ExoR–ExoS–ChvI regulatory cascade (Yao *et al.*, 2004; Wells *et al.*, 2007) and ExpR (Hoang *et al.*, 2008), which are known to activate *exo* genes, and positively by CbrA that represses EPS biosynthesis (Gibson *et al.*, 2007). Such opposite regulation was recently documented in pathogenic bacteria. There, it was governed by regulators that respond to the level of cyclic di-GMP, which is synthesized by a specialized phosphodiesterase (Hickman *et al.*, 2008; Nakhamchik *et al.*, 2008). It would be interesting to know if cyclic di-GMP could play this role in rhizobia as well.

S. meliloti motility gene expression is controlled by two principal regulators, VisNR and Rem (Sourjik *et al.*, 2000; Rotter *et al.*, 2006). Hoang *et al.* (2008) studied *S. meliloti* motility gene expression and flagellation at different growth phases and demonstrated that the ExpR/Sin quorum-sensing system negatively regulates motility gene expression through the VisN/VisR/Rem relay. In addition, the ExoS-dependent two-component system was shown to repress motility gene expression through VisN and Rem in parallel to quorum sensing (Hoang *et al.*, 2008). Bahlawane *et al.* (2008) have added MucR to the regulatory network for motility gene expression, showing that MucR binds to the promoter region of *rem* and represses its transcription.

PhoR–PhoB

One of the important environmental factors that affect EPS biosynthesis in *S. meliloti* is phosphorus availability. Soluble phosphate is scarce in the soil and therefore phosphorus starvation is a common stress for rhizobia. A response of *S. meliloti* to this stress is governed by the OmpR-type transcriptional regulator PhoB (Bardin and Finan, 1998). Under phosphate limitation, the expression of *exo* genes is enhanced and *exp* gene expression is activated in a *phoB*-dependent manner (Rüberg *et al.*, 1999; Mendrigal and Gonzalez, 2002). Our microarray studies have confirmed *phoB*-dependent activation of *exp* genes and partially *phoB*-dependent increase in *exo* gene transcription (Krol and Becker, 2004). Moreover, we found PhoB-activated transcription of *sinR* as well as transcription of *chvI*. Since both regulators are known to regulate *exo* and *exp* genes positively, they could be involved in a transmission of the phosphate limitation signal via PhoB. PhoB-dependent regulation of *sinR* points to a possible linkage between the phosphate-dependent and quorum sensing regulatory circuits. Repression of the motility genes under phosphate starvation correlated with activation of exopolysaccharide biosynthesis, although this effect was independent of *phoB* (Krol and Becker, 2004).

General stress response

The nutritional status of the cells can influence EPS production as well. Amino acid starvation induces the biosynthesis of the messenger molecule ppGpp by the ribosome-associated protein RelA. ppGpp is thought to associate directly with the RNA polymerase holoenzyme and to alter its specificity for various promoters. A *S. meliloti relA* mutant overproduces succinoglycan and shows an increased transcription of *exo* genes (Wells and Long, 2002), therefore it was suggested as a negative regulator of EPS biosynthesis. Another general stress response-related mechanism of EPS biosynthesis regulation was documented in *Rhizobium sp.* TAL1145 (Kaufusi *et al.*, 2004). In this strain, a mutation of the alternative sigma factor gene *rpoH2* resulted in a dramatic decrease in EPS production, *exo* gene expression and in the number of nitrogen-fixing nodules per plant.

Regulation of LPS and KPS biosynthesis

In comparison to the extensively studied regulation of EPS biosynthesis, our understanding of the genetic and physiological regulation of KPS and LPS biosynthesis in rhizobia is rather limited. The reason for this is a more complex sugar composition of these polysaccharides and a less obvious phenotypic appearance of the alterations in their expression, which makes it more difficult to study the subtle changes in their structures and production levels. In the last years, it became evident that changes of LPS structure induced during symbiotic interaction are an important part of bacteroid differentiation. In the course of symbiotic interaction or in response to flavonoids, added to the growth medium, the LPS methylation patterns of O-antigen and acylation of lipid A can change and even a new O-antigen can appear (Noel *et al.*, 2004; Reuhs *et al.*, 2005; D'Haeze *et al.*, 2007). In *Rhizobium* sp. NGR234, K-antigen production was found to decline after bacteria were exposed to plant flavonoids, and the decrease coincided with induction of a symbiotically active (bacteroid-specific) rhamnan-LPS, suggesting that an exchange of surface polysaccharides occurs during bacterial differentiation in the developing nodule (Le Quéré *et al.*, 2006).

The regulatory pathway of the flavonoid-dependent induction of the rhamnan O-antigen biosynthesis in *Rhizobium* sp. NGR234 was recently elucidated. It involved the regulatory genes *nodD1* and *ttsI*. *nodD1*, a known flavonoid-dependent transcriptional activator of Nod factor biosynthesis genes, was also responsible for activation of another transcriptional regulator, *ttsI* (Marie *et al.*, 2004). The latter acts on the promoters of the type three secretion system genes and rhamnose biosynthesis genes by binding to the consensus *tts* box sequences (Wassem *et al.*, 2008).

The symbiotic environment is able to trigger the addition of the VLCFA modification to the lipid A in an *acpXL* mutant of *R. leguminosarum*, which is unable to add this modification to its lipid A in the free-living state (Vedam *et al.*, 2006). However, the genetic mechanisms that control these alterations are not yet elucidated.

Conclusions

Bacterial surface polysaccharides are involved in rhizospheric root colonization and establishment of a symbiotic relationship with a host plant. EPS play an important role in the attachment of bacteria to soil particles and roots, as well as in improvement of the soil structure. EPS, KPS and LPS probably serve as signals during the initial phases of the symbiotic plant-bacterial interaction, whereas at the later stages an active remodelling of the cell surface takes place, which includes repression of EPS biosynthesis and structural changes in LPS. A strict control of surface polysaccharide biosynthesis is required to ensure optimal surface composition during both free-living growth and symbiotic interactions. This is especially well-characterized in *S. meliloti*, where a regulatory network that integrates stress- and population density-related physiological stimuli, was demonstrated to regulate EPS biosynthesis. Coupled opposite regulation of exopolysaccharide production and motility is a new feature that has been added to the regulation scheme in the last years. Changes in the rhizobial LPS structure *in planta* show features characteristic for mammalian intracellular pathogens. Although a significant number of the surface polysaccharides were already characterized in rhizobia, new structures are being resolved and new polysaccharide species are being discovered. Knowledge acquired in these fields will improve our understanding of the role of surface polysaccharides in rhizospheric and symbiotic fitness of nitrogen-fixing plant-associated bacteria.

References

Alami, Y., Achouak, W., Marol, C., and Heulin, T. (2000). Rhizosphere soil aggregation and plant growth promotion of sunflowers by an exopolysaccharide-producing

Rhizobium sp. strain isolated from sunflower roots. Appl. Environ. Microbiol. *66*, 3393–3398.

Amellal, N., Burtin, G., Bartoli, F., and Heulin, T. (1998). Colonization of wheat roots by an exopolysaccharide-producing pantoea agglomerans strain and its effect on rhizosphere soil aggregation. Appl. Environ. Microbiol. *64*, 3740–3747.

An, J., Carlson, R.W., Glushka, J., and Streeter, J.G. (1995). The structure of a novel polysaccharide produced by *Bradyrhizobium* species within soybean nodules. Carbohydr. Res. *269*, 303–317.

Bahat-Samet, E., Castro-Sowinski, S., and Okon, Y. (2004). Arabinose content of extracellular polysaccharide plays a role in cell aggregation of *Azospirillum brasilense*. FEMS Microbiol. Lett. *237*, 195–203.

Bardin, S.D., and Finan, T.M. (1998). Regulation of phosphate assimilation in *Rhizobium (Sinorhizobium) meliloti*. Genetics *148*, 1689–700.

Bartels, F.W., Baumgarth, B., Anselmetti, D., Ros, R., and Becker, A. (2003). Specific binding of the regulatory protein ExpG to promoter regions of the galactoglucan biosynthesis gene cluster of *Sinorhizobium meliloti* – a combined molecular biology and force spectroscopy investigation. J. Struct. Biol. *143*, 145–152.

Bahlawane, C., Baumgarth, B., Serrania, J., Rüberg, S., and Becker, A. (2008). Fine-tuning of galactoglucan biosynthesis in *Sinorhizobium meliloti* by differential WggR (ExpG)-, PhoB-, and MucR-dependent regulation of two promoters. J. Bacteriol. *190*, 3456–3466.

Bahlawane, C., McIntosh, M., Krol, E., and Becker, A. (2008). *Sinorhizobium meliloti* regulator MucR couples exopolysaccharide synthesis and motility. Mol. Plant–Microbe Interact. *21*, 1498–1509.

Barnett, M.J., Toman, C.J., Fisher, R.F., and Long, S.R. (2004). A dual-genome symbiosis chip for coordinate study of signal exchange and development in a prokaryote-host interaction. Proc. Natl. Acad. Sci. USA *101*, 16636–16641.

Baumgarth, B., Bartels, F.W., Anselmetti, D., Becker, A., and Ros, R. (2005). Detailed studies of the binding mechanism of the *Sinorhizobium meliloti* transcriptional activator ExpG to DNA. Microbiology *151*, 259–268.

Beck, S., Marlow, V.L., Woodall, K., Doerrler, W.T., James, E.K., and Ferguson, G.P. (2008). The *Sinorhizobium meliloti* MsbA2 protein is essential for the legume symbiosis. Microbiology. *154*, 1258–1270.

Becker, A., Bergès, H., Krol, E., Bruand, C., Rüberg, S., Capela, D., Lauber, E., Meilhoc, E., Ampe, F., de Bruijn, F.J., Fourment, J., Francez-Charlot, A., Kahn, D., Küster, H., Liebe, C., Pühler, A., Weidner, S., and Batut, J. (2004). Global changes in gene expression in *Sinorhizobium meliloti* 1021 under microoxic and symbiotic conditions. Mol. Plant–Microbe Interact. *17*, 292–303.

Becker, A., Fraysse, N., and Sharypova, L. (2005). Recent advances in studies on structure and symbiosis-related function of rhizobial K-antigens and lipopolysaccharides. Mol. Plant–Microbe Interact. *18*, 899–905.

Becker, A., Rüberg, S., Baumgarth, B., Bertram-Drogatz, P.A., Quester, I., and Pühler, A. (2002). Regulation of succinoglycan and galactoglucan biosynthesis in *Sinorhizobium meliloti*. J. Mol. Microbiol. Biotechnol. *4*, 187–190.

Becker, A., Rüberg, S., Küster, H., Roxlau, A.A., Keller, M., Ivashina, T., Cheng, H.P., Walker, G.C., and Pühler, A. (1997). The 32-kilobase *exp* gene cluster of *Rhizobium meliloti* directing the biosynthesis of galactoglucan: genetic organization and properties of the encoded gene products. J. Bacteriol. *179*, 1375–1384.

Becker, A., and Pühler, A. (1998). Specific amino acid substitutions in the proline-rich motif of the *Rhizobium meliloti* ExoP protein result in enhanced production of low-molecular-weight succinoglycan at the expense of high-molecular-weight succinoglycan. J. Bacteriol. *180*, 395–399.

Becker, B.U., Kosch, K., Parniske, M., and Müller, P. (1998). Exopolysaccharide (EPS) synthesis in *Bradyrhizobium japonicum*: sequence, operon structure and mutational analysis of an *exo* gene cluster. Mol. Gen. Genet. *259*, 161–171.

Bertram-Drogatz, P.A., Quester, I., Becker, A., and Pühler, A. (1998). The *Sinorhizobium meliloti* MucR protein, which is essential for the production of high-molecular-weight succinoglycan exopolysaccharide, binds to short DNA regions upstream of *exoH* and *exoY*. Mol. Gen. Genet. *257*, 433–441.

Bezzate, S., Aymerich, S., Chambert, R., Czarnes, S., Berge, O., and Heulin, T. (2000). Disruption of the *Paenibacillus polymyxa* levansucrase gene impairs its ability to aggregate soil in the wheat rhizosphere. Environ. Microbiol. *2*, 333–342.

Bhagwat, A.A., Mithöfer, A., Pfeffer, P.E., Kraus, C., Spickers, N., Hotchkiss, A., Ebel, J., and Keister, D.L. (1999). Further studies of the role of cyclic beta-glucans in symbiosis. An NdvC mutant of *Bradyrhizobium japonicum* synthesizes cyclodecakis-(1→3)-beta-glucosyl. Plant Physiol. *119*, 1057–1064.

Bhagwat, A.A., Tully, R.E., and Keister, D.L. (1992). Isolation and characterization of an *ndvB* locus from *Rhizobium fredii*. Mol. Microbiol. *6*, 2159–2165.

Bhat, U.R., Krishnaiah, B.S., and Carlson, R.W. (1991). Re-examination of the structures of the lipopolysaccharide core oligosaccharides from *Rhizobium leguminosarum* biovar *phaseoli*. Carbohydr. Res. *220*, 219–227.

Bianciotto, V., Andreotti, S., Balestrini, R., Bonfante, P., and Perotto, S. (2001). Extracellular polysaccharides are involved in the attachment of *Azospirillum brasilense* and *Rhizobium leguminosarum* to arbuscular mycorrhizal structures. Eur. J. Histochem. *45*, 39–49.

Bittinger, M.A., and Handelsman, J. (2000). Identification of genes in the RosR regulon of *Rhizobium etli*. J. Bacteriol. *182*, 1706–1713.

Bittinger, M.A., Milner, J.L., Saville, B.J., and Handelsman, J. (1997). *rosR*, a determinant of nodulation competitiveness in *Rhizobium etli*. Mol. Plant Microbe Interact. *10*, 180–186.

Borthakur, D., Barker, R.F., Latchford, J.W., Rossen, L., and Johnston, A.W. (1988). Analysis of *pss* genes of *Rhizobium leguminosarum* required for exopolysaccharide synthesis and nodulation of peas: their primary structure and their interaction with *psi* and other nodulation genes. Mol. Gen. Genet. *213*, 155–162.

Breedveld, M.W., Yoo, J.S., Reinhold, V.N., and Miller, K.J. (1994). Synthesis of glycerophosphorylated cyclic beta-(1,2)-glucans by *Rhizobium meliloti ndv* mutants. J. Bacteriol. *176*, 1047–1051.

Breedveld, M.W., and Miller, K.J. (1994). Cyclic beta-glucans of members of the family *Rhizobiaceae*. Microbiol. Rev. *58*, 145–161.

Broughton, W.J., Hanin, M., Relic, B., Kopcińska, J., Golinowski, W., Simsek, S., Ojanen-Reuhs, T., Reuhs, B., Marie, C., Kobayashi, H., Bordogna, B., Le Quéré, A., Jabbouri, S., Fellay, R., Perret, X., and Deakin, W.J. (2006). Flavonoid-inducible modifications to rhamnan O antigens are necessary for *Rhizobium sp.* strain NGR234-legume symbioses. J. Bacteriol. *188*, 3654–3663.

Castro, O.A., Zorreguieta, A., Ielmini, V., Vega, G., and Ielpi, L. (1996). Cyclic beta-(1,2)-glucan synthesis in *Rhizobiaceae*: roles of the 319-kilodalton protein intermediate. J. Bacteriol. *178*, 6043–6048.

Cedergren, R.A., Lee, J., Ross, K.L., and Hollingsworth, R.I. (1995). Common links in the structure and cellular localization of *Rhizobium* chitolipooligosaccharides and general *Rhizobium* membrane phospholipid and glycolipid components. Biochemistry. *34*, 4467–4477.

Chakravorty, A.K., Zurkowski, W., Shine, J., and Rolfe, B.G. (1982). Symbiotic nitrogen fixation: molecular cloning of *Rhizobium* genes involved in exopolysaccharide synthesis and effective nodulation. J. Mol. Appl. Genet. *1*, 585–596.

Chataigné, G., Couderc, F., and Poinsot, V. (2008). Polysaccharides analysis of sinorhizobial capside by on-line anion exchange chromatography with pulsed amperometric detection and mass spectrometry coupling. J. Chromatogr. A. *1185*, 241–250.

Cheng, H.P., and Walker, G.C. (1998a). Succinoglycan production by *Rhizobium meliloti* is regulated through the ExoS-ChvI two-component regulatory system. J. Bacteriol. *180*, 20–26.

Cheng, H.P., and Walker, G. C. (1998b). Succinoglycan is required for initiation and elongation of infection threads during nodulation of alfalfa by *Rhizobium meliloti*. J. Bacteriol. *180*, 5183–5191.

Chowdhury, S.P., Nagarajan, T., Tripathi, R., Mishra, M.N., Le Rudulier, D., and Tripathi, A. K. (2007). Strain-specific salt tolerance and osmoregulatory mechanisms in *Azospirillum brasilense*. FEMS Microbiol. Lett. *267*, 72–79.

Ciocchini, A.E., Guidolin, L.S., Casabuono, A.C., Couto, A.S., de Iannino, N.I., and Ugalde, R.A. (2007). A glycosyltransferase with a length-controlling activity as a mechanism to regulate the size of polysaccharides. Proc. Natl. Acad. Sci. USA. *104*, 16492–16497.

Cooper, J.E. (2007). Early interactions between legumes and rhizobia: disclosing complexity in a molecular dialogue. J. Appl. Microbiol. *103*, 1355–1365.

Cronan G.E., and Keating D.H. (2004). *Sinorhizobium meliloti* sulfotransferase that modifies lipopolysaccharide. J. Bacteriol. *186*, 4168–4176.

Dazzo, F.B., Truchet, G.L., Sherwood, J.E., Hrabak, E.M., Abe, M., and Pankratz, S.H. (1984). Specific phases of root hair attachment in the *Rhizobium trifolii*-clover symbiosis. Appl. Environ. Microbiol. *48*, 1140–1150.

Danhorn, T., and Fuqua, C. (2007). Biofilm formation by plant-associated bacteria. Annu. Rev. Microbiol. *61*, 401–422.

D'Antuono, A.L., Ott, T., Krusell, L., Voroshilova, V., Ugalde, R.A., Udvardi, M., and Lepek, V.C. (2008). Defects in rhizobial cyclic glucan and lipopolysaccharide synthesis alter legume gene expression during nodule development. Mol. Plant–Microbe Interact. *21*, 50–60.

D'Antuono A.L., Casabuono A., Couto A., Ugalde R.A., and Lepek V.C. (2005). Nodule development induced by *Mesorhizobium loti* mutant strains affected in polysaccharide synthesis. Mol. Plant–Microbe Interact. *18*, 446–457.

de Iannino, I.N., Briones, G., Kreiman, G., and Ugalde, R. (1996). Characterization of the biosynthesis of beta(1–2). cyclic glucan in *R. Fredii*. Beta(1–2) glucan has no apparent role in nodule invasion of Mc Call and Peking soybean cultivars. Cell Mol. Biol. (Noisy-le-grand). *42*, 617–629.

D'Haeze, W., Glushka, J., De Rycke, R., Holsters, M., and Carlson R.W. (2004). Structural characterization of extracellular polysaccharides of *Azorhizobium caulinodans* and importance for nodule initiation on *Sesbania rostrata*. Mol. Microbiol. 52, 485–500.

D'Haeze, W., Leoff, C., Freshour, G., Noel, K.D., and Carlson, R.W. (2007). *Rhizobium etli* CE3 bacteroid lipopolysaccharides are structurally similar but not identical to those produced by cultured CE3 bacteria. J. Biol. Chem. *282*, 17101–17113.

Dickstein, R., Bisseling, T., Reinhold, V.N., and Ausubel, F.M. (1988). Expression of nodule-specific genes in alfalfa root nodules blocked at an early stage of development. Genes Dev. *2*, 677–687.

Djordjevic, S.P., Batley, M., Redmond, J.W., and Rolfe, B. G. (1986). The structure of the exopolysacchride from *Rhizobium sp.* strain ANU280 (NGR234). Carbohydr. Res. *148*, 87–99.

Djordjevic, S.P., Chen, H., Batley, M., Redmond, J.W., and Rolfe, B.G. (1987). Nitrogen fixation ability of exopolysaccharide synthesis mutants of *Rhizobium sp.* strain NGR234 and *Rhizobium trifolii* is restored by the addition of homologous exopolysaccharides. J. Bacteriol. *169*, 53–60.

Doherty, D., Leigh, J.A., Glazebrook, J., and Walker, G.C. (1988). *Rhizobium meliloti* mutants that overproduce the *R. meliloti* acidic calcofluor-binding exopolysaccharide. J. Bacteriol. *170*, 4249–4256.

Drozański, W., Drozańska, D., and Lorkiewicz, Z. (1981). Structure of rigid-layer of *Rhizobium* cell wall. I. Purification on the peptidoglycan from the cellulose microfibrils. Acta Microbiol. Pol. *30*, 355–370.

Epple, G., vanDerDrift, K.M.G.M., Thomas-Oates, J.E., and Geiger, O. (1998). Characterization of a novel acyl carrier protein, RkpF, encoded by an operon involved in capsular polysaccharide biosynthesis in *Sinorhizobium meliloti*. J. Bacteriol. *180*, 4950–4954.

Ferguson, G.P., Jansen, A., Marlow, V.L., and Walker, G.C. (2006). BacA-mediated bleomycin sensitivity in *Sinorhizobium meliloti* is independent of the unusual lipid A modification. J. Bacteriol. *188*, 3143–3148.

Ferguson, G.P., Datta, A., Baumgartner, J., Roop, R.M. 2nd, Carlson, R.W., and Walker, G.C. (2004). Similarity to peroxisomal-membrane protein family reveals that *Sinorhizobium* and *Brucella* BacA affect lipid-A fatty acids. Proc. Natl. Acad. Sci. USA. *101*, 5012–5017.

Ferguson, G.P., Datta, A., Carlson, R.W., and Walker, G.C. (2005). Importance of unusually modified lipid A in *Sinorhizobium* stress resistance an legume symbiosis. Mol. Microbiol. *56*, 68–80.

Forsberg, L.S., Bhat, U.R., and Carlson, R. W. (2000). Structural characterization of the O-antigenic polysaccharide of the lipopolysaccharide from *Rhizobium etli* strain CE3. J. Biol. Chem. *275*, 18851–18863.

Forsberg L.S., and Carlson R.W. (2008). Structural characterization of the primary O-antigenic polysaccharide of the *Rhizobium leguminosarum* 3841 lipopolysaccharide and identification of a new 3-acetimidoylamino-3-deoxyhexuronic acid glycosyl component: a unique O-methylated glycan of uniform size, containing 6-deoxy-3-O-methyl-D-talose,n-acetylquinovosamine, and rhizoaminuronic acid (3-acetimidoylamino-3-deoxy-D-gluco-hexuronic acid). J. Biol. Chem. *283*, 16037–16050.

Forsberg, L.S., Noel, K.D., Box, J., and Carlson, R.W. (2003). Genetic locus and structural characterization of the biochemical defect in the O-antigenic polysaccharide of the symbiotically deficient *Rhizobium etli* mutant, CE166. Replacement of N-acetylquinovosamine with its hexosyl-4-ulose precursor. J. Biol. Chem. *278*, 51347–51359.

Fraysse, N., Lindner, B., Kaczynski, Z., Sharypova, L., Holst, O., Niehaus, K., and Poinsot, V. (2005). *Sinorhizobium meliloti* strain 1021 produces a low-molecular-mass capsular polysaccharide that is a homopolymer of 3-deoxy-D-manno-oct-2-ulosonic acid harboring a phospholipid anchor. Glycobiology. *15*, 101–108.

Fraysse, N., Jabbouri, S., Treilhou, M., Couderc, F., and Poinsot, V. (2002). Symbiotic conditions induce structural modifications of *SinoRhizobium* sp. NGR234 surface polysaccharides. Glycobiology. *12*, 741–748.

Fujishige, N.A., Kapadia, N.N., De Hoff, P.L., and Hirsch, A.M. (2006). Investigations of *Rhizobium* biofilm formation. FEMS Microbiol. Ecol. *56*, 195–206.

Gage, D.J. (2004). Infection and invasion of roots by symbiotic, nitrogen-fixing rhizobia during nodulation of temperate legumes. Microbiol. Mol. Biol. Rev. *68*, 280–300.

Gao, M., Chen, H., Eberhard, A., Gronquist, M.R., Robinson, J.B., Rolfe, B.G., and Bauer, W.D. (2005). *sinI*- and *expR*-dependent quorum sensing in *Sinorhizobium meliloti*. J. Bacteriol. *187*, 7931–7944.

Gibson, K.E., Barnett, M.J., Toman, C.J., Long, S.R., and Walker, G.C. (2007). The symbiosis regulator CbrA modulates a complex regulatory network affecting the flagellar apparatus and cell envelope proteins. J. Bacteriol. *189*, 3591–3602.

Gibson, K.E., Campbell, G.R., Lloret, J., and Walker, G.C. (2006). CbrA is a stationary-phase regulator of cell surface physiology and legume symbiosis in *Sinorhizobium meliloti*. J. Bacteriol. *188*, 4508–4521.

Gil-Serrano, A.M., Rodríguez-Carvajal, M.A., Tejero-Mateo, P., Espartero, J.L., Menendez, M., Corzo, J., Ruiz-Sainz, J.E., and Buendía-Clavería, A.M. (1999). Structural determination of a 5-acetamido-3,5,7,9-tetradeoxy-7-(3-hydroxybutyramido)-L-glycero-L-manno-nonulosonic acid-containing homopolysaccharide isolated from *Sinorhizobium fredii* HH103. Biochem. J. *342*, 527–535.

Glenn, S.A., Gurich, N., Feeney, M.A., and González, J.E. (2007). The ExpR/Sin quorum-sensing system controls succinoglycan production in *Sinorhizobium meliloti*. J. Bacteriol. *189*, 7077–7088.

Glucksmann, M.A., Reuber, T.L., and Walker, G.C. (1993a). Genes needed for the modification, polymerization, export, and processing of succinoglycan by *Rhizobium meliloti*: a model for succinoglycan biosynthesis. J. Bacteriol. *175*, 7045–7055.

Glucksmann, M.A., Reuber, T.L., and Walker, G.C. (1993b). Family of glycosyl transferases needed for the synthesis of succinoglycan by *Rhizobium meliloti*. J. Bacteriol. *175*, 7033–7044.

González, J.E., and Marketon, M.M. (2003). Quorum sensing in nitrogen-fixing rhizobia. Microbiol. Mol. Biol. Rev. *67*, 574–592.

González, J.E., and Keshavan, N.D. (2006). Messing with bacterial quorum sensing. Microbiol. Mol. Biol. Rev. *70*, 859–875.

González, J.E., Reuhs, B.L., and Walker, G.C. (1996). Low molecular weight EPS II of *Rhizobium meliloti* allows nodule invasion in *Medicago sativa*. Proc. Natl. Acad. Sci. USA. *93*, 8636–8641.

Goormachtig, S., Capoen, W., James, E.K., and Holsters, M. (2004). Switch from intracellular to intercellular invasion during water stress-tolerant legume nodulation. Proc. Natl. Acad. Sci. USA. *101*, 6303–6308.

Griffitts, J.S., and Long, S.R. (2008). A symbiotic mutant of *Sinorhizobium meliloti* reveals a novel genetic pathway involving succinoglycan biosynthetic functions. Mol. Microbiol. *67*, 1292–1306.

Hickman, J.W., and Harwood, C.S. (2008). Identification of FleQ from *Pseudomonas aeruginosa* as a c-di-GMP-responsive transcription factor. Mol. Microbiol. *69*, 376–389.

Hoang, H.H., Gurich, N., and González, J.E. (2008). Regulation of motility by the ExpR/Sin quorum-sensing system in *Sinorhizobium meliloti*. J. Bacteriol. *190*, 861–871.

Hoang, H.H., Becker, A., and González, J.E. (2004). The LuxR homolog ExpR, in combination with the Sin quorum sensing system, plays a central role in *Sinorhizobium meliloti* gene expression. J. Bacteriol. *186*, 5460–5472.

Hoang, H.H., Gurich, N., and González, J.E. (2008). Regulation of motility by the ExpR/Sin quorum-sensing system in *Sinorhizobium meliloti*. J. Bacteriol. *190*, 861–871.

Hotter, G.S., and Scott, D.B. (1997). The requirement for exopolysaccharide precedes the requirement for flavolan-binding polysaccharide in nodulation of *Leucaena leucocephala* by *Rhizobium loti* Arch. Microbiol. *167*, 182–186.

Hurek, T., and Reinhold-Hurek, B. (2003). *Azoarcus* sp. strain BH72 as a model for nitrogen-fixing grass endophytes. J. Biotechnol. *106*, 169–178.

Janczarek, M., and Skorupska, A. (2007). The *Rhizobium leguminosarum* bv. *trifolii* RosR: transcriptional regulator involved in exopolysaccharide production. Mol. Plant Microbe Interact. *20*, 867–881.

Jones, K.M., Kobayashi, H., Davies, B.W., Taga, M.E., and Walker, G.C. (2007). How rhizobial symbionts invade plants: the *Sinorhizobium-Medicago* model. Nat. Rev. Microbiol. *5*, 619–633.

Jones, K.M., Sharopova, N., Lohar, D.P., Zhang, J.Q., VandenBosch, K.A., and Walker, G.C. (2008). Differential response of the plant *Medicago truncatula* to its symbiont *Sinorhizobium meliloti* or an exopolysaccharide-deficient mutant. Proc. Natl. Acad. Sci. USA. *105*, 704–709.

Kaci, Y., Heyraud, A., Barakat, M., and Heulin, T. (2005). Isolation and identification of an EPS-producing *Rhizobium* strain from arid soil (Algeria): characterization of its EPS and the effect of inoculation on wheat rhizosphere soil structure. Res. Microbiol. *156*, 522–531.

Keating, D.H. (2007a). *Sinorhizobium meliloti* SyrA mediates the transcriptional regulation of genes involved in lipopolysaccharide sulfation and exopolysaccharide biosynthesis. J. Bacteriol. *189*, 2510–2520.

Keating, D.H. (2007b). The *Sinorhizobium meliloti* ExoR protein is required for the down-regulation of *lpsS* transcription and succinoglycan biosynthesis in response to divalent cations. FEMS Microbiol. Lett. *267*, 23–29.

Keller, M., Roxlau, A., Weng, W.M., Schmidt, M., Quandt, J., Niehaus, K., Jording, D., Arnold, W., and Pühler, A. (1995). Molecular analysis of the *Rhizobium meliloti mucR* gene regulating the biosynthesis of the exopolysaccharides succinoglycan and galactoglucan. Mol. Plant–Microbe Interact. *8*, 267–277.

Kiss, E., Reuhs, B., Kim, J., Kereszt, A., Petrovics, G., Putnoky, P., Dusha, I., Carlson, R.W., and Kondorosi, A. (1997). The rkpGHI and -J genes are involved in capsular polysaccharide production by *Rhizobium meliloti*. J. Bacteriol. *179*, 2132–2140.

Kiss, E., Kereszt, A., Barta, F., Stephens, S., Reuhs, B., Kondorosi, A., and Putnoky, P. (2001). The rkp-3 gene region of *Sinorhizobium meliloti* Rm41 contains strain-specific genes that determine K antigen structure. Mol. Plant–Microbe Interact. *14*, 1395–1403.

Krol, E., and Becker, A. (2004). Global transcriptional analysis of the phosphate starvation response in *Sinorhizobium meliloti* strains 1021 and 2011. Mol. Genet. Genomics. 272, 1–17.

Laus, M.C., Logman, T.J., Lamers, G.E., Van Brussel, A.A., Carlson, R.W., and Kijne, J.W. (2006). A novel polar surface polysaccharide from *Rhizobium leguminosarum* binds host plant lectin. Mol. Microbiol. *59*, 1704–1713.

Laus, M.C., Logman, T.J., Van Brussel, A.A., Carlson, R.W., Azadi, P., Gao, M.Y., and Kijne, J.W. (2004). Involvement of exo5 in production of surface polysaccharides in *Rhizobium leguminosarum* and its role in nodulation of *Vicia sativa* subsp. *nigra*. J. Bacteriol. *186*, 6617–6625.

Laus, M.C., van Brussel, A.A., and Kijne, J.W. (2005a). Role of cellulose fibrils and exopolysaccharides of *Rhizobium leguminosarum* in attachment to and infection of *Vicia sativa* root hairs. Mol. Plant Microbe Interact. *18*, 533–538.

Laus, M.C., van Brussel, A.A., and Kijne, J.W. (2005b). Exopolysaccharide structure is not a determinant of host-plant specificity in nodulation of *Vicia sativa* roots. Mol. Plant–Microbe Interact. *18*, 1123–1129.

Leigh, J.A., Signer, E.R., and Walker, G.C. (1985). Exopolysaccharide-deficient mutants of *Rhizobium meliloti* that form ineffective nodules. Proc. Natl. Acad. Sci. USA. *82*, 6231–6235.

Le Quéré, A.J., Deakin, W.J., Schmeisser, C., Carlson, R.W., Streit, W.R., Broughton, W.J., and Forsberg, L.S. (2006). Structural characterization of a K-antigen capsular polysaccharide essential for normal symbiotic infection in *Rhizobium* sp. NGR234: deletion of the *rkpMNO* locus prevents synthesis of 5,7-diacetamido-3,5,7,9-tetradeoxy-non-2-ulosonic acid. J. Biol. Chem. *281*, 28981–28992.

Lloret, J., Martín, M., Oruezabal, R.I., Bonilla, I., and Rivilla, R. (2002). MucR and mucS activate *exp* genes transcription and galactoglucan production in *Sinorhizobium meliloti* EFB1. Mol. Plant–Microbe Interact. *15*, 54–59.

Louch, H.A. and Miller, K.J. (2001) Synthesis of a low-molecular-weight form of exopolysaccharide by *Bradyrhizobium japonicum* USDA 110. Appl. Environ. Microbiol. *67*, 1011–1014.

Marie, C., Deakin, W.J., Ojanen-Reuhs, T., Diallo, E., Reuhs, B., Broughton, W.J., and Perret, X. (2004). TtsI, a key regulator of *Rhizobium* species NGR234 is required for type III-dependent protein secretion and synthesis of rhamnose-rich polysaccharides. Mol. Plant–Microbe Interact. *17*, 958–966.

Marketon, M.M., Gronquist, M.R., Eberhard, A., and González, J.E. (2002). Characterization of the *Sinorhizobium meliloti sinR/sinI* locus and the production of novel N-acyl homoserine lactones. J. Bacteriol. *184*, 5686–5695.

Martín, M., Lloret, J., Sánchez-Contreras, M., Bonilla, I., and Rivilla, R. (2000). MucR is necessary for galactoglucan production in *Sinorhizobium meliloti* EFB1. Mol. Plant–Microbe Interact. *13*, 129–135.

Mathis, R., Van Gijsegem, F., De Rycke, R., D'Haeze, W., Van Maelsaeke, E., Anthonio, E., Van Montagu, M., Holsters, M., and Vereecke, D. (2005). Lipopolysaccharides as a communication signal for progression of legume endosymbiosis. Proc. Natl. Acad. Sci. USA *102*, 2655–60.

McIntosh, M., Krol, E., and Becker, A. (2008). Competitive and co-operative effects in quorum sensing regulated galactoglucan biosynthesis in *Sinorhizobium meliloti*. J. Bacteriol. *190*, 5308–5317.

Mendrygal, K.E. and González, J.E. (2000). Environmental regulation of exopolysaccharide production in *Sinorhizobium meliloti*. J. Bacteriol. *182*, 599–606.

Mithöfer, A., Bhagwat, A.A., Keister, D.L., and Ebel, J. (2001). *Bradyrhizobium japonicum* mutants defective in cyclic beta-glucan synthesis show enhanced sensitivity to plant defense responses. Z. Naturforsch. [C]. *56*, 581–584.

Mora, P., Rosconi, F., Franco-Fraguas, L., and Castro-Sowinski, S. (2008). *Azospirillum brasilense* Sp7 produces an outer-membrane lectin that specifically binds to surface-exposed extracellular polysaccharide produced by the bacterium. Arch Microbiol. *189*, 519–524.

Mulligan, J.T., and Long, S.R. (1989). A family of activator genes regulates expression of *Rhizobium meliloti* nodulation genes. Genetics *122*, 7–18.

Nakhamchik, A., Wilde, C., and Rowe-Magnus, D.A. (2008). c-di-GMP regulates extracellular polysaccharide production, biofilm formation and rugose colony development in *Vibrio vulnificus*. Appl. Environ. Microbiol. *74*, 4199–4209.

Niehaus, K., Baier, R., Becker, A., and Pühler, A. (1996). Symbiotic suppression of the Medicago sativa defence system – the key of *Rhizobium meliloti* to enter the host plant? In Biology of Plant–Microbe Interactions, G. Stacey, B. Mullin and P.M. Gresshof, ed. (ISMPI, St Paul), pp. 349–352.

Noel, K.D., Box, J.M., and Bonne, V.J. (2004). 2-O-methylation of fucosyl residues of a rhizobial lipopolysaccharide is increased in response to host exudate and is eliminated in a symbiotically defective mutant. Appl. Environ. Microbiol. *70*, 1537–1544.

Noel, K.D., Forsberg, L.S., and Carlson, R.W. (2000). Varying the abundance of O antigen in *Rhizobium etli* and its effect on symbiosis with *Phaseolus vulgaris* J. Bacteriol. *182*, 5317–5324.

Nogales, J., Muñoz, S., Olivares, J., and Sanjuán, J. (2006). *Sinorhizobium meliloti* genes involved in tolerance to the antimicrobial peptide protamine. FEMS Microbiol. Lett. *264*, 160–167.

O'Neill, M.A., Darvill, A.G., and Albersheim, P. (1991). The degree of esterification and points of substitution by O-acetyl and O-(3-hydroxybutanoyl) groups in the acidic extracellular polysaccharides secreted by *Rhizobium leguminosarum* biovars *viciae*, *trifolii*, and *phaseoli* are not related to host range. J. Biol. Chem. *266*, 9549–9555.

Ormeño-Orrillo, E., Rosenblueth, M., Luyten, E., Vanderleyden, J., and Martínez-Romero, E. (2008). Mutations in lipopolysaccharide biosynthetic genes impair maize rhizosphere and root colonization of *Rhizobium tropici* CIAT899. Environ. Microbiol. *10*, 1271–1284.

Ozga, D.A., Lara, J.C., and Leig, J.A. (1994). The regulation of exopolysaccharide production is important at two levels of nodule development in *Rhizobium meliloti*. Mol. Plant–Microbe Interact. *7*, 758–765.

Parada, M., Vinardell, J.M., Ollero, F.J., Hidalgo, A., Gutiérrez, R., Buendía-Clavería, A.M., Lei, W., Margaret, I., López-Baena, F.J., Gil-Serrano, A.M., Rodríguez-Carvajal, M.A., Moreno, J., and Ruiz-Sainz, J.E. (2006). *Sinorhizobium fredii* HH103 mutants affected in capsular polysaccharide (KPS) are impaired for nodulation with soybean and *Cajanus cajan*. Mol. Plant–Microbe Interact. *19*, 43–52.

Parniske, M., Schmidt, P.E., Kosch, K., and Müller, P. (1994). Plant defence response of host plants with determinate nodules induced induced by EPS defective *exoB* mutants of *Bradyrhizobium japonicum*. Mol. Plant–Microbe Interact *7*, 631–638.

Pellock, B.J., Teplitski, M., Boinay, R.P., Bauer, W.D., and Walker, G.C. (2002). A LuxR homolog controls production of symbiotically active extracellular polysaccharide II by *Sinorhizobium meliloti*. J. Bacteriol. *184*, 5067–5076.

Petrovics, G., Putnoky, P., Reuhs, B., Kim, J., Thorp, T.A., Noel, K.D., Carlson, R.W., and Kondorosi, A. (1993). The presence of a novel type of surface polysaccharide in *Rhizobium meliloti* requires a new fatty acid synthase-like gene cluster involved in symbiotic nodule development. Mol. Microbiol. *8*, 1083–1094.

Philip-Hollingsworth, S., Hollingsworth, R.I., and Dazzo, F.B. (1989). Host-range related structural features of the acidic extracellular polysaccharides of *Rhizobium trifolii* and *Rhizobium leguminosarum*. J. Biol. Chem. *264*, 1461–1466.

Raffaele, S., Vailleau, F., Léger, A., Joubès, J., Miersch, O., Huard, C., Blée, E., Mongrand, S., Domergue, F., and Roby, D. (2008). A MYB transcription factor regulates very-long-chain fatty acid biosynthesis for activation of the hypersensitive cell death response in *Arabidopsis*. Plant Cell. *20*, 752–767.

Reeve, W.G., Dilworth, M.J., Tiwari, R.P., and Glenn, A.R. (1997). Regulation of exopolysaccharide production in *Rhizobium leguminosarum* biovar *viciae* WSM710 involves *exoR*. Microbiology. *143*, 1951–1958.

Reuber, T.L., Long, S., and Walker, G.C. (1991). Regulation of *Rhizobium meliloti* exo genes in free-living cells and *in planta* examined by using Tn*phoA* fusions. J. Bacteriol. *173*, 426–434.

Reuber, T.L., and Walker, G.C. (1993). Biosynthesis of succinoglycan, a symbiotically important exopolysaccharide of *Rhizobium meliloti*. Cell. *74*, 269–280.

Reuhs, B.L., Geller, D.P., Kim, J.S., Fox, J.E., Kolli, V.S.K., and Pueppke, S.G. (1998). *Sinorhizobium fredii* and *Sinorhizobium meliloti* produce structurally conserved lipopolysaccharides and strain-specific K antigens. Appl. Environ. Microbiol. *64*, 4930–4938.

Reuhs, B.L., Relić, B., Forsberg, L.S., Marie, C., Ojanen-Reuhs, T., Stephens, S.B., Wong, C.H., Jabbouri, S., and Broughton, W.J. (2005). Structural characterization of a flavonoid-inducible *Pseudomonas aeruginosa* A-band-like O antigen of *Rhizobium* sp. strain NGR234, required for the formation of nitrogen-fixing nodules. J. Bacteriol. *187*, 6479–6487.

Reuhs, B.L., Williams, M.N.V., Kim, J.S., Carlson, R.W., and Cote, F. (1995). Suppression of the Fix$^-$ phenotype of *Rhizobium meliloti exoB* mutants by *lpsZ* is correlated to a modified expression of the K-polysaccharide. J. Bacteriol. *177*, 4289–4296.

Rinaudi, L., Fujishige, N.A., Hirsch, A.M., Banchio, E., Zorreguieta, A., and Giordano, W. (2006). Effects of nutritional and environmental conditions on *Sinorhizobium meliloti* biofilm formation. Res. Microbiol. *157*, 867–875.

Rodriguez-Carvajal, M.A., Rodrigues, J.A., Soria-Diaz, M.E., Tejero-Mateo, P., Buendía-Clavería, A., Gutierrez, R., Ruiz-Sainz, J.E., Thomas-Oates, J., and Gil-Serrano, A. M. (2005). Structural analysis of the capsular polysaccharide from *Sinorhizobium fredii* HWG35. Biomacromolecules *6*, 1448–1456.

Rodríguez-Carvajal, M.A., Tejero-Mateo, P., Espartero, J.L., Ruiz-Sainz, J.E., Buendía-Clavería, A.M., Ollero, F.J., Yang, S.S., and Gil-Serrano, A.M. (2001). Determination of the chemical structure of the capsular polysaccharide of strain B33, a fast-growing soya bean-nodulating bacterium isolated from an arid region of China. Biochem. J. *357*, 505–511.

Rodríguez-Navarro, D.N., Dardanelli, M.S., and Ruíz-Saínz, J.E. (2007). Attachment of bacteria to the roots of higher plants. FEMS Microbiol. Lett. *272*, 127–136.

Rotter, C., Mühlbacher, S., Salamon, D., Schmitt, R., and Scharf, B. (2006). Rem, a new transcriptional activator of motility and chemotaxis in *Sinorhizobium meliloti*. J. Bacteriol. *188*, 6932–6942.

Rüberg, S., Pühler, A., and Becker, A. (1999). Biosynthesis of the exopolysaccharide galactoglucan in *Sinorhizobium meliloti* is subject to a complex control by the phosphate-dependent regulator PhoB and the proteins ExpG and MucR. Microbiology. *145*, 603–611.

Rüberg, S., Tian, Z.X., Krol, E., Linke, B., Meyer, F., Wang, Y., Pühler, A., Weidner, S., and Becker, A. (2003). Construction and validation of a *Sinorhizobium meliloti* whole genome DNA microarray: genome-wide profiling of osmoadaptive gene expression. J. Biotechnol. *106*, 255–268.

Rudrappa, T., Biedrzycki, M.L., and Bais, H.P. (2008). Causes and consequences of plant-associated biofilms. FEMS Microbiol. Ecol. *64*, 153–166.

Russo, D.M., Williams, A., Edwards, A., Posadas, D.M., Finnie, C., Dankert, M., Downie, J.A., and Zorreguieta, A. (2006). Proteins exported via the PrsD-PrsE type I secretion system and the acidic exopolysaccharide are involved in biofilm formation by *Rhizobium leguminosarum*. J. Bacteriol. *188*, 4474–4486.

Sadykov, M.R., Ivashina, T.V., Kanapin, A.A., Shliapnikov, M.G., and Ksenzenko, V.N. (1998). Structure–functional organization of exopolysaccharide biosynthetic genes in *Rhizobium leguminosarum* bv. viciae VF39. Mol. Biol. (Mosk). *32*, 797–804.

Sanchez-Contreras, M., Bauer, W.D., Gao, M., Robinson, J.B., and Downie J.A. (2007). Quorum-sensing regulation in rhizobia and its role in symbiotic interactions with legumes. Philos Trans R Soc Lond B Biol. Sci. *362*, 1149–1163.

Saravanan, V.S., Madhaiyan, M., Osborne, J., Thangaraju, M., and Sa, T.M. (2008). Ecological occurrence of *Gluconacetobacter diazotrophicus* and nitrogen-fixing *Acetobacteraceae* members: their possible role in plant growth promotion. Microb. Ecol. *55*, 130–140.

Sharypova, L.A., Chataigné, G., Fraysse, N., Becker, A., and Poinsot, V. (2006). Overproduction and increased molecular weight account for the symbiotic activity of the *rkpZ*-modified K polysaccharide from *Sinorhizobium meliloti* Rm1021. Glycobiology. *16*, 1181–1193.

Simsek, S., Ojanen-Reuhs, T., Stephens, S.B., and Reuhs, B.L. (2007). Strain-ecotype specificity in *Sinorhizobium meliloti-Medicago truncatula* symbiosis is correlated to succinoglycan oligosaccharide structure. J. Bacteriol. *189*, 7733–7740.

Smit, G., Swart, S., Lugtenberg, B.J.J., and Kijne, J.W. (1992). Molecular mechanisms of attachment of *Rhizobium* bacteria to plant roots. Mol. Microbiol. *6*, 2897–2903.

Sourjik, V., Muschler, P., Scharf, B., and Schmitt, R. (2000). VisN and VisR are global regulators of chemotaxis, flagellar, and motility genes in *Sinorhizobium* (*Rhizobium*) *meliloti*. J. Bacteriol. *182*, 782–788.

Stacey, G., Libault, M., Brechenmacher, L., Wan, J., and May, G.D. (2006). Genetics and functional genomics of legume nodulation. Curr. Opin. Plant Biol. *9*, 110–121.

Staehelin, C., Forsberg, L.S., D'Haeze, W., Gao, M.Y., Carlson, R.W., Xie, Z.P., Pellock, B.J., Jones, K.M., Walker, G.C., Streit, W.R., and Broughton, W.J. (2006). Exo-oligosaccharides of *Rhizobium sp.* strain NGR234 are required for symbiosis with various legumes. J. Bacteriol. *188*, 6168–6178.

Steenhoudt, O., and Vanderleyden, J. (2000). *Azospirillum*, a free-living nitrogen-fixing bacterium closely associated with grasses: genetic, biochemical and ecological aspects. FEMS Microbiol. Rev. *24*, 487–506.

Townsend, G.E. 2nd, Forsberg, L.S., and Keating, D.H. (2006). *Mesorhizobium loti* produces *nodPQ*-dependent sulfated cell surface polysaccharides. J. Bacteriol. *188*, 8560–8572.

Townsend, G.E. 2nd, and Keating, D.H. (2008). Identification and characterization of KpsS, a novel polysaccharide sulphotransferase in *Mesorhizobium loti*. Mol. Microbiol. *68*, 1149–1164.

Trent, M.S., Pabich, W., Raetz, C.R., and Miller, S.I. (2001) A PhoP/PhoQ-induced Lipase (PagL) that catalyzes 3-O-deacylation of lipid A precursors in membranes of *Salmonella typhimurium*. J. Biol. Chem. *276*, 9083–9092.

Turovskiy, Y., Kashtanov, D., Paskhover, B., and Chikindas, M.L. (2007). Quorum sensing: fact, fiction, and everything in between. Adv. Appl. Microbiol. *62*, 191–234.

Turska-Szewczuk, A., Lotocka, B., Kutkowska, J., Król, J., Urbanik-Sypniewska, T., and Russa, R. (2007). The incomplete substitution of lipopolysaccharide with O-chain prevents the establishment of effective symbiosis between *Mesorhizobium loti* NZP2213.1 and *Lotus corniculatus*. Microbiol. Res. doi:10.1016/j.micres.2006.11.011.

Turska-Szewczuk, A., Palusinska-Szysz, M., and Russa, R. (2008a). Structural studies of the O-polysaccharide chain from the lipopolysaccharide of symbiotically enhanced mutant Mlo-13 of *Mesorhizobium loti* NZP2213. Carbohydr. Res. *343*, 477–482.

Turska-Szewczuk, A., Pietras, H., Borucki, W., and Russa, R. (2008b). Alteration of O-specific polysaccharide structure of symbiotically defective *Mesorhizobium loti* mutant 2213.1 derived from strain NZP2213. Acta Biochim. Pol. *55*, 191–200.

Vanbleu, E., Choudhury, B.P., Carlson, R.W., and Vanderleyden, J. (2005). The *nodPQ* genes in *Azospirillum brasilense* Sp7 are involved in sulfation of lipopolysaccharides. Environ. Microbiol. *7*, 1769–1774.

van Workum, W.A.T., van Slageren, G., van Brussel, A.A.N., and Kijne, J.W. (1998). Role of exopolysacchrides of *Rhizobium leguminosarum* bv. *viciae* as host plant-specific molecules required for infection thread formation during nodulation of *Vicia sativa*. Mol. Plant–Microbe Interact. *11*, 1223–1241.

Vedam, V., Kannenberg, E., Datta, A., Brown, D., Haynes-Gann, J.G., Sherrier, D.J., and Carlson, R.W. (2006). The pea nodule environment restores the ability of a

Rhizobium leguminosarum lipopolysaccharide *acpXL* mutant to add 27-hydroxyoctacosanoic acid to its lipid A. J. Bacteriol. *188*, 2126–2133.

Vedam, V., Haynes, J.G., Kannenberg, E.L., Carlson, R.W., and Sherrier, D.J. (2004). A *Rhizobium leguminosarum* lipopolysaccharide lipid-A mutant induces nitrogen-fixing nodules with delayed and defective bacteroid formation. Mol. Plant–Microbe Interact. *17*, 283–291.

Vedam, V., Kannenberg, E.L., Haynes, J.G., Sherrier, D.J., Datta, A., and Carlson, R.W. (2003). A *Rhizobium leguminosarum* AcpXL mutant produces lipopolysaccharide lacking 27-hydroxyoctacosanoic acid. J. Bacteriol. *185*, 1841–1850.

Vriezen, J.A., de Bruijn, F.J., and Nüsslein, K. (2007). Responses of rhizobia to desiccation in relation to osmotic stress, oxygen, and temperature. Appl. Environ. Microbiol. *73*, 3451–3459.

Wang, L.X., Wang, Y., Pellock, B., and Walker, G.C. (1999). Structural characterization of the symbiotically important low-molecular-weight succinoglycan of *Sinorhizobium meliloti*. J. Bacteriol. *181*, 6788–6796.

Wang, P., Zhong, Z., Zhou, J., Cai, T., and Zhu, J. (2008). Exopolysaccharide biosynthesis is important for *Mesorhizobium tianshanense*: plant host interaction. Arch. Microbiol. *189*, 525–530.

Wassem, R., Kobayashi, H., Kambara, K., Le Quéré, A., Walker, G.C., Broughton, W.J., and Deakin, W.J. (2008). TtsI regulates symbiotic genes in *Rhizobium* species NGR234 by binding to *tts* boxes. Mol. Microbiol. *68*, 736–748.

Wells, D.H., Chen, E.J., Fisher, R.F., and Long, S.R. (2007). ExoR is genetically coupled to the ExoS-ChvI two-component system and located in the periplasm of *Sinorhizobium meliloti*. Mol. Microbiol. *64*, 647–664.

Wells, D.H., and Long, S.R. (2002). The *Sinorhizobium meliloti* stringent response affects multiple aspects of symbiosis. Mol. Microbiol. *43*, 1115–1127.

Wiedemann, G., and Müller, P. (2004). Use of Tn KPK2 for sequencing a 10.6-kb PstI DNA fragment of *Bradyrhizobium japonicum* and for the construction of *aspA* and *ndvA* mutants. Arch. Microbiol. *181*, 418–427.

Williams, A., Wilkinson, A., Krehenbrink, M., Russo, D., Zorreguieta, A., and Downie, J.A. (2008). Glucomannan-mediated attachment of *Rhizobium leguminosarum* to pea root hairs is required for competitive nodule infection. J. Bacteriol. *190*, 4706–4715.

Yang, C., Signer, E.R., and Hirsch, A.M. (1992). Nodules initiated by *Rhizobium meliloti* exopolysaccharide mutants lack a discrete, persistent nodule meristem. Plant Physiol. *98*, 143–151.

Yao, S.Y., Luo, L., Har, K.J., Becker, A., Rüberg, S., Yu, G.Q. Zhu, J.B., and Cheng, H.P. (2004). *Sinorhizobium meliloti* ExoR and ExoS proteins regulate both succinoglycan and flagellum production. J. Bacteriol. *186*, 6042–6049.

York, G.M., and Walker, G.C. (1998a). The *Rhizobium meliloti* ExoK and ExsH glycanases specifically depolymerize nascent succinoglycan chains. Proc. Natl. Acad. Sci. USA *95*, 4912–4917.

York, G.M., and Walker, G.C. (1998b). The succinyl and acetyl modifications of succinoglycan influence susceptibility of succinoglycan to cleavage by the *Rhizobium meliloti* glycanases ExoK and ExsH. J. Bacteriol. *180*, 4184–4191.

York, G.M., and Walker, G.C. (1997). The *Rhizobium meliloti exoK* gene and *prsD/prsE/exsH* genes are components of independent degradative pathways which contribute to production of low-molecular-weight succinoglycan. Mol. Microbiol. *25*, 117–134.

Levansucrase and Levan Formation in *Pseudomonas syringae* and Related Organisms

12

Abhishek Srivastava, Daria Zhurina and Matthias S. Ullrich

Abstract

Bacterial levan is formed wherever bacteria encounter moderate to high sucrose concentrations, i.e. in the plant environment, in the mammalian oral cavity, or during the fermentation of plant-borne substrates, i.e. for bioethanol production. Consequently, the transcriptional and translational regulation, synthesis, secretion, and enzymatic properties of levan-forming enzymes such as levansucrase (Lsc) from various bacterial species have raised substantial interest among researchers in plant pathology, dental medicine, food manufacturing, and renewable energy technologies. Experimental work with Lsc was accelerated by three major features, i.e. its intrinsically high protein stability as an extracellular enzyme, no need for reaction-catalysing co-factors, and an enormous ease of enzymatic detection due to its glucose releasing activity. Aside of many possible applied aspects, Lsc enzymatic activity as well as its regulation and secretion are exciting biochemical and molecular model systems for the in-depth understanding of global carbon utilization, protein transport processes, and biofilm formation.

Levan as a fructosyl polymer

Bacterial and plant-borne levans are typical fructan polymers. The term fructan covers both inulin and levan. Levan is made up of numerous fructosyl residues derived from sucrose and single glucose head groups. The linear 2,6-linked levan molecules are flexible and tend to be left-handed twist, whereas the twist of inulin is right-handed (French, 1989). Sucrose (β-D-fructofuranosyl-(2→1)-α-D-glucopyranoside) is a covalently (1→2-α, β-glycosidic bond) bonded compound consisting of glucose and fructose. Levan polymers can be several mega Dalton in size and levan polymers together become visible to the naked eye as a semi-translucent, whitish, homogenous mucus. Levan has the following interesting biochemical characteristics: It is biodegradable, water soluble, and forms liquid crystals being amorphously microcrystalline (Fuchs, 1956; Viney *et al.*, 1993). While small fructans have a rather sweet taste, longer fructan chains such as those found in levan form emulsions with a fat-like texture and have a rather neutral taste (Vijn and Smeekens, 1999). Below 20% solids, levan solutions exhibit typical Newtonian behaviour (Simms *et al.*, 1990). Levan has a tensile strength of 991 pounds per square inch and possesses flexible five-membered rings packed in a dense, spheroid structure. At low concentrations, the spheres roll smoothly. When the density of spheres passes a critical point and come into contact with each other, a rapid increase in viscosity is observed (Stivala and Zweig, 1981). At 50% solids, the levan polysaccharide resembles chewing gum (Simms *et al.*, 1990). Unlike most other polysaccharides, levan has a compact spherical structure and does not swell in water. The intrinsic viscosity of levan is only 0.14 dl/g (Kasapis *et al.*, 1994; Kasapis and Morris, 1994). Levan is resistant to organic solvents such as jet fuel and *d*-limonene (Stivala and Bahary, 1978) and is fairly heat stable with a melting point of 225°C.

Similar in structure to levan, fructan was first described by the German botanist, Rose, in boiling water extract from *Inula helenium* (Rose,

1804). This substance was later named inulin (Thompson, 1818). The term 'levulan' was first described as an undesirable by-product in sugar cane and liquor processing that caused increase in viscosity (Avigad, 1968; Lippmann, 1881). The term 'levan' was introduced as analogous to 'dextran' for a compound that was produced by an isolate of *Bacillus* sp. growing on sucrose (Greig-Smith, 1901). Gummy levan was later obtained from spores of *Aspergillus sydowi* (Kopeloff *et al.*, 1920). *Pseudomonas pruni* was described as the first levan/dextran-forming plant pathogen (Lyne *et al.*, 1940). In 1944, levan was first shown to be synthesized from sucrose by cell-free enzymes (Hestrin and Avineri-Shapiro, 1944). Shortly after this, levans from various micro-organisms were shown to be identical in terms of their immunological features (Hehre *et al.*, 1945). Levan was then described as fructofuranose-containing polysaccharide synthesized by plants and many aerobic bacteria from sucrose or raffinose (Evans and Hibbert, 1946). In 1957, levan from *Corynebacterium* species was described to possess β-2,1 and β-2,6 linkages (Avigad and Feingold, 1957). Finally, the interaction between the lectin, concanavalin A, and levan was first demonstrated (So and Goldstein, 1968).

Levan formation is of particular importance in three major areas. The plant-borne disaccharide, sucrose, is a readily available carbon source for (i) plant pathogens, (ii) bacteria of the oral cavity and (iii) microbes associated with bioethanol production. These microorganisms use the extracellular enzyme, levansucrase, to release glucose from sucrose and to polymerize fructosyl residues into the branched polymer levan. Glucose is then consumed inside the cell while the polymerized fructosyl residues along with single glucosyl head groups make up the extracellular mucus loosely associated with the cell envelope. Bacterial levan accumulation in form of dome-shaped mucoid structures is easily visualized when those microbes are grown on sucrose-containing agar media (Fig. 12.1).

Levansucrases: enzymatic properties and relatedness to other enzymes

Levansucrase (Lsc) (EC 2.4.2.10.) belonging to the superfamily of β-fructosidases and having a five-bladed β-propeller fold is found in various Gram-positive and Gram-negative bacteria. This extracellular enzyme cleaves sucrose into glucose and fructosyl residues, releases glucose, and polymerizes the fructosyl residues to form the β-(2,6) linked and β-(2,1) branched exopolysaccharide (EPS) levan, which has singular glucosyl head groups (Han, 1990). Additionally, Lsc is able to hydrolyse levan and to exchange glucosyl head groups of levan polymers as demonstrated by feeding bacterial cells with ^{14}C-labelled glucose (Kaspasi *et al.*, 1994).

The mirror reaction of Lsc's forward reaction is performed by dextransucrase (EC 2.4.1.5), a kind of glucansucrase from *Leuconostoc mesenteroides*, which forms dextran, a glucose-containing polymer (Bailey *et al.*, 1957). Alternansucrase (EC 2.4.1.140), another variety of glucansucrase from *L. mesenteroides*, can synthesize alternan with alternating α-(1,6) linkages and α-(1,3) linkages (Cote and Robyt, 1982). Another glycosyltransferase named mutansucrase from *Streptococcus*

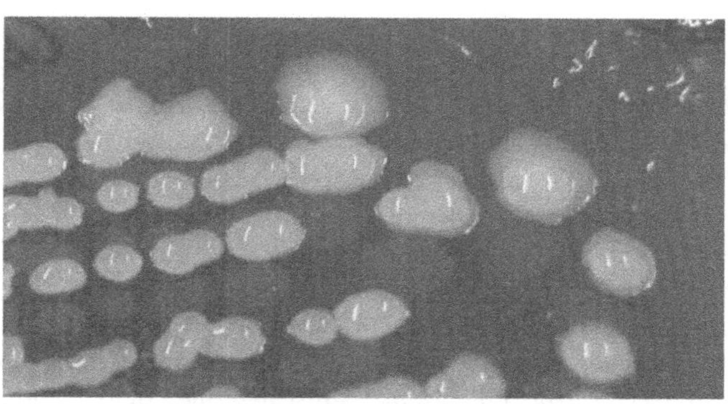

Figure 12.1 Typical colonies of *P. syringae* PG4180 grown on sucrose-containing minimal medium agar plates. The whitish mucus in form of dome-shaped colonies represents levan.

mutans (Wenham *et al.*, 1981) can synthesize mutans with mixed α-1,3 predominant, and α-1,6 linkages. A recently discovered glucansucrase-like enzyme termed reuteransucrase (EC 2.4.1.5) from *Lactobacillus reuteri* produces reuteran that contains mainly α-1,4 linkages together with α-(1,6) and α-(1,4,6) linkages (Kralj *et al.*, 2004). Inulosucrase (EC 2.4.1.9) of *Leuconostoc citreum* results in inulin β- 2,1 linear fructans (Olivares-Illana *et al.*, 2003). Amylosucrase (EC 2.4.1.4) from *Neisseria polysaccharea* has been shown to produce starch like amylan with α-1,4 linkages between glucosyl molecules (Sarcabal *et al.*, 2000). Protein-Nπ phosphohistidine:sugar Nπ-phosphotransferase (EC2.7.1.69), which is an enzyme II of the phosphor-transferase system can phosphorylate the fructosyl residue of sucrose thus generating sucrose 6-phosphate (Kornberg and Riordan, 1976; Postma and Roseman, 1976). In summary, one can clearly see that nature has evolved many different enzymatic reactions breaking down the most abundant plant transport sugar, sucrose.

Peptide sequences, enzymatic properties, protein location, and transcriptional regulation of Lsc differ most noticeably between Gram-positive and Gram-negative bacteria. In Gram-positive microorganisms, Lsc has been studied in detail in *Bacillus subtilis* (SacB), in different food-grade *Lactobacillus* and oral *Streptococcus* species, and in *Leuconostoc mesenteroides* (Morales-Arrieta *et al.*, 2006; Seibel *et al.*, 2006; Tieking *et al.*, 2005; van Hijum *et al.*, 2006). The three-dimensional structure of SacB of *B. subtilis* had been resolved revealing interesting aspects of the enzymatic activity of this enzyme (Meng and Futterer, 2003). Most notable is that (i) all studied Lsc of Gram-positive bacteria possess typical N-terminal signal peptides and are secreted via a classical type II protein transport mechanism across the membrane (Benyahia *et al.*, 1988); (ii) some were found to be membrane-bound mediated by C-terminal lipid anchors (Tieking *et al.*, 2005); and (iii) levan formation by *Streptococcus* sp. contributed to oral biofilm formation associated with dental plaque (Tentuta *et al.*, 2006).

For Gram-negative bacteria, Lsc and its transport have been studied to variable extent in *Pseudomonas syringae* (see below for details), *Erwinia amylovora, Zymomonas mobilis, Gluconacetobacter diazototrophicus, Rahnella aquatilis,* and *Acetobacter xylinum* (Martinez-Fleites *et al.*, 2005; Seo *et al.*, 2002; Song *et al.*, 1999; Tajima *et al.*, 2000; Vigants *et al.*, 2001). Generally, the Lsc proteins of Gram-negative bacteria share higher similarities among each other compared with those enzymes from Gram-positive organisms. The Lsc protein, LevU, from *Zymomonas mobilis* was found to be membrane-associated and mutational analysis had revealed that the polymerizing activity could be uncoupled from the sucrolytic activity of this enzyme (Song *et al.*, 1999). In contrast to Lsc enzymes of other Gram-negative bacteria, that of the plant endophyte, *G. diazotrophicus,* (LsdA) possesses a typical N-terminal signal peptide sequence and was shown to be transported by a type II protein secretion mechanism (Arrieta *et al.*, 2004). The crystal structure of LsdA was resolved suggesting a strong overall similarity with SacB from *B. subtilis* but noticeable structural differences in the transfructosylation domain (Martinez-Fleites *et al.*, 2005).

In the plant pathogen, *P. syringae*, at least two Lsc gene products, LscB and LscC, have been identified and analysed in detail (Hettwer *et al.*, 1998; Li and Ullrich, 2001; Li *et al.*, 2006). A third allele coding for Lsc, *lscA*, was shown to be not expressed but its gene product was fully functional upon heterologous expression under control of an inducible promoter (Hettwer *et al.*, 1998). The genes coding for LscB and LscC reside on an indigenous 60-kb plasmid and the chromosome, respectively. In this context it is notable that *P. syringae* seems to be the only bacterial organism with multiple Lsc iso-enzymes. Remarkably, the presence of 2–3 very similar genes for Lsc iso-enzymes is highly conserved in the genome of strain PG4180 (pv. *glycinea*) as well as in all three fully sequenced genomes of additional *P. syringae* strains, which are strain DC3000 (pv. *tomato*), strain B728 (pv. *syringae*), and strain 1449A (pv. *phaseolicola*) (www.pseudomonas.com). This conservation may reflect a particular importance of Lsc for the life style of this foliar and biotrophic plant pathogen. Even more excitingly, the two expressed gene products accumulated in either the periplasmic space or the extracellular space, respectively, making it tempting to speculate that both enzymes may possess specific functions (Li *et al.*, 2006). Differential localization of LscB and LscC was demonstrated in subcellular fractions derived from a reliable osmotic shock extraction procedure (see below). Sucrose acts as a major transport and storage sugar molecule in diverse crop plants. It is therefore tempting to speculate that the biotrophic plant pathogen, *P. syringae*, has evolved enzymatic mechanisms to particularly utilize this disaccharide with high efficiency.

Role of levan in biofilm formation and ecological fitness

Biofilm formation and EPS syntheses have always been tightly associated with each other due to the fact that the matrix surrounding cells in biofilms consists among others of diverse polysaccharides, (glyco-)proteins, and extracellular DNA (Branda *et al.*, 2005; Flemming *et al.*, 2007; Goller and Romeo, 2008). Numerous studies have shown that mutants defective in EPS synthesis show either no biofilm formation or significantly altered biofilm structures. For the current review, biofilms produced by pseudomonads and other bacteria occurring in sucrose-rich environments are of particular interest. Alginate, a poly-anionic EPS, is the best-investigated component of mucoid *P. aeruginosa* biofilms and its role in biofilm formation and infection has been established (Ryder *et al.*, 2007). However, recent reports demonstrated that a cellulose-like fibre structure encoded by the *psl* and *pel* gene clusters of non-mucoid *P. aeruginosa* is essential for biofilm formation (Ma *et al.*, 2007; Ryder *et al.*, 2007). Expression of *psl* seems to depend on the *wsp* locus encoding for a chemosensory system and mediated a wrinkly spreader phenotype in *P. fluorescens* (Spiers *et al.*, 2002) indicating the complexity of regulation of biofilm formation. Another important regulatory input for *P. aeruginosa* biofilms comes from the global two-component regulatory system, GacS/GacA (Goodman *et al.*, 2004). Similar results were found for biofilms formed by *P. putida* (Chang *et al.*, 2007) and those produced by *P. fluorescens* (Kives *et al.*, 2006).

For *P. syringae*, our recent results demonstrated that neither alginate nor levan or the alternative sigma factor, AlgT, were essential for biofilm formation indicating a role for a yet-to-be identified additional EPS for biofilm structure formation (Laue *et al.*, 2006; see below for more details). To elucidate this, mutants with defects in either, levan formation, alginate synthesis, or AlgT, were subjected to flow chamber experiments. Biofilms from the wild type of PG4180 as well as from levan- or alginate-deficient mutants showed indistinguishable formation patterns. However, alginate produced by *P. syringae* had an impact on initial cell attachment to artificial surfaces while levan accumulated in voids of bacterial colonies thus potentially representing a storage compound. Formation of biofilms on plant surfaces significantly enhances the epiphytic fitness of pathogenic *P. syringae* versus antagonistic epiphytes, such as *Pantoea agglomerans* (Danhorn and Fuqua, 2007). An H-NS-like negative regulator, MvaT, controlled the fimbrial *cup* gene clusters and biofilm formation as well as exo-product formation in *P. aeruginosa* (Vallet *et al.*, 2004) and in *P. fluorescens* (Baehler *et al.*, 2006), respectively. Not surprisingly, MvaT also played a role in quorum sensing-associated N-acylhomoserine lactone formation in *P. aeruginosa* (Diggle *et al.*, 2003) reflecting the complexity of regulatory effects mediated by this DNA binding protein.

The wild type of *P. syringae* strain PG4180 and its corresponding EPS mutants were analysed with respect to surface attachment, lectin binding, and biofilm formation in flow chambers (Laue *et al.*, 2006). Interestingly, neither alginate nor levan had a major impact on biofilm formation and its structure. In contrast to levan, presence of alginate seemed to improve cell-surface attachment. The lectin, concanavalin A (ConA), was identified to be specifically staining levan since its fluorescent signal was only detectable when Lsc was expressed and when sucrose was provided to the flow chamber medium. A third yet-to-be identified fibrous EPS structure was marked with a lectin from *Naja mossambica* (Laue *et al.*, 2006). Surprisingly, ConA-stainable levan accumulated in large cell-free voids in the centre of micro-colonies forming the biofilm. Since those centred voids are also formed when Lsc was absent or when no sucrose was provided, it may be speculated that these voids are the physical result of colony growth due to micro-aerobic and nutrient-deprived conditions inside colonies. We hypothesized that levan-filled voids in *P. syringae* biofilms might represent storage compartments being 'loaded' during sucrose presence and being utilized by the bacterial cells during periods of starvation (Laue *et al.*, 2006). This assumption was fuelled by the idea that the poly-fructan, levan, is a side-product of glucose utilization from sucrose via Lsc activity and its fructosyl residues could potentially being used diauxically when glucose is used up and no other carbon sources are available. Consequently, it is logical to next test sucrose-grown PG4180 biofilms under starvation conditions. Should the accumulated levan disappear and thus been used up subsequently, our hypothesis would be proven. A potential direct or indirect linkage of Lsc activities with the general glucose metabolism should then be analysed in detail.

The effects of sugars, particularly sucrose, on biofilm formation have been in-depth evaluated in oral (dental plaque) biofilms mainly formed by Gram-positive bacteria (Tentuta *et al.*, 2006; Vale *et al.*, 2007). Presence of sucrose significantly enhanced biofilm build-up in comparison to fructose, galactose, glucose, or lactose for diverse organisms, which all produce Lsc (Yang *et al.*, 2006). Furthermore, it was demonstrated that sucrose addition led to distinct population shifts when different streptococci and lactobacilli species were co-cultivated as dental plaque biofilms suggesting that this sugar gave advantage to lactobacilli (Tentuta *et al.*, 2006). The same laboratory showed that sucrose addition to oral biofilms led to increased EPS synthesis, acid accumulation, and subsequent enamel demineralization in patients (Vale *et al.*, 2007). When searching for differentially expressed genes during biofilm formation in *Staphylococcus aureus*, *S. mutans*, and *B. subtilis*, sugar uptake PTS genes were found (Sutrina *et al.*, 2007). In an unrelated study, numerous bacterial contaminants were identified during corn (sucrose)-based fuel ethanol production by yeasts (Skinner-Nemec *et al.*, 2007). Most of the identified bacterial contaminants were Gram-positive and all of them contained Lsc genes allowing the speculation that levan synthesis negatively impacted ethanol production yields. One can easily envisage that levan build-up in diverse production facilities such as bioreactors and tubes of any commercial (and thus non-sterile) high-throughput fermentation of sucrose-rich biofuel substrate will have a major negative impact on production yields.

Regulation and secretion of levansucrase

In the current review article most emphasis is brought to the transcriptional regulation and differential secretion of Lsc from *P. syringae* (see below). However, the extensive work by others with different bacterial organisms should be considered at first place. Expression of the *sacB* gene in various *Bacillus* species is sucrose- and salt-inducible via the SacY anti-terminator and the DegSU two-component system (Biedendieck *et al.*, 2007b; Li and Ferenec, 1997). In the plant pathogen, *E. amylovora*, which possesses a single *lsc* gene, overproduction of the regulatory proteins, RcsA and RcsB, strongly reduced Lsc synthesis (Bereswill and Geider, 1997). Moreover, *lsc* expression was found to be plant-inducible in this phytopathogenic organism (Zhao *et al.*, 2005). In contrast to the *lsc* genes of *P. syringae* and other Gram-negative bacteria, the Lsc-encoding *levU* gene of *Zymomonas mobilis* is transcribed as a bi-cistronic mRNA together with a sucrase gene (Song *et al.*, 1999). This might possibly be due to the lack of a phosphoenolpyruvate-dependent carbohydrate phosphotransferase (PTS) system in this organism. In the enteric bacterium, *R. aquatilis*, expression of the Lsc-encoding *lsrA* gene was shown to be stationary phase dependent but *rpoS* independent and was mediated by an activator (LsrS) and a repressor (LsrR) (Seo *et al.*, 2002). The gene encoding for LsdA of *G. diazotrophicus* is encoded on a constitutively synthesized bi-cistronic mRNA together with a levanase gene (Arrieta *et al.*, 2004). Interestingly, the *sacB* gene, its sucrose inducibility, and the export characteristics of its gene product from *B. megaterium* were successfully used as biotechnological tools for heterologous protein recovery from bacterial supernatants (Biedendieck *et al.*, 2007a,b; Malten *et al.*, 2006). It should also been noted that – for a long time – the signal peptide-deprived *sacB* gene of *B. subtilis* has been used as a lethal gene marker on plasmids used for homologous recombination experiments. This is due to the fact that intracellular Lsc activities lead to an accumulation of levan product inside the cell and subsequently to the bursting and disruption of cells expressing signal peptide-deprived *sacB*.

For years, our laboratory had been interested in the temperature-dependent gene expression of the plant pathogenic bacterium, *P. syringae* pv. *glycinea* PG4180 (for reviews see Braun *et al.*, 2007; Ullrich *et al.*, 2000). Several successful attempts were undertaken to identify low-temperature induced genetic loci or gene products. In this context, extracellular protein profiles were investigated and the 50-kDa protein, Lsc, was found to be secreted at the low and virulence-promoting temperature of 18°C in contrast to the optimal growth temperature of this bacterium, 28°C (Li and Ullrich, 2001). This phenotype could be confirmed by zymographic detection of Lsc in native polyacrylamide gel electrophoresis (Fig. 12.2). In contrast to the well-studied Lsc of *B. subtilis* (SacB), *P. syringae* Lsc was not cleaved during export and secretion and did not possess a recognizable signal peptide suggesting that it was not transported via the general secretory pathway. This was proven by N-terminal sequencing of the extracellular protein. A total of three functional *lsc* genes were cloned and mutagenized in PG4180 (Hettwer *et al.*, 1998; Li and Ullrich, 2001). The gene, *lscA*, was found to not be expressed under any condition tested (i.e. *in vitro, in planta,* or under numerous growth conditions) (Schenk *et al.*, 2008) although it could be expressed under control of the P*lac* promoter in *E. coli* and *P. syringae* (Hettwer *et al.*, 1998; Li and Ullrich, 2001). In contrast, simultaneous mutation of both *lscB* and *lscC* led to a levan-negative phenotype in mutant PG4180.M6 while single mutation of either gene rendered the resulting mutants levan-positive. Levan syntheses by the two single-locus mutants were indistinguishable from each other. These results suggested that both gene products were responsible for levan formation on sucrose-containing media. The predicted gene products of *lscB* and *lscC* showed 96% similarity towards each other indicating the potential of gene duplication. In line with this, *lscB* is located on an indigenous 60-kb plasmid while *lscC* has a chromosomal location (Li and Ullrich, 2001). The two gene products differ in only five amino acyl residues, most of which represent conservative replacements.

Both gene products could be extracted from the periplasmic space suggesting that they were not exported and secreted by either type I, type III or type IV transport systems (Li *et al.*, 2006). In that study it was furthermore shown that both *lscB* and *lscC* are expressed *in vitro* in a temperature-dependent manner with maximum mRNA accumulation during the early logarithmic growth phase at 18°C (Fig. 12.3). During later growth stages, the thermo-responsive difference remained while the overall amount of *lscB* and *lscC* mRNAs declined dramatically. The expression profiles of both genes were nearly identical. These results indicated a potentially particular role for Lsc during early logarithmic growth. Alternatively, it could be argued that early log phase synthesis of Lsc is sufficient for the bacterial cells since Lsc is an extremely stable protein. The respective *in vitro* results could be confirmed when *lsc* gene expression was analysed during the *in planta* infection process using a modified and optimized total RNA extraction procedure (Schenk *et al.*, 2008).

Excitingly, both gene products showed a differential secretion pattern: LscB was found in minor amounts in the periplasm but with large

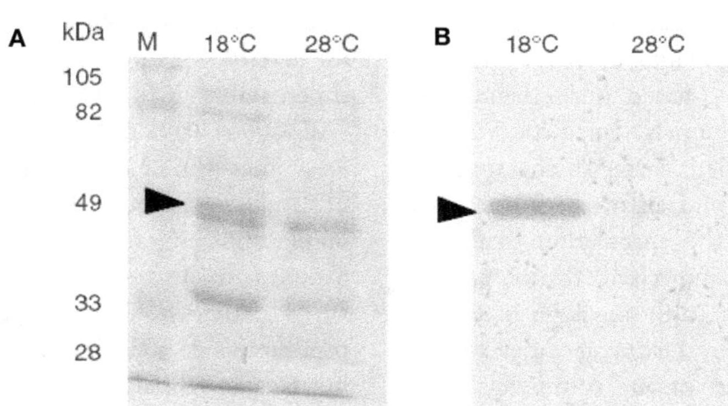

Figure 12.2 Extracellular protein profiles of P. syringae PG4180 as depicted by **A** SDS-PAGE and **B** zymographic detection of levansucrase in by native PAGE. Levansucrase activity corresponds to a ~50 kDa protein found predominantly during bacterial growth at 18°C.

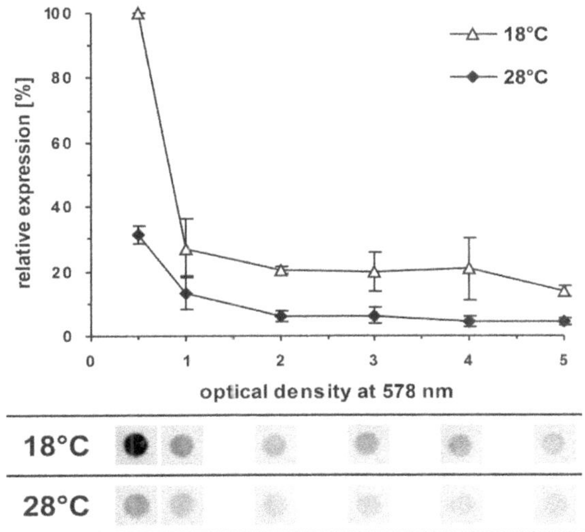

Figure 12.3 Transcriptional analysis of *lscB* and *lscC* by quantitation of transcript accumulation in P. syringae PG4180 during growth at 18°C and 28°C as determined by RNA dot blot analysis and as described by Li *et al*. (2006).

quantities in the supernatant while LscC mostly remained in the periplasm with only minor traces found in the supernatant (Li *et al.*, 2006). Interestingly, four of the five divergent amino acyl residues from the predicted protein sequences were conservative replacements while the fifth was a cysteine-to-serine exchange allowing speculations about different secondary structure formation (one versus two disulfide bridges). The quality of the subcellular fractionation technique was verified by expression of a MucA–PhoA hybrid, which clearly resided in the periplasm (Li *et al.*, 2006). A physiological rationale for this interesting differential secretion pattern remained to be elucidated but once again fuelled reasonable speculations about distinct enzymatic functions for either protein.

In absence of direct experimental evidence, i.e. mutation of the particular transport system, exported and secreted proteins can be assigned as substrates for a given transport system by three major criteria: (a) presence or absence in the periplasm of Gram-negative bacteria; (b) intrinsic amino acid sequence or mRNA sequence characteristics, which act as recognition signals for transport during or past translation; or (c) signal peptide cleavage during transport. Interestingly, the extracellular Lscs of *P. syringae* can not be clearly assigned to any of the known protein transport system since these enzymes lack a signal peptide, are found in the periplasm, and do not possess any assignable amino acid sequences characteristic for any of the systems. Whether or not their mRNAs harbour specific signals – such as postulated for Type III secreted proteins – is yet unknown. With these features, Lscs of *P. syringae* clearly differ from Lscs of *B. subtilis* and *G. diazotrophicus*, which both possess signal peptides and were shown to be transported via the general secretory pathway (Arrieta *et al.*, 2004; Benyahia *et al.*, 1988). In a preliminary study, *P. syringae* mutants with defects in the *hrp* type III protein secretion system and with a defect in the twin arginine transport (TAT) secretion pathway were checked for levan formation and extracellular as well as subcellular Lsc accumulation. Both mutants showed the normal wild-type phenotype for both assays suggesting that neither the type III secretion system nor the TAT pathway was responsible for Lsc export or secretion. It may be speculated that export and secretion of Lsc in *P. syringae* is mediated by an yet-to-be characterized novel protein translocation machinery or by a thus far unprecedented combination of known protein transport systems.

Currently ongoing work in our laboratory allowed the determination of the transcriptional start sites of *lscB* and *lscC* as well as the characterization of the respective upstream sequences in *P. syringae* (Srivastava *et al*, unpublished data). Not surprisingly, variable length of upstream sequence gave rise to variable expression levels for *lscB* suggesting the potential for several (independent) regulatory factors affecting transcription. Regulators such as those found in the enterobacterium, *E. amylovora*, i.e. RcsA and RcsB (Bereswill and Geider, 1997) could not be identified in the genome of *P. syringae*. Virulence assays with a *lsc*-deficient mutant of *P. syringae* indicated no particular role of levan formation for virulence or pathogenicity. However, the ecological fitness and the general survival efficiency were affected. Heterologous expression in a related levan-minus *Pseudomonas* strain with the help of a genomic library derived from *P. syringae* was used to identify a potential activator protein required for Lsc expression (Zhurina *et al.*, unpublished data). Finally, potential repressor proteins involved in the thermoregulation of Lsc expression have been identified using DNA affinity chromatography. These preliminary results will help to further elucidate the likely complex regulatory system involved in *lsc* gene expression in future studies.

Conclusions

The plant-borne disaccharide, sucrose, is a ubiquitously occurring and readily available carbon source for plant pathogens, bacteria of the oral cavity, and microbes associated with bioethanol production. Many of these microorganisms use the extracellular enzyme, levansucrase, to release glucose for consumption from sucrose and to polymerize fructosyl residues into the multiply branched polymer levan. Our previous studies showed that two levansucrases of the Gram-negative bacterium, *P. syringae* are expressed in a temperature- and growth phase-dependent manner with maximal levan formation at the virulence-promoting temperature of 18°C as opposed to the optimal growth temperature, 28°C. They showed distinct export and secretion patterns distinguishable from any classical protein secretion pathway. Expression of levansucrase led to an accumulation of levan in 'storage compartments' of biofilms. In our future work, it will have to be determined in detail how levansucrase gene expression is regulated, how the enzymes are transported via inner and outer membrane, and what the actual physiological role of levan formation is in context of carbon metabolism. The role of a newly identified regulator and its potential interaction with repressor proteins will be investigated. The export/secretion pathways of both levansucrases need to be identified and characterized. The hypothesis that levan formation might be a form of bacterial 'waste disposal' for later carbon recycling will be tested using an array of divergent and complementary experiments. Finally, the individual enzymatic activities of both iso-enzymes need to be elucidated in detail.

Acknowledgements

This work was supported by grants from the Deutsche Forschungsgemeinschaft and a European Community Marie Curie Early Trainings Stage grant.

References

Arrieta J.G., Sotolongo, M., Menendez, C., Alfonso, D., Trujillo, L.E., Soto, M., Ramirez, R., and Hernandez, L. (2004). A type II protein secretory pathway required for levansucrase secretion by *Gluconoacetobacter diazotrophicus*. J. Bacteriol. *186*, 5031–5039.

Avigad, G. and Feingold, D.S. (1957). Fructosides formed from sucrose by a *Corynebacterium*. Arch. Biochem. Biophys. *1*, 178–184.

Avigad, G. (1968). Levan. Encyclopedia of Polymer Science and Technology *8*, 711–718.

Baehler E., de Werra, P., Wick, L.Y., Pechy-Tarr, M., Mathys, S., Maurhofer, M., and Keel, C. (2006). Two novel MvaT-like global regulators control exoproduct formation and biocontrol activity in root-associated *Pseudomonas fluorescens* CHAO. Mol. Plant Microbe Interact. *19*, 313–329.

Bailey, R.W., Barker, S.A., Bourne, E.J. , and Stacey, M. (1957). The isolation and properties of the dextransucrase of Betacoccus arabinosaceous. J. Chem. Soc. (Lond.) Immunopolysaccharides. Part VI. pp. 3530–3536.

Benyahia F., Chambert, R., and Petit-Glatron, M.F. (1988). Levansucrase of *Bacillus subtilis*: Effects on the secretion process of single amino acid substitutions in the mature part of the protein. J. Gen. Microbiol. *134*, 3259–3268.

Bereswill S. and Geider, K. (1997). Characterization of the *rcsB* gene from *Erwinia amylovora* and its influence on exopolysaccharide synthesis and virulence of the fire blight pathogen. J. Bacteriol. *179*, 1354–1361.

Biedendieck R., Beine, R., Gamer, M., Jordan, E., Buchholz, K., Seibel, J., Dijkhuizen, L., Malten, M., and Jahn, D. (2007a). Export, purification, and activities of affinity tagged *Lactobacillus reuteri* levansucrase produced by *Bacillus megaterium*. Appl. Microbiol. Biotechnol. 74, 1062–1073.

Biedendieck R., Gamer, M., Jaensch, L., Meyer, S., Rohde, M., Deckwer, W.D., and Jahn, D. (2007b). A sucrose-inducible promoter system for the intra- and extracellular protein production in *Bacillus megaterium*. J. Biotechnol. *132*, 426–430.

Branda, S.S., Vik, A., Friedman, L., and Kolter, R. (2005). Biofilms: the matrix revisited. Trends Microbiol. *13*, 20–26.

Braun, Y., Smirnova, A.V., Weingart, H., Schenk, A., and Ullrich, M.S. (2007). A temperature-sensing histidine kinase – Function, genetics, and membrane topology. Meth. Enzymol. *423*, 222–249.

Chang, W.S., van de Mortel, M., Nielsen, L., de Guzman, G.N., Li, X., and Halverson, L.J. (2007). Alginate production by *Pseudomonas putida* creates a hydrated microenvironment and contributes to biofilm architecture and stress tolerance under water limiting conditions. J. Bacteriol. *189*, 8290–8299.

Cote, G.L. and Robyt, J.F. (1982). Isolation and partial characterization of an extracellular glucansucrase from Leuconostoc mesenteroides NRRL B-1355 that synthesizes an alternating (1 goes to 6), (1 goes to 3)-alpha-D-glucan. Carbohydr. Res. *101*, 57–74.

Danhorn, T. and Fuqua, C. (2007). Biofilm formation by plant-associated bacteria. Ann. Rev. Microbiol. *61*, 401–422.

Diggle, S.P., Winzer, K., Lazdunski, A., Williams, P., and Camara, M. (2002). Advancing the quorum in *Pseudomonas aeruginosa*: MvaT and the regulation of N-acylhomoserine lacton production and virulence gene expression. J. Bacteriol. *184*, 2567–2586.

Evans, T.H. and Hibbert, H. (1946). Bacterial polysaccharides. Adv. Carbohydr. Chem. *2*, 204.

Flemming, H.C., Neu, T.R., and Wozniak, D.J. (2007). The EPS matrix: The house of biofilm cells. J. Bacteriol. *189*, 7945–7947.

French, A. (1989). Chemical and physical properties of fructans. J. Plant Physiol. *134*, 125–136.

Fuchs, A. (1956). Synthesis of levan by pseudomonads. Nature *178*, 921.

Goller, C.C. and Romeo, T. (2008). Environmental influences on biofilm development. Curr. Top. Microbiol. Immunol. *322*, 37–66.

Goodman, A.L., Kulasekara, B., Rietsch, A., Boyd, D., Smith, R.S., and Lory, S. (2004). A signaling network reciprocally regulates genes associated with acute infection and chronic persistence in *Pseudomonas aeruginosa*. Dev. Cell *7*, 745–754.

Greig-Smith, R. (1901). The gum fermentation of sugar cane juice. Proc. Linnean Soc. New South Wales *26*, 589.

Han, Y.W. (1990). Microbial levan. Adv. Appl. Microbiol. *35*, 171–193.

Hehre, E.J., Genghof, D.S., and Neill, J.M. (1945). Serological reactions of two bacterial levans. J. Immunol. *51*, 5–13.

Hestrin, S. and Avineri-Shapiro, C. (1944). The mechanism of polysaccharide production from sucrose. J. Biochem. *38*, 2–10.

Hettwer, U., Jaeckel, F.R., Boch, J., Meyer, M., Rudolph, K., and Ullrich, M.S. (1998). Cloning, nucleotide sequence, and expression in *Escherichia coli* of levansucrase genes from the plant pathogens *Pseudomonas syringae* pv. *glycinea* and *P. syringae* pv. *phaseolicola*. Appl. Environ. Microbiol. *64*, 3180–3187.

Kasapis, S., Morris, E., Gross, M., and Rudolph, K. (1994). Solution properties of levan polysaccharide from *Pseudomonas syringae* pv. *phaseolicola*, and its possible primary role as a blocker of recognition during pathogenesis. Carbohyd. Polymers *23*, 55–64.

Kasapis, S. and Morris, E. (1994). Conformation and physical properties of two unusual microbial polysaccharides: *Rhizobium trifolii* CPS and Levan, in Food Hydrocolloids: Structures, Properties and Functions, ed by K. Nishinari, Plenum Press, New York, 97–103.

Kives, J., Orgaz, B., and SanJose, C. (2006). Polysaccharide differences between planktonic and biofilm-associated EPS from *Pseudomonas fluorescens* B52. Colloids and Surfaces B: Biointerfaces *52*, 123–127.

Kornberg, H.L. and Riordan, C. (1976). Uptake of galactose into *Escherichia coli* by facilitated diffusion. J. Gen. Microbiol. *94*, 75–89.

Kralj, S., van Geel-Schutten, G.H., van der Maarel, M. J. E. C., and Dijkhuizen L. (2004). Biochemical and molecular characterization of *Lactobacillus reuteri* reuteransucrase. Microbiol. *150*, 2099–2112.

Laue, H., Schenk, A., Li, H., Lambertsen, L., Neu, T.R., Molin, S., and Ullrich, M.S. (2006). Contribution of alginate and levan production to biofilm formation by *Pseudomonas syringae*. Microbiol. *152*, 2909–2918.

Li, H., Schenk, A., Srivastava, A., Zhurina, D., and Ullrich, M.S. (2006). Thermo-responsive expression and differential secretion of the extra-cellular enzyme levansucrase in the plant pathogenic bacterium *Pseudomonas syringae* pv. *glycinea*. FEMS Microbiol. Lett. *265*, 178–185.

Li, H. and Ullrich, M.S. (2001). Characterization and mutational analysis of three allelic *lsc* genes encoding levansucrase in *Pseudomonas syringae*. J. Bacteriol. *183*, 3282–3292.

Li, Y. and Ferenec, T. (1997). Gene organization and regulatory sequences in the sucrose utilization cluster of *Bacillus stearothermophilus* NUB36. Gene *195*, 195–200.

Lippmann, E.O. (1881). Ueber das Lavulan, eine neue, in der Melasse der Rubenzuckerfabriken vorkommende Gummiart. Chem. Berichte *14*, 1509.

Lyne, R.R., Peat, S., and Stacey, M.J. (1940). The constitution of certain levans formed by bacterial action. Polysaccharides, Part XXXIX. J. Chem. Soc. *12*, 237–241.

Ma, L., Jackson, K.D., Landry, R.M., Parsek, M.R., and Wozniak, D.J. (2006). Analysis of *Pseudomonas aeruginosa* conditional Psl variants reveals roles for the Psl polysaccharide in adhesion and maintaining biofilm structure postattachment. J. Bacteriol. *188*, 8213–8221.

Malten, M., Biedendieck, R., Gamer, M., Drews, A.C., Stammen, S., Buchholz, K., Dijkhuizen, L., and Jahn, D. (2006). A *Bacillus megaterium* plasmid system for the production, export, and one-step purification of affinity-tagged heterologous levansucrase from growth medium. Appl. Environ. Microbiol. 72, 1677–1679.

Martinez-Fleites, C., Ortiz-Lombardia, M., Pons, T., Tarbouriech, N., Taylor, E.J., Arrieta, J.G., Hernandez, L., and Davies, G.J. (2005). Crystal structure of levansucrase from the gram-negative bacterium *Gluconoacetobacter diazotrophicus*. Biochem. J. *390*, 19–27.

Meng, G. and Futterer, K. (2003). Structural framework of fructosyl transfer in *Bacillus subtilis* levansucrase. Nat. Struct. Biol. *11*, 935–941.

Morales-Arrieta, S., Rodriguez, M.E., Segovia, L., Lopez-Munguia, A., and Olvera-Carranza, C. (2006). Identification and functional characterization of *levS*, a gene encoding for a levansucrase from *Leuconostoc mesenteroides* NRRL B-512 F. Gene *376*, 59–67.

Olivares-Illana, V., López-Munguía, A., and Olvera, C. (2003). Molecular characterization of inulosucrase from *Leuconostoc citreum*: a fructosyltransferase within a glucosyltransferase. J. Bacteriol. *185*, 3606–3612.

Postma, P.W. and Roseman, S. (1976). The bacterial phosphoenolpyruvate: sugar phosphotransferase system. Biochim. Biophys. Acta *457*, 213–57.

Ryder, C., Byrd, M., and Wozniak, D.J. (2007). Role of polysaccharides in *Pseudomonas aeruginosa* biofilm development. Curr. Opin. Microbiol. *10*, 644–648.

Sarcabal, P., Remaud-Simeon, M., Willemot, R., Potocki de Montalk, G., Svensson, B., and Monsan, P. (2000). Identification of key amino acid residues in *Neisseria polysaccharea* amylosucrase. FEBS Lett. *474*, 33–37.

Schenk, A., Weingart, H., and Ullrich, M.S. (2008). Extraction of high-quality bacterial RNA from infected leaf tissue for bacterial *in planta* gene expression analysis by multiplexed fluorescent Northern hybridization. Mol. Plant Pathol. *9*, 227–235.

Seibel, J., Moraru, R., Götze, S., Buchholz, K., Naamnieh, S., Pawlowski, A., and Hecht, H.J. (2006). Synthesis of sucrose analogues and the mechanism of action of *Bacillus subtilis* fructosyltransferase (levansucrase). Carbohydr. Res. *341*, 2335–2349.

Seo, J.W., Jang, K.H., Kang, S.A., Song, K.B., Jang, E.K., Park, B.S., Kim, C.H., and Rhee, S.K. (2002). Molecular characterization of the growth phase-dependent expression of the *lsrA* gene, encoding levansucrase of *Rahnella aquatilis*. J. Bacteriol. *184*, 5862–5870.

Simms, P., Boyko, W., and Edwards, J. (1990). The structural analysis of a levan produced by *Streptococcus salivarius* SS2. Carbohydr. Res. *208*, 193–198.

Skinner-Nemec, K.A., Nichols, N.N., and Leathers, T.D. (2007). Biofilm formation by bacterial contaminants of fuel ethanol production. Biotechn. Lett. *29*, 379–383.

So, L.L. and Goldstein, I.J. (1968). Protein-Carbohydrate Interaction. XIII. The interaction of concanavalin A with α-mannans from a variety of microorganisms. J. Biol. Chem. *243*, 2003–2007.

Song, K.B., Seo, J.W., and Rhee, S.K. (1999). Transcriptional analysis of *levU* operon encoding saccharolytic enzymes and two apparent genes involved in amino acid biosynthesis in *Zymomonas mobilis*. Gene 232, 107–114.

Spiers, A.J., Kahn, S.G., Bohannon, J., Travisano, M., and Rainey, P.B. (2002). Adaptive divergence in experimental populations of *Pseudomonas fluorescens*. I. Genetic and phenotypic bases of wrinkly spreader fitness. Genet. *161*, 33–46.

Stivala, S. and Bahary, W. (1978). Some dilute-solution parameters of the levan of *Streptococcus salivarius* in various solvents. Carbohydr. Res. *67*, 17–21.

Stivala, S.S., Zweig, J.E. (1981). Physicochemical parameters of partially hydrolyzed *S. salivarius* levan fractions. Biopolymers. *20*, 605–619.

Sutrina, S.L. and McGeary, T., Bourne, C.A. (2007). The phosphotransferase system and biofilms in gram-positive bacteria. J. Mol. Microbiol. Biotechnol. *12*, 269–272.

Tajima, K., Tanio, T., Kobayashi, Y., Kohno, H., Fujiwara, M., Shiba, T., Erata, T., Munekata, M., and Takai, M. (2000). Cloning and sequencing of the levansucrase gene from *Acetobacter xylinum* NCI 1005. DNA Res. *7*, 237–242.

Tentuta, L.M.A., Ricomini-Filho, A.P., Del-Bel-Cury, A.A., and Cury, J.A. (2006). Effect of sucrose on the selection of *Mutans Streptococci* and *Lactobacilli* in dental biofilm formed *in situ*. Caries Res. *40*, 546–549.

Tieking, M., Ehrmann, M.A., Vogel, R.F., and Gänzle, M.G. (2005). Molecular and functional characterization of a levansucrase from the sourdough isolate *Lactobacillus sanfranciscensis* TMW 1.392. Appl. Microb. Biotechnol. *66*, 655–663.

Ullrich, M.S., Schergaut, M., Boch, J., and Ullrich, B. (2000). Temperature-responsive genetic loci in the plant pathogen *Pseudomonas syringae* pv. *glycinea*. Microbiol. *146*, 2457–2468.

Vale, G.C., Tabchoury, C.P.M., Arthur, R.A., Del-Bel Cury, A.A., and Paes Leme, A.F. (2007). Temporal relationship between sucrose-associated changes in dental biofilm composition and enamel demineralization. Caries Res. *41*, 406–412.

Vallet, I., Diggle, S.P., Stacey, R.E., Camara, M., Ventre, I., Lory, S., Lazdunski, A., Williams, P., and Filloux, A. (2004). Biofilm formation in *Pseudomonas aeruginosa*: Fimbrial cup gene clusters are controlled by the transcriptional regulator MvaT. J. Bacteriol. *186*, 2880–2890.

van Hijum. S.A.F.T., Kralj, S., Ozimek, L.K., Dijkhuizen, L., and van Geel-Schulten, I.G.H. (2006). Structure-function relationships of glucansucrase and fructansurcrase enzymes from lactic acid bacteria. Microbiol. Mol. Biol. Rev. *70*, 157–176.

Vigants, A., Hicke, H.G., and Marx, S.P. (2001). A simple and efficient method for the purification of membrane-bound levansucrase from *Zymomonas mobilis*. Curr. Microbiol. 42, 415–418.

Vijn, I. and Smeekens, S., (1999). Fructan: More Than a Reserve Carbohydrate? Plant Physiol. *120*, 351–360.

Viney, C., Huber, A., and Verdugo, P. (1993). Biodegradable polymers and packaging, eds, Ching, C., Kaplan, D.L., Thomas, E.L. Technomic, Lancaster, p. 209.

Wenham, D.G., Davies, R.M., and Cole, J.A. (1981). Insoluble glucan synthesis by mutansucrase as a determinant of the cariogenicity of *Streptococcus mutans*. J. Gen. Microbiol. *127*, 407–15.

Yang, Y., Sreenivasan, P.K., Subramanyam, R., and Cummins, D. (2006). Multiparameter assessments to determine the effects of sugars and antimicrobials on a polymicrobial oral biofilm. Appl. Environ. Microbiol. *72*, 6734–6742.

Zhao, Y., Blumer, S.E., and Sundin, G.W. (2005). Identification of *Erwinia amylovora* genes induced during infection of immature pear tissue. J. Bacteriol. *187*, 8088–8103.

Structure, Biosynthesis and Regulation of Capsular Exopolysaccharide of *Erwinia amylovora* and other *Erwinia* Species and Role in Pathogenicity 13

Klaus Geider

Abstract

The fire blight pathogen *E. amylovora* and related bacteria synthesize complex EPS with galactose, glucuronic acid and eventually glucose residues. For *E. amylovora*, the capsules of amylovoran have been investigated by staining techniques and visualized by light and electron microscopy. They protect the pathogen against plant defence recognition, provide a moist environment against desiccation and bind ions and nutrients to support the cell metabolism. Amylovoran consists of repeating units forming chains up to 5 MDa and can be detached from the cells by mild treatments. EPS-deficient mutants of *E. amylovora* and *E. pyrifoliae* are non-pathogenic. Synthesis requires at least 12 genes for assembly of the repeating units, transport and polymerization. Similar genes were found for other *Erwinia* species that produce EPS related to amylovoran. Expression of the EPS encoding gene clusters depends on specific or global regulators including siRNAs, such as RcsA, B, C or Hns and RsmA/*rsmB*, which may adjust the EPS level to actual needs of the bacteria. Degradation or absence of EPS capsules results in pathogen recognition by plant defence mechanisms and provides a strategy to control fire blight.

Taxonomy of the genus *Erwinia* now and before

Fire blight, caused by *Erwinia amylovora*, is a serious disease of apple, pear, quince and other rosaceous plants (Vanneste, 2000). Fire blight was described first in the eighteenth century for the north-east of North America (Denning, 1794). The genuine hosts may have been crab apple (*Malus* sp.), hawthorn (*Crataegus* sp.) and mountain ash (*Sorbus* sp.) (Thomson, 2000), and introduction of European apple and pear cultivars may have enhanced distribution of the disease and continuous outbreaks. A subtype of the pathogen affects mainly raspberry (Starr *et al.*, 1951). An Asian pear disease, bacterial shoot blight of pear (BSBP), which has been described to occur in Korea (Rhim *et al.*, 1999) and in Japan (Kim *et al.*, 2001), is caused by *Erwinia pyrifoliae*.

E. amylovora and *E. pyrifoliae* belong to the genus *Erwinia*, formerly defined to comprise Gram-negative bacteria, which are facultatively anaerobic. They had been classified as soft-rot erwinias, now the genus *Pectobacterium* with *E. carotovora* subsp. *carotovora* and *E. carotovora* subsp. *atroseptica*, and the genus *Dickeya* with *E. chrysantemi* as prominent species. The 'herbicola' group, now genus *Pantoea* with *P. agglomerans*, included *Erwinia herbicola*, and *Erwinia stewartii* with *P. stewartii* subsp. *stewartii* (here mostly called *P. stewartii*) as yellow Gram-negative bacteria. *Erwinia amylovora* is the type species of the genus *Erwinia* which includes pathogenic and non-pathogenic bacteria with white colonies from various hosts. Other *Erwinia* species have been reclassified into the genus *Brenneria* including *B. salicis*, *B. alni* or *B. quercina*.

The remaining genus *Erwinia* consists of *E. amylovora*, *E. pyrifoliae*, *E. psidii*, *E. toletana*, *E. papayae*, *E. mallotivora*, *E. tracheiphila* as pathogens and of *E. billingiae*, *E. tasmaniensis*, *E. persicina* and *E. rhapontici* as epiphytic bacteria, which are not or not convincingly associated with plant diseases (Fig. 13.1). A similar pattern for differentiation

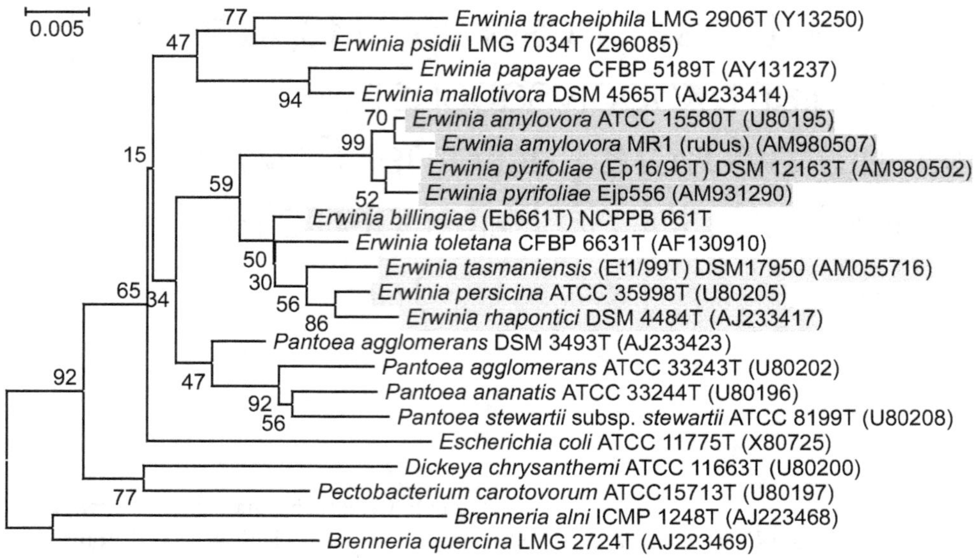

Figure 13.1 Relation of *Erwinia* species shown in a dendrogram from their 16S rRNA sequences. The nucleotide sequences (accession number at the right) were aligned with ClustalW and a dendrogram created with the program Mega 3.1 with boot straps for a series of *Erwinia* species, which have been in part reclassified into *Pantoea* and *Brenneria*. The remaining Erwinias species form clades for *E. amylovora* and *E. pyrifoliae* as well as *E. billingiae*, *E. toletana*, *E. tasmaniensis*, *E. persicina* and *E. rhapontici*. *E. tracheiphila/E. psidii* and *E. papayae/E. mallotivora* cluster separately. The pathogenic species *E. amylovora* and *E. pyrifoliae* are marked in grey, the other *Erwinia* species discussed in the text in light grey.

of *Erwinia* species has been obtained by protein analysis from whole cells applying MALDI-TOF (Sauer *et al.*, 2008). According to analysis of 16S rRNA sequence information, the genera *Brenneria* and *Pectobacterium/Dickeya* are separated from the genus *Erwinia* and members of the genus *Pantoea*. *E. persicina* and *E. rhapontici* form pink colonies on some agar plates. They do not cause HR in tobacco leaves (A. Wensing, unpublished) and primers of common HR-genes as from *hrpL*, *hrcC* or *hrcR* do not produce the correct PCR signals. *E. rhapontici*, which is sometimes found in clinical specimens, produces AHL (Jakovljevic *et al.*, 2008) and grows at 37°C unlike *E. persicina*. By us, disease symptoms on rhubarb (Letal, 1976) have not been observed in greenhouse experiments for the *E. rhapontici* type strain nor effects on beans (Gonzales *et al.*, 2005) for the *E. persicina* type strain. Based on their 16S rRNA, they are related to each other and to *E. tasmaniensis* and close to *E. toletana* and *E. billingiae*. *E. amylovora* and *E. pyrifoliae* form a similar clade, which comprises 'fruit tree' strains and raspberry strains for *E. amylovora* and for *E. pyrifoliae* pear pathogens from Korea and Japan. The diseases caused by these species have been well investigated. Spread of fire blight from North America to other countries was observed in 1916 to New Zealand and in 1956 to England (Vanneste, 2000). Events of long-distance distribution cannot be traced back, although genetic analysis may suggest pattern types which can be associated. European and Mediterranean strains are quite homogeneous (Jock *et al.*, 2002). Nevertheless, they can be differentiated by PFGE analysis. Accordingly, it can be concluded that *E. amylovora* with the PFGE pattern type Pt2 spread from Egypt (1965) to Turkey, Iran and the Balkans, and has now reached Hungary. Another wave spread from England to Central Europe and to the north, east and south to Norway, Sweden, Poland, Austria and Switzerland (Pt1). Other pattern types spread from western France to northern Spain (Pt4), and from Belgium/northern France to northern Italy and central Spain (Pt3). That distribution may be associated with trade of latently infected fire blight host plants. In contrast to strains from Europe and the Mediterranean region, North American strains are quite heterogeneous (Jock and Geider, 2004).

Single nucleotide polymorphisms (SNPs) show a coincidence of European/Mediterranean strains with those from New Zealand (I. Gehring and K. Geider, unpublished). Short distance spread is predominantly caused by insects visiting flowers, which become contaminated with the sticky bacterial slime.

E. pyrifoliae may face similar pathways of distribution. The limited information about the occurrence of the pathogen restricts strain differentiation on large scale. Nevertheless, strains from Korea and from Japan are related to each other, and taxonomically different from *E. amylovora* (Geider *et al.*, 2009; Waleron *et al.*, 2008). This relationship is also seen in the 16S rRNA sequences (Fig. 13.1). *E. pyrifoliae* does not produce levan. Most *E. amylovora* strains isolated from raspberry (*Rubus idaeus*) synthesize amylovoran with repeating units lacking the glucose side chain (Maes *et al.*, 2001) identical with the repeating units of pyrifolan (EPS from *E. pyrifoliae*). Nevertheless, at least one *Rubus* strain produces EPS identical with amylovoran from 'fruit tree' strains. The EPS structure may therefore not be associated with the host range.

Role of the capsular EPS amylovoran for *Erwinia amylovora*

Erwinia amylovora produces two types of EPS. Levan is a polyfructan and formed from sucrose by the secreted levansucrase, whereas amylovoran is more complex and consists of highly polymerized repeating units of five or six sugar residues. Synthesis of a gluco-oligosaccharide has been mentioned for *E. amylovora* (Smith *et al.*, 1995) and a *celA* gene for cellulytic activity was detected in the genome of the pathogen (Riekki *et al.*, 2000).

The capsular EPS amylovoran has important roles for *E. amylovora,* strikingly indicated by the fact that all non-producing mutants are non-virulent (Bugert and Geider, 1995). Symptoms on fire blight host plants including apple, pear, quince, hawthorn, cotoneaster or pyracantha are also caused by EPS and result in wilt and often droplets of ooze on the plant surface. The pathogen migrates mainly in the xylem, where are no barriers to slow down movement (Bogs *et al.*, 1998). On the other hand, this environment is poor in nutrients. A disease strategy is therefore clogging vessels by bacterial aggregates with EPS and extrusion into the parenchyma. Remigration into the xylem may enhance movement, but also prevent recognition by plant defence mechanisms. The dominant direction of spread in plants is from the shoots to the roots (Bogs *et al.*, 1998) shown with gfp-labelled bacteria, also found before with radioactively labelled cells (Suhayda and Goodman, 1981). Necrosis seems to be mainly caused by transfer of Hrp-proteins to plant cells (Bocsanczy *et al.*, 2008). Oozing cankers contribute to outbreak of the disease in spring. The sticky slime may also help to distribute the pathogen by insects visiting flowers. Inside plant tissue the EPS causes clogging of vessels by migrating bacteria and thereby symptoms of wilt (Bogs *et al.*, 1998). This property of isolated amylovoran is not restricted to fire blight host plants and can also be seen for other plants (Sjulin and Beer, 1978).

EPS synthesis on plates with minimal agar is enhanced in the presence of copper ions (MM2Cu) (Bereswill *et al.*, 1998). In addition to mucoid appearance the colonies become yellow, which is typical for *E. amylovora* forming white colonies on agar without copper salt, and is therefore a diagnostic feature. Many other bacteria do not show this colour shift or barely grow on minimal agar with copper salt. The gene *rcfA* is mainly responsible for yellow colour formation of *E. amylovora* (Zhang *et al.*, 2000). In lab assays with immature pears, this formation of ooze can also be used for diagnosis.

Amylovoran of *E. amylovora* forms loose capsules around the cells and apparently has multiple roles such as binding of ions and nutrients and most important, to shield the pathogen against recognition by plant defence mechanisms. Furthermore, the EPS provides a moist environment, protecting the cells against complete loss of water (Jock *et al.*, 2005).

Capsulated strains survive in dry habitats which was assayed on nitrocellulose filters and in HR lesions of non-host plants. The decrease of bacterial viability under dry conditions is more pronounced for EPS-mutants of *E. amylovora* than for wild-type cells. The capsules can be visualized by electron microscopy after fragile staining with ruthenium red (Bellemann *et al.*, 1994) or just by binding of FITC-labelled lectin from *Abrus precatorius* (Bereswill *et al.*, 1998), which is commercially available with severe restrictions in distribution due to its toxicity. The lectin binds to the terminal galactose residue of the side chain and is so far specific for *E. amylovora* (Bellemann *et al.*, 1994). In summary, the capsules may also be vital for long term survival of *E. amylovora.*

Structure of the repeating unit and its polymerization

Amylovoran can be isolated from suspension cultures in minimal medium with sorbitol (MM1, MM2C) (Nimtz *et al.*, 1996a) or separated from cells after growth on plates with minimal agar (Bernhard *et al.*, 1996). A slightly acidic pH enhances amylovoran production. To avoid contamination with agar, the surface should be covered with a sterile cellophane disk. In suspension cultures, the EPS is essentially detached from cell bodies, for isolation from cells on agar plates the EPS should be separated by vigorous shaking. Culture supernatants with amylovoran must be concentrated by pressure dialysis. Dialysis of amylovoran solutions against water is applied to remove salts, and the dialysate is then freeze-dried. These preparations can be kept at room temperature for a long time without detectable changes. The size of amylovoran from suspension cultures was determined by analytical ultracentrifugation and by migration in large pore HPLC columns to be approximately 1 MDa (Jumel *et al.*, 1997). When amylovoran is gently isolated from agar plates, the size is often considerably larger, in the range of 5 MDa corresponding to about 10,000 repeating units (M. Schollmeyer, C. Langlotz, A. Huber, and K. Geider, unpublished).

Further analysis of the EPS can be done with whole molecules or after cleavage into repeating units. Alkaline treatment results in removal of acyl-groups, which can be detected by changes of the IR spectra (Bernhard *et al.*, 1996). The sugar composition can be determined chemically, their linkages preliminary by methylation analysis and precise structure by NMR.

NMR-analysis of the repeating units requires enzymatic dissection of amylovoran. This is readily achieved with a viral EPS depolymerase (Kim and Geider, 2000). The repeating units can be further purified on agarose columns and are then ready to determine their molecular weight by electrospray (ESI) analysis (Kim *et al.*, 2002) and their linkages by NMR measurements (Nimtz *et al.*, 1996a).

Amylovoran consists of repeating units with galactose, glucuronic acid and some glucose (Fig. 13.2). The molar ratios are 4:1:0.6. The linkages have been analysed by 2D-NMR and a complete structure was established including decorations in the side chain (Nimtz *et al.*, 1994a). Amylovoran is partially substituted with a second side chain carrying a glucose in more than 50% of the repeating units.

Precursors of amylovoran synthesis and their genetics

According to the sugar composition of amylovoran, biosynthesis of the repeating units requires UDP-galactose, UDP-glucuronic acid and UDP-glucose. There is no evidence for use of other nucleotide diphosphates than UDP-activated sugar residues. Interconversion of UDP-galactose and UDP-glucose is catalysed by the epimerase GalE of *E. amylovora* (Metzger *et al.*, 1994). A *gal*-operon with *galETK* as in *E. coli* differs for *E. amylovora, E. billingiae,* and *E. tasmaniensis* with *galE* adjacent to the EPS encoding regions. The adjacent *galF* gene, also named *amsM,* does not encode for an described enzymatic activity and is supposed to participate as a subunit of GalU, the glucose-1-phosphate uridylyltransferase (UDP-glucose pyrophosphorylase) in conversion of galactose and UTP to UDP-galactose or to interact with GalE, the UDP-galactose/UDP-glucose epimerase (Fig. 13.3). The GalF proteins of several *Erwinia* species such as *E. amylovora, E. billingiae,* and *E. tasmaniensis,* but also *P. stewartii* show a high conservation at the N-terminus, which may indicate a specific topology for interaction with other proteins. AmsM/GalF-deficient mutants of *E. amylovora* were also unable to synthesize amylovoran (K. Geider, unpublished). The deficiency of *galE* mutants in amylovoran synthesis can be phenotypically complemented by addition of galactose to the growth medium (Metzger *et al.*, 1994). The GalU and GalF proteins of *E. amylovora, E. billingiae* and *E. tasmaniensis* are related (>70% similarity) and are located in about 0.3 Mb distance on the chromosomes.

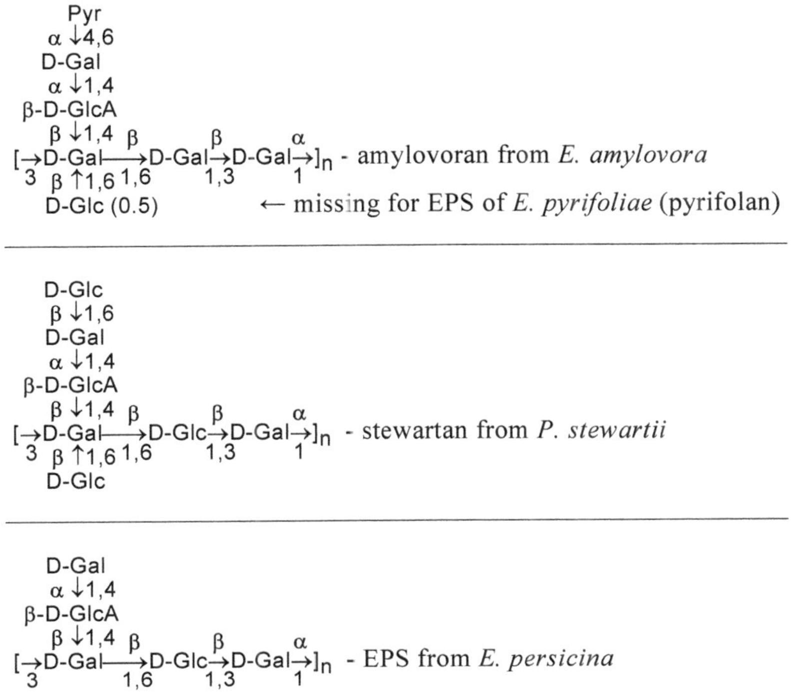

Figure 13.2 Structure of repeating units of EPS from *E. amylovora*, *E. pyrifoliae*, *P. stewartii* and *E. persicina*. The backbone consists of three sugars, which are differently linked galactose residues in case of amylovoran and pyrifolan. EPS from *P. stewartii* and *E. persicina* have a substitution of the second galactose by glucose. The third galactose carries a side chain with glucuronic acid and another galactose, which is terminated by pyruvate in case of amylovoran and pyrifolan and by glucose in case of stewartan. Decoration of the galactose such as acetyl groups in case of amylovoran and pyrifolan have not been described for EPS of *E. persicina*. A second side chain at the same branched galactose is glucose which is present in all repeating units of stewartan and absent for pyrifolan. Half of the repeating units of amylovoran are substituted by a glucose residue for 'fruit tree' strains. *E. amylovora* from raspberry may lack this glucose residue (Maes *et al.*, 2001), but not all of them do (unpublished).

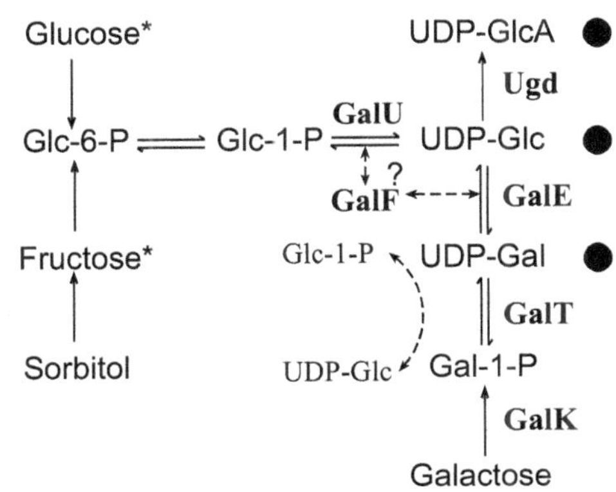

Figure 13.3 Genes involved in precursor synthesis for amylovoran. According to the sugar composition of amylovoran, UDP-gal, UDP-glcA and UDP-glc must be provided by the cells in order to synthesize the EPS. Due to low galactose in plants, these are mainly made from other carbon sources such as sorbitol, fructose and glucose including cleavage of sucrose. Glc-1 phosphate is converted to UDP-glc, a source for UDP-glcA and UDP-gal. Metabolism of sucrose is missing for *E. billingiae*, and the sorbitol operon is not present in the genome of *E. tasmaniensis* (see Fig. 13.4). *Also provided by cleavage of sucrose; •Precursors for amylovoran synthesis.

UDP-glucuronic acid is synthesized from UDP-glucose by the enzyme UDP-glucose dehydrogenase, encoded by the gene *ugd* adjacent to a *wcaG* homologue for an uridine diphosphate galacturonate 4-epimerase and the global regulator gene *hns* (Hildebrand *et al.*, 2006).

The energy in carbon metabolism is provided by glycolysis in the Embden–Meyerhof pathway. Since erwinias are facultatively anaerobe, they can also use the Entner–Douderoff pathway. EPS synthesis does not only require energy, it also weakens the cells. In tedious immuno-staining experiments, patches on the cell surface of *E. coli* indicates sites of secretion (Kröncke *et al.*, 1990). These extrusion sites may also be receptors for bacteriophages, which find their target as EPS coated cells, cleave amylovoran in case of *E. amylovora* and finally reach the cell walls. The EPS depolymerase gene of the *E. amylovora* phage φEa1h was fused with a His-tag and the purified enzyme applied for cleavage of various EPS species, which carry a 3-α galactose residue in the backbone (Nimtz *et al.*, 1996a) (Fig. 13.2).

Carbohydrates of fire blight host plants and sugar metabolism of *E. amylovora*

Sucrose and sorbitol are typical carbohydrates for host plants of *E. amylovora*. Sucrose is a general transport sugar for all green plants. It is an early product of photosynthesis and is distributed to all parts of a plant. It is also a component of nectar in flowers, which consists of sucrose, fructose and glucose (2:2:1). The viability of *E. amylovora* in high sucrose concentrations is remarkable in comparison to other bacteria (Bogs and Geider, 2000). Although levan synthesis can reduce the actual sucrose concentrations this property may be not connected to survival in high sucrose (Bogs and Geider, 2000). *E. amylovora* carries a sucrose uptake system, which comprises sensing, phosphorylation, transport and cleavage of sucrose (Fig. 13.4). Mutants in the sucrose operon of *E. amylovora* are non-virulent (Bogs and Geider, 2000) indicating the importance of sucrose for the pathogen.

Sorbitol is not as common as sucrose in plants. On the other hand, Rosaceae contain significant amounts which render sorbitol to a second transport carbohydrate. The content of many plants was investigated, and a relation of daily and seasonal changes was correlated to fire blight susceptibility (Suleman and Steiner, 1994).

The metabolism of sorbitol and sucrose differs for *E. amylovora, E. billingiae* and *E. tasmaniensis* (Fig. 13.4). *E. amylovora* can use both carbohydrates, whereas *E. billingiae* metabolizes sorbitol and *E. tasmaniensis* sucrose as an energy source. The sorbitol operon of *E. amylovora* consists of six genes involved carbohydrate uptake and cleavage including two regulators (Aldridge *et al.*, 1997). The *E. billingiae* strain Eb661 shows a similar arrangement for the sorbitol operon (Kube *et al.*, unpublished).

For *E. amylovora*, the sucrose operon (*scr*) comprises five genes, which encode uptake, phosphorylation and cleavage of sucrose in addition to regulation (Bogs *et al.*, 2000) (Fig. 13.4). For *E. tasmaniensis*, a similar operon is present (Kube *et al.*, 2008) and homologues of the central genes *scrA* and *scrB* are duplicated elsewhere in the genome. *E. tasmaniensis* may typically colonize flowers, where it is exposed to low amounts of sorbitol and not directly to plant defence mechanisms. The lacking synthesis of capsular EPS of *E. tasmaniensis* could be partially substituted by levan in this environment.

In contrast, *E. billingiae* is unable to utilize sucrose. The genetic information for sucrose metabolisms is missing (Kube *et al.*, unpublished). This includes absence of the *lsc* gene for levan synthesis. On the other hand, *E. billingiae* contains a sorbitol operon (Fig. 13.4) and efficiently metabolizes the carbohydrate. It is a powerful EPS-producer, but is unable to cause HR (Jakovlevic *et al.*, 2008). Since both properties are strictly required for virulence of plant pathogens, *E. billingiae* is restricted to epiphytic growth.

The ams region and functions in amylovoran synthesis

Due to different linkages in the backbone of the repeating unit of amylovoran and additional sugars in the side chain (Fig. 13.2), a complex set of glycosyl transferases and additional enzymes for translocation of the repeating units and their polymerization are required. The encoding genes are arranged in the *ams* region. Twelve genes are coding for functions in amylovoran followed by genes, *galF* and *galE* for precursor synthesis (Figs

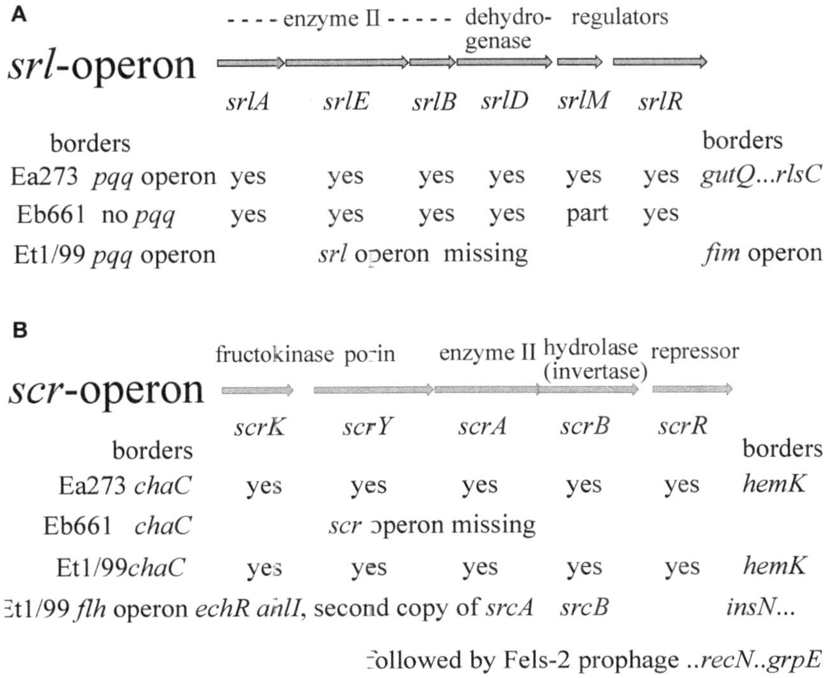

Figure 13.4 The operons for uptake and cell metabolism of **A** sorbitol and **B** sucrose in the genomes of *E. amylovora*, *E. billingiae* and *E. tasmaniensis*. The uptake of these carbohydrates is achieved by the phosphoenolpyruvate-dependent carbohydrate:phosphotransferase system (PTS). The responsible proteins are highly conserved with about 90% similarity between *E. billingiae* and *E. amylovora* for the *srl* operon and between *E. tasmaniensis* and *E. amylovora* for the *scr* operon. The insertion regions for *scr* of *E. amylovora* and *E. tasmaniensis* are related, but *E. tasmaniensis* carries a second copy of the *srcA* and *scrB* genes. The adjacent genes of a Fels-2 prophage may indicate horizontal gene transfer. The corresponding proteins for carbohydrate metabolism have approximately the same size indicating conservation during evolution, probably due to their specific functions in the catalytic steps.

13.3 and 13.5). All ORFs in the *ams* region were randomly mutagenized with transposon Tn5 and the insertion sites also localized by nucleotide sequencing (Bugert and Geider, 1995). An insertion in the intergenic region between genes *amsF* and *amsJ* had little influence on amylovoran synthesis. Mutants of Tn5-insertions within structural genes are unable to produce amylovoran (Bellemann and Geider, 1992; Menggad and Laurent, 1998). They are not virulent for fire blight host plants, but can still cause HR on non-host-plants. The *ams*-mutants are not only deficient in amylovoran synthesis, they are unable to multiply in host plant tissue (Bellemann and Geider, 1992). Apparently they are recognized by host plant defence mechanisms and often show necrotic lesions at the inoculations sites.

The biosynthesis of the repeating unit has been attempted in a subcellular system (C. Langlotz and K. Geider, unpublished). EDTA-treated cells of *ams* mutants were incubated with radioactively labelled UDP-galactose and the sugar intermediates extracted with the lipid carrier. In summary, no label was attached by cells of an *amsG* mutant, one galactose-residue was recovered from *amsD*-cells. Consequently, AmsG attaches the first galactose of the backbone to the undecaprenylphosphate carrier and AmsD the second galactose. AmsC has a remarkable topology due to several transmembrane domains and may differ in the cellular location from other glycosyl transferases. The protein may anchor the EPS chain in the membrane for attachment of the next repeating unit.

It is tempting to speculate the order for synthesis of repeating units of amylovoran by sequential involvement of AmsG, AmsD, AmsE (Fig. 13.6). The next sugar attachments are experimentally not well documented. It can be assumed that attachment to glycosyl residues in the

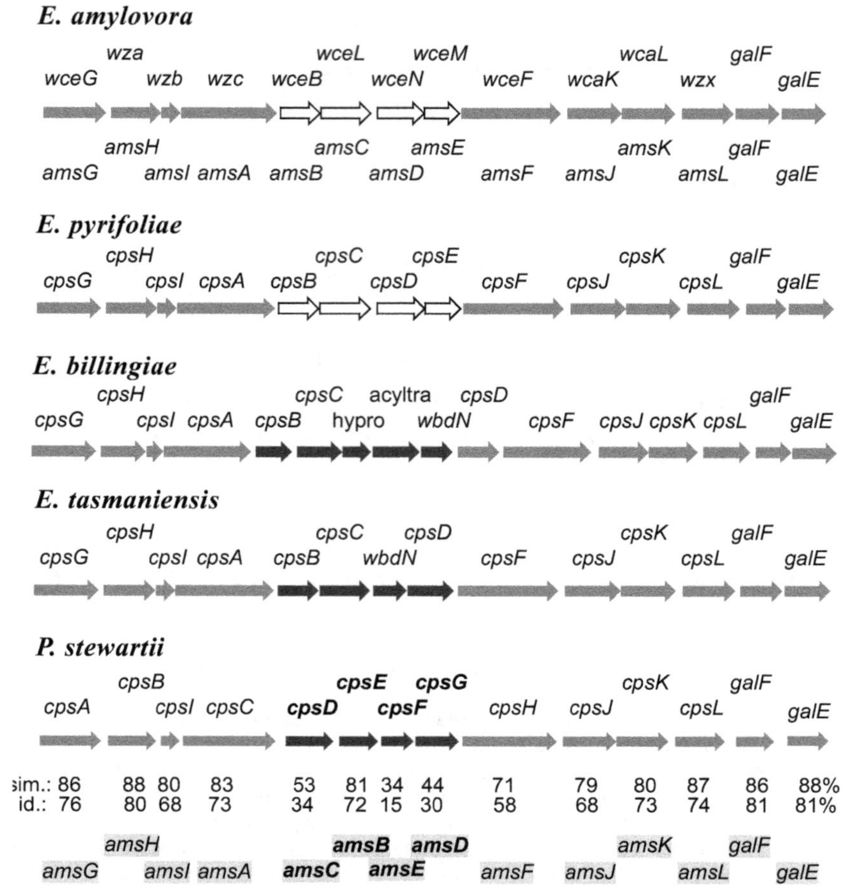

Figure 13.5 EPS encoding regions. The nucleotide sequences were derived from various sources (Kube *et al.*, in part unpublished; D. Coplin, unpublished) and have also been confirmed by partial and whole genomic sequences for the species *E. amylovora*, *E. pyrifoliae*, *E. billingiae*, *E. tasmaniensis* and *P. stewartii* subsp. *stewartii*. The nomenclature of the genes is under reconsideration (Reeves *et al.*, 1996), names associated with 'w' (for cell wall) to unify designations for genes from various bacteria with similar functions in polysaccharide synthesis. The new designations have been added to the *ams* genes of *E. amylovora* (first line). The corresponding genes of other *Erwinia* species with a *cps* region have been assigned to homologous *ams* genes. Also the *cps* genes of *P. stewartii* are mostly very similar to *ams* genes (shaded background, bottom) and values for protein similarity and identity have been added. It should be noted that the order of gene similarity and order of functions is broken in the centre region (*cpsD~amsC*; (except *cpsE/amsB* with high homology)); *cpsF/amsE*; *cpsG/amsD*) (dark arrows/white arrows in region from *E. amylovora* and *E. pyrifoliae*). The other proteins from the *cps/ams* clusters are highly related exceeding a similarity of 70%. The WbdN protein (similar to WbdN of *E. coli*) of *E. billingiae* and *E. tasmaniensis* is highly related to CpsF of *P. stewartii* (appr. 70%). sim., similarity with permissive amino acid exchanges; id., identical amino acids.

side chains is mediated by AmsB for glucuronic acid and the adjacent galactose by AmsK (C. Langlotz, M. Schollmeyer, M. Nimtz, D. Coplin and K. Geider, unpublished). ^{14}C-labelled UDP-glucuronic acid is incorporated in EDTA treated cells by *amsK* mutants, but not by *amsB* mutants. AmsJ is tentatively assigned to attach the pyruvyl residue terminating the side chain. In blast searches in nucleotide sequence data libraries, the amino acid sequence is related to putative, predicted pyruvyl transferases of *E. coli* and other bacteria (Johnson *et al.*, 2007).

Other genes in the *ams* region participate in processing of the repeating units. Little evidence is seen for a protein involved in polymerization. AmsF is a candidate, because it is related to the viral Dpo, which cleaves amylovoran into repeating units. It may be supported by AmsC

AmsG/WceG	**Gal**-P-P-LC
AmsD/WceN	**Gal**-Gal-P-P-LC
AmsE/WceM	**Gal**-Gal-Gal-P-P-LC
AmsB/WceB	**GlcA**-Gal-Gal-Gal-P-P-LC
AmsK/WceK	**Gal**-GlcA-Gal-Gal-Gal-P-P-LC
AmsJ/WceJ?	**pyr** Gal-GlcA-Gal-Gal-Gal-P-P-LC
	pyr Gal-GlcA-Gal*-Gal-Gal-P-P-LC
unknown protein	**Glc**

Figure 13.6 Gene functions for sequential synthesis of the repeating unit of amylovoran. As discussed in the text, the sequential attachment of sugars forming the repeating units is mediated by distinct Ams proteins, the glycosyl transferases. The site of initial start of attachment of a undecaprenylphosphate, lipid carrier (LC), which is recycled after polymerization of the repeating unit to the growing amylovoran chain. The components for carbohydrate attachment have been deduced from EDTA-treated cells and by mutant complementation and determination of the resulting repeating units by methylation analysis. The order presents the best evidence at present. The assignment of AmsJ circumstantial and will also be discussed in the text. *Attachment site for extension of existing amylovoran chain creating a new side chain.

with a flippase activity. AmsA has an activity of a tyrosine kinase (Ilan *et al.*, 1999). AmsH could be located in the outer membrane of *E. amylovora*. AmsL may be involved in pore formation (Bugert and Geider, 1995). After polymerization of the repeating units into the growing chain, a lipid carrier loaded with a diphosphate residue is released. The conversion into the active undecapenyl-monophosphate is most likely done by the AmsI phosphatase (Bugert and Geider, 1997), which could also be involved in other dephosphorylation steps.

Genetics of capsular EPS of other *Erwinia* species and a comparison with *P. stewartii*

Previously, *E. amylovora* and *E. stewartii* were classified into the same genus *Erwinia*. Mainly based on DNA/DNA hybridization kinetics, *E. stewartii* was transferred with other 'yellow Erwinias' into the genus *Pantoea* (Mergaert *et al.*, 1993). On the other hand, *P. stewartii* and *E. amylovora* also share many features for their EPS encoding genes. The alignment of the protein sequences from *ams*/*cps*-genes allows a good correlation of both EPS-encoding gene clusters of *E. amylovora* and *P. stewartii*. It is possible to transfer the gene cluster for stewartan synthesis to *E. amylovora*, when cloned in a broad host range plasmid, in order to restore EPS synthesis in this pathogen (Bernhard *et al.*, 1996). Several criteria confirmed these properties: (i) staining with FITC-labelled lectin from *A. precatorius*, specific for the galactose residue terminating the side chain of the repeating unit of amylovoran; (ii) sugar linkages determined by methylation analysis; (iii) sugar composition of isolated EPS. The merodiploids of *P. stewartii* regain virulence on corn, but *E. amylovora* merodiploids produce only weak symptoms on pear slices or apple seedlings.

The proteins AmsD/CpsC, AmsC/CpsD and AmsC/CpsF show low similarities as 15–35% identity compared with 70% and more for the other corresponding proteins from the EPS encoding regions of *E. amylovora* and *P. stewartii*. The best homology from alignments including protein sizes are given in Fig. 13.5. The *amsD* mutants can be complemented by the *P. stewartii* genes *cpsF*/*cpsG* in order to create a repeating unit of the amylovoran type with a glucose in the backbone (C. Langlotz, M. Schollmeyer, M. Nimtz, D. Coplin, and K. Geider, unpublished).

The EPS encoding regions have been sequenced for several *Erwinia* species. A striking homology to the *ams* region of *E. amylovora* is observed. In case of *E. amylovora* 'fruit tree' and raspberry strains, there are only differences on the nucleotide level, similar to the EPS encoding region of *E. pyrifoliae* compared to the *ams* cluster (Fig. 13.5). Changes for gene arrangements are observed for *E. billingiae* and *E. tasmaniensis* compared with *E. amylovora*. In particular, there is a rearrangement in the centre region as for *P. stewartii*. A gene related to *wbdN* of *E. coli* is similar to *cpsF* of *P. stewartii* (44%). CpsF has only a low homology to AmsE (34% similarity; 15% identity), indicating different catalytic specificities as discussed above.

Within the Erwinias mentioned in Fig. 13.1, *E. tasmaniensis* does not produce detectable amounts of EPS in various media. Apparently, expression of the EPS genes is down-regulated, because transformation with *rcsA* from *E. amylovora* results in EPS production of *E. tasmaniensis* (I. Müller and K. Geider, unpublished). Introduction of *rcsB* from *E. amylovora* or *P. stewartii* has little impact on EPS synthesis of *E. tasmaniensis*.

Stewartan, produced by *P. stewartii*, is highly related to amylovoran. In a similar approach as for amylovoran, the structure of stewartan was determined (Nimtz *et al.*, 1996b). A main difference is the substitution of the second galactose residue in the backbone of the repeating unit of amylovoran by glucose in stewartan (Fig. 13.2). The side chain of stewartan is terminated by glucose, whereas amylovoran carries pyruvate and acetyl residues at the galactose of the side chain. The sugar linkages, but not the composition, in the backbone of the repeating units are identical (Fig. 13.2). The backbone of EPS from *E. persicina* is structurally related to stewartan (Kiessling *et al.*, 2005), whereas pyrifolan, the EPS from *E. pyrifoliae* resembles amylovoran (Kim *et al.*, 2002). In EPS profiles by electrospray (ESI), presence or absence of a glucose residue as a second side chain was measured. Accordingly, this residue is present in all repeating units of stewartan and totally absent for pyrifolan of *E. pyrifoliae* (Kim *et al.*, 2002).

A puzzle is the mechanisms how cells count their EPS polymerization steps for a unique chain length. A narrow size range as for the K1 antigen of *E. coli* with 160 to 230 residues suggests an active process with terminal modifications (Whitfield, 2006). The considerably larger size of amylovoran and pyrifolan with up to 10,000 repeating units may require a network of fine tuning during chain elongation. Considering an inhomogeneous size distribution around 1 MDa for amylovoran from suspension cultures, it can be assumed that the chains grow synchronously with the cell. Elongation will be terminated in the stationary phase, when cell metabolism slows down, resulting in a size of 5 MDa. The chain will then stick to the cells permanently to provide protection as an EPS capsule.

Regulation of ams gene expression

EPS synthesis of *E. amylovora* depends on properties of individual strains and on the environment. The polysaccharide production is well visible on plates with minimal agar and sorbitol as a carbon source (MM2C) (Bereswill *et al.* 1998). Freshly isolated strains may slow down during propagation on plates and storage, but even isolates from the same plant can show a different level of mucoidy. Hignett (1987) has therefore postulated that colonization of plant tissue needs bacteria with 'normal' and with low amylovoran production. The latter could be recognized by plant defence mechanisms to induce an HR-like reaction in the host tissue providing nutrients for the 'normal' unrecognized population with EPS capsules. On the other hand, not all natural populations contain both types of EPS producers and there is apparently a cross-feeding of mutants. This was shown for mixed populations of *E. amylovora*, where both produce complementary components, i.e. amylovoran and the DspE protein, respectively (Bellemann and Geider, 1992). Even a capsule from complementation may prevent pathogen recognition by the plant tissue.

Bacterial sensing is often achieved by a protein in the outer membrane, which transfers a phosphate group to an activator protein. EPS synthesis of *E. amylovora* is also regulated by a two-component system with a sensor protein and activator proteins (Fig. 13.7). This system was described in detail for T-DNA transfer by *Agrobacterium tumefaciens* (Bodman *et al.*, 2003) and for colanic acid of *E. coli* (Majdalani and Gottesman, 2006). *A. tumefaciens* is sensing the environment for damage of plant cells in order to transfer the T-DNA to its target. Integration into the plant genome causes growth of a tumour. The VirA protein is the sensor for plant signals such as release of acetosyringon or organic acids and phosphorylates VirG protein. This activator binds to specific promoters in the *vir*-region in order to activate expression of proteins encoded by the *virB* cluster, *virC*, *virD*, *virE* and *virG* itself, thereby enhancing its own expression. Consequently, the transfer of T-DNA is initiated with steps comprising pore formation and complexing of the single-stranded DNA intermediate. Colanic acid synthesis of *E. coli* depends on several proteins for expression of the responsible operon. RcsC/RscD is the sensor in the bacterial membrane with activities of a phosphorylase and a kinase. The phosphate group is transferred to RcsD (formerly YojN), which activates RcsB. RcsB forms a complex with RcsA in order to enhance colanic acid synthesis. For *E. amylovora*, a gene analogous to *rcsD* of *E. coli* is also present in the genome, so synthesis of amylovoran and of colanic acid may follow similar schemes, where RcsB is strictly re-

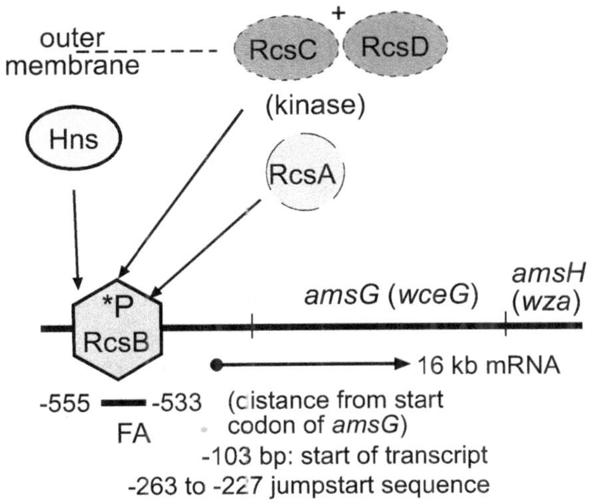

Figure 13.7 Regulation of amylovoran synthesis. The sensor protein RcsC is located in the outer cell membrane and transfers a phosphate group to the adjacent RcsD protein. This activation is used for phosphorylation of RcsB. The activated RcsB protein can act alone, but is more efficient in a complex with RcsA. RcsB attaches upstream of the *ams* region (Wehland *et al.*, 1999), and its binding is enhanced by AmsA. The recognition site and structural properties of RcsB were further investigated. The main transcript of the *ams* region starts about 100 bp upstream of *amsG* and is terminated downstream of *amsL* (Bugert and Geider, 1995). A jumpstart sequence is often found upstream of EPS encoding operons of bacteria (Hobbs and Reeves, 1994). For the *ams* operon of *E. amylovora*, the consensus motif 'MASYRYAYTGGTAGCTGWNRAGCCARGGGCGGTAGCGTR' occurs between the positions –263 to –227 bp upstream of *amsG*. Binding of the global regulator H-NS in the *ams* can also affect gene expression.

quired. The sensor RcsC would activate RcsB by phosphorylation, and the phosphorylated RcsB is complexed by RcsA to bind in about 550 bp distance to *amsG* enhancing transcription of the *ams* region (Kelm *et al.*, 1997). The binding has been further narrowed to a position from –555 to –533 bp upstream of *amsG* (Wehland *et al.*, 1999). A central control motif of 13 nucleotides was defined by specific interaction of the RcsB/RcsA complex. The 23-bp DNA was mutagenized at various positions. Some alterations even resulted in a significantly increased retardation in band shift assays. The binding domain of RcsB was characterized by NMR spectroscopy to contain four α-helices, similar to structures of regulatory proteins (Pristovsek *et al.*, 2003). The binding site upstream of *amsG* is also recognized by RcsB alone. In this case, the binding constant of RcsB was approximately one order of magnitude higher than in the presence of RcsA. It was concluded that RcsA does not the alter DNA binding specificity, but stabilizes the complex of RcsB with the DNA binding sequence. RcsB may also interact with other coactivators to control expression of additional operons in bacteria. Expression of *rcsA* is regulated by several factors. Binding of Hns at the promoter site of *rcsA* reduces the level of protein expression, which is further controlled by binding of a small RNA (Sledjeski and Gottesman, 1995). A dominant role has the *lon* protease for posttranscriptional regulation of *rcsA* in *E. amylovora* (Eastgate *et al.*, 1995). Cleavage of RcsA is low in *lon*-mutants resulting in high EPS synthesis.

A large 15-kb mRNA is transcribed for the *ams* region starting upstream of *amsG* and covers the genes *amsG* to *amsL* (*wceG* to *wzx*) (Bugert and Geider, 1995). Additional internal promoters are likely, but have not been defined yet. An RcsA-like protein, RcsV, can substitute to some extent *rcsA* mutants and influences viscosity of stewartan synthesized by *P. stewartii* (Aldridge *et al.*, 1998).

Mutations in *rcsA* and *rcsB* reduce amylovoran synthesis (Bernhard *et al.*, 1990; Bereswill *et al.*, 1997). Overexpression of RcsB causes high levels of amylovoran, which can interfere with the viability of cells. On the other hand, a lack of RcsA can be partially compensated by overexpression of RcsB. The effect of *rcsC* and *rcsD* on EPS synthesis of *E. amylovora* is still open. Among *E. amylovora*, *E. pyrifoliae*, *E. billingiae* and *E. tasmaniensis*, RcsB is highly conserved with 90–100% similarity, whereas RcsA and RcsC are more variable within these *Erwinia* species with 80–95% similarity of the proteins.

In addition to specific regulators, global regulators can also interfere with amylovoran synthesis. A complex regulation system has been described for *E. carotovora* subsp. *carotovora* (syn. *Pectobacterium carotovorum*), which shares homology with *E. amylovora* (Ma *et al.*, 2001). The posttranscriptional regulator of secondary metabolism (RSM) RsmA affects extracellular enzymes and harpin synthesis of *P. carotovora*. The RNA binding protein RsmA lowers the half life of many mRNAs, whereas *rsmB* is an untranslated mRNA neutralizing RsmA effects. High expression of *rsmB* in *E. amylovora* elevates synthesis of amylovoran as it was judged by the mucoid appearance of colonies. On the other hand, a plasmid with *rsmA* decreases EPS synthesis of *E. amylovora* and symptom formation in apple shoots. The small *rsmB* transcript belongs to the siRNAs (small interfering RNA) silencing genes in microorganisms and in eukaryotic cells (Vogel and Papenfort, 2006).

The DNA binding protein Hns of *E. amylovora* interferes with levan and amylovoran synthesis of *E. amylovora*. An *hns* gene is present on the chromosome and a related gene on plasmid pEA29 (Hildebrand *et al.*, 2006). Attempts to create double mutants were unsuccessful. This emphasizes an important role of Hns for the cell metabolism of *E. amylovora*. Mutants in one *hns* gene show an enhanced level of amylovoran synthesis, whereas overexpression of the gene decreases amylovoran production (Hildebrand *et al.*, 2006). Other *Erwinia* species such as *E. tasmaniensis* (Kube *et al.*, 2008) or *E. billingiae* (Kube *et al.*, unpublished) carry at least two *hns*-like genes in their genome, which are related to either the chromosomal *hns* gene of *E. amylovora* or *hns* on plasmid pEA29.

Autoinducers may have a pleiotrophic role for gene regulation in bacteria and can also effect EPS synthesis. Acyl homoserine lactones (AHL, AI-1) were first described for quorum sensing of soil and marine bacteria (Bodman *et al.*, 2003). *Agrobacterium tumefaciens* senses via the concentration of AHL its population density in order to initiate transfer of the Ti-plasmid, and *Vibrio harveyi* initiates bioluminescence at high bacterial densities.

Repression of EPS synthesis by the regulator EsaR has been observed for *P. stewartii* (Bodman *et al.*, 1998), which is relieved by binding of AHL. In consequence, several properties of the corn pathogen are affected (Koutsoudis *et al.* 2006). Mutants in *esaI*, which are defective in AHL synthesis, form dense biofilms, which can be relieved by addition of exogenous AHL. Surface attachment is decreased for an *esaR* mutant (regulator), and both mutants are localized at the site of infection rather then migrating efficiently into the xylem as a wild-type strain or do not show a typical colonization pattern. An interference with expression of the activator protein RcsA explained this negative regulation (Carlier and von Bodman, 2006). The EsaR quorum-sensing repressor binds to a site centred between the two promoters of the EPS synthesis activator gene *rcsA*, blocking transcription elongation from the regulated promoter under non-inducing conditions.

Among the *Erwinia* species discussed for EPS synthesis in this review most do not produce detectable AHL. Exceptions are *E. billingiae* and *E. rhapontici*. On the other hand, a second autoinducer (AI-2), furanoyl borate in case of *V. harveyi* (Federle and Bassler, 2003) has also been studied for effects on EPS synthesis of *E. amylovora* (A. Wensing *et al.*, unpublished). Mutations in a central gene for AI-2 synthesis, *luxS*, reduces the production of amylovoran. The effect does not influence formation of fire blight symptoms due to an endogenous AI-2-like compound in pear tissue and presumably also in apple.

Degradation of amylovoran by a viral EPS depolymerase and implications for control of fire blight

Bacteriophages need specific receptors on the cell surface in order to infect their host. They have also been characterized to be eventually used for control of fire blight (Gill *et al.*, 2003). In case of *E. amylovora*, the EPS capsules are a barrier shielding proteinaceous receptors at the surface of the bacteria. Many *E. amylovora* phages have therefore the ability to degrade amylovoran. A protein in the viral coat serves as an EPS depolymerase. The encoding gene has been sequenced (Kim and Geider, 2000) and the EPS degrading enzyme purified. The nucleotide sequence is highly conserved in many related bacteriophages. On the other hand, PCR primers from *dpo* of phage φEa1h do not produce an amplicon for other *E. amylovora* phages such as φEa116 (I. Müller and K. Geider, unpublished). It can be concluded that the *dpo* gene is not present in those phages or carries a different nucleotide sequence.

The gene from φEa1h has been cloned and expressed in bacteria. The enzyme cleaves amylovoran, stewartan and EPS of other *Erwinia* species (Nimtz *et al.*, 1996b; Kim *et al.*, 2002; I. Müller and K. Geider, unpublished). The cleavage site is located at the galactose residue in the backbone, which had been attached to the lipid carrier during synthesis of the repeating unit. Fused to a his-tag, the enzyme was purified. It can be applied as a tool for cleavage of amylovoran into repeating units, and also to degrade the EPS capsules of *E. amylovora* (Kim and Geider, 2000). This results in recognition of the pathogen by host defence mechanisms and thus to inactivation of the bacteria in plant tissue. Expression of the gene in pear and apple resulted in decrease of symptom formation of transgenic plants (Malnoy *et al.*, 2005; Flachowski *et al.*, 2008).

Mutants of *E. amylovora* in *ams* genes and in the *hrp* region have been investigated for competition of wild-type cells for symptom formation on fire blight host plants (Tharaud *et al.*, 1997). Their release for disease control appears risky due to a potential of reversions to virulent cells. On the other hand, the application of epiphytic *Erwinia* species related to *E. amylovora* in several properties including a similar EPS may provide an efficient tool for control of fire blight.

Conclusions

Exopolysaccharides surround many plant-associated bacteria and can also float freely into their environment. They are supposed to protect the cells against desiccation and may bind ions and nutrients. Pathogenic erwinias may also be protected against recognition by plant defence mechanisms. In rare cases as for *E. tasmaniensis*, EPS seems to be not readily produced, despite the presence of a complete set of synthesis genes and their regulators.

Amylovoran is strictly required for pathogenicity of *E. amylovora*. Enzymatic degradation *in planta* reduces fire blight symptoms. The structure of EPS from various *Erwinia* species and the responsible regions for biosynthesis are related and may be derived from common ancestors.

The *rcs* two-component system seems to regulate EPS synthesis of several *Erwinia* species and of *P. stewartii*, although it has been investigated in detail only for *E. amylovora* and *P. stewartii* subsp. *stewartii*. Hns encoding genes are also present in these *Erwinia* species and have been shown to participate in expression of genes for synthesis of amylovoran. Global regulation includes the interference of siRNAs (small interfering RNAs), which can modulate many steps in the actions of regulatory proteins. It is not surprising that *Erwinia* species often share many properties such as EPS synthesis, HR induction or carbohydrate metabolism. The ability to produce EPS and to cause HR seems to be a basic property for virulence of bacteria. The pathogenic species *E. amylovora* and *E. pyrifoliae* share both properties, whereas most other *Erwinia* species lack HR induction and *E. tasmaniensis* is deficient in synthesis of capsular EPS.

References

Aldridge, P., Bernhard, F., Bugert, P., Coplin, D.L., and Geider, K. (1998). Characterization of a gene locus from *Erwinia amylovora* with regulatory functions in exopolysaccharide synthesis of *Erwinia* spp. Can. J. Microbiol. *44*, 657–666.

Aldridge, P., Metzger, M., and Geider, K. (1997). Genetics of sorbitol metabolism by *Erwinia amylovora* and its influence on bacterial virulence. Mol. Gen. Genet. *256*, 611–619.

Bellemann, P., Bereswill, S., Berger, S., and Geider, K. (1994). Visualization of capsule formation by *Erwinia amylovora* and assays to determine amylovoran synthesis. Int. J. Biol. Macromol. *16*, 290–296.

Bereswill, S., and Geider, K. (1997). Characterization of the *rcs*B gene from *Erwinia amylovora* and its influence on exopolysaccharide synthesis and virulence of the fire blight pathogen. J. Bacteriol. *179*, 1354–1361.

Bereswill, S., Jock, S., Bellemann, P., and Geider, K. (1998). Identification of *Erwinia amylovora* by growth morphology on agar containing copper sulfate and by capsule staining with lectin. Plant Disease *82*, 158–164.

Bernhard, F., Poetter, K., Geider, K., and Coplin, D. (1990). The *rcs*A gene from *Erwinia amylovora*: Identification, nucleotide sequence, and regulation of exopolysaccharide biosynthesis. Molecular Plant–Microbe Interactions *3*, 429–437.

Bernhard, F., Schullerus, D., Bellemann, P., Nimtz, M., Coplin, D.L., and Geider, K. (1996). Genetic transfer of amylovoran and stewartan synthesis between *Erwinia amylovora* and *Erwinia stewartii*. Microbiology *142*, 1087–1096.

Bocsanczy, A.M., Nissinen, R.M., Oh, C.-S., and Beer, S.V. (2008). HrpN of *Erwinia amylovora* functions in the translocation of DspA/E into plant cells. Molecular Plant Pathol. *9*, 425–434.

Bogs, J., Bruchmüller, I., Erbar, C., and Geider, K. (1998). Colonization of host plants by the fire blight pathogen *Erwinia amylovora*, marked with genes for bioluminescence and fluorescence. Phytopathology *88*, 416–421.

Bogs, J., and Geider, K. (2000). Molecular analysis of sucrose metabolism of *Erwinia amylovora* and influence on bacterial virulence. J. Bacteriol. *182*, 5351–5358.

Bugert, P., and Geider, K. (1995). Molecular analysis of the *ams* operon required for exopolysaccharide synthesis of *Erwinia amylovora*. Mol. Microbiol. *15*, 917–33.

Bugert, P., and Geider, K. (1997). Characterization of the *amsI* gene product as a low molecular weight acid phosphatase controlling exopolysaccharide synthesis of *Erwinia amylovora*. FEBS Letters *400*, 252–256.

Carlier, A.L., and von Bodman, S.B. (2006). The *rcsA* Promoter of *Pantoea stewartii* subsp. *stewartii* features a low-level constitutive promoter and an EsaR Quorum-Sensing-regulated promoter. J. Bacteriol. *188*, 4581–4584.

Denning, W. (1794). On the decay of apple trees. New York Society for the promotion of Agricultural Arts and manufacturers Transaction *2*, 219–222.

Eastgate, J.A., Taylor, N., Coleman, M.J., Healy, B., Thompson, L., and Roberts, I.S. (1995). Cloning, expression, and characterization of the *lon* gene of *Erwinia amylovora* – evidence for a heat shock response. J. Bacteriol. *177*, 932–937.

Federle, M.J., and Bassler, B.L. (2003). Interspecies communication in bacteria. J. Clin. Investig *112*, 1291–1299.

Flachowski, H., Richter, K., Kim, W.-S., Geider, K., and Hanke, M.-V. (2008). Expression of a viral EPS-depolymerase induces fire blight resistance in apple (*Malus domestica* Borkh.). Ann. Appl. Biol., *153*, 345–355.

Geider, K., Auling, G., Jakovljevic, V., and Völksch, B. (2009). A polyphasic approach assigns the pathogenic *Erwinia* strains from diseased pear trees in Japan to *Erwinia pyrifoliae*. Lett. Appl. Microbiol. 48, 324–330.

Gill, J.J., Svircev, A.M., Smith, R., and Castle, A.J. (2003). Bacteriophages of *Erwinia amylovora*. Appl. Environ. Microbiol. *69*, 2133–2138.

González A.J., Tello J.C., and de Cara, M. (2005). First report of *Erwinia persicina* from *Phaseolus vulgaris* in Spain. Plant Disease *89*,109.

Hignett, R. (1987). Inhibition of fireblight disease by injection of D-galactose or L-fucose into susceptibel apple palnts. Phys. Mol. Plant Pathol. *30*, 131–138.

Hildebrand, M., Aldridge, P., and Geider, K. (2006). Characterization of *hns* genes from *Erwinia amylovora*. Mol. Gen Genomics *275*, 310–319.

Hobbs, M., and Reeves, P.R. (1994). The JUMPstart sequence: a 39 bp element common to several polysaccharide gene clusters. Mol. Microbiol. *12*, 855–856.

Ilan, O., Bloch, Y., Frankel, G., Ullrich, H., Geider, H., and Rosenshine, I. (1999). Protein tyrosine kinases associated with production of exopolysaccharide and virulence of animal and plants bacterial pathogens. EMBO J. *18*, 3241–3248.

Jakovljevic, V., Jock, S., Du, Z., and Geider, K. (2008). Hypersensitive response and acyl-homoserine lactone production of the fire blight antagonists *Erwinia tasmaniensis* and *E. billingiae*. Microbial Biotechnology *1*, 416–424.

Jock, S., and Geider, K. (2004). Molecular differentiation of *Erwinia amylovora* strains from North America and of two Asian pear pathogens by analyses of PFGE patterns and *hrpN* genes. Env. Microbiol. *6*, 480–490.

Jock, S., Kim, W.-K., Donat, V., Lopez, M.M., Bazzi, C., and Geider, K. (2002). Following spread of fire blight in Western, Central and Southern Europe by molecular differentiation of *Erwinia amylovora* strains with PFGE analysis. Env. Microbiol. *4*, 106–114.

Jock, S., Langlotz, C., and Geider, K. (2005). Survival and possible spread of *Erwinia amylovora* and related plant-pathogenic bacteria exposed to environmental stress conditions. J. Phytopathol. *153*, 87–93.

Johnson, T.J., Kariyawasam, S., Wannemuehler, Y., Mangiamele, P., Johnson, S.J., Doetkott, C., Skyberg, J.A., Lynne, A.M., Johnson, J.R., and Nolan, L.K. (2007). O1:K1:H7 Shares strong similarities with human extraintestinal pathogenic *E. coli* genomes. J. Bacteriol. *189*, 3228–3236.

Jumel, K., Geider, K., and Harding, S.E. (1997). The solution molecular weight and shape of the bacterial exopolysaccharides amylovoran and stewartan. Int. J. Biol. Macromol. *20*, 251–258.

Kelm, O., Kieker, C., Geider, K., and Bernhard, F. (1997). Interaction of the regulator proteins RcsA and RcsB with the promoter region of the operon for amylovoran biosynthesis by *Erwinia amylovora*. Mol. Gen. Genet. *256*, 72–83.

Kiessling, P., Senchenkova, S.O.N., Ramm, M., and Knirel, Y.A. (2005). Structural studies on the exopolysaccharide from *Erwinia persicina*. Carbohyr. Res. *340*, 1761–1765.

Kim, W.-S., and Geider, K. (2000). Characterization of a viral EPS-depolymerase as a potential tool for control of fire blight. Phytopathology *90*, 1263–1268.

Kim, W.-S., Hildebrand, M., Jock, S., and Geider, K. (2001). Molecular comparison of pathogenic bacteria from pear trees in Japan and the fire blight pathogen *Erwinia amylovora*. Microbiology/SGM *147*, 2951–2959.

Kim, W.-S., Schollmeyer, M., Nimtz, M., Wray, V., and Geider, K. (2002). Genetics of biosynthesis and structure of the capsular exopolysaccharide from the Asian pear pathogen *Erwinia pyrifoliae*. Microbiology/SGM *148*, 4015–4024.

Koutsoudis, M.D., Tsaltas, D., Timothy, D., Minogue, T.D., and von Bodman, S.B. (2006). Quorum-sensing regulation governs bacterial adhesion, biofilm development, and host colonization in *Pantoea stewartii* subspecies *stewartii*. Proc. Natl. Acad. Sci. USA *103*, 5983–5988.

Kröncke, K.-D., Boulnois, G., Roberts, I., Bitter-Suermann, D., Golecki, J.R., Jann, B., and Jann, K. (1990). Expression of the *Escherichia coli* K5 capsular antigen: immunoelectron microscopic and biochemical studies with recombinant *E. coli*. J. Bacteriol. *172*, 1085–1091.

Kube, M., Migdoll, A., Müller, I., Kuhl, H., Beck, A., Reinhardt, R., and Geider, K. (2008). The genome of *Erwinia tasmaniensis* strain Et1/99, a novel non-pathogenic bacterium in the genus *Erwinia*. Environ. Microbiol. *10*, 2211–2222.

Letal, J.R. (1976). Crown rot of rhubarb. Can. J. Plant Pathol. *12*, 445–448.

Ma, W., Cui, Y., Liu, Y., Dumenoy, C.K., Mukherjee, A., and Chatterjee, A.K. (2001). Molecular characterization of global regulatory RNA species that control pathogenicity factors in *Erwinia amylovora* and *Erwinia herbicola* pv. pypsophilae. J. Bacteriol. *183*, 1870–1880.

Maes, M., Orye, K., Bobev, S., Devreese, B., Van Beeumen, J., De Bruyn, A., Busson, R., Herdewijn, P., Morreel, K., and Messens, E. (2001). Influence of amylovoran production on virulence of *Erwinia amylovora* and a different amylovoran structure in *E. amylovora* isolates from Rubus. Eur. J. Plant Pathol. *107*, 839–844.

Majdalani, N., and Gottesman, S. (2006). The Rcs phosphorelay: A complex signal transduction system. Annu. Rev. Microbiol. *59*, 379–405.

Malnoy, M., Faize, M., Venisse, J.-S., Geider, K., and Chevreau, E. (2005). Expression of viral EPS-depolymerase reduces fire blight susceptibility in transgenic pear. Plant Cell Rep. *23*, 632–638.

Menggad, M., and Laurent, J. (1998). Mutations in *ams* genes of *Erwinia amylovora* affect the interactions with host plants. Eur. J. Plant Pathol. *10*, 313–322.

Mergaert, J., Verdonck, L., and Kersters, K. (1993). Transfer of *Erwinia ananas* (synonym, *Erwinia uredovora*) and *Erwinia stewartii* to the genus *Pantoea* emend. as *Pantoea ananas* (Serrano 1928) comb. nov. and *Pantoea stewartii* (Smith 1898) comb. nov., respectively, and description of *Pantoea stewartii* subsp. *indologenes* subsp. nov. Int. J. Syst. Bacteriol, *43*, 162–173.

Metzger, M., Bellemann, P., Bugert, P., and Geider, K. (1994). Genetics of galactose metabolism of *Erwinia amylovora* and its influence on polysaccharide-synthesis and virulence of the fire blight pathogen. J. Bacteriol. *176*, 450–459.

Nimtz, M., Mort, A., Domke, T., Wray, V., Zhang, Y., Qiu, F., Coplin, D., and Geider, K. (1996a). Structure of amylovoran, the capsular exopolysaccharide from the fire blight pathogen *Erwinia amylovora*. Carbohydr. Res. *287*, 59–76.

Nimtz, M, Mort, A., Wray, V., Domke, T., Zhang, Y., Coplin, D.L., and Geider, K. (1996b). Structure of stewartan, the capsular exopolysaccharide from the corn pathogen *Erwinia stewartii*. Carbohydr. Res. *288*, 189–201.

Pristovsek, P., Sengupta, K., Lohr, F., Schafer, B., von Trebra, M.W., Ruterjans, H., and Bernhard, F. (2003). Structural analysis of the DNA-binding domain of the *Erwinia amylovora* RcsB protein and its interaction with the RcsAB box. J. Biol. Chem. *278*, 17752–17759.

Reeves, P.R., Hobbs, M., Valvano, M.A., Skurnik, M., Whitfield, C., Coplin, D., Kido, N., Klena, J., Maskell, D., Raetz, C.R.H., and Rick, P.D. (1996). Bacterial polysaccharide synthesis and gene nomenclature. Trends Microbiol. *4*, 495–503.

Rhim, S.-L., Völksch, B., Gardan, L., Paulin, J.-P., Langlotz, C., Kim, W.-S., and Geider, K. (1999). *Erwinia pyrifoliae*, an *Erwinia* species different from *Erwinia amylovora*, causes a necrotic disease of Asian pear trees. Plant Pathol. *48*, 514–520.

Riekki, R., Palomaki, T., Virtaharju, O., Kokko, H., Romantschuk, M., and Saarilahti, H.T. (2000). Members of the *amylovora* group of *Erwinia* are cellulolytic and possess genes homologous to the type II secretion pathway. Mol. Gen. Genet. *263*, 1031–1037.

Sauer, S., Freiwald, A., Maier, T., Kube, M., Reinhardt, R., Kostrzewa, M., and Geider, K. (2008). Classification and identification of bacteria by mass spectrometry and computational analysis. PLoS ONE 3: e2843. doi:10.1371/journal.pone.0002843.

Sjulin, T.M., and Beer, S.V. (1978). Mechanisms of wilt induction by amylovorin in cotoneaster shoots and its relation to wilting of shoots infected by *Erwinia amylovora*. Phytopathology *68*, 89–94.

Sledjeski, D., and Gottesman, S. (1995). A small RNA acts as an antisilencer of the H-NS-silenced *rcsA* gene of *Escherichia coli*. Proc. Natl. Acad. Sci USA *92*, 2003–20037.

Smith, A.R.W., Rastall, R.A., Blake, P., and Hignett, R.C. (1995). Gluco-oligosaccharide production by a strain of *Erwinia amylovora*. Microbios *83*, 27–39.

Starr, M.P., Carona, C., and Folsom, D. (1951) Bacterial blight of raspberry. Phytopathology *41*, 915–919.

Suhayda, C.G., and Goodman. R.N. (1981). Infection courts and systemic movement of ^{32}P-labeled *Erwinia amylovora* in apple petioles and stems. Phytopathology 7, 656–660.

Suleman, P., and Steiner, P.W. (1994). Relationship between sorbitol and solute potential in apple shoots realitve to fire blight symptoms development after infection by *Erwinia amylovora*. Phytopathology *84*, 1244–1250.

Tharaud, M., Laurent, J., Faize, M., and Paulin, J.F. (1997). Fire blight protection with avirulent mutants of *Erwinia amylovora*. Microbiology *143*, 625–632.

Thomson, S.V. (2000). Epidemiology of fire blight. In: Fire Blight – The Disease and its causal Agent, *Erwinia amylovora*, ed. Vanneste JL (New York: CABI Publishing), pp. 9–39.

Vanneste, J.L. (2000). (Ed) Fire Blight – The disease and its causal agent, *Erwinia amylovora*. CABI Publishing, New York.

Vogel, J., and Papenfort, K. (2006). Small non-coding RNAs and the bacterial outer membrane. Curr. Opin. Microbiol. *9*, 605–611.

von Bodman, S.B., Majerczak, D.R., and Coplin, D.L. (1998). A negative regulator mediates quorum-sensing control of exopolysaccharide production in *Pantoea stewartii* subsp. *stewartii*. Proc. Natl. Acad. Sci. USA *95*, 7687–7692.

von Bodman, S.B., Bauer, W.D., and Coplin, D.L. (2003). Quorum sensing in plant-pathogenic bacteria. Annu. Rev. Phytopathol. *41*, 455–82.

Waleron, M., Waleron, K., Geider, K., and Lojkowska, E. (2008). Application of RFLP analysis of *recA*, *gyrA* and *rpoS* gene fragments for rapid differentiation of *Erwinia amylovora* from *Erwinia* strains isolated in Korea and Japan. Eur. J. Plant. Pathol. *121*, 161–172.

Wehland, M., Kiecker, C., Coplin, D.L., Kelm, O., Saenger, W., and Bernhard, F. (1999). Identification of an RcsA/RcsB recognition motif in the promoters of exopolysaccharide biosynthetic operons from *Erwinia amylovora* and *Pantoea stewartii* subspecies *stewartii*. J. Biol. Chem. *274*, 3300–3307.

Whitfield, C. (2006). Biosynthesis and assembly of capsular polysaccharides in *Escherichia coli*. Annu. Rev. Biochem. *75*, 39–68.

Zhang. Y., Jock, S., and Geider, K. (2000). Genes of *Erwinia amylovora* involved in yellow color formation and release of a low-molecular weight compound during growth in the presence of copper ions. Mol. Gen. Genet. *264*, 233–240.

Osmoregulated Periplasmic Glucans (OPGs), Alginate and Biofilm Formation in *Pseudomonas syringae* 14

Alejandro Penaloza-Vazquez, Christina M. Baker and Carol L. Bender

Abstract

In recent years it became increasingly clear that the cell envelope, loosely attached, or tightly fixed exopolymeric substances might play an important role for plant pathogenic bacteria such as *Pseudomonas syringae*. Herein, the role and synthesis of the osmoregulated periplasmic glucans (OPGs) in *P. syringae* are described in detail. OPGs are produced by many proteobacteria, and are important for the bacterial envelope and bacterial–host interactions. Mutants defective in production of OPGs in *P. syringae* exhibit a pleiotropic phenotype including the overproduction of the exopolysaccharide (EPS) alginate, decrease in cell viability under low osmolarity, low epiphytic fitness on leaves, and are unable to form biofilms on plant leaves. Results for the plant pathogen are compared with those for the opportunistic human pathogen, *P. aeruginosa*. In addition, we discussed the possibility of molecular interactions between OPG production, the physical integrity of the membrane, and the associated EPS regulatory systems.

Introduction

Pseudomonas syringae is a Gram-negative plant pathogenic bacterium and is related to other important pseudomonads, including the human pathogen *P. aeruginosa*, and the environmentally important *P. fluorescens* and *P. putida*. Strains of *Pseudomonas syringae* are divided into pathogenic variants (pathovars; pvs), which are named according to their host range. This species produces a wide variety of symptoms in plants including blights (extensive tissue death), wilting, and leaf spots (Hirano and Upper, 2000). *P. syringae* strains are diverse and interact with their host plants in specific interactions. Diseases produced by strains of *P. syringae* include bacterial blight of soybean (*P. syringae* pv. *glycinea*), halo blight of beans (*P. syringae* pv. *phaseolicola*), leaf spot of *Brassica* spp. (*P. syringae* pv. *maculicula*), brown spot disease of common bean *Phaseolus vulgaris* (*P. syringae* pv. *syringae*), and bacterial speck of tomato (*P. syringae* pv. *tomato*). The infection cycle of *P. syringae* involves epiphytic (surface) colonization, entry into the intercellular space (apoplast), multiplication, and the production of visible symptoms.

In non-host plants, *P. syringae* elicits a reaction known as the hypersensitive response or HR, which is characterized by plant cell necrosis 12–24 h post-inoculation. A gene cluster known as the *hrp* (hypersensitive response and pathogenicity) region is conserved in phytopathogenic bacteria, and is required for *P. syringae* to cause disease on susceptible host plants, and elicit the HR on non-host plants (de Torres *et al.*, 2006; Gopalan and He, 1996; He *et al.*, 2007; Oguiza and Asensio, 2005). This gene cluster encodes the type III secretion system (T3SS), which is conserved in Gram-negative plant and animal pathogens and is used to transport virulence proteins into host cells (Galan and Collmer, 1999; Grant *et al.*, 2006). The T3SS and its associated secreted virulence effector proteins (T3SEs) play a major role in the invasion and extraction of nutrients by bacteria from their plant hosts (McCann and Guttman, 2008).

Among the genes associated with the T3SS and the HR is the *hrpM* locus of *P. syringae* pv. *syringae,* which was reported previously (Mukhopadhyay *et al.,* 1988). The gene product of *hrpM* was found to be related to proteins required for the production of osmoregulated periplasmic glucans (OPGs) of *Escherichia coli* encoded by the *mdoGH/opgGH* operons (Hanoulle *et al.,* 2004; Loubens *et al.,* 1993). In *P. syringae* pv. *tomato* DC3000 the gene corresponding to *hrpM* is annotated as *mdoH* (PSPTO_1524) (Buell *et al.,* 2003).

Osmoregulated periplasmic glucans

Biosynthesis of osmoregulated periplasmic glucans

The envelope of Gram-negative bacteria is characterized by a specialized cell compartment, the periplasm, which is located between the inner and outer cell membranes (Oliver, 1996). The periplasm forms a concentrated gel-like matrix containing the murein sacculus and a variety of specialized proteins found exclusively in this compartment. Osmoregulated periplasmic glucans (OPGs) are located in the periplasm and share several common characteristics but also exhibit structural diversity among various bacterial species (Bohin, 2000). In proteobacteria, OPGs are involved in osmoprotection, biofilm formation, virulence, resistance to antibiotics (Hanoulle *et al.,* 2004), and are critical for proper organization of the cell envelope (Lequette *et al.,* 2007a).

OPGs of *E. coli* (formerly also known as membrane-derived oligosaccharides (MDOs)) are representative members of a family of glucans found in the periplasmic space of Gram-negative bacteria (Bohin, 2000; Schulman and Kennedy, 1979). In *E. coli, mdoG* (*opgG*) and *mdoH* (*opgH*) compose the *mdoGH/opgGH* operon (Hanoulle *et al.,* 2004; Loubens *et al.,* 1993). Both genes are required for the synthesis of OPGs (Lacroix *et al.,* 1991; Loubens *et al.,* 1993). The production of OPGs is under osmotic control and is inversely correlated to the osmolarity of the external milieu (Lequette *et al.,* 2004; Miller *et al.,* 1986).

The structure of OPGs consists of 6–13 glucose units joined by β-1,2 and β-1,6 linkages, and are variously substituted with succinic acid O-ester, phosphoethanolamine, and *sn*-1-phosphoglycerol (Schulman and Kennedy, 1979). The OPGs from *P. syringae* pv. *syringae* belong to Family I, which are acyclic, neutral glucans that are similar in size to the OPGs of *E. coli* (5–12 glucose residues) (Talaga *et al.,* 1994).

The *hrpM* locus from *P. syringae* pv. *syringae* strain R32 consists of 5,490 nucleotides and contains two genes, orf1 and *hrpM* (M23555), which are presumably arranged as an operon with transcription initiating upstream of orf1 (Shimizu *et al.,* 2003). Orf1 encodes MdoG (GenBank accession no. cl01208), which is necessary for the synthesis of OPGs. The exact function of MdoG remains unclear; however, it has been suggested that it may catalyse the addition of branches to a linear glucan backbone (Bohin, 2000).

The HrpM protein is a membrane glycosyltransferase, which belongs to glycosyl transferase family 2. Proteins belonging to this family transfer sugar from UDP-glucose, UDP-N-acetylgalactosamine, GDP-mannose, or CDP-abequose to a range of substrates including cellulose, dolichol phosphate and teichoic acids (Unligil and Rini, 2000). The MdoG and HrpM proteins are highly conserved in phytopathogenic bacteria (Tables 14.1 and 14.2).

Regulation and function of OPGs in plant pathogenic bacteria

OPG synthesis is subject to osmoregulation at the transcriptional and/or the post-translational level depending on the bacterial species. In *E. coli,* transcription of the *mdoGH* operon followed the same osmotically controlled response as OPG synthesis. When *E. coli* was grown in low osmolarity medium, OPGs represented about 5% of the cellular dry weight, but the OPG content dropped dramatically when cells were grown in media with higher osmolarities, even if an osmoprotectant such as betaine was present (Bohin, 2000; Lequette *et al.,* 2007b).

OPG synthesis is very important for the virulence of phytopathogenic bacteria, and OPG mutants show reduced disease symptoms on their plant hosts and fail to elicit an HR on non-host plants (Minsavage *et al.,* 2004; Mukhopadhyay *et al.,* 1988; Page *et al.,* 2001). Several regions of the bacterial genome that are involved in the pathogenicity of *P. syringae* pv. *syringae* R32 were identified by Tn5 mutagenesis (Mukhopadhyay *et

Table 14.1 Results of BLASTX analysis of the MdoG protein encoded by orf1 (gb|M23555; from 28 to 1155 bp)

Strain	orf1 Accession	% Identity	% Similarity
P. syringae pv. *syringae* B728a	Psyr_0377	99	100
P. syringae pv. *phaseolicola*	PSPPH_0359	98	99
P. syringae pv. *tomato* DC3000	PSPTO_5162	95	97
P. fluorescens Pf-5	Pfl_0413	74	86
P. fluorescens PfO-1	PflO1_0373	73	85
P. putida KT2440	PP_5026	73	86
P. aeruginosa PAO1	PA5078	71	84
E. coli 0157:H7	ECs1426	68	83

Table 14.2 Results of BLASTX analysis of the HrpM protein encoded by *hrpM* (gb|M23555; from 1363 to 2839 bp)

Strain	*hrpM* Accession	% Identity	% Similarity
P. syringae pv. *syringae* B728a	Psyr_0378	100	100
P. syringae pv. *phaseolicola*	PSPPH_0360	99	100
P. syringae pv. *tomato* DC3000	PSPTO_5161	97	98
P. fluorescens Pf-5	Pfl_0414	87	93
P. fluorescens PfO-1	PflO1_0374	86	93
P. putida KT2440	PF_5025	83	91
P. aeruginosa PAO1	PA5077	77	87
E. coli 0157:H7	ECs1427	63	77

al., 1988). One of these mutants, PS9021, failed to induce disease symptoms on host bean and to cause a HR on non-host tobacco (Mukhopadhyay *et al.*, 1988). The mutant did not grow *in planta*, but grew on minimal agar medium and exhibited a mucoid colony morphology, unlike the parental strain which had firm and non-mucoid colonies. Sequence analysis of the PS9021 mutant revealed a Tn5 insertion in the *hrpM* gene (Mukhopadhyay et al., 1988).

OPGs are also important in the virulence of other phytopathogenic species. *Xanthomonas campestris* pv. *vesicatoria* mutants deficient in OPG biosynthesis showed an attenuated virulence in tomato (Minsavage *et al.*, 2004), and *Erwinia chrysanthemi* OPG mutants are non-pathogenic when inoculated into potato tubers or chicory leaves (Page *et al.*, 2001). Proteomic analysis of an OPG-defective mutant of *E. chrysanthemi* showed that in addition to the loss of virulence, the OPG mutant exhibited pleiotropic phenotypes that affected motility, salt tolerance, exoenzyme secretion, exopolysaccharide (EPS) synthesis, and membrane lipid composition. In addition to these effects, the synthesis of catabolic enzymes was enhanced, and there was also a greater abundance of proteins required for various stress responses (chaperones involved in folding and degradation). Thus, in the OPG mutant of *E. chrysanthemi*, loss of *mdoGH* expression activated many cellular processes involved in response to environmental stress, suggesting that the mutant strain is impaired in the perception of its surrounding environment. This was found to be the result of changes in both envelope structure and cellular metabolism (Bouchart *et al.*, 2007). Their results emphasized the importance of the physical integrity of the cell membrane, and its importance during bacteria-surface interactions with the environment. For instance, one general characteristics of the OPG mutants is the overproduction of EPS, which is well known to protect cells from environmental injury (Bohin, 2000). This was demonstrated in OPG mutants of *E. chrysanthemi* as well as in other Gram-negative bacteria (Bohin, 2000).

The exopolysaccharide alginate

Arrangement of the alginate gene clusters in *P. syringae*

The genetic basis of pathogenicity and virulence in *P. syringae* is complex and includes global regulators such as the products of the *gacA* and *gacS* genes, the *Pseudomonas* sigma factor, PsrA (Chatterjee *et al.*, 2007; Chatterjee *et al.*, 2003; Hrabak and Willis, 1992; Kang and Gross, 2005; Kitten *et al.*, 1998; Quinones *et al.*, 2004), products of the *hrp* cluster (He *et al.*, 2007), phytotoxins (Bender *et al.*, 1999; Strange, 2007), and exopolysaccharides (Laue *et al.*, 2006; Schenk *et al.*, 2006; Yu *et al.*, 1999). The EPS alginate, a copolymer of O-acetylated β-1,4 linked D-mannuronic acid and its C-5 epimer L-guluronic acid, is produced by a variety of pseudomonads, including *P. syringae*. In *Pseudomonas* the alginate regulatory and biosynthetic genes are chromosomally encoded. The biosynthetic gene cluster from *algD* to *algA* in *P. aeruginosa* and *P. syringae* are similar (60–70% identity at the nucleotide level).

Alginate biosynthesis has been most extensively studied in *P. aeruginosa*, and a number of structural and regulatory genes from this species have been cloned and characterized. A comparison between the genome of *P. aeruginosa* PAOI and the genomes of *P. syringae* pv. *tomato* DC3000, *P. syringae* pv. *phaseolicola* 1448A, and *P. syringae* pv. *syringae* B728a showed that the order and arrangement of the alginate structural gene cluster from *algD* to *algA* are virtually identical in these *P. syringae* strains to that previously described for *P. aeruginosa* PAO1 (Fig. 14.1). *P. syringae* pv. *syringae* FF5 also contains orthologues of the alginate structural genes *algD*, *alg8*, *alg44*, *algG*, *algX* (*alg60*), *algL*, *algF*, and *algA* (Penaloza-Vazquez *et al.*, 1997). Not surprisingly, the order and arrangement of the gene cluster was virtually identical to *P. aeruginosa* (Penaloza-Vazquez *et al.*, 1997). In addition, it was reported that the alginate biosynthetic genes are very well conserved throughout *Pseudomonas* spp. (Muhammadi and Ahmed, 2007).

Experiments were designed to investigate if the alginate biosynthetic gene clusters from *P. aeruginosa* 8830 and *P. syringae* pv. *syringae* FF5 were functionally interchangeable. These studies involved the use of FF5.31, a mutant of *P. syringae* pv. *syringae* FF5 containing a Tn5 insertion in *algL* with polar affects on the expression of *algA*. Two derivatives of *P. aeruginosa* 8830 containing mutations in *algD* (designated 8835) and *algA* (designated 8853) were also utilized. The alginate biosynthetic gene cluster from *algD* to *algA* from *P. syringae* pv. *syringae* FF5 was cloned in a construct designated pSK2 and this was mobilized into the two *P. aeruginosa* mutants under control of the native *P. syringae* promoter (Kidambi *et al.*, 1995). Furthermore, a plasmid construct designated pAD501, containing the *P. aeruginosa* 8830 alginate biosynthetic genes under control of the *P. aeruginosa* promoter, was mobilized into FF5.31. Normally these plasmids restore their respective host strains for alginate production *in vitro*. However, when pSK2 and pAD501 were introduced into the *P. aeruginosa* mutants and FF53.1, respectively, alginate production was not restored. These results indicated that the alginate structural gene clusters from *P. aeruginosa* 8830 and *P. syringae* pv. *syringae* FF5 were not functionally interchangeable when expressed from their native promoters (Kidambi *et al.*, 1995; Penaloza-Vazquez *et al.*, 1997).

Regulation of alginate biosynthesis in *P. syringae* and *P. aeruginosa*

The genes involved in alginate regulation and production in *P. aeruginosa* are grouped into three main loci. The regulatory gene locus consists of *algB*, *algC*, *algR3*, *algR* (*algR1*), which have a positive regulatory effect on transcription of the *algD* promoter. Another gene cluster is involved in the switch from the non-mucoid to the mucoid mode of growth, and consists of *algT* and the *mucABCD* operon. The *muc* genes have a negative regulatory role in alginate production in both *P. aeruginosa* and *P. syringae* (Keith and Bender, 1999; Schenk *et al.*, 2008a).

Activation of alginate production in response to environmental signals involves two promoters, P*algC* and P*algD*, which map at two distinct locations in *P. aeruginosa* and *P. syringae* (Chitnis and Ohman, 1993; Penaloza-Vazquez *et al.*, 1997). In both species, the DNA region spanning *algD* to *algA* is transcribed as an operon under the control of the *algD* promoter. AlgD encodes GDP-mannose dehydrogenase (Chitnis and Ohman,

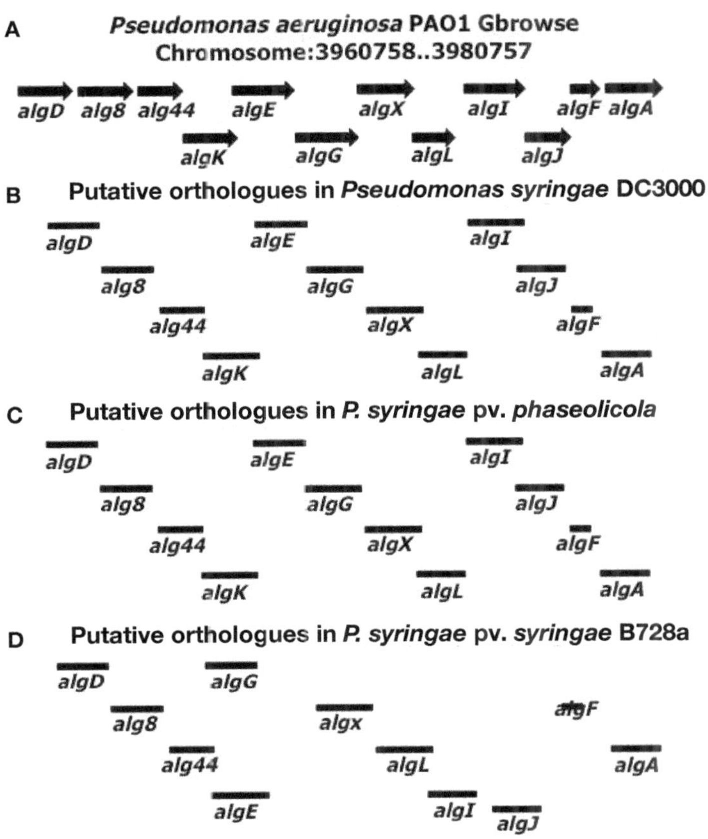

Figure 14.1 Localization of the alginate gene clusters in genomes of **A** *P. aeruginosa* PAO1, **B** *P. syringae* pv. *tomato* DC3000, **C** *P. syringae* pv. *phaseolicola*, and **D** *P. syringae* pv. *syringae* B728a. Chromosome region 3960758 to 3980757 of *P. aeruginosa* PAO1 was used to fetch gene clusters from *P. syringae* genome sequences.

1993; Deretic *et al.*, 1987a; Penaloza-Vazquez *et al.*, 1997). Using transcriptional fusions, a region upstream of *algD* was shown to contain the transcriptional start site for alginate biosynthesis in *P. aeruginosa* (Deretic *et al.*, 1987a) and *P. syringae* (Penaloza-Vazquez *et al.*, 1997).

Despite the similarities in the biosynthetic pathway mentioned previously, there are transcriptional and regulatory variations between *P. aeruginosa* and *P. syringae*. In *P. aeruginosa*, AlgR (AlgR1) positively controls transcription at the *algD* promoter (Deretic *et al.*, 1987b; Kato and Chakrabarty, 1991; Mohr *et al.*, 1992) and *algC* promoters (Zielinski *et al.*, 1991; Zielinski *et al.*, 1992). However, in *P. syringae* pv. *syringae* FF5, AlgR is not required for the activation of *algD* (Fakhr *et al.*, 1999), but it is required for transcriptional activation of *algC* (Penaloza-Vazquez *et al.*, 2004).

One of the key regulatory genes controlling alginate biosynthesis in *P. aeruginosa* and *P. syringae* is *algT*, which encodes the alternate sigma factor, σ^{22} (Hershberger *et al.*, 1995). σ^{22} is required for the transcription of *algD*, *algR*, and *algT* in *P. aeruginosa* (Wozniak and Ohman, 1994). σ^{22} recognition sequences were found upstream of *algR1* and *algD* in *P. syringae*, suggesting that *algT* is required for transcription of these genes in *P. syringae* as well (Fakhr *et al.*, 1999) (Fig. 14.2). Alginate biosynthesis is tightly controlled by proteins encoded by the *algT-muc* regulatory gene cluster in *P. aeruginosa*. In *P. aeruginosa*, this cluster consists of *algT* and the *mucABCD* operon (Wood and Ohman, 2006). The function of *mucC* is unknown but it is thought to have a negative regulatory role in alginate production in *P. aeruginosa* (Boucher *et al.*, 1997). In contrast, a homologue of the *mucC* gene is not present in *P. syringae* (Keith and Bender, 2001). Gene products of *mucA*, *mucB* (*algN*) and *mucD* negatively impact the *algD* promoter by inhibiting the activity of σ^{22}, as mutations in these genes lead to conversion to mucoidy (Boucher *et al.*, 1997; Goldberg *et al.*, 1993; Martin *et al.*, 1993; Schurr *et al.*, 1994; Wood and Ohman, 2006; Xie *et al.*, 1996). A polar insertion in *algT* did not block production of *mucD* in *P. aeruginosa* (Wood and Ohman, 2006), and indicated that another promoter (P*mucD*) allowed for *mucD* transcription which was not *algT*-dependent. Analysis of an *algT* mutant of *P. syringae* pv. *glycinea* revealed that AlgT activates *mucA, mucB, algD,* and *algT,* whereas *mucD* transcription is not dependent on AlgT *in vitro.* Further experiments showed that *mucD* in *P. syringae* pv. *glycinea* is not co-transcribed with the *algT-muc* operon, but instead is transcribed as a monocistronic mRNA (Schenk *et al.*, 2006).

Environmental factors are signals for alginate gene expression

In *P. aeruginosa* and *P. syringae* the alginate biosynthetic genes are normally silent (not expressed) (Goldberg *et al.*, 1993). The switch from a non-mucoid to a mucoid phenotype depends on the presence of environmental factors, which activate expression of the alginate regulatory genes, leading to alginate production. Environmental factors such as oxidative stress, heat shock, and high osmolarity induce alginate production by *P. aeruginosa in vitro* and presumably also *in vivo* (Shankar *et al.*, 1995).

Dehydration and osmotic stress are conserved signals for alginate gene expression and production. Addition of NaCl and ethanol to a sucrose-containing medium led to increased production of alginate by several *P. syringae* pathovars (Singh and Fett, 1995). In *P. aeruginosa* AlgR is known to activate *algD* (Kato and Chakrabarty, 1991; Kimbara and Chakrabarty, 1989) and *algC* (Zielinski *et al.*, 1992) under high osmolarity (Fujiwara *et al.*, 1993). At the transcriptional level, high osmolarity is a signal for enhanced *algD* expression in *P. aeruginosa* (Berry *et al.*, 1989), and *P. syringae* pv. *syringae* FF5 (Penaloza-Vazquez *et al.*, 1997). More recently, microarray analysis indicated that many alginate regulatory genes were up-regulated in *P. aeruginosa* PA14 during high osmolarity, including *algT, algR, mucA,* and *mucB* (Aspedon *et al.*, 2006). Additionally, an *algT* (*algU*) deletion mutant of *P. fluorescens* was sensitive to osmotic stress (Schnider-Keel *et al.*, 2001).

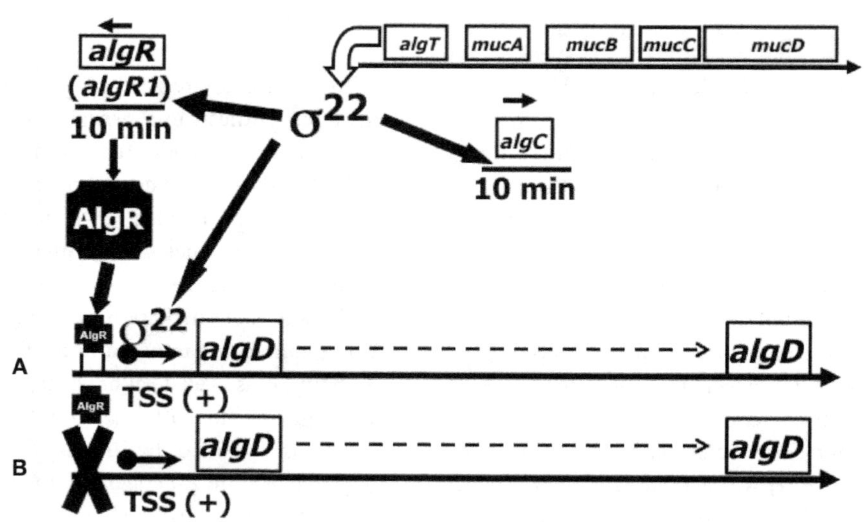

Figure 14.2 Regulation of the *algD* promoter in *Pseudomonas*. **A** Sequence analysis of the upstream region of the *algD* gene in *Pseudomonas aeruginosa* showed the presence of the alternate sigma factor, σ^{22} which is encoded by *algT* (Deretic *et al.*, 1987b; Hershberger *et al.*, 1995; Kato and Chakrabarty, 1991; Mohr *et al.*, 1992). **B** Sequence analysis of the *algD* region in *P. syringae* pv. *syringae* FF5 showed the presence of the sigma 22 binding site but not the presence of the *algR* consensus binding site [shown by ×] (Fakhr *et al.*, 1999, Penaloza-Vazquez *et al.*, 2004).

Heat shock is another common signal for alginate production. The *algR* promoter of *P. aeruginosa* PA01 was induced by heat shock (Schurr *et al.*, 1995). Although the optimal temperature for cultivating *P. syringae* FF5 is 28°C, transcription of the *algD* promoter in this strain (FF5) was significantly higher at 32°C, indicating that elevated temperature is also a signal for alginate expression in some strains of *P. syringae* (Penaloza-Vazquez *et al.*, 1997). However, in *P. syringae* pv. *glycinea* PG4180, expression of both *algT* and *algD* was temperature-dependent *in planta*, with higher expression at 18 than 28°C, and thermoregulation was not observed *in vitro* (Schenk *et al.*, 2008b). *algT* transcription in *P. aeruginosa* PAO1 (Schurr *et al.*, 1995) and *P. syringae* pv. *syringae* FF5 was activated by heat shock (Keith and Bender, 1999), and *algT* mutants in both organisms displayed increased sensitivity to heat (Keith and Bender, 1999; Martin *et al.*, 1994).

Despite many common signals, some of the environmental signals for alginate expression in *P. syringae* differ from those in *P. aeruginosa* (Kidambi *et al.*, 1995; Penaloza-Vazquez *et al.*, 1997). Alginate production in *P. syringae* was stimulated when copper sulphate was added to the medium *in vitro*, but copper was not a signal for alginate synthesis in *P. aeruginosa* (Kidambi *et al.*, 1995). *P. syringae* is commonly exposed to copper in the form of bactericidal compounds, which are applied to reduce and control disease in the field, whereas *P. aeruginosa* does not normally encounter this in the human lung (Kidambi *et al.*, 1995). In a subsequent study, copper sulphate was shown to stimulate transcription from the *algD* promoter from *P. syringae* pv. *syringae* FF5 (Penaloza-Vazquez et al., 1997).

P. aeruginosa algT (*algU*) mutants defective in alginate production displayed increased sensitivity to paraquat, which induces the production of reactive oxygen species (ROS) and causes oxidative stress (Martin *et al.*, 1994). Similarly, in *P. syringae* pv. *syringae* FF5, *algT* is expressed in response to paraquat, hydrogen peroxide (H_2O_2), and copper sulphate. The *algT* gene product, σ^{22}, activates expression of *algD*, leading to alginate production under oxidative stress (Keith and Bender, 1999). During colonization of plant tissue, *P. syringae* may be exposed to H_2O_2, so both this compound and copper sulphate, may be unique signals for *algT* activation in *P. syringae* (Keith and Bender, 1999). Additionally, in *P. syringae* pv. *tomato* DC3000, *algD* was expressed 8–12 h post inoculation during the hypersensitive response (HR) in tobacco plants (Keith *et al.*, 2003). During the HR, the plant produces cytotoxic quantities of H_2O_2 and O_2^-; thus these results suggest that the production of alginate by *P. syringae* would be stimulated by exposure to these stresses.

Role of alginate in plant pathogenesis of *P. syringae*

EPSs are involved in invasion, multiplication, and survival of bacteria within host plant tissues, and allow the bacterium to avoid plant defence responses. In plant pathogenesis, EPS can plug pit membranes between vessels or block xylem vessels thus interfering with water transport in the xylem leading to disease symptoms such as wilting (Denny, 1995; Leigh and Coplin, 1992). A scarcity of water in the apoplast of plant tissue can hinder bacterial multiplication. It is assumed that EPS remains closely associated with the bacterial cell as a capsule or slime layer, so that moisture is retained and bacterial cells are protected from desiccation (Denny, 1995). During pathogenesis, EPSs are also thought to be beneficial for the pathogen by minimizing the contact between bacteria and plant cells thus promoting colonization (Denny, 1995). On the surface of plant leaves and inside the plant tissue, EPS production increases the dissemination and epiphytic fitness of bacteria, and protects bacteria from desiccation stress and ROS (Keith *et al.*, 2003; Yu *et al.*, 1999).

Other fluorescent plant-pathogenic *Pseudomonas* spp. that produce alginate *in vitro* include *P. cichorii*, *P. savastanoi*, and *P. viridiflava* (Fett *et al.*, 1986). *P. syringae* pathovars known to produce alginate include pvs. *glycinea*, *lachrymans*, *pisi*, *morsprunorum*, *phaseolicola*, *tabaci*, *tomato*, *syringae*, *aptata*, *tagetis*, and *papulans* (Fett *et al.*, 1986; Gross and Rudolph, 1987b). Strains of *P. syringae* also produce alginate *in planta*, which has multiple functions in bacteria–host interactions. Alginate has been isolated from soybean leaves infected with *P. syringae* pv. *glycinea* (Osman *et al.*, 1986) and from bean plants (*Phaseolus vulgaris*) infected by *P. syringae* pv. *phaseolicola* (Gross and Rudolph, 1987a). Alginate is the major EPS produced in water-soaked lesions of susceptible plant leaves inoculated with *P. syringae* pathovars (Fett and Dunn, 1989; Gross and Rudolph, 1987a). These studies suggest that alginate is at least partially responsible for the water-soaked appearance of disease lesions (Denny, 1995; Gross and Rudolph, 1987a; Leigh and Coplin, 1992).

Alginate clearly functions in the virulence and epiphytic fitness of some *P. syringae* strains. The first genetic evidence supporting this contention was provided by Yu *et al.* (1999). In their study, an *algL* (encoding alginate lyase) mutant was generated in *P. syringae* pv. *syringae* 3525, which causes bacterial brown spot of bean (Yu *et al.*, 1999). This alginate-defective mutant was significantly impaired in its ability to colonize the surface of tomato leaves, which is a non-host for this pathogen. On leaves of the host (common bean), the mutant retained the ability to form lesions; however, the symptoms were less severe, and the bacterial population was significantly reduced compared with the wild type. Thus alginate contributes to the epiphytic fitness and virulence of *P. syringae* pv. *syringae* 3525 *in planta* (Yu *et al.*, 1999).

Interestingly, an alginate regulatory gene, *mucD*, was shown to function in the virulence of *P. aeruginosa* on *Arabidopsis thaliana*. The *mucD* mutant of *P. aeruginosa* strain PA14 was significantly impaired in its ability to cause disease in *Arabidopsis*, while an *algD* mutant of *P. aeruginosa* PA14 was not impaired in pathogenicity. A *mucD–algD* double mutant of *P. aeruginosa* PA14 showed the same phenotype as the *mucD* mutant *in planta* (Yorgey *et al.*, 2001).

Additionally, gene expression studies have indicated that the alginate biosynthetic genes are expressed during the *P. syringae*-plant interaction. For example, *in vivo* expression technology (IVET) was used to show that *algA* was induced during infection of *A. thaliana* with *P. syringae* pv. *tomato* DC3000 (Boch *et al.*, 2002). In another study, an *algD* transcriptional fusion was constructed and used to monitor alginate gene expression in plants inoculated with *P. syringae* pv. *tomato* DC3000 (Keith *et al.*, 2003). In leaves of the host, the expression of *algD* was activated within 72 h post inoculation and correlated with the development of water-soaked lesions. In leaves of resistant or non-host plants, *algD* expression was also observed very quickly after inoculation. Collectively, these results showed that *algD* is induced when *P. syringae* pv. *tomato* DC3000 is inoculated to susceptible, resistant, and non-host plants (Keith *et al.*, 2003).

A role for alginate in the virulence and epiphytic fitness of *P. syringae* pv. *syringae* FF5, causal agent of rapid dieback on ornamental pear trees, was established (Penaloza-Vazquez *et al.*, 2004). An *algR* mutant of strain FF5 was impaired in pathogenicity and produced small, localized non-spreading lesions on pear host plants and a reduction in epiphytic fitness was observed with the *algR* mutant on non-host plants (Penaloza-Vazquez *et al.*, 2004). When the *algR* mutant was complemented, epiphytic fitness was restored, and lateral spreading of necrotic lesions from the main vein similar to the wild-type was observed. These results indicate that *algR* gene contributes to both virulence and epiphytic fitness in *P. syringae* pv. *syringae* FF5. Furthermore, the results of this study suggest that a functional copy of *algR* is required for *P. syringae* pv. *syringae* FF5 to move systemically in leaf tissue (Penaloza-Vazquez *et al.*, 2004).

The alginate biosynthetic genes also contribute to the *P. syringae* pv. *glycinea*–plant interaction. In comparison to an alginate-deficient *algT* mutant, an alginate-producing strain of *P. syringae* pv. *glycinea* PG4180 with a functional copy of *algT* exhibited increased epiphytic fitness and enhanced symptom development *in planta* (Schenk *et al.*, 2006). A subsequent study showed that AlgT promotes the survival, epiphytic fitness, virulence and growth of *P. syringae* pv. *glycinea* 4180 *in planta* (Schenk *et al.*, 2008b). AlgT has a role in the stress response, and in this work the authors suggested that AlgT may act as a global regulator of traits associated with virulence and fitness of *P. syringae in planta* in addition to its role in alginate biosynthesis (Schenk *et al.*, 2008b).

Biofilms

The traditional view of bacteria as planktonic, unicellular organisms is giving way to a different perspective, in which the planktonic state is only part of the life cycle of a bacterial population (Espinosa-Urgel, 2004). The establishment of sessile microbial populations or biofilms has been known to occur in both natural habitats and artificial settings. However, the biofilm mode of growth is now acknowledged as a key survival strategy for bacteria in all environments where an interface is present (Davey and O'Toole, 2000; Stoodley *et al.*, 2002). In nature, a bacterial biofilm is a structured community of bacterial cells enclosed in a self-produced polymeric matrix and adherent to an inert or living surface, which constitutes a protected mode of growth that allows survival in hostile environments (Chang *et al.*, 2007; Costerton *et al.*, 1995; Prakash *et al.*, 2003).

Biofilms are important environmental reservoirs for pathogens, and the biofilm mode of growth provides organisms with survival advantages that increase their virulence (Coenye *et al.*, 2007; Parsek and Singh, 2003). The role of biofilms in the pathogenesis of some chronic human infections has been well studied (Parsek and Singh, 2003; Schaber *et al.*, 2007). Bacteria present in biofilms differ from free-living cells in growth rate, chemical composition of EPS (Beech *et al.*, 1999), and resistance to antibiotics and antibodies (Stewart, 2002; Szomolay *et al.*, 2005). Additionally, studies using DNA microarray technology have suggested that biofilms may have a unique pattern of gene expression (An and Parsek, 2007; Lazazzera, 2005).

Biofilm formation in plant–bacteria interactions

Plant surfaces are complex and dynamic environments, and bacteria can colonize leaves, roots, and internal vasculature (Morris and Monier, 2003). Each tissue type has unique chemical and physical properties that present challenges and opportunities for microbial colonists (Ramey *et al.*, 2004). Most plant-associated bacteria are found in the phyllosphere or the rhizosphere, both of which are nutrient-poor (Fujishige *et al.*, 2006; Lindow and Brandl, 2003; Morris and Monier, 2003; Wilson *et al.*, 1999). The leaf surface is a hostile environment and bacteria are subject to stresses such as extreme temperature, fluctuations in relative humidity (RH), desiccation, wind, and UV light (Hirano and Upper, 2000; Lindow and Brandl, 2003; Wilson *et al.*, 1999). Bacteria are known to survive in protected sites on leaves such as the substomatal chamber (Roos and Hattingh, 1983), the base of glandular trichomes, and depressions along veins and between epidermal cells (Mansvelt and Hattingh, 1989; Monier and Lindow, 2004; Schneider and Grogan, 1977). Phytopathogenic bacteria often survive and multiply as epiphytes on the aerial parts of plants without causing any visible symptoms (Hirano and Upper, 2000; Jacques *et al.*, 2005; Morris *et al.*, 2007). This asymptomatic phase allows bacterial populations to attain sizes permitting disease development when conditions are favourable (Hirano and Upper, 1983).

In addition to surviving epiphytically on diverse plant species, *P. syringae* is also present in non-agricultural habitats, such as those associated with native, indigenous plants, and water (e.g. rain, snow, alpine streams and lakes) (Morris *et al.*, 2008). The ability of this bacterium to survive for long periods on plant leaves and on rocks is due in part to (i) the ability of *P. syringae* to cause disease on diverse species of plants, and (ii) the ability of *P. syringae* to form biofilms on plant leaves and river epilithon (rock-attached biofilms composed of algae, diatoms, rotifers, bacteria and nematodes) (Dulla and Lindow, 2008; Laue *et al.*, 2006; Morris *et al.*, 2007; Ramey *et al.*, 2004). *P. syringae* produces diffusible quorum-sensing molecules that aid in the formation of aggregates or biofilms on the plant leaf surface (Dulla and Lindow, 2008; Quiniones *et al.*, 2005; Von Bodman *et al.*, 2003).

Other phytopathogenic bacteria associated with plants have been observed to form aggregates or biofilms (Morris and Monier, 2003). In the case of *X. axonopodis* pv. *phaseoli*, planktonic population sizes decreased dramatically when bean plants were exposed to desiccation stress, whereas biofilm population sizes did not. However, returning plants to high RH allowed planktonic population sizes to increase *de novo* (Jacques *et al.*, 2005). Hence, aggregation in biofilms could offer protection from external stress to epiphytic cells of *X. axonopodis* pv. *phaseoli*, which has been shown for *P. syringae* (Monier and Lindow, 2003).

Under field conditions, solitary population sizes of *P. syringae* are sensitive to rain and temperature fluctuations. However, *P. syringae* biofilm populations remain stable under those environmental variations. Hence, protection against harsh environmental conditions could arise either from endophytic localization (Beattie and

Lindow, 1994; Wilson *et al.*, 1999) or from aggregation in biofilms. A stable population size in a variable environment and a population size close to the threshold necessary for symptom induction led to the hypothesis that biofilms constitute a reservoir of cells for *X. axonopodis* pv. *phaseoli* in the field (Jacques *et al.*, 2005). Additionally, plant-associated biofilms undergo chromosomal rearrangements and are hot spots for conjugative plasmid transfer, favoured by the close proximity between cells and the constant supply of nutrients coming from the plant in the form of exudates or leachates (Espinosa-Urgel, 2004).

Biofilm formation also plays a major role in the virulence of other phytopathogens, for example *Xylella fastidiosa*, a xylem-limited bacterium that depends on insect vectors for transmission. *X. fastidiosa* causes Pierce's disease of grapevine and diseases of other important plants (almond, tobacco, avocado, citrus) and forms an aggregated biofilm inside the plant (Newman *et al.*, 2004). Vessels become occluded by dense colonization, and blocked vessels are associated with disease symptoms such as wilting. A diffusible signal molecule (DSF) is required for biofilm formation in the insect vector and for vector transmission to plants and may regulate expression of traits required for interactions between bacteria, plant and insect (He *et al.*, 2006; Newman *et al.*, 2004).

Role of biofilms in colonization of plants by human enteric pathogens

The formation of biofilms on seeds, sprouts, and other plant material used for human consumption is a health concern as they may harbour pathogens. Microbial aggregates on commercial sprouts exhibited biofilms with a thickness ranging from 5 to 12 μm (Fett and Cooke, 2005). Most of the illnesses linked to fresh produce are due to post-harvest microbial contamination rather than contamination in the field. Epidemics of food-borne illness linked to fresh produce increased in the USA from 1992 to 1993 (2178 cases per year average) (Brandl, 2006). With the recognition of pre-harvest contamination, new studies regarding the fitness of enteric pathogens on crop plants are starting to emerge. Enteric pathogens come in contact with plants in the field through contaminated irrigation water, manure, bioaerosols, flies, or insects on manure (Brandl, 2006).

However, some of the main factors contributing to the introduction of human pathogenic bacteria are related to the poor hygiene practices by field-workers. The lack of on-site sanitation facilities can easily result in outbreaks of shigellosis, as a low infectious dose of *Shigella* is easily contracted from human faeces (Brandl, 2006). There are studies that also support the hypothesis that pre-harvest contamination of crops by *Salmonella enterica* plays a role in outbreaks linked to fresh fruits and vegetables (Brandl and Mandrell, 2002). *E. coli* O157:H7 is the causal agent of one of the largest outbreaks of bacterial enteric disease, with over 6000 cases epidemiologically linked to contaminated radish sprouts in Japan in 1996 (Brandl, 2006).

Human pathogens are aided in their colonization of plants by naturally occurring bacteria. Fett (2000) examined and observed naturally occurring biofilms consisting of environmental bacteria on different parts of alfalfa, broccoli, cloves and sunflower sprouts. He concluded that naturally occurring biofilms would afford protected colonization sites for human pathogens. Estimates of biofilm abundance showed that bacteria in biofilms constitute 10–40% of the bacterial population on broad-leaf endive and parsley (Morris *et al.*, 1998).

Removal of pathogenic *E. coli* O157:H7 and *Salmonella* from alfalfa and mung bean seeds had limited success after treatment with commercially available antibacterial products (Fett and Cooke, 2003). The products utilized were Citrex, Pangermex, Citricidal, Citrobio, and Environne. These products were tested at concentrations of up to 20,000 ppm in sterile tap water and compared with buffered chlorine (at 16,000 ppm). Alfalfa seeds were inoculated with cocktails of *Salmonella* and *E. coli* O157:H7 to give final initial concentrations of 9.0 and 7.0 CFU/g, respectively. The seeds were treated with Citrex, Pangermex, and Citricidal at 20,000 ppm and chlorine at 16,000 ppm, and incubated for 10 min. All treatments produced similar log reductions for alfalfa and mung bean seeds inoculated with *E. coli* O157:H7 and *Salmonella* (3.42 to 3.46 log CFU/g and 3.56 to 3.74 log CFU/g, respectively), and all four treatments were significantly ($P < 0.05$) more effective than the control treatment (a buffer wash). Citrobio at 20,000 ppm was as effective as the other three products and chlorine against *Salmonella* but not against *E. coli* O157:H7. Environne was less effective (producing reductions of 2.2–2.9 log CFU/g) than the control treatment (which produced reductions of 2.1–2.3 log CFU/g) against either pathogen. None of the treatments reduced seed germination (Fett and Cooke, 2003).

Biofilm formation also plays a role in the persistence of *Salmonella* after chlorination treatment of parsley, as biofilms enhance the resistance of plant-associated pathogens to disinfectants or biocides. To test the hypothesis that biofilm formation would enhance the adhesion and survival of *Salmonella* of leafy vegetables, the adhesion and persistence of *S. typhimurium* strain ATCC14028 and its biofilm-deficient isogenic mutant MAE52 were compared. The biofilm-defective mutant does not express the gene products of *bcsA* (bacterial cellulose synthesis A) and *agfBA* (also called *csgBA* for curli subunit genes A and B). Thus, it does not produce the two main extracellular matrix compounds, cellulose and curli, and therefore a biofilm is not formed (Lapidot *et al.*, 2006). Following inoculation of parsley and rinsing with water or chlorine solution, both the wild-type and biofilm-defective mutant had similar survival properties, and up to 3-log population reductions were observed in both strains, this indicates that the biofilm matrix of *Salmonella* probably does not play a significant role in initial adhesion and survival after disinfection. However, after a one-week storage period, the biofilm-producing strain survived chlorination significantly better than the biofilm-deficient mutant (Lapidot *et al.*, 2006). Although the biofilm matrix promotes the persistence of *Salmonella* after chlorination of parsley, the authors cite other mechanisms that are likely to be important, such as the ability to enter and survive in plant tissue to evade the chlorine treatment and the presence of pre-existing biofilms (Lapidot *et al.*, 2006).

Microscopic observations revealed that *Salmonella* can survive near the stomata and within cracks in the cuticle of parsley leaves (Duffy *et al.*, 2005). Gandhi *et al.* observed bacteria at a depth of 12 μm within intact alfalfa sprout tissue, indicating that bacteria are able to enter and survive in alfalfa tissue (Gandhi *et al.*, 2001). Brandl and Mandrell investigated the adhesion of *S. enterica* serovar Thompson on cilantro plants (Brandl and Mandrell, 2002). They observed that *S. enterica* Thompson developed lesions in cilantro and was present at extremely high densities in some lesions. Bacterial cell counts revealed that despite extensive washing, *Salmonella* was residing in niches on the cuticular surface of healthy leaves. Using microscopy, bacteria were also found to internalize cilantro tissue after gaining access through damaged regions of the cuticle (Brandl and Mandrell, 2002).

Barak *et al.* (2005) found that some of the virulence factors that are important for mammalian pathogenicity by *S. enterica* function during attachment to plant tissues as well. The *S. enterica* genes important for virulence in animal systems are *agfB* (encoding the surface-exposed aggregative fimbria curli nucleator), *agfD* (encoding transcriptional regulator of the LuxR superfamily) and *rpoS* (stationary-phase sigma factor). Both AgfD and RpoS have been reported to regulate fimbria and cellulose production, and RpoS regulates other adhesins such as pili, All these genes were also required for colonization of various plants, and the authors stated that plants can serve as a vector of *S. enterica* from animal to animal (Barak *et al.*, 2005).

In addition to animal pathogens that are merely transmitted through plants, there are those that cross host kingdoms and infect both plants and animals, such as *Enterococcus faecalis* and *P. aeruginosa*. *E. faecalis* is a ubiquitous Gram-positive bacterium that is usually found in soil and water (Muller *et al.*, 2001). Some *E. faecalis* strains (FA2-2–, V583, and OG1RF) cause disease in *Arabidopsis* via a sequential array of events, including attachment to leaves and roots, congregation of bacteria in stomata or wounds, colonization in intercellular spaces, formation of communities on root surfaces, disruption of membrane structures, and maceration and rotting of petioles (Jha *et al.*, 2005). The ubiquitous bacterium *P. aeruginosa* is the quintessential opportunistic pathogen, and certain isolates infect a broad range of host organisms, from plants to humans. The pathogenic promiscuity of particular variants may reflect an increased virulence gene repertoire beyond the core *P. aeruginosa* genome. He and co-workers (He *et al.*, 2004) have characterized two *P. aeruginosa* pathogenicity islands designated PAPI-1 and PAPI-2, which span 108 and 11 kb, respectively, in the genome of PA14, a highly virulent clinical isolate. The large PAPI-1 island is absent from the *P. aeruginosa* reference strain PAO1, and only part of the smaller PAPI-2 element is found in this isolate. Although PA14 and PAO1 infect similar hosts, PA14 is considerably more virulent in several model organisms (Hendrickson *et al.*, 2001; Rahme *et al.*, 1995). Most of the predicted proteins encoded by the PAPI genes exhibit no homology with any proteins of known function, and mutation of several of these genes reveals their relevance in both plant and animal pathogenicity(He *et al.*, 2004).

The virulence of *P. aeruginosa* PA14 is facilitated by the acquisition of pathogenicity islands via horizontal gene transfer (HGT), and this occurs when cells are in intimate contact within biofilms (Espinosa-Urgel, 2004). Mobile genetic elements can be exchanged promiscuously between a broad spectrum of bacteria and contribute to bacterial genome plasticity. As components of the horizontal gene pool, mobile genetic elements include insertion sequences, transposons, integrons, bacteriophages, genomic islands (such as pathogenicity islands), plasmids and combinations of these elements. Driven in part by the accelerating pace of whole-genome sequencing, increasing evidence is emerging to support of significant gene shuffling by HGT among members of the domains Bacteria, Archaea and Eukarya (Sorensen *et al.*, 2005).

EPS regulatory systems

The Wsp system

Recently the Wsp chemosensory signal transduction system has been identified, and genetic analyses show that the regulation and function of this pathway is analogous to the chemotaxis (Che) pathway from *E. coli* (Bantinaki *et al.*, 2007; Guvener and Harwood, 2007). The Wsp system (Fig. 14.3A) includes a membrane-bound receptor protein, WspA, which is similar to the methylation-dependent chemotaxis receptors. WspR, a response regulator GGDEF protein, catalyses c-di-GMP synthesis when phosphorylated (Bantinaki *et al.*, 2007; Guvener and Harwood, 2007). Other Wsp system components include CheW-like scaffolding proteins (WspB and WspD), a CheA/Y hybrid histidine sensor kinase with a receiver domain (WspE), a methyltransferase CheR homologue (WspC), and a methylesterase CheB homologue (WspF) (Bantinaki *et al.*, 2007).

The biosynthesis of the novel regulatory nucleotide bis-(3′–5′)-cyclic dimeric guanosine monophosphate (cyclic-di-GMP or c-di-GMP) is catalysed by the protein WspR, which has been found in bacterial pathogens. This unusual compound is a bacterial intracellular secondary messenger, and has been shown to regulate several functions in diverse bacteria including cellulose synthesis (Amikam and Benziman, 1989; Ross *et al.*, 1990), EPS biosynthesis, biofilm formation (Hickman *et al.*, 2005), cellular adhesion components (Jenal, 2004), and motility (Amikam and Galperin, 2006).

Cellulose synthase, which is involved in regulating cellulose synthesis, was found in bacterial membrane preparations and specifically binds to c-di-GMP (Weinhouse *et al.*, 1997). Genetic data indicated that proteins containing a conserved GGDEF domain (referring to a conserved central sequence pattern) possess diguanylate cyclase activity (Ausmees *et al.*, 2001). This GGDEF domain was detected in many proteins, and most sequenced bacterial genomes contain several different GGDEF proteins (Ausmees *et al.*, 2001). c-di-GMP transmits environmental stimuli into a range of different cellular functions (Jenal and Malone, 2006), and proteins with GGDEF domains are predicted to regulate cell surface adhesiveness (D'Argenio and Miller, 2004) and motility (Simm *et al.*, 2004) by controlling c-di-GMP levels.

In addition to cellulose synthase, other proteins have been proposed to bind c-di-GMP. The PilZ ('pills') domain (Pfam domain PF07238) was found at the C terminus of cellulose synthetase (Amikam and Galperin, 2006), and bioinformatic analysis suggested that the PilZ domain binds to c-di-GMP. The physical interaction between the PilZ domains and c-di-GMP has been confirmed in the protein VCA0042/

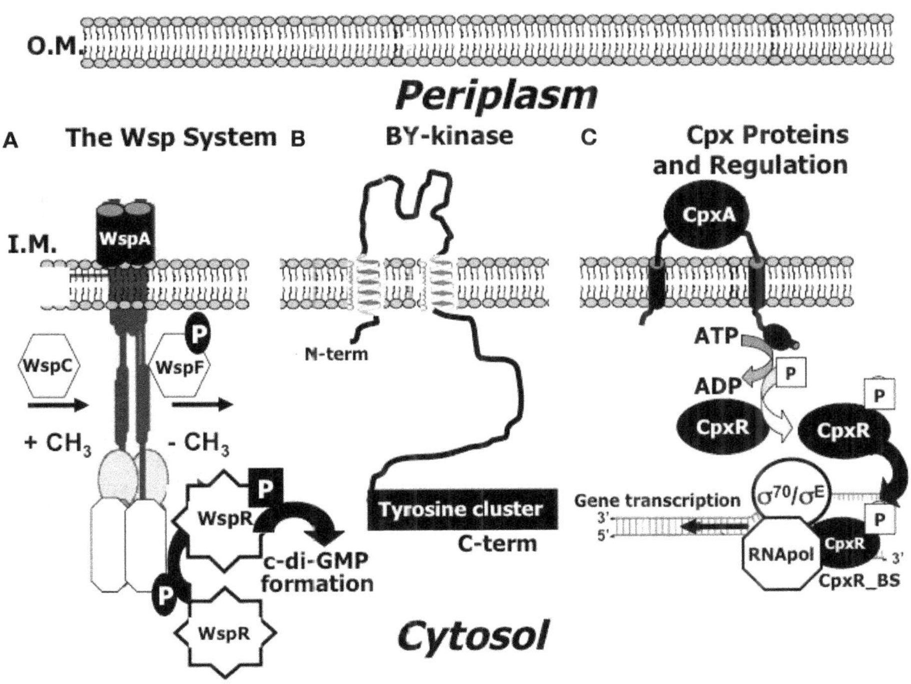

Figure 14.3 Bacterial regulatory systems involved in EPS production. **A** Recently the *Wsp* system has been studied, which is associated with the bis-(3′-5′)-cyclic dimeric guanosine monophosphate (cyclic-di-GMP or c-di-GMP). The c-di-GMP is a universal bacterial secondary messenger that regulates biofilm formation, motility, production of extracellular polysaccharide and multicellular behavior in diverse bacteria (Amikam and Galperin, 2006; Jenal and Malone, 2006). **B** BY-Kinases (Wzc), Protein kinases, the enzymes that catalyse protein phosphorylation, belong to one large protein family defined by Hanks and Hunter (Hanks and Hunter, 1995) on the basis of structural and sequence features. Although there is a rich diversity of structures, regulation modes, and substrate specificities among the protein kinases, there are also common structural features like the tyrosine cluster at the C-terminal end. These conserved structural motifs provide clear indications as to how these enzymes manage to transfer the δ-phosphate of a purine nucleotide triphosphate to the hydroxyl groups of their protein substrates. **C** The two-component signal transduction system of CpxA-CpxR, controls the expression of genes encoding cell envelope proteins involved in protein folding and degradation (Pogliano *et al.*, 1997; Shimohata *et al.*, 2002). CpxA is a sensor in the cell envelope, whereas CpxR is the corresponding response regulator of the CpxA sensor (Raivio and Silhavy, 1997). When CpxR is phosphorylated (CpxR~P) is able to bind a putative CpxR consensus-binding site that is found upstream of a number of genes involve in response to a variety of stresses including membrane-protein damage, starvation, and high osmolarity (De Wulf *et al.*, 2000; De Wulf *et al.*, 2002).

PlzD of *Vibrio cholerae* by crystallographic studies. The binding of c-di-GMP to VCA0042/PlzD protein causes a major change in the interdomain interactions in crystal structures of VCA0042/PlzD. Also, fluorescence-resonance energy transfer experiments demonstrate reorientation of the PilZ domains in VCA0042/PlzD upon binding c-di-GMP in solution (Benach *et al.*, 2007).

Alg44, which is involved in the biosynthesis of alginate, also contains a PilZ domain (Amikam and Galperin, 2006; Merighi *et al.*, 2007). Recent studies have shown that increasing c-di-GMP levels by overexpression of highly active diguanylate cyclases (which are enzymes that produce c-di-GMP from two molecules of GTP), enhanced formation of alginate in mucoid strains (Merighi *et al.*, 2007). However, overexpression of phosphodiesterases (PDEs), which are enzymes involved in c-di-GMP hydrolysis, reduced formation of alginate in mucoid strains (Merighi *et al.*, 2007). In addition, Merighi and coworkers engineered substitutions in several conserved residues of the PilZ domain of Alg44, and these substitutions resulted in simultaneous loss of c-di-GMP binding and abrogated alginate production by *P. aeruginosa* (Merighi *et al.*, 2007). It has been proposed that the Alg44 binds c-di-GMP, and this event controls polymerization or transport of the polysaccharide alginate (Merighi *et al.*, 2007; Remminghorst and Rehm, 2006).

It is likely that c-di-GMP binds to other protein domains that have yet to be described. For instance, *X. campestris* pv. *campestris (Xcc)*, the causal agent of black rot disease of cruciferous crops (Berg *et al.*, 2006; Zhao *et al.*, 2000), produces the RpfG protein with an HD-GYP domain, which is involved in the cyclic di-GMP turnover (Ryan *et al.*, 2006). The RpfG-HD-GYP protein domain regulates virulence, biofilm formation, and the synthesis of extracellular enzymes in *Xcc* (Ryan *et al.*, 2007). Very recently, the FleQ protein in *P. aeruginosa* was shown to be a c-di-GMP-responsive transcription factor, but does not have amino acid motifs that resemble previously defined c-di-GMP binding domains (Hickman and Harwood, 2008).

c-di-GMP transmits environmental stimuli into a range of different cellular functions (Jenal and Malone, 2006), and several model systems for c-di-GMP signalling have been described in *Pseudomonas, Vibrio, Caulobacter* and *Salmonella* (Guvener and Harwood, 2007; Jenal and Malone, 2006).

The bacterial tyrosine kinases (BY-kinases)

Protein kinases, the enzymes that catalyse protein phosphorylation, exhibit diverse structure, regulation, and substrate specificity. The conserved structural motifs in protein kinases provide clear indications as to how these enzymes manage to transfer the δ-phosphate of a purine nucleotide triphosphate to the hydroxyl groups of their protein substrates. The comparison of the sequences in the catalytic fragment of the protein kinases has been used to arrange these enzymes in evolutionary trees that group subfamilies of closely related enzymes (Hanks and Hunter, 1995). Protein kinases use ATP as a phosphate donor and phosphorylate proteins exclusively on serine, threonine or tyrosine. Phosphorylation is an enzymatically reversible modification in which the phosphoryl group is removed by phosphoprotein phosphatases, hence its remarkable regulatory potential. Phosphoprotein phosphatases also exhibit specificity for phospho-amino acid residues (Grangeasse *et al.*, 2007).

Initially, protein kinases similar to those found in eukaryotes were not identified in bacteria, and this protein modification was largely ignored in bacteria until the mid-1990s. Since then, several tyrosine kinases exhibiting unexpected features have been identified in a variety of bacteria. These enzymes use homologues of Walker motifs of nucleotide-binding proteins for their catalytic mechanism (Soulat *et al.*, 2007), thus defining an idiosyncratic type of bacterial tyrosine kinases. Recently, bacterial tyrosine-kinases (BY) have been found to phosphorylate an increasing list of endogenous protein substrates. This discovery contributes to the emerging picture that bacterial tyrosine phosphorylation constitutes an important regulatory arsenal of bacterial physiology in addition to the classical serine/threonine kinases and the 'two-component' and phosphotransferase systems (Grangeasse *et al.*, 2007).

The BY-kinases are inner membrane components and participate in the regulation and export of EPS and capsular polysaccharides (CPS). The molecular mechanism responsible for this regulation varies with the bacterial species analysed (Obadia *et al.*, 2007). Special attention was previously paid to the BY-kinase Wzcca of *E. coli* K-12, which is involved in the production of the EPS colanic acid commonly found in Enterobacteriaceae, and autophosphorylates by using a cooperative two-step process (Obadia *et al.*, 2007) (Fig. 14.3B).

Cpx proteins and regulation

The two-component signal transduction system of CpxA-CpxR controls the expression of genes encoding cell envelope proteins involved in protein folding and degradation (Pogliano *et al.*, 1997; Shimohata *et al.*, 2002). The disruption of normal protein trafficking in the *E. coli* cell envelope can activate the Cpx two-component system, which modulates the expression of gene products that fold or degrade the mislocalized proteins (Raivio and Silhavy, 1997) (Fig. 14.3C). CpxA is the environmental sensor in the cell envelope and is widely conserved among Pseudomonads (Table 14.3), whereas CpxR is the corresponding response regulator (Raivio and Silhavy, 1997). When CpxR is phosphorylated (CpxR~P), bind-

Table 14.3 Results of BLAST/blastp analysis of the CpxA protein from *P. syringae* pv. *syringae* B728a encoded by the gene Psyr_3792

Strain	Accession	% Identity	% Similarity
P. syringae pv. *syringae* B728a	Psyr_3792	100	100
P. syringae pv. *tomato* DC3000	PSPTO_1803	87	93
P. syringae pv. *phaseolicola* 1448A	PSPPH_1461	87	93
P. aeruginosa PAO1	PA3206	79	89
P. fluorescens Pf-5	PFL_4474	75	87
P. putida KT2440	PP_3371	58	70

ing occurs to a putative CpxR consensus site located upstream of multiple stress-response genes (De Wulf *et al.*, 2000; De Wulf *et al.*, 2002). These findings indicate the presence of an extensive Cpx regulon including genes transcribed by σ^{70} and the alternative sigma factor σ^{e} (encoded by *rpoE*) (Dorel *et al.*, 2006; Raivio and Silhavy, 1997).

The Cpx pathway has attracted broad interest because of its impressive involvement in the regulation of bacterial physiology. CpxR~P has been shown to modify the transcription of over 20 genes, and 100 target operons have been predicted to be modulated by CpxR~P based on bioinformatic analysis (De Wulf *et al.*, 2002). The activation of the Cpx pathway appears to result in the shutdown of adherence structure biogenesis that becomes unnecessary in biofilm-embedded cells, and in the dormancy of such immobilized bacteria. Therefore, Cpx is a strategic signalling pathway for facing adverse conditions and for settling biofilm communities (Dorel *et al.*, 2006). Currently, the regulatory mechanisms of the CpxR response (transcriptional and post-transcriptional) and the interactions between CpxR and bacterial regulatory systems are dependent on the expression of *rpoE* (Redford and Welch, 2006).

Due to the importance of the Cpx pathway in stress adaptation and pathogenicity in *E. coli*, we explored the presence and distribution of the *cpxA*/*cpxR* system in different *Pseudomonas* genomes. The CpxA protein encoded by the gene *cpxA* (gi|157156338) and *cpxR* from *P. syringae* pv. *syringae* B728a encoded by the gene Psyr_3793 were utilized. NCBI/BLAST/blastp analysis and the results from this query are shown in Tables 14.3 and 14.4. The results from both analysis showed that both genes are highly conserved in *Pseudomonas*.

Conclusions and perspectives

EPS such as alginate and the osmoregulated periplasmic glucans (OPGs) play an important role in the biology and pathogenicity of *P. syringae* and other bacteria. At the same time, EPS biosynthesis requires a significant expenditure of energy, which may explain why EPS biosynthesis is highly regulated and complex. One example of this tightly controlled regulation is the regulation of alginate biosynthesis.

The production of OPGs is required for *P. syringae* and other bacteria to respond to different environmental challenges. It is possible that the absence of OPGs changes the cell membrane

Table 14.4 Results of BLAST/blastp analysis of the CpxR protein from *P. syringae* pv. *syringae* B728a encoded by the gene Psyr_3793

Strain	Accession	% Identity	% Similarity
P. syringae pv. *syringae* B728a	Psyr_3793	100	100
P. syringae pv. *tomato* DC3000	PSPTO_1806	96	98
P. syringae pv. *phaseolicola* 1448A	PSPPH_3547	95	98
P. fluorescens Pf-5	Pfl_4475	81	88
P. putida KT2440	PP_3372	72	83
P. fluorescens PfO-1	Pfl_1483	40	58

conformation, thus affecting other membrane-bound protein receptors such as WspA, BY-kinases, and Cpx. To answer these questions, it is necessary to perform studies by electronic microscopy to evaluate the physical integrity of the cell membrane. Furthermore, monoclonal antibodies (mAbs) that recognize the membrane epitopes could be used. Conformational changes in the cell membrane epitopes may occur when OPGs are absent; if this happens then the native mAbs will not be able to recognize the native epitopes. These results could suggest the importance of OPGs in the cell membrane.

To obtain a better understanding about the OPGs role in the cell membrane, genomic and transcriptomic comparison of whole-genome microarrays, proteomic analyses, and quantitative PCR could be used to characterize the differences between *P. syringae* wild-type strains (like DC3000) and to evaluate gene expression and proteomic changes caused by the *P. syringae* DC3000 *hrpM* mutant.

Biofilms are microniches of conditions different from the ambient environment and impart microbial communities with functions not available outside of the biofilm environment. At the same time biofilms play an important role in bacterial ecology and biology and protect cells from desiccation, UV, predation and bactericides. Biofilms play an important role in the colonization of plants by enteric pathogens and enhance HGT and synergistic interactions. More studies are needed to evaluate the beneficial and detrimental effects of microbial biofilms in the plant–microbe interaction.

Acknowledgements

C.L.B. acknowledges support from the National Science Foundation, and the Oklahoma Center for Advancement of Science and Technology. This work was also supported by the Oklahoma Agricultural Experiment Station (OAES). C.M.B. acknowledges support from an Oklahoma State University Graduate College Research Fellowship.

References

Amikam, D., and Benziman, M. (1989). Cyclic diguanylic acid and cellulose synthesis in *Agrobacterium tumefaciens*. J. Bacteriol. *171*, 6649–6655.

Amikam, D., and Galperin, M.Y. (2006). PilZ domain is part of the bacterial c-di-GMP binding protein. Bioinformatics *22*, 3–6.

An, D., and Parsek, M.R. (2007). The promise and peril of transcriptional profiling in biofilm communities. Curr. Opin. Microbiol. *10*, 292–296.

Aspedon, A., Palmer, K., and Whiteley, M. (2006). Microarray analysis of the osmotic stress response in *Pseudomonas aeruginosa*. J. Bacteriol. *188*, 2721–2725.

Ausmees, N., Mayer, R., Weinhouse, H., Volman, G., Amikam, D., Benziman, M., and Lindberg, M. (2001). Genetic data indicate that proteins containing the GGDEF domain possess diguanylate cyclase activity. FEMS Microbiol. Lett. *204*, 163–167.

Bantinaki, E., Kassen, R., Knight, C.G., Robinson, Z., Spiers, A.J., and Rainey, P.B. (2007). Adaptive divergence in experimental populations of *Pseudomonas fluorescens*. III. Mutational origins of wrinkly spreader diversity. Genetics. *176*, 441–453.

Barak, J.D., Gorski, L., Naraghi-Arani, P., and Charkowski, A.O. (2005). *Salmonella enterica* virulence genes are required for bacterial attachment to plant tissue. Appl. Environ. Microbiol. *71*, 5685–5691.

Beattie, G.A., and Lindow, S.E. (1994). Comparison of the behavior of epiphytic fitness mutants of *Pseudomonas syringae* under controlled and field conditions. Appl. Environ. Microbiol. *60*, 3799–3808.

Beech, I., Hanjagsit, L., Kalaji, M., Neal, A.L., and Zinkevich, V. (1999). Chemical and structural characterization of exopolymers produced by *Pseudomonas sp* NCIMB 2021 in continuous culture. Microbiology *145*, 1491–1497.

Benach, J., Swaminathan, S.S., Tamayo, R., Handelman, S.K., Folta-Stogniew, E., Ramos, J.E., Forouhar, F., Neely, H., Seetharaman, J., and Camilli, A. (2007). The structural basis of cyclic diguanylate signal transduction by PilZ domains. EMBO J. *26*, 5153–5166.

Bender, C.L., Alarcon-Chaidez, F., and Gross, D.C. (1999). *Pseudomonas syringae* phytotoxins: Mode of action, regulation, and biosynthesis by peptide and polyketide synthetases. Microbiol. Mol. Biol. Rev. *63*, 266–292.

Berg, T., Tesoriero, L., and Hailstones, D.L. (2006). A multiplex real-time PCR assay for detection of *Xanthomonas campestris* from brassicas. Lett. Appl. Microbiol. *42*, 624–630.

Berry, A., DeVault, J.D., and Chakrabarty, A.M. (1989). High osmolarity is a signal for enhanced *algD* transcription in mucoid and nonmucoid *Pseudomonas aeruginosa* strains. J. Bacteriol. *171*, 2312–2317.

Boch, J., Joardar, V., Gao, L., Robertson, T.L., Lim, M., and Kunkel, B.N. (2002). Identification of *Pseudomonas syringae* pv. *tomato* genes induced during infection of *Arabidopsis thaliana*. Mol. Microbiol. *44*, 73–88.

Bohin, J.P. (2000). Osmoregulated periplasmic glucans in Proteobacteria. FEMS Microbiol. Lett. *186*, 11–19.

Bouchart, F., Delangle, A., Lemoine, J., Bohin, J.P., and Lacroix, J.M. (2007). Proteomic analysis of a non-virulent mutant of the phytopathogenic bacterium *Erwinia chrysanthemi* deficient in osmoregulated periplasmic glucans: change in protein expression is not restricted to the envelope, but affects general metabolism. Microbiology *153*, 760–767.

Boucher, J.C., Schurr, M.J., Yu, H., Rowen, D.W., and Deretic, V. (1997). *Pseudomonas aeruginosa* in cystic fibrosis: role of *mucC* in the regulation of alginate production and stress sensitivity. Microbiology *143*, 3473–3480.

Brandl, M.T. (2006). Fitness of human enteric pathogens on plants and implications for food safety. Annu. Rev. Phytopathol. *44*, 367–392.

Brandl, M.T., and Mandrell, R.E. (2002). Fitness of *Salmonella enterica* serovar Thompson in the cilantro phyllosphere. Appl. Environ. Microbiol. *68*, 3614–3621.

Buell, C.R., Joardar, V., Lindeberg, M., Selengut, J., Paulsen, I.T., Gwinn, M.L., Dodson, R.J., Deboy, R.T., Durkin, A.S., and Kolonay, J.F. (2003). The complete genome sequence of the *Arabidopsis* and tomato pathogen *Pseudomonas syringae* pv. *tomato* DC3000. Proc. Natl. Acad. Sci. USA *100*, 10181–10186.

Chang, W.S., van de Mortel, M., Nielsen, L., Nino de Guzman, G., Li, X., and Halverson, L.J. (2007). Alginate production by *Pseudomonas putida* creates a hydrated microenvironment and contributes to biofilm architecture and stress tolerance under water-limiting conditions. J. Bacteriol. *189*, 8290–8299.

Chatterjee, A., Cui, Y., Hasegawa, H., and Chatterjee, A.K. (2007). PsrA, the *Pseudomonas* sigma regulator, controls regulators of epiphytic fitness, quorum-sensing signals, and plant interactions in *Pseudomonas syringae* pv. *tomato* strain DC3000. Appl. Environ. Microbiol. *73*, 3684–3694.

Chatterjee, A., Cui, Y.Y., Yang, H.L., Collmer, A., Alfano, J.R., and Chatterjee, A.K. (2003). GacA, the response regulator of a two-component system, acts as a master regulator in *Pseudomonas syringae* pv. *tomato* DC3000 by controlling regulatory RNA, transcriptional activators, and alternate sigma factors. Mol. Plant–Microbe Interact. *16*, 1106–1117.

Chitnis, C.E., and Ohman, D.E. (1993). Genetic analysis of the alginate biosynthetic gene cluster of *Pseudomonas aeruginosa* shows evidence of an operonic structure. Mol. Microbiol. *8*, 583–593.

Coenye, T., Peeters, E., and Nelis, H.J. (2007). Biofilm formation by *Propionibacterium acnes* is associated with increased resistance to antimicrobial agents and increased production of putative virulence factors. Res. Microbiol. *158*, 386–392.

Costerton, J.W., Lewandowski, Z., Caldwell, D.E., Korber, D.R., and Lappinscott, H.M. (1995). Microbial biofilms. Annu. Rev. Microbiol. *49*, 711–745.

D'Argenio, D.A., and Miller, S.I. (2004). Cyclic di-GMP as a bacterial second messenger. Microbiol. *150*, 2497–2502.

Davey, M.E., and O'Toole, G.A. (2000). Microbial biofilms: from ecology to molecular genetics. Microbiol. Mol. Biol. Rev. *64*, 847–867.

de Torres, M., Mansfield, J.W., Grabov, N., Brown, I.R., Ammouneh, H., Tsiamis, G., Forsyth, A., Robatzek, S., Grant, M., and Boch, J. (2006). *Pseudomonas syringae* effector AvrPtoB suppresses basal defence in *Arabidopsis*. Plant J. *47*, 368–382.

De Wulf, P., Akerley, B.J., and Lin, E.C.C. (2000). Presence of the Cpx system in bacteria. Microbiology *146*, 247–248.

De Wulf, P., McGuire, A.M., Liu, X.Q., and Lin, E.C.C. (2002). Genome-wide profiling of promoter recognition by the two-component response regulator CpxR-P in *Escherichia coli*. J. Biol. Chem. *277*, 26652–26661.

Denny, T.P. (1995). Involvement of bacterial polysaccharides in plant pathogenesis. Annu. Rev. Phytopathol. *33*, 173–197.

Deretic, V., Gill, J.F., and Chakrabarty, A.M. (1987a). Gene *algD* coding for GDPmannose dehydrogenase is transcriptionally activated in mucoid *Pseudomonas aeruginosa*. J. Bacteriol. *169*, 351–358.

Deretic, V., Gill, J.F., and Chakrabarty, A.M. (1987b). *Pseudomonas aeruginosa* infection in cystic fibrosis: nucleotide sequence and transcriptional regulation of the *algD* gene. Nucleic Acids Res. *15*, 4567–4581.

Dorel, C., Lejeune, P., and Rodrigue, A. (2006). The Cpx system of *Escherichia coli*, a strategic signaling pathway for confronting adverse conditions and for settling biofilm communities? Res. Microbiol. *157*, 306–314.

Duffy, E.A., Cisneros-Zevallos, L., Castillo, A., Pillai, S.D., Ricke, S.C., and Acuff, G.R. (2005). Survival of *Salmonella* transformed to express green fluorescent protein on Italian parsley as affected by processing and storage. J. Food Prot. *68*, 687–695.

Dulla, G., and Lindow, S.E. (2008). Quorum size of *Pseudomonas syringae* is small and dictated by water availability on the leaf surface. Proc. Natl. Acad. Sci. USA *105*, 3082–3087.

Espinosa-Urgel, M. (2004). Plant-associated Pseudomonas populations: molecular biology, DNA dynamics, and gene transfer. Plasmid *52*, 139–150.

Fakhr, M.K., Penaloza-Vazquez, A., Chakrabarty, A.M., and Bender, C.L. (1999). Regulation of alginate biosynthesis in *Pseudomonas syringae* pv. *syringae*. J. Bacteriol. *181*, 3478–3485.

Fett, W.F. (2000). Naturally occurring biofilms on alfalfa and other types of sprouts. J. Food Prot. *63*, 625–632.

Fett, W.F., and Cooke, P.H. (2003). Reduction of *Escherichia coli* O157:H7 and *Salmonella* on laboratory-inoculated alfalfa seed with commercial citrus-related products. J. Food Prot. *66*, 1158–1165.

Fett, W.F., and Cooke, P.H. (2005). A survey of native microbial aggregates on alfalfa, clover and mung bean sprout cotyledons for thickness as determined by confocal scanning laser microscopy. Food Microbiol. *22*, 253–259.

Fett, W.F., and Dunn, M.F. (1989). Exopolysaccharides produced by phytopathogenic *Pseudomonas syringae* pathovars in infected leaves of susceptible hosts. Plant Physiol *89*, 5–9.

Fett, W.F., Osman, S.F., Fishman, M.L., and Siebles, T.S. (1986). Alginate production by plant-pathogenic Pseudomonads. Appl. Environ. Microbiol. *52*, 466–473.

Fujishige, N.A., Kapadia, N.N., and Hirsch, A.M. (2006). A feeling for the micro-organism: structure on a small scale. Biofilms on plant roots. Botan. J. Linn Soc. *150*, 79–88.

Fujiwara, S., Zielinski, N.A., and Chakrabarty, A.M. (1993). Enhancer-like activity of AlgR1-binding site in alginate gene activation: positional, orientational, and sequence specificity. J. Bacteriol. *175*, 5452–5459.

Galan, J.E., and Collmer, A. (1999). Type III secretion machines: bacterial devices for protein delivery into host cells. Science *284*, 1322–1328.

Gandhi, M., Golding, S., Yaron, S., and Matthews, K.R. (2001). Use of green fluorescent protein expressing *Salmonella* Stanley to investigate survival, spatial location, and control on alfalfa sprouts. J. Food Prot. *64*, 1891–1898.

Goldberg, J.B., Gorman, W.L., Flynn, J.L., and Ohman, D.E. (1993). A mutation in *algN* permits trans activation of alginate production by *algT* in *Pseudomonas* species. J. Bacteriol. *175*, 1303–1308.

Gopalan, S., and He, S.Y. (1996). Bacterial genes involved in the elicitation of hypersensitive response and pathogenesis. Plant Dis. *80*, 604–610.

Grangeasse, C., Cozzone, A.J., Deutscher, J., and Mijakovic, I. (2007). Tyrosine phosphorylation: an emerging regulatory device of bacterial physiology. Trends Biochem. Sci. *32*, 86–94.

Grant, S.R., Fisher, E.J., Chang, J.H., Mole, B.M., and Dangl, J.L. (2006). Subterfuge and manipulation: Type III effector proteins of phytopathogenic bacteria. Annu. Rev. Microbiol. *60*, 425–449.

Gross, M., and Rudolph, K. (1987a). Demonstration of levan and alginate in bean plants (*Phaseolus vulgaris*) infected by *Pseudomonas syringae* pv *phaseolicola*. J. Phytopathol. *120*, 9–19.

Gross, M., and Rudolph, K. (1987b). Studies on the extracellular polysaccharides (EPS) produced *in vitro* by *Pseudomonas phaseolicola*. 1. Indications for a polysaccharide resembling alginic acid in 7 *Pseudomonas syringae* pathovars. J. Phytopathol. *118*, 276–287.

Guvener, Z.T., and Harwood, C.S. (2007). Subcellular location characteristics of the *Pseudomonas aeruginosa* GGDEF protein, WspR, indicate that it produces cyclic-di-GMP in response to growth on surfaces. Mol. Microbiol. *66*, 1459–1473.

Hanks, S.K., and Hunter, T. (1995). Protein kinases. 6.The eukaryotic protein-kinase superfamily-kinase (catalytic) domain-structure and classification. FASEB J. *9*, 576–596.

Hanoulle, X., Rollet, E., Clantin, B., Landrieu, I., Odberg-Ferragut, C., Lippens, G., Bohin, J.P., and Villeret, V. (2004). Structural analysis of *Escherichia coli* OpgG, a protein required for the biosynthesis of osmoregulated periplasmic glucans. J. Mol. Biol. *342*, 195–205.

He, J.X., Baldini, R.L., Deziel, E., Saucier, M., Zhang, Q.H., Liberati, N.T., Lee, D., Urbach, J., Goodman, H.M., and Rahme, L.G. (2004). The broad host range pathogen *Pseudomonas aeruginosa* strain PA14 carries two pathogenicity islands harboring plant and animal virulence genes. Proc. Natl. Acad. Sci. USA *101*, 2530–2535.

He, P., Shan, L., and Sheen, J. (2007). Elicitation and suppression of microbe-associated molecular pattern-triggered immunity in plant–microbe interactions. Cell Microbiol. *9*, 1385–1396.

He, Y.W., Xu, M., Lin, K., Ng, Y.J.A., Wen, C.M., Wang, L.H., Liu, Z.D., Zhang, H.B., Dong, Y.H., and Dow, J.M. (2006). Genome scale analysis of diffusible signal factor regulon in *Xanthomonas campestris* pv. campestris: identification of novel cell-cell communication-dependent genes and functions. Mol. Microbiol. *59*, 610–622.

Hendrickson, E.L., Plotnikova, J., Mahajan-Miklos, S., Rahme, L.G., and Ausubel, F.M. (2001). Differential roles of the *Pseudomonas aeruginosa* PA14 *rpoN* gene in pathogenicity in plants, nematodes, insects, and mice. J. Bacteriol. *183*, 7126–7134.

Hershberger, C.D., Ye, R.W., Parsek, M.R., Xie, Z.D., and Chakrabarty, A.M. (1995). The *algT* (*algU*) gene of *Pseudomonas aeruginosa*, a key regulator involved in alginate biosynthesis, encodes an alternative sigma factor (sigma E). Proc. Natl. Acad. Sci. USA *92*, 7941–7945.

Hickman, J.W., and Harwood, C.S. (2008). Identification of FleQ from *Pseudomonas aeruginosa* as a c-di-GMP-responsive transcription factor. Mol. Microbiol. *69*, 376–389.

Hickman, J.W., Tifrea, D.F., and Harwood, C.S. (2005). A chemosensory system that regulates biofilm formation through modulation of cyclic diguanylate levels. Proc. Natl. Acad. Sci. USA *102*, 14422–14427.

Hirano, S.S., and Upper, C.D. (1983). Ecology and epidemiology of foliar bacterial plant-pathogens. Annu. Rev. Phytopathol. *21*, 243–269.

Hirano, S.S., and Upper, C.D. (2000). Bacteria in the leaf ecosystem with emphasis on *Pseudomonas syringae*-a pathogen, ice nucleus, and epiphyte. Microbiol. Mol. Biol. Rev. *64*, 624–653.

Hrabak, E.M., and Willis, D.K. (1992). The lemA gene required for pathogenicity of *Pseudomonas syringae* pv syringae on bean is a member of a family of 2-component regulators. J. Bacteriol. *174*, 3011–3020.

Jacques, M.A., Josi, K., Darrasse, A., and Samson, R. (2005). *Xanthomonas axonopodis* pv. *phaseoli* var. *fuscans* is aggregated in stable biofilm population sizes in the phyllosphere of field-grown beans. Appl. Environ. Microbiol. *71*, 2008–2015.

Jenal, U. (2004). Cyclic di-guanosine-monophosphate comes of age: a novel secondary messenger involved in modulating cell surface structures in bacteria? Curr. Opin. Microbiol. *7*, 185–191.

Jenal, U., and Malone, J. (2006). Mechanisms of cyclic-di-GMP signaling in bacteria. Annu. Rev. Genet. *40*, 385–407.

Jha, A.K., Bais, H.P., and Vivanco, J.M. (2005). *Enterococcus faecalis* mammalian virulence-related factors exhibit potent pathogenicity in the *Arabidopsis thaliana* plant model. Infect. Immun. *73*, 464–475.

Kang, H.J., and Gross, D.C. (2005). Characterization of a resistance-nodulation-cell division transporter system associated with the *syr-syp* genomic island of *Pseudomonas syringae* pv. *syringae*. Appl. Environ. Microbiol. *71*, 5056–5065.

Kato, J., and Chakrabarty, A.M. (1991). Purification of the regulatory protein AlgR1 and its binding in the far upstream region of the *algD* promoter in *Pseudomonas aeruginosa*. Proc. Natl. Acad. Sci. USA *88*, 1760–1764.

Keith, L.M., and Bender, C.L. (1999). AlgT (sigma22) controls alginate production and tolerance to environmental stress in *Pseudomonas syringae*. J. Bacteriol. *181*, 7176–7184.

Keith, L.M., and Bender, C.L. (2001). Genetic divergence in the *algT-muc* operon controlling alginate biosynthesis and response to environmental stress in *Pseudomonas syringae*. DNA Seq. *12*, 125–129.

Keith, R.C., Keith, L.M., Hernandez-Guzman, G., Uppalapati, S.R., and Bender, C.L. (2003). Alginate gene expression by *Pseudomonas syringae* pv. *tomato* DC3000 in host and non-host plants. Microbiology. *149*, 1127–1138.

Kidambi, S.P., Sundin, G.W., Palmer, D.A., Chakrabarty, A.M., and Bender, C.L. (1995). Copper as a signal for alginate synthesis in *Pseudomonas syringae* pv. *syringae*. Appl. Environ. Microbiol. *61*, 2172–2179.

Kimbara, K., and Chakrabarty, A.M. (1989). Control of alginate synthesis in *Pseudomonas aeruginosa*: regulation of the *algR1* gene. Biochem. Biophys. Res. Commun. *164*, 601–608.

Kitten, T., Kinscherf, T.G., McEvoy, J.L., and Willis, D.K. (1998). A newly identified regulator is required for virulence and toxin production in *Pseudomonas syringae*. Mol. Microbiol. *28*, 917–929.

Lacroix, J.M., Loubens., I., Tempete., M., Menichi., B., and Bohin., J.P. (1991). The mdoA locus of *Escherichia coli* consists of an operon under osmotic control. Mol. Microbiol. *5*, 1745–1753.

Lapidot, A., Romling, U., and Yaron, S. (2006). Biofilm formation and the survival of *Salmonella* Typhimurium on parsley. Int. J. Food Microbiol. *109*, 229–233.

Laue, H., Schenk, A., Li, H., Lambertsen, L., Neu, T.R., Molin, S., and Ullrich, M.S. (2006). Contribution of alginate and levan production to biofilm formation by *Pseudomonas syringae*. Microbiology. *152*, 2909–2918.

Lazazzera, B.A. (2005). Lessons from DNA microarray analysis: the gene expression profile of biofilms. Curr. Opin. Microbiol. *8*, 222–227.

Leigh, J.A., and Coplin, D.L. (1992). Exopolysaccharides in plant-bacterial interactions. Annu. Rev. Microbiol. *46*, 307–346.

Lequette, Y., Odberg-Ferragut, C., Bohin, J.P., and Lacroix, J.M. (2004). Identification of *mdoD*, an *mdoG* paralog which encodes a twin-arginine-dependent periplasmic protein that controls osmoregulated periplasmic glucan backbone structures. J. Bacteriol. *186*, 3695–3702.

Lequette, Y., Rollet, E., Delangle, A., Greenbergi, E.P., and Bohin, J.P. (2007). Linear osmoregulated periplasmic glucans are encoded by the *opgGH* locus of *Pseudomonas aeruginosa*. Microbiology. *153*, 3255–3263.

Lindow, S.E., and Brandl, M.T. (2003). Microbiology of the phyllosphere. Appl. Environ. Microbiol. *69*, 1875–1883.

Loubens, I., Debarbieux, L., Bohin, A., Lacroix, J.M., and Bohin, J.P. (1993). Homology between a genetic-locus (*mdoA*) involved in the osmoregulated biosynthesis of periplasmic glucans in *Escherichia coli* and a genetic-locus (*hrpM*) controlling pathogenicity of *Pseudomonas syringae*. Mol. Microbiol. *10*, 329–340.

Mansvelt, E.L., and Hattingh, M.J. (1989). Scanning electron microscopy of invasion of apple leaves and blossoms by *Pseudomonas syringae* pv. *syringae*. Appl. Environ. Microbiol. *55*, 533–538.

Martin, D.W., Schurr, M.J., Mudd, M.H., and Deretic, V. (1993). Differentiation of *Pseudomonas aeruginosa* into the alginate-producing form: inactivation of *mucB* causes conversion to mucoidy. Mol. Microbiol. *9*, 497–506.

Martin, D.W., Schurr, M.J., Yu, H., and Deretic, V. (1994). Analysis of promoters controlled by the putative sigma factor AlgU regulating conversion to mucoidy in *Pseudomonas aeruginosa*: relationship to sigma E and stress response. J. Bacteriol. *176*, 6688–6696.

McCann, H.C., and Guttman, D.S. (2008). Evolution of the type III secretion system and its effectors in plant–microbe interactions. New Phytol. *177*, 33–47.

Merighi, M., Lee, V.T., Hyodo, M., Hayakawa, Y., and Lory, S. (2007). The second messenger bis-(3′-5′)-cyclic-GMP and its PilZ domain-containing receptor Alg44 are required for alginate biosynthesis in *Pseudomonas aeruginosa*. Mol. Microbiol. *65*, 876–895.

Miller, K.J., Kennedy, E.P., Reinhold, V.N., and Kennedy, E.P. (1986). Osmotic adaptation by gram-negative bacteria: possible role for periplasmic oligosaccharides. Science *231*, 48–51.

Minsavage, G.V., Mudgett, M.B., Stall, R.E., and Jones, J.B (2004). Importance of *opgH*(Xcv) of *Xanthomonas campestris* pv. vesicatoria in host-parasite interactions Mol. Plant–Microbe Interact. *17*, 152–161.

Mohr, C.D., Leveau, J.H., Krieg, D.P., Hibler, N.S., and Deretic, V. (1992). AlgR-binding sites within the *algD* promoter make up a set of inverted repeats separated by a large intervening segment of DNA. J. Bacteriol *174*, 6624–6633.

Monier, J.M., and Lindow, S.E. (2003). Differential survival of solitary and aggregated bacterial cells promotes aggregate formation on leaf surfaces. Proc. Natl Acad. Sci. USA *100*, 15977–15982.

Monier, J.M., and Lindow, S.E. (2004). Frequency, size, and localization of bacterial aggregates on bean leaf surfaces. Appl. Environ. Microbiol. *70*, 346–355.

Morris, C.E., Kinkel, L.L., Xiao, K., Prior, P., and Sands, D.C. (2007). Surprising niche for the plant pathogen *Pseudomonas syringae*. Infect Genet. Evol. *7*, 84–92.

Morris, C.E., and Monier, J.M. (2003). The ecological significance of biofilm formation by plant-associated bacteria. Annu. Rev. Phytopathol. *41*, 429–453.

Morris, C.E., Monier, J.M., and Jacques, M.A. (1998). A technique to quantify the population size and composition of the biofilm component in communities of bacteria in the phyllosphere. Appl. Environ. Microbiol. *64*, 4789–4795.

Morris, C.E., Sands, D.C., Vinatzer, B.A., Glaux, C., Guilbaud, C., Buffiere, A., Yan, S.C., Dominguez, H., and Thompson, B.M. (2008). The life history of the plant pathogen *Pseudomonas syringae* is linked to the water cycle. ISME J. 2, 321–334.

Muhammadi, and Ahmed, N. (2007). Genetics of bacterial alginate: Alginate genes distribution, organization and biosynthesis in bacteria. Curr. Genomics *8*, 191–202.

Mukhopadhyay, P., Williams, J., and Mills, D. (1988). Molecular analysis of a pathogenicity locus in *Pseudomonas syringae* pv. *syringae*. J. Bacteriol. *170*, 5479–5488.

Muller, T., Ulrich, A., Ott, E.M., and Muller, M. (2001). Identification of plant-associated enterococci. J. Appl. Microbiol. *91*, 268–278.

Newman, K.L., Almeida, R.P.P., Purcell, A.H., and Lindow, S.E. (2004). Cell-cell signaling controls *Xylella fastidiosa* interactions with both insects and plants. Proc. Natl. Acad. Sci. USA *101*, 1737–1742.

Obadia, B., Lacour, S., Doublet, P., Baubichon-Cortay, H., Cozzone, A.J., and Grangeasse, C. (2007). Influence of tyrosine-kinase Wzc activity on colanic acid production in *Escherichia coli* K12 cells. J. Mol. Biol. *367*, 42–53.

Oguiza, J.A., and Asensio, A.C. (2005). The VirPphA/AvrPtoB family of type III effectors in *Pseudomonas syringae*. Res. Microbiol. *156*, 298–303.

Oliver, D.B. (1996). Perplasm. In: *Escherichia coli* and *Salmonella*. In Cellular and Molecular Biology F.C. Neidhardt, Curtiss III, R., Ingraham, J.L., Lin, E.C.C., Low, K.B., Maganasik, B., Reznikoff, W.S., M., Schaechter, M., and Umbarger, H.E., eds. (American Society for Microbiology, Washington, DC.), pp. 88–103.

Osman, S.F., Fett, W.F., and Fishman, M.L. (1986). Exopolysaccharides of the phytopathogen *Pseudomonas syringae* pv. *glycinea*. J. Bacteriol. *166*, 66–71.

Page, F., Altabe, S., Hugouvieux-Cotte-Pattat, N., Lacroix, J.M., Robert-Baudouy, J., and Bohin, J.P. (2001). Osmoregulated periplasmic glucan synthesis is required for *Erwinia chrysanthemi* pathogenicity. J. Bacteriol. *183*, 3134–3141.

Parsek, M.R., and Singh, P.K. (2003). Bacterial biofilms: An emerging link to disease pathogenesis. Annu. Rev. Microbiol. *57*, 677–701.

Penaloza-Vazquez, A., Fakhr, M.K., Bailey, A.M., and Bender, C.L. (2004). AlgR functions in *algC* expression and virulence in *Pseudomonas syringae* pv. *syringae*. Microbiology. *150*, 2727–2737.

Penaloza-Vazquez, A., Kidambi, S.P., Chakrabarty, A.M., and Bender, C.L. (1997). Characterization of the alginate biosynthetic gene cluster in *Pseudomonas syringae* pv. *syringae*. J. Bacteriol. *179*, 4464–4472.

Pogliano, J., Lynch, A.S., Belin, D., Lin, E.C.C., and Beckwith, J. (1997). Regulation of *Escherichia coli* cell envelope proteins involved in protein folding and degradation by the Cpx two-component system. Genes & Development *11*, 1169–1182.

Prakash, B., Veeregowda, B.M., and Krishnappa, G. (2003). Biofilms: A survival strategy of bacteria. Curr. Sci. *85*, 1299–1307.

Quiniones, B., Dulla, G., and Lindow, S.E. (2005). Quorum sensing regulates exopolysaccharide production, motility, and virulence in *Pseudomonas syringae*. Mol. Plant Microbe Interact. *18*, 682–693.

Quinones, B., Pujol, C.J., and Lindow, S.E. (2004). Regulation of AHL production and its contribution to epiphytic fitness in *Pseudomonas syringae*. Mol. Plant–Microbe Interact. *17*, 521–531.

Rahme, L.G., Stevens, E.J., Wolfort, S.F., Shao, J., Tompkins, R.G., and Ausubel, F.M. (1995). Common virulence factors for bacterial pathogenicity in plants and animals. Science *268*, 1899–1902.

Raivio, T.L., and Silhavy, T.J. (1997). Transduction of envelope stress in *Escherichia coli* by the Cpx two-component system. J. Bacteriol. *179*, 7724–7733.

Ramey, B.E., Koutsoudis, M., von Bodman, S.B., and Fuqua, C. (2004). Biofilm formation in plant–microbe associations. Curr. Opin. Microbiol. *7*, 602–609.

Redford, P., and Welch, R.A. (2006). Role of sigma E-regulated genes in *Escherichia coli* uropathogenesis. Infect. Immun. *74*, 4030–4038.

Remminghorst, U., and Rehm, B.H.A. (2006). Alg44, a unique protein required for alginate biosynthesis in *Pseudomonas aeruginosa*. FEBS Lett. *580*, 3883–3888.

Roos, I.M.M., and Hattingh, M.J. (1983). Scanning electron-microscopy of *Pseudomonas syringae* pv. *morsprunorum* on sweet cherry leaves. J. Phytopathol. *108*, 18–25.

Ross, P., Mayer, R., Weinhouse, H., Amikam, D., Huggirat, Y., Benziman, M., de Vroom, E., Fidder, A., de Paus, P., Sliedregt, L.A., *et al.* (1990). The cyclic diguanylic acid regulatory system of cellulose synthesis in *Acetobacter xylinum*. Chemical synthesis and biological activity of cyclic nucleotide dimer, trimer, and phosphothioate derivatives. J. Biol. Chem. *265*, 18933–18943.

Ryan, R.P., Fouhy, Y., Lucey, J.F., Crossman, L.C., Spiro, S., He, Y.W., Zhang, L.H., Heeb, S., Camara, M., and Williams, P. (2006). Cell-cell signaling in *Xanthomonas campestris* involves an Hd-GYP domain protein that functions in cyclic di-GMP turnover. Proc. Natl. Acad. Sci. USA *103*, 6712–6717.

Ryan, R.P., Fouhy, Y., Lucey, J.F., Jiang, B.L., He, Y.Q., Feng, J.X., Tang, J.L., and Dow, J.M. (2007). Cyclic di-GMP signalling in the virulence and environmental adaptation of *Xanthomonas campestris*. Mol. Microbiol. *63*, 429–442.

Schaber, J.A., Hammond, A., Carty, N.L., Williams, S.C., Colmer-Hamood, J.A., Burrowes, B.H., Dhevan, V., Griswold, J.A., and Hamood, A.N. (2007). Diversity of biofilms produced by quorum-sensing-deficient clinical isolates of *Pseudomonas aeruginosa*. J. Med. Microbiol. *56*, 738–748.

Schenk, A., Berger, M., Keith, L.M., Bender, C.L., Muskhelishvili, G., and Ullrich, M.S. (2006). The *algT* gene of *Pseudomonas syringae* pv. *glycinea* and new insights into the transcriptional organization of the *algT-muc* gene cluster. J. Bacteriol. *188*, 8013–8021.

Schenk, A., Weingart, H., and Ullrich, M.S. (2008). The alternative sigma factor AlgT, but not alginate synthesis, promotes *in planta* multiplication of *Pseudomonas syringae* pv. *glycinea*. Microbiology. *154*, 413–421.

Schneider, R.W., and Grogan, R.G. (1977). Tomato leaf trichomes, a habitat for resident populations of *Pseudomonas tomato*. Phytopathol. *67*, 898–902.

Schnider-Keel, U., Lejbolle, K.B., Baehler, E., Haas, D., and Keel, C. (2001). The sigma factor AlgU (AlgT) controls exopolysaccharide production and tolerance towards desiccation and osmotic stress in the biocontrol agent *Pseudomonas fluorescens* CHA0. Appl. Environ. Microbiol. *67*, 5683–5693.

Schulman, H., and Kennedy, E.P. (1979). Localization of membrane-derived oligosaccharides in the outer envelope of *Escherichia coli* and their occurrence in other gram-negative bacteria. J. Bacteriol. *137*, 686–688.

Schurr, M.J., Martin, D.W., Mudd, M.H., and Deretic, V. (1994). Gene cluster controlling conversion to alginate-overproducing phenotype in *Pseudomonas aeruginosa*: functional analysis in a heterologous host and role in the instability of mucoidy. J. Bacteriol. *176*, 3375–3382.

Schurr, M.J., Yu, H., Boucher, J.C., Hibler, N.S., and Deretic, V. (1995). Multiple promoters and induction by heat shock of the gene encoding the alternative sigma factor AlgU (sigma E) which controls mucoidy in cystic fibrosis isolates of *Pseudomonas aeruginosa*. J. Bacteriol. *177*, 5670–5679.

Shankar, S., Ye, R.W., Schlictman, D., and Chakrabarty, A.M. (1995). Exopolysaccharide alginate synthesis in *Pseudomonas aeruginosa*: enzymology and regulation of gene expression. Adv. Enzymol. Relat. Areas Mol. Biol.*70*, 221–255.

Shimizu, R., Russell, B.W., and Mills, D. (2003). Genetic organisation and proposed function of the *hrpM* locus of *Pseudomonas syringae* pv. *syringae*. In Presentations from the 6th International Conference on *Pseudomonas syringae* pathovars and related pathogens, Maratea, Italy, September 1–519, 2002, pp. 157–164.

Shimohata, N., Chiba, S., Saikawa, N., Ito, K., and Akiyama, Y. (2002). The Cpx stress response system of *Escherichia coli* senses plasma membrane proteins and controls HtpX, a membrane protease with a cytosolic active site. Genes to Cells *7*, 653–662.

Simm, R., Morr, M., Kader, A., Nimtz, M., and Romling, U. (2004). GGDEF and EAL domains inversely regulate cyclic di-GMP levels and transition from sessility to motility. Mol. Microbiol. *53*, 1123–1134.

Singh, S., and Fett, W.F. (1995). Stimulation of exopolysaccharide production by fluorescent Pseudomonads in sucrose media due to dehydration and increased osmolarity. FEMS Microbiol. Lett. *130*, 301–306.

Sorensen, S.J., Bailey, M., Hansen, L.H., Kroer, N., and Wuertz, S. (2005). Studying plasmid horizontal transfer *in situ*: A critical review. Nature Rev. Microbiol. *3*, 700–710.

Soulat, D., Jault, J.M., Geourjon, C., Gouet, P., Cozzone, A.J., and Grangeasse, C. (2007). Tyrosine-kinase Wzc from *Escherichia coli* possesses an ATPase activity regulated by autophosphorylation. FEMS Microbiol. Lett. *274*, 252–259.

Stewart, P.S. (2002). Mechanisms of antibiotic resistance in bacterial biofilms. Int. J. Med. Microbiol. *292*, 107–113.

Stoodley, P., Sauer, K., Davies, D.G., and Costerton, J.W. (2002). Biofilms as complex differentiated communities. Annu. Rev. Microbiol. *56*, 187–209.

Strange, R.N. (2007). Phytotoxins produced by microbial plant pathogens. Nat. Prod. Rep. *24*, 127–144.

Szomolay, B., Klapper, I., Dockery, J., and Stewart, P.S. (2005). Adaptive responses to antimicrobial agents in biofilms. Environ. Microbiol. *7*, 1186–1191.

Talaga, P., Fournet, B., and Bohin, J.P. (1994). Periplasmic glucans of *Pseudomonas syringae* pv syringae. J. Bacteriol. *176*, 6538–6544.

Unligil, U.M., and Rini, J.M. (2000). Glycosyltransferase structure and mechanism. Curr. Opin. Struct. Biol. *10*, 510–517.

Von Bodman, S.B., Bauer, W.D., and Coplin, D.L. (2003). Quorum sensing in plant-pathogenic bacteria. Annu. Rev. Phytopathol. *41*, 455–482.

Weinhouse, H., Sapir, S., Amikam, D., Shilo, Y., Volman, G., Ohana, P., and Benziman, M. (1997). c-di-GMP-binding protein, a new factor regulating cellulose synthesis in *Acetobacter xylinum*. FEBS Lett. *416*, 207–211.

Wilson, M., Hirano, S.S., and Lindow, S.E. (1999). Location and survival of leaf-associated bacteria in relation to pathogenicity and potential for growth within the leaf. Appl. Environ. Microbiol. *65*, 1435–1443.

Wood, L.F., and Ohman, D.E. (2006). Independent regulation of MucD, an HtrA-like protease in *Pseudomonas aeruginosa*, and the role of its proteolytic motif in alginate gene regulation. J. Bacteriol. *188*, 3134–3137.

Wozniak, D.J., and Ohman, D.E. (1994). Transcriptional analysis of the *Pseudomonas aeruginosa* genes *algR*, *algB*, and *algD* reveals a hierarchy of alginate gene expression which is modulated by *algT*. J. Bacteriol. *176*, 6007–6014.

Xie, Z.D., Hershberger, C.D., Shankar, S., Ye, R.W., and Chakrabarty, A.M. (1996). Sigma factor-anti-sigma factor interaction in alginate synthesis: inhibition of AlgT by MucA. J. Bacteriol. *178*, 4990–4996.

Yorgey, P., Rahme, L.G., Tan, M.W., and Ausubel, F.M. (2001). The roles of *mucD* and alginate in the virulence of *Pseudomonas aeruginosa* in plants, nematodes and mice. Mol. Microbiol. *41*, 1063–1076.

Yu, J., Penaloza-Vazquez, A., Chakrabarty, A.M., and Bender, C.L. (1999). Involvement of the exopolysaccharide alginate in the virulence and epiphytic fitness of *Pseudomonas syringae* pv. *syringae*. Mol. Microbiol. *33*, 712–720.

Zhao, Y.F., Damicone, J.P., Demezas, D.H., and Bender, C.L. (2000). Bacterial leaf spot diseases of leafy crucifers in Oklahoma caused by pathovars of *Xanthomonas campestris*. Plant Dis. *84*, 1008–1014.

Zielinski, N.A., Chakrabarty, A.M., and Berry, A. (1991). Characterization and regulation of the *Pseudomonas aeruginosa algC* gene encoding phosphomannomutase. J. Biol. Chem. *266*, 9754–9763.

Zielinski, N.A., Maharaj, R., Roychoudhury, S., Danganan, C.E., Hendrickson, W., and Chakrabarty, A.M. (1992). Alginate synthesis in *Pseudomonas aeruginosa*: environmental regulation of the *algC* promoter. J. Bacteriol. *174*, 7680–7688.

Part VI

Polysaccharides of Lactic Acid Bacteria

Ecology of Exopolysaccharide Formation by Lactic Acid Bacteria: Sucrose Utilization, Stress Tolerance, and Biofilm Formation

15

Michael Gänzle and Clarissa Schwab

Abstract

Lactic acid bacteria (LAB) synthesize a wide variety of exopolysaccharides (EPS); these polysaccharides are synthesized extracellularly from sucrose by glycansucrases, or intracellularly by glycosyltransferases from sugar nucleotide precursors. This chapter provides an overview on emerging concepts related to the ecological significance of EPS production by LAB. Biofilm formation, stress resistance and sucrose utilization are clearly linked to the formation of EPS in individual species of LAB. The high frequency of homopolysacharide (HoPS) and heteropolysaccharide (HePS) producing LAB in the oral cavity and intestinal ecosystems argues in favour for an important role of EPS formation for the persistence of LAB in these habitats. The intricate regulatory network controlling the expression of glycansucrases in oral streptococci is in keeping with the contribution of HoPS and extracellular glycansucrases to biofilm formation and persistence in the oral cavity. EPS production by intestinal lactobacilli may play a comparable role. Glycansucrases in *Lb. reuteri* are regulated in response to stress sensed by the cytoplasmic membrane. The products of glycansucrases improve survival of lactobacilli in a scenario characterized by strong fluctuations in water activity, temperature, pH, and nutrient supply, and the presence of natural inhibitors. Because the expression of glycansucrases in many strains of lactobacilli and *Leuconostoc* species is induced by sucrose, the contribution of glycansucrases to sucrose catabolism may be their main ecological role in some strains.

Introduction

Lactic acid bacteria (LAB) synthesize polysaccharides which are utilized as cell wall constituents (peptidoglycan) or secreted from the cell. The latter remain attached to the outer cell wall and form a capsule (capsular polysaccharide, CPS) and/or are released into the environment as exopolysaccharides (EPS) (Cerning, 1990). The description of EPS formation by lactic acid bacteria dates back to Pasteur, who observed that ropy spoilage of wine, the wine disease 'graisse', was caused by microbial activity (Pasteur 1861, as cited in Lonvaud-Funel, 1999). Orla-Jensen (1943) described the ability of *Leuconostoc* spp. (Betacoccus), mesophilic lactobacilli (Betabacteria), pediococci, and oral streptococci to form polysaccharides from sucrose, and indicated the role of polysaccharide formation in the spoilage of apple cider and beer as well as the formation of dextran by *Leuconostoc* spp. during the extraction of sucrose from sugar beets. Dextran from *Leuconostoc* species has been produced on an industrial scale since the 1940s and finds a broad range of applications (Leathers, 2002). Compared with slimy exopolysaccharides, the formation of CPS by non-pathogenic LAB attracted attention only sporadically (Brooker, 1977; Mozzi *et al.*, 2006; van der Meulen *et al.*, 2007).

With respect to biosynthetic pathways, two classes of EPS from lactic acid bacteria can be distinguished, polysaccharides that are synthesized from sucrose by extracellular glycansucrases, and polysaccharides with (ir)regular repeating units that are synthesized intracellularly by glycosyltransferases from sugar nucleotide precursors.

The repeating units are polymerized extracellularly. The terms 'heteropolysaccharides' (HePS) and 'homopolysaccharides' (HoPS) are generally used to differentiate products of glycosyltransferase and glycansucrase activities, respectively. Exopolysaccharides were described, however, that are synthesized intracellularly by glycosyltransferases and exhibit uniform monomer composition and linkage types, and glycansucrases are capable of formation of heterooligosaccharides.

The differences in the biosynthetic pathways have ecological implications. HePS synthesis requires metabolic energy in the form of activated sugar nucleotides that are derived from central carbon metabolism. Competition of glycosyltransferases for sugar nucleotide precursors as well as the capacity of the lipid carrier for EPS assembly limit HePS formation to levels that are generally less than 0.4 g/l (van der Meulen *et al.*, 2007; Mozzi *et al.*, 2006). In contrast, extracellular HoPS synthesis by glycansucrases requires no expense of metabolic energy by the producer organism because the energy of the osidic bond of sucrose is conserved to synthesize oligo- and polysaccharides, and heterooligosaccharides in the presence of additional acceptor sugars (Korakli and Vogel, 2006). The activity of the glycansucrases is therefore uncoupled from growth. Glycansucrases enable high sucrose turnover (more than 100 g/l sucrose in batch culture) and high yields of polysaccharides (3–15 g/l or g/kg DM) (Korakli *et al.*, 2001; Tieking *et al.*, 2003). So far, no lactic acid bacterium has been described which concurrently utilizes both EPS biosynthesis systems.

Several recent reviews outline the current state of the art on application, diversity, biosynthesis, and genetics of HePS and HoPS formation by LAB (Korakli and Vogel, 2006; van Hijum *et al.*, 2006; Degeest *et al.*, 2001; Laws and Marshall, 2001). This chapter provides an overview on emerging concepts concerning the ecological significance of EPS production by lactic acid bacteria. On the basis of the frequency of EPS-producing lactic acid bacteria in various ecosystems, the genetic organization of EPS biosynthetic genes and their regulation, emphasis is placed on the discussion of those studies that demonstrated an ecological role of EPS formation.

Occurrence of exopolysaccharide producing lactic acid bacteria

EPS formation has been described for LAB associated with different gastrointestinal ecosystems, food fermentations and spoilage, and with plants. It is not possible to draw a line between habitats populated with HePS or HoPS producing LAB only, but in some niches one type of EPS formation seems to be more dominant than the other.

The presence of HePS or HoPS forming LAB varies within the different sections of the gastrointestinal tract. In the mouth, glucans and fructans synthesized by oral streptococci of the species *Streptococcus mutans, S. downei, S. sobrinus* and *S. salivarius* are involved in persistence and biofilm formation on the tooth enamel (Cvitkovitch *et al.*, 2003). Lactobacilli are autochthonus to the upper gastrointestinal regions of ducks (crop), pigs (esophagus) and rodents (forestomach) (Brooker and Fuller, 1975; Mitsuoka, 1992; Tannock, 1992). HoPS producers are frequent among the lactobacilli associated with those habitats, i.e. *Lactobacillus acidophilus, Lb. animalis, Lb. crispatus, Lb. johnsonii, Lb. mucosae, Lb. panis, Lb. pontis,* and *Lb. reuteri* (Tieking *et al.*, 2003; Kralj *et al.*, 2004b). Tieking *et al.* (2005b) screened all lactobacilli (42 strains) isolated from intestinal samples of ducks and pigs for the presence of glycansucrase genes and production of glucans and fructans. More than 25% of the isolates produced HoPS and oligosaccharides from sucrose amd genes coding for glycansucrases were detected in most isolates. Especially in the species *Lb. reuteri*, virtually all strains are capable of HoPS production and the presence of several glycansucrases in a single strain is the rule rather than the exception (Schwab and Gänzle, 2005; Table 15.1).

In contrast, a screening of human fecal and mucosal isolates from the genera *Bifidobacterium* and *Lactobacillus* indicated that these strains produce HePS rather than HoPS (Ruas-Madiedo *et al.*, 2007). Out of 365 isolates, 17% of the strains were identified as EPS producers.

The high proportion of HePS producing strains among dairy starter cultures reflects the relevance of EPS formation for the quality of fermented milk products and cheese. HePS production has been frequently described for strains of the species *S. thermophilus, Lactococcus lactis,*

Table 15.1 Glycansucrase genes and polysaccharide formation in *Lb. reuteri* (adapted from Gänzle and Schwab, 2005)

Origin	Strain	Inulosucrase	Reuteransucrase	*ftfA*	*ftfB*	Levansucrase	Polysaccharides
Rodents	100–23	+	+	+	+	–	Levan
	FUA 3041	–	+	–	–	–	Glucan
	FUA 3042	–	–	+	–	–	Fructan
	FUA 3044	–	–	+	+	–	Fructan
Swine	TMW1.272	+	+	–	n.d.	–	Glucan
	TMW1.279	+	–	–	n.d.	–	–
	TMW1.285	+	–	–	n.d.	–	–
	TMW1.146	+	–	–	n.d.	–	–
	TMW1.194	+	+	–	n.d.	–	–
	TMW1.138	+	+	–	n.d.	–	Glucan
Ducks	TMW1.967	+	–	–	n.d.	–	–
	TMW1.972	+	–	–	n.d.	–	–
	TMW1.973	+	–	–	n.d.	–	–
	TMW1.974	+	+	–	n.d.	–	Glucan
	TMW1.975	+	+	–	n.d.	+	–
	TMW1.976	+	+	–	n.d.	–	Glucan
	TMW1.980	+	–	+	n.d.	–	–
Humans	DSM20016	+	+	–	n.d.	–	Fructan
	LTH5794	–	–	–	+	–	Levan
	LTH5795	–	–	–	+	–	Fructan
	LTH5796	–	–	–	+	–	Fructan
	LTH5797	–	–	–	+	–	Fructan
Sourdough	LTH2584	–	–	+	–	–	–
	TMW1.106	+	+	–	–	–	Reuteran
	LTH5448	–	–	+	–	–	Levan

+, present; –, absent; n.d., not determined.

Lb. delbrueckii, and *Lb. helveticus,* but other species of LAB also harbour HePS producing strains (de Vuyst *et al.,* 2001; Broadbent *et al.,* 2003; Mozzi *et al.,* 2006). Polysaccharide formation is a requirement for the appropriate texture in traditional fermented dairy products such as vilii and kefir (Oba *et al.,* 1999; Kontusaari and Forsén, 1987; Wszolek *et al.,* 2006). In contrast, HePS producing strains are rarely isolated from cereal fermentations (Tieking *et al.,* 2003; van der Meulen *et al.,* 2007; Sanni *et al.,* 2002). Glucan or fructan formation was frequently observed in strains of the cereal-associated species *Lb. sanfranciscensis, Lb. sakei, Lb. parabuchneri,* and *Lb. curvatus* (Tieking *et al.,* 2003; Kralj *et al.,* 2004b).

EPS production of *Pediococcus parvulus* is implicated in the ropy spoilage of wine. A plasmid-coded glycosyltransferase which is highly conserved among wine and cider spoiling bacteria confers the ropy phenotype in pediococci, oenococci, and *Lb. diolivorans* (Gindreau *et al.,* 2001; Walling *et al.,* 2001; Dols-Lafargue *et al.,* 2008).

Leuconostoc mesenteroides and *Weissella confusa* frequently harbour strains that produce dextran, alternan or levan from sucrose. A screening of more than 150 food isolates of lactic acid bacteria revealed that 8 out of 10 EPS-positive strains were glucan-producing *Leuconostoc* or *Weissella* species, which may reflects the association of these organisms with plant material that is rich in sucrose as main carbon source (van der Meulen *et al.*, 2007).

Genetic organization and transcriptional regulation of exopolysaccharide formation

Genetic organization of HePS formation

The biosynthesis of HePS occurs by combined activity of housekeeping as well as *eps* specific enzymes. Housekeeping genes as part of the central carbon metabolism control the formation of sugar nucleotides and are regulated independently of HePS synthesis, whereas genes responsible for the assembly, transportation and polymerization of HePS are specific and clustered on one operon (de Vuyst *et al.*, 2001; Stingele and Jolly, 2001). An unusual gene arrangement exists in *Lb. rhamnosus*, where genes involved in the formation of sugar nucleotide precursors are located within the EPS operon. These genes possess their own promoter but can also be transcribed together with the *eps* genes (Péant *et al.*, 2005). In streptococci, *eps* operons are frequently chromosomally located, whereas plasmid encoded *eps* operons are frequently found in lactococci (de Vuyst *et al.*, 2001; Dabour and LaPointe, 2005).

The general organization and transcriptional control of EPS operons is highly conserved (de Vuyst *et al.*, 2001; Stingele and Jolly, 2001). Regulatory genes are located at the 5′ end, glycosyltransferases represent the core region which is surrounded by genes coordinating polymerization, chain length and export. The regulatory gene cluster owns similarities to genes regulating the capsular biosynthesis of Gram-positive pathogens, e.g. *S. pneumonia*, and is highly conserved among lactic acid bacteria (Stingele and Jolly, 2001). Similar to *Streptococcus* spp., HePS formation of *Lc. lactis* is under control of a phosphoregulatory system (Nierop-Groot and Kleerebezem, 2007). In *S. pneumonia*, the regulatory genes are also involved in linking the CPS formation to cell wall synthesis (Morona *et al.*, 2006); in LAB these genes have a comparable function. The genetic structure of the priming glycosyltransferases is highly conserved among Gram-positive lactic acid bacteria, while the structure of the other glycosyltransferases is more variable. Replacement of the priming glycosyltransferase of *Lc. lactis* with priming glycosyltransferases of other *Lc. lactis* strains or capsule producing *S. pneumonia* maintains EPS biosynthesis (van Kranenburg *et al.*, 1999).

At the start of HePS biosynthesis, sugar nucleotides, building blocks of HePS, are derived from de novo synthesis or from the general precursor glucose-1-phosphate. The first sugar nucleotide is linked to lipid carrier located in the cytoplasmic membrane by the priming glycosyltransferase, and sugar nucleotides are subsequently added to form repeating units by the activity of glycosyltransferases. Repeating units are composed of three to eight monosaccharides (de Vuyst *et al.*, 2001). These repeating units can be used in cell wall synthesis or for HePS formation. During HePS synthesis, the repeating units are transported across the membrane to the outer layer and polymerized to form the final HePS (de Vuyst *et al.*, 2001). HePS synthesized by different strains varies in monosaccharide composition, linkage type, degree of polymerization and branching. The biosynthesis of HePS is an energy demanding process and is closely linked to growth. Cells growing at suboptimal conditions synthesized higher amounts of HePS than cells at optimal conditions, probably due to lower demands for cell wall synthesis (Degeest *et al.*, 2001; Sánchez *et al.*, 2006).

Beta-glucan biosynthesis observed in lactobacilli, pediococci and oenococci resembles heteropolysaccharide formation (Werning *et al.*, 2006; Velasco *et al.*, 2007; Ibaruru *et al.*, 2007). Beta-glucan formation occurs from the precursor UDP-glucose which is assembled by the single membrane bound β-glucan synthase Gtf (Velasco *et al.*, 2007). The Gtf is plasmid encoded in lactobacilli and pediococci but located on the chromosome of *Oenococcus oeni* (Werning *et al.*, 2006). The composition of the synthesized β-glucan is independent of the carbohydrate source (Velasco *et al.*, 2007; Ibaruru *et al.*, 2007).

Genetic organization of HoPS synthesis

Genes coding for HoPS synthesizing glycansucrases form single operons, for example the monocistronic operon of the dextransucrase of

Lu. mesenteroides NRRL B512-F (Quirasco *et al.*, 1999). *S. mutans* harbours the glucosyltransferases GtfB, GtfC and GtfD and the fructosyltransferase Ftf (Shimamura *et al.*, 1983; Shiroza and Kuramitsu, 1988). The genes *gtfB* and *gtfC* are successively arranged on the genome and can be either independently transcribed from different promoters or tandemly transcribed from the *gtfB* promoter. *gtfD* is located further downstream as monocistronic operon (Fujiwara *et al.*, 1996; Smorawinska and Kuramitsu, 1995; Chen *et al.*, 2006). In *S. mutans*, EPS formation is regulated by the two-component systems VicRK (formerly termed CovRS) (Senadheera *et al.*, 2005) which is located on the tricistronic operon *vicKRX*. Additional transcriptional regulators include the pleiotropic transcriptional regulator *frp* in *S. mutans* and the positive transcriptional regulator *rgg*, which is located immediately upstream of the sole glycosyltransferases *gtfG* of *S. gordonii* (Wang and Kuramitsu, 2006; Sulavik *et al.*, 1992).

The presence of silent and chimeric glycansucrases provides evidence for their genetic instability and variability. A transposase and an IS-element, were found adjacent to the levansucrase genes of *Lb. reuteri* 121 and *Lb. sanfranciscensis* TMW1.392, respectively (van Hijum *et al.*, 2004; Tieking *et al.*, 2005a) and these mobile genetic elements may account loss of glycansucrase expression in *Lb. reuteri* 121 after several generations of growth (van Geel Schutten *et al.*, 1999). Several glycansucrases from lactobacilli or *Leuconostoc* spp. were described that may be considered natural chimeric proteins. For example, *Lb. reuteri* LTH5448 harbours *ftfA*; the N-terminal variable and C-terminal domains of *ftfA* are 98% and 96% identical to *inu* but the catalytic domain shares only 72% identical amino acids with *inu* (Schwab and Gänzle, 2006). The inulosucrase IslA from *Lu. citreum* harbours a Ftf catalytic domain flanked by N- and C-terminal domains that are similar to an alternansucrase (Olivares-Illana *et al.*, 2003).

Transcriptional regulation of EPS synthesis

Substrate regulation

The transcription level of the EPS operon responsible of HePS formation is independent of sugar source. Increasing amounts of sugar nucleotides of the same type does not affect synthesis of EPS (Boels *et al.*, 2001 and 2003a). Decreased EPS formation of for example *Lc. lactis* NIZOB40 or *P. parvulus* 2.6 during growth on fructose is hypothesized to be caused by limiting activity of FBPase which transforms fructose-6-P in fructose-1-P or by different activities of internalizing systems for glucose and fructose (Looijesteijn *et al.*, 1999; Velasco *et al.*, 2007). Increased amounts of EPS were obtained by homologous overexpression of the entire *eps* operon of *Lc. lactis* NIZO B40 operon, indicating that expression of the *eps* operon and not availability of precursors constitutes the limiting factor in EPS biosynthesis (Boels *et al.*, 2003b).

In contrast, the synthesis of HoPS is related to the presence of the substrate. Sucrose is the sole substrate for glucosyltransferases, whereas fructosyltransferases use raffinose and sucrose. In *Lu. mesenteroides*, the expression of dextransucrase in is relatively low in the presence of carbon sources other than sucrose and is increased by the addition of sucrose (Neely and Nott, 1962; Quirasco *et al.*, 2007). A sucrose concentration of 20 g/l stimulates dextransucrase DsrD expression in *Lu. mesenteroides* Lcc4 and increased sucrose levels do not result in a further increase in enzyme synthesis (Neubauer *et al.*, 2003). The Ftf of *S. mutans* is up-regulated in the presence of sucrose or excess glucose (Kiska and Macrina, 1994; Shemesh *et al.*, 2006). Remarkably, increased transcription of *ftf* in the presence of sucrose was verified *in vivo* on the tooth of rats (Grey *et al.*, 1997). The two component regulatory system *vicKR* mediates the upregulation of *ftfA* in the presence of sucrose (Lee *et al.*, 2004; Shemesh *et al.*, 2006) and is furthermore involved in biofilm formation, competence, and stress tolerance (Senadheera *et al.*, 2005, 2007). In contrast to Ftf, substrate regulation of the Gtfs of *S. mutans* is strain dependent. The presence of sucrose increased *gtfD* transcription and decreased *gtfB* and *gtfC* transcription in the oral cavity isolate *S. mutans* MT8148 (Fujiwara *et al.*, 2002). In another human oral isolate, *S. mutans* GS5, sucrose increased expression of *gtfB* and *gftC* and only exert a limited affect on expression of *gtfD* (Shemesh *et al.*, 2006). Glycansucrases in *S. mutans* furthermore underlie regulation of RegM, which has homology to Ccpa of *Bacillus subtilis*, a global regulator of sugar metabolism and catalyst of carbon catabolite repression (Browngardt *et al.*, 2004).

Strains of *Lb. reuteri* maintain sucrose phosphorylases and glycansucrases as sucrose utilization systems and at least one of these enzymes is expressed constitutively (Årsköld *et al.*, 2007; Schwab *et al.*, 2007). Expression of *ftfA* in *Lb. reuteri* LTH5448, which also harboured a sucrose inducible sucrose phosphorylase (Data not shown) and possibly another active sucrose utilizing mechanism, was induced by sucrose. In contrast, *inu* and *gtfA* were constitutively expressed in *Lb. reuteri* TMW 1.106; however, *inu* and *gtfA* knock down mutants of *Lb. reuteri* TMW 1.106 were capable of growing on sucrose as the sole carbohydrate source only after adaption (Schwab *et al.*, 2007). *Lb. reuteri* ATCC55730 harbours a sucrose inducible sucrose phosphorylase and a constitutively expressed glucosyltransferase *gtfO* (Årsköld *et al.*, 2007).

Physiological state and environmental conditions

In lactobacilli, transcription of glycansucrases is regulated by environmental conditions. Elevated transcription of the fructosyltransferase genes *ftfA* and *inu* of *Lb. reuteri* LTH5448 and TMW1.106, and increased formation of levan and FOS was observed during exposure to environmental stressors. The common denominator of inducing agents was their interaction with, or their effect on bacterial membranes which resulted in increased membrane lateral pressure (Fig. 15.1). Accordingly, fructosyltransferase expression in *Lb. reuteri* may be mediated through alterations of the membrane (Schwab and Gänzle, 2006).

Ecological relevance of EPS

Glycansucrases and sucrose metabolism in HoPS-forming LAB

Sucrose phosphorylases, sucrose-phosphate transport systems and glycansucrases are alternative pathways for sucrose metabolism in LAB. In *Streptococcus* and *Leuconostoc* species, the major proportion of sucrose is converted to HoPS by glycansucrases and only a portion of sucrose is transported into the cell for phosphorylysis by sucrose phosphorylase and conversion to lactate, acetate and ethanol (Dols *et al.*, 1997; Hiratsuka *et al.*, 1998). In *Lb. sanfranciscensis* TMW1.392, levansucrase expression is not regulated by sucrose but deletion of the levansucrase abolished the ability of strain to grow with sucrose as the sole carbon source, indicating that levansucrase is the sole active sucrose utilization system (Tieking *et al.*, 2005a). In contrast, *Lb. reuteri* TMW1.106 and LTH5448 express sucrose phosphoporylases and were still able to grow after the inactivation of glycansucrase genes (Schwab *et al.*, 2007). Inactivation of *ftfA* had no effect on metabolite formation of *Lb. reuteri* LTH5448; similar amounts of lactate, acetate, ethanol and mannitol were formed independent of the presence of an active FtfA (Schwab *et al.*, 2007).

EPS and EPS formation byproducts as storage polymers and substrates

The majority of EPS positive LAB do not degrade the polymers they produce (Czerning, 1990; Kim and Fogler, 1999), making a role of EPS as substrate for the producer strain unlikely. The enzymatic degradation of HePS by the producer strain has only been reported for *S. thermophilus* and *Lb. rhamnosus* upon prolonged fermentation (Degeest *et al.*, 2002; Pham *et al.*, 2000). Pham *et al.* (2000) identified an array of glycohydrolases involved in the degradation process, whose activity released fermentable sugars into the growth medium and resulted in decreased polymer size and viscosity. Nevertheless, EPS may act as substrates for other organism in complex ecosystems. HePS produced by intestinal bifidobacteria stimulated short chain fatty acid production during *in vitro* cultivation of faecal microbiota (Salazar *et al.*, 2008).

HoPS produced by streptococci and lactobacilli in the oral cavity or the upper intestinal tract of humans and animals may serve as a carbon source to bacteria in the same ecological niche or in the distal parts of the digestive tract. Dextran and inulin formed by oral streptococci act as substrate of other groups of organism associated with the dental caries biofilm community, for example the family *Bifidobacteriacae* (Birkhead *et al.*, 1979; Sato and Kuramitsu, 1996). Strains of bifidobacteria have been isolated from the oral cavity which possess extracellular dextranase activity and are thus able to degrade the polymer (Crociani *et al.*,

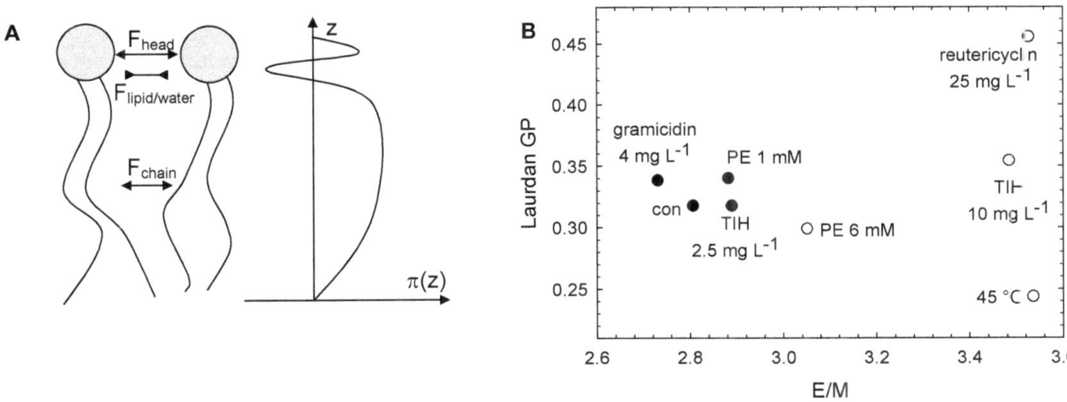

Figure 15.1 Diagram of the lateral pressure profile in biological membranes and effect of membrane fluidity and lateral pressure on fructosyltransferase expression of *Lb. reuteri*. **A** Distribution of the lateral pressure within a lipid monolayer in bacterial membranes. Shown are forces related to headgroup repulsion (F_{head}), interfacial tension ($F_{lipid/water}$) and the outward chain pressure (F_{chain}). To the right is plotted the resulting lateral pressure π as a function of depth into the monolayer, z. (Adapted from Templer *et al.* 1998). **B** Plot of the fluidity of membranes as measured by the laurdan general polarization (GP) of *Lb. reuteri* TMW1.106 against the membrane lateral pressure as measured by the eximer/monomer (E-M) ratio of d-pyrenyl phosphatidylcholine. Data was acquired at conditions which do not induce *inu* overexpression (•) and conditions which induce *inu* overexpression (○). Con, reference conditions (37°C, no additives), PE, phenylethanol, TIH, trans-isohumulone (Schwab and Gänzle, 2006).

1996; Kaster and Brown 1983; Modesto *et al.*, 2006). Oral isolates of *S. mutans* and *S. salivarius* possess extracellular fructan hydrolases (exo-3-D fructosidase) (Burne *et al.*, 1987; Takahashi *et al.*, 1985) and thus degrade inulin and levan type fructans (van Houte and Jansen, 1968; Walker *et al.*, 1983). Likewise, bifidobacteria are capable of levan and inulin metabolism (Marx *et al.*, 2000; Korakli *et al.*, 2002). Formation of these polymers by *Lb. reuteri* in the forstomach of mice or the crop of chicken may serve as carbon source for bifidobacteria and other members of the intestinal microbiota in the distal parts of the intestinal tract.

The microflora of traditional sourdoughs is typically composed of two to four strains of lactobacilli, and the dominant species in traditional sourdoughs is *Lb. sanfranciscensis* (Hammes *et al.*, 1996; Vogel *et al.*, 1999). Sucrose is the major fermentable carbohydrate in wheat and rye grains but the majority of strains of *Lb. sanfranciscensis* are not capable of sucrose hydrolysis (Hammes *et al.*, 1996). Remarkably, each sourdough harbours at least one strain capable of EPS formation via extracellular glycansucrases (Tieking *et al.*, 2003). Sucrose turnover by extracellular glucansucrases and fructansucrases invariably results in the (extracellular) liberation of frucrose and glucose, respectively, from sucrose. These products become available for EPS-negative and sucrose-negative lactobacilli that are present in the same dough.

Bacteriophage resistance and stress tolerance of EPS-producing lactic acid bacteria

Reports considering the sensitivity of EPS forming LAB towards bacteriophages are not consistent. Several reports have linked exopolysaccharide formation by LAB to phage resistance. HePS produced by *Lc. lactis* moderately protected against infection by phages ML3, ø712 and øc2; protection from the latter two phages was attributed to interference with phage adsorption to cell surface receptors (Looijesteijn *et al.*, 2001; Forde and Fitzgerald, 2003). Other reports, however, did not indicate a role of EPS in phage resistance of *Lc. lactis* (Deveau *et al.*, 2002). Capsular polysaccharides produced by *S. thermophilus* increased the sensitivity to phage infection (Rodriguez *et al.*, 2008). These inconsistent results may be attributed to the heterogenous monosaccharide composition of HePS as well as differences in cell surface recognition by individual phages (Broadbent *et al.*, 2003).

HePS and HoPS production of LAB improves their survival at adverse environmental conditions. Formation of a HePS by *Lc. lactis* protected the producing bacteria against copper, nisin, and lysozyme (Looijesteijn *et al.* 2001) and an encapsulated strain of *S. thermophilus* exhibited improved survival in ice-cream compared to non-encapsulated mutants (Hong and Marshall, 2001). The comparison of isogenic gtf$^+$ and gtf$^-$ variants of *Lc. lactis* and *P. parvulus* demonstrated that the formation of β-glucans by wine-spoiling pediococci and oenoccci improves the survival at low pH as well as in the presence of ethanol and SO_2, stress factors occurring during wine making (Dols-Lafargue *et al.*, 2008).

In oral streptococci, cells grown in a high cell-density biofilm are more resistant to acid and oxidative stress than planktonic-phase cells (Cvitkovitch *et al.*, 2003). Dextran, levan and FOS added at 50 g/l improved survival *Lb. reuteri* towards the membrane-active inhibitors nisin, reutericyclin, and CCCP (Schwab and Gänzle, 2006). Likewise, dextran formation by *Lu. mesenteroides* provided protection during long-term survival at acidic conditions (Kim *et al.*, 2000).

Fructans and FOS protect biological membranes during dehydration or freezing (Schwab *et al.*, 2007; Crowe *et al.*, 2001; Vereyken *et al.*, 2003a and 2003b). This protective effect can be attributed to specific interactions between fructans and the lipid headgroups of biological membranes which maintain the membrane in the physiological liquid-crystalline phase under adverse conditions. Comparable effects of glucans and gluco-oligosaccharides have been reported (Crowe *et al.*, 2001). Protection of bacterial membranes by fructo-oligosaccharides can be exploited in commercial applications for the protection of starter cultures and probiotic cultures during spray drying, freeze drying, or frozen storage (Corcoran, *et al.*, 2004; Schwab *et al.*, 2007). The protective effect of fructo-oligosaccharides on survival of freeze dried *Lb. reuteri* was superior when compared with sucrose, indicating that the conversion of sucrose to fructans and fructo-oligosaccharides by bacterial fructosyltransferases improves survival of the producer strain (Fig. 15.2) (Schwab *et al.*, 2007).

Bacterial EPS and biofilm formation

Biofilms are bacterial communities whose cells are attached to a surface or to each other. Biofilm are generally characterized by the presence of a polymeric extracellular matrix, which can be supplied by polysaccharides formed by the biofilm community itself (Donlan and Costerton, 2002). Biofilm formation proceeds by the attachment of individual cells to a surface, followed by the formation of microcolonies and the production of an extracellular polymeric matrix that allows maturation to a three-dimensional structure. Bacterial biofilm formation is instrumental for the degradation of solid substrates and the persistence of bacteria in natural aquatic ecosystems. Moreover, biofilms are involved in bacterial pathogenesis (O'Toole *et al.*, 2000) and the interaction of animals and plants with commensal or symbiotic bacteria (Rudrappa *et al.*, 2008; Macfarlane and Dillon, 2007).

Several exopolysaccharides of lactic acid bacteria were shown to contribute to the formation of biofilms *in vitro* (Leeber *et al.*, 2007; Leather and Côté, 2007) and β-glucan formation by wine-spoiling pediococci and oenococci increased adhesion to surfaces as a prerequisite for biofilm formation (Dols-Lafargue *et al.*, 2008). The ecological role of biofilm formation and the contribution of EPS to biofilms of lactic acid bacteria is best understood for the adherence and persistence of streptococci to tooth surfaces, the persistence of lactobacilli in the proximal parts of the gastrointestinal tract of animals, and the microbial communities within kefir grains. All three biofilm communities have in common that the EPS and/or glycansucrases of lactic acid bacteria play a major role in maintaining and forming the biofilm.

Streptococci colonization on tooth surfaces: a biofilm formed by HoPS

Bacterial HoPS, extracellular glucosyltransferase and glucan binding proteins (Gbp) are integral components of *S. mutans* biofilms on tooth surfaces. Soluble and insoluble glucans synthesized by glycansucrases form the main component of the extracellular polysaccharide matrix; accordingly, biofilm formation is strongly enhanced by the presence of sucrose (Tam *et al.*, 2007). The glucosyltransferases GtfB and GtfC are cell

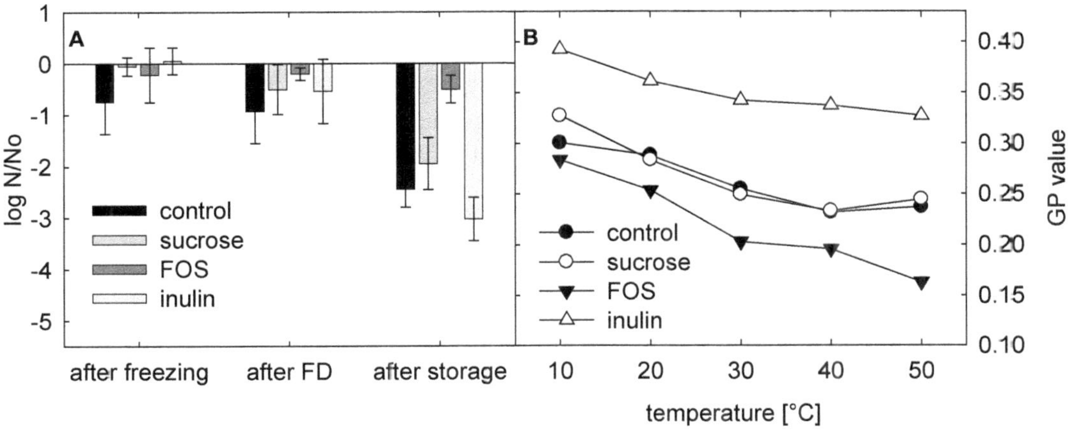

Figure 15.2 Effect of sucrose, fructo-oligosaccharides, and inulin on dry survival of *Lb. reuteri* TMW1.106 (Panel A) and the fluidity of the membrane (Panel B). **A** Survival of stationary phase *Lb. reuteri* TMW1.106 after freezing, freeze-drying and storage in phosphate buffer (PB) or PB with 7.5%, sucrose, FOS, or inulin. **B** Membrane fluidity of *Lb. reuteri* as measured by the laurdan general polarization (GP) at temperatures ranging from 10 to 50 °C. Cells of *L. reuteri* TMW1.106 were incubated in PB (control) or in PB containing 7.5% sucrose, FOS, or inulin (Schwab *et al.*, 2007).

surface associated and produce mainly water-insoluble glucans rich in α-(1–3) linkages (>85%, mutan). GtfD is released in cell supernatant and forms water-soluble, mainly α-(1–6) linked glucan (>70%, dextran) (Monchois *et al.*, 1999). Inactivation of either of the enzymes and thus loss of the respective polysaccharide decreases attachment. Earlier results indicated that the polysaccharide formed by GtfC contributes the most to the attachment process (Cross *et al.*, 2007; Fukushima *et al.*, 1992; (Tsumori and Kuramitsu, 1997). In *S. gordonii*, the single GtfG takes on the duties of the three *S. mutans* Gtfs: it produces α-(1–6) and α-(1–3) linked glucans and it possesses glucan binding capabilities. Loss of GtfG does not affect initial colonization but reduces the persistence on the tooth surface (Tanzer *et al.*, 2008).

The role of the fructan formed by Ftf in oral streptococci is mainly that of a storage polysaccharide, but it also seems to take part in the adhesion process (Rozen *et al.*, 2001). The fructosyltransferase Ftf of *S. mutans* shows affinity for the glucans formed by *S. mutans* and may therefore be involved in formation of the network (Rozen *et al.*, 2004). Inactivation of Ftf results in reduced biofilm depth and cariogenicity (Schroeder *et al.*, 1989).

Four different groups of membrane-bound glucan binding proteins (Gbp) have been identified. Secreted or cell bound Gbps are involved in virulence and biofilm architecture. Their role is to link bacteria and extracellular glucans or to cross-link glucans for the formation of a network (Banas and Vickerman, 2003; Lynch *et al.*, 2007). The cell bound GbpC acts as the primary binder of glucan, and its loss strongly affects biofilm structure and biofilm depth (Lynch *et al.*, 2007). Glucosyltransferases and some glucan binding proteins possess homologous glucan-binding domains, indicating that glucosyltransferases themselves contribute to cellular adhesion to the extracellular matrix (Banas and Vickerman, 2003; Cross *et al.*, 2007).

Biofilms formed by *Lb. reuteri* within proximal gastrointestinal tract: the role of glycansucrases

Lactobacilli dominate the microflora in the proximal digestive tracts of rodents, pigs and birds (Brooker and Fuller, 1975; Mitsuoka, 1992; Tannock, 1992). Autochthonous lactobacilli directly adhere to the non-secretory stratified squamous epithelium which lines the forestomach of rodents, the oesophagus of pigs and the crops of birds, and form stable biofilms which persist throughout the life of the host animal (Tannock, 2004; Walter, 2006; Walter, 2008). In contrast to the colonization of tooth surfaces of mutans streptococci, biofilm formation of lactobacilli occurs in the virtual absence of sucrose, implying different functions of bacterial glycansucrases (Walter *et al.*, 2008). Nevertheless, the relevance of glycansucrases of *Lb. reuteri* epithelium in maintenance of the strain in the gastrointestinal tract has been shown (Walter *et al.*, 2008). In vivo expression of the glycansucrases of *Lb. reuteri* was a first indicator of the ecological functinality of the enzymes. The strains *Lb. reuteri* LTH5448 and TMW1.106 possess the levansucrase FtfA and the reuteransucrase GtfA and inulosucrase Inu, respectively, and establish stable populations in the forestomach of ex-lactobacilli mice. Transcription studies verified that *ftfA* and *gtfA* are expressed *in vivo* (Fig. 15.3) (Schwab, 2006). Transcription of *inu* was not detectable probably due to the detection limit of real-time PCR. However, *ftfA* and *gtfA* were transcribed in five of five host animals (Fig. 15.3) by bacteria scraped from the forestomach epithelium or present in the luminal contents. Expression levels varied between host animals but were not different in relation to *in vitro* expression. Mice were fed with commercially available rat chow which contained mainly maltose and glucose and only traces of raffinose and sucrose (Walter *et al.*, 2008), excluding sucrose induction of glycansucrases. Inactivation of *gtfA* and *inu* of *Lb. reuteri* TMW1.106 reduced the competitiveness of the *inu* mutant when inoculated together with wild-type *Lb. reuteri* TMW1.106, and decreased competitiveness of the *gtfA* mutant when grown with the EPS-negative *Lb. johnsonii* #21 (Walter *et al.*, 2008). Deletion of *gtfA* or *inu* strongly affected *in vitro* biofilm formation in the presence of sucrose. Lack of *inu* or *gtfA* decreased the sucrose dependent autoaggregation of the mutant strains compared with the wild type and resulted in impaired biofilm formation (Walter *et al.*, 2008). Based on those observations, Walter *et al.* (2008) proposed different roles of Inu and GtfA as glucan binding protein and glucan synthesizer, respectively (Fig. 15.4).

The kefir grain: a polysaccharide matrix provided by the heteropolysaccharide kefiran

Microbial grains, often referred to as kefir grains, are used as starters for the production of fermented alcoholic milk beverages. The grains are added to goat or cows milk and are fermented at mesophilic conditions to enable a lactic acid and an alcoholic fermentation. Kefir originates from the Caucasus region and is now available world wide; similar products exist in Mexico (Bulgaros) or

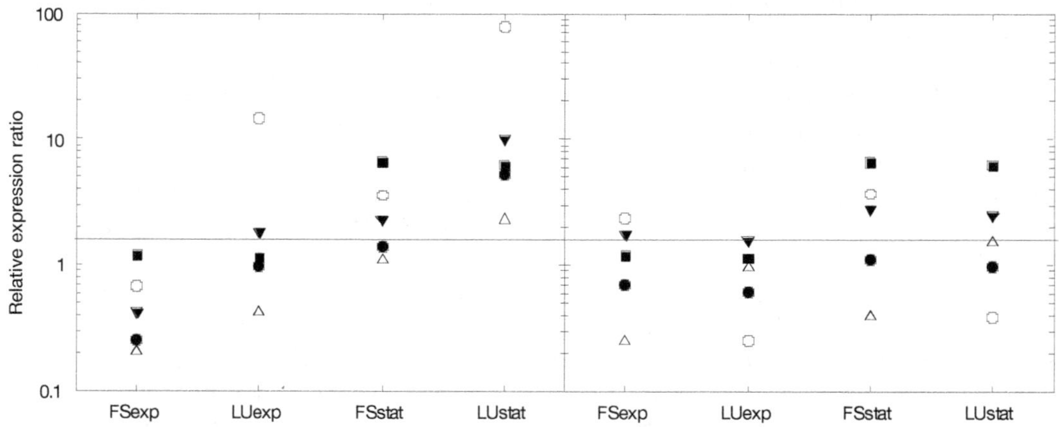

Figure 15.3 Relative expression ratios of **A** *gtfA* and **B** *ftfA* of *Lb. reuteri* TMW1.106 and LTH5448, respectively, isolated from forestomach epithelium (FS) and luminal (LU) stomach contents of five reconstituted lactobacilli-free mice in relation to expression of glycansucrase genes in *Lb. reuteri* strains grown to exponential (exp) or early stationary (stat) growth state in MRS. The 16SrRNA gene was used as reference gene. Identical symbols represent expression ratios within the same animal. mRNA was isolated, reverse transcribed and genes of interest were amplified by PCR (Primer: 16SrRNA gene Lac1 5′-AGCAGTAGGGAATCTTCCA-3′, Lac2 5′-ATTCCACCGCTACACATG-3′ (Walter *et al.* 2001), *ftfA* FTF-rt-forward 5′-AGCCGACAATGATGGTGT C-3′, FTFrevTEST 5′-GGCATTTGTACAACATGTCC, *gtfA* DexforTEST 5′-AATTAAACTGGTTATACTATCTC-3′, Dex-rt-reverse 5′-GAGTTCATACCATCTGCAGC-3′). Specificity of primers and absence of chromosomal DNA was confirmed (Schwab, 2006).

Brazil (Wszolek *et al.*, 2006). Motaghi *et al.* (1997) showed that it is possible to grow kefir grains on the skin of a sheep stomach bag. Pasteurized milk and sheep intestinal microflora were incubated at mesophilic temperature. The coagulate was frequently replaced with fresh milk. After a period of three months, the polysaccharide-protein matrix had developed. The polysaccharide – kefiran- is composed of equimolar amounts of glucose and galactose (Kooiman, 1968). In *Lb. kefiranofaciens,* a strain frequently associated with kefiran formation, the polysaccharide forms a capsule around the bacterial cell (Mukai, 1990); Frengova *et al.* (2002) isolated kefiran producing *Lb. delbrueckii* subsp. *bulgaricus* from Bulgarian kefir grains.

Within the grain lactic acid bacteria (lactobacilli, leuoconostoc and lactococci), acetic acid bacteria (*Acetobacter* spp.) and yeasts (*Zygosaccharomyces, Candida,* and *Saccharomyces*) are present. The microbial flora is dominated by lactic acid bacteria, with *Lb. kefiri, Lb. kefiranofaciens, Lb. delbruecki* subsp. *bulgaricus, Lu. mesenteroides* subsp. and *Lc. lactis* subsp. being prominent representatives, however, the cultivation method of the kefir grains and the regional origin clearly affect the microflora composition (Witthuhn *et al.*, 2005; Chen *et al.*, 2008; Frengova *et al.*, 2002; Pintado *et al.*, 1996; Witthuhn *et al.*, 2004; Motaghi *et al.*, 1997; Seydim-Guzel *et al.*, 2005). LAB, acetic acid bacteria and yeast live in a symbiotic community. EPS forming lactobacilli provide the polysaccharide matrix. Lactose utilization by lactic acid bacteria results in a pH drop, which enhances yeast growth. In return, the yeasts form alcohol, deliver peptides through hydrolysis of milk proteins, and generate an anaerobic atmosphere. Yeasts as well as *Acetobacter* spp., are capable of synthesizing vitamin B (Wszolek *et al.*, 2006). Increased kefiran formation in mixed culture of *Lb. kefiranofaciens* and *Saccharomyces cerevisiae* has been related to physical contact between yeast and lactobacilli cells (Cheirsilp *et al*, 2003; Frengova *et al.*, 2002).

Conclusion

A high proportion of EPS-forming strains in isolates from a given ecosystem likely reflects a competitive advantage of EPS-positive strains over non-producing strains. The high frequency of EPS-producing LAB in the oral cavity and intestinal ecosystems thus argues in favour for an important role of EPS formation for the persistence of LAB in those habitats. Moreover, LAB involved in food fermentation and food spoilage appear to benefit from EPS formation, e.g. in specific dairy fermentations and in wine spoilage. Additional clues for the ecological relevant of EPS formation can be derived form the regulatory network of EPS biosynthetic genes. The intricate regulatory network controlling the expression of glycansucrases in oral streptococci clearly is

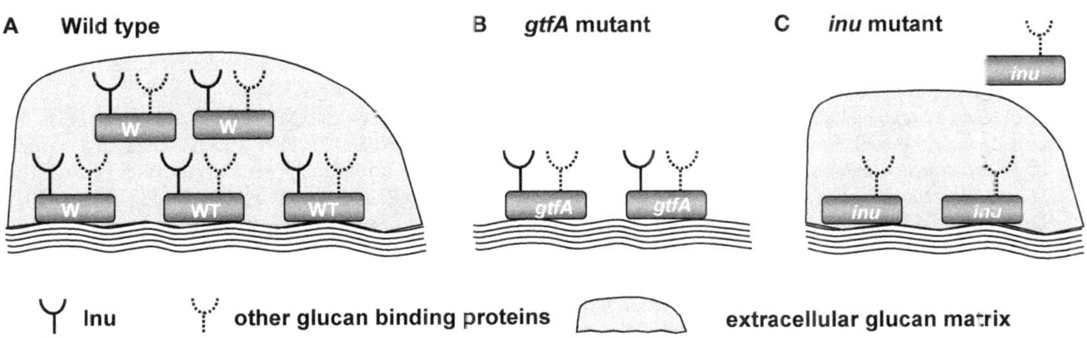

Figure 15.4 Model for biofilm formation by *Lb. reuteri* in the forestomach of mice. The drawing is based on studies of *in vitro* and *in vivo* colonization experiments with reconstituted lactobacilli-free mice (Walter *et al.*, 2008). **A** GtfA (reuteransucrase) activity in the wild-type strain contributes to the formation of an extracellular matrix, the inulosucrase Inu acts as one of the glucan-binding proteins. **B** The *gtfA* mutant strain lost the ability for formation of an extracellular matrix and exhibits delayed colonization when administered alone but retains the ability to adhere to the biofilm when the glucan-forming wild-type strain is present. **C** The *inu* mutant strain has a reduced ability to adhere to the extracellular matrix and is washed out when the glucan-forming wild-type strain is present.

in keeping with the contribution of EPS and extracellular glycansucrases to biofilm formation and persistence in the oral cavity. The regulation of glycansucrases in *Lb. reuteri* argues in favour of HoPS and FOS production as a response of lactobacilli towards stress sensed by the cytoplasmic membrane. The products of glycansucrases are likely to improve the survival of lactobacilli in a scenario characterized by strong fluctuations in water activity, temperature, pH, and nutrient supply, and the presence of natural inhibitors as they are likely to occur in decaying plants upon injury of the plant tissue or during gastrointestinal transit. Because the expression of glycansucrases in many strains of lactobacilli and *Leuconostoc* species is induced by sucrose, the contribution of glycansucrases to sucrose catabolism may be their main ecological role in some strains.

Taken together, there is no single answer to the question why lactic acid bacteria produce EPS. Biofilm formation, stress resistance as well as sucrose utilization are clearly linked to the formation of HePS or HoPS in individual strains. In many cases, EPS producing enzymes, particularly glycansucrases, exhibit multiple functionality, e.g. biofilm formation, stress resistance, and sucrose utilization in *Lb. reuteri*.

References

Årsköld, E., Svensson, M., Grage, H., Roos, S., Rådström, P., and van Niel, E.W.J. (2007). Environmental influences on exopolysaccharide formation in *Lactobacillus reuteri* ATCC55730. Int. J. Food Microbiol. *116*, 159–167.

Banas, J.A., and Vickerman, M.M. (2003). Glucan-binding proteins of the oral strepcocci. Crit. Rev. Oral Biol. Med. *14*, 89–99.

Birkhead, D., Rosell, K.G., and Granath, K. (1979). Structure of extracellular water soluble polysaccharides synthesized from sucrose by oral strains of *Streptococcus mutans*, *Streptococcus salivarius*, *Streptococcus sanguis*, and *Actinomyces viscosus*. Arch. Oral Biol. *24*, 53–61.

Boels, I.C., Kleerebezem, M., and de Vos, W. (2003a). Engineering of carbon distribution between glycolysis and sugar nucleotide biosynthesis in *Lactococcus lactis*. Appl. Environm. Microbiol. *69*, 1129–1135.

Boels, I.C., van Kranenburg, R., Kanning, M.W., Fong Chong, B., de Vos, W., and Kleerebezem, M. (2003b). Increased exopolysaccharide production in *Lactococcus lactis* due to increased levels of expression of the NIZO B40 *eps* gene cluster. Appl. Environm. Microbiol. *69*, 5029–5031.

Boels, I.C., Ramos, A., Kleerebezem, M., and de Vos, W. (2001). Functional analysis of the *Lactococcus lactis* galU and galE genes and their impact on sugar nucleotide and exopolysaccharide biosynthesis. Appl. Environm. Microbiol. *67*, 3033–3040.

Broadbent, J.R., McMahon, D.J., Welker, D.L., Oberg, C.J., and Moineau, S. (2003). Biochemistry, genetics, and applications of exopolysaccharide production in *Streptococcus thermophilus*: A review. J. Dairy Sci. *86*, 407–423.

Brooker, B.E., and Fuller, R. (1975). Adhesion of lactobacilli to the chicken crop epithelium. J. Ultrastruct. Res. *52*, 21–31.

Brooker, B.E. (1977). Ultrastructural surface changes associated with dextran synthesis by *Leuconostoc mesenteroides*. J. Bact. *131*, 288–292.

Browngardt, C.M., Wen, Z.R., and Burne, R.A. (2004). RegM is required for optimal fructosyltransferase and glucosyltransferase expression in *Streptococcus mutans*. FEMS Microbiol. Lett. *240*, 75–79.

Burne, R.A., Schilling, K., Bowen, W., and Yasbin, R.E. (1987). Expression, purification, and characterization of an exo-β-D-fructosidase of *Streptococcus mutans*. J. Bacteriol. *169*, 4907–4911.

Cerning, J. (1990). Exocellular polysaccharides produced by lactic acid bacteria. FEMS Microbiol. Rev. *87*, 113–130.

Cheirsilp, B., Shoji, H., Shimizu, H., and Shioya, S. (2003). Interactions between *Lactobacillus kefiranofaciens* and *Saccharomyces cerevisiae* in mixed culture for kefiran production. J. Biosc. Bioeng. *96*, 279–284.

Chen, H.-C., Wang, S.-Y., and Chen, M.-J. (2008). Microbiological study of lactic acid bacteria in kefir grains by culture-dependent and culture-independent methods. Food Microbiol. *25*, 492–501.

Chen, P.-M., Chen, J.-Y., and Chia, J.-S. (2006). Differential regulation of *Streptococcus mutans gtfBCD* genes in response to copper ions. Arch. Microbiol. *185*, 127–135.

Corcoran, B.M., R.P. Ross, G.F. Fitzgerald, and Stanton, C. (2004). Comparative survival of probiotic lactobacilli spray-dried in the presence of prebiotic substances, J. Appl. Microbiol. *96*, 1024–1039.

Crociani, F., Biavati, B., Alessandrini, A., Chiarini, C., and Scardovi, V. (1996). *Bifidobacterium inopinatum* sp. *nov.* and *Bifidobacterium denticolens* sp. *nov.* two new species isolated from human dental caries. Int. J. Syst. Bacteriol. *46*, 564–571.

Crowe, J.H., Crowe, L.M., Oliver, A.E., Tsvetkova, M., Wolkers, W., and Tablin, F. (2001). The trehalose myth revisited: introduction to a symposium on stabilization of cells in the dry state. Cryobiology *43*, 89–105.

Cross, S.E., Kreth, J., Zhu, L., Sullivan, R., Shi, W., Qi, F., and Gimzewski, J.K. (2007). Nanomechanical properties of glucans and associated cell-surface adhesion of *Streptococcus mutans* probed by atomic force microscopy under *in situ* conditions. Microbiology *153*, 3124–3132.

Cvitkovitch, D.G., Li, Y.H., and Ellen, R.P. (2003). Quorum sensing and biofilm formation in streptococcal infections. J. Clin. Invest. *112*, 1626–1632.

Dabour, N., and LaPointe, G. (2005). Identification and molecular characterization of the chromosomal exopolysaccharide biosynthesis gene cluster from *Lactococcus lactis* subsp. *cremoris* SMQ-461. Appl. Environm. Microbiol. *71*, 7414–7425.

de Vuyst, L., de Vin, F., Vaningelem, F., and Degeest, B. (2001). Recent developments in the biosynthesis and applications of heteropolysaccharides from lactic acid bacteria. Int. Dairy J. *11*, 687–707.

Degeest, B., Mozzi, F., and de Vuyst, L. (2002). Effect of medium composition and temperature and pH changes on exopolysaccharide yields and stability during *Streptococcus thermophilus* LY03 fermentations. Int. J. Food Microbiol. 79 (3) 161–174.

Degeest, B., Vaningelgem, F., and de Vuyst, L. (2001). Microbial physiology, fermentation kinetics, and process engineering of heteropolysaccharide production by lactic acid bacteria. Int. Dairy J. *11*, 747–757.

Deveau, H., van Calsteren, M.-R., Moineau, S. (2002). Effect of Exopolysaccharides on Phage-Host Interactions in Lactococcus lactis. *Appl. Environ. Microbiol. 68*, 4364–4369.

Dols-Lafargue, M., Lee, H.Y., Le Marrec, C., Heryaud, A., Chambat, G., and Lonvaud-Funel, A. (2008). Characterization of gtf, a glucosyltransferase gene in the genomes of *Pediococcus parvulus* and *Oenococcus oeni*, two bacterial species commonly found in wine. Appl. Environ. Microbiol. *74*, 4079–4090.

Dols, M., Chraibi, W., Remaud-Simeon, Lindley, N.D., and Monsan, P.F. (1997). Growth and energetics of *Leuconostoc mesenteroides* NRRL B-1299 during metabolism of various sugars and their consequences for dextransucrase production. Appl. Environm. Microbiol. *63*, 2159–2165.

Dols, M., Remaud-Simeon, M., Willemot, R.-M., Vignon, M.R., and Monsan, P.F. (1998). Structural characterization of the maltose acceptor-products synthesized by *Leuconostoc mesenteroides* NRRL B-1299 dextransucrase. Carbohydr. Res. *305*, 549–559.

Donlan, R.M., and Costerton, J.W. (2002). Biofilms: Survival mechanism of clinically relevant microorganisms. Clin. Microbiol. Rev. *15*, 167–193.

Forde, A., and Fitzgerald, G.F. (1999) Analysis of exopolysaccharide (EPS) production mediated by the bacteriophage adsorption blocking plasmid pCI658, isolated from Lactococcus lactis subsp. cremoris HO2. Int. Dairy J. *9*, 465–472.

Frengova, G.I., Simova, E., Beshkova, D.M., and Simov, Z.I. (2002). Exopolysaccharides produced by lactic acid bacteria of kefir grains. Z. Naturforsch. *57c*, 805–810.

Fujiwara, T., Hashino, T., Oshima, T., and Hamada, S. (2002). Differential and quantitative analysis of mRNA expression of glucosyltransferases from *Streptococcus mutans* MT8148. J. Dent. Res. *81*, 109–113.

Fujiwara, T., Tamesada, M., Bian, Z., Kawabata, S., Kimura, S., and Hamada, S. (1996). Deletion and reintroduction of glucosyltransferase genes of *Streptococcus mutans* and role of their gene products in sucrose dependent cellular adherence. Microb. Pathog. *20*, 225–233.

Fukushima, K., Ikeda, T., and Kuramitsu H.K. (1992). Expression of *Streptococcus mutans* gtf genes in *Streptococcus milleri*. Infect. Immun. *60*, 2815–2822.

Gänzle, M.G., and Schwab, C. (2005). Exopolysaccharide production by intestinal lactobacilli. In Probiotics and prebiotics: scientific aspects, G.W. Tannock, ed (Norfolk, UK, Caister Academic Press), pp. 83–96.

Gindreau, E., Walling, E., and Lonvaud-Funel, A. (2001). Direct polymerase chain reaction detection of ropy *Pediococcus damnosus* strains in wine. J. Appl. Microbiol. *90*, 535–542.

Grey, W.T., Curtiss, R., III and Hudson, M.C. (1997) Expression of the *Streptococcus mutans* fructosyltransferase gene within a mammalian host. Infect. Immun. *65*, 2488–2490.

Hammes, W.P., Stolz, P., and Gänzle, M.G. (1996). Metabolism of lactobacilli in traditional sourdoughs. Adv. Food Sci. *18*, 176–184.

Hiratsuka, K., Wang, B., Sato, Y., and Kuramitsu, H. (1998). Regulation of sucrose-6 phosphate hydrolase activity in *Streptococcus mutans*: Characterization of the *scrR* gene. Infect. Immun. *66*, 3736–3743.

Hong, S.H., and Marshall, R.T. (2001). Natural exopolysaccharides enhance survival of lactic acid bacteria in frozen dairy dessert. J. Dairy Sci. *84*, 1367–1374.

Ibarburu, I., Soria-Díaz, M.E., Rodríguez-Carvajal, M.A., Velasco, S.E., Tejero-Mateo, P., Gil-Serrano, A.M., Irastorza, A., and Dueñas, M.T. (2007). Growth and exopolysaccharide (EPS) production by *Oenococcus oeni* I4 and structural characterization of their EPSs. J. Appl. Microbiol. *103*, 477–486.

Kaster, A.G., and Brown, L.R. (1983). Extracellular dextranase activity produced by human oral strains of the genus *Bifidobacterium*. Infect. Immun. *42*, 716–720.

Kim, D.S., Thomas, S., and Fogler, H.S. (2000). Effects of pH and trace minerals on long-term starvation of *Leuconostoc mesenteroides*. Appl. Environ. Microbiol. *66*, 976–981.

Kim, S.S., and Fogler, H.S. (1999). The effects of exopolymers on cell morphology and culturability of *Leuconostoc mesenteroides* during starvation. Appl. Microbiol. Biotechnol. *52*, 839–844.

Kiska, D.L., and Macrina, F.L. (1994). Genetic regulation of fructosyltransferase in *Streptococcus mutans*. Infect. Immun. *62*, 1241–1251.

Kooiman, P. (1968). The chemical structure of kefiran, the watersoluble polysaccharide of the kefir grain. Carbohydr. Res. *7*, 200–211.

Korakli, M., and Vogel, R.F. (2006). Structure/function relationship of homopolysaccharide producing glycansucrases and therapeutic potential of their synthesised glycans. Appl. Microbiol. Biotechnol. *71*, 790–803.

Korakli, M., Gänzle, M.G., and Vogel, R.F. (2002). Metabolism by bifidobacteria and lactic acid bacteria of polysaccharides from wheat and rye, and exopolysaccharides produced by *Lactobacillus sanfranciscensis*. J. Appl. Microbiol. 92, 958–965.

Korakli, M., Rossmann, A., Gänzle, M., and Vogel, R.F. (2001). Sucrose metabolism and exopolysaccharide production in wheat and rye sourdoughs by *Lactobacillus sanfranciscensis*. J. Agric. Food Chem. *49*, 5194–5200.

Kralj, S., van Geel-Schutten, G.H., van der Maarel, M.J.E.C., and Dijkhuizen, L. (2004a). Biochemical and molecular characterization of *Lactobacillus reuteri* 121 reuteransucrase. Microbiol. *150*, 2099–2112.

Kralj, S., van Geel-Schutten, G.H., Dondorff, M.M.G., Kirsanovs, S., van der Maarel, M.H E.C., and Dijkhuizen, L. (2004b). Glucan synthesis in the genus

Lactobacillus: isolation and characterization of glucansucrase genes, enzymes and glucan products from six different strains. Microbiology *150*, 3681–3690.

Laws, A., and Marshall, V.M. (2001). The relevance of exopolysaccharides to the rheological properties in milk fermented with ropy strains of lactic acid bacteria. Int. Dairy J. *11*, 709–721.

Leather, T.D., and Côté, G.L. (2008). Biofilm formation by exopolysaccharide mutants of *Leuconostoc mesenteroides* strain NRRL B-1355. Appl. Microbiol. Biotechnol. *78*, 1025–1031.

Leathers, T. (2002). Dextran. In Biopolymers, Steinbüchel, A., Baets, S.D., and Vandamme, E.H, eds. (Weinheim, Germany, Wiley VCH,), pp. 299–322.

Lee, S.F., Delaney, G.D., and Elkhateeb, M. (2004). A two-component *covRS* regulatory system regulates expression of fructosyltransferase and a novel extracellular carbohydrate in *Streptococcus mutans*. Infect. Immun. *72*, 3968–3973.

Leeber, S., Verhoeven, T.L.A., Perea Vélez, M., Vanderleyden, J., and de Keersmaecker, S.C.J. (2007). Impact of environmental and genetic factors on biofilm formation by the probiotic strain *Lactobacillus rhamnosus* GG. Appl. Environm. Microbiol. *73*, 6768–6775.

Lonvaud-Funel, A. (1999). Lactic acid bacteria in the quality improvement and depreciation of wine. Antonie van Leeuwenhoek *76*, 317–334.

Looijesteijn P.J., Trapet, L., de Vries, E., Abee, T., and Hugenholtz, J. (2001) Physiological function of exopolysaccharides produced by *Lactococcus lactis*. Int. J. Food Microbiol. *64*, 71–80.

Looijesteijn, P.J., Boels, I.C., Kleerebezem, M., and Hugenholtz, J. (1999). Regulation of exopolysaccharide production by *Lactococcus lactis* subsp. *cemoris* by the sugar source. Appl. Environm. Microbiol. *65*, 5003–5008.

Lopitz-Ostoa, F., Rementeria, A., Elguzabal, N., and Garaizar, J. (2006). Kefir: A symbiotic yeasts-bacteria community with alleged healthy capabilities. Rev. Iberoam. Micol. *23*, 67–74.

Lynch, D.J., Fountain, T.L., Mazurkiewicz, J.E., and Banas, J.A. (2007). Glucan-binding proteins are essential for shaping *Streptococcus mutans* biofilm architecture. FEMS Microbiol. Lett. *268*, 158–165.

Macfarlane, S., and Dillon, J.F. (2007). Microbial biofilm in the human gastrointestinal tract. J. Appl. Microbiol. *102*, 1187–1196.

Marx, S.P., Winkler, S., and Hartmeier, W. (2000). Metabolization of β-(2,6)-linked fructose-oligosaccharides by different bifidobacteria. FEMS Microbiol. Lett. *182*, 163–169.

Mitsuoka, T. (1992). The human gastrointestinal tract. In The Lactic Acid Bacteria, vol. 1. The Lactic Acid Bacteria in Health and Disease, Wood, B.J.B., ed. (London, UK, Elsevier Appl. Sci.), pp. 69–114.

Modesto, M., Biavati, B., and Mattarelli, P. (2006). Occurrence of the family of *Bifidobacteriaceae* in human dental caries and plaque. Caries Res. *40*, 271–276.

Monchois, V., Willemot, R.M., and Monsan, P. (1999). Glucansucrases: mechanism of action and structure-function relationships. FEMS Microbiol. Rev *23*, 131–151.

Morona, J.K., Morona, R., and Paton, J.C. (2006). Attachment of capsular polysaccharide to the cell wall of *Streptococcus pneunomiae* type 2 is required for invasive disease. Proc. Natl. Aca. Sci. USA, *103*, 8505–8510.

Motaghi, M., Mazaheri, M., Moazami, N., Farkhondeh, A., Fooladi, M.H., and Goltapeh, E.M. (1997). Kefir production in Iran. World J. Microbiol. Biotechn. *13*, 579–581.

Mozzi, F., Vaningelgem, F., Hébert, E.M., van der Meulen, R., Remedios Fouliqué Moreno, M., Font de Valdez, G., and de Vuyst, L. (2006). Diversity of heteropolysaccharide-producing lactic acid bacterium strains and their biopolymers. Appl. Environm. Microbiol. *72*, 4431–4435.

Mukai, T., Toba, T., Itoh, T., and Adachi, S. (1990). Structural investigation of the capsular polysaccharide from *Lactobacillus kefiranofaciens*. Carbohydr. Res. *204*, 227–232.

Neely, W., and Nott, J. (1962). Dextransucrase, an induced enzyme from *Leuconostoc mesenteroides*. Biochemistry *1*, 1136–1140.

Neubauer, H., Bauche, A., and Mollet, B. (2003). Molecular characterization and expression analysis of the dextransucrase DsrD of *Leuconostoc mesenteroides* Lcc4 in homologous and heterologous *Lactococcus lactis* cultures. Microbiol. *149*, 973–982.

Nierop-Groot, M.N., and Kleerebezem, M. (2007). Mutational analysis of the *Lactococcus lactis* NIZO B40 exopolysaccharide (EPS) gene cluster: EPS biosynthesis correlates with unphosphorylated EpsB. J. Appl. Microbiol. *103*, 2645–2656.

O'Toole, G., Kaplan, H.B., and Kolter, R. (2005). Biofilm formation as microbial development. Ann. Rev. Microbiol. *54*, 49–79.

Oba, T., Higashimura, M., Iwasaki, T., Matser, A.M., Steeneken, P.A.M., Robijin, G.W., and Sikkema, J. (1999). Viscoelastic properties of aqueous solutions of the phosphopolysaccharide 'viilian' from *Lactococcus lactis* subsp. *cremoris* SBT 0495. Carbohydr. Pol. *39*, 275–281.

Olivares-Illana, V., López-Munguia, A., and Olvera, C. (2003). Molecular characterization of inulosucrase from *Leuconostoc citreum*: a fructosyltransferase within a glucosyltransferase. J. Bacteriol. *185*, 3606–3612.

Orla-Jensen, S. (1943), The lactic acid bacteria. In Biologiske Skrifter Bind II Nr. 3, Kopenhagen, Denmark, Det Kongelige Danske Videnskabernes Selskab.

Péant, B., LaPointe, G., Gilbert, C., Atlan, D., Ward, P., and Roy, D. (2005). Comparative analysis of the exopolysaccharide biosynthesis gene clusters from four strains of *Lactobacillus rhamnosus*. Microbiol. *151*, 1839–1851.

Pfaffl, M.W., Horgan, G.W., and Dempfle, L. (2002). Relative expression software tool (REST) for group-wise comparison and statistical analysis of relative expression results in real-time PCR. Nuc. Acids. Res. *30*, e36.

Pham, P.L., Dupont, I., Roy, D., Lapointe, G., and Cerning, J. (2000). Production of exopolysaccharide by *Lactobacillus rhamnosus* R and analysis of its enzymatic degradation during prolonged fermentation. Appl. Environm. Microbiol. *66*, 2302–2310.

Pintado, M.E., Lopes Da Silva, J.A., Fernandes, P.B., Xavier Malcata, F., and Hogg, T.A. (1996). Microbiological and rheological studies on Portugese kefir grains. Int. J. Food Sci. Tech. *31*, 15–26.

Quirasco, M., López-Munguía, A., Remauld-Simeon, M., Monsan, P., and Farrés, A. (1999). Induction and transcription studies of the dextransucrase gene in *Leuconostoc mesenteroides* NRRL B-512F. Appl. Environm. Microbiol. *65*, 5504–5509.

Rodríguez, C., Van der Meulen, R., Vaningelgem, F., Font de Valdez, G., Raya, R., and de Vuyst, L. (2008). Sensitivity of capsular-producing *Streptococcus thermophilus* strains to bacteriophage adsorption. Lett. Appl. Microbiol. *46*, 462–468.

Rozen, R., Bachrach, M., Bronshteyn, M, Gedalia, I., and Steinberg, D. (2001). The role of fructans on dental biofilm formation by *Streptococcus sorbinus*. *Streptococcus gordonii*, *Streptococcus mutans*, and *Actinomyces viscosus*. FEMS Microbiol. Lett. *195*, 205–210.

Rozen, R., Steinberg, D., and Bachrach, G. (2004). *Streptococcus mutans* fructosyltransferase interactions with glucans. FEMS Microbiol. Lett. *232*, 39–43.

Ruas-Madiedo, P., Moreno, J.A., Salazar, N., Delgado, S., Mayo, B., Margolles, A., and de los Reyes-Gavilán, C.G. (2007). Screening of exopolysaccharide-producing *Lactobacillus* and *Bifidobacterium* strains isolated from the human intestinal microbiota. Appl. Environm. Microbiol. *73*, 4385–4388.

Rudrappa, T., Biedrzycki, NM., and Bais, H.P. (2008). Causes and consequences of plant-associated biofilms. FEMS Microbiol. Ecol. *64*, 153–166.

Salazar, M., Gueimonde, M., Hernandez-Barranco, A.M., Ruas-Dadieto, P.K., de low Reyer-Gavilán, C.G. (2008). Exopolysaccharides produced by intestinal *Bifidobacterium* strains act as fermentable substrates for human intestinal bacteria. Appl. Environ. Microbiol. *74*, 4737–4745.

Sánchez, J.-I., Martínez, B., Guillén, R., Jiménez-Díaz, A., and Rodríguez, A. (2006). Culture conditions determine the balance between two-different exopolysaccharides produced by *Lactobacillus pentosus* LPS26. Appl. Environ. Microbiol. *72*, 7495–7502.

Sanni, A.I., Onilude, A.A., Ogunbanwo, S.T., Fadahunsi, I.F., and Afolabi, R.O. (2002). Production of exopolysaccharides by lactic acid bacteria isolated from traditional fermented foods in Nigeria. European Food Res. Tech. *214*, 405–407.

Sato, S., and Kuramitsu, H.K. (1986). Isolation and characterization of a fructosyltransferase gene from *Streptococcus mutans* GS-5. Infect. Immun. *52*, 166–170.

Schroeder, V.A., Michalek, S.M., and Macrina, F.L. (1989). Biochemical characterization and evaluation of virulence of a fructosyltransferase-deficient mutant of *Streptococcus mutans* V403. Infect. Immun. *57*, 3560–3569.

Schwab, C. (2006). Ecological relevance and regulation of fructosyltransferases in *Lactobacillus reuteri*. Doctoral thesis at the TU München, Faculty Centre of Life Sciences Weihenstephan.

Schwab, C., and Gänzle, M.G. (2006). Effect of membrane lateral pressure on the expression of fructosyltransferases in *Lactobacillus reuteri*. Syst. Appl. Microbiol. *29*, 89–99.

Schwab, C., Walter, J., Tannock, G.W., Vogel, R.F., and Gänzle, M.G. (2007). Sucrose utilization and impact of sucrose on glycosyltransferase expression in *Lactobacillus reuteri*. Syst. Appl. Microbiol. *30*, 433–443.

Senadheera, M.D., Guggenheim, B., Spatafora, G.A., Huang, Y.C., Choi, J., Hung, D.C., Treglown, J.S., Goodman, S.D. Ellen, R.P., and Cvitkovitch, D. G. (2005). A VicRK signal transduction system in *Streptococcus mutans* affects *gtfBCD*, *gbpB*, and *ftf* expression, biofilm formation, and genetic competence development. J. Bacteriol. *187*, 4064–4076.

Senadheera, M.D., Lee, A.W. , Hung, D.C., Spatafora, G.A., Goodman, S.D., and Cvitkovitch, D.G. (2007) The *Streptococcus mutans vicX* gene product modulates *gtfB/C* expression, biofilm formation, genetic competence, and oxidative stress tolerance. J. Bacteriol. *189*, 1451–1458.

Seydim-Guzel. Z., Wyffels, J.T., Sydim, A.C., and Green, A.K. (2005). Turkish kefir and kefir grains: microbial enumeration and electron microscobic observation. Int. J. Dairy Tech. *1*, 25–29.

Shemesh, M, Tam, A., Feldman, M., and Steinberg, D. (2006). Differential expression profiles of *Streptococcus mutans* ftf, gtf and vicR genes in the presence of dietary carbohydrates at early and late exponential growth phases. Carbohydr. Res. *341*, 2090–2097.

Shimamura, A., Tsumori, H., and Mukasa, H. (1983). Three kinds of extracellular glucosyltransferases from *Streptococcus mutans* 6715 (serotype g). FEBS Lett. *157*, 79–84.

Shiroza, T., and Kuramitsu, H.K. (1988). Sequence analysis *Streptococcus mutans* fructosyltransferase gene and flanking regions. J.Bacteriol. *170*, 810–816.

Smorawinska, M., and Kuramitsu, H.K. (1995). Primer extension analysis of *Streptococcus mutans* promoter structures. Oral Microbiol. Immunol. *10*, 183–192.

Stingele, F., and Jolly, L. (2001). Molecular organization and functionality of exopolysaccharide gene clusters in lactic acid bacteria. Int. Dairy J. 733–745.

Sulavik, M.C., Tardif, G., and Clewell, D.B. (1992). Identification of a gene, *rgg*, which regulates expression of glucosyltransferase and influences the Spp phenotype of *Streptococcus gordonii*. J. Bacteriol. *174*, 3577–3586.

Takahashi, N., Mizuno, F., and Talamori, K. (1985). Purification and preliminary characterization of exo-β-D-fructosidase in *Streptococcus salivarius* KTA-19. Infect. Immun. *47*, 271–276.

Tam, K., Kinsinger, N., Ayala, P., and Qi, F. (2007). Real-time monitoring of *Streptococcus mutans* biofilm formation using a quartz crystal microbalance. Caries Res. *41*, 474–483.

Tannock, G.W. (1992). Lactic microbiota of pigs, mice and rats. In The Lactic Acid Bacteri, vol. 1. The Lactic Acid Bacteria in Health and Disease, B.J.B. Wood , ed. (London, UK, Elsevier Appl. Sci), pp.21–48.

Tannock, G.W. (2004). A special fondness for lactobacilli. Appl. Environ. Microbiol. 79, 3189–3194.

Tanzer, J.M., Thompson, A.M., Grant, L.P., Vickerman, M.M., and Scannapieco, F.A. (2008). *Streptococcus gordonii's* sequences strain CH1 glucosyltransferase determines persistent but not initial colonization of teeth on rats. Arch. Oral Biol. 53, 133–140.

Templer, R.H., Castle, S.J., Curran, R., Tumbles, G., and Klug, D.R. (1998). Sensing isothermal changes in the lateral pressure in model membranes using di-pyrenyl phosphatidylcholine. Faraday Discuss. *111*, 41–53.

Tieking, M., Ehrmann, M.A., Vogel, R.F., and Gänzle, M.G. (2005a). Molecular and functional characterization of a levansucrase from the sourdough isolate *Lactobacillus sanfranciscensis* TMW1.392. Appl. Environm. Microbiol. *66*, 655–663.

Tieking, M., Kaditzky, S., Valcheva, R., Korakli, M., Vogel, R.F., and Gänzle, M.G. (2005b). Extracellular homopolysaccharides and oligosaccharides from intestinal lactobacilli. J. Appl. Microbiol. *99*, 692–702.

Tieking, M., Korakli, M., Ehrmann, M.A., Gänzle, M.G., and Vogel, R.F. (2003). In situ production of exopolysaccharides during sourdough fermentation by cereal and intestinal isolates of lactic acid bacteria. Appl. Environm. Microbiol. *69*, 945- 952.

Torino, M.I., Hébert, E.M., Mozzi, F., and Font de Valdez, G. (2005). Growth and exopolysaccharide production by *Lactobacillus helveticus* ATCC15807 in an adenine-supplemented chemically defined medium. J. Appl. Microbiol. *99*, 1123–1129.

Tsumori, H., and Kuramitsu, H. (1997). The role of the *Streptococcus mutans* glucosyltransferases in the sucrose-dependent attachment to smooth surfaces: essential role of the GtfC enzyme. Oral Microbiol. Immunol. *12*, 274–280.

Van der Meulen, R., Groscu-Tudor, S., Mozzi, F., Veningelgem, F., Zamfir, M., Font de Valdez, G., and de Vuyst, L. (2007). Screening of lactic acid bacteria isolates from dairy and cereal products for exopolysaccharide production and genes involved. Int. J. Food Microbiol. *118*, 250–258.

van Geel-Schutten, G.Y., Faber, E.J., Smit, E., Bonting, K., Smith, M.R., ten Brink, B., Kamerling, J.P., Vliegenthart, J.F., and Dijkhuizen, L. (1999). Biochemical and structural characterization of the glucan and fructan exopolysaccharides synthesized by the *Lactobacillus reuteri* wild-type strain and by mutant strains. Appl. Environ. Microbiol. *65*, 3008–3014.

Van Hijum, S.A.F.T., Kralj, S., Ozimek, L.K., Dijkhuizen, L., and van Geel-Schutten, I.G.H. (2006). Structure-function relationships of glucansucrase and fructansucrase enzymes from lactic acid bacteria. Microbiol. Molec. Biol. Rev. *70*, 157–176.

Van Hijum, S.A., Szalowska, E., van der Maarel, M.J., and Dijkhuizen, L. (2004). Biochemical and molecular characterization of a levansucrase from *Lactobacillus reuteri*. Microbiol. *150*, 621–630.

Van Houte, J., and Jansen, H. M. (1968). Levan degradation by streptococci isolated from human dental plaque. Arch. Oral Biol. *13*, 827–830.

Van Kranenburg, R., Vos, H.R., van Swam, I., Kleerebezem, M., and W. de Vos (1999). Functional analysis of glycosyltransferase genes from *Lactococcus lactis* and other gram-positive cocci: complementation, expression and diversity. J. Bacteriol. *181*, 6347–6353.

Velasco, S.E., Yebra, M.J., Monedero, V., Ibarburu, I., Dueñas, M.T., and Irastorza, A. (2007). Influence of the carbohydrate source on β-glucan production and enzyme activities involved in sugar metabolism in *Pediococcus parvulus* 2.6. Int. J. Food Microbiol. *115*, 325–334.

Vereyken, I.J., Chupin, V., Islamov, A., Kuklin, A., Hincha, D.K., and Kruijff, B. (2003a). The effect of fructan on the phospholipid organization in the dry state. Biophys. J. *85*, 3058–3065.

Vereyken, I.J., Chupin, V., Demel, R.A., Smeekens, S.C.M., and De Kruijff, B. (2003b). Fructans insert between the headgroups of phospholipids, Biochim. Biophy. Acta *1510*, 307–320.

Vogel, R.F., Knorr, R., Müller, M.R.A. Steudel, U., Gänzle, M.G., and Ehrmann, M.A. (1999). Non-dairy lactic fermentations: The cereal world. Antonie van Leeuwenhoek *76*, 403–411.

Walker, G.J., Hare, M.D., and Morry-Jones, J.G. (1983). Activity of fructanase in batch cultures of oral streptococci. Carbohydr. Res. *113*, 101–112.

Walling, E., Gindreau, E., Lonvaud-Funel, A. (2001). La biosynthèse d'exopolysaccharide par des souches de *Pediococcus damnosus* isolèes du vin: mise au point d'outils moléculaires de détection. Lait *81*, 289–300.

Walter, J. (2006). The microecology of lactobacilli in the gastrointestinal tract. In: Probiotics and Prebiotics-Scientific Aspects. G.W. Tannock, ed. (Norfolk, UK, Caister Academic Press), pp. 51–82.

Walter, J. (2008). The ecological role of lactobacilli in the gastrointestinal tract: Implications for fundamental and biomedical research. Appl. Environm. Microbiol. *e-pub ahead of print*.

Walter, J., Schwab, C., Loach, D.M., Gänzle, M.G., and Tannock, G.W. (2008). Glucosyltransferase A (GtfA) and inulosucrase (Inu) of *Lactobacillus reuteri* TMW1.106 contribute to cell aggregation, *in vitro* biofilm formation, and colonization of the mouse gastrointestinal tract. Microbiol. *154*, 72–80.

Wang, B., and Kuramitsu, H.K. (2006). A pleiotropic regulator, Frp, affects exopolysaccharide synthesis, biofilm formation and competence development in *Streptococcus mutans*. Infect. Immun. *74*, 4581–4589.

Werning, M.L., Ibarburu, I., Dueñas, M.T., Irastorza, A., Navas, J., and López, P. (2006). *Pediococcus parvulus gtf* gene encoding the GTF glycosyltransferase and its application for specific PCR detection of β-D-glucan producing bacteria in foods and beverages. J. Food Prot. *69*, 161–169.

Witthuhn, R.C., Schoeman, T., and Britz, T.J. (2004). Isolation and characterisation of the microbial population of different South African grains. Int. J. Dairy Tech. *57*, 33–37.

Witthuhn, R.C., Schoeman, T., and Britz, T.J. (2005). Characterisation of the microbioal population at different stages of kefir production and kefir grain mass cultivation. Int. Dairy J. *15*, 383–389.

Wouters, J.T.M., Ayad, E.H.E., Hugenholtz, J., and Smit, G. (2002). Microbes from raw milk for fermented dairy products. Int. Dairy J. *12*, 91–109.

Wszolek, M., Kupiec-Teahan, B., Skov Guldager, H., and A.Y. Tamine (2006). Production of kefir, koumiss and other related products. In: Fermented milks, A.Y. Tamine, ed. (Oxford, UK, Blackwell Science Ltd), pp.174–216.

Biosynthesis and Chemical Composition of Exopolysaccharides Produced by Lactic Acid Bacteria

16

Patricia Ruas-Madiedo, Nuria Salazar and Clara G. de los Reyes-Gavilán

Abstract

Exopolysaccharides (EPS) produced by lactic acid bacteria (LAB) can be classified according to the chemical composition and biosynthesis mechanism as homopolysaccharides (HoPS) and heteropolysaccharides (HePS). HoPS are composed of a single monosaccharide type, either glucose or fructose, and they are polymerized extracellularly by means of glycansucrases, sucrose being the donor of the corresponding sugar moiety. These α- or β-D-glucans and β-D-fructans are polymers of high molar mass ($\geq 1\times10^6$ Da) which can have different degrees of branching and they are usually produced in amounts higher than 1 g/l, although this value varies according to the strain and the production conditions. The glucansucrase and fructansucrase enzymes present a conserved functional structure in which four domains can be recognized, although enzymes produced by different strains do not share similar sequences. The HePS are built up of repeating unit structures of two or more types of monosaccharides, substituted monosaccharides and other organic and inorganic molecules. To date, around 42 different unique structures comprising from di- to octasaccharides have been described. Galactose, glucose, rhamnose and to a lower extent N-acetyl-glucosamine and N-acetyl-galactosamine are present in their composition. The HePS biosynthesis is a complex process not fully understood, that involves the activity of several specific, as well as housekeeping, enzymes which also participate in the metabolism of carbohydrates and in the synthesis of some components of the cell-wall. The genes involved in HePS synthesis are organized in *eps* clusters that share a common structural organization in which genes involved in regulation, export, polymerization and chain length determination, as well as glycosyltransferases responsible for the intracellular assembly of the repeated units, can be recognized. This report summarizes an overview of the chemical composition, structure and physical properties, as well the genetics and biosynthesis of EPS produced by LAB and some bifidobacteria and propionibacteria strains.

Introduction

Lactic acid bacteria (LAB) are GRAS (Generally Recognized as Safe) or QPS (Qualified Presumption of Safety) bacteria according to FDA and EFSA definitions, respectively. From a long time ago, they have been employed as culture starters in the manufacture of a wide diversity of fermented foods due to their acidifying capability and their participation in flavour and aroma development. In addition, in the dairy industry some LAB producing exopolysaccharides (EPS) are being used because of their ability to improve the rheological and textural properties of fermented dairy products. Besides, some LAB, *Bifidobacterium* and *Propionibacterium* strains with probiotic characteristics are able to produce EPS. Some of the beneficial effects have been attributed to the presence of an EPS surrounding the bacteria surface, although this evidence has been obtained by *in vitro* or with animal models and there is a lack of human intervention studies. Regarding the ecological role of these polymers for the producing bacteria, it seems that they could be involved in protection against adverse factors and could interact with the environment, being responsible for the adhesion to, and colonization of, diverse ecological niches.

LAB are a group of Gram-positive, non-spore forming, cocci, coccobacilli and rods bacteria having a G+C content lower than 50% (Axelsson, 2004; Pot, 2008). In general they are catalase negative, non mobile, non pigmented, oxygen-tolerant anaerobic and, after glucose fermentation, they form lactic acid as the main final product (homofermentatives) or lactic acid, acetic acid, ethanol and CO_2 (heterofermentatives). LAB are involved in food fermentation and they are also isolated from mucoses of healthy humans and animals. Thereby genera *Bifidobacterium* and *Propionibacterium,* which are present in similar ecosystems, are often associated with LAB even if their G+C content is higher than 50% and they belong to the phylum Actinobacteria (BXIV) (Garrity *et al.*, 2002). LAB, belonging to phylum Firmicutes (BXIII) and class Bacilli (Garrity *et al.*, 2002), comprise of the genera *Aerococcus, Alloiococcus, Carnobacterium, Enterococcus, Lactobacillus, Lactococcus, Leuconostoc, Oenococcus, Pediococcus, Streptococcus, Tetragenococcus, Vagococcus* and *Weissella* (Fig. 16.1). Traditionally, some strains of these genera have been employed as starters for food production due to the anti-microbial capability against food borne pathogens and the flavour and aroma production. Additionally, LAB strains able to produce EPS have been selected for their technological application in the dairy industry (Hassan, 2008). More recently, specific LAB strains have been included in functional foods due to their health or probiotic benefits; EPS being one of the bacterial molecules which could be involved in a such functionality (Ruas-Madiedo *et al.*, 2008).

The EPSs are exocellular polymers that can form a cohesive layer or capsule covalently linked to the cell surface, for example that of the pathogen *Streptococcus pneumoniae.* Also, EPS can form a slime layer which is loosely (non-covalently) attached to the cell surface or is secreted to the environment (Fig. 16.2). Two phenotypes, 'ropy' and 'mucoid' can be distinguished for EPS-producing strains; the first is detected by the formation of a long filament when an inoculation

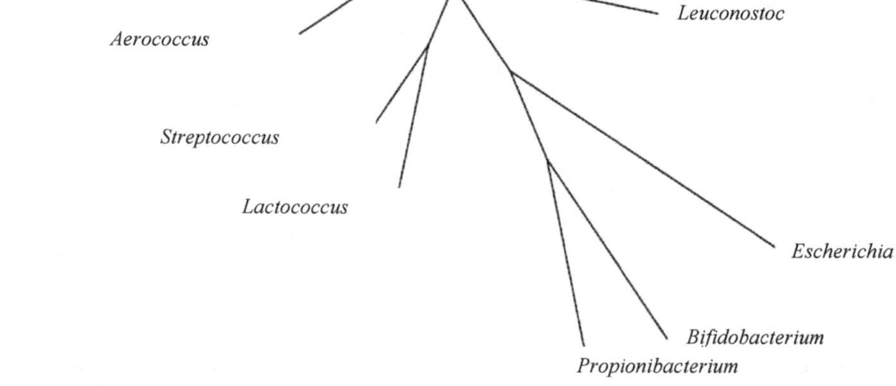

Figure 16.1 Unrooted phylogenetic tree of lactic acid bacteria genera and the high G+C content *Bifidobacterium* and *Propionibacterium* genera. The structure was generated from the annotated bacterial 16S rRNA sequences of the type species for each genus (Garrity *et al.*, 2002) obtained from the Ribosomal Database Project (http://rdp.cme.msu.edu). Sequence alignment was done with the Multiple Sequence Alignment by Clustalw (http://align.genome.jp) using the Neighbour Joining algorithm.

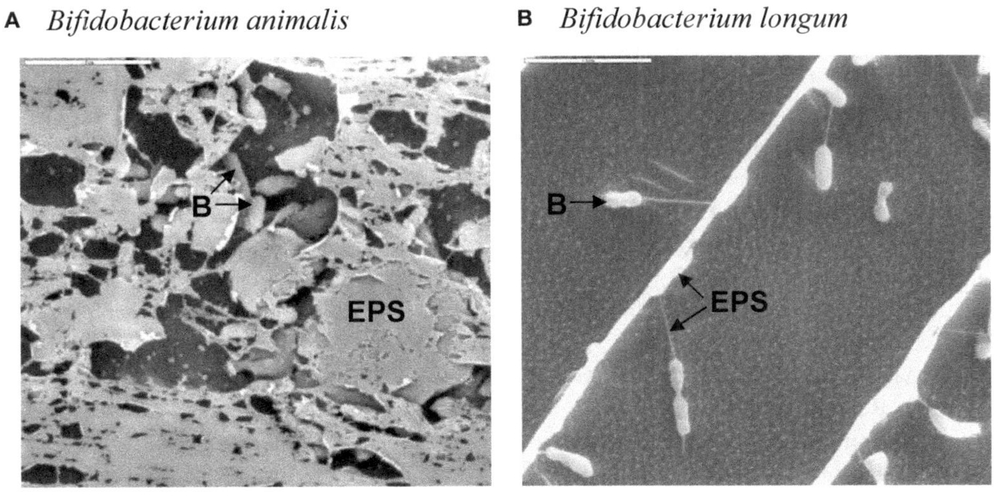

Figure 16.2 Cryo-scanning electron microscopy visualization of two EPS producing **A** 'ropy' and **B** 'mucoid' strains. Bacteria were grown in MRS-cysteine for 24 h at 37°C in anaerobic conditions. After washing twice with PBS buffer, the 50-times concentred cell suspensions were frozen in liquid N_2 and directly visualized in a Low Temperature Scanning Electron Microscopy (LTSEM) coupled with an Oxford CT-1500 HF cryo-preparation system (Oxford Instruments, Cambridge, UK) at CIB-CSIC (Madrid). Bars represent 5 μm, EPS = exopolysaccharide matrix or filaments, B, bacteria.

loop is introduced into the colony, and the second by the glistening and smooth appearance of the colonies growing in agar plates (Ruas-Madiedo *et al.*, 2005b). Depending on their monosaccharide composition and biosynthesis mechanism, EPS produced by LAB can be divided into two categories, homopolysaccharides (HoPS), which contain only one type of monosaccharide, and heteropolysaccharides (HePS) which are composed of repeating units of two or more types of monosaccharides, substituted monosaccharides and other organic and inorganic molecules (De Vuyst *et al.*, 2001; Laws *et al.*, 2001b; Monsan *et al.*, 2001).

The ecological function of EPS in producing LAB has not yet been clearly determined although it might be related to better adaptation to, and recognition of the environment. It has been suggested that these biopolymers could act as protective agents against desiccation, bacteriophage attack, antimicrobial-compound challenges, and escaping phagocytosis. They can be also involved in adhesion to surfaces like the human intestinal mucosa. In this way, we have shown that EPS produced by LAB bacteria could interact with human mucus and thereby modify the adhesion of probiotic and enteropathogen bacteria to this glycoprotein polymer that covers our intestinal mucosa (Ruas-Madiedo *et al.*, 2006a,b). EPS may exert beneficial effects on human health, such as prebiotic and serum lowering cholesterol abilities, immune modulation and anticarcinogenic activity. However, these benefits have been suggested from *in vitro* results or using animal models, and there is a lack of clinical evidence by means of human intervention studies. Finally, EPS-producing strains are employed in the dairy industry to reduce the syneresis in yoghurt production, to improve the rheological and textural properties of fermented milks and as fat replacement in the manufacture of low-fat cheeses. However, the presence of EPS-producing LAB is not a desirable characteristic in fermented beverages, such as wine, cider or beer, or in processed meats (Ruas-Madiedo *et al.*, 2008).

Homopolysaccharides

HoPS are composed of glucose or fructose, named glucans and fructans, respectively, with residues with varying glycosidic linkages, degrees of branching and molecular weights. Most of these HoPS are synthesized from sucrose through the action of extracellular enzymes called glycansucrases. However, recently it has been reported that β-glucans produced by some LAB bacteria are synthesized by means of glycosyltransferases, the mechanism of synthesis still remains unclear.

Composition, structure and yield

According to their primary structure (linkage type), glucans can be classified as α- and β-D-glucans whereas, only β-D-fructans have been described in LAB (Table 16.1) (Monsan *et al.*, 2001; Ruas-Madiedo *et al.*, 2002a; Korakli and Vogel, 2006; van Hijum *et al.*, 2006). In addition, a polygalactan composed of a pentameric repeating unit of galactose has been described (Gruter *et al.*, 1992). Regarding the carbon involved in the linkage, glucans are subdivided into dextran [α-(1→6)], mutan [α-(1→3)], reuteran [α-(1→4)] and alternan [alternating α-(1→6)/α-(1→3)]. In general, this classification was made based on the fact that the main linkage type represents more than 50% of the total linkages. But other linkages can be present in the molecule due to the branching pattern of these biopolymers, which is variable depending on the producing strain (van Hijum *et al.*, 2006). In addition, the strain *Leuconostoc mesenteroides* B-1355 is able to synthesize an α-glucan containing α-(1→2) linkages (Smith *et al.*, 1998). α-glucans have been isolated from several strains of the genera *Lactobacillus*, *Leuconostoc* and *Streptococcus* (Table 16.1). As a recent example, the EPS produced by *Lactobacillus reuteri* 180 is a (1→3, 1→6) α-D- glucan with no repeating units present, as analysed by 1D/2D ^{1}H and ^{13}C NMR spectroscopy (van Leeuwen *et al.*, 2008). Recently, the production of glucan by the *Lactococcus lactis* strain 1.8 has also been described for the first time (van der Meulen *et al.*, 2007). The production of β-glucans (Table 16.1) is not a common feature in LAB and only a few strains have been reported belonging to the genera *Lactobacillus*, *Oenococcus* and *Pediococcus*, isolated from fermented beverages, and *Propionibacterium* isolated from dairy products (Werning *et al.*, 2006; Deutch *et al.*, 2008; Dols-Lafargue *et al.*, 2008). The linkage type present in the β-glucan produced by *Lactobacillus diolivorans* G-77 and *Pediococcus parvulus* 2.6 (previously identified as *Pediococcus damnosus* 2.6) is β-(1→3) as determined by NMR techniques (Dueñas-Chasco *et al.*, 1997, 1998).

Two types of β-fructans have been described in LAB, levan and inulin-like (Table 16.1) which contain β-(2→6) and β-(2→1) osidic bonds, respectively. These fructose HoPS are associated with genera *Streptococcus*, *Lactobacillus* and *Leuconostoc* (Monsan *et al.*, 2001; Tieking *et al.*, 2005b; Korakli and Vogel, 2006; van Hijum *et al.*, 2006). Recent examples of β-fructans characterized by NMR techniques are the high molar mass inulin-like produced by the probiotic *Lactobacillus johnsonii* NCC533 (Anwar *et al.*, 2008) and the levan polymer produced by *Leuconostoc mesenteroides* ATCC8293 (Olvera *et al.*, 2007). In addition, other fructans with non-determined structures have been described in strains of *Weissella confusa*, *Lactobacillus panis*, *Lactobacillus pontis* and *Lactobacillus frumenti* (Tieking *et al.*, 2003).

The molar mass (MM) of both glucan and fructan HoPS, is commonly higher than 10^6 Da and varied according to the producing strain. The MM has been measured by size exclusion chromatography (HPLC), although some values could be overestimated due to the presence of aggregates formed in aqueous solutions (Ruas-Madiedo *et al.*, 2002a). The average MM of the dextran produced by *L. mesenteroides* NRRL B-25F varies from 6.2 to 7.1 × 10^6 Da (Cerning, 1990) and the strain *L. reuteri* Lb121 produces an α-glucan of 3.5 x10^6 Da (van Geel-Schutten *et al.*, 1999). The MM of fructans also show large variations from 2 × 10^4 to 50 × 10^6 Da, as reported by van Hijum and co-workers (2006). The levans produced by several *Streptococcus mutans* strains have MMs ranging from 2.7 to 21.6 × 10^6 Da and the strain *S. mutans* JC2 produces an inulin-like fructan of 12 × 10^6 Da (Cerning, 1990). In addition to the producing strain, the MM of the HoPS varies depending on culture media composition and culture conditions. These parameters, together with the method employed for the isolation of the EPS, also have influence on the yield reported for these biopolymers (Degeest *et al.*, 2001; Ruas-Madiedo and de los Reyes-Gavilán, 2005b). In general, the amount of HoPS produced by LAB is higher than 1 g/l with a few strains reaching values of up to 10 g/l. This is the case of *L. reuteri* Lb121, which simultaneously produces two glucan and fructan polymers (van Geel-Schutten *et al.*, 1999). Only dextran, the α-glucan synthesized by *L. mesenteroides* is produced at industrial level for food applications (yield lower than 20 g/l). In

Table 16.1 Types of homopolysaccharides produced by LAB strains and their predominant linkages; modified from Ruas-Madiedo *et al.* (2002a), Korakli and Vogel (2006) and van Hijum *et al.* (2006)

HoPS	Predominant linkage (≥ 50%)	Species
α-Glucan		
Dextran	α-D-Glc*p* (1→6)	*Lactobacillus reuteri*
		Lactobacillus sakei
		Lactobacillus fermentum
		Lactobacillus parabuchneri
		Leuconostoc mesenteroides
		Streptococcus mutans
		Streptococcus downei
		Streptococcus sobrinus
		Streptococcus salivarius
		Streptococcus gordonii
Mutan	α-D-Glc*p* (1→3)	*Lactobacillus reuteri*
		Streptococcus mutans
		Streptococcus downei
		Streptococcus sobrinus
Aternan	α-D-Glc*p* (1→6)/α-D-Glc*p* (1→3)	*Leuconostoc mesenteroides*
Reuteran	α-D-Glc*p* (1→4)	*Lactobacillus reuteri*
Others	α-D-Glc*p* (1→2)	*Leuconostoc mesenteroides* B-1355
	Unknown	*Lactococcus lactis* 1.8
β-Glucan		
	β-D-Glc*p* (1→3)	*Lactobacillus diolivorans* G77
		Pediococcus parvulus 2.6[a]
	Unknown	*Oenococcus oeni*
		Pediococcus parvulus
		Propionibacterium freudenreichii
Fructan		
Levan	β-D-Fru*p* (2→6)	*Lactobacillus reuteri*
		Lactobacillus sanfranciscensis
		Streptococcus sobrinus
		Streptococcus salivarius
Inulin-like	β-D-Fru*p* (2→1)	*Lactobacillus reuteri*
		Leuconostoc citreum
		Streptococcus mutans
Others	Unknown	*Lactobacillus frumenti*
		Lactobacillus panis
		Lactobacillus pontis
		Weisella confusa
Polygalactan	α-D-Gal*p*/β-D-Gal*p*	*Lactococcus lactis* subsp. *lactis* H414

Glu, glucose; Gal, galactose; Fru, fructose.

[a]Previously *Pediococcus damnosus* 2.6.

general, the yield of EPS (both HoPS and HePS) produced by LAB is low compared with other bacterial (hetero)-polymers such as gellan or xanthan synthesized at industrial level for several applications (in range higher than 20 g/l) by the Gram-negative bacteria, *Sphingomonas paucimobilis* and *Xanthomonas campestris*, respectively (Sutherland, 1998).

HoPS-synthesizing enzymes

The synthesis of HoPS requires the presence of sucrose as the main substrate employed by glycansucrases to catalyse the transfer of a corresponding glycosyl moiety to a polymer in formation. According to the Nomenclature Committee of the International Union of Biochemistry and Molecular Biology, the glycansucrase enzymes are classified based on the reaction catalysed and the product specificity. Enzymes synthesizing glucan HoPS are named glycansucrases (GS) and are limited to LAB, whereas those synthesizing fructans, named fructansucrase (FS), are present in both Gram-positive and Gram-negative bacteria. Some authors have referred to GS as glucosyltransfereses and FS as fructrosytransferases. However, the glycosyltransferases, which also catalyse the transfer of different sugar units, are not descriptive for the substrate and product specificity of sucrose-type enzymes (van Hijum *et al.*, 2006). These glycoside hydrolases (GH) are classified into different families based on their amino acid sequence. GS have been included in the family GH70 and FS in the family GH68 [more detailed information is given in the CAZY 'Carbohydrate-Active Enzymes' web page (http://www.cazy.org)].

GS are grouped in dextransucrase (EC.2.4.1.5) and alternansucrase (EC.2.4.1.140) enzymes, whereas the mutansucrase and reuteransucrase enzymes are classified together with dextransucrases. Also, two FS enzymes have been described: inulosucrase (EC.2.4.1.9) and levansucrase (EC.2.4.1.10) (van Hijum *et al.*, 2006). Both glycansucrase enzymes are large extracellular proteins with average MM of 160 kDa for GS and, as an example, 140 kDa for the levansucrase described for *Streptococcus salivarius* ATCC13419 (Monchois *et al.*, 1999; Korakli and Vogel, 2006; van Hijum *et al.*, 2006). Based on the amino acid sequences of GS and FS, four distinct structural domains could be distinguished although, even if both catalyse similar reactions and use sucrose as substrate, they do not share similar sequences. The four structural domains (Fig. 16.3) are (i) a signal peptide that targets these extracellular enzymes for secretion, (ii) a N-terminal variable domain, (iii) a conserved catalytic domain, and (iv) a C-terminal binding domain (Korakli and Vogel, 2006; van Hijum *et al.*, 2006). The four conserved regions found in GS enzymes were first identified in α-amylase enzymes (family GH13). The signal peptide of GS contains 36 to 40 amino acids that are well conserved in Gram-positive bacteria. In the N-terminal region, the stretch of amino acids between the signal peptide and the catalytic region is highly variable in composition and length (200–700 amino acids) and several repeating units (123–129 amino acids) have

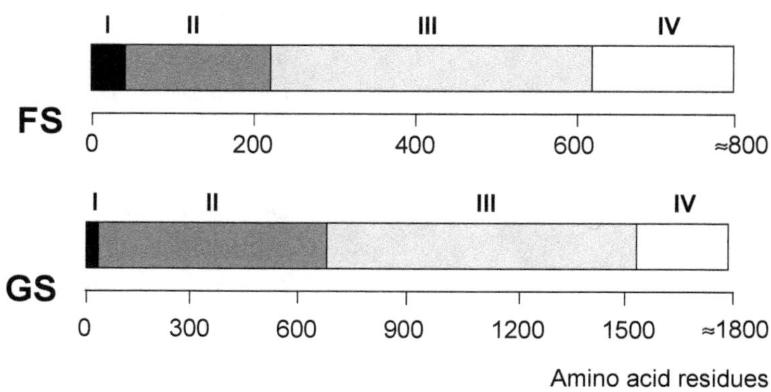

Figure 16.3 Schematic representation of the structure of glucansucrases (GS) and fructansucrases (FS) involved in the synthesis of HoPS. Maps are not scaled. I: signal peptide, II: N-terminal variable domain, III: catalytic domain, IV: C-terminal binding domain (Modified from van Hijum *et al.* (2006)).

been found, although its function still remains unclear. The highly conserved catalytic region, with about 1000 amino acids, comprises the sucrose-binding and sucrose-cleaving domain. The detailed structure of GS is not known, however secondary-structure prediction studies of the catalytic domain have shown similar structures to enzymes of the GH13 family. These analyses have resulted in the identification of three amino acid residues involved in catalysis in both GH70 and GH13 families. In the sequence of the GS from *L. reuteri* Lb121 (Kralj *et al.*, 2002) these amino acids are Asp1024, Glu1061 and Asp1133 and play a crucial role in catalysis (van Hijum *et al.*, 2006). Finally, the C-terminal domain of GS, of about 400 amino acid residues, is composed of a series of tandem repeats and it is involved in the binding to the glucan chain in formation. For a few GS enzymes, it has been shown that this domain is also involved in the determination of the glucan structure and it can be important for the polymer chain length. However, its precise role in catalysis still remains unclear (Monchois *et al.*, 1999; Korakli and Vogel, 2006; van Hijum *et al.*, 2006). Although about 60 GS have been classified in the GH70 family, the first three-dimensional structure determined by biochemical and crystallographic studies has recently been reported and it corresponds with the GS produced by the strain *L. reuteri* 180 (Pijning *et al.*, 2008). This characterization may provide further insight into the structure-function relationship of GS enzymes from LAB.

Regarding the structural organization of FS, the N-terminal signal peptide is composed of 36 to 39 amino acids. The N-terminal variable region varies in size between different FS and it is constituted by direct repeats of 4–7 × 14 amino acids (Korakli and Vogel, 2006). To date, no clear function has been assigned for this domain. The conserved catalytic domain is formed with about 500 amino acids which is shared among all members of FS family GH68. The studies of structure-function of this region have been carried out in FS from non-LAB, and only some of the results have been corroborated for FS of LAB origin. It seems that some acidic residues (two Asp and Glu) are present in highly conserved sequence motifs of FS from GH13 family, as previously indicated for GS of the same family. This has been proven for the inulosucrase and levansucrase of the strain *L. reuteri* Lb121 (van Hijum *et al.*, 2006). The sucrose-binding domain has been identified, among others, in the levansucrase produced by *Bacillus subtilis* (family GH13). Besides, the residue Asp339 from the levansucrase of this strain is involved in binding calcium, ion that is needed for the enzymatic activity. A similar role has been demonstrated for the residues Asp520Asn and Asp500Asn of inulosucrase and levansucrase, respectively, synthesized by the strain *L. reuteri* Lb121 (Ozimek *et al.*, 2005). Sequence alignment of the FS GH68 family indicated that residues involved in the binding of calcium are conserved in most enzymes from Gram-positive bacteria, but are absent in those of Gram-negative bacteria (Ozimek *et al.*, 2005; van Hijum *et al.*, 2006). Finally, the C-terminal domain of different FS presents variable length and it might affect the fructan size and/or enzyme specificity, as occurs for GS. However, this domain is also involved in the attachment of the FS to the cell wall of the producing bacteria. A C-terminal LPXTG cell wall-anchoring motif is found in FS from *L. reuteri* strains Lb121 (van Hijum *et al.*, 2006), LTH5448 and TMW1.106 (Korakli and Vogel, 2005) and *L. sanfranciscensis* TMW1.392 (Korakli and Vogel, 2006). A similar motif is also present in the cell associated levansucrase produced by *S. salivarius* ATCC25975 (Rathsam and Jacques, 1998).

In addition to HoPS synthesized from sucrose by means of glycansucrases, in the last few years glycosyltransferases (GTF) able to synthesize β-glucans have been described in LAB (Walling *et al.*, 2005; Werning *et al.*, 2006; Doels-Lafargue *et al.*, 2008) and *Propionibacterium* (Deutsch *et al.*, 2008) (Table 16.2). In *P. parvulus* and *L. diolivorans gtf* genes are located on plasmids, whereas for *Oenococcus oeni* and *Propionibacterium freudenreichii* they are located in the chromosome. These *gtf* sequences are placed near to mobile elements which could explain the integration of this gene in chromosomes. In *O. oeni* IOEB0205 the *gtf* is surrounded by two open reading frames (ORF). The ORF1 located upstream shows high homology with transposases of *Lactobacillus casei* and the bacteriophage phiAT3, as well as with transposases of the family IS30 detected in several LAB (Doels-Lafargue *et al.*, 2008). The downstream ORF2 has a protein sequence identical to that of

Table 16.2 Glycosyltransferase (GTF) genes involved in the synthesis of β-glucans in LAB and *Propionibacterium* strains

Strain	Location	Accession no. (bp)	Amino acid GTF	Reference
Pediococcus parvulus IOEB880[a]	Plasmid	AF196967 (5558)	567	Walling *et al.* (2005)
P. parvulus 2.6	Plasmid	AY999683 (4535)	567	Werning *et al.* (2006)
Oenococcus oeni IOEB0205	Chromosome	EU556433 (3528)	567	Doels-Lafargue *et al.* (2008)
O. oeni I4	Chromosome	AY999685 (1704)	567	Werning *et al.* (2006)
Lactobacillus diolivorans G77	Plasmid	AY999684 (1704)	567	Werning *et al.* (2006)
Propionibacterium freudenreichii subsp. *shermanii* TL34	Chromosome	AM850120 (1536)	511	Deutsch *et al.* (2008)

[a]Previously identified as *Pediococcus damnosus* IOEB880 (Deutsch *et al.*, 2008)

putative IS30 transposase from *L. casei* ATCC334 (Doels-Lafargue *et al.*, 2008). In the strain *P. parvulus* 2.6 the *tnp* positioned downstream of *gtf*, also encodes for a putative transposase of IS30 family. In addition, the *traA* sequence located upstream from *gtf* is putatively involved in conjugation, with high homology to its counterparts in *Lactobacillus* and *L. lactis* species (Werning *et al.*, 2006). The sequences of *gtf* from *P. parvulus*, *O. oeni* and *L. diolivorans* show high identity among them (from 97.8 to 99.1%) but this homology decreases with respect to that of *Pr. freudenreichii* subsp. *shermanii* TL34 (around 32% identity and 51% similarity). Besides, the predicted protein products from *gtf* produced by *P. parvulus* 2.6 and *Pr. freudenreichii* subsp. *shermanii* TL34 align with the polysaccharide synthase of *St. pneumoniae* serotype 37 (33% and 52% identity, respectively) (Werning *et al.*, 2006; Deutsch *et al.*, 2008). This polysaccharide synthase (GenBank accession no. CAB51329) is responsible for the production of a β-glucan capsular polysaccharide in the pneumococci. By means of agglutination experiments using *S. pneumoniae* serotype 37 antisera, the presence of (1→3, 1→2)-β-D-glucan capsule was demonstrated for seven strains of *Pr. freudenreichii* subsp. *shermanii* (Deutsch *et al.*, 2008), for *P. parvulus* 2.6 (Werning *et al.*, 2006), and for *P. parvulus* IOEB8801 (Walling *et al.*, 2005). Finally, preliminary studies of the GTF produced by *P. parvulus* 2.6 indicated that this protein belongs to the COG1215 membrane-bound GTF family. Topology predictions show that a conserved cytosolic domain is flanked by two and four transmembrane segments (Werning *et al.*, 2006). The mechanism of β-glucan synthesis in LAB by means of these GTF still remains to be elucidated.

Synthesis of HoPS

Depending on the acceptor molecule, and using sucrose as the substrate, the glycansucrases (both GS and FS) described above are able to catalyse three types of reactions (Fig. 16.4): (i) polymerization of α-glucans or β-fructans, respectively, (ii) sucrose hydrolysis and (iii) glucosylation or fructosylation of acceptor molecules to render different low MM oligosaccharides and, in some cases, hetero-oligosaccharides (Monchoi *et al.*, 1999; Monsan *et al.*, 2001; Korakli and Vogel, 2006; van Hijum *et al.*, 2006).

The catalytic mechanism of GS for the synthesis of α-glucans has not been totally clarified. The key step of the transfer of the D-glucosyl unit is the formation of a covalent glucosyl-enzyme intermediate. This procedure, as previously indicated for GS of LAB and α-amylases of family GH13, involves a triad consisting of two aspartic and one glutamic residue (Monsan *et al.*, 2001). α-glucan is synthesized by autopolymerization and is a single chain mechanism, which does not involve the formation of initial oligosaccharides. The synthesis could be separated into three different steps: initiation, elongation and termination, the last one corresponding to the dissociation of the α-glucan from the enzyme (Tsuchiya *et al.*, 1953). As well, two alternative mechanisms have been proposed for the elongation step: occurring at the non-reducing or at the reducing end of the

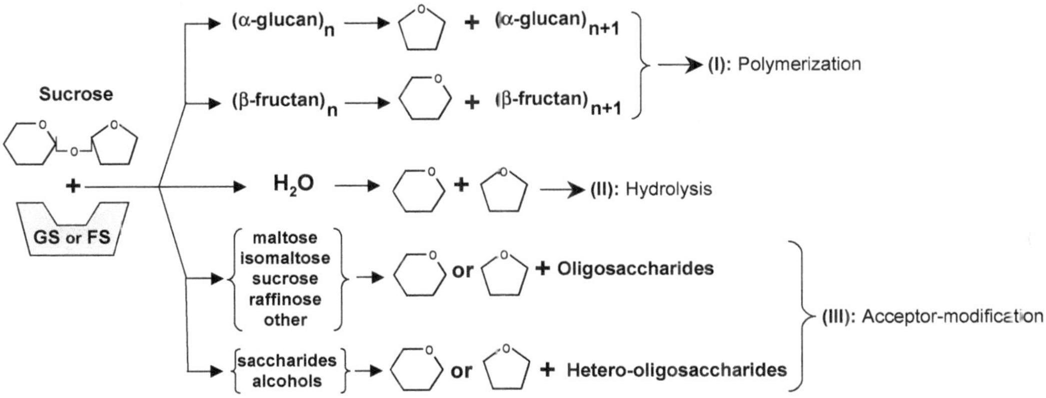

Figure 16.4 Schematic representation of reactions catalysed by glycansucrases: GS (glucansucrase) and FS (fructansucrase) (Modified from Korakli and Vogel (2006) and van Hijum *et al.* (2006)).

α-glucan chain (Monchois *et al.*, 1999). Also, little is known about the catalytic mechanism of FS in the synthesis of β-fructans. A multiple chain elongation mechanism in which the fructose residues are added randomly to a fructan acceptor has been proposed (Chambert *et al.*, 1974). The catalytic transfructosylation mechanism could be done in two-steps involving an acidic and a nucleophilic group. The nucleophilic (Asp86) group responsible for binding to the fructose moiety was determined from the three-dimensional crystal structure of *B. subtilis* levansucrase (Meng and Futterer, 2003). Another Asp residue located in the conserved region of the catalytic domain has been shown to be involved in the stabilization of the transition state. Mutation of this Asp in levansucrases from *S. salivarius* and *L. reuteri* Lb121 resulted in a dramatic decrease of the enzymatic activity (Song and Jacques, 1999; Ozimek *et al.*, 2004).

The GS and FS are able to synthesize oligosaccharides (OS) with different linkage types when another molecule, different from α-glucan or β-glucan, is employed as the acceptor. Dextransucrases from LAB are able to employ numerous sugars as acceptors, which could be classified into two groups according to their ability to compete with α-glucan or their effect on the reaction speed: (i) the strong acceptors, such as maltose or isomaltose, which strongly inhibit the glucan synthesis, and (ii) the weak acceptors, such as fructose or melibiose (Monchoi *et al.*, 1998). The synthesis of OS occurs by the successive transfer of glucosyl units to the OS in formation which is alternately product and substrate. These strong acceptors allow the formation of a variety of different OS. However, weak acceptors like fructose only produce one disaccharide called leucrose [α-D-glucopyranosyl-(1→5)-β-D-fructofuranoside] where the glucose from sucrose is transferred to the fructose. This reaction begins when fructose is accumulated at the end of the α-glucan synthesis (Korakli and Vogel, 2006). Leucrose synthesis has been reported from the glucansucrase of *Streptococcus downei* MFe28 (Monchois *et al.*, 2000) and by the reuteransucrase of *L. reuteri* Lb121 (Kralj *et al.*, 2004). The trisaccharides, panose [α-D-glucopyranosyl-(1→6)-α-D-glucopyranosyl-(1→4)-D-glucose] and maltotriose [α-D-glucopyranosyl-(1→4)-α-D-glucopyranosyl-(1→4)-D-glucose] are produced by the same GS when maltose acts as acceptor (Kralj *et al.*, 2004). Panose synthesis has been reported as well for the glucansucrase of *L. mesenteroides* NRR B-512F (Monchois *et al.*, 1997). Fructooligosaccharides with different linkage types can be synthesized by means of FS when the fructose from sucrose is transferred to acceptor molecules, other than β-fructan polymer, such as sucrose or maltose. Several strains of *Lactobacillus* from food and animal origin are able to produce fructooligosaccharides. This is the case of the trisaccharide 1-kestose [β-D-fructofuranoside-(2→1)-β-D-fructofuranoside-(2→1)-α-D-glucopyranoside] produced by two *Lactobacillus acidophilus* strains (Tieking *et al.*, 2005b) or the inulin-oligosaccharide from *L. reuteri* (van Hijum *et al.*, 2002). Besides, the tetrasaccharide nystose [β-D-fructofuranoside-(2→1)-β-D-fructofuranoside-(2→1)-β-D-fructofuranoside-(2→1)-α-D-glucopyranoside] is produced together with 1-kestose by the activity of the FS from *L. sanfranciscencis* (Tieking *et al.*, 2005a) and *L. reuteri* (van Hijum *et al.*, 2002). In addition to OS containing glucose or fructose previously described, the production of galactooligosaccharides has been reported for the strain *L. reuteri* NCIMB44152 due to the activity of an α-galactosidase whose synthesis is induced when raffinose is employed as the carbon source (Tzortzis *et al.*, 2003). Finally, the synthesis of heterooligosaccharides (HeOS) could take place when fructosyl putative acceptors such as raffinose, maltotriose, maltose, xylose, or arabinose are available. Using these acceptors, the levansucrase obtained from *L. sanfranciscensis* TMW1.392 produces the corresponding HeOS: 1-β-fructosylraffinose, 1-β-fructofuranosylmaltotriose, 1-β-fructofuranosylmaltose (or erlose), xylsucrose, 1-β-fructosylxylsucrose and arabsucrose. The identification of these HeOS was performed by HPAEC-PAD which also showed the formation of further tri-, tetra- and pentasaccharides (Tieking *et al.*, 2005c). The synthesis of erlose using maltose as the acceptor was also reported for the recombinant levansucrase from *L. mesenteroides* B-512 FMC expressed in *Escherichia coli* (Kang *et al.*, 2005).

The role of HoPS, as well as of OS and HeOS, in the environment niches of the producing LAB still remains to be elucidated. Cell associated HoPS may be involved in the adherence of the producing bacteria to a surface, such as the teeth and the intestinal mucosa (Rozen *et al.*, 2001). Besides, the short-chain polysaccharides could have functional applications given that they could be employed as prebiotic substrates for the beneficial microbiota of the intestinal ecosystem (Dal Bello *et al.*, 2001; Korakli *et al.*, 2002).

Heteropolysaccharides

HePS are constructed of a backbone of repeated units which could be unbranched or branched at different carbon positions. These units are composed of monosaccharides, derivatives monosaccharides and of organic- and inorganic-substituted monosaccharides. The synthesis of HePS is a complex process in which several genes organized in *eps* clusters are involved. The function of most of these genes, as well as the mechanism of HePS synthesis still needs to be fully understood. However, it seems that the repeating units are formed intracellularly, exported, and polymerized outside the cell.

Composition, structure and yield

The most abundant monosaccharides present in the HePS repeating units are galactose and glucose followed by rhamnose and, into a lower extent, N-acetyl-glucosamine and N-acetyl-galactosamine. Other monosaccharides such as fucose, ribose and mannose have been described in a few HePS. Organic and inorganic molecules such as acetyl, glucuronic acid, glycerol, glycerol-phosphate and phosphate are present in several polymers. To date, a wide variety of different HePS repeating units composed of from two to eight monosaccharides have been determined by NMR spectroscopy techniques. An updated review of the composition of the repeating units forming HePS from LAB, as well as from the few described bifidobacteria and propionibacteria, are collected in Table 16.3. The new NMR-repeating unit structures, which were not collected in former reviews (De Vuyst *et al.*, 2001; Laws *et al.*, 2001a,b; Ruas-Madiedo *et al.*, 2002a; Broadbent *et al.*, 2003), are summarized in Table 16.4.

At the time of writing, the NMR spectra for 6 *L. lactis* subsp. *cremoris*, 28 *Lactobacillus*, 23 *Streptococcus*, 3 *Bifidobacterium*, and 5 *Propionibacterium* strains have been reported. Numerous other articles have described the monosaccharide composition of HePS using diverse techniques such as TLC, HPLC and/or GC. From these NMR spectra, 42 unique structures could be found (Table 16.3) and, as far as is possible to note, they are not shared among strains of the different genera. From these structures, including the newest, few generalizations can be indicated (Laws *et al.*, 2001b). The monosaccharides are present as the α- or β-anomer in pyranose or furanose ring conformation, but the first seems to be the most abundant given that only galactose adopted the furanose form. Without exception, glucose and galactose have D-absolute configuration whereas rhamnose has L-absolute configuration. For rhamnose there is a clear preference for the α-anomer and for glucose

Table 16.3 Chemical composition of the repeating units forming HePS synthesized by LAB, bifidobacteria and propionibacteria whose structures have been determined by NMR spectroscopy; updated from De Vuyst *et al*. (2001), Laws *et al*. (2001a,b), Ruas-Madiedo *et al*. (2002a) and Broadbent *et al*. (2003)

Genus	Strain	Monosaccharide ratio						Subunit	Reference
		Glc	Gal	Rha	NAcGlc	NAcGal	Others	Group[a]	
Lactococcus	*Lc. lactis* subsp. *cremoris* H414[b]	–	5	-	–	–	–	1	Gruter *et al*. (1992)
	Lc. lactis subsp. *cremoris* STB0495	2	2	1	–	–	PO_4	2	Nakajima *et al*. (1992)
	Lc. lactis subsp. *cremoris* NB40	2	2	1	–	–	PO_4	2	van Casteren *et al*. (1998)
	Lc. lactis subsp. *cremoris* AHR53	2	2	1	–	–	PO_4	2	Yang *et al*. (1999)
	Lc. lactis subsp. *cremoris* NB39	2	3	2	–	–	–	3	van Casteren *et al*. (2000b)
	Lc. lactis subsp. *cremoris* NB891	3	2	–	–	–	Acetyl	4	van Casteren *et al*. (2000a)
Lactobacillus	*Lb. acidophilus* LMG9433	3	1	–	1	–	–	5	Robijn *et al*. (1996a)
	Lb. acidophilus 5e2	3	3	–	1	–	–	6	Laws *et al*. (2008)
	Lb. diolivorans G77[c]	3	–	–	–	–	–	7	Dueñas–Chasco *et al*. (1998)
	Lb. delbrueckii subsp. *bulgaricus* rr	1	5	1	–	–	–	8	Gruter *et al*. (1993)
	Lb. delbrueckii subsp. *bulgaricus* LY03, 24, 25	1	5	1	–	–	–	8	Marshall *et al*. (2001b)
	Lb. delbrueckii subsp. *bulgaricus* 291	3	2	–	–	–	–	9	Faber *et al*. (2001a)
	Lb. delbrueckii subsp. *bulgaricus* NCFB2074	4	3	–	–	–	–	10	Harding *et al*. (2005)
	Lb. delbrueckii subsp. *bulgaricus* EU23	3	–	3	–	–	–	11	Harding *et al*. (2003)
	Lb. delbrueckii subsp. *bulgaricus* LBB.B332	1	2	2	–	–	–	12	Sánchez-Medina *et al*. (2000b)
	Lb. delbrueckii subsp. *bulgaricus* LBB.B26	2	3	–	–	–	–	13	Sánchez-Medina *et al*. (2000a)
	Lb. helveticus TY1–2	3	3	–	1	1	–	14	Yamamoto *et al*. (1994)
	Lb. helveticus NCDO 766	4	2	–	–	–	–	15	Robijn *et al*. (1995a)
	Lb. helveticus TN–4	3	3	–	–	–	–	16	Yamamoto *et al*. (1995)
	Lb. helveticus Lh59	3	3	–	–	–	–	16	Stingele *et al*. (1997)

Table 16.3 *continued*

Genus	Strain	Monosaccharide ratio						Subunit	Reference
		Glc	Gal	Rha	NAcGlc	NAcGal	Others	Group[a]	
Lactobacillus (cont.)	*Lb. helveticus* NCC2745	3	3	–	–	–	–	17	Jolly *et al.* (2002)
	Lb. helveticus 2091	2	4	–	–	–	–	18	Staaf *et al.* (1996)
	Lb. helveticus Lb161	5	2	–	–	–	–	19	Staaf *et al.* (2000)
	Lb. helveticus K16	4	2	–	–	–	Glycerol-PO_4	20	Yang *et al.* (2000)
	Lb. kefiranofaciens WT–2B	3	3	–	–	–	–	21	Maeda *et al.* (2004)
	Lb. paracasei 34–1	–	3	–	1	–	–	22	Robijn *et al.* (1996b)
	Lb. rhamnosus C83	2	3	–	–	–	–	23	Vanhaverbeke *et al.* (1998)
	Lb. rhamnosus GG	–	4	1	1	–	–	24	Landersjö *et al.* (2002)
	Lb. rhamnosus KL37–C	3	2	–	–	–	–	25	Lipinski *et al.* (2003)
	Lb. rhamnosus RW9595–M, R	2	1	4	–	–	Pyruvate-acetal	26	van Calsteren *et al.* (2002)
	Lb. sakei 0–1	2	–	3	–	–	Acetyl/ glycerol-PO_4	27	Robijn *et al.* (1995b)
Streptococcus	*St. macedonicus* SC136	3	2	–	1	–	–	28	Vincent *et al.* (2001)
	St. thermophilus CNCMI733, 734, 735	1	2	–	–	1	–	28	Doco *et al.* (1990)
	St. thermophilus EU21	1	2	–	–	1	–	28	Marshall *et al.* (2001b)
	St. thermophilus SFi6	1	2	–	–	1	–	28	Stingele *et al.* (1996)
	St. thermophilus IMDO01, 2, 3; NCFB859	1	2	–	–	1	–	28	Marshall *et al.* (2001b)
	St. thermophilus SFi20	1	2	–	–	1	–	29	Navarini *et al.* (2001)
	St. thermophilus SFi12	1	3	2	–	–	–	30	Lemoine *et al.* (1997)
	St. thermophilus SFi39	2	2	–	–	–	–	30	Lemoine *et al.* (1997)

	St. thermophilus SY89, SY102	2	2	–	–	–	–	31	Marshall *et al*. (2001b)
	St. thermophilus OR901	–	5	2	–	–	–	31	Bubb *et al*. (1997)
	St. thermophilus Rs, Sts	–	5	2	–	–	–	32	Faber *et al*. (1998)
	St. thermophilus MR–1C	–	5	2	–	–	Fucose	33	Low *et al*. (1998)
	St. thermophilus S3	–	4	2	–	–	Acetyl	34	Faber *et al*. (2001b)
	St. thermophilus EU20	2	3	2	–	–	–	35	Marshall *et al*. (2001a)
	St. thermophilus 8S	1	2	–	–	1	Ribose	36	Faber *et al*. (2002)
	St. thermophilus THS	3	2	–	–	–	–	37	Nordmark *et al*. (2005a)
Bifidobacterium	*Bif. infantis* ATCC15697 [d]	–	2	–	–	–	–	38	Tone-Shimokawa *et al*. (1996)
	Bif. catenulatum YIT4016 [d]	–	3	–	–	–	–	39	Nagaoka *et al*. (1996)
	Blf. longum YIT4028	–	3	2	–	–	–	40	Nagaoka *et al*. (1995)
Propionibacterium	*Pr. freudenreichii* subsp. *shermanii* JS [e]	2	–	–	–	–	–	41	Nordmark *et al*. (2005b)
	Pr. freudenreichii subsp. *shermanii* TL34 [e]	2	–	–	–	–	–	41	Deutsch *et al*. (2008)
	Pr. freudenreichii 109 [f], 111 [f]	2	–	–	–	–	Man/Gluc. acid	42	Dobruchoska *et al*. (2008)
	Pr. thoenii 126 [f]	2	–	–	–		Man/Gluc. acid	42	Dobruchoska *et al*. (2008)

[a]Groups of EPS repeating subunit that share unique NMR structure.

[b]The EPS H414 is a polygalactan composed of a pentasaccharide repeated unit.

[c]The EPS G–77 is a β-glucan composed of a trisaccharide repeated unit, with a backbone of (1→3) linkages an a side glucose group linked (1→2).

[d]The EPS ATCC15697 and YIT4016 are considered as cell–wall polysaccharides.

[e]The EPS JS is a β-glucan composed of a disaccharide repeated unit (see Table 16.4).

[f]The EPS 109, 111 and 126 are composed of a pentasaccharide repeated unit with a glucose: mannose: glucuronic acid ratio of 2:2:1.

Table 16.4 Structures of the repeating units forming HePS synthesized by LAB, bifidobacteria and propionibacteria which were not summarized in previous reviews (De Vuyst *et al*. 2001; Laws *et al*. 2001a,b; Ruas-Madiedo *et al*. 2002a; Broadbent *et al*. 2003)

Genus	Strain	Structure[a,b]	Reference
Lactobacillus	*Lb. acidophilus* 5e2	β-D-Glc*p* 1 ↓ 4 →6)-α-D-Glc*p*-(1→3)-β-D-Glc*p*-(1→3)-β-D-Glc*p***NAc**-(1→3)-α-D-Gal*p*-(1→ 6 ↑ 1 β-D-Gal*p*-(1→4)-β-D-Glc*p*	Laws *et al*. (2008)
	Lb. delbrueckii subsp. *bulgaricus* NCFB2074	α-D-Gal*p*-(1→3)-α-D-Glc*p* α-D-Gal*p* 1 1 ↓ ↓ 3 6 →4)-α-D-Glc*p*-(1→3)-α-D-Gal*p*-(1→ 2 ↑ 1 β-D-Gal*p*-(1→4)-β-D-Glc*p*	Harding *et al*. (2005)
	Lb. delbrueckii subsp. *bulgaricus* EU23	α-L-Rha*p* 1 ↓ 3 →2)-α-L-Rha*p*-(1→4)-α-D-Glc*p*-(1→3)-β-L-Rha*p*-(1→4)-β-D-Glc*p*-(1→4)-α-D-Glc*p*-(1→	Harding *et al*. (2003)
	Lb. delbrueckii subsp. *bulgaricus* LBB.B332	→3)-α-D-Glc*p*-(1→3)-α-D-Gal*p*-(1→3)-α-L-Rha*p*-(1→2)-α-L-Rha*p*-(1→2)-α-D-Glc*p*-(1→	Sánchez-Medina *et al*. (2007b)
	Lb. delbrueckii subsp. *bulgaricus* LBB.B26	α-D-Glc*p* 1 ↓ 6 →3)-α-D-Gal*p*-(1→3)-β-D-Gal*p*-(1→4)-β-D-Glc*p*-(1→3)-β-D-Gal*f*-(1→	Sánchez-Medina *et al*. (2007a)

	Lb. helveticus NCC2745	β-D-Gal*p*-(1→6)-α-D-Glc*p* 1 ↓ 6 →3)-β-D-Gal*p*-(1→3)-α-D-Gal*p*-(1→6)-α-D-Glc*p*-(1→3)-β-D-Glc*p*-(1→	Jolly *et al*. (2002)
	Lb. kefiranofaciens WT-2B	→6)-β-D-Glc*p*-(1→6)-β-D-Gal*p*-(1→4)-α-D-Gal*p*-(1→3)-β-D-Gal*p*-(1→4)-β-D-Glc*p*-(1→ 2 ↑ 1 β-D-Glc*p*	Maeda *et al*. (2004)
	Lb. rhamnosus GG	→3)-β-D-Gal*f*-(1→3)-β-D-Gal*p*-(1→4)-α-D-Glc*p*NAc-(1→3)-α-L-Rha*p*-(1→3)-α-D-Gal*p*-(1→ 6 ↑ 1 β-D-Gal*f*	Landersjö *et al*. (2002)
	Lb. rhamnosus KL37-C	→3)-α-D-Glc*p*-(1→2)-β-D-Gal*f*-(1→6)-α-D-Gal*p*-(1→6)-α-D-Glc*p*-(1→3)-β-D-Gal*f*-(1→	Lipinski *et al*. (2003)
	Lb. rhamnosus RW9595-M, R	**Py**-α-D-Gal*p* 1 ↓ 2 →3)-α-L-Rha*p*-(1→3)-β-D-Glc*p*-(1→3)-α-L-Rha*p*-(1→3)-α-L-Rha*p*-(1→3)-α-L-Rha*p*-(1→2)-α-D-Glc*p*-(1→	van Calsteren *et al*. (2002)
Streptococcus	*St. thermophilus* THS	**Py**-α-D-Gal*p* 1 ↓ 2 →3)-α-L-Rha*p*-(1→3)-β-D-Glc*p*-(1→3)-α-L-Rha*p*-(1→3)-α-L-Rha*p*-(1→3)-α-L-Rha*p*-(1→2)-α-D-Glc*p*-(1→	Nordmark *et al*. (2005a)

Table 16.4 *continued*

Genus	Strain	Structure[a,b]	Reference
Bifidobacterium	*Bif. infantis* ATCC15697 [c]	β-D-Gal*p*-(1→4)-β-D-Glc*p*-(1→6)-α-D-Glc*p* 1 ↓ 4 →3)-β-D-Gal*p*-(1→4)-β-D-Glc*p*-(1→	Tone-Shimokawa *et al*. (1996)
	Bif. catenulatum YIT4016	→3)-β-D-Gal*f*-(1→3)-α-D-Glc*p*-(1→	Nagaoka *et al*. (1996)
	Bif. longum YIT4028	→6)-β-D-Gal*f*-(1→5)-β-D-Gal*f*-(1→ 6 ↑ 1 α-D-Gal*p*	Nagaoka *et al*. (1995)
Propionibacterium	*Pr. freudenreichii* subsp. *shermanii* JS	→2)-α-L-Rha*p*-(1→3)-α-D-Gal*p*-(1→2)-α-L-Rha*p*-(1→3)-α-D-Gal*p*-(1→ 6 ↑ 1 α-D-Gal*p*	Nordmark *et al*. (2005b)
	Pr. freudenreichii 109, 111 and *Pr. thoenii* 126	→3)-β-D-Glc*p*-(1→2)-β-D-Glc*p*-(1→	Dobruchoska *et al*. (2008)

[a]Glc, glucose; Gal, galactose; Rha, rhamnose; Man, mannose; GlcNAc, N-acetyl-glucosamine; GlcA, glucuronic acid; Py, pyruvate acetal.

[b]*p*, pyranose ring conformation; *f*, furanose ring conformation.

[c]The monosaccharides of this backbone are partially substituted at hydroxyl of C6 with β-D-Glc*p*.

and galactose there is a slight abundance of the β-anomer.

The predominant linkages in the backbone chain for glucose and galactose are made through the hydroxyls of C3 and C4. Less frequently these monosaccharides are attached in positions C6 (Table 16.3) (groups 5, 6, 14, 15, 17, 18, 21, 22, 23, 25, and 35) and C2 (groups 3, 8, 12, 19, 23, 25, 26, 30, 32, 34, and 41). Rhamnose is always attached via either the hydroxyl of C2 or C3. Exceptionally, a C5 linkage is present in the main chain of the polymers produced by the strains *Lactobacillus helveticus* TN4 and Lh59 (group 16) and *Bifidobacterium catenulatum* YIT4016 (group 39) in the three cases belonging to a β-D-galactofuranose. It is interesting to point out that in the backbone of the repeating unit of the HePS produced by three *Propionibacterium* strains (group 42) mannose is present linked at C3. This is the first HePS reported including mannose in its structure as determined by NMR spectroscopy (Dobruchoska *et al.*, 2008). Also, it has been indicated that one of the two EPS produced by *Lactobacillus pentosus* LPS26 has mannose and glucose (ratio 1:3) in its composition (Sánchez *et al.*, 2006) although no NMR structure is available to date.

Finally, the repeating units of the polymers produced by *Bifidobacterium infantis* ATCC15697 and *Pr. freudenreichii* subsp. *shermanii* JS and TL34 are formed by only two monosaccharides. The first case (Table 16.4) is a glucogalactan with a backbone with (1→3) linkages (Tone-Shimmokawa *et al.*, 1996) which could be partially substituted at the hydroxyl of C6 with β-D-Glc*p*. The second structure (Table 16.3, group 41) is a disaccharide composed exclusively of glucose [→3)-β-D-Glu*p*-(1→2)-β-D-Glc*p*-(1→] (Nordmark *et al.*, 2005b; Deutsch *et al.*, 2008). Given that the EPS JS was isolated from a whey-permeated medium and sucrose was not present in its formulation, it seems that the synthesis could be similar to that of β-glucan formerly indicated for the strain TL34 (Deutsch *et al.*, 2008).

The linkage types present in the backbone of the HePS, as well as the presence of side groups, are of special interest given that they confer different degrees of chain stiffness and water solubility properties and thereby, they produce variable viscosifying capabilities in aqueous systems. For example, monosaccharides connected by β-(1→4) bonds result in stiffer chains compared with α-(1→4) or β-(1→3) bonds (Bianchi *et al.*, 1986). In this way, the EPS produced by *L. lactis* subsp. *cremoris* B40 whose main backbone is linked with β-(1→4) bonds presented greater stiffness than the EPS B39 which contains α-(1→3) linkages (Ruas-Madiedo *et al.*, 2002b). A high content of (1→6) linkages could render polymers with a lower solubility degree (Laws *et al.*, 2001b). On the other hand, the branched linkage and presence of side groups also play an important role in the technological functionality of HePS. Tuinier *et al.* (2001) showed that the removal of terminally linked galactosyl residue on the side chain of EPS produced by the strains *L. lactis* subsp. *cremoris* B39 and B891 tended to decrease the chain stiffness and consequently the thickening efficiency. The presence of negative charge in the HePS containing phosphate (Table 16.3: groups 2, 20 and 27) causes, depending on the ionic strength, an increase in the intramolecular repulsion forces in the polymer chain. And hence, leads to an increase of the hydrodynamic volume and the intrinsic viscosity of the polymer in solution (Ruas-Madiedo *et al.*, 2002a, 2002b).

In addition to the primary structure of the repeating units, the molar mass of HePS play a key role in the viscosity intensifying ability of these polymers. The MM of HePS is generally lower than that of HoPS and the reported values varied among LAB species. The ranges indicated in literature oscillate from 4×10^4 to 9×10^6 Da for *St. thermophilus*, from 1×10^5 to 2×10^6 for *L. lactis* subsp. *cremoris*, and from 2.5×10^4 to 1.4×10^5 for some *Lactobacillus* strains (Ruas-Madiedo and de los Reyes-Gavilán, 2005b). Mozzi and co-authors (2006) have recently shown a broad MM range among the HePS produced by thermophilic and mesophilic LAB strains from dairy origin. They established an arbitrary classification between strains able to produce HePS with high MM (> 1×10^6 Da) and low MM (< 1×10^6 Da) and, noticeably, most of the strains produced low MM polymers (ranging between 8×10^3 and 9.5×10^5 Da). As far as we know, no data are available in literature reporting the MM of EPS from *Propionibacterium*. The values of *Bifidobacterium* strains showed a wide variability in the size of their EPS. The strain *Bifidobacterium adolescentis*

M101-4 seems to produce a water-soluble HePS fraction with an average MM estimated at 1.5×10^6 Da (Hosono *et al.*, 1997). The MM of the EPS produced by *Bifidobacterium longum* BB-79 is 3.5×10^5 Da (Roberts *et al.*, 1995) and that of *Bif. longum* YIT4028 is 1.4×10^4 Da (Nagaoka *et al.*, 1995). *Bif. longum* ATCC15707 produced two different MM EPS fractions of 3.5×10^5 and 1.2×10^6 Da (Abbad-Andaloussi *et al.*, 1995). Similarly, *Bif. infantis* ATCC15697 produced two EPS of different size: 9.4×10^3 and 1×10^5 Da (Tone-Shimokawa *et al.*, 1996). Results obtained in our group (P. Ruas-Madiedo, N. Salazar and C.G. de los Reyes-Gavilán, unpublished data) about the MM of EPS produced by *Bifidobacterium* strains from human microbiota origin also showed that the same strain could produce more than one EPS fraction with different M. This fact has already been described for some EPS from LAB. The culturing conditions, mainly media composition and temperature, could induce the production of more than one EPS (Sánchez *et al.*, 2006). Degeest and De Vuyst (1999) showed that the production level of two EPS with different MM, but having the same repeating unit composition, was affected by the composition of the culture medium. In the strain *St. thermophilus* LY03 a shift from a high MM HePS (1.8×10^6 Da) to a lower MM HePS (4.1×10^5 Da) was detected in presence of different nitrogen sources. In addition, the production of more than one type of EPS varying in size and chemical composition has also been reported, as occurred for a strain *L. delbrueckii* subsp. *bulgaricus* (Petry *et al.*, 2000).

The MM of the EPS can be measured by size exclusion chromatography (GPC-HPLC) using either standard (currently dextran or pullulan) calibration curves or using a MALLS (multi angle laser light scattering) detector (Ruas-Madiedo and de los Reyes-Gavilán, 2005b). By using this detector, the MM (kg/mol) and the radius of gyration [R_g (nm)] of the polymer can be determined and from that, the intrinsic viscosity (η) can be calculated [$\eta \approx 3.1 \times 10^{24} R_g{}^3/M$ (m^3/kg)] (Tuinier *et al.*, 1999a,b). For random coil EPS, R_g (≈ size) depends on MM (≈ length) and also on the flexibility (or thickening efficiency Ø) of the molecule ($R_g{}^2 \approx MM/Ø$). These parameters, as previously indicated, are related to the chemical composition, linkage type and presence of side chains in the repeating units that build the polymer. That means that in order to obtain high intrinsic viscosity, the MM of EPS should be high, thereby R_g will increase, and the repeating unit of EPS should be relatively stiff (Ruas-Madiedo *et al.*, 2002a,b). This could be illustrated by the work of Faber *et al.* (1998). They observed that milks fermented with the strains *St. thermophilus* Rs and Sts had different posthumus viscosity (36 and 126 s, respectively), even if both EPS were produced in equal amounts and their chemical composition was identical. However, the MM of the HePS Rs was lower (2.6×10^6 Da) than that of the HePS Sts (3.7×10^6 Da). In the same way, Tuinier *et al.* (2001) found that the release of the galactose located in the side chain of the EPS from *L. lactis* subsp. *cremoris* B39 by using a β-galactosidase from *Aspergillus aculeatus* (van Casteren *et al.*, 2000b) reduced the MM and the R_g of the resulting polymer, which reduced its intrinsic viscosity by half.

In spite of the high viscosifying capability at low concentrations of some HePS, due to their intrinsic physicochemical properties, these polymers have not been used until now as food additives because of their low yield. In general, HePS are produced in lower amounts (the yield does not exceed 0.6 g/l) than HoPS and its production is regulated by the composition of the culture medium (carbon and nitrogen sources, and other nutrients), growth conditions (pH, temperature, oxygen concentration, etc.) and incubation time (Degeest *et al.*, 2001). Also, as previously indicated, the isolation and measurement procedures (Ruas-Madiedo and de los Reyes-Gavilán, 2005b) as well as the presence of other polysaccharides (mainly glucomannans) in rich culture media (Vaningelgem *et al.*, 2004) could influence the reported EPS production of some strains. The latter authors found higher production of EPS by 26 *St. thermophilus* strains when the milk medium was supplemented with yeast extract and peptone, although, EPS isolated from the enriched non-fermented medium already yielded from 100 to 120 mg of PDM (powder dry material) per litre. However, the presence of yeast extract and peptone could enhance the growth and hence, the growth-associated EPS production. EPS yield also varies according to the producing LAB strain and values collected from literature oscillate from 25 to 600 mg/l (Ruas-Madiedo and de los Reyes-Gavilán, 2005b). Only a few exceptional strains are able to produce higher amounts of HePS under optimized culturing conditions. This is the case of *Lactobacillus rhamnosus* RW-9595M which produces around 2 g/l of polymer (Bergmaier *et al.*, 2005). Scarce articles report the yield of EPS produced by *Bif. longum* strains. For strain ATCC15707, the yield is around 140 mg/l (Abbad-Andaloussi *et al.*, 1995) and for strain BB-7 near 160 mg/l (Roberts *et al.*, 1995).

In short, the effect of EPS on the physical properties of dairy products depends not only on the amount of EPS produced during milk fermentation but also on the physico-chemical characteristics of the polymers. In addition, the interactions among EPS and other components of the food matrix, such as the milk protein network, play an important role in the development of the rheological and textural characteristics of the fermented milk-products (Laws and Marshall, 2001b; Ruas-Madiedo *et al.*, 2005a, 2008).

HePS clusters and related proteins

The genes involved in the synthesis of HePS are organized in complex clusters that show a highly conserved structural organization (van Kranenburg *et al.*, 1999a; Jolly and Stingelle, 2001; Boadbent *et al.*, 2003). The first *eps* cluster described for LAB from food origin was that of *St. thermophilus* Sfi6 (Stingelle *et al.*, 1996) consisting of a 14.4-kb region which comprises 13 genes. At present, several *eps* cluster from LAB have been characterized and their sequences have been deposited in gene bank databases (Table 16.5). Most studied species were the thermophilic *L. delbrueckii* subsp. *bulgaricus* and *St. thermophilus* due to their application as starters in yoghurt manufacturing to improve the viscosity and texture of the products. Also *L. lactis* subsp. *cremoris* strains, isolated from Scandinavian fermented milks, have received special attention due to the high viscosity intensifying capability of their HePS. In addition, *eps* clusters from other species, such as *L. helveticus, L. johnsonii* and *L. rhamnosus* are being characterized because of their putative probiotic interest. In addition, the analysis of the available *Bifidobacterium* genomes also reveals the presence of putative *eps* (also named *cps*) clusters. This is the case of *Bif. adolescentis* ATCC 15703 (Suzuki *et al.*, 2006), *Bifidobacterium dentium* ATCC 27678 (Sudarsanam *et al.*, 2008) and *Bif. longum* NCC2705 (Schell *et al.*, 2002). In addition, our group has recently shown the presence of HePS-producing *Bifidobacterium* strains among those isolated from the human microbiota using a primer targeting for the priming-GTF (Provencher *et al.*, 2003), which initiates the synthesis of the HePS repeating units. The predicted amino acid sequences belonging to priming-GTF of several *Bif. longum* and *Bifidobacterium pseudocatenulatum* strains showed identity (83 to 95%) with two putative priming-GTF (RfbP and CpsD) from *Bif. longum* NCC2705 (Ruas-Madiedo *et al.*, 2007). However, as far as we know, neither data are available on the gene data bases, nor are articles reporting genes involved in HePS production from *Propionibacterium* genus. The EPS-producing phenotype is mainly plasmid-associated in *Lactococcus* strains (van Kranenburg *et al.*, 1997, 1999b; Forde and Fitzgerald, 2003; Knoshaug *et al.*, 2007) with the exception of the strain *L. lactis* subsp. *cremoris* SMQ-461 (Dabour and LaPointe, 2003). In this strain, a 13.2-kb chromosomal region consisting of 15 ORFs is flanked by *tnp* genes at the 5′ and 3′ ends, which could explain the possible transfer of the *eps* cluster from a plasmid. In contrast, for *Lactobacillus* and *Streptococcus* strains, EPS production is found to be encoded by the chromosome (Stingele *et al.*, 1996; Almirón-Roig *et al.*, 2000; Lamothe *et al.*, 2002; Péant *et al.*, 2005). The organization of the *eps* clusters of these LAB are highly comparable with *cps* clusters in *St. pneumoniae* (García and López, 1997; Kolkman *et al.*, 1997; Morona *et al.*, 1997) and *Streptococcus agalactiae* (Yamamoto *et al.*, 1999) which are involved in the synthesis of capsular polysaccharides (CPS) in these bacteria.

All *eps* clusters from LAB have an operon structure with a high coding density and most genes, which are usually oriented in one direction, are transcribed as a single mRNA. Although the function of most EPS-related genes is only inferred from sequence homologies, the structural organization of *eps* clusters from LAB is strikingly maintained (Fig. 16.5). These genes encode enzymes that are involved in regulation, biosynthesis of the repeating units, chain length determination, polymerization, and export (Jolly

Table 16.5 Clusters of *eps* genes found in the GeneBank database (http://www.ncbi.nlm.nih.gov) as involved in the synthesis of HePS from LAB

Genus	Strain	*eps* cluster			Accession no.	Reference
		Location[a]	Size (kb)	No. of genes		
Lactococcus	*Lc. lactis* subsp. cremoris NIZO B40	Plasmid	12	14	AF036485	van Kranenburg *et al*. (1997, 2000)
	Lc. lactis subsp. cremoris NIZO B35	Plasmid	5.4	10	AF100297	van Kranenburg *et al*. (1999b)
	Lc. lactis subsp. cremoris NIZO B891	Plasmid	2.9	8	AF100298	van Kranenburg *et al*. (1999b)
	Lc. lactis subsp. cremoris HO2	Plasmid	21.8	23	AF142639	Forde and Fitzgerald (2003)
	Lc. lactis subsp. cremoris SMQ-461	Chromosome	13.2	15	AY741550	Dabour and LaPointe (2005)
	Lc. lactis subsp. cremoris Ropy352	Plasmid	22.4	19	EF192213	Knoshaug *et al*. (2007)
Lactobacillus	*Lb. delbrueckii* subsp. bulgaricus Lfi5	Chromosome	18	14	AF267127	Lamothe *et al*. (2002)
	Lb. helveticus NCC2745	Genome	14	11	AX009415.1	Jolly *et al*. (2002)
	Lb. johnsonii ATCC33200	Genome	22,2	15	EF138833	Berger *et al*. (2007)
	Lb. johnsonii ATCC11506	Genome	36.5	29	EF138834	Berger *et al*. (2007)
	Lb. johnsonii NCC2767	Genome	24.6	23	EF138835	Berger *et al*. (2007)
	Lb. rhamnosus ATCC9505	Chromosome	18.5	17	AY659976	Péant *et al*. (2005)
	Lb. rhamnosus RW6541M	Chromosome	18.5	17	AY659977	Péant *et al*. (2005)
	Lb. rhamnosus R	Chromosome	18.5	17	AY659978	Péant *et al*. (2005)
	Lb. rhamnosus RW9595M	Chromosome	18.5	17	AY659979	Péant *et al*. (2005)
Streptococcus	*St. macedonicus* Sc136	Genome	NI	18	-	Jolly and Stingelle (2001)
	St. thermophilus Sfi6	Chromosome	14.5	13	U40830	Stingelle *et al*. (1996)
	St. thermophilus NCFB 2393	Chromosome	11.2	12	Y17900	Almirón-Roig, *et al*. (2000)
	St. thermophilus Sfi39	Genome	20	15	AF373595	Germond *et al*. (2001)
	St. thermophilus MR-2C	Genome	23.2	17	AY061649	Broadbent *et al*. (2003)
	St. thermophilus MR-310	Genome	27.8	14	AF448502	Broadbent *et al*. (2003)
	St. thermophilus MR-360	Genome	23.2	17	AF434993	Broadbent *et al*. (2003)
	St. thermophilus MR-1C	Genome	44.4	19	AF448249	Broadbent *et al*. (2003)
	St. thermophilus MTC330	Genome	23.2	14	AF430847	Welker *et al*., (2004)
	St. thermophilus LY0	Genome	14.7	15	DQ393658	de Vin *et al*. (2006)

[a]The location of *eps* cluster in the genome means that the reference did not specifically indicate whether it was plasmidic or chromosomic.

NI, not indicated.

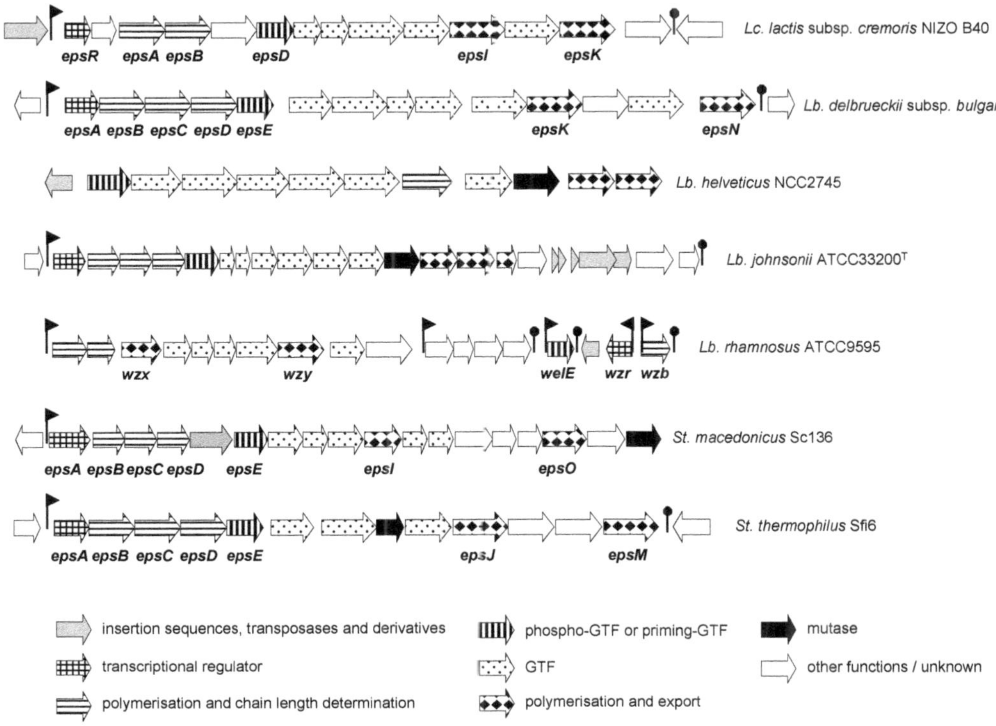

Figure 16.5 Physical maps of the *eps* clusters involved in the synthesis of HePS in some representative LAB species. Putative or established functions for individual genes are presented in different grey and black marks. Maps are not scaled, flags represent potential ORF and pins represent putative terminators; (Modified from van Kranenburg *et al.* (1999a), Jolly and Stingelle (2001), and Broadbent *et al.* (2003)).

and Stingele, 2001; Broadbent *et al.*, 2003). The core of the *eps* clusters is occupied by genes involved in the synthesis of the repeating units, which are built-up by the sequential activity of GTF (glycosyltransferases). These enzymes catalysed the transfer of sugar moieties from activator donor molecules to specific acceptor molecules, thereby forming a glycosidic bond. GTF using nucleoside diphospho-sugars, nucleoside monophospho-sugars and sugar phosphates can be classified into distinct sequence-based families (EC.2.4.1.X., http://www.cazy.org) and were first described by Campbell *et al.* (1997). The initial step in the synthesis of the HePS repeating units is the transfer of a sugar-1-phosphate to a lipophilic carrier molecule that is anchored in the membrane. This is achieved by the priming-GTF or undecaprenyl-glycosyl-1-phosphate transferase, which, unlike other GTF recognizes the lipid carrier as well as the sugar residue and does not catalyse the glycosidic linkage.

In general, *gtf* genes have little similarity to each other (Jolly and Stingele, 2001; Provencher *et al.*, 2003) but *priming-gtf* genes are similar in Gram-positive and Gram-negative bacteria, particularly in the C-terminus (Wang *et al.*, 1996; Provencher *et al.*, 2003). The C-terminal aminoacidic sequence of priming-GTF defines a conserved region showing high similarity with the regions Bactransf (bacterial sugar transferases of the Pfam family, accession number PF02397) and WcaJ (sugar transferases involved in the lipopolysaccharide synthesis, accession number COG2148). The EpsE from *St. thermophilus* Sfi6 and CNRZ1066 are phospho-galactosyltransferases (Stingelle *et al.*, 1996; Minic *et al.*, 2007) while other transfer glucose, such as CpsE from *St. thermophilus* NCFB2393 (Almirón-Roig *et al.*, 2000), EpsD from *L. lactis* subsp. *cremoris* NIZO B40 (van Kranenburg *et al.*, 1997) and WelE from *L. rhamnosus* RW-9595M (Péant *et al.*, 2005). Interestingly, the EPS gene clusters reported to date (Fig. 16.5) present the *priming-gtf* gene just upstream of all other *gtf* genes. But in the four *L. rhamnosus* strains examined by Péant *et al.* (2005) the *welE* encoding the priming-GTF is located after the other GTF and is also transcribed from its own promoter. This suggests an independent expression of the priming-GTF that may lead to higher amounts of mRNA, and thus to an improved initiation of the HePS repeated-units biosynthesis.

The functionality of *gtf* genes has been studied by homologous and heterologous expression. Specifically, the functional analysis of *epsE* from *St. thermophilus* Sfi6 (Stingele *et al.*, 1999) and *cpsE* from *St. thermophilus* NCFB2393 (Almirón-Roig *et al.*, 2000) has demonstrated the priming-GTF activity of these enzymes. Recently, it has also been demonstrated that the expression of *epsABCD* from *L. lactis* subsp. *cremoris* B40 is necessary for the heterologous production of the HePS from *St. pneumoniae* serotype 14 in the lactococci. The *epsD* was the encoding for the priming-GTF of strain B40 (Nierop-Groot *et al.*, 2008). With respect to GTF other than priming-GTF, a summary of putative *gtf* genes described from LAB encoding a variety of α- and β-GTF were collected (Jolly and Stingele, 2001). Their sugar specificity was mainly postulated from protein sequence comparisons and few functional studies have been undertaken. The assignment of GTF activities has been carried out mainly with strains *St. thermophilus* Sfi6 (Stingele *et al.*, 1999) and *L. lactis* subsp. *cremoris* strains (van Kranenburg *et al.*, 1999b). In the last, EpsE and EpsF link glucose to lipid-linked glucose and EpsG links galactose to lipid-linked cellobiose. In *St. thermophilus* Sfi6 EpsG is able to transfer N-acetyl-glucosamine on a lipid-linked precursor harbouring a β-galactose. The cloning and expression of GTF appear to be very difficult and dependent on the expression system, which could account for the low level of success in functionality studies (Jolly and Stingele, 2001). Finally, as far as is known, the three dimensional structure of GFT from LAB has not been reported, which is an essential requirement for a better understanding of the mode of action of these enzymes.

The 5'region of each *eps* cluster (Fig. 16.5) appears to encode proteins involved in the regulation of HePS synthesis. In *St. thermophilus* Sfi6 the gene product of *epsA* has significant homology with the LytR protein from *B. subtilis* which is involved in the regulation of the autolysin *lytABC* operon (Stingelle, *et al.*, 1996). It seems that the *epsA* regulatory gene is conserved in other LAB with homology values varying from 29%, for *St. macedonicus* Sc136, to 95%, for the *epsR* from *L. lactis* subsp. *cremoris* B40 (Jolly and Stingele, 2001). Similarly, *L. johnsonii* strains (ATCC33200^T, ATCC11506, NCC533, and NCC2767) and *Lactobacillus gasseri* (ATCC33323^T) conserved the putative transcriptional regulator gene in 5'-end position (Berger *et al.*, 2007). But in the strain *L. rhamnosus* RW9595-M the putative regulator is located in the 3'-end region. In addition, the *wzr* regulator gene, whose predicted product shows 31% identity with EpsA from *L. delbrueckii* subsp. *bulgaricus* Lfi5, is organized in the opposite transcriptional direction (Péant *et al.*, 2005). These regulator proteins contain three putative transmembrane segments at their N-terminal end. They are also predicted to have a signal anchor and contain most likely extracellular domains (Jolly and Stingelle, 2001). But currently there are no further indications about the detailed regulation mechanism of HePS syntheses.

Also located in the 5'-end region of each *eps* cluster are the genes involved in polymerization and chain length determination of HePS. The products of the *eps* gene clusters are homologues to those of Gram-negative bacteria such as the *O*-antigen polymerization and export proteins Wzx, Wzy and Wzz. Indeed, Wzz proteins in *E. coli* are thought to be involved in the chain length determination of the *O*-antigen, because *wzz* mutants have random length distribution of this antigen. Also, the ExoP-like proteins from *Rhizobium meliloti* as well as CpsC and CpsD from *St. pneumoniae* putatively have the same function. These proteins share 21–61% identity with the EpsC and EpsD from *St. thermophilus* Sfi6 and Sfi39 and *St. macedonicus* Sc136, with the EpsB and EpsC from *L. delbrueckii* subsp. *bulgaricus* Lfi5 and with the EpsA and EpsB from *L. lactis* subsp. *cremoris* B40 (Jolly and Stingelle, 2001). Similarly, *wzz* putative homologues are present *in L. johnsonii* and *L. gasseri* strains (Berger *et al.*, 2007). These ExoP-like proteins have a transmembrane domain, containing two putative transmembrane helices, and a domain containing a nucleotide-binding motif. Both domains are present in a single protein (ExoP) in *R. meliloti* but it is split into two proteins in the case of the LAB, as previously indicated. The EpsC (*St. thermophilus* Sfi6 and Sfi39 and *St. macedonicus* Sc136), EpsB (*L. delbrueckii* subsp. *bulgaricus* Lfi5) and EpsA (*L. lactis* subsp. *cremoris* NIZO B40) are homologues of ExoP (*R. meliloti*), Wzz (*E. coli*), CpsB (*St. agalactiae*) and CpsC (*St. pneumoniae*) proteins. The Wzz proteins have two potential transmembrane domains and are located in the cytoplasmic membrane with the central domain exposed to the periplasm (Whitfield and Roberts, 1999). They share their hydrophobicity profile and the central domain, which is predicted to be extracellular, with the LAB homologues (Jolly and Stingele, 2001). The homologous nucleotide binding motif of ExoP from *R. meliloti* and Wzc from *E. coli* in LAB is encoded by *espD* and *cpsD* (*Streptococcus*), *epsC* (*Lactobacillus*) and *espB* (*Lactococcus*) (Stingelle *et al.*, 1996; van Kranenburg *et al.*, 1999b; Lamothe *et al.*, 2002).

Protein Wzc and the homologous ExoP have been identified as an autophophorylating protein-tyrosine kinase associated with a specific phosphotyrosine-protein phosphatase Wzb (Vincentet al., 1999). In *St. pneumoniae* the homologous proteins for Wzc are CpsC and CpsD and for Wzb is CpsB (Morona *et al.*, 2000). It has been reported that several HePS-producing strains of *L. rhamnosus* present *wzb* gene, whose protein shows high sequence identity with that of *L. bulgaricus* Lfi5 (51%) and with CpsB of *St. pneumoniae* (Peant *et al.*, 2005). Also, most *eps* clusters from LAB contain a gene product that shares homology with the CpsB of the pneumococci (Jolly and Stingele, 2001). In *St. pneumoniae* the activity of the CpsC, CpsD and CpsB proteins is needed for the CPS formation (Morona *et al.*, 2000). However, it is not clear how the LAB homologous could interact for the synthesis of HePS. Recently, Minic *et al.* (2007) have shown in *St. thermophilus* CNRZ1066 that the HePS synthesis took place in mutants lacking EpsB but not lacking EpsC or EpsD. An EpsD phosphorylated form was found in all strains except the *epsC* mutant, indicating that EpsC is necessary for EpsD phosphorylation. Moreover, the phosphorylated form of EpsD was found to be associated with the plasmatic membrane, possibly due to the interaction with EpsC. However, the most interesting result was that it seems that EpsC and EpsD are required for the function of EpsE (the priming-GTF). The mutation of Tyr200, which is predicted to be a phosphorylation site and is present as conserved motif in several priming-GTF, leads to EpsE inactivation. This suggests that EpsD (a tyrosine-kinase) controls the EpsE activity. A mechanism of EpsE regulation has been proposed taking into account this fact and the well-established data for streptococci (Minic *et al.*, 2007). EpsE could form a high-molecular-weight complex lacking priming-GTF activity. When EpsC is present in the membrane, it recruits EpsD which is then phosphorylated. The phosphorylated EpsD, possibly in conjunction with EpsC, could attract EpsE into a complex and activate the priming-GTF activity through the phosphorylation of the Tyr200 residue. The role of EpsB could be to dephosphorylate either EpsD or EpsE, as has been demonstrated for its homologous CpsD for *St. pneumoniae* (Bebder and Yother, 2001). Nevertheless, the mechanism of HePS chain-length regulation is still far from being totally elucidated.

Regarding the *eps* cluster region involved in polymerization and export, it is generally located downstream (3′ region) of the central *eps* cluster core (Fig. 16.5). The proteins Wzx (RfbX) from *E. coli* and ExoT from *R. meliloti* are thought to be related to the export of bacterial polysaccharides. Wzx, also called 'flippase', seems to translocate the undecaprenyl-linked *O*-antigen subunits across the membrane to show the carbohydrate moiety in the periplasmic side (Whitfield and Roberts, 1999). The homologous LAB proteins, showing moderate homology with these export proteins, are EpsK in *L. lactis* subsp. *cremoris* B40, EspN in *L. delbrueckii* subsp. *bulgaricus* Lfi5, Eps10 in *L. helveticus* NCC2745, EspO in *St. macedonicus* Sc136, EpsM in *St. thermophilus* Sfi6 (Jolly and Stingele, 2001) and Wzx in *L. rhamnosus* strains (Péant *et al.*, 2005). Remarkably, in the *L. rhamnosus* strains the *wzx* gene is located in the 5′ region of the *eps* cluster. A putative *wzx* flippase gene has been described for *L. johnsonii* and *L. gasseri* strains (Berger *et al.*, 2007), all situated in the 3′ region of the *eps* cluster core.

The protein Wzy from *E. coli* is the polymerase of the *O*-antigen synthesis (Drummelsmith and Whitfield, 2000) and has similar hydrophobicity plots to that of the LAB homologous (Jolly and Stingele, 2001). EpsI from *L. lactis* subsp.

cremoris B40 and *St. macedonicus* Sc136, EpsJ from *St. thermophilus* Sfi6 and Sfi39 and EpsK from *L. delbrueckii* subsp. *bulgaricus* Lfi5 are similar to this polymerase (Stingele *et al.*, 1996; van Kranenburg *et al.*, 1997; Lamothe *et al.*, 2002). Also, Wzy from *L. rhamnosus* RW-9595M shares a protein-sequence identity of 20% with those of *St. thermophilus* (Péant *et al*, 2005) and *wzy* is present also in the *eps* clusters of *L. johnsonii* and *L. gasseri* (Berger *et al.*, 2007). It is still not known whether the export of the HePS repeating units in LAB occurs through the Wzy homologous. However, the *O*-antigens synthesized by this mechanism are also heteropolymers and the undecaprenyl-phosphate lipid carrier (C_{55}) is employed to link the repeating units (Withfield and Roberts, 1999), as it seems to occur in HePS synthesis. Other mechanisms could also be possible given that the protein Eps11 from *L. helveticus* NCC2745 does not share homology with Wzy, but with the inner membrane protein of *St. pneumoniae* (Jolly and Stingele, 2001).

The presence of mobile elements bordering the *eps* clusters is also frequent (Fig. 16.5). In the case of *L. lactis* subsp. *cremoris* SMQ-461 several genes are potentially involved in DNA recombination and mobility functions. Some of these sequences show amino acid homology with transposase IS1077E and yuil, and transposon-related ygcE from *L. lactis* subsp. *lactis* IL-1403 and IS1253-like transposase protein from *St. thermophilus* (Dabour and LaPointe, 2005). Many strains of *Lactococcus* have been shown to carry copies or partial sequences of insertion elements (van Kranenburg *et al.*, 1997), some of them identified as belonging to ISS1, IS981 and IS982 families (Forde and Fitzgerald, 2003; Dabour and LaPointe, 2005; Knosaugh *et al.*, 2007). Within the region encoding for HePS biosynthesis, *orf1* encodes a truncated transposase of the IS1165 family in *L. rhamnosus* strains (Péant *et al.*, 2005). Also, by means of diverse genetic tools as well as *in silico* analysis of several genomes from *L. johnsonii* strains, the presence of genes associated with mobile DNA was very often found close to EPS gene clusters (Berger *et al.*, 2007). Mobile elements are well known for enabling genetic horizontal exchange among different genera. At the same time, the presence of the mobile elements could also explain the instability of HePS phenotype in some strains.

HePS synthesis mechanisms

The synthesis of HePS involves the employment of intermediate sugar nucleotides (nucleoside phosphorylated-sugars) which are formed from intermediates of the central carbon metabolism. A key intermediate linking the anabolic pathways of HePS production and the catabolic pathways of sugar degradation appears to be the glucose-1-phosphate which is derived from glucose-6-phosphate by the activity of the phosphoglucomutase. At this point, the flux of carbon bifurcates towards the glycolysis for ATP formation and cell component synthesis, and towards the biosynthesis of sugar nucleotides for HePS formation. Several studies have demonstrated a correlation between HePS synthesis and the activity of sugar nucleotide biosynthetic enzymes (Boels *et al.*, 2001a,b; De Vuyst *et al.*, 2001). Other intermediate molecules are shared for the synthesis of the cell wall-component, such as peptidoglycan, teichoic and lipoteichoic acids, and for the synthesis of HePS. An example is the undecaprenyl-phosphate lipid carrier, which is located in the cellular membrane. Consequently, sugar catabolism, cell wall formation, and synthesis of HePS are cellular processes that are competing in EPS-producing LAB and the reactions related to these metabolic pathways require the activity of several housekeeping enzymes (Broadbent *et al.*, 2003).

The mechanism of biosynthesis of HePS and HePS-like molecules (such as *O*-antigen) is relatively well-defined for pathogenic Gram-negative bacteria (such as *E. coli* and *Salmonella*) given that these polymers play an important role in virulence. In these bacteria, the assembly of the basic repeating unit occurs at the cytoplasmatic membrane and involves sequential transfer of diphospho-sugar nucleotides to the lipid carrier C_{55}. Once the basic repeating unit is assembled, the lipid-linked intermediates are usually translocated across the membrane and polymerized outside the cell by means of extracellular polymerases. Then, the HePS may be covalently linked to the cell to form a capsule, or released into the media as loosly associated EPS slime (Sutherland, 2001). Several authors have studied the HePS synthesis of pathogenic Gram-positive bacteria such as *St. pneumoniae, Streptococcus pyogenes* and *St. agalactiae* (Broadbent *et al.*, 2003). From this work it has been established that streptococci produced HePS by at least two different mechanisms. Syntheses of simple HePS, such as the hyaluronic acid of *St. pyogenes* or the linear HePS (consisting of a disaccharide repeating unit) from *St. pneumoniae* serotype 3, do not include the formation of lipid-linked intermediates, whereas synthesis of more complex HePS does (Yother, 1999). Functional studies and homology comparisons described in previous sections, mainly for *St. thermophilus* Sfi6 and *L. lactis subsp. cremoris* B40 strains, strongly suggest that the HePS synthesis in LAB requires the formation of lipid-linked intermediates (Stingele *et al.*, 1996, 1999; van Kranenburg *et al.*, 1997, 1999b; Almirón-Roig *et al.*, 2000; Germond *et al.*, 2001).

From the work with these strains a HePS putative mechanism of biosynthesis could be proposed (Fig. 16.6). The first step of assembly of the repeating units is carried out by the priming-GTF (EpsE for strain Sfi6 and EpsD for B40) which attaches the corresponding nucleotide diphosphorylated-sugar (UDP-galactose for Sfi6 and UDP-glucose for B40) to the undecaprenyl-phosphate lipid carrier C_{55}. This reaction is followed by the action of other GTF that sequentially add the monosaccharides from their corresponding UDP-sugar precursor to form the monosaccharide sequence. For *St. thermophilus* Sfi6 three GTF (EpsG, EpsI and EpsF) attach N-acetyl-galactosamine, glucose and galactose to complete the tetrasaccharide repeated unit. In *L. lactis* subsp. *cremoris* B40, glucose, galactose, rhamnose and galactose-phosphate (*epsE, epsF, epsG, epsH,* and *epsJ*) are branched to build the pentasaccharide repeating unit. Afterwards the repeated units are predicted to translocate across the membrane through an unknown mechanism which could involve the activity of a 'flippase'-like protein (EpsM for Sfi6 and EpsK for B40). The polymerization of the repeating units will take place by means of a polymerase (EpsJ for Sfi6 and EpsI for B40) and several enzymes will be involved in the chain-length determination (EpsC/Eps D/EpsB for Sfi6 and EpsA/EpsB for B40). Finally, the undecaprenyl-phosphate lipidic carried will be re-translocated and dephosphorylated (Boels *et al.*, 2001b; Broadbent *et al.*, 2003). This synthesis mechanism is very similar to that proposed for the production of xanthan from *X. campestris* which has been more extensively studied (van Kranenburg *et al.*, 1999a). However, the export, polymerization and chain-length de-

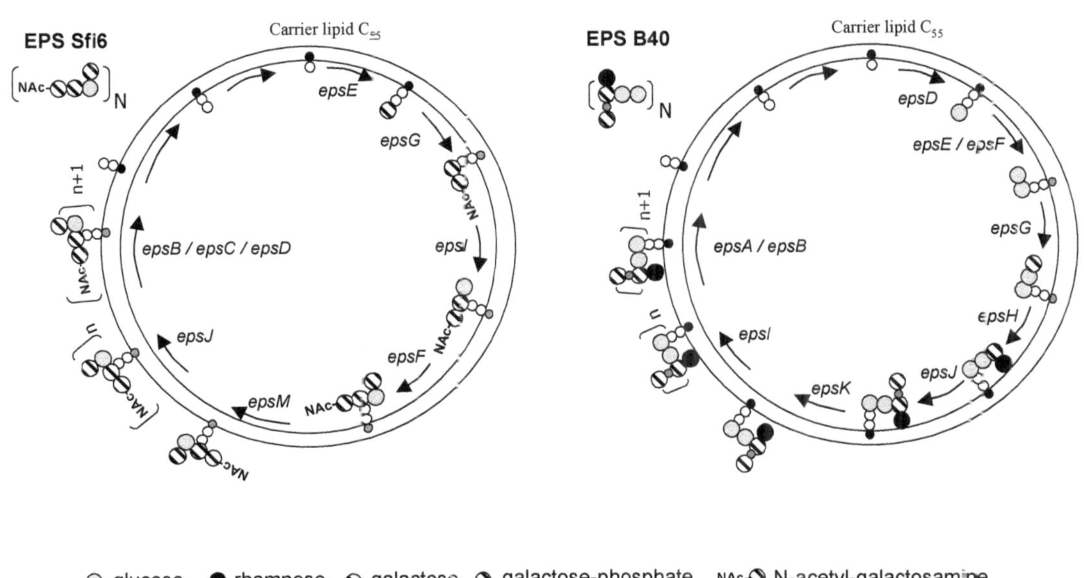

Figure 16.6 Model for the synthesis of HePS from *Streptococcus thermophilus* Sfi6 and *Lactococcus lactis* subsp. *cremoris* B40 (Modified from Boels *et al.* (2001b) and Broadbent *et al.* (2003)).

termination, as well as the overall HePS-synthesis regulation process is still far from being totally understood in LAB.

Conclusions

LAB are able to produce a wide variety of EPS varying in chemical composition, structure and physical properties. Contrary to this, the highly conserved structure of the genetic determinants for EPS synthesis, either HoPS or HePS, suggests a common mechanism of biosynthesis, which is still far from being totally elucidated, mainly for HePS. The rapid development of genetic techniques and analysis of whole or partial genome sequences will be valuable tools to better understand this process. However, the purification and structural characterization of enzymes involved in EPS synthesis from LAB still needs to be undertaken. The knowledge of biosynthesis mechanism is the first step towards a successful enhancement of EPS production. This could be achieved either by the optimizing the yield under defined fermentation conditions, or by genetic engineering towards overexpressing EPS production or producing polymers with different compositions, branching patterns and/or molar mass designed for specific applications. However, it is important to point out that genetically modified bacteria and their products still require legal approval and need to be accepted by the consumer, at least in the European frame, before being applied at industrial level. Finally, in addition to the technological or functional applications of the EPS produced by LAB, a future challenge will be to gain insight into the role of these polymers for the producing bacteria. EPS synthesis is a complex and energy consuming process which directly competes with the metabolism of carbohydrates and the bacterial cell wall formation. But, cell-associated EPS production seems to be an important and stable phenotype for some bacteria to be able to attach and to colonize certain environments, such as diverse human mucosa mucoses.

Acknowledgements

The authors are grateful for the financial support of their research work in projects AGL2006-03336 and AGL2007-62736/ALI by the Spanish Ministry of Education and Science (MEC). N. Salazar also acknowledges the Spanish MEC for her FPI pre-doctoral fellowship.

References

Abbad-Andalousii, S., Talbaoui, H., Marczak, R., and Bonaly, R. (1995). Isolation and characterization of exocellular polysaccharide produced by *Bifidobacterium longum*. Appl. Microbiol. Biotechnol. *43*, 995–1000.

Almirón-Roig, E., Mulholland, F., Gasson, M.J., and Griffin, A.M. (2000). The complete *cps* gene cluster from *Streptococcus thermophilus* NCFB 2393 involved in the biosynthesis of a new exopolysaccharide. Microbiology *146*, 2793–2802.

Anwar, M.A., Kralj, S., van der Maarel, J.E.C., and Dijkhuizen, L. (2008). The probiotic *Lactobacillus johnsonii* NCC 533 produces high-molecular-mass inulin from sucrose by using an inulosucrase enzyme. Appl. Environ. Microbiol. *74*, 3426–3433.

Axelsson, L. (2004). Lactic acid bacteria: classification and physiology. In Lactic Acid Bacteria. Microbial and Functional Aspects, S. Salminen, A. von Wright, and A. Ouwehand, eds. (New York: Marcel Dekker, Inc.), pp. 1–66.

Bebder, M.H., and Yother, J. (2001). CpsB is a modulator of capsule-associated tyrosine kinase activity in *Streptococccus pneumoniae*. J. Biol. Chem. *276*, 47966–47974.

Berger, B., Pridmore, R.D., Barretto, C., Delmas-Julien, F., Schereiber, K., Arigoni, F., and Brussow, H. (2007). Similarity and differences in the *Lactobacillus acidophilus* group identified by polyphasic analysis and comparative genomics. J. Bacteriol. *189*, 1311–1321.

Bergmaier, D., Champagne, C.P., and Lacroix, C. (2005). Growth and exopolysaccharide production during free and immobilized cell chemostat culture of *Lactobacillus rhamnosus* RW-9595M. J. Appl. Microbiol. *98*, 272–284.

Bianchi, E., Ciferri, A., Conio, G., Lanzavecchia, L., and Terbojevich, M. (1986). Mesophase formation and chain rigidity in cellulose and derivates. 5. Cellulose acetate in N, N-dimethylacetaminde. Macromolecules *19*, 630–636.

Boels, I.C., Ramos, A., Kleerebezem, M., and de Vos, W.M. (2001a). Functional analysis of the *Lactococcus lactis galU* and *galE* genes and their impact on sugar nucleotide and exopolysaccharide biosynthesis. Appl. Environ. Microbiol. *67*, 3033–3040.

Boels, I.C., van Kraneburg, R., Hugenholtz, J., Kleerebezem, M., and de Vos, W.M. (2001b). Sugar catabolism and its impact oh the biosynthesis and engineering of exopolysaccharide production in lactic acid bacteria. Int. Dairy J. *11*, 723–732.

Bubb, W.A., Urashima, T., Fujiwara, R., Shinnai, T., and Ariga, H. (1997). Structural characterisation of the exocellular polysaccharide produced by *Streptococcus thermophilus* OR901. Carbohyd. Res. *301*, 41–50.

Broadbent, J.R., McMahon, D.J., Welker, D.L., Oberg, C.J., and Moineau, S. (2003). Biochemistry, genetics, and applications of exopolysaccharide production in

Streptococcus thermophilus: a review. J. Dairy Sci. *85*, 407–423.

Campbell, J.A., Davies, G.J., Bulone, V., and Henrissat, B. (1997). A classification of nucleotide-diphospho-sugar glycosyltransferases based on amino acid sequence similarities. Biochem. J. *326*, 929–939. [Erratum published in Biochem. J. (1998) *329*, 719].

Cerning, J. (1990). Exocellular polysaccharides produced by lactic acid bacteria. FEMS Microbiol. Rev. *87*, 113–130.

Chambert, R., Treboul, G., and Dedonder, R. (1974). Kinetic studies of levansucrase of *Bacillus subtilis*. Eur. J. Biochem. *41*, 285–300.

Dabour, N., and LaPointe, G. (2005). Identification and molecular characterization of the chromosomal exopolysaccharide biosynthesis gene cluster from *Lactococcus lactis* subsp. *cremoris* SMQ-461. Appl. Environ. Microbiol. *71*, 7414–7425.

Dal Bello, F., Walter, J., Hartel, C., and Hammes, W.P. (2001). *In vitro* study of prebiotic properties of levan-type exopolysaccharides from lactobacilli and non-digestible carbohydrates using denaturing gradient gel electrophoresis. Syst. Appl. Microbiol. *24*, 232–237.

Degeest, B., and De Vuyst, L. (1999). Indication that the nitrogen source influences both amount and size of exopolysaccharides produced by *Streptococcus thermophilus* LY03 and modelling of the bacterial growth and exopolysaccharide production in a complex medium. Appl. Environ. Microbiol. *65*, 2863–2870.

Degeest, B., Vaningelgem, F., and De Vuyst, L. (2001). Microbial physiology, fermentation kinetics, and process engineering of heteropolysaccharides production by lactic acid bacteria. Int. Dairy J. *11*, 747–757.

Deutsch, S.M., Falentin, H., Dols-Lafargue, M., LaPointe, G., and Roy, D. (2008). Capsular exopolysaccharides biosynthesis gene of *Propionibacterium freudenreichii* subsp. *shermanii*. Int. J. Food Microbiol. *125*, 352–358.

de Vin, F., Herman, L., and De Vuyst, L. (2006). Biodiversity of *eps* gene clusters and their glycosyltransferase in *Streptococcus thermophilus*. Published in the GeneBank database (http://www.ncbi.nlm.nih.gov) Accession number DQ393658.

De Vuyst, L., de Vin, F., Vaningelgem, F., and Degeest, B. (2001) Recent developments in the biosynthesis and applications of heteropolysaccharides from lactic acid bacteria. Int. Dairy J. *11*, 687–707.

Dobruchowska, J.M., Gerwig, G.J., Babuchowski, A., and Kamerling, J.P. (2008). Structural studies on exopolysaccharides produced by three different propionibacteria strains. Carbohyd. Res. *343*, 726–745.

Doco, T., Wieruzeski, J.M., Fournet, B., Carcano, D., Ramos, P., and Loones, A. (1990). Structure of an exocellular polysaccharide produced by *Streptococcus thermophilus*. Carbohyd. Res. *198*, 313–321.

Doels-Lafargue, M., Lee, H.Y., Le Marrec, C., Heyraud, A., Chambat, G., and Lonvaud-Funel, A. (2008). Characterization of *gtf*, a glucosyltransferase gene in the genomes of *Pediococcus parvulus* and *Oenococcus oeni*, two bacterial species commonly found in wine. Appl. Environ. Microbiol. *74*, 4079–4090.

Drummelsmith, J., and Whitfield, C. (2000). Translocation of group 1 capsular polysaccharide to the surface of *Escherichia coli* requires a multimetic complex in the outer membrane. EMBO J. *19*, 57–66.

Dueñas-Chasco, M.T., Rodríguez-Carvajal, M.A., Tejero-Mateo, P., Franco-Rodríguez, G., Espartero, J.L., Irastorza-Iribas, A., and Gil-Serrano, A.M. (1997). Structural analysis of the exopolysaccharide produced by *Pediococcus damnosus* 2.6. Carbohyd. Res. *303*, 453–458.

Dueñas-Chasco, M.T., Rodríguez-Carvajal, M.A., Tejero-Mateo, P., Espartero, J.L., Irastorza-Iribas, A., and Gil-Serrano, A.M. (1998). Structural analysis of the exopolysaccharides produced by *Lactobacillus* sp. G-77. Carbohyd. Res. *307*, 125–133.

Faber, E.J., Zoon, P., Kamerling, J.P., and Vliegenthart, J.F.G. (1998). The exopolysaccharides produced by *Streptococcus thermophilus* Rs and Sts have the same repeating unit but differ in viscosity of their milk cultures. Carbohyd. Res. *310*, 269–276.

Faber, E.J., Kamerling, J.P., and Vliegenthart, J.F.G. (2001a). Structure of the extracellular polysaccharide produced by *Lactobacillus delbrueckii* subsp. *bulgaricus* 291. Carbohyd. Res. *331*, 183–194.

Faber, E.J., van den Haak, M.J., Kamerling, J.P., and Vliegenthart, J.F.G. (2001b). Structure of the exopolysaccharide produced by *Streptococcus thermophilus* S3. Carbohyd. Res. *331*, 173–182.

Faber, E.J., van Haaster, D.J., Kamerling, J.P., and Vliegenthart, J.F.G. (2002). Characterization of the exopolysaccharide produced by *Streptococcus thermophilus* 8S containing an open chain nononic acid. Eur. J. Biochem. *269*, 5590–5598.

Forde, A., and Fitzgerald, G.F. (2003). Molecular organization of exopolysaccharide (EPS) encoding genes on the lactococcal bacteriophage adsorption blocking plasmid, pCI658. Plasmid *49*, 130–142.

García, E., and López, R. (1997). Molecular biology of capsular genes of *Streptococcus pneumoniae*. FEMS Microbiol. Lett. *149*, 1–10.

Garrity, G.M., Johnson, K.L., Bell, J., and Searles, D.B. (2002). Taxonomic outline of the prokaryotes. Bergey's Manual of the Systematic Bacteriology, Release 3.0, July, Bergey's Manual Trust (New York, USA: Springer-Verlag New York Inc.).

Germond, J.E., Delley, M., D'Amico, N., and Vincent, S.J. (2001). Heterologous expression and characterization of the exopolysaccharide from *Streptococcus thermophilus* Sfi39. Eur. J. Biochem. *268*, 5149–5156.

Gruter, M., Leeflang, B.R., Kuiper, J., Kamerling, J.P., and Vliegenthart, J.F.G. (1992). Structure of the exopolysaccharide produced by *Lactococcus lactis* subsp. *cremoris* H414 grown in a defined medium or skimmed-milk. Carbohyd. Res. *231*, 273–291.

Gruter, M., Leeflang, B.R., Kuiper, J., Kamerling, J.P., and Vliegenthart, J.F.G. (1993). Structural characterisation of the exopolysaccharide produced by *Lactobacillus delbrueckii* subspecies *bulgaricus* rr grown in skimmed milk. Carbohyd. Res. *239*, 209–226.

Harding, L.P., Marshall, V.M., Elvin, M., Gu, Y., and Laws, A.P. (2003). Structural characterisation of a perdeuteriomethylated exopolysaccharide by NMR spectroscopy: characterisation of the novel exopolysaccharide produced by *Lactobacillus delbrueckii* subsp. *bulgaricus* EU23 Carbohyd. Res. *338*, 61–67.

Harding, L.P., Marshall, V.M., Hernandez, Y., Gu, Y., Maqsood, M., McLay, N., and Laws, A.P. (2005). Structural characterization of a highly branched exopolysaccharide produced by *Lactobacillus delbrueckii* subsp. *bulgaricus* NCFB2074. Carbohyd. Res. *340*, 1107–1111.

Hassan, A.N. (2008). Possibilities and challenges of exopolysaccharides-producing lactic cultures in dairy foods. J. Dairy Sci. *91*, 1282–1298.

Hosono, A., Lee, J., Ametani, A., Natsume, M., Hirayama, M., Adachi, T., and Kaminogawa, S. (1997). Characterization of a water-soluble polysaccharide fraction with immunopotentiating activity from *Bifidobacterium adolescentis* M101–4. Biosci. Biotechnol. Biochem. *61*, 312–316.

Jolly, L., and Stingele, F. (2001). Molecular organization and functionality of exopolysaccharide gene clusters in lactic acid bacteria. Int. Dairy J. *11*, 733–745.

Jolly, L., Newell, J., Porcelli, I., Vincent, S.J.F., and Stingelle, F. (2002). *Lactobacillus helveticus* glycosyltransferases: from genes to carbohydrate synthesis. Glycobiology *12*, 319–327.

Kang, H.K., Seo, M.Y., Seo, E.S., Kim, D., Chung, S.Y., Kimura, A., Day, D.F., and Robyt, J.F. (2005). Cloning and expression of levansucrase from *Leuconostoc mesenteroides* B-512 FMC in *Escherichia coli*. BBA-Gene Struct. Expr. *1727*, 5–15.

Knoshaug, E.P., Ahlgren, J.A., and Trempy, J.E. (2007). Exopolysaccharide expression in *Lactococcus lactis* subsp. *cremoris* Ropy352: evidence for novel gene organization. Appl. Environ. Microbiol. *73*, 897–905.

Kolkman, M.A.B., Wakarchuk, W., Nuijten, P.J.M., and van der Zeijst, B.A.M. (1997). Capsular polysaccharide synthesis in *Streptococcus pneumoniae* serotype 14: Molecular analysis of the complete *cps* locus and identification of genes encoding glycosyltransferases required for the biosynthesis of the tetrasaccharide subunit. Mol. Microbiol. *26*, 197–208.

Korakli, M., Gänzle, M.G., and Vogel, R.F. (2002). Metabolism by bifidobacteria and lactic acid bacteria of polysaccharides from wheat, rye, and exopolysaccharides produced by *Lactobacillus sanfranciscensis*. J. Appl. Microbiol. *92*, 958–965.

Korakli, M., and Vogel, R.F. (2006). Structure/function relationship of homopolysaccharide producing glycansucrases and therapeutic potential for their synthesised glycans. Appl. Microbiol. Biotechnol. *71*, 790–803.

Kralj, S., van Geel-Schutte, G.H., Rahaoui, H., Leer, R.J., Faber, E.J., van der Maarel, M.J.E.C., and Dijkhuizen, J. (2002). Molecular characterization of a novel glucosyltransferase from *Lactobacillus reuteri* strain Lb121 synthesizing a unique, highly branched glucan with α-(1→4) and α-(1→6) glucosidic bonds. Appl. Environ. Microbiol. *68*, 4283–4291.

Kralj, S., van Geel-Schutte, van der Maarel, M.J.E.C., and Dijkhuizen, J. (2004). Biochemical and molecular characterization of *Lactobacillus reuteri* 121 reuteransucrase. Microbiology *150*, 2099–2112.

Lamothe, G.T., Jolly, L., Mollet, B., and Stingele, F. (2002). Genetic and biochemical characterization of exopolysaccharide biosynthesis by *Lactobacillus delbrueckii* subsp. *bulgaricus*. Arch. Microbiol. *178*, 218–228.

Ländersjo, C., Yang, Z., Huttunen, E., and Widmalm, G. (2002). Structural studies of the exopolysaccharides produced by *Lactobacillus rhamnosus* strain GG (ATCC 53103). Biomacromolecules *3*, 880–884.

Laws, A., Gu, Y., and Marshall, V. (2001a). Biosynthesis, characterisation, and design of bacterial exopolysaccharides from lactic acid bacteria. Biotechnol. Adv. *19*, 597–625.

Laws, A.P, and Marshall, V.M. (2001b). The relevance of exopolysaccharides to the rheological properties in milk fermented with ropy strains of lactic acid bacteria. Int. Dairy J. *11*, 709–721.

Laws, A.P., Chadha, M.J., Chacon-Romero, M., Marshall, V.M., and Maqsood, M. (2008). Determination of the structure and molecular weights of the exopolysaccharides produced by *Lactobacillus acidophilus* 5e2 when grown on different carbon feeds. Carbohyd. Res. *343*, 301–307.

Lemoine, J., Chirat, F., Wieruszeski, J.M., Strecker, G., Favre, N., and Neeser, J.R. (1997). Structural characterization of the exocellular polysaccharides produced by *Streptococcus thermophilus* SFi39 and SFi12. Appl. Environ. Microbiol. *63*, 3512–3518.

Lipinski, T., Jones, C., Lemercinier, X., Korzeniowska-Kowal, A., Strus, M., Rybka, J., Gamian, A., and Heczko, P.B. (2003). Structural analysis of the *Lactobacillus rhamnosus* strain KL37-C exopolysaccharides. Carbohyd. Res. *338*, 605–609.

Low, D., Ahlgren, J.A., Horne, D., McMahon, D.J., Oberg, C.J., and Broadbent, J.R. (1998). Role of *Streptococcus thermophilus* MR-1C capsular exopolysaccharide in cheese moisture retention. Appl. Environ. Microbiol. *64*, 2147–2151.

Maeda, H., Zhu, X., Suzuki, S., Suzuki, K., and Kitamura, S. (2004). Structural characterization and biological activities of an exopolysaccharide kefiran produced by *Lactobacillus kefiranofaciens* WT-2B. J. Agric. Food Chem. *52*, 5533–5538.

Marshall, V.M., Dunn, H., Elvin, M., McLay, N., Gu, Y., and Laws, A.P. (2001a). Structural characterisation of the exopolysaccharide produced by *Streptococcus thermophilus* EU20. Carbohyd. Res. *331*, 413–422.

Marshall, V.M., Laws, A.P., Gu, Y., Levander, F.R., Rådström, P., De Vuyst, L., Degeest, B., Vaningelgem, F., Dunn, H., and Elvin, M. (2001b). Exopolysaccharide-producing strains of thermophilic lactic acid bacteria cluster into groups according to their EPS structure. Lett. Appl. Microbiol. *32*, 433–437.

Meng, G., and Futterer, K. (2003). Structural framework of fructosyl transfer in *Bacillus subtilis* levansucrase. Nat. Struct. Biol. *10*, 935–941.

Minic, Z., Delorme, C.M.C., Faurie, J.M., Mecier, G., Ehrlich, D., and Renault, P. (2007). Control of EpsE, the phosphoglycosyltransferase initiating exopolysaccharide synthesis in *Streptococcus thermophilus*, by EpsD tyrosine kinase. J. Bacteriol. *189*, 1351–1357.

Monchois, V., Remaud-Simeon, M., Russell, R.R.B., Monsan, P., and Willemont, R.M. (1997). Characterization of *Leuconostoc mesenteroides* NRRL B-512F dextransucrase (DSRS) and identification of amino-acid residues playing a key role in enzyme activity. Appl. Microbiol. Biotechnol. *48*, 465–472.

Monchois, V., Willemot, R.M., and Monsan, P. (1999). Glucansucrases: mechanism of action and structure-function relationships. FEMS Microbiol. Rev. *23*, 131–151.

Monchois, V., Vignon, M., and Russell, R.R.B. (2000). Mutagenesis of Asp-569 of glucosyltransferase I glucansucrase modulates glucan and oligosaccharides synthesis. Appl. Environ. Microbiol. *66*, 1923–1927.

Monsan, P., Bozonnet, S., Albenne, C., Joucla, G., Willemot, R.M., and Remaud-Siméon, M. (2001). Homopolysaccharides from lactic acid bacteria. Int. Dairy J. *11*, 675–685.

Morona, J.K., Morona, R., and Paton, P.J. (1997). Characterization of the locus encoding the *Streptococcus pneumoniae* type 19F capsular polysaccharide biosynthetic pathway. Mol. Microbiol. *23*, 751–763.

Morona, J.K., Paton, J.C., Miller, D.C., and Morona, R. (2000). Tyrosine phosphorylation of CpsD negatively regulates capsular polysaccharide biosynthesis in *Streptococcus pneumoniae*. Mol. Microbiol. *35*, 1431–1442.

Mozzi, F., Vaningelgem, F., Hébert, E.M., van der Meulen, R., Foulquié-Moreno, M.R., Font de Valdez, G., and De Vuyst, L. (2006). Diversity of heteropolysaccharide-producing lactic acid bacterium strains and their biopolymers. Appl. Environ. Microbiol. *72*, 4431–4435.

Nagaoka, M., Shibata, H., Kimura, I., Hashimoto, S., Kimura, K., Sawada, H., and Yokokura, T. (1995). Structural studies on a cell wall polysaccharide from *Bifidobacterium longum* YIT4028. Carbohyd. Res. *274*, 245–249.

Nagaoka, M., Hashimoto, S., Shibata, H., Kimura, I., Kimura, K., Sawada, H., and Yokokura, T. (1996). Structure of a galactan from cell walls of *Bifidobacterium catenulatum* YIT4016. Carbohyd. Res. *281*, 285–291.

Nakajima, H., Hirota, T., Toba, T., Itoh, T., and Adachi, S. (1992). Structure of the extracellular polysaccharide from slime-forming *Lactococcus lactis* subsp. *cremoris* STB 0495. Carbohyd. Res. *224*, 245–253.

Navarini, L., Abatangelo, A., Bertocchi, C., Conti, E., Bosco, M., and Picotti, F. (2001). Isolation and characterization of the exopolysaccharide produced by *Streptococcus thermophilus* SFi20. Int. J. Biol. Macromol. *28*, 219–226.

Nierop-Groot, M.N., Godefrooij, J., and Kleerebezem, M. (2008). Heterologous expression of the pneumococcal serotype 14 polysaccharide in *Lactococcus lactis* requires lactococcal *epsABC* regulatory genes. Appl. Environ. Microbiol. *74*, 912–915.

Nordmark, E.L., Yang, Z., Huttunen, E., and Wildmalm, G. (2005a). Structural studies of and exopolysaccharide produced by *Streptococcus thermophilus* THS. Biomacromolecules *6*, 105–108.

Nordmark, E.L., Yang, Z., Huttunen, E., and Wildmalm, G. (2005b). Structural studies of an exopolysaccharide produced by *Propionibacterium freudenreichii ssp. shermanii* JS. Biomacromolecules *6*, 521–523.

Olvera, C., Centeno-Leija, S., and López-Munguía, A. (2007). Structural and functional features of fructansucrases present in *Leuconostoc mesenteroides* ATCC 8293. Antonie van Leeuwenhoek *92*, 11–20.

Ozimek, L.K., van Hijum, S.A.F.T., van Koningsveld, G.A., van der Maarel, M.J.E.C., van Geel-Schutten, G.H., and Dijkhuizen, L. (2004). Site directed mutagenesis study of the three catalytic residues of the fructosyltransferases from *Lactobacillus reuteri*. FEBS Lett. *560*, 131–133.

Ozimek, L.K., Euverink, G.J., van der Maarel, M.J.E.C., and Dijkhuizen, L. (2005). Mutational analysis of the role of calcium ions in the *Lactobacillus reuteri* strain Lb121 fructosyltransferase (levansucrase and inulosucrase) enzymes. FEBS Lett. *579*, 1124–1128.

Péant, B., LaPointe, G., Atlan, G.D., Ward, P., and Roy, D. (2005). Comparative analysis of the exopolysaccharide biosynthesis gene clusters from four strains of *Lactobacillus rhamnosus*. Microbiology *151*, 1839–1851.

Petry, S., Furlan, S., Crepeau, M.A., Cerning, J., and Desmazeaud, M. (2000). Factors affecting the exocellular polysaccharide production by *Lactobacillus delbrueckii* subsp. *bulgaricus* grown in a chemically defined medium. Appl. Environ. Microbiol. *66*, 342–3431.

Pijning, Y., Vujicic-Zagar, A., Kralj, S., Eeuwema, W., Dijkhuizen, L., and Dijkstra, B.W. (2008). Biochemical and crystallographic characterization of a glucansucrase from *Lactobacillus reuteri* 180. Biocatal. Biotransform. *26*, 12–17.

Pot, B. (2008). The taxonomy of lactic acid bacteria. In Bactéries lactiques. De la génétique aux ferments, G. Corrieu and F.M. Luquet, eds. (Paris, France: Lavoisier), pp. 1–152.

Provencher, C., LaPointe, G., Sirois, S., van Calsteren, M.R., and Roy, D. (2003). Consensus-generate hybrid oligonucleotide primers for amplification of priming glycosyltransferase genes of the exopolysaccharide locus in strains of *Lactobacillus casei* group. Appl. Environ. Microbiol. *69*, 3299–3307.

Rathsam, C., and Jacques, N.A. (1998). Role of C-terminal domains in surface attachment of the fructosyltransferase of *Streptococcus salivarius* ATCC25975. J. Bacteriol. *180*, 6400–6403.

Roberst, C.M., Fett, W.F., Osman, S.F., Wijey, C., O'Connor, J.V., and Hoover, D.G. (1995). Exopolysaccharides production by *Bifidobacterium longum* BB-79. J. Appl. Bacteriol. *78*, 463–468.

Robijn, G.W., Thomas, J.R., Haas, H., van den Berg, D.J.C., Kamerling, J.P., and Vliegenthart, J.F.G. (1995a). The structure of the exopolysaccharides produced by *Lactobacillus helveticus* 766. Carbohyd. Res. *276*, 137–154.

Robijn, G.W., van den Berg, D.J.C., Haas, H., Kamerling, J.P., and Vliegenthart, J.F.G. (1995b). Determination of the structure of the exopolysaccharides produced by *Lactobacillus sake* 0–1. Carbohyd. Res. *276*, 117–136.

Robijn, G.W., Gutierrez-Gallego, R., van den Berg, D.J.C., Haas, H., Kamerling, J.P., and Vliegenthart, J.F.G. (1996a). Structural characterization of the exopolysaccharides produced by *Lactobacillus acidophilus* LMG9433. Carbohyd. Res. *288*, 203–218.

Robijn, G.W., Wienk, H.L.J., van den Berg, D.J.C., Haas, H., Kamerling, J.P., and Vliegenthart, J.F.G. (1996b). Structural studies of the exopolysaccharides produced by *Lactobacillus paracasei* 34–1. Carbohyd. Res. *285*, 129–139.

Rozen, R., Bachrach, G., Bronshteyn, M., Gedalia, I., and Steinberg, D. (2001). The role of fructans on dental biofilm formation by *Streptococcus sobrinus*, *Streptococcus mutans*, *Streptococcus gordonii* and *Actinomyces viscous*. FEMS Microbiol. Lett. *195*, 205–210.

Ruas-Madiedo, P., Hogenholtz, J., and Zoon, P. (2002a). An overview of the functionality of exopolysaccharides produced by lactic acid bacteria. Int. Dairy J. *12*, 163–171.

Ruas-Madiedo, P., Tuinier, R., Kanning, M., and Zoon, P. (2002b). Role of exopolysaccharides produced by *Lactococcus lactis* subsp. *cremoris* on the viscosity of fermented milks. Int. Dairy J. *12*, 689–695.

Ruas-Madiedo, P., Alting, A., and Zoon, P. (2005a). Effect of exopolysaccharides and proteolytic activity of *Lactococcus lactis* subsp. *cremoris* strains on the viscosity and structure of fermented milks. Int. Dairy J. *15*, 155–164.

Ruas-Madiedo, P., and de los Reyes-Gavilán, C.G. (2005b). Methods for the screening, isolation and characterization of exopolysaccharides produced by lactic acid bacteria. J. Dairy Sci. *88*, 843–856.

Ruas-Madiedo, P., Gueimonde, M., Margolles, A., de los Reyes-Gavilán, C.G., and Salminen, S. (2006a). Effect of exopolysaccharide isolated from 'viili' on the adhesion of probiotics and pathogens to intestinal mucus. J. Dairy Sci. *89*, 2355–2358.

Ruas-Madiedo, P., Gueimonde, M., de los Reyes-Gavilán, C.G., and Salminen, S. (2006b). Exopolysaccharides produced by probiotic strains modify the adhesion of probiotics and enteropathogens to human intestinal mucus J. Food Prot. *69*, 2011–2015.

Ruas-Madiedo, P., Moreno, J.A., Salazar, N., Delgado, S., Mayo, B., Margolles, A., and de los Reyes-Gavilán, C. (2007). Screening of exopolysaccharide producing *Lactobacillus* and *Bifidobacterium* strains isolated from the human intestinal microbiota. Appl. Environ. Microbiol. *73*, 4385–4388.

Ruas-Madiedo, P., Abraham, A., Mozzi, F., and de los Reyes-Gavilán, C.G. (2008). Functionality of exopolysaccharides produced by lactic acid bacteria. In Molecular aspects of lactic acid bacteria for traditional and new applications, B. Mayo, P. López, and G. Pérez-Martínez, eds. (Kerala, India: Research Signpost), pp. 137–166.

Sánchez, I., Martínez, B., Guillén, R., Jiménez-Díaz, R., and Rodríguez, A. (2006). Culture conditions determine the balance between two different exopolysaccharides produced by *Lactobacillus pentosus* LPS26. Appl. Environ. Microbiol. *72*, 495–7502.

Sánchez-Medina, I., Gerwig, G.J., Urshev, Z.L., and Kamerling, J.P. (2007a). Structure of a neutral exopolysaccharide produced by *Lactobacillus delbrueckii* ssp. *bulgaricus* LBB.B26. Carbohyd. Res. *342*, 2430–2439.

Sánchez-Medina, I., Gerwig, G.J., Urshev, Z.L., and Kamerling, J.P. (2007b). Structural determination of a neutral exopolysaccharides produced by *Lactobacillus delbrueckii* ssp. *bulgaricus* LBB.B332. Carbohyd. Res. *342*, 2735–2744.

Schell, M.A., Karmirantzou, M., Snell, B., Vilanova, D., Berger, B., Pessi, G., Zwahlen, M.C., Desiere, F., Bork, P., Delley, M., Pridmore, R.D., and Arigoni, F. (2002). The genome sequence of *Bifidobacterium longum* reflects its adaptation to the human gastrointestinal tract. Proc. Natl. Acad. Sci. USA *99*, 14422–14427.

Smith, M.R., Zahanley, J.C., Wong, R.Y., Lundin R.E., and Ahlgren J.A. (1998). A mutant strain of *Leuconostoc mesenteroides* B-1355 producing a glucosyltransferase synthesising β-(1→2) glucosidic linkages. J. Ind. Microbiol. Biotechnol. *21*, 37–45.

Song, D.D., and Jacques, N.A. (1999). Mutation of aspartic acid residues in the fructosyltransferase of *Streptococcus salivarius* ATCC 25975. Biochem J. *344*, 259–264.

Staaf, M., Widmalm, G., Yang, Z., and Huttunen, E. (1996). Structural elucidation of an extracellular polysaccharide produced by *Lactobacillus helveticus*. Carbohyd. Res. *291*, 155–164.

Staaf, M., Yang, A., Huttunen, E., and Widmalm, G. (2000). Structural elucidation of the viscous exopolysaccharides produced by *Lactobacillus helveticus* Lb161. Carbohyd. Res. *326*, 113–119.

Stingele, F., Neeser, J.R., and Mollet, B. (1996). Identification and characterization of the *eps* (exopolysaccharide) gene cluster from *Streptococcus thermophilus* Sfi6. J. Bacteriol. *178*, 1680–1690.

Stingele, F., Lemoine, J., and Neeser, J.R. (1997). *Lactobacillus helveticus* Lh59 secretes and exopolysaccharide that is identical to the one produced by *Lactobacillus helveticus* TN-4, a presumed spontaneous mutant of *Lactobacillus helveticus* TY1-2. Carbohyd. Res. *302*, 197–202.

Stingele, F., Newell, J.W., and Neeser, J.R. (1999). Unraveling the function of glycosyltransferases in *Streptococcus thermophilus* Sfi6. J. Bacteriol. *181*, 6354–6360.

Sudarsanam, P., Ley, R., Guruge, J., Turnabaugh, P.J., Mahwald, M., Liep, D., and Gordon, J. (2008). Draft genome sequence of *Bifidobacterium dentium* (ATCC 27678). Published in the GeneBank database (http://www.ncbi.nlm.nih.gov) Accession no. NZ_ABIX02000002.

Sutherland, I.W. (1998). Novel and established applications of microbial polysaccharides. Trends Biotechnol. *16*, 41–46.

Sutherland, I.W. (2001). Microbial polysaccharides from Gram-negative bacteria. Int. Dairy J. *11*, 663–674.

Suzuki, T., Tsuda, Y., Kanou, N., Inoue, T., Kumazaki, K., Nagano, S., Hirai, S., Tanaka, K., and Watanabe, K. (2006). *Bifidobacterium adolescentis* complete genome sequence. Published in the GeneBank database (http://www.ncbi.nlm.nih.gov) Accession no. AP009256.

Tieking, M., Korakli, M., Ehrmann, M.A., Gänzle, M.G., and Vogel, R.F. (2003). *In situ* production of exopolysaccharides during sourdough fermentation by cereal and intestinal isolates of lactic acid bacteria. Appl. Environ. Microbiol. *69*, 945–952.

Tieking, M., Ehrmann, M.A. Vogel, R.F., and Gänzle, MG. (2005a). Molecular and functional characterization of a levansucrase from the sourdough isolated *Lactobacillus sanfranciscensis* TMW 1.392. Appl. Microbiol. Biotechnol. *66*, 655–663.

Tieking, M., Kaditzky, S., Valcheva, R., Korakli, M., Vogel, R.F., and Gänzle, MG. (2005b). Extracellular homopolysaccharides and oligosaccharides from intestinal lactobacilli. J. Appl. Microbiol. *99*, 692–702.

Tieking, M., Kühnl, W., and Gänzle, M.G. (2005c). Evidence for formation of heterooligosaccharides by *Lactobacillus sanfranciscensis* during growth in wheat sourdough. J. Agric. Food Chem. *53*, 2456–2461.

Tone-Shimokawa, Y., Toida, T., and Kawashima, T. (1996). Isolation and structural analysis of polysaccharide containing galactofuranose from the cell walls of *Bifidobacterium infantis*. J. Bacteriol. *178*, 317–320.

Tsuchiya, H.M, Hellman, N.N., and Koepsell, H.J. (1953). Factors affecting molecular weight of enzymatically synthesized dextran. J. Am. Chem. Soc. *75*, 757–758.

Tuinier, R., Zoon, P., Cohen-Stuart, M.A., Fleer, G., and de Kruif, C.G. (1999a). Concentration and shear-rate dependence of an exocellular polysaccharide. Biopolymers *50*, 641–646.

Tuinier, R., Zoon, P., Olieman, C., Cohen-Stuart, M.A., Fleer, G., and de Kruif, C.G. (1999b). Isolation and physical characterisation of an exocellular polysaccharide. Biopolymers *49*, 1–9.

Tuinier, R., van Casteren, W.H.M., Looijesteinj, P.J., Schols, H.A., Voragen, A.G.J., and Zoon, P. (2001). Effects of structural modifications on some physical characteristics of exopolysaccharides from *Lactococcus lactis*. Biopolymers *59*, 160–166.

Tzortzis, G., Jay, A.J., Baillon, M.L.A., Gibson, G.R., and Rastall, R.A. (2003). Synthesis of α-galactooligosaccharides with α-galactosidase from *Lactobacillus reuteri* of canine origin. Appl. Microbiol. Biotechnol. *63*, 286–292.

van Calsteren, M.R., Pau-Roblot, C., Be, A., and Roy, D. (2002). Structure determination of the exopolysaccharides produced by *Lactobacillus rhamnosus* strains RW-9595M and R. Biochem. J. *363*, 7–17.

van Casteren, W.H.M., Dijkema, C., Schols, H.A., Beldman, G., and Voragen, A.G.J. (1998). Characterisation and modification of the exopolysaccharides produced by *Lactococcus lactis* subsp. *cremoris* B40. Carbohydr. Polym. *37*, 123–130.

van Casteren, W.H.M., de Waard, P., Dijkema, C., Schols, H.A., and Voragen, A.G.J. (2000a). Structural characterisation and enzymic modification of the exopolysaccharide produced by *Lactococcus lactis* subsp. *cremoris* B891. Carbohyd. Res. *327*, 411–422.

van Casteren, W.H.M., Dijkema, C., Schols, H.A., Beldman, G., and Voragen, A.G.J. (2000b). Structural characterization and enzymic modification of the exopolysaccharide produced by *Lactococcus lactis* subsp. *cremoris* B39. Carbohyd. Res. *324*, 170–181.

van der Meulen, R., Grosu-Tudor, S., Mozzi, F., Vaningelgem, F., Zamfir, M., Font de Valdez, G., and De Vuyst, L. (2007). Screening of lactic acid bacteria isolates from dairy and cereal products for exopolysaccharides production and genes involved. Int. J Food Microbiol. *118*, 250–258.

van Geel-Schutten, G.H., Faber, E.J., Smit, E., Monting, K., Smith, M.R., Ten Brink, B., Kamerling, J.P., Vliegenthart, J.F.G., and Dijkhuizen, L. (1999). Biochemical and structural characterization of the glucan and fructan exopolysaccharides synthesised by the *Lactobacillus reuteri* wild-type strains and by mutant strains. Appl. Environ. Microbiol. *65*, 3008–3014.

Vanhaverbeke, C., Bosso, C., Colin-Morel, P., Gey, C., Gamar-Nourani, L., Blondeau, K., Simonet, J.M., and Heyraud, A. (1998). Structure of an extracellular polysaccharide produced by *Lactobacillus rhamnosus* strain C83. Carbohyd. Res. *314*, 211–220.

van Hijum, S.A.F.T., van Geel-Schutten, G.H., Rahaoui, H., van der Maarel, M.J., and Dijkhuizen, M.J. (2002). Characterization of novel fructosyltransferase from *Lactobacillus reuteri* that synthesized high molecular weight inulin and inulin-oligosaccharides. Appl. Environ. Microbiol. *68*, 4390–4398.

van Hijum, S.A.F.T., Kralj, S., Ozimek, L.K., Dijkhuizen, L., and van Geel-Schutten, I.G.H. (2006). Structure-function relationships of glucansucrase and fructansucrase enzymes from lactic acid bacteria. Microbiol. Mol. Biol. Rev. *70*, 157–176.

Vaningelgem, F., Zamfir, M., Mozzi, F., Adriany, T., Vancanneyt, M., Swings, J., and De Vuyst, L. (2004). Biodiversity of exopolysaccharides produced by *Streptococcus thermophilus* strains is reflected in their production and their molecular and functional characteristics. Appl. Environ. Microbiol. *70*, 900–912.

van Kranenburg, R., Marugg, J.D., van Swan, I.I., Willem, N.J., and de Vos, W.M. (1997). Molecular characterization of the plasmid-encoded *eps* gene essential for exopolysaccharide biosynthesis in *Lactococcus lactis*. Mol. Microbiol. *24*, 387–397.

van Kranenburg, R., Boels, I.C., Kleerebezem, M., and de Vos, W.M. (1999a). Genetics and engineering of microbial exopolysaccharides for food: approaches for the production of existing and novel polysaccharides. Cur. Opin. Biotechnol. *10*, 498–504.

van Kranenburg, R., Vos, H.R., van Swan, I.I., Kleerebezem, M., and de Vos, W.M. (1999b). Functional analysis of glycosyltransferase genes from *Lactococcus lactis* and other gram-positive cocci: complementation, expression, and diversity. J. Bacteriol. *181*, 6347–6353.

van Kranenburg, R., Kleerebezem, M., and de Vos, W.M. (2000). Nucleotide sequence analysis of the lactococcal EPS plasmid pNZ4000. Plasmid *43*, 130–136.

van Leeuwen, S.S., Kralj, S., ven Geel-Schutten, I.H., Gerwing, G.J., Dijkhuizen, L., and Kamerling, J.P. (2008). Structural analysis of the alpha-D-glucan (EPS180) produced by the *Lactobacillus reuteri* strain 180 glucansucrase GTF180enzyme Carbohyd. Res. *343*, 1237–1250.

Vincent, C., Doublet, P., Grangeasse, C., Vaganay, E., Cozzone, A.J., and Duclos, B. (1999). Cells of *E. coli* contain a protein-tyrosine kinase, Wzc, and a phophotyrosine-protein phosphatase, Wzb. J. Bacteriol. *181*, 3472–3477.

Vincent, S.J., Faber, E.J., Neeser, J.R., Stingele, F., and Kamerling, J.P. (2001). Structure and properties of the exopolysaccharide produced by *Streptococcus macedonicus* Sc136. Glycobiology *11*, 131–139.

Walling, E., Gindreau, E., and Lonvaud-Funel, A. (2005). A putative glucan syntase gene *dps* detected in exopolysaccharides-producing *Pediococcus damnosus* and *Oenococcus oeni* strains isolated from wine and cider. Int. J. Food Microbiol. *98*, 53–62.

Wang, L., Liu, D., and Reeyes, P.R. (1996). C-terminal half of *Salmonella enterica* WbaP (RfbP) is the galactosyl-1-

phosphate transferase domain catalyzing the first step of *O*-antigen synthesis. J. Bacteriol. *178*, 2598–2604.

Welker, D.L., Jensen, R.L., and Broadbent, J.R. (2004). The exopolysaccharide synthesis gene cluster of *Streptococcus thermophilus* strain MTC330. Published in the GeneBank (http://www.ncbi.nlm.nih.gov) Accession no. AF430847.

Werning, M.L., Ibarburu, I., Dueñas, M.T., Irastorza, A., Navas, J., and López, P. (2006). *Pediococcus parvulus gtf* gene encoding the GTF glycosyltransferase and its application for specific PCR detection of β-D-glucan-producing bacteria in foods and beverages. J. Food Prot. *69*, 161–169.

Whitfield, C., and Roberts, I.S. (1999). Structure, assembly and regulation of expression of capsules in *Escherichia coli*. Mol. Microbiol. *31*, 1307–1319.

Yamamoto, Y., Morosaki, S., Yamauchi, R., Kato, K., and Sone, Y. (1994). Structural study on an exocellular polysaccharide produced by *Lactobacillus helveticus* TY1-2. Carbohyd. Res. *261*, 67–78.

Yamamoto, Y., Nunome, T., Yamauchi, R., Kato, K., and Sone, Y. (1995). Structure of an exocellular polysaccharide *of Lactobacillus helveticus* TN-4, a spontaneous mutant strain of *Lactobacillus helveticus* TY1-2. Carbohyd. Res. *275*, 319–332.

Yamamoto, Y., Miyake, K., Koike, M., Watanabe, M., Machida, M., Ohta, M., and Iijima, S. (1999). Molecular characterization of type specific capsular polysaccharide biosynthesis genes of *Streptococcus agalactiae*. J. Bacteriol. *181*, 5176–5184.

Yang, Z., Huttunen, E., Staaf, M., Widmalm, G., and Tenhu, H. (1999). Separation, purification and characterisation of extracellular polysaccharides produced by slime-forming *Lactococcus lactis* ssp. *cremoris* strains. Int. Dairy J. *9*, 631–638.

Yang, Z., Staaf, M., Huttunen, E., and Widmalm, G. (2000). Structure of a viscous exopolysaccharide produced by *Lactobacillus helveticus* K16. Carbohyd. Res. *329*, 465–469.

Yother, J. (1999). Common themes in the genetics of streptococcal capsular polysaccharides. In Genetics of Bacterial Polysaccharides, J.B. Goldberg, ed. (New York, USA: CRC Press), pp. 161–184.

Part VII

Commercial Applications of Microbial Polysaccharides

Commercial Exploitation of Homo-exopolysaccharides in Non-dairy Food Systems

17

Florian W. Waldherr and Rudi F. Vogel

Abstract

Lactic acid bacteria play a crucial role in various food fermentations. Several strains can produce long-chain sugar polymers called exopolysaccharides (EPS). These can be classified according to whether they are composed of different sugar monomers or just one kind, i.e. as hetero- or homopolysaccharides. Whereas heteropolysaccharides are intensively used as additives in milk products, homopolysaccharides can be introduced into sourdough products, influencing structural quality and baking ability and reducing bread staling factors. Additionally, beneficial effects on enteral health are discussed. An example of the industrial use of EPS in bakery products is the addition of dextran in panettone and other breads. Also, the addition of non-bacterial hydrocolloids is well established in industrial baking. Several studies of the effects of replacing these additives by bacterial EPS have been carried out, providing data regarding such aspects of dough and bread as textural factors, water retention and moisture and specific bread volume. Practically no information is available on the effects of bacterial EPS in other fermented foods such as fermented meat products, sauerkraut or vinegar. EPSs can be defined as additives or, in the case of *in situ* production, as starter cultures. The addition of purified EPS has to be labelled on the end product, which is a disadvantage since consumers demand for fewer additives in foods. On the other hand, *in situ* production appears to be less effective in traditional wheat and rye dough systems due to strain-dependent acid formation, which may be required but counteracts positive EPS effects. Forthcoming chances of EPS applications may therefore lie within special applications as in gluten-free breads in which both reduced pH and EPS should have synergistic positive effects on structure.

EPS-producing bacteria in food

Bacteria in traditional fermentations are generally considered to be safe for use in human nutrition. As humans started to ferment different foods very early, utilization for thousands of years has proven the safety of the traditional fermentation floras. New, enhanced starter cultures are mostly based on the strains common in traditional fermentations exploiting the growing knowledge on the special properties of specific strains.

Remarkably, lactic acid bacteria (LAB) playing a natural role in food fermentations can often be also found as part of (healthy) human intestinal flora. Despite their low numbers compared with other commensals, they are believed to play crucial roles in the intestinal ecosystem and impact of the immune system. LAB isolated from naturally fermented foods and the intestinal ecosystem are widely used as fermentation starters or probiotics. For others the production of prebiotic factors, supporting desirable intestinal floras is described.

The gut as well as different food systems are very complex environments requiring different survival strategies for microorganisms. An adaptation to available nutrients and substrates and protective mechanisms against negative environmental influences like low pH, high salt concentrations, low water activity, bile salts and other chemical components must be developed. The production of long chain exopolysaccharides (EPS) can be part of a microbial defence system. Apart from the effects mediated by the EPS itself, potential roles in the persistence of EPS producers in the intestinal tract have been dedicated to involvement in cellular adhesion of EPS producing enzymes (Walter at al, 2008). Nevertheless, the impact of these substances for the producing bacteria strains is not fully understood yet. Pathogens like oral streptococci responsible for development of dental caries use such molecules for biofilm formation. On the other hand, similar sugar chains are produced by food starter bacteria, showing prebiotic properties.

Since bacterial EPS show comparable characteristics as variable additives used in food industry to improve different process and product properties, replacement of these chemical supplements by food born substances might be a desirable ambition.

A first success in this field has already been achieved in the use of heteropolysaccharides of bacterial origin to optimize the rheological and textural properties of fermented milk products, which is reviewed in another chapter. On the other hand, homopolysaccharides produced by LAB have useful properties for applications especially in non-dairy food systems. EPS producers are easily detected upon growth on sucrose, and sometimes naturally produce large amounts of EPS. Fig. 17.1 shows an example for a strain of *Lactocabillus hilgardii.*

Types and structures of EPSs produced by fermenting bacteria

EPS can be distinguished by different criteria. One possibility is to divide into capsular EPS which remain ultimately attached to the cell surface of the producing microorganism forming a capsule, and slime polysaccharides which are just loosely bound to the cell wall or are completely secreted into the medium (Kumar *et al.* 2007). Hassan *et al.* (2007) further distinguished the producing bacteria into ropy and nonropy strains, which does not correlate with the capsular or unattached appearance of EPS. Since not all unattached EPSs produce ropiness, he introduced four groups: group I as capsule-forming ropy strains that produce capsules and unattached ropy EPS; group II representing capsule-forming non-ropy strains that produce capsules and possibly unattached EPS; group III including noncaspule-forming ropy strains; and group IV, which are strains producing no or undetectable EPS (Hassan *et al.* 2007).

An alternative possibility to differentiate EPS-producing bacteria is to consider the two main types of EPS composition: heteropolysaccharids (HePSs) and homopolysaccharides (HoPSs). HePSs consist of regular or irregular repeating units that are composed of 3–8 carbohydrate moieties. These are produced by several enzymes intracellularly from sugar nucleotide precursers. The subunits are brought to the cell surface where they are linked to long chains or networks (De

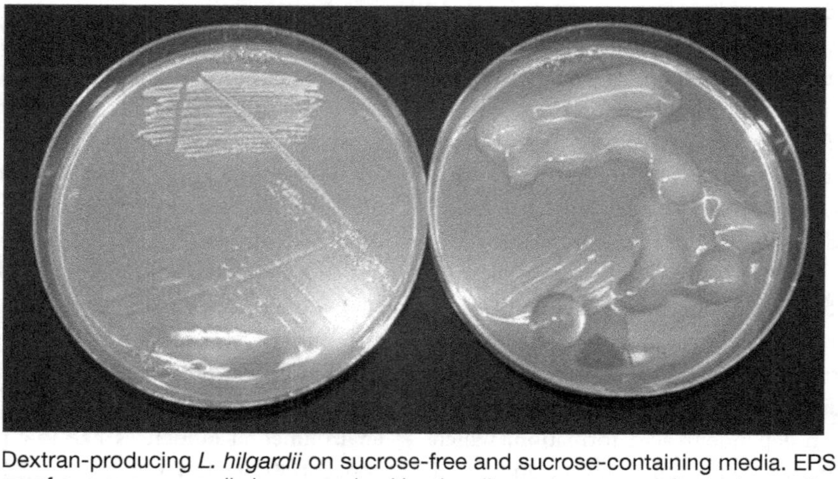

Figure 17.1 Dextran-producing *L. hilgardii* on sucrose-free and sucrose-containing media. EPS production in the presence of sucrose can easily be recognized by the slimy appearance of colonies on the agar plate.

Vuyst *et al.* 2001; Tieking and Gänzle 2005). The synthesis is similar to the cell wall synthesis and is energy-dependent. Two to eight different monosaccharides can be contained (mostly glucose, galactose, rhamnose and fructose), and amounts mostly do not exceed 2 g/L. Arabinan was reported to be produced in final amounts up to 5 g/l. (De Vuyst and Degeest 1999; Bergmaier *et al.* 2005; Korakli and Vogel 2006).

In this chapter we focus on HoPSs. They are result of their sucrose metabolism, the reactions of which re summarized in Fig. 17.2. HoPS consist of one monosaccharide, mostly glucose or fructose, with the resulting EPS designated glucans or fructans respectively. Monomers are connected with one type of bond; however, branching is possible. Depending on the dominant osidic bondages in the EPS glucans and fructans can be divided in more detail. Up to date four different glucans have been described. The most common is dextran with α-(1→6) linked glucose subunits. In the linear molecule branching is possible at positions 2, 3 or 4. Glucans with mainly α-(1→3) linked subunits are called mutan. A glucan with alternating α-(1→6) and α-(1→3) glucosidic linkages was called alternan. Finally there is reuteran with a majority of α-(1→4) linkages, also containing α-(1→6) glucosidic bonds and α-(1→4,6) branching points. In contrast to the more variable glucans just two types of fructans have been described yet. Mainly β-(2→6) linked fructose monomers are called levan. They can casually contain β-(2→1) branches. In inulins β-(2→1) are the dominating linkage, β-(2→6)

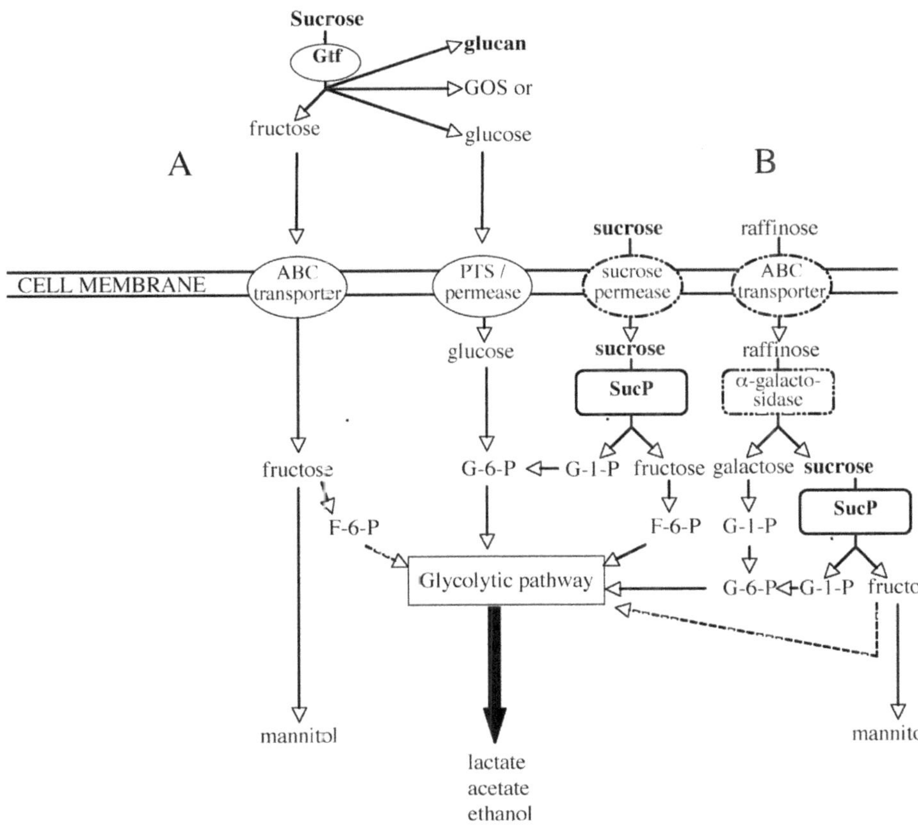

Figure 17.2 Sucrose metabolism of lactic acid bacteria. Extracellular glycosyltransferases (here a gtf) split sucrose molecules and transfer one of the basic sugar molecules onto accessible aceptors resulting in free fructose or glucose monomers, oligosaccharides or EPS. Free glucose is transported into the cell by various systems and can intracellularly be metabolized to lactate, acetate and ethanol. Some strains are capable to use fructose as electron acceptor and produce mannitol to gain redox equivalents or produce sucrose phosphorylase competing with EPS formation. The latter enzyme is also involved in the interlinked raffinose pathway.

branching is possible (van Hijum *et al.* 2006). HoPS are formed extracellularly by a one enzyme reaction. In most cases the substrate for that reactions is sucrose, which provides the necessary energy for the reaction in its osidic bondage between the fructose and glucose monomer. HoPSs are often formed in relatively large amounts up to 40 g/l (Korakli and Vogel 2006).

Enzymes involved in homo-exopolysaccharide biosynthesis

In contrast to the more complex synthesis of HePSs, HoPSs are built up by a single enzyme. Corresponding to the division in glucans and fructans, two classes of glykosyltransferases can be distinguished: fructosyltransferases or fructansucrases (FTFs) for the synthesis of fructans and glucosyltransferases or glucansucrases (GTFs) for the synthesis of glucans, respectively.

Fructosyltransferases (FTFs)

FTFs are extracellular enzymes. They transfer fructose monomers to acceptor molecules. The possible reactions are summarized in Fig. 17.3. The necessary energy and the fructose subunit are gained by the cleavage of the glycosidic bond of the substrate which usually is sucrose. The use of raffinose as substrate has also been described. Different acceptor molecules are possible, resulting in three different product groups. Coupling of fructose units to a growing fructan chain results in an elongated chain. The reaction is called polymerization. By the use of alternative sugar molecules, e.g. sucrose or maltose a so called acceptor reaction produces fructooligosaccharides

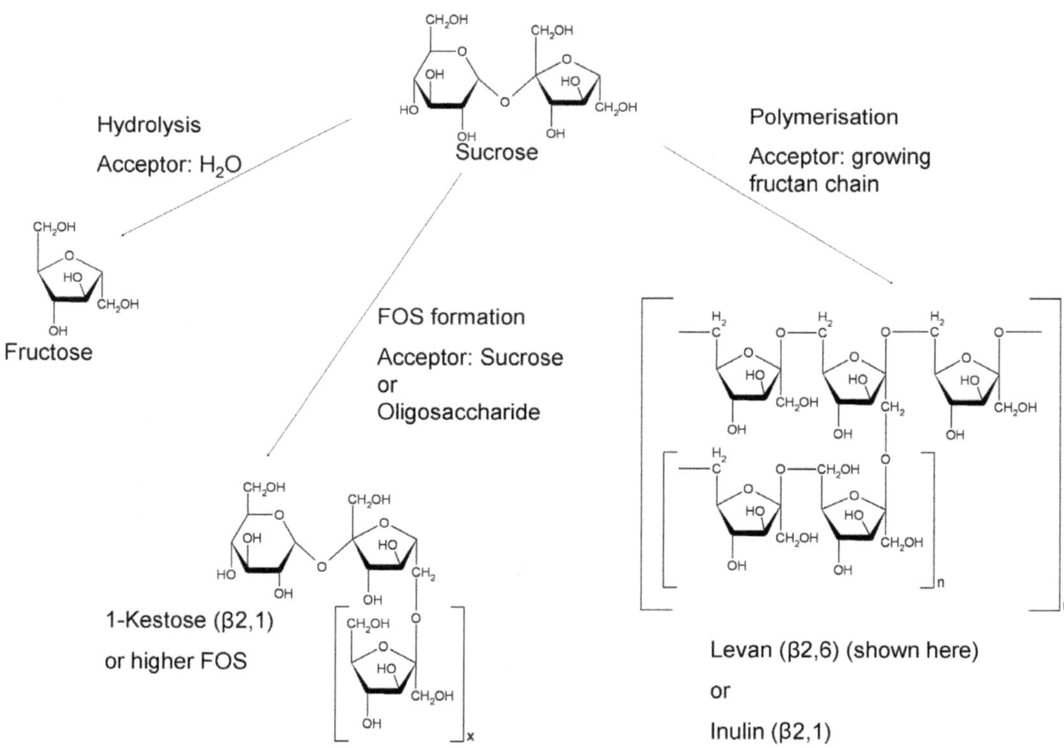

Figure 17.3 Sucrose utilization by fructosyltransferases. The basic fructose transfer reaction can result in three different classes of products depending on the acceptor molecule. When water is used as an acceptor, sucrose is hydrolysed in a glucose and a fructose monomer. When fructose is transferred to another sucrose molecule, the oligosaccharide kestose is the result. Depending on the enzyme, addition of further fructose units and therewith the formation of higher oligosaccharides is possible. Polymerization results in EPS by transferring fructose monomers to a growing fructan chain. Two types of fructans are described: β-2,6 linked levan (shown here) and β-2,1 linked inulin.

(FOS). Examples therefore are sucrose-based FOSs. Sucrose with an additional linked fructose is 1-kestose (GF2) followed by 1,1-nystose (GF3) and 1,1,1-kestopentaose (GF4). The third type of acceptor molecule is water. The transfer of the activated fructose to a water molecule results in a hydrolysis of the substrate sugar (Korakli and Vogel 2006).

Two structural different fructans can be produced by bacterial FTFs: Levan is a mainly β-(2→6) linked chain of fructose monomers and occasionally occurring β-(2→1) linked branches. The producing enzymes are called levansucrases. Alternatively inulin, which is a dominantly β-(2→1) linked polyfructose molecule with β-(2→6) branches can be synthesized by inulinsucrases. The molecular mass of fructans produced by bacteria varies from 2×10^4 to 50×10^6 daltons.

Levansucrases are widely found in both Gram-positive and Gram-negative bacteria. In terms of food fermentations it is worth mentioning the levansucrases found in lactobacilli, leuconostocs, streptococci and *Gluconoacetobacter* species. Such enzymes are also described and very well characterized in *Bacillus* and *Zymomonas* species (van Hijum *et al.* 2006). Several fructosyltransferases are described in lactobacilli. Inulinsucrases have been found in *L. reuteri* (van Hijum *et al.* 2002) and *L. johnsonii* (Anwar *et al.* 2008), while lactobacilli levansucraes are known in *L. reuteri* (van Hijum *et al.* 2001), *L. sanfranciscensis* (Tieking *et al.* 2005) and *L. panis* (Waldherr *et al.* 2008). Uncharacterized lactobacillus produced fructans have been described for *L. frumenti* and *L. pontis* (Tieking *et al.* 2003b). To date, inulosucrases have been found only in LAB such as streptococci, lactobacilli and leuconostocs.

All known bacterial FTFs share a common basic structure. All of them are extracellular enzymes so they carry an N-terminal signal peptide (36–39 amino acids) followed by a variable domain. This domain varies greatly in length and sequence. In some FTFs direct repeats can be found. The core of the enzymes is a relatively conserved catalytic domain of approximately 500 amino acids including highly conserved amino acid residues shared with the members of the GH68 family enzymes. The C-terminal domain consists of a proline-rich (PXXP-repeats), a spacer region and, in lactobacilli and streptococci, a LPXTG cell wall anchoring motif (Korakli and Vogel 2006).

The three-dimensional structures of *Bacillus* levansucrases are known, revealing a rare fivefold β-propeller topology including a deep, negatively charged central pocket. Fig. 17.4 shows a three-dimensional structure model of *L. panis* TMW 1.648 levansucrase obtained by computional molecular modelling process on basis on X-ray structure of *Bacillus* fructosyltransferases. The amino acid residudes important for substrate binding and transfer reaction have been identified by mutational analysis in different levansucrases. The activity of several FTFs has been described to be dependent on Ca^{2+} ions. Three-dimensional structure and mutational analysis revealed a binding site with corresponding asparagic acid residues. These residues are highly conserved in FTFs of Gram-positive origin but absent in corresponding enzymes from Gram-negative organisms.

Glucosyltransferases (GTFs)

Analogous to the FTFs, GTFs are extracellular enzymes cleaving their substrate sucrose and transferring glucose monomers to an acceptor using the energy of the cleaved glycosidic bond. Different groups of acceptor molecules can be used by GTFs as well carrying out a hydrolysis reaction if the acceptor is a water molecule or tranglycosylation reactions resulting in a growing glucan chain or glucooligosaccharides with different types of bondages. GTFs are large enzymes with an avarage molecular mass of 160 kDa.

The glucans produced by bacterial GTFs are exclusivley α-glucans that can be differentiated in four groups. Glucan residues in dextrans are linked α-(1→6) with branching points at position 2, 3 or 4. The corresponding enzymes are called dextransucrases. In mutan-type glucans α-(1→3) are dominating (up to 90%) they are produced by so called mutansucrases. Alternan consists of glucose subunits alternating linked with α-(1→3) and α-(1→6) glycosidic bonds. The producing enzyme was named alternansucrase. The latest type of glucans is reuteran, characterized by mainly α-(1→4) glycosidic linkages. Following the used nomenclature the producing enzyme is called reuteransucrase. In this case α-(1→6) linkages beside the dominating α-(1→4) ones and α-(1→4,6)

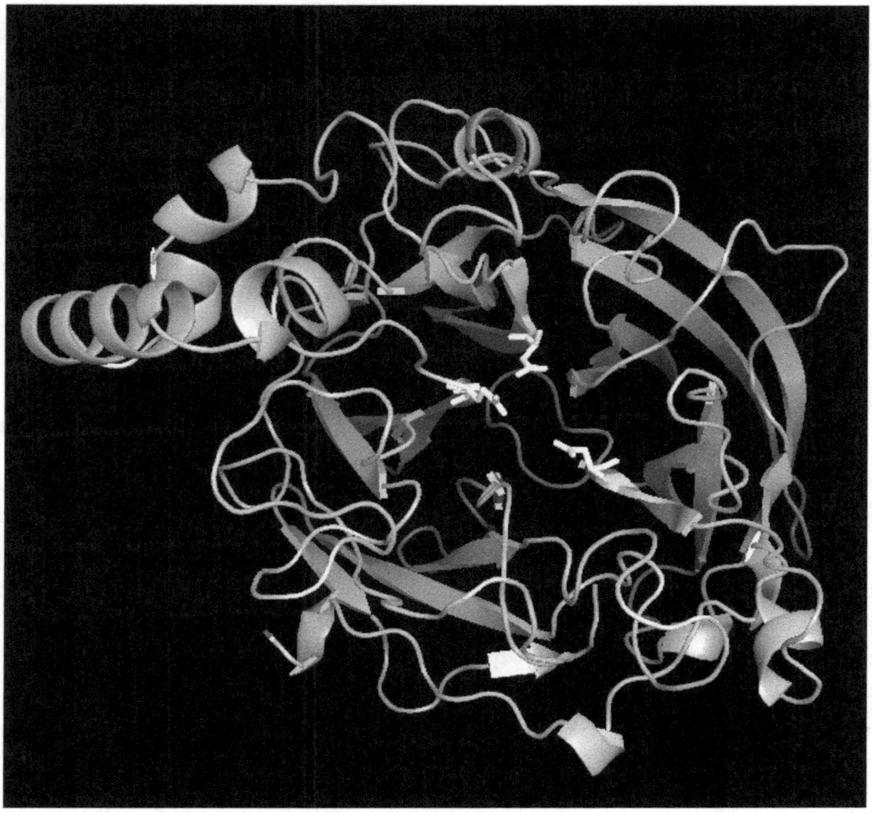

Figure 17.4 Three-dimensional structure model of *Lactobacillus panis* TMW 1.648 levansucrase obtained by computional molecular modelling process on basis on X-ray structure of *Bacillus* fructosyltransferases. The catalytic centre is based inside a five-bladed β propeller topology.

branchings are described. In addition to the main types, a glucan with mainly α-(1→6) linkages and a remarkable number of α-(1→2) bonds between the glucose monomers is reported. The type of glucan produced is determined by the enzyme. The glucans produced can, beside the type of glucose linkages, vary in chain length, number of branches and solubility.

In contrast to the widespread distribution of bacterial FTFs throughout different bacterial groups, GTFs are found only in the group of LAB. The reason for this remains unknown. Thus, GTF genes are found in lactobacilli, leuconostocs and streptococci.

Like the FTFs, GTFs share a common basic structure of four domains, which is shown in Fig. 17.5. At the N-terminus there is a signal peptide responsible for the extracellular location of the enzyme. Its sequence is highly conserved, 32 to 34 amino acids long and typical for Gram-positive bacteria. It is followed by an unconserved variable domain, which varies in sequence and length. The function of that part of the enzymes is not known yet. The largest part is about 1000 amino acids in length and has been identified as catalytic region including the sucrose-binding domain and the catalytic amino acid residues of the active centre of the GTF. Owing to its functionality it is highly conserved. C-terminally located is a glucan binding domain of approximately 400 amino acids. This domain is composed of a series of repeats that can be classified and is characteristic for each GTF.

There is no three-dimensional structure model of a bacterial GTF. Nevertheless, mutational analysis revealed data about the reaction mechnisms and substrate binding sites. It has been shown that exchanging amino acid residues, the solubility of the produced glucan can be influenced and the linkage type of the glucan can be changed (Korakli and Vogel 2006; van Hijum *et al.* 2006).

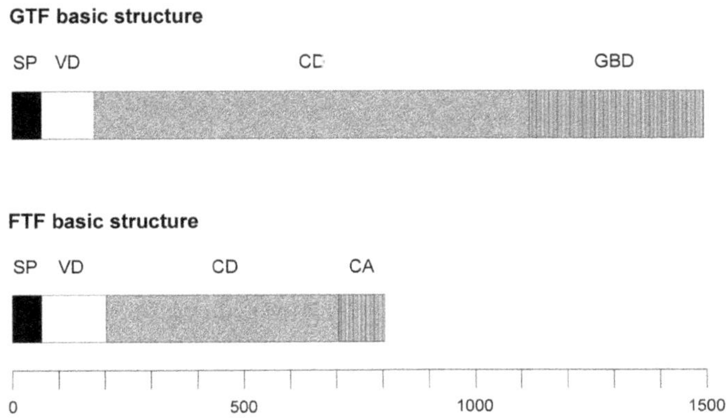

Figure 17.5 Schematic basic structure of glucan and fructan sucrases. Scala represents length in amino acids. Both protein classes have and N-terminal signal peptide (SP) for extracellular location followed by a variable region (VD) that can also vary in length. For both enzyme groups the core of the protein is the enzymatic active catalytic domain (CD). Therein important regions are highly conserved. Fructosyltransferases C-terminally carry conserved cell wall anchor motif (CA). In glycosyltransferases the C-terminus forms a glucan-binding domain (GBD).

Applications

EPS when produced within or added to food can have enormous and very different effects on the properties of the product. Examples are modification of the water retention in the food and influences on mouthfeel properties. However, glucans and fructans are practically tasteless and thus can serve as supplements in all kinds of food without affecting product taste.

EPS can be contained in practically all groups of fermented food, so all these foods are targets for the optimization of characteristics by directed application of EPS.

Dairy applications

Fermented dairy products are strongly influenced by EPS-producing starter cultures. Due to the substrate milk not only homo- but also hetero-EPS play a role in cheese and yoghurt fermentations. There are effects on the structure formation of fermented milk, e.g. by modifying the formation of casein aggregates. Other factors affected by EPS production during fermentation are rheology, texture and syneresis.

An example of a beneficial application of EPS in dairy food technology is the increase in moisture retention resulting from water binding and thus an enhancement of the textural and functional properties of reduced-fat cheese. EPSs are introduced in these products using EPS-producing starter cultures.

EPS impacts in fermented dairy products are complex and has been examined and reviewed for a long time (Hassan *et al.* 2007). In this book that field has its own chapter.

EPS applications in sourdough bread

Sourdough principles

A further important group of fermented foods is sourdoughs. Such doughs spontaneously develop by the fermentation of a mixture of cereal with water by the natural occurring flora. Consequently, such methods belong to the oldest food biotechnological processes in the world. Today, sourdoughs are added in low percentages to the final bread doughs, which are further leavened by the addition of bakers yeast *Saccharomyces cerevisiae*. The use of sourdough in bread dough lowers the pH thus inhibiting flour amylase and activating proteases with strong impacts in texture and flavour, respectively, of the bread produced by the use of sourdough. In wheat breads, gluten protein is responsible for forming a stable structure by linking gluten proteins to a stable network during the baking process. With rye flours that network formation is insufficient due to an different protein composition. Here, a low pH is required to achieve bakeability. The dropping pH during sourdough fermentation enables pentosanes to take over structure building function during baking. In rye breads, but also in wheat, enhancement of taste compared with breads without sourdough is an important reason to use sourdoughs since consumers demand more tasty breads and bakery products.

The sourdough flora consists mainly of LAB of many types as recently reviewed (Vogel *et al.* 2002; De Vuyst and Vancanneyt 2007). Still, typical autochthonous bacteria dominate the fermentations mainly as dictated by fermentation temperature, time and refreshment cycles of these preparations. Dominant species are namely heterofermentative ones (e.g. *L. sanfranciscensis, L. pontis, L. reuteri, L. fermentum*) and some (facultative) homofermentative species (e.g. *L. plantarum, L. amylovorus*). In addition to bacteria, yeasts like acid tolerant *Candida milleri* or *Saccharomyces cerevisiae* can be found, the latter being mostly contaminants from the bakeries using it as a major leavening agent in the bread doughs. Typically a sourdough flora consists of two to five strains of LAB only.

Traditionally, sourdough initially is started with a spontaneous flora. If the fermentation has the desired effects, a part of the old dough is used to restart the fermentation with fresh substrate. This backslopping procedure is applied over long periods of time (years) resulting in a stable flora. Such sourdoughs are called type I sourdoughs. The evolved process has several steps and can not be integrated in industrial production of bread. For large scale and facilitated application procedures other types of sourdoughs designated Type II doughs have been developed. These are started by using adapted strains to ferment a liquid and thus pumpable dough. Type III sourdoughs can be dried (spray or drum drying) or liquid doughs stabilized by pasteurization or cooling. Type II and III doughs have much easier one step processing and enable a stable end product quality and convient handling. On the one hand, time is saved as a result of the usually higher temperatures and one step fermentations while on the other hand the fermentative flora changes. These sourdoughs allow a quick and flexible production of bakery products with typical sourdough properties in industrial scale.

Potential and prevalence of bacterial EPSs in sourdoughs

A persistent problem is bread staling when storing the end products for a long time. To extend the shelf-life and improve structural properties, the addition of plant polysaccharides is a common practice and the addition of fructans and fructooligosaccharides to improve rheological properties of wheat doughs and bread quality has been patented.

Nevertheless, EPSs influencing doughs are also produced during dough fermentation by the LAB of the starter flora. Experiments by Tieking and Gänzle (2005) found that approximately 20% of 140 tested sourdough strains were homo-EPS producers (19 strains produced fructans, eight strains produced glucan and three produced EPS of unknown composition). Since EPS production by sourdough organisms is so common, Tieking and Gänzle (2005) postulated that the probability of any sourdough flora containing at least one EPS-producing strain is high. Initial studies have shown that bacterial EPS can have significant effects on dough and bread. Potential beneficial effects on the following technological properties has been postulated: (i) water absorption of the dough, (ii) dough rheology and machinability, (iii) dough stability during frozen storage, (iv) loaf volume, (v) bread structure and (vi) bread staling and shelf-life. Additional added value functions of bacterial EPS and oligosaccharides are possible (Decock and Cappelle 2005; Tieking and Gänzle 2005; Arendt *et al.* 2007).

Dextran in sourdough

Practically applied is the use of bacterial dextran in sourdough fermentations. Two patents claim the enhancement of dough and bread by the addition of this glucose based polysaccharide. The use of *Leuconostoc mesenteroides* dextran is well studied and a good example for the improvement of dough and product properties by bacterial EPS. *Leuconostoc mesenteroides* is a sourdough related LAB species since strains can be found in panettone dough. Panettone is a typical Italian sweet bread, consumed mainly during the Christmas period. Since the high consumption is concentrated in a very short time frame of some weeks, production has to start already some months before the selling period. However, stored for a long time period before consumption, panettone stays fresh. Dextran from the sourdough organism *L. mesenteroides* has been identified as cause of that storage stability and the typical crumb structure with large lengthy holes. High amounts of EPS can be formed in the dough since in the traditional panettone process (type I dough) the dough is refreshed up to seven times with increasing amounts of sugar. Apparently, the microflora is able to adapt to high sucrose concentration and *L. mesenteroides* can produce large amounts of dextran from sugar substrate. Due to the long glucan chains, a clear structure build up was proven. The company Puratos® holds a patent for that technology. The produced EPS rates by a particular *L. mesenteroides* strain could be optimized resulting in approximately 25% dextran in dry matter of the sourdough. This dough can be conserved and stabilized with different methods to obtain Type II sourdough. Active cooling or pasteurizing does not affect the functionality of dextran. Also drying without a loss of dextran functional properties is possible and accordingly the application of the dough as type III sourdough preparation. Such ready-to-use dextran containing products enable the baker to easier organize his production in comparison with the time consuming traditional sourdough process without any loss in end product quality. The use of such type III sourdoughs can also bring advantages for rye bread and toast bread. Such existing doughs can also be easily used for the development of new products (Decock and Cappelle, 2005; Lacaze *et al.*, 2007).

Although dextran is defined as an α-(1→6) linked glucan chain with a distinct number of α-(1→3) branches, nothing is specified about the chain length of the polysaccharide. In fact, dextran fractions with different molecular weights (MW) exist and Ross *et al.* showed a dependence of the effect on dough rheology from the average MW of the glucan. *L. mesenteroides* dextrans of lower MW decreased the amount of water bound while higher molecular weight fractions increased the parameter. The texture of the dough was increased with higher MW dextrans while the development time of the dough showed a decreasing tendency. However it has to be mentioned that the time was increased compared with dextran free controls. Influences on dough strength (increase) and extensibility (decrease) also varied with the MW of the added dextran with lower MW dextrans having grater effects than bigger ones. This effects on dough resulted in a decreased loaf volume in pan breads of fermented dough when added lower molecular weight dextran fractions. Unfermented doughs were less affected by the additives. As basic functionality of the dextrans in dough, Ross *et al.* (1992) stated alteration of water binding capacity of the dough and influences on the formation of a gluten network by hydrogen bonds or steric interactions.

Levan application in sourdough

Although the application of bacterial dextrans is widely examined and established in industry less is known about the effects of the bacterial fructans levan and inulin. However, patents claim an improvement in dough rheology and bread quality. First experiments concerning the influence of *L. sanfranciscensis* levan on sourdough parameters has been performed. The effects appeared to be similar to that of dextrans; however, levan turned out to be less effective than glucan addition (Korakli *et al.* ,2000; Brandt, 2001; Kaditzky, 2008).

Effects of hydrocolloids and EPS in dough systems

Bacterial HoPS act in sourdough systems as hydrocolloids. Hydrocolloids are commonly called gums and are extensively used in the food industry. Hydrocolloids have been shown to be able to influence rheology and texture of aqueous systems by stabilizing emulsions, suspensions and foams and to modify starch gelatination. They have been reported to extend the overall quality of a product during time and are applicated as fat replacers, gluten substitutes in the production of gluten-free bread and as therapeutic fibre source. In the baking industry their potential as bread making improvers is of growing importance.

The effects of bacterial EPS and other hydrocolloids on dough and bread can be traced back to two main molecular functionalities: (i) the alteration of water retention and (ii) interactions with other dough components as gluten and flour proteins as well as starch and therewith influencing the formation of structural networks during fermentation and baking. Dextran has been shown to constitutively build up a structure and so increase viscosity of a 1% solution when a shear force is applied. A lineup of long linear dextran chains and a stabilization by hydrogen bonds is proposed as molecular base of that phenomenon (Lacaze *et al.*, 2007). The level of ability to form hydrogen bonds and the possible steric interactions of EPS and structure forming substances in dough are crucial factors in effectivity. So it is essential if EPS chains are flexible or have less degrees of freedom.

Regarding the dough and bread systems, this affects such different dough parameters as water content, dough consistency, development time, stability, strength, elasticity and retaining ability. In bread moisture, crumb and crust firmness and elasticity are influenced. In addition to that, the staling of bread can be modified. Examples for such additives are guar gum, locust bean gum, alginate, xanthan gum, κ-carragenan, carboxymethylcellulose and hydroxypropylmethylcellulose (HPMC).

Most work has been done on the effects of the usually added hydrocolloids in food which are not of bacterial origin but isolated from plants or algae. There are different types of commercially applied hydrocolloid additives concerning their chemical properties. An important factor is the presence of polar groups (e.g. hydroxyl groups). In contrast to (HPMC) or alginate, xanthan and κ-carragenenan are neutral hydrocolloids. Also bacterial glucans and fructans do not contain hydroxyl or similar groups and so can be compared with that group of food additives concerning their functionality in respective food systems.

Three representative studies show the effects of additives on rheological properties of sourdough system. Influences on dough parameters has been examined for the non bacterial hydrocolloids HPMC, alginate, κ-carragenan and xanthan independently from other additives as enzymes by Rosell *et al.* (2001) and for different molecular weight fractions of *L. mesenteroides* dextran (Ross *et al.*, 1992; Guarda *et al.*, 2004) as well as for *L. sanfranciscensis* levan (Kaditzky, 2008).

Farinograph measurements give information about the water binding potential and consistency parameters of the dough. Water binding capacity was determined as amount of water to reach a consistency of 500 Brabender units (BU), a value permitting high water content and comparability. All of the non-bacterial hydrocolloids have been shown to increase water binding capacity of the dough up to 4%. Following expectations, the effect was bigger using hydrocolloids with polar hydroxyl groups (alginate, HPMC). The effect of dextran depends on the MW of the used glucan. High MW fractions (around 500 kDa) increased the Farinograph water absorption (FWA) while smaller molecules (10–70 kDa fractions) decreased the parameter 3–4%. *In situ* produced levan showed a positive effect of water absorption in sourdough resulting in an increase of about 1% compared with control dough. The dough developing time (DDT) is the time required to reach a dough consistency of 500 BU. It was determined for the four non-bacterial substances and turned out to be altered in a different manner by each hydrocolloid. While xanthan and alginate prolonged DDT, κ-carragenan and HPMC had no significant effect. Also dextrans of different MW and levan showed tendency to increase the DDT.

The stability of the dough, defined as time during which dough consistency is at 500 BU, was determined for the non bacterial hydrocolloids and levan. While HPMC and κ-carragenan lowered stability in comparison to control, alginate, xanthan and *in situ* produced levan was capable of elongation of dough stability. In case of hydrocolloids, the effect was dependent on the amount of additive added. Xanthan, HPMC and alginate was able to increase stability when added in 0.5% but not in 0.1% dosage. κ-Carragenan enforced its destabilizing effect when contained in higher amounts. However, the influence of levan merely was significant. For dextran comparable data are not available but Ross *et al.* (1992) therefore reported that the farinograph maximal consistency (FMC) (centre of farinograph curve at the time of peak dough development in BU) was reduced with decreasing molecular weight of dextran. When increasing amounts of 40-kDa dextran are added the parameter is further decreased.

Elasticity parameters can be measured with extensiograph experiments to obtain information about viscoelastic behaviour of dough. Extensibility and resistance to extension are the respective parameters. Desirable is dough with a combination of good resistance and good extensible results. The initial resistance to deformation predicts the dough handling properties and the fermentation tolerance. Extensiograph measurements are repeated after 45, 90 and 135 minutes for xanthan, alginate, HPMC, κ-carragenan and different MW dextrans. When non-bacterial hydrocolloids were added resistance was increased in comparison to control, but stabilized over time with a tendency to increase, while control resistance was reduced with resting time. High MW dextran fraction showed comparable results with a significant increase of resistance with elongated resting time. In contrast, lower MW dextran immediately enforced and additionally increased dough resistance during time. These results was determined with additional water (compared with control) in supplemented doughs to adjust a consistency of 500 BU. Unadjusted supplemented doughs with reduced farinograph consistency showed lowered resistance values but reached the control level after a resting time of 135 minutes. It has to be mentioned that Rosell *et al.* (2001) (non-bacterial hydrocolloids) reported decreasing resistance in control dough while the control doughs of the same study (dextrans of different MW) documented an increasing resistance of both of their control doughs.

Non-bacterial hydrocolloids increased dough extensibility but without an effect of time while particularly low MW dextran fractions decreased that parameter and further reduced it during resting time with a comparable rate to control doughs. To display the overall effect of the additives the quotient of resistance and extensibility was calculated. Non-bacterial hydrocolloids showed a decreased ratio compared with the control but positive effects concerning the stability of viscoelastic parameters during resting time. All dextran fractions showed better over all effects than the control at all time points although in contrast to the control dough of the hydrocolloid experiments the according control doughs also tended to increase their viscoelastic behaviour. Extensiograph data for levan containing doughs did not show a significant influence of the fructan. However, a trend to reduced resistance in levan containing doughs was observed. In addition to that, a clear stabilizing effect of levan on the elasticity parameters resistance and extensibility during resting time compared with a more sensitive control could be detected. That matches with the results of the other additives, particularly dextrans. Nevertheless dextran turned out to have more influence on viscosity than levan.

Rossell *et al.* (2001) additionally performed alveograph studies to get information about the gas retaining potential of hydrocolloid supplemented dough. Experiments revealed a significant increase of gas retaining properties when xanthan was added to the dough. Alginate, HPMC and κ-carragenan had a similar but less effective influence. Unfortunately, comparable data are not available for EPS-supplemented dough.

Influences on rheological, elasticity and water binding factors have, in addition to their technical relevance, an influence on the microflora and therewith on the fermentation behaviour of the dough system. Relevant parameters can be controlled by a rheofermenter concerning dough development and gaseous release. Kaditzky *et al.* (unpublished data) could not find an influence of EPS on the results of those measurements. Appearing changes could rather be traced back to effects of organic acids. In contrast to that an influence of non-bacterial hydrocolloids on the fermentation was reported. Dough height was decreased by xanthan and alginate, probably due to the interactions of that molecules and wheat flour proteins limiting the free expansion of the dough during fermentation. Except alginate the used hydrocolloids increased the time of maximum dough development, alginate decreased it significantly. All of the additives decreased the loss of dough volume and so can support processes which require both slower and longer fermentation. Concerning gas retention, results support the use of HPMC as improving agent. Comparable data have not yet been reported for the application of dextran in sourdough systems.

Concerning the dough rheological and viscoelastic properties of sourdough, it can be concluded that the addition of certain hydrocolloids can have beneficial influences. These factors are of interest for industrial machinability of sourdoughs. On the other hand, they do also affect the fermentation behaviour. As expected, bacterial EPS have been shown to have similar effects and can be classified as neutral hydrocolloids with comparable effects. Dextran turned out to be more effective in altering rheological and elastic parameters than fructans (Tieking *et al.*, 2003a). Due to their origin in sourdough organisms EPS application to sourdoughs would correspond with consumers' demand for less ('chemical') additives in food.

Effects of hydrocolloids and EPS on baking and bread quality parameters and staling

Hydrocolloids and therewith bacterial EPS, based on their molecular functionality, affect not only the rheological properties of sourdough. Modified water binding capacity and interactions with structure forming proteins, glutens and starch are also influencing baking and bread qualities. An important economical parameter in this context is the volume of bread after baking. An improvement of specific volume of bread was reported for the non-bacterial hydrocolloids κ-carragenan, HPMC and xanthan, but not for alginate, and could partially be reproduced in a following study. But only HPMC turned out in several studies to constantly improve bread volume in economic scales. Bacterial levan resulted in an increase of loaf volumes of 1% to 2% (Rosell *et al.*, 2001; Guarda *et al.*, 2004). Dextrans of different MW were supplemented in pan breads with rapid and fermented doughs using two different flours. Only in some conditions a slight volume extension could be noticed, moreover there was a decrease of volume compared with dextran-free control in a majority of cases. In contrast to that, a specific volume extension of 12% and, in combination with application of an enzyme mix, 25% was reported recently (Lacaze *et al.*, 2007). In comparable studies of bacterial EPS in sourdough, dextran was reported to be more efficient in influencing the loaf volume than the fructans levan and inulin (Tieking *et al.*, 2003a; Tieking and Gänzle, 2005).

Data of parameters of bread quality like firmness and moisture of crumb are less comparable, but positive effects by the addition of dextran have been reported and are claimed by a patent. Namely a significantly improved oven jump due to better dough stability and gas retention and an altered crumb structure should be mentioned. Additionally a better because moister mouthfeel could be achieved by dextran supplementation. This dextran effect is described to become clearer with a longer conservation time of the bread. This would be especially interesting concerning the staling and firming of bread and bakery products. The extension of shelf-life is a desirable aim for improvements in baking technology since bakery products with few exceptions like panettone loose their freshness in short time. Aside from microbial spoilage physicochemical changes shorten the shelf live. Stored bread looses its palatability due to staling and firming. Staling is a complex process resulting in retrogradation of starch to a more crystalline form and migration of the moisture from starch granules to interstitial spaces. Staling is independent of environmental moisture and proceeds fastest at temperatures just above freezing. Staled bread appears dry and leathery. The decisive factor in staling can be the gluten–starch interactions and variations in water activity and moisture content in and near the crust. Sourdough has been reported to have positive effects on bread staling, whereas the anti-staling effects are dependent on the particular starter strains. Responsible for the beneficial effects of sourdough flora is obviously the acidification, hydrolysis of starch and proteolytic activity. Different additives has been tried to delay staling, among them organic acids, α-amylases and hydrocolloids like HPMC (Corsetti *et al.* 1998; Corsetti *et al.* 2000). Since bacterial homopolysaccharides can have similar effects in dough like non bacterial hydrocolloid additives and are of dough relevant microbial source they are good alternatives as anti-staling agents. An example is the patent 'use of a polysaccharide as bread improver' of van Geel-Schutten (2006), which claims the application of different LAB homopolysaccharides (e.g. *L. reuteri* dextran and reuteran) in sourdough and baking. They describe beside positive effects on dough rheological parameters a delayed staling compared with reference breads without the addition of polysaccharides. They evaluated the crumb softness by hand and found a similar hardening to a reference bread containing a not named anti-staling additive in polysaccharide breads. Calorimetric measurements showed no significant difference in amylopectin crystallization compared with the blank (van Geel-Schutten 2006). Kaditzky documented an anti-staling effect when *L. sanfranciscensis* levan was added to wheat dough. Therefore texture profile analysis (TPA) was performed to assess the hardness of breads and differential scanning calorimetry (DSC) was used to display the rate of starch retrogradadtion. Both parameters was positively affected by the fructan addition (Kaditzky 2008). For the addition of dextran a significant influence on the quality conservation of milk bread has been reported, too. A sourdough-free reference was compared with a 5% dextran-based sourdough containing sample regarding hardness, pH and water activity. After 2 weeks, water activity was not different in sample and control and pH was only slightly lower (0.3 units) in sourdough containing milk bread, but hardness was significantly reduced. A panel test showed that the dextran sourdough containing milk bread was significantly preferred. A similar consumer reaction could be achieved in a ranking experiment with sourdough containing and sourdough free butter brioches. Dextran content appeared to be also effective in conservation of rye mixed bread that stayed softer when dextran based sourdough was used. Firming of these breads was considerably delayed compared with a dextran free sourdough control. Tasting the bread revealed an improvement of crumb softness and moistness. A beneficial effect on the specific volume of the dextran containing rye mixed breads completed the conserving effects (Lacaze *et al.*, 2007).

Beneficial health aspects of homoexopolysaccharides (added value functions)

A growing tendency in consumers demand for foods with additional nutrition and health functions is noted in the last years. Examples are yoghurts containing probiotic lactobacillus strains and margarine lowering cholesterol level. The principal function of pro- and prebiotic additives in food is to boost gut bacteria populations with positive effects on human intestinal tract. The beneficial effects may either be due to advantageous metabolics and metabolic products of probiotic strains. Another desirable component is the suppressing and blocking of pathogenic bacteria by occupation of ecological niches and receptors on human intestinal cells.

Probiotic properties of bifidobacteria are widely accepted and also certain LAB are reported to have health-promoting effects. Amongst others, cereal fermentations can be a source for such probiotic LAB. Some of them have been shown to be able to adhere to human intestinal cells (Modler, 1994; Lee and Salminen, 1995; Corsetti and Settanni, 2007). It is important to distinguish pro- and prebiotic additives in foods. The term probiotic always constitutes a viable micro organism in the product which is designated to reach the human gut alive and establish in that environment to develop its beneficial properties. In contrast to that, prebiotic additives are substances supporting existing advantageous strains in the gut, e.g. carbohydrates that can be exclusively used by bifidobacteria. For prebiotics particularly enforcing bifidobacterial flora the term bifidogenic has been used (Corsetti and Settanni, 2007).

Naturally, pre- and probiotics must not be affected or destroyed during food processing and ingestion. That postulates at least stability against stomach acid and, depending of food processing steps, a temperature resistance. Therefore it is not possible to directly introduce probiotics in heat treated foods like baked goods.

Although sourdough can contain a spectrum of potentially probiotic LAB and respective strains have been isolated from cereal fermentations as mentioned above, bakery products in contrast to fermented milk products, due to their inevitable baking process step are not suitable as primary carrier for probiotics. On the other hand, fructooligosaccharides of the inulintype (GF2–1-kestose, GF3–1,1-nystose) and the fructans levan and inulin are known to selectively stimulate growth and metabolism of bifidobacteria and so belong to the prebiotic substances (Dal Bello *et al.* 2001). The metabolism of levan produced by sourdough organism *L. sanfranciscensis* could be shown (Korakli *et al.* 2002). For the fructan forming enzymes, particularly inulinsucrases, of cereal lactobacilli the potential of forming frutctooligosaccharides (FOS) was reported. Since fructans and FOS are tolerant against high temperatures during baking and sufficiently resistant against low pH, they can add an additional value function to sourdough breads when applicated to dough or produced by selected starter strains during fermentation.

Further, bacterial EPS should have useful effects in the field of gluten-free bread making. Patients with coeliac disease suffer a pathological alteration of gut epithelial cells when consuming gluten-containing bread. Since the gluten network, as mentioned above, is the most important structure-forming factor and gluten proteins are included in most cereal flours, technological inventions must be found to produce gluten-free breads. The use of alternative flours like brown rice results in gluten-free breads with poor textural properties. The use of sourdough fermentations can plainly increase bread quality. A special selection of effectively EPS-producing sourdough starter strains or the additional supplementation with bacterial homopolysaccharides such as dextran can bring a further progress in gluten-free bread production.

Further groups of food

Today applications of bacterial HoPS in food industry concentrate on dairy and sourdough technologies and some patents claim their use. Nevertheless, LAB play crucial roles in further fields of fermented foods and potential polysaccharide producers are among them. Although there is no established application to date, a potential use of EPS as product improver for meat fermentations, sauerkraut and vinegar production can be discussed.

Salami is a group of sausages consisting of raw meat fermented by strains of *L. sakei*, *L. curvatus*, *L. plantarum* and *Micrococcaceae*. The soft structure of the fresh sausage is altered by acidification and the resulting denaturation of meat proteins and the concomitant drying process. Acidification and drying are the main conservatory mechanisms in salami production and a quick lowering of pH is essential to achieve product safety in the early fermentation stage. Although the starter cultures used are potential EPS producers no data about the role of EPS in meat fermentations has been reported yet. This may be referred to lesser expected positive influences of EPS on endproduct quality. Still, modifications in spreadability of slightly fermented raw sausages or mouth feel is thinkable as well as introduction of pre- or symbiotics in fermented meat products.

An importnt fermentation organism in sauerkraut is *Leuconostoc mesteroides*. Strains of that species have been found to be effective dextran producers. In consequence it can be considered that there are certain amounts of glucan in the fermentation, limited by substrate availability. In all probability there is an influence on mouth feel. However, there is no apparent demand for an alteration of existing sauerkraut fermentations. Fermentations are stable and used starter strains produce desired sensoric and textural properties. Sauerkraut also is not yet a food with a strong demand on added value functions.

Vinegar is fermented from whine by acetic acid bacteria such as *Acetobacter* and *Gluconobacter* in an aerobic process. The ethanol of whine is converted to acetate. Industrial vinegar fermentations have, in contrast to traditional ones, a risk to be delayed by bacteriophages. The critical point concerning this is a monoculture of a high performance starter strain in a liquid and mixed fermentation environment. Since phages often are very strain specific, a rolating system of different starter strains with similar fermentation properties can avoid loss by phage attacks. However, capsular EPS has been reported to have protectional functions including a shelter to phage infection, probably by steric prohibition of phage docking. Since certain gluconobacter strains are capable of production of EPS, a selection of those strains might be helpful to obtain phage resistant vinegar starters. Nevertheless, an influence on the product can not be excluded and a decrease of fermentation rates due to worse substrate and product transports throughout the EPS capsule is thinkable and an economic problem.

Beside from applications in fermented foods, which might be source of EPS itself, bacterial EPS could replace plant and seaweed derived polysaccharides. These additives are widely used as thickening, gelling, stabilizing and water-binding agents. Dextrans are already used in confectionary for improvement of moisture retention, viscosity and inhibition of sugar crystallization. Also an application as gelling agent in gum and jelly candies is known. Crystallization inhibition in ice-cream and use for achievement of desirable body and mouth feel in pudding mixtures are further examples (Khan *et al.*, 2007).

Addition of EPS versus *in situ* production

In fermented food, especially in sourdough systems, two basically different ways of applications of EPS additives are possible: Either the *ex situ* production and addition of the EPS as pure substance in distinct amounts at a defined time point or the *in situ* production by the starter cultures during the fermentation. Therefore effective

EPS producing starters can be chosen or even be engineered by genetic methods. Naturally the latter option does not exist when supplementing non-fermented food products with EPS.

Most studies concentrate on direct addition of EPS, for example dextran in certain amounts. Others just report the effect of sourdough containing EPS producing strains and in most cases do not determine amounts and kind of EPS produced in the fermentation. Only few data are available that allow a comparison of the effects of added EPS and *in situ* production of polysaccharides.

Beside of effects on product properties, addition of EPS preparations in the sourdough has some advantages. The amount of EPS added is defined and can be optimized, as well as the time point of addition. This allows a strict control of the EPS impact on dough and bread quality. More critical is the fact that this kind of application makes EPS an additive that has to be mentioned on the additive list on the packaging whereas the consumer increasingly demands for fewer additives in food products and so the acceptance of the product can be negatively influenced. In addition to that, EPS preparations must be obtained separately and thus increase production costs.

In vitro produced EPS can enhance dough and bread properties without having to be mentioned on ingredients lists. It does not have to be bought and so does not cause additional costs compared with conventional processes. The potential and possibilities of development of starter lactobacilli derived from sourdough ecosystems concerning various metabolic properties has been reported (Vogel *et al.*, 2002). Nevertheless some aspects have to be considered: EPS production is dependent on different environmental factors. Influences of dough yield (DY), pH, substrate concentration and fed batch processing has been examined (Kaditzky and Vogel, 2008). Glucan and fructan production of an *L. reuteri* strain and fructan production of a *L. sanfranciscensis* strain was studied. It turned out that more liquid doughs (higher DY) had higher EPS concentrations at comparable conditions. A reason for that could be a better diffusion of substrates and extracellular glycosyltransferases. An optimal pH was found at optimum growth level for EPS producing strains although the enzyme pH optimum was lower. Further, an increase was obtained by continuous addition of sucrose over fermentation time. Based on these data an optimization of EPS production according to other process criteria can result in significantly enhanced dough and bread quality if appropriate EPS forming starter cultures are chosen. However, similar amounts of *in situ* produced EPS compared to added EPS may not result to the same effect level since *in situ* production is a dynamic process progressing in parallel to fermentation. In sourdough systems, it was observed that the production of acids can counterbalance the positive EPS effects, and that it is crucial to select an EPS producing fermentation organisms not solely along the properties of its EPS, but also along its fermentation performance and product ratio in mixed acid fermentations. (Kaditzky and Vogel, 2008).

The use of genetically engineered starter strains with enhanced EPS-forming versus fermentation properties is thinkable but could be in conflict with legal guidelines and consumers' demand for more natural food.

Conclusions

In conclusion, HoPS applications focus on non-dairy food applications namely (sour)dough systems. Water retention, baking abilities and staling can be beneficially influenced. Independently from the class of EPS added, these effects could be seen when applied pure EPS in dough before baking. Nevertheless this kind of application would have to be mentioned on the food label, which is undesired since consumers' demand is for more natural and unmodified food with fewer additives.

Labelling can be avoided when strains with a long traditional and safe use are tested towards their ability to produce EPS and an effective EPS producing strain is chosen and introduced in the fermentation process. However, it has to be taken in account that the metabolism of the fermentation flora (e.g. acetate production) can negatively influence EPS yields and counteract beneficial EPS effects. Still these metabolic functions erve other food properties and may even be required, e.g. to achieve bakeability in rye breads. In that case choosing a weak acidifying but strong EPS producing strain would not be solely advantageous. Forthcoming chances of EPS applications may lie within special applications as in gluten-free breads where both reduced pH and EPS should have synergistic positive effects on structure.

References

Anwar, M.A., Kralj, S., van der Maarel, M.J., and Dijkhuizen, L. (2008). The probiotic *Lactobacillus johnsonii* NCC 533 produces high-molecular-mass inulin from sucrose by using an inulosucrase enzyme. Appl. Environ. Microbiol. 74, 3426–3433.

Arendt, E.K., Ryan, L.A., and Dal Bello, F. (2007). Impact of sourdough on the texture of bread. Food Microbiol. *24*, 165–174.

Bergmaier, D., Champagne, C.P., and Lacroix, C. (2005). Growth and exopolysaccharide production during free and immobilized cell chemostat culture of *Lactobacillus rhamnosus* RW-9595M. J. Appl. Microbiol. *98*, 272–284.

Brandt, M.J. (2001) Mikrobiologische Wechselwirkungen von technologischer Bedeutung in Sauerteigen. In: Institut für Lebensmitteltechnologie. Universität Hohenheim, Hohenheim, p. 134.

Corsetti, A., Gobbetti, M., Balestrieri, F., Paoletti, F., Russi, L., and Rossi J. (1998). Sourdough lactic acid bacteria effects on bread firmness and staling. J. Food Sci. *63*, 347–351.

Corsetti, A. (2000). Combined effect of sourdough lactic acid bacteria and additives on bread firmness and staling. J. Agric. Food Chem. *48*, 3044–3051.

Corsetti, A., and Settanni, L (2007). Lactobacilli in sourdough fermentation. Food Res. Int. *40*, 539–558.

Dal Bello, F., Walter, J., Hertel, C., and Hammes, W.P. (2001) In vitro study of prebiotic properties of levan-type exopolysaccharides from Lactobacilli and non-digestible carbohydrates using denaturing gradient gel electrophoresis. Syst Appl. Microbiol. *24*, 232–237.

De Vuyst, L., De Vin, F., Vaningelgem, F., and Degeest, B (2001). Recent developments in the biosynthesis and applications of heteropolysaccharides from lactic acid bacteria. Int. Dairy J. *11*, 687–707.

De Vuyst, L., and Degeest, B. (1999). Heteropolysaccharides from lactic acid bacteria. FEMS Microbiol. Rev. *23*, 153–177.

De Vuyst, L., and Vancanneyt, M. (2007). Biodiversity and identification of sourdough lactic acid bacteria. Food Microbiol. *24*, 120–127.

Decock, P., and Cappelle, S. (2005). Bread technology and sourdough technology. Trends Food Sci. Technol. *16*, 113–120.

Guarda, A., Rosell, C.M., Benedito, C., and Galotto, M.J. (2004). Different hydrocolloids as bread improvers and antistaling agents. Food Hydrocolloids *18*, 241–247.

Hassan, A.N., Awad, S., and Mistry, V.V. (2007). Reduced fat process cheese made from young reduced fat Cheddar cheese manufactured with exopolysaccharide-producing cultures. J. Dairy Sci. *90*, 3604–3612.

Kaditzky, S., and Vogel, R.F. (2008). Optimization of exopolysacchride yields in sourdoughs fermeted by lactobacilli. Eur. Food Res. Technol. (in press).

Kaditzky, S.B. (2008). Sucrose metabolism in lactobacilli and bifidobacteria. In: Lehrstuhl für Technische Mikrobiologie. Technische Universität München, Freising, p. 137.

Khan, T., Park, J.K., and Kwon, J.-H. (2007). Functional biopolymers produced by biochemical technology considering applications in food engineering. Korean J. Chem. Eng. *24*, 816–826.

Korakli, M., Gänzle, M.G., and Vogel, R.F. (2002). Metabolism by bifidobacteria and lactic acid bacteria of polysaccharides from wheat and rye, and exopolysaccharides produced by *Lactobacillus sanfranciscensis*. J. Appl. Microbiol. *92*, 958–965.

Korakli, M., Schwarz, E., Wolf, G., Hammes, WP (2000) Produciton of mannitol by *Lactobacillus sanfranciscensis*. Advanc. Food Sci. *22*, 1–4.

Korakli, M., and Vogel, R.F. (2006). Structure/function relationship of homopolysaccharide producing glycansucrases and therapeutic potential of their synthesised glycans. Appl. Microbiol. Biotechnol. *71*, 790–803.

Kumar, A.S., Mody, K., and Jha, B. (2007). Bacterial exopolysaccharides – a perception. J. Basic Microbiol. *47*, 103–117.

Lacaze, G., Wick, M., and Cappelle, S. (2007). Emerging fermentation technologies: development of novel sourdoughs. Food Microbiol. *24*, 155–160.

Lee, Y.-K., and Salminen, S. (1995) .The coming of age of probiotics. Trends Food Sci. Technol. *6*, 241–245.

Modler,, H.W. (1994). Bifidogenic factors – sources, metabolism and application. Int. Dairy J. *4*, 383–407.

Rosell, CM, Rojas, JA, Benedito de Barber, C (2001). Influence of hydrocolloids on dough rheology and bread quality. Food Hydrocolloids *15*, 75–81.

Ross, A., S, McMaster, GJ, Tomlinson, DJ, Cheetham, NWH (1992). Effects of dextrans of differing molecular weights on the rheology of wheat flour doughs and the quality chracteristics of pan and arabic breads. J. Sci. Food Agric. 60:91–98.

Tieking, M., Ehrmann, MA, Vogel, RF, Gänzle, MG (2005). Molecular and functional characterization of a levansucrase from the sourdough isolate Lactobacillus sanfranciscensis TMW 1.392. Appl. Microbiol. Biotechnol. *66*, 655–663.

Tieking, M., Gänzle MG (2005) Exopolysaccharides from cereal-associated lactobacilli. Trends Food Sci. Technol. *16*, 79–84.

Tieking, M., Kaditzky, S.B., Gänzle, M.G., and Vogel, R.F. (2003a). Biodiversity and potential for baking applications of glycosyltransferases in lactobacilli for use in sourdough fermentation. In: L. dV (ed.) Sourdough, from Fundamentals to Applications. Brussels: Vrije Universiteit Brussel (VUB), IMDO, pp. 58–59.

Tieking, M., Korakli, M., Ehrmann, M.A., Gänzle, M.G., and Vogel, R.F. (2003b). In situ production of exopolysaccharides during Sourdough fermentation by cereal and intestinal isolates of lactic acid bacteria. Appl. Environ. Microbiol. *69*, 945–952.

van Geel-Schutten, GH (2006) Use of a polysaccharide as bread improver. Patent no. WO 2006/062410 A1.

van Hijum, S.A., Bonting, K.,van der Maarel, M.J., and Dijkhuizen, L. (2001) Purification of a novel fructosyltransferase from *Lactobacillus reuteri* strain 121 and characterization of the levan produced. FEMS Microbiol. Lett. *205*, 323–328.

van Hijum, S.A., Kralj, S., Ozimek, L.K., Dijkhuizen, L., and van Geel-Schutten, I.G. (2006) Structure-function relationships of glucansucrase and fructansucrase enzymes from lactic acid bacteria. Microbiol. Mol. Biol. Rev. *70*, 157–176.

van Hijum, S.A., van Geel-Schutten, G.H., Rahaoui, H., van der Maarel, M.J., and Dijkhuizen, L. (2002) Characterization of a novel fructosyltransferase from *Lactobacillus reuteri* that synthesizes high-molecular-weight inulin and inulin oligosaccharides. Appl. Environ. Microbiol. *68*, 4390–4398.

Vogel, R.F., Ehrmann, MA, and Gänzle, M.G. (2002) Development and potential of starter lactobacilli resulting from exploration of the sourdough ecosystem. Antonie Van Leeuwenhoek *8*, 631–638.

Waldherr, F.W., Meissner, D., and Vogel, R.F. (2008) Genetic and functional characterization of *Lactobacillus panis* levansucrase. Arch Microbiol. *190*, 497–505.

Walter, J., Schwab, C., Loach, D.M., Gänzle, M.G., and Tannock, G.W. (2008) Glucosyltransferase A (GtfA) and inulosucrase (Inu) of *Lactobacillus reuteri* TMW1.106 contribute to cell aggregation, *in vitro* biofilm formation, and colonization of the mouse gastrointestinal tract. Microbiology *154*, 72–80.

Exploitation of Exopolysaccharides from Lactic Acid Bacteria: Nutritional and Functional Benefits

18

Alan D. Welman

Abstract

The quest to find food ingredients with valuable bioactive properties has encouraged interest in exopolysaccharides (EPSs) from lactic acid bacteria (LAB). Functional food products that offer health and sensory benefits beyond their nutritional composition are becoming progressively more important in the food supply chain. Informed consumers are becoming increasingly more health conscious and more aware that disease prevention can be effected through dietary means and, more specifically, by particular components of foods. Although the sensory benefits of EPSs are now well established, more knowledge to fully understand how these polysaccharides can impact on human health and nutrition is needed. Because of the wide variation in molecular structures of EPSs, and the complexity of the mechanisms by which physical changes in foods and bioactive effects are elicited, the future commercial development of EPS-producing cultures and ingredients will depend heavily upon this platform of knowledge. In this chapter, evidence for the health properties that are attributable to LAB EPSs is reviewed, linking some of the very first findings in this area to current discoveries. Because the sensory attributes of foods are critical to their acceptance, the basis for the texture-imparting properties of EPSs is also reviewed; this section describes what is known about the basis of the structure–function relationships of EPSs, and their applications in a dairy food context. It is hoped that this chapter will stimulate interest in an area in which far more fundamental biochemical and clinical understanding is required.

Commercial importance

Historically, EPS-producing cultures (EPS$^+$ cultures) have been exploited, particularly in Europe, to produce mainly fermented dairy foods with novel physical properties (Cerning, 1995). By including EPS$^+$ cultures into these foods, improved viscosity and water binding can result, imparting enhanced sensory attributes such as 'mouthfeel', textural qualities and a longer residence time in the mouth (and hence taste perception) (Duboc and Mollet, 2001; Font de Valdez *et al.*, 2003). EPSs also have the effect of reducing syneresis. For example, there is a high demand for smooth and creamy yoghurt products; these textural characteristics are conventionally supplied by the addition of fat, sugars and proteins (e.g. pectin, starch or gelatin) (Jolly *et al.*, 2002). In more recent times, consumers have become more discriminating –not only do they desire these sensory attributes, but also they want a product that is low in additives (Van der Meulen *et al.*, 2007) or devoid of fat and sugar and a product that carries a 'natural' label. Functional food manufacturers, in turn, now want to connect their products with 'all-natural' and 'free-from' messages.

Regulatory directives also influence the swing to more natural ingredients. In Europe, for example, food regulations stipulate that a natural-type yoghurt cannot contain artificial stabilizers or hydrocolloids. These factors, combined with the need to lower costs of production, render EPSs from LAB an attractive alternative for inclusion into foods if they can mimic the sensory qualities elicited by conventional additives. Some suppliers of stock starter cultures have already exploited these opportunities, producing EPS$^+$ cultures that they claim will enhance the viscosity of yoghurt by up to 20%. Although these EPS$^+$ starter cultures often produce unique flavours and taste profiles in yoghurt, the EPSs that they produce have no taste of their own.

Apart from these sensory effects, evidence for the potential health benefits of LAB EPSs, such as anti-tumour effects, immunostimulatory activity and the lowering of cholesterol, has been accumulating (Welman and Maddox, 2003). Although much of this research is still in its infancy, the possibility for incorporating selected health-promoting EPSs into foods remains a distinct possibility. Successful exploitation of these attributes has the ability to expand the functional footprint of foods and their resultant benefits.

The nature of EPSs from LAB

EPSs are naturally produced by some LAB. They are long-chain polysaccharides that are made up of branched, repeating units of sugars, substituted sugars or sugar derivatives, most commonly glucose, galactose and rhamnose, occurring in various ratios (De Vuyst and Degeest, 1999; De Vuyst *et al.*, 2001). Microbial EPSs can be usefully classified into two main groups: 'homopolysaccharides', which have only one monosaccharide type, either D-glucose or D-fructose (these can be subclassified as either glucans or fructans respectively), and 'heteropolysaccharides', with structurally dissimilar monomer units. LAB EPSs characteristically have the heteropolysaccharide structure, and are assembled by LAB as multiple copies of oligosaccharides, each consisting of between three and eight monomers. Each of the oligo-units has more than one monomer type linked together at different positions. These variations result in a variety of compositional and structural variants, which also differ in their size (from 4.0×10^4 to 6.0×10^6 Da) and the space that they occupy. The polysaccharides are secreted outside of the cell and, depending on the bacterial strain, either remain attached to the cell (forming a capsular EPS) or become unattached in the surrounding medium, often generating a 'ropy' property (Hassan, 2008).

Apart from the influence of their conformational structure on functionality, the amount of EPS that is produced by LAB is relatively low when compared with other polysaccharides, such as xanthan or gellan. Because of the broad range of different structures and compositions of EPSs produced by LAB, combined with their various functional effects, there is no way of establishing a universal benchmark for an effective EPS titre or usage level in foods (Welman and Maddox, 2003). For example, one of the highest EPS titres reported is 2.8 g/l in *Lactobacillus rhamnosus* RW-9595M (Macedo *et al.*, 2002), whereas the majority of publications reflect titres below 1.0 g/l. Strategies that have been used to produce higher levels of EPS have mostly centred on understanding how metabolic flux can be engineered to overproduce these metabolites (Welman *et al.*, 2006).

Physiological nutrition

In more recent times, a shift in consumer perceptions about food and its effects has been fuelled by increasing awareness of health concerns related to obesity, cancer, atherosclerosis, elevated blood pressure and immune health. The more educated consumers pay attention to health claims on a food label, and expect the ingredients to offer a preventative clinical benefit, even when they are healthy. This trend has become pervasive and regulatory authorities have begun to exercise control over the level of claims made so that the consumer is not misled. Effectively, this means that clinical nutrition claims have to be supported by solid scientific evidence.

To date, although there have been few reports of possible proven clinical nutrition applications for EPSs from LAB, there is a growing body of evidence for the potential for this type of use. Table 18.1 summarizes the reported effects and benefits of LAB EPSs, as well as their proposed mechanisms of action.

Cholesterol-lowering effects

Hypercholesterolaemia is generally considered to be one of the risk factors for atherosclerosis. There is the potential for EPSs to disrupt the absorption of cholesterol in the intestines by binding it in a similar fashion to that achieved by plant-based polysaccharides or dietary fibre (Pigeon *et al.*, 2002).

Nakajima *et al.* (1992) reported that the EPS from *Lactococcus lactis* subsp. *cremoris* SBT 0495 enhanced the metabolism of serum cholesterol

Table 18.1 Reported health benefits associated with exopolysaccharides from Lactic Acid Bacteria and their proposed mechanisms of action

Examples of EPS-producing LAB	Specific effect of EPS	Postulated mechanisms of action	Ultimate health benefit
Lactococcus lactis subsp. *cremoris* *Lactobacillus kefiranofaciens*	Cholesterol-lowering	EPS binds to cholesterol, diminishing cholesterol absorption in the intestine EPS binds to bile acids, inducing a compensatory conversion of cholesterol to the acids, and reducing the free cholesterol pool	Cardiovascular – reduces risk of atherosclerosis
Lactococcus lactis subsp. *cremoris* *Lactobacillus delbrueckii* subsp. *bulgaricus* *Lactobacillus rhamnosus* *Lactobacillus helveticus*	Immunogenic effects	Induction of cytokines Increased T-lymphocytes Activation of macrophages	Stimulation of immune system
Lactobacillus delbrueckii subsp. *bulgaricus* *Lactobacillus plantarum* *Lactobacillus sanfranciscensis*	Gastrointestinal health	Assists in anchoring probiotics to intestinal mucus Inhibits pathogen binding to intestinal walls Bifidogenic (prebiotic) effect	Enhances gut health; protection against intestinal pathogens
Lactococcus lactis subsp. *cremoris* *Lactobacillus delbrueckii* subsp. *bulgaricus* *Lactobacillus kefiranofaciens*	Anticarcinogenic	Induction of mitogenic response Enhancement of T-cells	Anti-carcinogenic effects
Lactobacillus kefiranofaciens	Reduction in blood pressure	Unknown	Cardiovascular health

in rats. This strain was included in a basic diet in conjunction with 0.5% cholesterol, and the diet was compared with the same diet with and without the non-EPS-forming variant of that strain. The lowest serum cholesterol levels and the highest serum high-density lipoprotein (HDL) cholesterol/total cholesterol ratios were found in rats that were administered the diet with the SBT 0495 strain. The authors postulated that the EPS acted like dietary fibre, diminishing the absorption of cholesterol and bile acids in the intestine. The ability of yoghurt LAB cultures to bind bile acids is the basis of a possible mechanism whereby cholesterol is reduced indirectly by EPS. The postulate is that, because cholesterol is used to synthesize free bile acid, a compensatory increase in the production of bile acid to replace that excreted from the body with EPS will reduce the pool of cholesterol (Maeda *et al.*, 2004a).

An EPS that has attracted attention is kefiran (galactoglucan), produced by *Lactobacillus kefiranofaciens*. In a study by Maeda *et al.* (2004b), groups of rats fed kefiran were found to have lower serum levels of total cholesterol, low-density lipoprotein (LDL) cholesterol and triglycerides, and also lower levels of cholesterol, triglycerides and phospholipids in the liver. Advanced plaque with lipid deposition in the aortic arch was also diminished in the kefiran-fed rats. The mechanism for cholesterol reduction by kefiran is not entirely clear, but the observation that dose-dependent increases in faecal cholesterol, bile acid (a cholesterol metabolite) and total sterol are associated with kefiran ingestion may throw some light on this. Because the rate of excretion of food cholesterol was unaltered, it was suggested that kefiran does not inhibit food absorption, but rather binds with enterohepatic cholesterol in the intestine

(Maeda *et al.*, 2005). However, the mechanism of liver cholesterol reduction remains unclear. Not all investigations using kefiran have been reported as being successful. In a human trial (n = 13), dairy kefiran was reported to have no effect on total cholesterol, LDL cholesterol, HDL cholesterol and triglyceride concentrations in mildly hypercholesterolaemic male subjects, after 4 weeks of dietary supplementation (St-Onge *et al.*, 2002).

More recently, Liu *et al.* (2006) demonstrated a lowering of serum triacylglycerol and total cholesterol and an accumulation of cholesterol in the liver of cholesterol-fed hamsters when also fed milk kefir (or soya milk kefir).

Immunopotentiating effects

The role of carbohydrates in the recognition of antigens by the adaptive immune system has become an accepted fact (Cobb and Kasper, 2005). The option of enhancing the immune status (and hence the health status) of individuals by manipulating carbohydrate constituents of the normal diet has been advanced by successive findings that demonstrate an immunopotentiating activity of LAB EPSs.

Some of the earlier studies into the activation of the immune system by LAB EPSs showed that EPSs could cause an increase in the levels of T-lymphocytes (Forsén *et al.*, 1987) and induce the production of cytokines and activate macrophages; this induction was thought to be due to a phosphopolysaccharide (Kitazawa *et al.*, 1996). Nakajima *et al.* (1995) reported the enhancement of specific antibody production in mice when the EPS of *Lactococcus lactis* subsp. *cremoris* was administered via an intraperitoneal route. The acidic polysaccharide (APS) component of *Lactobacillus delbrueckii* subsp. *bulgaricus* OLL1073R-1 was shown to demonstrate mitogenic activity for murine B-lymphocytes; the presence of a phosphate group on this EPS was thought to play an important role in this regard (Kitazawa *et al.*, 1998).

Chabot *et al.* (2001) isolated EPS fractions from *Lactobacillus rhamnosus* RW-9595M on the basis of molecular weight and tested them *in vitro* for their possible modulating properties on mouse splenocytes, on a macrophage-like cell line (murine RAW 264.7) and on peripheral blood mononuclear cells (PBMC) from 14 healthy human donors. The largest molecular weight fraction (> 1000 kg/mol), as well as the unfractionated (total) EPS isolate, elicited a cytokine response in both the mouse cells and the human cells. In the human PBMC, no interferon-γ production was seen even though IL-12p40 secretion was enhanced. These results pointed to the possibility of using LAB EPS to enhance the immune system of certain individuals who might respond to such a challenge. Whereas it is an accepted fact that the EPSs exert their effects via cell membrane receptors, it was not apparent as to which receptors were involved. However, it was speculated that a mannose receptor may be implicated, because of the rich rhamnose content of *Lactobacillus rhamnosus* EPS.

The immunostimulatory effects of *Lactobacillus delbrueckii* subsp. *bulgaricus* (strain OLL1073R-1) were investigated further by Nishimura-Uemura *et al.* (2003), who studied the stimulatory effect of a neutral polysaccharide (NPS) and a phosphopolysaccharide (APS) from this strain on macrophages (J774.1 murine cell line). Neither of these fractions induced nitrite release, but both effected morphological changes, and to a greater extent with the APS fraction than the NPS fraction. These effects were enhanced by a lipopolysaccharide (LPS) fraction. The differences in effect were thought to be related to differences in the amount of phosphorylation of these structurally similar fractions and the nature of the binding via these groups to LPS receptors on the macrophages. The effect of the NPS and APS fractions on induction of the host immune defence system by cytokine release was also investigated. Interleukin 1α mRNA expression (as well as the expression of a number of other cytokines) was strongly induced by 12 h of stimulation with the APS fraction, but the same effect was not achieved with NPS. EPSs from *Lactobacillus helveticus* R389 were also reported to effect mucosal immunomodulation, causing an increase in the number of IgA+, IL-10+, IL-2+ and IL-6+ cells when fed to BALB/c mice (Vinderola *et al.*, 2007).

Natural killer (NK) cell activity is thought to play an important part in preventing tumour development. A study in humans has demonstrated that the high cytotoxic activity of peripheral blood lymphocytes is associated with a reduced risk of cancer (Imai *et al.*, 2000). In a separate study, the APS component of the OLL1073R-1 strain was further separated into a high-molecular weight fraction (phosphorylated) and a low-molecular weight fraction. Oral administration of the high-molecular weight fraction to mice significantly increased interferon-γ production by mouse splenocytes, and enhanced NK cell activity. However, these effects were not replicated with administration of yoghurt fermented with the starter combination of *Lactobacillus bulgaricus* OLL1256 and *Streptococcus thermophilus* OLS3295 (Makino *et al.*, 2006). In summary, it would appear to be possible that it is the conformational structure of the EPS, or the presence of a phosphate group, that has an important influence on its immunostimulatory effect (Kitazawa *et al.*, 2000).

A fermented milk drink that has attracted significant interest is kefir, which is produced by a consortium of LAB, yeasts and acetic acid bacteria, trapped within a network of kefir grains, EPSs and proteins. This complex matrix has been shown to enhance phagocytic activity of peritoneal and pulmonary macrophages and to modulate the mucosal response at remote sites (Vinderola *et al.*, 2005). Oral administration of the kefir EPS kefiran to mice induced a gut mucosal response, increasing the number of IgA+ cells in both the small intestine and the large intestine, and enhancing systemic immunity by releasing cytokines into the blood circulation (Vinderola *et al.*, 2006).

However, not all studies have proved to be successful. EPS from *Bifidobacterium lactis* Bb12 did not significantly stimulate lymphocyte proliferation or increase interferon-γ and IL-10 secretion in mice splenocytes (Amrouche *et al.*, 2006).

Effects on intestinal health

Inflammatory bowel disease (IBD) is a group of inflammatory conditions of the large intestine and, in some cases, of the small intestine; these conditions are thought to be caused by autoimmune states. One postulate is that IBD stems from a poorly regulated immune reaction to intestinal flora. Probiotic therapy has been suggested to assist in the treatment of this condition. Sengül *et al.* (2006) evaluated the effects of an EPS$^+$ *Lactobacillus delbrueckii* subsp. *bulgaricus* strain (B3) and a strain producing lower amounts of EPS (A13) on a rat experimental colitis model. Compared with a colitis control, histological scores of damage to the colon and myeloperoxidase activity (indicators of the disease) were lower in both treated groups, but were significantly lower when the higher EPS producer was used. A therapeutic (and possibly dose-related) role for EPS$^+$ probiotics is suggested.

EPSs could potentially also play a role in enhancing the health benefits of probiotics. Combinations of probiotics have been shown to synergistically cause a significant enhancement of their adhesion to intestinal mucus (Collado *et al.*, 2007) and EPSs could play a role in modifying this adhesion (Ruas-Madiedo *et al.*, 2006a,b). Suppressor mutations in the *Lactobacillus rhamnosus luxS* mutant CMPG5413, resulting in a defect in adherence, biofilm formation and EPS production, showed a shorter residence time in the murine gastrointestinal tract. This suggested a role for biofilm formation and EPSs in the ability of *Lactobacillus rhamnosus* GG to persist in the gastrointestinal tract for longer time periods (Lebeer *et al.*, 2008).

Cell-surface-bound EPSs may also play a role in inhibiting pathogen binding to the intestinal wall. Surface EPSs of *Lactobacillus plantarum* Lp6 were found to play a role in inhibiting adhesion of bacteria to the intestine (Sun *et al.*, 2007). The EPS from *Lactobacillus rhamnosus* ATCC 9595 has also been shown to have cholera-toxin-binding activities, rendering it a potential candidate for use as a functional food ingredient (Kim *et al.*, 2006). Interestingly, cell wall polysaccharides have been implicated in anti-ulcer effects. In acetic-acid-induced gastric ulcer and ethanol-induced erosion models in rats, oral administration of *Bifidobacterium breve* YIT4014 and 4043 and *Bifidobacterium bifidum* YIT4007, or the polysaccharide fractions from these strains, assisted in ulcer healing by enhancing host repair mechanisms (Nagaoka *et al.*, 1994). Administration of polysaccharides, in particular those with a rhamnose content exceeding 60%, caused an increase in the levels of epidermal growth factor and basic fibroblast growth factor in gastric epithelial cells. Macrophage production of

6-ketoprostaglandin $F_1\alpha$ was also increased by the polysaccharide fraction.

Anti-carcinogenic effects

A number of reports of LAB EPSs having anti-carcinogenic effects exist. A polysaccharide isolated from *Lactobacillus helveticus* var. *jugurti* by Oda *et al.* (1983) was demonstrated to significantly increase the life span of mice with ascites Sarcoma-180 when injected intraperitoneally. This effect could not be repeated *in vitro*, suggesting a host-mediated effect.

Kitazawa *et al.* (1991) discovered that *Lactococcus lactis* subsp. *cremoris* KVS 20 (injected via an intraperitoneal route) also inhibited the growth of Sarcoma-180 tumours in mice; however, this result could not be replicated *in vitro*, suggesting that the mechanism of prevention of tumour growth was mediated via the immune system. Their postulate that the anti-carcinogenic effect was caused principally by the EPS from the organism was supported by a subsequent study in which B-cell-dependent mitogenic activity was induced by its EPS (Kitazawa *et al.*, 1992). The cells of *Lactobacillus delbrueckii* subsp. *bulgaricus* (OLL1073R-1) were demonstrated to effect a host-mediated activity in mice (Ebina *et al.*, 1995). The EPS isolated from that strain was separated into an NPS and an APS, with the APS (a phosphopolysaccharide) shown to have mitogenic activity against murine B-lymphocytes. The phosphate group was shown to be the inducer of the mitogenic properties of this EPS (Maeda *et al.*, 2004a). Kefiran has also been implicated in anti-tumour effects. EPS from kefir grains (KGP) activated Peyer's patch cells from tumour-bearing mice, causing these cells to secrete a soluble factor that increased the mitogenic response of thymocytes, and splenocytes from normal mice (Furukawa *et al.*, 1996). The KGP also had a significant protective effect against the pulmonary metastasis and growth of Lewis lung carcinoma when mice were treated twice with KGP, either prior to or after injection of the carcinoma cells (Furukawa *et al.*, 2000).

If the eventual aim of this line of research was to incorporate EPSs into functional foods, then efficacy via an oral route would have to be demonstrated. EPSs from kefir (KGF-C) delivered orally were found to slow the growth of tumours by immune enhancement of T-cells (Maeda *et al.*, 2004b).

Cardiovascular effects

EPSs have also been reported to influence blood pressure. A hexasaccharide produced by *Lactobacillus kefiranofaciens* (kefiran) has been demonstrated to significantly suppress blood pressure increases in stroke-prone spontaneously hypertensive SHRSP/Hos rats (Maeda *et al.*, 2004b).

When three fractions (LCP1, LCP2 and LCP3) of EPSs from *Lactobacillus casei* strain LC2W were fed to spontaneously hypertensive rats (SHR) and normotensive Wistar rats, at a daily dosage of 15 mg/kg body weight for 7 days, one of the fractions (LCP1) was shown to significantly decrease the systolic blood pressure of SHR with no change in the heart rate. LCP1 did not have any effect on the blood pressure or heart rate of the Wistar rats. These results suggested an anti-hypertensive effect for the LCP1 fractions in rats (Lianzhong *et al.*, 2008).

Sensory properties

Despite the fact that research on the possible health benefits of EPSs is at an early stage, knowledge of the structural and functional relationships that result in the textural changes caused by these polymers is more well established. EPSs, although not imparting any taste of their own, are increasingly being used to enhance the appeal of functional foods to consumers. The physical means by which they effect these enhancements is complex; a brief review of what is known about these structural and functional relationships follows.

EPSs – structural and functional relationships

The design of EPS food polymers requires an understanding of their chemical composition and structure. Knowledge of how these characteristics determine the nature of the physical interactions with other components of the food into which they will be incorporated is also important.

The reason why EPSs impart viscosity to aqueous solutions is multi-factorial. One of the main contributing factors is related to the nature of the EPS polymers and their ability to occupy a significant amount of space per unit mass in the matrix, while remaining suspended in the aqueous phase. Two factors that define the dimensions of the EPS polymers have been identified: the molar mass (M) and the radius of gyration (R_g) (Tuinier *et al.*, 2001). The molar mass provides a measure of the length of the molecule and the radius of gyration gives an indication of its size. Together, the terms define the intrinsic viscosity ($[\eta]_0$) of the molecule (or how able it is to make the solution viscous) using the equation (Tuinier *et al.*, 1999a; Ruas-Madiedo *et al.*, 2005):

$$[\eta]_0 = 3.1 \times 10^{24} \times R_g^{\ 3} \times M^{-1} \text{ (units: m}^3\text{/kg)}$$

Other factors that also contribute to the texture of the product include the interaction between the EPSs and proteins, the bacterial cells themselves with attached EPS and the presence of hydration water 'trapped' by the polymers. The last property has the effect of increasing the concentration of EPSs in the aqueous phase (Duboc and Mollet, 2001).

Translating the results from a viscometer to the sensory effects experienced in the consumer's mouth is complicated by the fact that fermented milk products often decrease in viscosity when exposed to higher shear forces. The shear rates generated by the act of chewing and swallowing are approximately 30–50/s and this has implications for the functional effects of EPSs (Hassan, 2008). An additional property of EPSs is that of 'thixotropy', which is the ability of the polymer to return to its original viscosity upon cessation of the shear forces – an 'elasticity'. When an oscillatory stress is applied, the response of the polymer in the matrix can be measured by two factors: the storage modulus (G'), which relates to elasticity, and the loss modulus (G''), corresponding to viscosity. The overall charge on the EPS molecule exerts an influence on this elastic behaviour. For example, linear NPSs (produced by *Lactobacillus helveticus*) were found to increase viscosity, but had no influence on elasticity, whereas negatively charged polysaccharides (*Lactobacillus sakei* and *Lactobacillus paracasei*) contributed towards elasticity, but not viscosity. This was ascribed to electrostatic interactions between the EPS and the positively charged casein particles, resulting in a poor dispersion in the aqueous phase and hence imparting a very poor viscosifying effect (van den Berg *et al.*, 1995). The nature of the chemical bond between the individual sugars has also been found to influence viscosity. Higher viscosities have also been related to a greater polysaccharide chain stiffness (Faber *et al.*, 1998); for example, β(1,4) bonds render stiffer chains than α(1,4) or β(1,3) linkages (Tuinier *et al.*, 1999a). This chain stiffness can also be influenced by the presence of side chains; removal of terminally linked galactosyl residues from EPS (from two *Lactococcus lactis* subsp. *cremoris* strains) can decrease chain stiffness and hence the resultant viscosity (Tuinier *et al.*, 2001).

Apart from these factors, the physical amount of polysaccharide can have a positive effect on viscosity generation. However, because of the wide diversity of structural and compositional EPS types from LAB and their different functional effects, increasing only the quantity of EPS in a dairy product will produce varying results. In general, heteropolysaccharides from LAB have been reported to be produced in the range 50–600 mg/l, whereas as much as 10 g/l of homopolysaccharides have been produced, e.g. glucan and fructan exopolysaccharides produced by *Lactobacillus reuteri* (Van Geel-Schutten *et al.*, 1999).

Applications

Fermented milk applications

The widest application of EPSs has been in fermented milks. Historically, the genesis of these fermented milks occurred in the Middle East and Europe and these foods became part of the traditional diet. The commercial production of yoghurt was first started in Europe (Madrid, Spain) by Danone (Shah, 2006).

Consumers are attracted to thick, smooth and creamy textures, and EPS$^+$ starter cultures and adjunct cultures have been able to provide these characteristics. Generally, these thickened fermented milks are often characterized by a 'ropiness', in that a thread-like feature can be pulled from the surface of the milk. However, some thick yoghurts are not ropy, and hence that relationship does not always apply (Van der Meulen *et al.*, 2007). Generally, fermented milks containing LAB that have EPS located in capsules closely associated with the bacteria (capsular EPS) have been found to be less ropy than those with EPS 'free' in the surrounding matrix (Hassan *et al.*, 1996). EPSs from LAB have the additional effect of enhancing the perception of taste in the mouth as they increase the residence time of the fermented milk, thereby prolonging the time of contact with the taste receptor cells.

The most cost-effective way of introducing these beneficial properties of EPS into fermented drinks is through the use of LAB cultures that produce these EPSs, either as starters or as adjuncts. This has sponsored screening programmes for EPS$^+$ LAB that produce EPSs that enhance the textural properties of yoghurt. Folkenberg *et al.* (2006) undertook a sensory evaluation (gel firmness, ropiness, creaminess and mouth thickness) and a rheological analysis (complex modulus 'G^*', phase angle 'PA', hysteresis loop area and syneresis volume) of combinations of seven EPS$^+$ and four non-EPS$^+$ strains (39 yoghurts in total). It was found that some aspects of the sensory analysis correlated well with the analytical results: G^* correlated with sensory gel firmness, viscosity (at 241/s) correlated with sensory mouth thickness and a relationship between sensory ropiness and hysteresis loop area was established. Ropiness, although a good indicator of EPS production and texturizing ability, was not found to be essential for EPS$^+$ cultures to generate more creaminess or gel firmness – a commercially desirable attribute. Interestingly, greater ropiness has also been shown through confocal laser scanning microscopy to be related to EPS associating with protein in the network (Folkenberg *et al.*, 2006). Mozzi *et al.* (2006) evaluated 31 EPSs (heteropolysaccharides) out of 201 candidate strains, classifying the polymers into nine groups, based upon their monomer composition, finding as well that ropiness and capsular polysaccharide formation were strain dependent. Others have screened large numbers of strains for homopolysaccharide production (Van Geel-Schutten *et al.*, 1998). The physical basis of the texture improvement in yoghurt is not entirely clear cut, but texture is linked to EPS intrinsic viscosity, which, in turn, is determined by molecular mass and stiffness (Faber *et al.*, 1998; Kleerebezem *et al.*, 1999; Doleyres *et al.*, 2005).

In addition, it appears that viscosity is also associated with attractive interactions (depletion interactions) that occur between aggregated whey protein colloid particles or between casein micelles in the presence of EPS (Tuinier *et al.*, 1999b, 2000). Confocal laser microscopy studies by Hassan *et al.* (2003) have shed further light on this. Yoghurts made with EPS$^+$ cultures were observed to have large pores associated with EPS as well as a protein network consisting of thick strands of densely aggregated protein particles. 'EPS channels' formed in the serum were suggested to create a more polymer-like rheological behaviour in the continuous phase, resulting in an increased viscosity. More recently, it has been concluded that EPSs produced by *Lactococcus lactis* subsp. *cremoris* (JFR1) impact upon textural characteristics, not only by their molecular properties, but also by virtue of their ability to interact with whey proteins at acidic pH (Ayala-Hernández *et al.*, 2008).

This ability to improve texture in yoghurts, *in situ*, and to diminish cost in use by decreasing the use of added texturizers, has been exploited by some companies through the supply of selected combinations of cultures.

EPS$^+$ starter cultures are also employed to control syneresis in yoghurt (Wacher-Rodarte *et al.*, 1993), especially in regions where the use of stabilizers derived from plants or animals is not allowed (Cerning, 1995). The shear-induced microstructure in yoghurt made with an EPS$^+$ culture, and observed to be made up of compartmentalized protein aggregates situated between EPS-containing channels (see above), is thought to diminish syneresis as well (Hassan *et al.*, 2003). EPSs also have the ability to bind water (Hassan *et al.*, 1996; Rawson and Marshall, 1997), which plays a role in this regard. Mechanical pumping of yoghurt is known to increase syneresis, but this effect can be diminished by the use of EPS$^+$ cultures (Duboc and Mollet, 2001).

Protective effects related to the freeze–thaw cycle have also been reported. Capsular EPSs have also been found to have greater survival rates in frozen dairy desserts, after storage at −29°C for a period of 16 days (Hong and Marshall, 2001).

Cheese applications

The application of LAB EPSs in cheese has not enjoyed the same profile as that in fermented milks. This is partially because of different perceived values for the respective applications. In fermented milks, a thickening effect has been the principal proposition for its value for many years. In cheese applications, however, the reasons for the incorporation of EPSs have more to do with a relatively modern and growing consumer trend of a preference for low-fat foods, and the ability of LAB EPSs to replace palatability in reduced-fat cheeses. Making a low-fat cheese product is not as simple as it would appear. Commercially, large volumes of Mozzarella cheese are produced for the pizza industry and the fat in Mozzarella plays an important role in generating texture, as well as meltability. One of the problems associated with the removal of fat in cheese is the resultant loss of desirable organoleptic properties (McMahon *et al.*, 1996) as well as poor meltability in such applications (Mistry and Anderson, 1993). Removal of the fat reduces the amount of moisture and renders the cheese harder, which impedes the ability of the cheese to stretch, melt and flow. These cheeses then become scorched during pizza baking, and show incomplete shred fusion (Fife *et al.*, 1996). The cheese tends to undergo a loss of pliability, hence requiring more heat input for melting (McMahon and Oberg, 1998).

This induced defect can be explained at a structural level. In Mozzarella cheese, fat becomes trapped within 'channels' in the protein matrix, thus providing the consistency associated with this cheese. Removing the fat causes a narrowing of these channels, leaving less space for water (Mozzi *et al.*, 2006), resulting in poor melt and stretch characteristics and defective texture (Doleyres *et al.*, 2005). However, it is known that these adverse properties can be significantly reduced if more water is brought into the cheese structure. LAB EPSs, and particularly capsular EPSs, have been found to raise the moisture levels of cheeses (Perry *et al.*, 1997, 1998). For example, EPS produced by *Streptococcus thermophilus* MR-1C (an octasaccharide repeating unit consisting of galactose, rhamnose and fucose) was found to significantly enhance moisture retention in low-fat Mozzarella cheese (Low *et al.*, 1998). Similarly, both capsular and ropy EPS$^+$ *Streptococcus thermophilus* cultures increased moisture retention in low-fat Mozzarella cheese. The cheeses made with EPS had softer properties, and showed less springiness and chewiness and improved stretchability and meltability, when compared with control cheeses (Zisu and Shah, 2006).

The commercial interest in using EPS$^-$ LAB starters in cheese has been tempered by the fact that the EPS can accumulate in the cheese whey, hence reducing efficiencies of membrane processing and whey protein concentration and drying. It would seem that this problem can be alleviated by using LAB that produce non-ropy, capsular EPS, because the polysaccharide is bound to the cell wall (Broadbent *et al.*, 2003). This is supported by the prior studies of Petersen *et al.* (2000), who demonstrated that the capsular but not ropy EPSs produced by *Streptococcus thermophilus* strains are much less viscous than their ropy counterparts. They are hence more suitable candidates for the purpose of raising the moisture levels of Mozzarella cheese.

A physical method of making a softer cheese with better stretch and flow properties is to pre-acidify the cheese milk. The result is that calcium is removed from the casein micelles, resulting in fewer cross-linkages between the casein polymers. Zisu and Shah (2007) investigated the effects of pre-acidification, in combination with encapsulated ropy and encapsulated non-ropy strains, on moisture retention and textural characteristics in low-fat Mozzarella cheeses. Pre-acidification, alone, had no effect on moisture retention. However, the use of EPS cultures (an encapsulated strain that could produce a ropy EPS – *Streptococcus thermophilus* 1275, and an encapsulated strain – *Streptococcus thermophilus* 285) in conjunction with pre-acidification did elicit an increase in moisture. When the ropy strain was used, the water-holding capacity of the cheese was poor when heated during the first 28 days of storage and the cheese was also coated with a layer of slime. However, the capsular strain retained water and did not have a sticky layer.

Reduced-fat applications in Cheddar cheese have also been investigated. Dabour *et al.* (2006) studied the effect of including ropy or capsular EPS$^+$ *Lactococcus lactis* subsp. *cremoris* strains, and the combination thereof, into Cheddar cheese starter culture. They found that the strains could assist with an increase in moisture retention by up to 4.8%, and could modify the texture in freshly pressed reduced-fat Cheddar cheese.

Ropy EPS cultures were found to significantly enhance rennet coagulation properties, decrease manufacturing time and elevate cheese moisture by 1.8%, and to raise milk fat recovery and cheese yield. Interestingly, the capsular EPS culture had little or no effect on these parameters (Rynne *et al.*, 2007).

A positive effect by EPSs on the texture of reduced-fat Cheddar cheese is also evident. Reduced-fat Cheddar cheeses made with EPS^+ cultures were softer and exhibited lower viscoelastic moduli than those made with non-EPS^+ cultures (Hassan *et al.*, 2007). Similarly, reduced-fat Cheddar cheese made with ultrafiltered milk and EPS^+ cultures had higher body and texture scores than those made with non-EPS^+ cultures (Agrawal and Hassan, 2008). It has been noted that an adverse side effect of using specific EPS^+ cultures could be that the particular strains used lack the debittering enzymes needed to generate a suitable flavour profile, and that it may be necessary to incorporate adjunct cultures with these enzyme systems.

The use of a ropy strain of *Lactobacillus plantarum* (162RM) in combination with the starter consortium *Lactococcus lactis* subsp. *lactis, Lactococcus lactis* subsp. *cremoris* and *Lactococcus lactis* subsp. *diacetylactis* also increased the moisture content and improved the hard rubbery texture of Kareish cheese, which has a low fat content (Zambou *et al.*, 2004).

Summary and conclusions

An increasing body of evidence has revealed that certain dietary EPSs from LAB have the potential to directly influence some aspects of human health. Most of the potential benefits probably arise from immunomodulatory effects, and the positive effects on gut health. The cholesterol-lowering ability of EPSs also remains a fertile area for further research, because of the dietary prevalence of this problem and its acceptance as a major risk factor for cardiovascular disease. If these types of ingredients can be developed on a substantial scientific and clinical foundation, the remaining major challenges for commercialization will be regulatory challenges and the need to be able to convey to the consumer a realistic and understandable message as to their benefits.

Consumer acceptance of food ingredients also depends significantly on sensory perception. The use of LAB EPSs as agents to improve the physical functionality of foods, particularly in fermented dairy products, is supported by the proven sensory and textural qualities that they have demonstrated. Mainly because only relatively low levels of functional polysaccharides can be produced within these systems, *in situ* use of EPS^+ cultures, either as adjuncts or as part of consortia of starter cultures, remains the most practical way to deliver these textural qualities. Consumer preferences for natural products are likely to encourage the use of these polysaccharides, as will the drive to leverage more textural and sensory value from them. EPSs from LAB can also be applied in reduced-fat formats, where texture loss most often occurs.

References

Agrawal, P., and Hassan, A.N. (2008). Characteristics of reduced fat Cheddar cheese made from ultrafiltered milk with an exopolysaccharide-producing culture. J. Dairy Res. *75*, 182–188.

Ai, L.-Z., Guo, B.-H., Zhang, H., Wu, Z.J., Chen, W., Wang, Y.Y., and Tang, J.A. (2008). Isolation and antihypertensive effect of exopolysaccharides from *Lactobacillus casei* LC2W. Milchwissenschaft *63*, 3–6.

Amrouche, T., Boutin, Y., Prioult, G., and Fliss, I. (2006). Effects of bifidobacterial cytoplasm, cell wall and exopolysaccharide on mouse lymphocyte proliferation and cytokine production. Int. Dairy J. *16*, 70–80.

Awad, S., Hassan, A.N., and Halaweish, F. (2005). Application of exopolysaccharide-producing cultures in reduced-fat Cheddar cheese: composition and proteolysis. J. Dairy Sci. *88*, 4195–4203.

Ayala-Hernández, I., Hassan, A., Goff, H.D., Mira de Orduña, R., and Corredig, M. (2008). Production, isolation and characterization of exopolysaccharides produced by *Lactococcus lactis* subsp. *cremoris* JFR1 and their interaction with milk proteins. Article in press. Int. Dairy J. doi:10.1016/j.idairyj.2008.06.008.

Broadbent, J.R., McMahon, D.J., Welker, D.L., Oberg, C.J., and Moineau, S. (2003). Biochemistry, genetics and applications of exopolysaccharide production in *Streptococcus thermophilus*: a review. J. Dairy Sci. *86*, 407–423.

Cerning, J. (1995). Production of exocellular polysaccharides by lactic acid bacteria and dairy propionibacteria. Lait *75*, 463–472.

Chabot, S., Yu, H.-L., De Léséleuc, L., Cloutier, D., Van Calsteren, M.-R., Lessard, M., Roy, D., Lacroix, M., and Oth, D. (2001). Exopolysaccharides from *Lactobacillus rhamnosus* RW-9595M stimulate TNF, IL-6 and IL-12 in human and mouse cultured immunocompetent cells, and IFN-γ in mouse splenocytes. Lait *81*, 683–697.

Cobb, B.A., and Kasper, D.L. (2005). Coming of age: carbohydrates and immunity. Eur. J. Immunol. 35, 352–356.

Collado, M.C., Meriluoto, J., and Salminen, S. (2007). Development of new probiotics by strain combinations: is it possible to improve the adhesion to intestinal mucus? J. Dairy Sci. 90, 2710–2716.

Dabour, N., Kheadr, E., Benhamou, N., Fliss, I., and LaPointe, G. (2006). Improvement of texture and structure of reduced-fat Cheddar cheese by exopolysaccharide-producing lactococci. J. Dairy Sci. 89, 95–110.

Dal Bello, F., Walter, J., Hertel, C., and Hammes, W.P. (2001). *In vitro* study of prebiotic properties of levan-type exopolysaccharides from lactobacilli and non-digestible carbohydrates using denaturing gradient gel electrophoresis. Syst. Appl. Microbiol. 24, 232–237.

De Vuyst, L., and Degeest, B. (1999). Exopolysaccharides from lactic acid bacteria. Technological bottlenecks and practical solutions. Macromol. Symp. 140, 31–41.

De Vuyst, L.F., De Vin, F., Vaningelgem, F., and Degeest, B. (2001). Recent developments in the biosynthesis and applications of heteropolysaccharides from lactic acid bacteria. Int. Dairy J. 11, 687–707.

Doleyres, Y., Schaub, L., and Lacroix, C. (2005). Comparison of the functionality of exopolysaccharides produced *in situ* or added as bioingredients on yogurt properties. J. Dairy Sci. 88, 4146–4156.

Duboc, P., and Mollet, B. (2001). Applications of exopolysaccharides in the dairy industry. Int. Dairy J. 11, 759–768.

Ebina, T., Ogama, N., and Murata, K. (1995). Antitumor effects of *Lactobacillus bulgaricus* 878R. Biotherapy 9, 65–70.

Faber, E.J., Zoon, P., Kamerling, J.P., and Vliegenthart, J.F.G. (1998). The exopolysaccharides produced by *Streptococcus thermophilus* Rs and Sts have the same repeating unit but differ in viscosity of their milk cultures. Carbohydr. Res. 310, 269–276.

Fife, R.L., McMahon, D.J., and Oberg, C.J. (1996). Functionality of low fat Mozzarella cheese. J. Dairy Sci. 79, 1903–1910.

Folkenberg, D.M., Dejmek, P., Skriver, A., and Ipsen, R. (2005). Relation between sensory texture properties and exopolysaccharide distribution in set and in stirred yoghurts produced with different starter cultures. J. Texture Stud. 36, 174–189.

Folkenberg, D.M., Dejmek, P., Skriver, E., Guldager, H.S., and Ipsen, R. (2006). Sensory and rheological screening of exopolysaccharide producing strains of bacterial yoghurt cultures. Int. Dairy J. 16, 111–118.

Font de Valdez, G., Torino, M.I., De Vuyst, L., and Mozzi, F. (2003). Applications review: Food-grade heteropolysaccharides: ongoing research and future trends of biopolymers from lactic acid bacteria. Appl. Biotechnol. Food Sci. Policy 1, 223–234.

Forsén, R., Heiska, E., Herva, E., and Arvilommi, H. (1987). Immunobiological effects of *Streptococcus cremoris* from cultured milk 'viili'; application of human lymphocyte culture techniques. Int. J. Food Microbiol. 5, 41–47.

Furukawa, N., Takahashi, T., and Yamanaka, Y. (1996). Effects of supernatant of Peyer's patch cell culture with kefir grain components on the mitogenic response of thymocyte and splenocyte in mice. Anim. Sci. Technol. 67, 153–159.

Furukawa, N., Matsuoka, A., Takahashi, T., and Yamanaka, Y. (2000). Anti-metastatic effect of kefir grain components on Lewis lung carcinoma and highly metastatic B16 melanoma in mice. J. Agric. Sci. 45, 62–70.

Hassan, A.N. (2008). ADSA Foundation Scholar Award: Possibilities and challenges of exopolysaccharide-producing lactic cultures in dairy foods. J. Dairy Sci. 91, 1282–1298.

Hassan, A.N., Frank, J.F., Schmidt, K.A., and Shalabi, S.I. (1996). Textural properties of yogurt made with encapsulated nonropy lactic cultures. J. Dairy Sci. 79, 2098–2103.

Hassan, A.N., Ipsen, R., Janzen, T., and Qvist, K.B. (2003). Microstructure and rheology of yogurt made with cultures differing only in their ability to produce exopolysaccharides. J. Dairy Sci. 86, 1632–1638.

Hassan, A.N., Awad, S., and Mistry V.V. (2007). Reduced fat process cheese made from young reduced fat Cheddar cheese manufactured with exopolysaccharide-producing cultures. J. Dairy Sci. 90, 3604–3612.

Hong, S.H., and Marshall, R.T. (2001). Natural exopolysaccharides enhance survival of lactic acid bacteria in frozen dairy desserts. J. Dairy Sci. 84, 1367–1374.

Imai, K., Matsuyama, S., Miyake, S., Suga, K., and Nakachi, K. (2000). Natural cytotoxic activity of peripheral-blood lymphocytes and cancer incidence: an 11-year follow-up study of a general population. Lancet 356, 1795–1799.

Jolly, L., Vincent, S.J.F., Duboc, P., and Neeser, J.-R. (2002). Exploiting exopolysaccharides from lactic acid bacteria. Antonie van Leeuwenhoek 82, 367–374.

Kim, J.-U., Kim, Y., Han, K.-S., Oh, S., Whang, K.-Y., Kim, J.-N., and Kim, S.-H. (2006). Function of cell-bound and released exopolysaccharides produced by *Lactobacillus rhamnosus* ATCC 9595. J. Microbiol. Biotechnol. 6, 939–945.

Kitazawa, H., Toba, T., Itoh, T., Kumano, N., Adachi, S., and Yamaguchi, T. (1991). Antitumoral activity of slime-forming encapsulated *Lactococcus lactis* subsp. *cremoris* isolated from Scandinavian ropy sour milk, 'viili'. Anim. Sci. Technol. 62, 277–283.

Kitazawa, H., Yamaguchi, T., and Itoh, T. (1992). B-cell mitogenic activity of slime products produced from slime-forming, encapsulated *Lactococcus lactis* ssp. *cremoris*. J. Dairy Sci. 75, 2946–2951.

Kitazawa, H., Itoh, T., Tomioka, Y., Mizugaki, M., and Yamaguchi, T. (1996). Induction of IFN-gamma and IL-1 alpha production in macrophages stimulated with phosphopolysaccharide produced by *Lactococcus lactis* spp. *cremoris*. Int. J. Food Microbiol. 31, 99–106.

Kitazawa, H., Harata, T., Uemura, J., Saito, T., Kaneko, T., and Itoh, T. (1998). Phosphate group requirement for mitogenic activation of lymphocytes by an extracellular phosphopolysaccharide from *Lactobacillus delbrueckii* spp. *bulgaricus*. Int. J. Food Microbiol. 40, 169–175.

Kitazawa, H., Ishii, Y., Uemura, J., Kawai, Y., Saito, T., Kaneko, T., Noda, K., and Itoh, T. (2000). Augmentation of macrophage functions by an extracellular phosphopolysaccharide from *Lactobacillus delbrueckii* ssp. *bulgaricus*. Food Microbiol. *17*, 109–118.

Kleerebezem, M., van Kranenburg, R., Tuinier, R., Boels, I.C., Zoon, P., Looijesteijn, E., Hugenholtz, J., and de Vos, W.M. (1999). Exopolysaccharides produced by *Lactococcus lactis*: from genetic engineering to improved rheological properties? Antonie van Leeuwenhoek *76*, 357–365.

Korakli, M., Gänzle, M.G., and Vogel, R.F. (2002). Metabolism by bifidobacteria and lactic acid bacteria of polysaccharides from wheat and rye, and exopolysaccharides produced by *Lactobacillus sanfranciscensis*. J. Appl. Microbiol. *92*, 958–965.

Lebeer, S., Claes, I.J.J., Verhoeven, T.L.A., Shen, C., Lambrichts, I., Ceuppens, J.L., Vanderleyden, J., and De Keersmaecker, S.C.J. (2008). Impact of *luxS* and suppressor mutations on the gastrointestinal transit of *Lactobacillus rhamnosus* GG. Appl. Environ. Microbiol. *74*, 4711–4718.

Liu, J.-R., Wang, S.-Y., Chen, M.-J., Chen, H.-L., Yueh, P.-Y., and Lin, C.-W. (2006). Hypocholesterolaemic effects of milk-kefir and soyamilk-kefir in cholesterol-fed hamsters. Br. J. Nutr. *95*, 939–946.

Low, D., Ahlgren J.A., Horne, D., McMahon, D.J., Oberg, C.J., and Broadbent, J.R. (1998). Role of *Streptococcus thermophilus* MR-1C capsular exopolysaccharide in cheese moisture retention. Appl. Environ. Microbiol. *64*, 2147–2151.

Macedo, M.G., Lacroix, C., Gardner, N.J., and Champagne, C.P. (2002). Effect of medium supplementation on exopolysaccharide production by *Lactobacillus rhamnosus* RW-9595M in whey permeate. Int. Dairy J. *12*, 419–426.

Maeda, H., Zhu, X., Omura, K., Suzuki, S., and Kitamura, S. (2004a). Effects of an exopolysaccharide (kefiran) on lipids, blood pressure, blood glucose, and constipation. Biofactors *22*, 197–200.

Maeda, H., Zhu, X., Suzuki, S., Suzuki, K., and Kitamura, S. (2004b). Structural characterization and biological activities of an exopolysaccharide kefiran produced by *Lactobacillus kefiranofaciens* WT-2B^{T}. J. Agric. Food Chem. *52*, 5533–5538.

Maeda, H., Mizumoto, H., Suzuki, M., and Tsuji, K. (2005). Effects of kefiran-feeding on fecal cholesterol excretion, hepatic injury and intestinal histamine concentration in rats. Biosci. Microflora *24*, 35–40.

Makino, S., Ikegami, S., Kano, H., Sashihara, T., Sugano, H., Horiuchi, H., Saito, T., and Oda, M. (2006). Immunomodulatory effects of polysaccharides produced by *Lactobacillus delbrueckii* ssp. *bulgaricus* OLL1073R-1. J. Dairy Sci. *89*, 2873–2881.

McMahon, D.J., and Oberg, C.J. (1998). Influence of fat, moisture and salt on functional properties of mozzarella cheese. Aust. J. Dairy Technol. *53*, 98–101.

McMahon, D.J., Alleyne, M.C., Fife, R.L., and Oberg, C.J. (1996). Use of fat replacers in low fat Mozzarella cheese. J. Dairy Sci. *79*, 1911–1921.

Mistry, V.V., and Anderson, D.L. (1993). Composition and microstructure of commercial full-fat and low-fat cheeses. Food Struct. *12*, 259–266.

Mozzi, F., Vaningelgem, F., Hébert E.M., Van der Meulen, R., Moreno, M.R.F., Font de Valdez, G., and De Vuyst, L. (2006). Diversity of heteropolysaccharide-producing lactic acid bacterium strains and their biopolymers. Appl. Environ. Microbiol. *72*, 4431–4435.

Nagaoka, M., Hashimoto, S., Watanabe, T., Yokokura, T., and Mori, Y. (1994). Anti-ulcer effects of lactic acid bacteria and their cell wall polysaccharides. Biol. Pharm. Bull. *17*, 1012–1017.

Nakajima, H., Suzuki, Y., Kaizu, H., and Hirota, T. (1992). Cholesterol lowering activity of ropy fermented milk. J. Food Sci. *57*, 1327–1329.

Nakajima, H., Toba, T., and Toyoda, S. (1995). Enhancement of antigen-specific antibody production by extracellular slime products from slime-forming *Lactococcus lactis* subsp. *cremoris* SBT 0495 in mice. Int. J. Food Microbiol. *25*, 153–158.

Nishimura-Uemura, J., Kitazawa, H., Kawai, Y., Itoh, T., Oda, M., and Saito, T. (2003). Functional alteration of murine macrophages stimulated with extracellular polysaccharides from *Lactobacillus delbrueckii* ssp. *bulgaricus* OLL1073R-1. Food Microbiol. *20*, 267–273.

Oda, M., Hasegawa, H., Komatsu, S., Kambe, M., and Tsuchiya, F. (1983). Anti-tumor polysaccharide from *Lactobacillus* sp. Agric. Biol. Chem. *47*, 1623–1625.

Perry, D.B., McMahon, D.J., and Oberg, C.J. (1997). Effect of exopolysaccharide-producing cultures on moisture retention in low fat Mozzarella cheese. J. Dairy Sci. *80*, 799–805.

Perry, D.B., McMahon, D.J., and Oberg, C.J. (1998). Manufacture of low fat Mozzarella cheese using exopolysaccharide-producing starter cultures. J. Dairy Sci. *81*, 563–566.

Petersen, B.L., Dave, R.I., McMahon, D.J., Oberg, C.J., and Broadbent, J.R. (2000). Influence of capsular and ropy exopolysaccharide-producing *Streptococcus thermophilus* on Mozzarella cheese and cheese whey. J. Dairy Sci. *83*, 1952–1956.

Pigeon, R.M., Cuesta, E.P., and Gilliland, S.E. (2002). Binding of free bile acids by cells of yogurt starter culture bacteria. J. Dairy Sci. *85*, 2705–2710.

Rawson, H.L., and Marshall, V.M. (1997). Effect of 'ropy' strains of *Lactobacillus delbrueckii* ssp. *bulgaricus* and *Streptococcus thermophilus* on rheology of stirred yogurt. Int. J. Food. Sci. Technol. *32*, 213–220.

Ruas-Madiedo, P., and de los Reyes-Gavilán, C.G. (2005). Invited review: Methods for the screening, isolation, and characterization of exopolysaccharides produced by lactic acid bacteria. J. Dairy Sci. *88*, 843–856.

Ruas-Madiedo, P., Gueimonde, M., Margolles, A., de los Reyes-Gavilán, C.G., and Salminen, S. (2006a). Exopolysaccharides produced by probiotic strains modify the adhesion of probiotics and enteropathogens to human intestinal mucus. J. Food Prot. *69*, 2011–2015.

Ruas-Madiedo, P., Gueimonde, M., de los Reyes-Gavilán, C.G., and Salminen, S. (2006b). Short communication: Effect of exopolysaccharide isolated from 'viili' on the adhesion of probiotics and pathogens to intestinal mucus. J. Dairy Sci. *89*, 2355–2358.

Rynne, N.M., Beresford, T.P., Kelly, A.L., Tunick, M.H., Malin, E.L., and Guinee, T.P. (2007). Effect of exopolysaccharide-producing adjunct starter cultures on the manufacture, composition and yield of half-fat cheddar cheese. Aust. J. Dairy Technol. *62*, 12–18.

Sengül, N., Aslím, B., Uçar, G., Yücel, N., Işik, S., Bozkurt, H., Sakaoğullari, Z., and Atalay, F. (2006). Effects of exopolysaccharide-producing probiotic strains on experimental colitis in rats. Dis. Colon Rectum *49*, 250–258.

Shah, N.P. (2006). Health benefits of yogurt and fermented milks. In Manufacturing Yogurt and Fermented Milks, R.C. Chandan, ed. (Ames, Iowa: Blackwell Publishing Professional), pp. 327–340.

St-Onge, M.-P., Farnworth, E.R., Savard, T., Chabot, D., Mafu, A., and Jones, P.J.H. (2002). Kefir consumption does not alter plasma lipid levels or cholesterol fractional synthesis rates relative to milk in hyperlipidemic men: a randomized controlled trial. BMC Complement. Altern. Med. *2*, 1–7.

Sun, J., Le G.-W., Shi, Y.-H., and Su, G.-W. (2007). Factors involved in binding of *Lactobacillus plantarum* Lp6 to rat small intestinal mucus. Lett. Appl. Microbiol. *44*, 79–85.

Tuinier, R., ten Grotenhuis, E., Holt, C., Timmins, P.A., and de Kruif C.G. (1999a). Depletion interaction of casein micelles and an exocellular polysaccharide. Phys. Rev. E. *60*, 848–856.

Tuinier, R., Zoon, P., Cohen-Stuart, M.A., Fleer, G.J., and de Kruif, C.J. (1999b). Concentration and shear-rate dependence of an exocellular polysaccharide. Biopolymers *50*, 641–646.

Tuinier, R., Dhont, J.K.G., and de Kruif C.G. (2000). Depletion-induced phase separation of aggregated whey protein colloids by an exocellular polysaccharide. Langmuir *16*, 1479–1507.

Tuinier, R., van Casteren, W.H.M., Looijesteijn, P.J., Schols, H.A., Voragen, A.G.J., and Zoon, P. (2001). Effects of structural modifications on some physical characteristics of exopolysaccharides from *Lactococcus lactis*. Biopolymers *59*, 160–166.

van den Berg, D.J.C., Robijn, G.W., Janssen, A.C., Giuseppin, M.L.F, Vreeker, R., Kamerling, J.P., Vliegenthart, J.F.G., Ledeboer, A.M., and Verrips, C.T. (1995). Production of a novel extracellular polysaccharide by *Lactobacillus sake* 0-1 and characterization of the polysaccharide. Appl. Environ. Microbiol. *61*, 2840–2844.

Van der Meulen, R., Grosu-Tudor, S., Mozzi, F., Vaningelgem, F., Zamfir, M., Font de Valdez, G., and De Vuyst, L. (2007). Screening of lactic acid bacteria isolates from dairy and cereal products for exopolysaccharide production and genes involved. Int. J. Food Microbiol. *118*, 250–258.

Van Geel-Schutten, G.H., Flesch, F., Ten Brink, B., Smith, M.R., and Dijkhuizen, L. (1998). Screening and characterization of Lactobacillus strains producing large amounts of exopolysaccharides. Appl. Microbiol. Biotechnol. *50*, 697–703.

Van Geel-Schutten, G.H., Faber, E.J., Smit, E., Bonting, K., Smith, M.R., Ten Brink, B., Kamerling, J.P., Vliegenthart, J.F.G., and Dijkhuizen, L. (1999). Biochemical and structural characterization of the glucan and fructan exopolysaccharides synthesized by the *Lactobacillus reuteri* wild-type strain and by mutant strains. Appl. Environ. Microbiol. *65*, 3008–3014.

Vincent, S.J.F., Faber, E.J., Neeser, J.-R., Stingele, F., and Kamerling, J.P. (2001). Structure and properties of the exopolysaccharide produced by *Streptococcus macedonicus* Sc136. Glycobiology *11*, 131–139.

Vinderola, C.G., Duarte J., Thangavel, D., Perdigón, G., Farnworth, E., and Matar, C. (2005). Remote-site stimulation and duration of the immune response by kefir. Eur. J. Inflammation *3*, 63–74.

Vinderola, G., Perdigón, G., Duarte, J., Farnworth, E., and Matar, C. (2006). Effects of the oral administration of the exopolysaccharide produced by *Lactobacillus kefiranofaciens* on the gut mucosal immunity. Cytokine *36*, 254–260.

Vinderola, G., Matar, C., Palacios, J., and Perdigón, G. (2007). Mucosal immunomodulation by the non-bacterial fraction of milk fermented by *Lactobacillus helveticus* R389. Int. J. Food Microbiol. *115*, 180–186.

Wacher-Rodarte, C., Galvin, M.V., Farres, A., Gallardo, F., Marshall, V.M.E., and Garcia-Garibay, M. (1993). Yogurt production from reconstituted skim milk powders using different polymer and non-polymer forming starter cultures. J. Dairy Res. *60*, 247–254.

Welman, A.D., and Maddox, I.S. (2003). Exopolysaccharides from lactic acid bacteria: perspectives and challenges. Trends Biotechnol. *21*, 269–274.

Welman, A.D., Maddox, I.S., and Archer, R.H. (2006). Metabolism associated with raised metabolic flux to sugar nucleotide precursors of exopolysaccharides in *Lactobacillus delbrueckii* subsp. *bulgaricus*. J. Ind. Microbiol. Biotechnol. 33, 391–400.

Zambou, N.F., Ahmed, N., Tchouanguep, T.F., and El-Soda, M. (2004). Effect of ropy and capsular exopolysaccharides producing strain of *Lactobacillus plantarum* 162RM on characteristics and functionality of fermented milk and soft Kareish type cheese. Afr. J. Biotechnol. *3*, 512–518.

Zisu, B., and Shah, N.P. (2006). Role of microbial exopolysaccharides on moisture retention, texture and functionality of low-fat mozzarella cheeses. Aust. J. Dairy Technol. *61*, 253–260.

Zisu, B., and Shah, N.P. (2007). Texture characteristics and pizza bake properties of low-fat Mozzarella cheese as influenced by pre-acidification with citric acid and use of encapsulated and ropy exopolysaccharide producing cultures. Int. Dairy J. *17*, 985–997

Synthesis of Bacterial Polysaccharides as a Limiting Factor for Biofuel Production

19

Karuppasamy Geetha, Narayanan Rajnish, Jeyaprakash Rajendhran, and Paramasamy Gunasekaram

Abstract

Bioethanol has been considered as a potential alternative energy source to the conventional petroleum-based fuel. *Zymomonas mobilis* is an efficient ethanol-producing bacteria and it is advantageous over *Saccharomyces cerevisiae* with respect to ethanol productivity and tolerance. *Z. mobilis* possesses nearly 100% theoretical ethanol conversion rate on glucose-based media, while it has only about 70% theoretical yield from sucrose. The reduced ethanol yield has been attributed to the formation of by-products such as levan, fructooligosaccharides and sorbitol. The efficiency of ethanol production by *Z. mobilis* from sucrose-based substrates can be improved by limiting the formation of by-products either by optimizing the fermentation conditions or genetic improvements of *Z. mobilis*. Recent strategies developed for the improved production of ethanol, while limiting the formation of levan and *vice versa* are discussed.

Introduction

Fructans are one of the most important groups of components in human nutrition. These are fructose-containing sugars comprising several fructose units linked to glucose unit. Different types of fructans are present in nature, which are classified based on the glycosidic linkages by which the fructose residues are linked to each other. Fructans can broadly be divided into three groups. The first group comprises of the inulins, which are linear fructans, where the fructose units are linked via a β-(2,1) bond (Pontis and del Campillo, 1985). 1-Kestose is the shortest fructan of the inulin type (Fig. 19.1a). The second group of fructans are the levans, which are also linear fructans, but the fructose units are linked via a β-(2,6) bond (Suzuki and Pollock, 1986). This type of fructans is found in most of the bacterial sources of fructans. The 6-kestose is the shortest fructan of the levan type (Fig. 19.1b). The third group is the fructans of mixed type, which is also referred as graminan type (Carpita *et al.*, 1989; Lewis, 1993). These fructans have both β-(2,1) and β-(2,6) linkage bonds between fructose units, and thus contain branches. These fructans are found in grasses and 6,6&1 kestopentaose is simplest fructan of this type (Fig. 19.1c).

Small fructans have a sweet taste, whereas longer fructan chains form emulsions with a fat-like texture and a neutral taste. The human digestive tract does not contain enzymes able to degrade fructans; therefore, there is strong interest from the food industry to use them as low-calorie food ingredients. Plants also store fructans as reserve carbohydrates. However, fructan synthesis in bacteria is simpler than plant fructan biosynthesis because of the use of a single biosynthetic enzyme. Enzymes that have been studied extensively are levansucrase and fructosyltransferase. Levansucrase, which produces levans, was isolated from *Bacillus subtilis* (Dedonder, 1966) and fructosyltransferase, which produces inulins, was isolated from *Streptococcus mutans* (Carlsson, 1970).

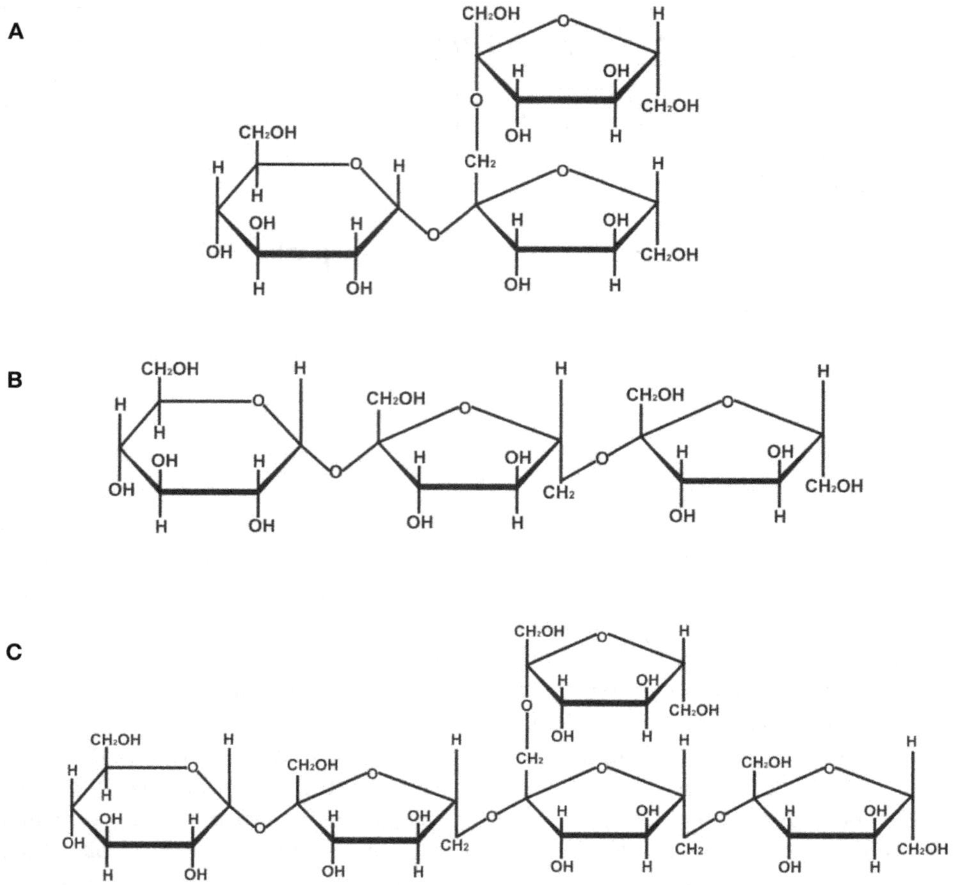

Figure 19.1 Types of fructans. **A** 1-Kestose; **B** 6-kestose; **C** 6,6&1 kestopentaose.

Fructopolymer: important components of functional food

A food can be regarded functional if it is demonstrated to affect beneficially one or more target functions in the body in a way relevant to either improved state of well being or reduction in disease risk. A functional food could be normally consumed in the diet for the desirable effects (Roberfroid, 1996). The interest in developing multifunctional additives with extra nutritional properties resulted in the identification of polysaccharides with prebiotic attributes. The human intestinal microbiota comprises a wide range of metabolically active microorganisms that play an important role in influencing the health. Prebiotics are defined as non-digestible food ingredients that benefit the host by selective modification of the growth and activity of selected number of bacteria in the colon. Bacterial species belonging to the group bifidobacteria are generally considered to be health promoting and recognized for the prevention of disease and maintenance of health. Therefore, strategies are developed for promoting the growth of bifidobacteria using prebiotics.

To be effective, the prebiotic compounds should be undigested and unabsorbed in the upper gastrointestinal tract; then they should reach the colon and be selectively metabolized by the intestinal bifidobacteria. The nondigestible oligosaccharides (NDO) such as FOS, xylooligosaccharides (XOS), galactooligosaccharides (GOS), and isomaltooligosaccharides (IMO) are fermented by bifidobacteria and thus demonstrate prebiotic effects in humans. FOS and inulin are used as prebiotics. The degree of polymerization of FOS can vary from 2 to 35 (Sangeetha *et al.*, 2002). FOS mainly consists of 1-kestose, nystose and 1-fructofuranosyl nystose which have one to three units of fructose, bound at the β-(2,1) position of sucrose, respectively. These FOS can be enzymatically synthesized from sucrose by using fructosyltransferase (EC 2.4.1.9) (Park *et al.*, 1999).

The human GI tract harbours a wide repertoire of bacteria, which constitutes the intestinal microbiota of an individual. This is established at birth and the distribution of the microorganisms varies according to the diet. The β-(1,4) linkage in fructans are resistant to digestion by digestive enzymes of the upper GI tract and therefore serve as substrates for utilization by intestinal microflora (Gibson and Roberfroid, 1995). The FOS undergo fermentation in the colon by the colonic microflora leading to the production of acetate, lactate and short chain fatty acids, which benefit the host. An increased concentration of short-chain fatty acids facilitates the colonic absorption of minerals such as magnesium and calcium leading to the improvement of health. Studies with human volunteers have demonstrated that a diet composed of inulin and FOS increased the bifidobacterial counts in faecal microflora (Gibson, 1999). Mice fed with a diet comprised of inulin and FOS were more resistant to colorectal cancer, systemic enteric infection and systemic candidiasis than the control mice fed with a cellulose diet (Buddington *et al.*, 2002). The increased resistance of mice towards pathogens and tumour biogenesis is attributed to the enhanced levels of immune response elicited by the colonic microflora in the presence of inulin-type fructans.

The fructopolymers have also been shown to have an anti diabetic effect and increased secretion of glucagon like peptide (GLP), a peptide that induces insulin secretion. The dietary supplementation of FOS lowers the plasma lipid concentrations, and increases the levels of GLP. Cani *et al.* (2006) have reported that FOS administration improved glucose tolerance and stimulation of insulin secretion. Thus, the dietary intake of inulin and FOS result in improved health. In view of these, several prebiotic foods are marketed worldwide with the aim of prevention of intestinal infections and improvement of human health.

Microbial strains producing fructopolymer

Fructan-producing bacteria can be found in a wide range of taxa, including plant pathogens and the bacteria present in oral and gut flora of animals and humans. Fructan-producing bacteria were reported from the following genera, *Tolypothrix, Pseudomonas, Xanthomonas, Zymomonas, Azotobacter, Erwinia, Streptococcus, Bacillus, Actinomyces, Rothia* and *Arthrobacter* (Hendry and Wallace, 1993). Fructans from the strains of *S. mutans* are known for their involvement in dental caries (Uchiyama, 1993). Saito *et al.* (1997) have reported the production of oligofructans from levan using fructosyltransferase of *Arthrobacter nicotinovorans* GS-9 cloned in *Escherichia coli.* Inulin production has been described in several strains of *Streptococcus mutans* and *Leuconostoc citreum* (Olivares-Illana *et al.*, 2002).

Several exopolysaccharide (EPS)-producing lactic acid bacteria (LAB) such as *Lactobacillus delbrueckii* subsp. *bulgaricus* and *Lactococcus lactis* have been isolated from fermented dairy products, in particular, from yogurt. The majority of EPS synthesized by LAB are heteropolysaccharides composed of repeating units of two or more monosaccharides, mainly galactose, glucose, rhamnose and fructose. *Lactobacillus reuteri* strains are known to produce homopolysaccharides of the fructan type (Hijum *et al.*, 2002). Korakli *et al.* (2001) have reported the production of EPS of the fructan type by *Lactobacillus sanfranciscensis.* Similarly, Tieking *et al.* (2003) have described the ability of several lactobacilli of sourdough and intestinal origin to produce EPS of the fructan and glucan types. Levan is one of the fructose biopolymer mainly linked by β-(2,6)-glycosidic bonds, with some β-(2,1)-linked branch chains. Several bacteria such as *Z. mobilis, Bacillus circulans, Pseudomonas* sp., *Xanthmonas* sp., and *Streptococcus* sp. *Arthrobacter ureafaciens* and *Microbacterium* produce levan. Bacterial fructans produced and their corresponding enzymes are listed in Table 19.1.

Industrial applications of levan

Industrial application of microbial polysaccharides is highly dependent on their physicochemical and physiological properties, such as viscosity, solubility, suspending properties, rheological features, stability, and water-holding capacity. Levan has potential applications in various industries (Table 19.2). It is used as sweetener and thickening agent in food industry. In the pharmaceutical

Table 19.1 Microbial sources of fructopolymers

Fructopolymer	Enzyme	Source	Reference
1-Kestose (GF_2) and Nystose	fructosyltransferase	*Bacillus marcerans* EG-6	Park *et al.* (2001)
FOS (GF_2,GF_3 and GF_4)	Extracellular fructosyltransferase	*Aspergillus oryzae* CFR 202	Sangeetha *et al.* (2002)
Levan	levansucrase	*Zymomonas mobilis*	Ananthalakshmy *et al.* (1999a)
Levan	fructosyltransferase	*Actinomyces naeslundii* WVU45	Bergeron *et al.* (2000)
Levan	fructosyltransferase	*Clostridium acetobutilicum*	Nolling *et al.* (2001)
Levan	Levan fructotransferase (lftA)	*Actinomyces ureafaciens* K2032	Song *et al.* (1996)
Levan	Extracellular levansucrase	*Bacillus subtilis* NRC 33a	Abdel-Fattah *et al.* (2005)
Levan	Extracellular levansucrase	*Pseudomonas syringae pv. glycinea* PG4180	Laue *et al.* (2006)
Fructan	fructosyltransferase	*Lactobacillus sanfranciscensis*	Korakli *et al.* (2002)
Inulin	fructosyltransferase	*Lactobacillus reuteri* strain 121	Hijum *et al.* (2002)
Inulin	Inulin-producing enzyme (IPE)	*Bacillus* sp. 217C-11	Wada *et al.* (2003)

Table 19.2 Industrial applications of levan

Application	Usage	Source	Reference
Pharmaceutical industry	Blood plasma extender	*Z. mobilis*	Kim *et al.* (1998)
	Modification of cell membrane (tumour cells)	*Z. mobilis*	Shih and Yu (2005)
	Tablet binder	*Bacillus* sp. TH4–2	Ammar *et al.* (2002)
Food industry	Bulking agent	*Z. mobilis*	Kim *et al.* (2005)
	Add sweetness to foods	*B. polymyxa*	Han (1989)
	Fillers	–	Kirk *et al.* (1991)
	Substitute for gum Arabic	*B. subtilis*	Kim *et al.* (2005)
Others	Selective plugging agent	*B. licheniformis*	Ramsay *et al.* (1989)
	Encapsulating agent	*B. polymyxa*	Han (1989)
	Emulsifier	–	Han (1990)
	Stabilizer and thickener	–	Han (1990)
	Soil sealant	*B. licheniformis*	Ghaly *et al.* (2007)
	Formulation aid	*Bacillus* sp. TH4–2	Ammar *et al.* (2002)
	Provide viscosity	*B. subtilis* NRC 33a	Ahmed (2005)
	Carrier for flavour and fragrances	*Z. mobilis*	Kim *et al.* (2005)
	Cosmetic	*Z. mobilis*	Kim *et al.* (2005)

industry, levan could be used as a plasma extender (Kazap *et al.*, 1980). Like the dextrans, levan can be transfused to maintain fluid volume of the blood in event of great necessity, supplemental to the use of whole blood and plasma. It is also shown to exert excellent cell-proliferating, skin moisturizing, and skin irritation-alleviating effects, and is thus used as a blending component in cosmetics (Kim *et al.*, 2005). Levan derivatives such as sulphated, phosphated or acetylated levans are asserted to be anti-AIDS agents (Earl and Per, 1998). Levan has been used as a coating material in drug delivery formulation (Earl and Per, 1998). Further, it exhibits tumour inhibitory activity as evidenced by the inhibition of Sarcoma 180 and Ehrlich carcinoma (Lopez *et al.*, 1997).

In addition, levan has a number of other applications such as a surfactant for household use due to its excellent surface-active properties, a glycol/levan aqueous two-phase system for the partitioning of proteins, etc. It is incorporated in photographic emulsions to improve silver granularity and preparation of molecular sieves for gel filtration (Han, 1989). It is a suitable substitute for gum Arabic in food, pharmaceutical, medicine, cosmetics and textile emulsifier, formulation aid, stabilizer, and surface finishing agent (Kim *et al.*, 2005).

Synthesis of levan by *Zymomonas mobilis*

Production of microbial levan as EPS is largely affected by the concentration of nutrients in culturing medium and the environmental conditions. *Z. mobilis* produces an extracellular levansucrase (SacB), an extracellular sucrase (SacC) and an intracellular sucrase (SacA). The extracellular levansucrase synthesizes levan by transferring fructose moiety of sucrose to pre-existing acceptor molecules (Johnes *et al.*, 1991). The genes encoding these three enzymes (*SacA*, *SacB* and *SacC*) have been cloned, sequenced and characterized. The extracellular SacC has higher specific activity on sucrose than SacB. The SacB and the SacC contribute 30% and 70%, respectively, to the extracellular sucrose hydrolysing activity of *Z. mobilis*. Thus, the SacC is the main enzyme responsible for sucrose hydrolysis during sucrose fermentation (Ananthalakshmy and Gunasekaran, 1999a). However, fructose molecules formed by the action of SacC are not converted into levan.

Z. mobilis SacB is naturally expressed as an extracellular protein but has no putative signal peptide and it has been proposed that it could be secreted through a signal-peptide-independent pathway (Song *et al.*, 1993). The SacB catalyses the hydrolysis of sucrose into fructose and glucose resulting in formation of levan by the transfer of fructose moiety to pre-existing acceptor at β-2, 6-linkage. The reaction is :

$$\underset{\text{sucrose}}{\text{GF}} + \underset{\text{levan}(n)}{\text{G(F)}n} \rightarrow \underset{\text{glucose}}{\text{G}} + \underset{\text{levan}(n+1)}{\text{G(F)}n+1} \quad (19.1)$$

The levansucrase also has invertase (hydrolase) activity, in which it catalyses the following reaction:

$$\underset{\text{sucrose}}{\text{GF}} + \underset{\text{water}}{H_2O} \rightarrow \underset{\text{glucose}}{\text{G}} + \underset{\text{fructose}}{\text{F}} \quad (19.2)$$

Z. mobilis ferment glucose and fructose to ethanol anaerobically via the Entner–Doudoroff (ED) pathway (Fig. 19.2).

The levan produced by *Z. mobilis* has been shown to be typical bacterial levans containing only fructose in the furanose ring form, linked via *p*(2–6) with *p*(2–1) branch points and a very high molecular mass (approximately 10 MDa). *Z. mobilis* has several advantages over other levan producing bacteria such as higher sugar uptake, lower biomass production and without requirement of controlled addition of oxygen during the fermentation (Gunasekaran *et al.*, 1994). In addition to levan, FOS are also produced by *Z. mobilis* in sucrose medium (Viikari, 1984). At low sucrose concentrations (10–15% at 25°C), ethanol is produced while at higher concentration of sucrose (50–70% at 45°C), FOS are synthesized.

Biosynthesis of levan and FOS by *Z. mobilis* has been improved in the presence of high salt concentrations (0.6M NaCl) during sucrose fermentation (Bekers *et al.*, 2000). Levan produced by *Z. mobilis* is stable for 120 h in a culture medium at temperature 25 or 30°C and pH not lower than 3.7. Levan synthesis in water miscible organic solvents has also been reported. 2-Methyl-2-propanol had positive effect on levan production (Castillo and Mungiua, 2004). Senthilkumar and Gunasekaran (2005) have reported a mutant strain of *Z. mobilis* CT2 with defective *sacC*, for production of higher levels of levan (27.2 g/l) from 200 g/l of sucrose at 25°C and pH 5.0.

Genetic improvements for levan production

The levan production can be improved by the optimization of fermentation conditions or genetic manipulation of the producer organism. *Z. mobilis* B4286 produced a low concentration of levan (5.7 g/l) during sucrose fermentation.

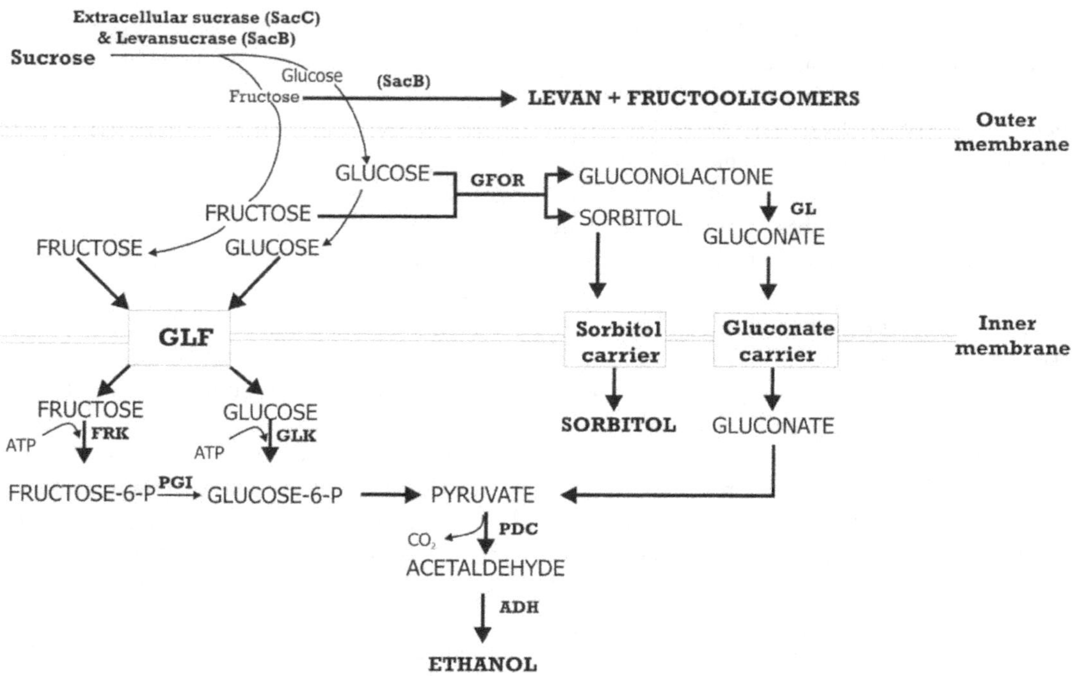

Figure 19.2 Synthesis of ethanol and by-products, levan, FOS and sorbitol in *Z. mobilis*. The key steps in the pathway of sugar transport and its conversion to ethanol are presented. GFOR, glucose fructose oxidoreductase; GL, gluconolactonase; GLF, glucose transport protein; FRK, fructokinase; GLK, glucokinase; PGI, phosphoglucoisomerase; PDC, pyruvate decarboxylase; ADH, alcohol dehydrogenase.

The fermentation conditions such as substrate concentration, pH and temperature etc. have been optimized and the levan production was improved to 12.0 g/l, a twofold higher concentration (Gunasekaran and Raj, 1999). Using chemical mutagenesis, levan overproducing mutants of *Z. mobilis* ZML1 and ZML2 were obtained from the strain *Z. mobilis* B4286 (Ananthalakshmy and Gunasekaran, 1999a). These mutants produced levan to a concentration of 20 g/l and the improved production was associated with the overproduction of SacB.

In order to overproduce levan, the levansucrase (*sacB*) gene, from *Z. mobilis* has been cloned and expressed efficiently in *E. coli* and the overproduced levansucrase was used for the enzymatic synthesis of levan (Song *et al.*, 1996). Further, the *SacB* gene dosage in *Z. mobilis* was increased by cloning and expression of cloned *sacB* gene. The *sacB* gene was cloned in an *E. coli-Z. mobilis* shuttle vector pZA22 and resultant recombinant plasmid pLSD19 was transferred to a sucrose mutant of Z. *mobilis* Zsuc1, defective in production of extracellular levansucrase and sucrase and its parent strain B14023 (Ananthalakshmy and Gunasekaran, 1999b). The *sacB* gene expressed efficiently in Zsuc1 and B14023. The plasmid was stable in these strains, and the *sacB* gene expression improved the sucrose fermentation efficiency of Zsuc1 to a level equal to that of the parent strain. The *sacB* gene expression in B14023 (pLSD19) was increased, which resulted in an increased levan production. Thus, the recombinant *Z. mobilis* strain with multiple copies of *SacB* is considered as an efficient candidate for the enhanced level of levan production.

In *Z. mobilis*, the product of *zliE* gene activates the synthesis of extracellular SacB and SacC while the product of *zliS* gene stimulates the secretion of both enzymes. The overexpression of cloned *zliE* gene in *Z. mobilis* strain Z6 and Z6C resulted in a fivefold increased production of SacB and SacC (Kondo *et al.*, 1994). The knock-out of *sacC* gene in *Z. mobilis* improved the levan production. Using insertional inactivation with tetracycline resistance gene, a *sacC* defective *Z. mobilis* strain CT2 was constructed. This recombinant strain produced higher level of levansucrase during sucrose fermentation. The *sacC* deletion in *Z. mobilis* alters the transcript length of the bi-cistronic transcript and increases the stability of *sacB* messenger. Due to the increased stability of the messenger, SacB synthesis is increased (Senthilkumar *et al.*, 2004). The increased SacB activity results in overproduction of levan than from the wild-type strains (Fig. 19.3).

Production of bioethanol from *Zymomonas mobilis*

The natural energy resources such as fossil fuel, petroleum and coal are being utilized at a rapid rate and these resources have been estimated to last over a few years. Biofuels such as bioethanol are becoming a viable alternative to fossil fuels. Some biological processes have rendered possible

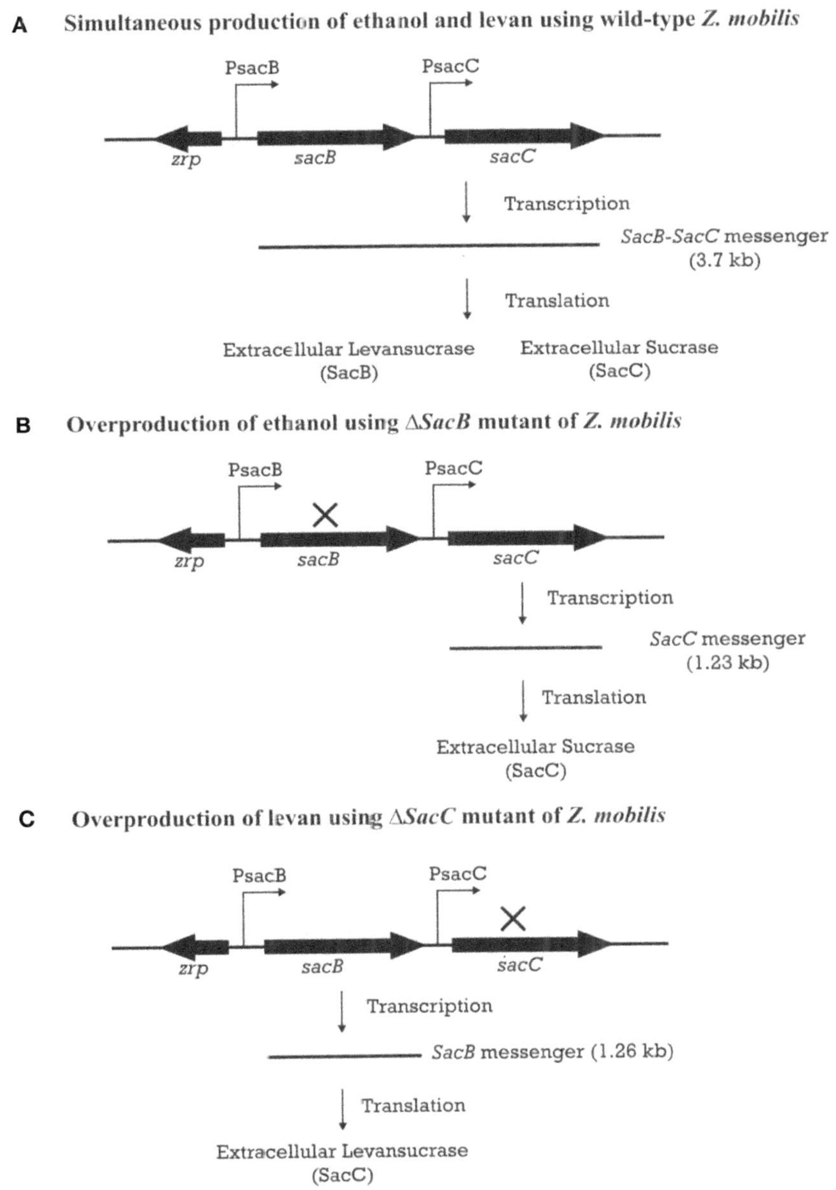

Figure 19.3 Genetic engineering strategies for improved ethanol or levan production from *Z. mobilis*. **A** Wild-type strains are used for simultaneous production of ethanol and levan. **B** Deletion of *SacB* gene enhances the ethanol production due to the inhibition of levan and FOS synthesis; **C** Deletion of *sacC* gene enhances the levan synthesis, while decreasing ethanol production.

routes for producing ethanol and methane as alternative energy sources. This has stimulated studies on the cost and efficiency of industrial processes for bioethanol production (Gunasekaran and Raj, 1999). Traditionally, ethanol has been produced in batch fermentation with yeast strains. However, yeast stains do not tolerate high concentration of ethanol, which affects the efficiency of ethanol conversion from sugars. This necessitated the strain improvement programme for obtaining alcohol-tolerant strains for fermentation process. *Z. mobilis* is considered as an alternative organism for the large-scale fuel ethanol production due to its ability to grow at higher sugar concentrations (>25% glucose) and to tolerate ethanol at concentrations up to 13% (w/v). Further, it has been reported to provide near theoretical yield of ethanol from glucose and fructose (Baratti and Bu' Lock, 1986). The crude sucrose substrates such as sugar beet, baggasse and molasses have also shown promise as substrates for direct fermentation to ethanol. However, the limitation of *Z. mobilis* is its utilizable substrate range, which is restricted to glucose, fructose and sucrose.

Co-production of levan and bioethanol by *Zymomonas mobilis*

Z. mobilis has also been used for the simultaneous production of ethanol and levan in sucrose medium. During ethanol production by *Z. mobilis*, FOS and levan can be produced as valuable by-products at ethanol distilleries using sucrose substrates as raw material. Sucrose concentration and temperature are the most important factors, which regulate activity of *Z. mobilis* SacB mediated FOS synthesis and levan formation. Crittenden and Doelle (1994) have reported that at low sucrose concentrations (10–15%) and 45–50°C, ethanol production is favoured, while at higher sucrose concentration (50–70%) and 25°C, levan synthesis is favoured. Ananthalakshmy and Gunasekaran (1999b) have reported that at 25°C, the levan production was higher (15 g/l) while no levan formation was observed at 35–40°C. In addition, ethanol production was also higher (48.2–57 g/l) at 25–35°C and the ethanol yield declined (47 g/l) at 40°C. The loss of levan production by *Z. mobilis* during fermentation at 35 and 40°C can be due to either complete inactivation of SacB or the loss of its ability to synthesize levan without affecting sucrose hydrolysing activity. Crittenden and Doelle (1994) have demonstrated that levan formation was thermolabile, levansucrase is inactivated irreversibly at 35°C. However, its sucrose hydrolytic activity was not affected. Therefore, at high temperature, levan forming activity of SacB is abolished.

For simultaneous levan and ethanol production, sugar beet juice has been used as a suitable raw material (Bekers *et al.*, 1999). The accumulation of monomeric sugars in the fermentation medium lowers the levan production (Viikari and Gisler, 1986). Addition of glucose or fructose (50 g/l to 150 g/l) into the medium reduced the levan production from 12.6 g/l to 2.8 g/l (on glucose addition) and 6.7 g/l (on fructose addition). In contrast, ethanol production is increased with added monomeric sugars in the medium (Ananthalakshmy and Gunasekaran, 1999a).

Influence of levan synthesis on bioethanol production

Z. mobilis can efficiently utilize glucose and fructose to produce ethanol and carbon dioxide as the sole fermentation products (Baratti and Bu' Lock, 1986). However, the ethanol yield is always low in sucrose containing medium. Formation of by-products such as levan, FOS and sorbitol results in more than 20% reduction of ethanol yield when compared with glucose fermentation (Viikari, 1984). Influence of intra- and extracellular sucrases of *Z. mobilis* on the ethanol production in presence of various by-products has been studied (Kannan *et al.*, 1997). The sucrose fermentation kinetics of two mutant strains LS1 (*sacB*–) and LS1A (*sacA*–), which are phenotypically levan non-producers, were studied. The ethanol productions by these mutants were significantly higher (44.2 g/l) than that of the parent strain B-806 (37.5 g/l). The SacA did not have any role in ethanol fermentation ability of the mutants LS1 and LS1A as the yield parameters such as cell yield and ethanol yield of these strains were same. Thus, the improved ethanol production by the mutants in comparison with their parent strain could be due to the loss of levan production (Kannan *et al.*, 1997).

Immobilization of *Z. mobilis* cells increased the ethanol production while the levan formation was considerably reduced (Bekers *et al.*, 2001). When stainless steel wire spheres (WS), alumina (Al_2O_3) and calcium alginate were used as carriers, the ethanol concentrations were 27.2 g/l, 26.4 g/l and 28.3 g/l respectively and levan concentrations were 14.4, 14.2 and 13.6 respectively after three cycles of fermentation. Although the system productivity was low, the values of the ratio of ethanol to levan concentrations were higher when Al_2O_3 immobilized cells were used in fermentation medium (Bekers *et al.*, 2001). Kannan *et al.*, (1998) have reported similar results by immobilizing the mutant strain LS1A (*sacA*–) and its parent strain *Z. mobilis* B-806 in Ca-alginate beads. Since the mutant strain was non-levan producer, the immobilized cells were stable in sucrose fermentations with 150 g/l sucrose and therefore produced higher amounts of ethanol (73.5 g/l) than the parent strain (65.2 g/l).

Genetic improvements for bioethanol production

The efficiency of ethanol production by *Z. mobilis* from sucrose-based substrates can be improved by limiting the formation of by-products either by optimizing the fermentation conditions or developing mutants of *Z. mobilis* that do not produce levan. Kannan *et al.* (1993) reported the isolation of levansucrase mutants of *Z. mobilis* and their improved ethanol production from sucrose. Two mutant strains LS1 and LS2 of *Z. mobilis*, isolated by NTG mutagenesis, were not able to synthesize active levansucrase, leading to the loss of both the sucrase- and levan-forming activity of the enzyme. However, these mutants still possessed the invertases SacA and SacC. Due to the presence of these two enzymes, the mutants could efficiently ferment sucrose. The loss of levan-forming ability of the mutants led to improvement in the bioethanol production.

Recombinant DNA technology has been exploited to construct microbial strains to produce ethanol from these low-cost agricultural substrates. To obtain ethanologenic strains utilizing the above-mentioned substrates, either genes coding for hydrolases of other organisms are transferred to *Z. mobilis* (Song *et al.*, 1998) or genes of *Z. mobilis* involved in ethanol synthesis are transferred to other organisms with hydrolytic enzymes (Kannan *et al.*, 1995). The cellulase gene from *Erwinia chrysanthemi* coding for endoglucanase Z has been subcloned into a broad-host range plasmid pGSS33 and successfully expressed in *Z. mobilis* ATCC 10988 (Brestic-Goachet *et al.*, 1989). The endoglucanase gene *celZ* was expressed efficiently and the level of expression was higher in *Z. mobilis* than in *E. coli*. In another report, the gene from *Bacillus licheniformis* coding for a thermostable α-amylase was subcloned into the broad-host range plasmid pKT210 in *E. coli*. This recombinant plasmid was then transferred into *Z. mobilis* ATCC31821 by conjugation (Brestic-Goachet *et al.*, 1990). The amylase gene was expressed, although at a lower level than in *E. coli* and the amylase activity was immunologically identical in *Z. mobilis*, *E. coli* and *B. licheniformis*.

Similarly, *Z. mobilis* does not ferment the xylose, which is commonly found in lignocellulosic feedstocks. Consequently, metabolic engineering of xylose fermentation in *Z. mobilis* has been employed for ethanol production from the feedstocks. Considerable research has been conducted to develop a xylose-fermenting *Z. mobilis*. The xylose isomerase and xylulokinase genes from *Xanthomonas campestris* were expressed in *Z. mobilis*. However, the recombinant strain could not grow on xylose as the sole carbon source (Linden and Hahn-Hfigerdal, 1989). However, a strain of *Z. mobilis* capable of growing on xylose and converting it to ethanol was successfully developed through the coordinate expression of the *E. coli* xylose isomerase, xylulokinase, transketolase, and transaldolase genes (Hinman *et al.*, 1989).

In addition, other host organisms were also explored through a range of genetic engineering approaches. In this regard, many Gram positive bacteria and yeasts have been genetically engineered for bioethanol production. The alcohol dehydrogenase and the pyruvate decarboxylase genes from *Z. mobilis* were cloned and expressed in *E. coli* and *Lactobacillus casei* (Gold *et al.*, 1996). Although the recombinant strains could produce higher amounts of ethanol (6%), these recombinant strains were sensitive to the high concentration of ethanol. In order to ferment pentose sugars to bioethanol, the genes encoding xylulokinase, xylose reductase and xylitol dehydrogenase were cloned and overexpressed in *S. cerevisiae* H158. Fermentation studies using media containing pentose or lignocelluloses hydrolysates indicated increased ethanol yield by the recombinant yeast strains (Johansson *et al.*, 2001). Fujita *et al.* (2002) constructed novel *S. cerevisiae* strains for the fermentation of cellulose to ethanol by cell surface codisplay of *Aspergillus aceulatus* β-glucosidase 1 (BGL 1) and *T. reesei* endoglucanase 2 (EG 2). The novel yeast strains could convert β-glucan to ethanol and a high yield (93.3% of theoretical yield) and high productivity (0.53 g/l/h) were achieved for direct ethanol production from β-glucan. Recently, Yanase *et al.* (2007) constructed engineered *Zymobacter palmae* strains with *E. coli* genes encoding the xylose catabolic enzymes. Fermentation with recombinant *Z. palmae* using media containing xylose, a maximum of 95% of the theoretical yield of ethanol was obtained with no by-product formation.

Conclusions

Z. mobilis coproduces levan, fructooligomers and ethanol during sucrose fermentation. Since the production of levan is disadvantageous in bioethanol production with respect to sugar conversion to ethanol, strain that produces very little or no levan would be useful for bioethanol production. The levansucrase negative mutants produce more ethanol. However, the by-products such as levan, FOS and sorbitol are also commercially important. Several technologies have been developed for levan, FOS and ethanol production through recombinant strain construction, process optimization strategies and immobilization of enzyme or whole cells. Metabolic engineering of *Z. mobilis* holds strong potential for the improvement of ethanol production and by-product synthesis.

Acknowledgements

The authors gratefully acknowledge the Department of Biotechnology, New Delhi, for providing financial support (Project No. BT/PR-7451/PID/20/284/2006). Central facility at Centre for Excellence in Genomic Sciences, Madurai Kamaraj University is gratefully acknowledged.

References

Abdel-Fattah, A.F., Mahmoud, D.A., and Esawy, M.A. (2005). Production of levansucrase from *Bacillus subtilis* NRC 33a and enzymic synthesis of levan and fructooligosaccharides. Curr. Microbiol. *51*, 402–407.

Ahmed, F. (2005). Production of levansucrase from *Bacillus subtilis* NRC 33a and enzymic synthesis of levan and fructo-oligosaccharides. Microbiology *51*, 402–407.

Ammar, B.Y. (2002). Characterization of a thermostable levansucrase from *Bacillus* sp. TH4-2 capable of producing high molecular weight levan at high temperature. J. Biotechnol. *99*, 111–119.

Ananthalakshmy, V.K., and Gunasekaran, P. (1999a). Isolation and characterization of mutants from levan producing *Zymomonas mobilis*. J. Biosci. Bioeng. 87, 214–217.

Ananthalakshmy, V.K., and Gunasekaran, P. (1999b). Overproduction of levan in *Zymomonas mobilis* by using clone *sacB* gene. Enzyme Microb. Technol. 25, 109–115.

Baratti, J. ,and Bu' Lock, J.D. (1986). *Zymomonas mobilis*: a bacterium for ethanol production. Biotechnol. Adv. 4, 95–115.

Bekers, M., Laukevics, J., Upite, D., Kaminska E and Linde R (1999). Flame burned stainless steel wire spheres as carrier of *Zymomonas mobilis* cells and extracellular levansucrase. Acta Biotechnol. *19*, 341–348.

Bekers, M., Vigants, A., Laukevics, J., Toma, M., Rapoports, A,., and Zikmanis, P. (2000). The effect of osmo-induced stress on product formation by *Zymomonas mobilis* on sucrose. Int. J. Food Microbiol. 55, 147–50.

Bekers, M., Laukevics, J., Karsakevich, A., Ventina, E., Kaminska, E., Upite, D., Vina, I., Linde, R., and Scherbaka, R, (2001). Levan-ethanol biosynthesis using *Zymomonas mobilis* cells immobilized by attachment and entrapment. Process Biochem. *36*, 979–986.

Bergeron, L.J., Morou-Bermudez, E., and Burne, R.A. (2000). Characterization of the fructosyltransferase gene of *Actinomyces naeslundii* WVU45. J. Bacteriol. *182*, 3649–3654.

Brestic-Goachet, N., Gunasekaran, P., Cami, B., and Baratti J. (1989). Transfer and expression of an *Erwinia chrysanthemi* cellulase gene in *Zymomonas mobilis*. J. Gen. Microbiol. *135*, 893–901.

Brestic-Goachet, N., Gunasekaran, P., Cami, B., and Baratti, J. (1990). Transfer and expression of a *Bacillus licheniformis* α-amylase gene in *Zymomonas mobilis*. Arch. Microbiol. *153*, 219–225.

Buddington, K.K., Donahoo, J.B. and Buddinigton, R.K. (2002). Dietary oligofructose and inulin protect mice from enteric and systemic pathogens and tumor inducers. J. *Nutr. 132, 472–477.*

Cani, P.D., Knauf, C., Iglesiaus, M.A., Drucker, D.J., Delzennae, N.M., and Burcelin, R. (2006). Improvement of glucose tolerance and hepatic insulin sensitivity by oligofructose requires a functional glucagon like peptide 1 receptor. Diabetes 55, 1484–1490.

Carlsson J. (1970). A levansucrase from *Streptococcus mutans*. Caries Res. *4*, 97–113.

Carpita, N.C., Kanabus J., and Housley, T.L. (1989). Linkage structure of fructans and fructan oligomers from *Triticum aestivum* and *Festuca arundinacea* leaves. *J. Plant Physiol.* 134, 162–168.

Castillo E and Mungiua, A.L. (2004). Synthesis of levan in water miscible organic solvents. J. Biotechnol. *114*, 207–217.

Crittenden, R., and Doelle, H. (1994). Identification and characterization of the extracellular sucrases of *Zymomonas mobilis* UQM-2716 (Atcc-39676). Appl. Microbiol. Biotechnol. *41*, 302–308.

Dedonder, R. (1966). Levansucrase from *Bacillus subtilis*. Methods Enzymol. *86*, 500–505.

Earl, RJ., and Per, J.G (1998). Levan derivatives, their preparation, composition and applications including medical and food application. Patent No.Wo/1998/003184.

Fujita, Y., Takahashi, S., Ueda, M., Tanaka, A., Okada, H., Morikawa, Y., Kawaguchi, T., Arai, M., Fukuda, H., and Kondo, A. (2002). Direct and efficient production of ethanol from cellulosic material with a yeast strain displaying cellulolytic enzymes. Appl. Environ. Microbiol. *68*, 5136–5141.

Ghaly, A.E., Arab, F., Mahmoud, N.S., and Higgins J. (2007). Production of levan by *Bacillus licheniformis* for use as a soil sealant in earthen manure storage structures. Am. J. Biotechnol. Biochem. *3*, 47–54.

Gibson, G.R. and Roberfroid M (1995). Dietary modulation of the human colonic microbiota: introducing the concept of prebiotics. J. Nutr. 125, 1401–1412.

Gibson, G.R. (1999). Dietary modulation of the human gut microflora using the prebiotics oligofructose and inulin. J. *Nutr. 129*, 1438S-1441S.

Gold, R., Meagher, M.M., Tong, S., Hutkins, R.W., and Conway T (1996). Cloning and expression of the *Zymomonas mobilis* production of ethanol genes in *Lactobacillus casei*. Curr. Microbiol. 33, 256–260.

Gunasekaran, P., Kannan, T.R., Ananthalakshmy, V.K., and Amutha R (1994). Molecular biology and genetics of *Zymomonas mobilis*. Indian J. Microbiol. *34*, 159–185.

Gunasekaran P and Raj, K.C. (1999). Ethanol fermentation technology – *Zymomonas mobilis*. Curr. Sci. 77, 56–68.

Han, Y.M. (1989). Levan production by *Bacillus polymyxa*. J. Ind. Microbiol. *4*, 447–452.

Han, Y.M. (1990). Production and characterization of microbial levan. J. Agric. Food Chem. 38, 393–396.

Hendry, G.A. F and Wallace, R.K. (1993). The origin, distribution, and evolutionary significance of fructans. *In* M Suzuki, N.J. Chatterton, eds, Science and Technology of Fructans. CRC Press, Boca Raton, FL, pp 119–139.

Hinman, N.D., Wright, J.D., Hoagland W and Wyman, C.E. (1989). Xylose fermentation: an economic analysis. Appl. Biochem. Biotechnol. *20/21*, 391–410.

Hijum, V.S. A F T, Van Geel-Schutten, G.H., Rahaoui, H., Van der Maarel M J., and Dijkhuizen L (2002). Characterization of a novel fructosyltransferase from *Lactobacillus reuteri* that synthesizes high-molecular-weight inulin and inulin oligosaccharides. Appl. Environ. Microbiol. *68*, 4390–4398.

Johansson, B., Christensson, C., Hobley, T. and Hägerdal, B.H. (2001). Xylulokinase overexpression in two strains *of Saccharomyces cerevisiae* also expressing xylose reductase and xylitol dehydrogenase and its effect on fermentation of xylose and lignocellulosic hydrolysate. Appl. Environ. Microbiol. *67*, 4249–4255.

Johnes, M.R., Greenfield, P.F., and Dolle, H.W. (1991). Byproducts from *Zymomonas mobilis*. Adv. Biochem. Eng. Biotechnol. *44*, 97–121.

Kannan, T.R., Mukundan, A.G., and Gunasekaran P (1993). Fermentation characteristics of levansucrase mutants of *Zymomonas mobilis*. J. Ferment. Bioeng. 75, 265–270.

Kannan, T.R., Kannan, G., Nadra Ait-Abdelkader, Augier-Magro, V., Baratti J., and Gunasekaran, P. (1995). Molecular cloning and characterization of the extracellular sucrase gene (sacC) of *Zymomonas mobilis*. Arch. Microbiol. 163, 195–204.

Kannan, T.R., Sangiliyandi, G. and Gunasekaran, P. (1997). Influence of intra- and extracellular sucrases of *Zymomonas mobilis* on the ethanol production and by-product formation. Biotechnol. Lett. *19*, 661–664.

Kannan, T.R., Sangiliyandi, G., and Gunasekaran, P. (1998). Improved ethanol production from sucrose by a mutant of *Zymomonas mobilis* lacking sucrases in immobilized cell fermentation. Enzyme Microb. Technol. 22, 179–184.

Kazap, I., Leibovici J., and Wolman, M. (1980). Blood levels of high-molecular levan in mice as a function of the route and duration of administration. Arzneimittelforschung *30*, 459–462.

Kim, C.H., Lee, J.Y., Kim, M.G., Song, K.B., Seo J.W., Chung, B.H., Chang S.J., and Rhee, S.K. (1998). Fermentation strategy to enhance plasmid stability during the cultivation of *Escherichia coli* for the production of recombinant levansucrase. J. Ferment. Bioeng. *86*, 391–394.

Kim, K.H., Chunq, C.B., Kim, Y.H., Kim, K.S., Han C S., and Kim C. H (2005). Cosmeceutical properties of levan produced by *Zymomonas mobilis*. J. Cosmet. Sci. 56, 395–406.

Kirk, R.E., Othmer, D.F., Kroschwitz J. I., and HoweGrant H (1991). Kirk-Othmer Encyclopedia of Chemical Technology. (Vol.16). Wiley Inc., New York, New York.

Kondo, Y., Toyada, A., Fukushi, H., Yanase, H., Tonomura, K., Kawasaki H., and Sakai T. (1994). Cloning and characterization of a pair of genes that stimulate the production and secretion of *Zymomonas mobilis* extracellular levansucrase and invertase. Biosci. Biotechnol. Biochem. *58*, 526–530.

Korakli, M., Rossman, A., Gänzle M. G and Vogel, R.F. (2001). Sucrose metabolism and exopolysaccharide production in wheat and rye sourdoughs by *L. sanfranciscensis*. J. Agri. Food Chem. *49*, 5194–5200.

Korakli, M., Gänzle, M.G., and Vogel, R.F. (2002). Metabolism by bifidobacteria and lactic acid bacteria of polysaccharides from wheat and rye, and exopolysaccharides produced by *L. sanfranciscensis*. J. Appl. Microbiol. 92, 958–965.

Laue, H., Schenk, A., Li, H., Lambertsen, L., Neu, T.R., Molin S., and Ullrich, M.S. (2006). Contribution of alginate and levan production to biofilm formation by *Pseudomonas syringae*. Microbiology *152*, 2909–2918.

Lewis, D.H. (1993). Nomenclature and diagrammatic representation of oligomeric fructans – a paper for discussion. New Phytol. *124*, 583–594.

Linden T., and Hahn-Hfigerdal B. (1989). Fermentation of lignocellulose hydrolysates with yeasts and xylose isomerase. Enzyme Microb. Technol. *11* 583–589.

Lopez, C.E., Lima, R.M.O.C., and Defranca, F.P. (1997). Antitumor activity of levan produced by *Zymomonas mobilis*. Biotechnol. Lett. *19*, 19–21.

Nolling, J., Breton, G., Omelchenko, M.V., Makarova, K.S., Zeng, Q., Gibson, R., Lee, H.M., Dubois, J., Qiu, D., Hitti, J. Wolf, Y.I., Tatusov, R.L., Sabathe, F., Doucette-stamm, L., Soucaille, P., Daly, M.J., Bennett, J.N., Koonin, E.V., and Smith, D.R. (2001). Genome sequence and comparative analysis of the solvent-producing bacterium *Clostridium acetobutylicum*. J. Bacteriol. *183*, 4823–38.

Olivares-Illana, V., Wacher-Rodarte, C., Le Borgne S., and López-Munguía A. (2002). Characterization of a cell-associated inulosucrase from a novel source: A *Leuconostoc citreum* strain isolated from Pozol, a fermented corn beverage from Mayan origin. J. Ind. Microbiol. Biotechnol. *28*, 112–117.

Park, J.P., Bae, J.T., and Yun, J.W. (1999). Critical effect of ammonium ions on the enzymatic reaction of a novel transfructosylating enzyme for fructooligosaccharide production from sucrose. Biotechnol. Lett. *21*, 987–990.

Park, J.P., Oh, T.K., and Yun, J.W. (2001). Purification and characterization of a novel transfructosylating enzyme from *Bacillus marcerans* EG-6. Process Biochem. *37*, 471–476.

Pontis, H.G., and Del Campillo, E. (1985). *Fructans: In Biochemistry of Storage Carbohydrates in Green Plants*, P. M. Dey and R. A. Dixon, eds., Academic Press, London, pp. 205–227.

Ramsay, J.A., Cooper, D.G., and Neufeld, R.J. (1989). Effects of oil reservoir conditions on the production of water-insoluble levan by *Bacillus licheniformis*. Geomicrobiol. J. *7*, 155–165.

Roberfroid, M.B. (1996). Functional effects of food components on the gastrointestinal system. Nutr. Rev. 54: S38-S42.

Saito, K., Yokota, A., and Tomita, F. (1997). Molecular cloning of levan fructotransferase gene from *Arthrobacter nicotinovorans* GS-9 and its expression in *Escherichia coli*. Biosci. Biotechnol. Biochem. *61*, 2076–2079.

Sangeetha P.T., Ramesh M.N., and Prapulla S.G. (2002). Influence of media components and reaction parameters on the production of fructosyl transferase and fructooligosaccharides. Sciences des Aliments. *22*, 277–287.

Senthilkumar, V., Rameshkumar, N., Busby S.J., and Gunasekaran P (2004). Disruption of the *Zymomonas mobilis* extracellular sucrase gene (*sacC*) improves levan production. J. Appl. Microbiol. 96, 671–6.

Senthilkumar, V., and Gunasekaran, P. (2005). Influence of fermentation conditions on levan production by *Zymomonas mobilis* CT2. Indian J. Biotechnol. 4, 491–496.

Shih, I. and Yu, Y. (2005). Simultaneous and selective production of levan and poly(γ-glutamic acid) by *Bacillus subtilis*. Biotechnol. Lett. 27, 103–106.

Song, K.B., Joo, H.K., and Rhee, S.K. (1993). Nucleotide sequence of levansucrase gene (*levU*) of *Zymomonas mobilis* ZM1 (ATCC 10988). Biochim. Biophys. Acta. *1173*, 320–4.

Song, K.B., Belghith, H., and Rhee, S.K. (1996). Production of levan, a fructose polymer, using an overexpressed recombinant levansucrases. Ann. N. Y. Acad. Sci. *799*, 601–607.

Song, K.B., Seo, J.W., Kim M.G., and Rhee, S.K. (1998). Levansucrase of *Rahnella aquatilis* ATCC33071: Gene cloning, expression, and levan formation. Ann. NY Acad. Sci. *864*, 506–511.

Suzuki, M. and Pollock, C.J. (1986). Extraction and characterization of the enzymes of fructan biosynthesis in timothy grass (*Phleum pratense* L.). Can. J. Bot. *64*, 1884–1887.

Tieking, M., Korakli, M., Ehrmann, M.A., Gänzle, M.G., and Vogel, R.F. (2003). In situ production of exopolysaccharides during sourdough fermentation by cereal and intestinal isolates of lactic acid bacteria. Appl. Environ. Microbiol. 69. 945–952.

Uchiyama, T. (1993). Metabolism in microorganisms, Part II. Biosynthesis and degradation of fructans by microbial enzymes other than levansucrase. In M Suzuki, N.J. Chatterton, eds, Science and Technology of Fructans. CRC Press, Boca Raton, FL, pp. 169–190.

Viikari, L. (1984). Formation of levan and sorbitol from sucrose by *Zymomonas mobilis*. Appl. Microbiol. Biotechnol. *19*, 252–255.

Viikari, L. and Gisler, R. (1986). By-products in the fermentation of sucrose by different *Zymomonas* strains. Appl. Microbiol. Biotechnol. 23, 240–244.

Wada, T., Ohguchi, M. and Iwai, Y. (2003). Novel enzyme of *Bacillus sp.* 217C-11 that produces inulin from sucrose. Biosci. Biotechnol. Biochem. *67*, 1327–34.

Yanase, H., Sato, D., Yamamoto, K., Matsuda, S., Yamamoto, S., and Okomoto, K. (2007). Genetic engineering of *Zymobacter palmae* for ethanol production from xylose. Appl. Environ. Microbiol. 73, 2592–2599.

Index

Other Books of Interest

Neisseria

Molecular Mechanisms of Pathogenesis

Edited by: **Caroline Genco and Lee Wetzler**
c. 300, January 2010
ISBN: 978-1-904455-51-6 $310/£150

With contributions from 40 neisseria experts, this comprehensive volume reviews the latest research, theory and clinical significance of molecular mechanisms in meningococcal disease. Topics include: gene expression, genomics, biofilms, denitrification, adhesion strategies and mechanisms of cellular invasion. A section on the host response to neisserial infection covers inate immunity, complement, apoptosis, and acquired immunity while a section devoted to clinical correlation deals with vaccine development, epidemiology and antibiotic resistance. Essential reading!

Molecular Oral Microbiology

Edited by: **Anthony H. Rogers**
x + 292 February 2008
ISBN: 978-1-904455-24-0 $300/ £150

This excellent volume is aimed at students and researchers in the field of oral biology and oral microbiology. Topics include oral microbial taxonomy, identification and typing, applied genomics, horizontal gene transfer, cell-cell communication, cariogenic bacteria, *Aggregatibacter actinomycetemcomitans*, *Porphyromonas gingivalis*, *Treponema denticola*, host-pathogen interaction, host defense mechanisms, and vaccination against oral infections.

Reviews:
"well-referenced and nicely illustrated" from SciTech Book News (March 2008) pp. 71.

"... a broad and comprehensive text covering the molecular aspects of dental caries and periodontal disease. .... of value to all microbiologists with an interest in molecular microbiology ... " from Microbiology Today (2008)

"All of the chapters are well written, up to date and thoroughly referenced. There is also an extensive index. The book deals with the subject areas in depth and provides a wealth of information for the reader. It will be an important text for any clinician or researcher with an interest in oral microbiology." from ACM News (2008) 2: 18

Lactobacillus Molecular Biology

From Genomics to Probiotics

Edited by: **Åsa Ljungh and Torkel Wadström**
x + 206 January 2009
ISBN: 978-1-904455-41-7 $310/£150

This major new work focuses on recent research on the molecular biology and genomics of *Lactobacillus*. Topics covered include phylogenetics, taxonomy, comparative genomics, functional genomics, the intestinal microflora, surface proteins, stress responses, interaction with the immune system, probiotics, anti-cancer potential, and much more. Essential reading for all scientists involved with lactic acid bacteria or probiotic research and a recommended book for all microbiology laboratories.

Reviews:
"... contributions from respected international scientists, many of which are leaders in their respective fields, this book constitutes an authoritative resource about both fundamental research and applications of lactobacilli. ... essential and up to date information for anyone interested in the biology of lactobacilli. The book will clearly be of interest to microbiologists, nutritionists, food scientists, and medical practitioners alike, and it is a valuable contribution to the probiotic literature." from Biotechnol. J. (2009)

".....an essential reference for medical researchers, microbiologists and biotechnologists, especially, dairy technologists." from Int. Microbiol. (2008) 11:

Real-Time PCR

Current Technology and Applications

Edited by: **J. Logan, K. Edwards and N. Saunders**
x + 284 pp., January 2009
ISBN: 978-1-904455-39-4 $310 / £150

Essential manual providing a comprehensive guide to the most up-to-date technologies and applications as well as providing an overview of the theory of this increasingly important technique.

Reviews:
"... a comprehensive overview of the RT-PCR technology, which is as up-to-date as a book can be ..." Mareike Viebahn in Current Issues in Molecular Biology (2009)

Mycobacterium
Genomics and Molecular Biology

Edited by: **Tanya Parish and Amanda Brown**
viii + 214 January 2009
ISBN: 978-1-904455-40-0 $300/ £150

Expert scientists critically review the current and most recent advances in the genomics and molecular biology of mycobacteria. Topics covered include strain variation and evolution, hypervirulent strains, electron transport and respiration, lipid biosynthesis, DNA repair, oxygen signaling, sulphur metabolism, protein secretion, the protein kinase family, and much more.

Reviews:
"This very timely book reviews the current knowledge about the genetics and cell biology of the mycobacterium species. It is clear that this information will lead to new treatment options and preventive strategies that may cure or prevent the serious disease of tuberculosis." from Doodys (2009)

Lab-on-a-Chip Technology
Fabrication and Microfluidics

Edited by: **K. E. Herold and A. Rasooly**
c. 420 pp., August 2009
ISBN: 978-1-904455-46-2 $310 / £150

This comprehensive volume presents the current technologies in the field and includes theoretical and technical information to enable both the understanding of the technology and the reproduction of experiments. The book aims to help the reader to understand current LOC technologies, to perform similar experiments, to design new LOC systems and to develop new methodologies and applications.

Lab-on-a-Chip Technology
Biomolecular Separation and Analysis

Edited by: **K. E. Herold and A. Rasooly**
c. 320 pp., August 2009
ISBN: 978-1-904455-47-9 $310 / £150

A comprehensive volume focusing on the applications of LOC technology in the biomedical and life sciences. Includes biomolecule separation, electrophoresis, chromatography, protein and cell separation, genetic and transcriptome analysis, PCR, cell viability analysis and microorganism capturing. A volume of exceptional importance to current science.

ABC Transporters in Microorganisms

Edited by: **Alicia Ponte-Sucre**
c. 220 pp., August 2009
ISBN: 978-1-904455-49-3 $310 / £150

The structure, physiology and evolution of ABC transporters and their special characteristics in bacteria, yeast, trypanosomes and malaria parasites. Recent research on the role of ABC transporters in the circumvention of drug resistance in microorganisms.

Real-Time PCR in Microbiology
From Diagnosis to Characterization

Edited by: Ian M. Mackay
x + 454 pp., Sept. 2007
ISBN 978-1-904455-18-9 $300 / £150

A background for the novice, a theoretical reference for the experienced user, and useful discussions of future developments. Includes: oligonucleotide design, target preparation, standardisation, quantification, various applications, and future challenges. The final chapter is a roundtable discussion providing an insightful, topical and interesting discourse with contributions from over 30 experts.

Reviews:
"I am impressed with the amount of valuable information in this book - information that is difficult to find in a single source. I have purchased three additional copies for each laboratory and teaching area. It is clear that I will use this book many times ..." from Rebecca T. Horvat, University of Kansas Medical Center (from Doody.com)

Bacterial Secreted Proteins
Secretory Mechanisms and Role in Pathogenesis

Edited by: **Karl Wooldridge**
xii + 512 pp., April 2009
ISBN: 978-1-904455-42-4 $310 / £150

Extensive publication on bacterial secreted proteins, secretory systems and their vital role in bacterial pathogenesis. Immense value to all microbiologists, molecular biologists, public health scientists and other researchers and professionals.

Also of Interest

Aspergillus: Molecular Biology and Genomics	2010
Environmental Molecular Microbiology	2010
Frontiers in Dengue Virus Research	2010
Pili and Flagella: Current Research and Future Trends	2009
Microbial Toxins: Current Research and Future Trends	2009
Acanthamoeba: Biology and Pathogenesis	2009
Clostridia: Molecular Biology in the Post-genomic Era	2009
Plant Pathogenic Bacteria: Genomics and Molecular Biology	2009
Microbial Production of Biopolymers and Polymer Precursors	2009
Pathogenic Fungi: Insights in Molecular Biology	2008
Staphylococcus: Molecular Genetics	2008
Plasmids: Current Research and Future Trends	2008
Corynebacteria: Genomics and Molecular Biology	2008
Leishmania: After The Genome	2008
Archaea: New Models for Prokaryotic Biology	2008
RNA and the Regulation of Gene Expression	2008
Legionella Molecular Microbiology	2008
Epigenetics	2008
Animal Viruses: Molecular Biology	2008
Segmented Double-Stranded RNA Viruses	2008
Acinetobacter Molecular Microbiology	2008
Pseudomonas: Genomics and Molecular Biology	2008
Microbial Biodegradation: Genomics and Molecular Biology	2008
The Cyanobacteria: Molecular Biology, Genomics and Evolution	2008
Coronaviruses: Molecular and Cellular Biology	2007
Bacteriophage: Genetics and Molecular Biology	2007
Candida: Comparative and Functional Genomics	2007
Bacillus: Cellular and Molecular Biology	2007
AIDS Vaccine Development: Challenges and Opportunities	2007
Alpha Herpesviruses: Molecular and Cellular Biology	2007
Pathogenic *Treponema*: Molecular and Cellular Biology	2007
PCR Troubleshooting: The Essential Guide	2006
Influenza Virology: Current Topics	2006
Microbial Subversion of Immunity: Current Topics	2006
Cytomegaloviruses: Molecular Biology and Immunology	2006
Papillomavirus Research: From Natural History To Vaccines and Beyond	2006

NOTES

NOTES

www.ingramcontent.com/pod-product-compliance
Lightning Source LLC
LaVergne TN
LVHW062112200926
848484LV00007B/1645

* 9 7 8 1 9 0 4 4 5 5 4 5 5 *